Marine Biology

A flash of lightning:
Into the gloom
Goes the heron's cry
—MATSUO BASHŌ

What goes out, and comes in
Doesn't end or begin
It ebbs and flows
And there it goes again
—TOM CHAPIN

FIFTH EDITION

Marine Biology

Function, Biodiversity, Ecology

JEFFREY S. LEVINTON
Stony Brook University

New York Oxford
OXFORD UNIVERSITY PRESS

Oxford University Press is a department of the University of Oxford.
It furthers the University's objective of excellence in research, scholarship,
and education by publishing worldwide. Oxford is a registered trademark of
Oxford University Press in the UK and certain other countries.

Published in the United States of America by Oxford University Press
198 Madison Avenue, New York, NY 10016, United States of America.

Library of Congress Cataloging-in-Publication Data

Names: Levinton, Jeffrey S.
Title: Marine biology : function, biodiversity, ecology / Jeffrey S.
   Levinton, Stony Brook University.
Description: Fifth edition. | New York : Oxford University Press, [2017] |
   Includes bibliographical references and index.
Identifiers: LCCN 2017013346| ISBN 9780190625276 (pbk.: alk. paper)
Subjects: LCSH: Marine biology. | Marine biology—Textbooks.
Classification: LCC QH91 .L427 2017 | DDC 577.7—dc23 LC record available at https://lccn.loc.gov/2017013346

9 8 7 6 5 4 3

Printed by Marquis, Canada

For Joan, Nathan, Andy, and all others in the choir

BRIEF CONTENTS

CONTENTS

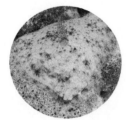

MARINE ORGANISMS: FUNCTION AND ENVIRONMENT

■ **CHAPTER 5 The Chemical and Physical Environment   74**

■ **CHAPTER 6 Life in a Fluid Medium   98**

■ **CHAPTER 7 Reproduction, Dispersal, and Migration   109**

ORGANISMS OF THE OPEN SEA

PATTERNS AND PROCESSES IN THE WATER COLUMN

■ CHAPTER 12 Productivity, Food Webs, and Global Climate Change 239

ORGANISMS OF THE SEABED

■ CHAPTER 13 Benthic Microorganisms, Seaweeds, and Sea Grasses 256

■ CHAPTER 14 The Diversity of Benthic Marine Invertebrates 268

■ **CHAPTER 15   Benthic Life Habits   297**

COASTAL BENTHIC ENVIRONMENTS

■ **CHAPTER 16   The Tidelands: Rocky Shores, Soft-Substratum Shores, Marshes, Mangroves, Estuaries, and Oyster Reefs   317**

■ **CHAPTER 17   The Shallow Coastal Subtidal: Sea Grass Beds, Rocky Reefs, Kelp Forests, and Coral Reefs   365**

FROM THE SHELF TO THE DEEP SEA

HUMAN IMPACT ON THE SEA

Welcome to the fifth edition of *Marine Biology: Function, Biodiversity, Ecology*. We marine biology instructors are lucky to have exciting creatures, great adventures, and continuing discoveries that manage to enchant and stimulate students. My greatest challenge is to find an organization that combines student inspiration with the principles and current practice they need to understand the subject at a high level. Many students find marine biology a little daunting. No surprise to me, because it is always a challenge to organize the information in a useful text that informs and challenges without swamping the students with so many new concepts and facts. I believe, though, that every student can acquire the broad spectrum of biological knowledge included in marine biology while appreciating the organismal diversity of the marine realm. The goal of this text is to appeal to a wide range of students, while also preparing future specialists with the knowledge and tools needed to conduct their research. I continue to be convinced that students *must learn concepts* along with the facts and begin to *think and reason like scientists*. Just as important, they must feel the pulse of current happenings.

That is why the text addresses three major principles and process-driven themes: *functional biology*, *ecological processes*, and *biodiversity*. It is why the text uses full color throughout in drawings that depict marine biological processes and a large number of photographs to connect students with marine environments and organisms. I have added yet more photographs to this edition, thanks to my own journeys and my wonderful friends. It is also why I have paid much attention to recent advances and include a series of essays called *Hot Topics in Marine Biology*. These features bring current exciting research to the students, with a diverse range of marine systems. In both the text and the Hot Topics, I try to connect the students to some of the most important recent research with an extensive literature section that is now online so that students can get to the best work for term papers and projects without too much distraction in the text. In my own classes I am convinced that a good teacher must show students the way to good science because reference databases are not user-friendly or self-explanatory, where finding excellent science research is concerned.

The Philosophy of This Text

This text is designed for a one-semester course at the sophomore to senior level. Some students will have already taken a college-level biology course with coverage of organismal diversity, and they will benefit greatly. A small number may even have taken a course in marine invertebrates or vertebrates and an introductory course in oceanography.

That said, I have successfully taught from this text for many years, and many students had no background in organismal biology or ecology. If the book is supplemented with journal articles—many of which are cited and recommended in the text or can be gleaned from the extensive online reference list—it can also be used in a more advanced undergraduate course in marine ecology. The fifth edition contains many new references to the primary literature, which are now accessible online with a simple link mentioned at the end of each text chapter. This is a valuable resource to get students started on term papers and essays. The Marine Biology Web page (ww.oup.com/us/levinton), which I founded a number of years ago, links students to many more views of marine biology and to a greater diversity of organisms. My career advice page had over 1.3 million visits as of 2016.

I have taught Marine Biology for over 40 years and have always been amazed at the diversity of students who take the course. Biology majors, marine science majors, geology majors, psychology majors, and even some humanities majors sit side by side. At my university, marine science has become a separate undergraduate discipline in recent years, and these students have truly learned the interdisciplinary nature of marine biology. All learn a great deal, and all seem to come away with a love for the ocean. You don't have to convince them to be there: They *want* to learn about marine biology. I do my best to keep that excitement alive, and I find that field trips and the use of color photographs and other illustrations throughout the text help a great deal. So do the online resources accompanying this text. *Marine Biology Explorations* includes hundreds of photographs from habitats discussed in the text and still more. As mentioned, an expanded and updated reference list is also available online. In class, I keep a large map of the oceans on the wall for the geographic context of our discussions. I have taken many of my students with me to marine labs, and they have launched careers in science or used their backgrounds to enter other areas. I hope the text will help a wider audience to get excited about marine life. I hope, too, that they will understand how the ocean works and why our marine realm is so threatened.

A Principles-Driven Approach

Marine biology applies the principles of ecology and evolution, using the crucial tools of cell biology, biomechanics, and molecular biology to a wide range of marine biological studies. These fields of study and their interactions govern the text's three overarching themes: functional biology, biodiversity, and ecological processes. **Function** refers to the way organisms solve problems and how physical and chemical factors constrain and select the solutions. What shape should a maneuvering fish have relative to a continuously swimming fish? How does a small peptide manage to be such an effective poison when injected by a snail into a prey? How does this specific biochemical adaptation feed into an understanding of biodiversity? I believe this textbook is unique in combining effectively functional biology with ecological thinking. **Biodiversity** is an essential part of marine biology, and I introduce the topic both through introductions to the principles used to study and explain biodiversity and to the factors that strongly affect marine biodiversity. A separate chapter discusses biodiversity and the processes that regulate it, both ecological and evolutionary. It is crucial that the student see the historical roots of many current distributions, which are affected by processes ranging from plate tectonics to climate change. This edition adds a great deal of coverage of recent advances, including molecular tools used to study marine biodiversity (especially in difficult situations like the microbes in the plankton), dispersal, speciation, and the rise of marine adaptations. This edition includes a **bonus chapter on molecular tools in marine biology**, available online. Your students can see how molecular tools, old and new, are used in a wide range of marine biological applications.

Ecology examines the interactions of organisms with their environment and tries to understand the distribution and abundance of organisms. It involves a series of processes, which I introduce in the context of a hierarchy—from *individual populations* to *ecosystems*. It also involves a discussion of important ecological processes along with accounts of major marine habitats and communities. This edition pays special attention to modern concepts of populations and species interactions, including connectivity, metapopulations, regional genetic differentiation, large-scale control of dispersal, biological invasions, and alternative stable states of communities. My background in geology, ecology, and evolutionary biology allows me to frequently combine ecological, evolutionary, and geological thinking in discussing marine biology problems.

Organization

We begin with a brief historical background. Marine biology has a history that is worth understanding, but it is also crucial to introduce the student to how science works. Chapter 1 therefore discusses framing and testing hypotheses, as well as making tests practical enough that they can be put to direct use. From the very beginning, I introduce the student not only to the scientific method, but also to how it translates into an intellectual tool with real-world applications. Chapters 1 through 7 introduce basic principles of how the ocean works in a physical, chemical, and ecological context and how marine organisms function with these constraints. The second chapter gives the student a comprehensive introduction to oceanography and the important

properties of seawater that might affect marine organisms. A new Chapter 3 covers climate change and the interaction of climate oscillations, such as El Niño, with longer-term trends. In Chapter 4, I introduce ecological and evolutionary principles so that students can work their way through concepts using marine examples. This allows all students to be brought up to a level of ecological thinking and an understanding of oceanographic processes. They will see this "big picture" as they read the rest of the text. A chapter on the physical-chemical environment discusses how temperature and other important physical variables affect marine organismal function and survival. The book extensively discusses both macro- and microscale effects of climate change, including ocean acidification, range changes, and ecosystem effects. A crucial chapter then introduces students to how the physics of fluids shapes the constraints and adaptations of marine organisms. As far as I know, this crucial subject is missing in all other marine biology texts, and allows a connection to a complete understanding of how the marine organisms function in the rather complex fluid environment. This subject is absolutely essential to see how the ocean works and how the same seawater environment has drastically different impacts on organisms' different sizes and shapes. These chapters conclude with a comprehensive introduction to reproductive strategies, larval dispersal, and migration, which sets up the big picture of the geographic distribution of marine species, down to the microscale of how mobile marine larvae succeed in finding a place to live in a turbulent and stressful world.

Chapters 8 through 12 cover the *organisms* and *processes* that are important in the water column of the open sea, including coverage of the major organisms from plankton to whales and the latest ideas on the rise and demise of phytoplankton blooms. This organism-process approach is essential so that students will understand the overall economy of the marine realm, while not forgetting the major players on the ecological stage. A new Chapter 10 emphasizes adaptations and processes in the water column, ranging from bioluminescence to diel vertical migrations. This leads to a detailed discussion of the processes that cause the genesis of phytoplankton blooms, the major drivers of global productivity and often local ecologically harmful blooms. Chapter 12 uses a global-scale approach to show how biological studies of the ocean lead to an understanding of the world's potential for fisheries and the global biological impact on the ocean of climate change.

Bottom organisms and habitats are covered in Chapters 13–18, which depend both directly and indirectly on the water world above. In Chapters 13 and 14, I cover benthic creatures and then go on to discuss the principles necessary to understand the ecology of marine bottom organisms (Chapter 15) and the major near-shore marine bottom habitats (Chapters 16–19). By necessity, I have been selective. I emphasize those habitats that are not only important and interesting, but also those in which important principles can be illustrated to their best advantage. This is a major reason why so much attention is paid to the tidelands, our ecologically best-known marine habitats. Community-level interactions are emphasized, as is global climate change as it relates to major changes in habitats such as coral reefs. I discuss a range of geographic locations so that the instructor will find local examples in many instances. Crucial habitats such as the intertidal, seagrasses, coral reefs, mangroves, estuaries, salt marshes, kelp forests, and others are discussed both from the points of habitat distinctions and ecological processes and the impacts of biological invasions and climate change. Oyster reefs are highlighted because of their great worldwide importance as foci for biodiversity and their role in ecosystem services. I discuss the drivers of coral reef ecology but also discuss the great problems they face from disease, ocean warming, and acidification.

Chapter 18 then looks at the important gradient from the continental shelf to the deep sea, paying special attention to some of the fascinating discoveries about biological function and fascinating habitats, from hot vents to deep-water coral mounds to the recently discovered subsurface bacterial realm over 500 m beneath the sea floor. In a new chapter, 19, on polar biology, I have greatly expanded coverage of Arctic and Antarctic environments, which are the front lines of climate change effects. I incorporate a wide range of discussions from the organismal to the ecosystem level, including the decline of sea ice, ice algae, and the crucial resource of krill in Antarctic food webs. Chapter 20 focuses on and summarizes what we know about marine biodiversity geographic patterns, including sections on invasive species, conservation of biodiversity, and conservation genetics. More and more, students and researchers have focused their attention to the deteriorating conditions of the ocean, and conservation is a major field of emphasis.

Finally, Chapters 21 and 22 tackle other human interactions with the sea, as both a source of food and, unfortunately, a waste receptacle. I cover human effects on the ocean. Throughout the text, the effects of climate change are brought up in many contexts, and how those effects are related to chemical issues such as acidification and facilitation of biological invasions. I also place strong emphasis on the reorganization of communities that has been initiated by the interaction of human activities and strong ecological interactions found in natural communities and in food webs. The impact of overfishing on populations and trophic cascades is a crucial part of a complete chapter on fisheries and mariculture. I include in Chapter 21 a section on drug discovery in the ocean, because of the great student interest in this subject and the connections between biodiversity and the new sources of compounds to combat pain and disease, such as cancer. It is a revelation to me at least how a dangerous animal like cone snails can synthesize toxins that hold great hope for reduction of pain without the side effect of drug addiction. The roles of toxic substances, eutrophication, and hypoxia are discussed clearly and in depth. I discuss the fascinating topic of evolutionary responses to stress and novel toxic substances introduced into the ocean. I also cover developing problems such as introduction of microplastic particles into the ocean.

A Refined Learning Package

This text has a series of pedagogical features designed to help students absorb a wide range of information and concepts by engaging their imaginations, helping them organize and prioritize important principles, and keeping them focused on the big picture, without getting lost in the details. **Hot Topics in Marine Biology** essays throughout the text introduce students to recent advances in the understanding of marine biology and discuss current issues, especially marine–biological debates and discoveries. Instructors can use these essays to kick off discussion, to expand a student's horizons, for course assignments, or as topics for term papers. A new section on the amazing discoveries of homing of sharks to highly localized reproductive sites is found in Chapter 7. The great strides made in reducing the scourge of shark finning is covered in an essay in Chapter 21. A Hot Topics essay addresses exciting new molecular techniques used to identify the virus that caused the recent catastrophic sea star wasting disease on the U.S. west coast (Chapter 16). I discuss in Chapter 10 the exciting studies that show how blue whales balance food and energy needs by highly sophisticated optimization of foraging dive times. On the individual level, I show the connection between the crystal eyes of some living chitons and their striking resemblance to those of trilobites hundreds of millions of years ago (Chapter 5). I also focus in on the exciting discoveries on how sea turtles keep their feet warm in a cold ocean (Chapter 5). Some Hot Topics have been retained from the fourth edition because they are still "hot," such as the use of dogs to locate whale scat for molecular and hormonal data, and the possible adaptation of microbes to breaking down oil in the Gulf of Mexico.

Key Concept full-sentence summary statements begin nearly every section of the text to help students identify central points of discussion and to foreshadow what's to come. These headings allow students to discern the forest from the trees and to quickly scan the basic progression of material by looking ahead through the chapter. Each chapter ends with a bulleted **Chapter Summary** and a variety of **Review Questions**. Instructors and students can use these to follow up on important issues in marine biology. The combination of these features and the Key Concept heading sentences successfully guides the student through a complex subject. **Going Deeper** boxes explain equations and related concepts in marine biology. Especially in early chapters, they will help students learn often-difficult material or refresh their memory of elementary courses (e.g., photosynthesis). They also allow instructors who choose to omit them to press on with no interruptions. An example is the discussion of Leslie matrices in Chapter 21, which give the student an idea of how age-structured population models help to understand impacts of various factors on fisheries and management decisions. Extensive **References** lists of classic and contemporary scholarship that instructors may assign as reading and that can lead students to further assignments are linked online from the text. These help students see that marine biology is a living field of research, not just a static textbook of "known" facts, without interrupting the flow of the text. A comprehensive **Glossary** of marine biology at the end of the text provides students access to get a quick definition of important concepts, processes, and terms. A **list of journals** is included as a resource for students in writing term papers and for further research.

What Is New and Noteworthy in the Fifth Edition?

Expanded illustration program. We continue the fifth edition using a rich color presentation in order to better demonstrate marine biological principles and introduce organismal diversity in a vivid and captivating visual presentation. The new edition includes many new photos and line drawings (many of the photos generously contributed by colleagues), and I believe students will benefit greatly from having the color photos integrated directly into the relevant textual discussion at hand.

More applications. To engage students with the diversity of marine biology today and to highlight the real-world applications of what they are learning, I've written many new in-text examples, including seven new **Hot Topics in Marine Biology**. Students will see how molecular tools can be used to study the origin of a major disease, how conservation efforts have succeeded and will continue to succeed in reducing shark finning, and how ocean acidification is now a major danger to shellfisheries.

Current and expanded topics maintain the excitement that underlies my philosophy of teaching and have been carefully selected to bring the text up-to-date while still remaining focused on the most important *principles* students need to learn.

- *Evidence and effects of climate change.* I have greatly expanded coverage of climate change and have added a new chapter that delves into the important issues, especially the difference between climate oscillations and protracted climate trends. I pay special attention to temperature change in the global ocean and in coastal areas (Chapter 3); the ocean acidification (Chapters 2, 8, 12, 17, 21); and the role of climate change in changing species distributions, facilitating biological invasions, and causing thermal stress (Chapters 2, 3, 5, several others).
- *New ideas that challenge us all.* It is hard to accept sometimes that textbook accounts are incomplete or even wrong. But our field is rapidly changing with new discoveries and outlooks. I pay special attention to an incipient revolution in our thinking about the advent of the spring phytoplankton bloom, which for many years has been explained using the classic Sverdrup model. This is about to change, and I attempt to show the problems with the classic model and how we might build a new approach, based on recent research. I look forward to hearing how students react to this.

- *Ecological interactions.* Strong attention is paid to major ecological interactions that are relevant to ecosystem structure, such as trophic cascades (Chapters 16 and 17), ecological reorganizations in New England and elsewhere (Chapter 16), molecular approaches to ecology and evolution (Chapters 4, 5, 6, 7, and others, including a new **bonus chapter** online), natural and human-induced phase shifts (Chapters 3 and 17), biological invasions (Chapters 3, 7, 8, 16, 17, 18, 19, 20), and climate change (Chapters 2, 3, 4, 5, 7, 8, 10, 11, 12, 16, 17, 19, 20).
- *Methods of environmental assessment, from remote sensing to the molecular level.* I have also expanded coverage of the latest methods for remote sensing, estimating world productivity, and assessing the stress on and change of ecosystems, including satellite methods and ocean observatories (Chapters 1 and 12), acoustic detection of fish and marine mammals (Chapter 7, 10), genetic and molecular studies of population differentiation (Chapters 5, 7, 9, 20), the shifting baseline concept (Chapter 20), diversity gradients and the tropical origins of biodiversity (Chapter 20), and molecular methods to assay the cause of disease and the diversity of microorganisms in the water column (Chapters 11 and 16).
- *Human impact on biodiversity.* This edition expands coverage of the decline of coral reefs and adds insights on other biological impacts such as the increase of sponges at the expense of corals (Chapter 17), overfishing and the issue of relating management decisions to management of the basis of ecosystem function (Chapter 21), the declines of sharks and other apex predators (Chapter 21), and the effects of pollution, especially with coverage of the Deepwater Horizon well blowout and recent expansion of inputs of plastics into the ocean (Chapter 22).
- *Emphasis on polar biology.* We notice right away from the new accomplishments of polar ecologists how much there is to learn about polar food webs and how climate change is rapidly changing the nexus of sea ice, productivity in the nearby ocean, and especially the changing nutrient supplies and productivity of crucial food species such as krill. I have established a separate chapter to discuss these issues and to continue to discuss the dangers ahead for polar communities in the face of climate change.
- *Molecular approaches.* I continue to emphasize molecular studies because they are becoming so important in the study of environmental stress, identification of genetic differentiation of species, and identification of difficult groups of microorganisms. As mentioned, there is a free online bonus chapter on molecular methods that your students can use as a resource.

Supplements

Marine Biology, Fifth Edition, is accompanied by a wealth of electronic resources for both students and instructors, including a FREE Companion Website (www.oup.com/us/levinton) and FREE access to the **Instructor's Resource** Ancillary Resource Center (www.oup.com/us/levinton/resources).

Companion Website: Maintained by the author, this Companion Website (www.oup.com/us/levinton) provides a multitude of resources for both students and instructors.

- **Student Resources**
  - **Marine Biology Explorations.** Explore the ocean's biodiversity that will take you through nine different marine habits; including over 450 photos with annotations!
  - **Marine Biology in the News.** Frequently updated current breakthroughs in marine biology research.
  - **Extensive web links** to marine biology topics and research literature. You will also find information on careers in marine biology and worldwide marine laboratories.
  - **Bonus molecular tools chapter.** An overview of how molecular tools, old and new, are used in a wide range of marine biological applications.
  - **Hot Topics Archive.** Hot Topics may cool off, but their relevance to marine biology is lasting. The student website will now feature a full archive of past Hot Topics.

- **Instructor Resources** (available to adopters of the text and password-protected)
  - **Electronic Images.** All illustrations from the text available in electronic format for download for lecture presentations.
  - **PowerPoint Lecture Notes.** Over 400 lecture notes slides organized by chapter.
  - **Video guide.** New to the fifth edition, the online instructors resources will now include a guide to video and multimedia most relevant to marine biology topics.
  - **Test Bank.** This comprehensive resource includes approximately 400 questions written by the author himself in editable Word files for easy customization (available on the Ancillary Resource Center: contact your Oxford University Press sales representative for details).

Acknowledgments

Many people have helped me—too many to mention all individually. I am especially grateful to the many individuals who shared their photographs and research experiences with me. Many sent me preprints, photographs, data, and just plain interesting discussions, which helped me greatly to understand fields unfamiliar to me. I am deeply grateful to the late Bob Guillard, who was so instrumental in making my first book on marine ecology readable and hopefully interesting. I am grateful to Debra Abercrombie, The American Museum of Natural History (Mammal Department), Michael Beherenfeld, Kelly Benoit Bird, John Dolan, Sonny Gruber, Drew Harvell, Hyemi Kim, Bruce Robison, Carl Safina, George Waldbusser, and Haikun Xu. As usual, my wife Joan was supportive and helpful with suggestions.

For the current edition, Jason Noe served as editor and continually moved this project forward. Barbara Mathieu worked as an excellent liaison between me and copy editors and artists to bring the project to completion. I would also like to thank Benjamin Olcott, assistant editor; Patrick Lynch, editorial director; David Jurman, marketing manager; Frank Mortimer, director of marketing; Michele Laseau, Art Director; and Lisa Grzan, managing editor. I also benefited a tremendous amount from careful review of the manuscript and additional reviewer recommendations based on the fourth edition. I had a wonderful panel of advisors for the fourth edition but have continued to be lucky with the truly excellent panel that helped me with revisions for the fifth. I could not have done better with such an excellent panel:

Chantale Bégin, University of South Florida

John Berges, University of Wisconsin–Milwaukee

Bopi Biddanda, Grand Valley State University

Susan Bratton, Baylor University

Virginia Dudley, Grossmont College

Tara Duffy, Northeastern University

Michael Franklin, California State University Northridge

Aaren Freeman, Adelphi University

Richard Grippo, Arkansas State University

Gail B. Hartnett, University of New Haven

Catherine Hurlbut, Florida State College at Jacksonville

David Kirchman, University of Delaware

Elizabeth Lacey, The Richard Stockton College of New Jersey

Dean Lauritzen, City College of San Francisco

Annie Lindgren, Portland State University

Sue Lowery, University of San Diego

Tim McLean, Tulane University

Xiaozhen Mou, Kent State University

Antonios Pappantoniou, Housatonic Community College

Clayton A. Penniman, Central Connecticut State University

Kristin Pollizzotto, Kingsborough Community College

Michael Robinson, Barry University

Santiago Salinas, Kalamazoo College

Erik P. Scully, Towson University

David Tapley, Salem State University

Ione Hunt von Herbing, University of North Texas

Mary K. Wicksten, Texas A&M University

Lawrence Wiedeman, University of Saint Francis

John Timothy Wootton, University of Chicago

I also again thank reviewers whose insights contributed to past editions: Jelle Atema, Boston University; Susan S. Bell, University of South Florida; Larry E. Brand, University of Miami; Christopher Brown, Florida International

University; James E. Byers, University of New Hampshire;
Edward J. Carpenter, San Francisco State University;
Gerardo Chin-Leo, Evergreen State College; Paul Dayton,
Scripps Institute of Oceanography; Chris D'Elia, Louisiana
State University; Sean Patrick Grace, Southern Connecti-
cut State University; Larry G. Harris, University of New
Hampshire; William W. Kirby-Smith, Duke University;
Alan J. Kohn, University of Washington; Derek R. Lavoie,
Cuesta College; Larry R. McEdward, University of Flor-
ida; George McManus, University of Connecticut; Amy
Moran, Clemson University; Stephen Norton, East Carolina
University; Jan A. Pechenik, Tufts University; Kathleen A.
Reinsel, Wittenberg University; David Scheel, Alaska
Pacific University; Eric P. Scully, Towson University;
Jayson Smith, California State University, Fullerton;
Alan E. Stiven, University of North Carolina; Philip Sze,
Georgetown University; Keith Walters, Coastal Carolina
University; and Judith S. Weis, Rutgers University.

Jeffrey Levinton
Stony Brook University, Stony Brook, New York

Marine Biology

CHAPTER 1

Sounding the Deep

Marine Biology as a Discipline

On every coast of the world, scientists work in field locations and in marine stations ranging from multimillion-dollar structures to small shacks with fanciful paintings of lobsters and crabs above the door. Some put out to sea in large ships, whereas others scarcely wet their knees (**Figure 1.1**).

The purpose of this textbook is to give you an organized way of turning a fascination for the sea into an appreciation of the principles of marine biology that reflect the function and ecology of marine life. Snorkel on a coral reef and you will see coexisting schools of large numbers of species of fishes. But why do so many species coexist in a very limited space? How do all these creatures interact to form the seascape? Such questions require an organized approach to a complex and somewhat foreign world. By the time you have finished your course and this textbook, you will be more familiar with that world. You will also be familiar with many human activities that put this world in peril, which makes an understanding of marine biology extremely important.

■ **Marine biology is a subject mixing functional biology and ecology.**

Marine biology is a diverse subject, but its main elements are functional biology and ecology. **Functional biology** is the study of how an organism carries out basic functions such as reproduction, locomotion, feeding, and the cellular and biochemical processes related to digestion, respiration, and other aspects of metabolism. Problems relating to function are quite varied. They might deal with questions such as: When a whale dives for food to very great depths, how does it conserve oxygen? **Ecology**, on the other hand, is the study of the interaction of organisms with their physical and biological environments and how these interactions determine the distribution and abundance of the organisms. For example, why do so many species requiring limited space coexist on a coral reef? Why doesn't one superior species win out and displace the others?

Because ecology is an environmental subject, the field of marine biology must cover the **basic aspects of marine**

FIG. 1.1 A fascination with marine creatures led the late Howard Sanders first to make major contributions to our understanding of the ecology of intertidal and shallow marine bottom communities. Later, he pioneered American research in the deep sea, discovered marine animals previously unknown to science, and unlocked the secret of the deep-sea bottom's great biodiversity. (Photograph courtesy of the Woods Hole Oceanographic Institution)

habitats. We shall therefore spend considerable space explaining the various seascapes that are important to marine life.

■ Biodiversity is the third major factor in marine biological studies.

Marine environments can be very rich in species but vary tremendously in the number of species, or **biodiversity**. Coral reefs may contain thousands of species, but a rocky shore in high latitudes may contain fewer than 50. How do species arise? Why is there variation in species numbers from habitat to habitat and from time to time within a given locality? What is the consequence of living in a very diverse community? How can so many different species make a living in a coral reef? Functional biology, ecology, and biodiversity, as you will see, are very interactive components.

Historical Background of Marine Biology

■ Marine biology began with simple observations of the distribution and variety of marine life.

A native lore of the biology of the sea has accumulated over thousands of years by those living near the shore and by fishing peoples. The earliest formal studies in marine biology date back to a time when there was little distinction among scientific specialties. Early biologists were "natural philosophers" who made general observations about anatomy and life habits. We owe the beginning of this tradition of natural philosophy to Aristotle (384–327 B.C.) and his Greek contemporaries, who recorded their observations on the distribution and habits of shore life. Aristotle described the anatomy of the octopus and other marine creatures, noticed that some sharks give birth to live young, and observed that some whales have structures that resemble hog bristles instead of teeth.

The next major steps forward took place in the eighteenth century, when a number of Europeans began to observe and classify living creatures. Most prominent among these was Linnaeus (1707–1778), who developed the modern means of naming species. He described hundreds of marine animal and plant species and developed larger-scale classifications. The great French biologist Georges Cuvier (1769–1832) classified all animals into four major classes of body plans: Articulata, Radiata, Vertebrata, and Mollusca.

The eighteenth century was an important era of oceanic exploration. A number of expeditions circumnavigated the globe, bringing glory to explorers and new Pacific territory to European nations. Many of these expeditions had scientific components as well, and scientific staff was charged with collecting terrestrial and marine plants and animals. The voyage of French captain Nicolas Thomas Baudin explored the tropical Pacific, and numerous marine mollusks were returned to France. Captain James Cook supervised the mapping of eastern Australia in 1770, and his scientific staff collected biological specimens all over the Pacific Ocean, which became the foundation of large collections in Great Britain.

Until the nineteenth century, most marine biology consisted of the description of anatomy and the naming and classification of species. Little was known about function and ecology. The only knowledge of open-ocean life was confined to experience with animals that were fished or observed (or in the case of mermaids and sea monsters, imagined) in the open sea. By the early 1800s, however, the study of **natural philosophy** had become popular, and a number of brilliant individuals devoted their lives to studying the ocean and its denizens.

■ In the nineteenth century, marine biology developed into a science involving ecology and hypothesis testing.

Edward Forbes (1815–1854) of the Isle of Man was the first of the great English-speaking marine biologists. After failing at art and abandoning his medical school studies, he set out to sea and participated in a number of expeditions in which a bottom dredge was used to dig into the seabed and collect organisms. He was the naturalist on the *Beacon*, a ship that sailed on the Mediterranean Sea. Forbes found that the number of creatures decreased with increasing depth and then proposed what was probably the first marine biological **hypothesis**, or testable statement about the world of the sea: the **azoic theory**, which stated that no life existed on seafloors deeper than 300 fathoms (1,800 ft).

Forbes also discovered that different species live at different depths, and he proposed that the broader the depth zone of a species, the wider its geographic extent. Forbes opened up the ocean to scientific research and was appointed to the most prestigious post in natural philosophy of those times at the University of Edinburgh, Scotland. He published maps of geographic distributions of organisms along with a natural history of European seas. He inspired countless followers to an interest in natural science.

During this time, many great pioneers from a number of European countries joined Forbes. In 1850, Norwegian marine biologist Michael Sars disproved the azoic theory by collecting and describing 19 species that live deeper than 300 fathoms in Norwegian fjords. His work inspired a new interest in deep-sea biology. The first plankton net was used during this period, and crude submersibles were developed. Marine biology was on its way.

Although he is usually remembered for his theory of evolution by means of natural selection, Charles Darwin (1809–1881) is the other great English father of marine biology (**Figure 1.2**). As a young man, he worked as naturalist on the H.M.S. *Beagle*, which sailed around the world in the years 1831–1836. He later wrote *The Voyage of the Beagle*, one of the best-selling travel books of the nineteenth century. Darwin made extensive collections of marine animals and concentrated his own later efforts on the classification of barnacles.

While on the *Beagle*, Darwin (1842) formed a theory of the development of coral reefs. He pictured the growth of coral reefs as a balance between the growth of corals upward and the sinking of the seafloor. If Forbes's azoic theory was the first important marine biology hypothesis, then Darwin's coral reef theory was the second. This subsidence theory was published in Darwin's first scientific book, and its brilliance was immediately recognized. Previously, most

FIG. 1.2 Charles Darwin is best remembered for his theory of natural selection, but he made many important contributions to marine biology, including a book on coral reefs and a classification of barnacles that remains essentially unchanged to the present.

had believed that coral reefs in the open Pacific developed from the colonization and growth of corals on submerged extinct oceanic volcanoes. In contrast, Darwin argued that coral reefs developed around emergent rock that was slowly sinking, and this downward motion was balanced by upward growth of the corals. In this subsidence theory as applied to the development of atolls (horseshoe-shaped rings of coral islands), Darwin was proven correct. About 100 years after the theory was developed, scientists drilled a hole in Eniwetok Atoll in the Marshall Islands of the Pacific and bored through hundreds of meters of coral rock before hitting the volcanic rock basement below. Since reef corals can grow only in very shallow water, this finding proved that the reef had been growing upward for millions of years as the island was sinking. Darwin was not completely right about coral reefs, however, insofar as he theorized that all reefs in the world are stages of subsidence leading to atolls. This has proven to be wrong; many reefs are not subsiding, and atolls are special cases of reefs on volcanoes rising from oceanic crust (see Chapter 18).

Fisheries research began in earnest in the nineteenth century and became central in marine biological research. Such research was also the beginning of **applied marine biology** and was necessitated by a need to understand how to find and manage populations of fish. England was first at this activity in 1863. Many nations began research efforts later in the century (see Chapter 22). In the United States, the Fish Commission sought to relate characteristics of the oceanic environment to the life history of fishes. Marine ecology became synonymous with fisheries research, and Canada used a fisheries emphasis to develop distinguished laboratories on both the Atlantic and Pacific coasts.

W. B. Carpenter and C. Wyville Thomson led a major expedition in 1868–1869 that foreshadowed the later great *Challenger* expedition. Both had a passion for marine biology, and they convinced the British government to outfit the *Lightning*, a steam- and sail-powered ship that dredged the seabed of the northern waters of the British Isles. Like Norwegian biologist Michael Sars, they found marine life deeper than 300 fathoms and thus also helped disprove Edward Forbes's azoic theory. The deep maintained its allure to marine biologists, as it was thought to be a museum of living fossils because many animals such as stalked crinoids were found in ancient fossil deposits and in deep water in the living ocean as well. They also found distinct bodies of water with different temperatures, which was an early discovery of the distinctness of some oceanic water masses.

■ The voyage around the world of the H.M.S. *Challenger* gave us the first global-scale view of marine biology.

These expeditions set the stage for the great *Challenger* expedition (1872–1876) that circumnavigated the globe and provided the first global perspective on the ocean's biotic diversity (**Figure 1.3**). The voyage was led by C. Wyville Thomson and by the great naturalist John Murray. The *Challenger* sampled the waters and bottoms of all seas except the Arctic. After the expedition, 50 volumes were needed to describe the tremendous number of organisms that were recovered.

On this expedition, chemist John Buchanan was able to disprove the existence of a so-called primordial slime on the

FIG. 1.3 The H.M.S. *Challenger* at St. Paul's Rocks, a remote equatorial mid-Atlantic Island.

seafloor, called *Bathybius*, which was supposed to be capable of giving rise to higher forms of life. The famous zoologist Thomas H. Huxley had published a well-known paper claiming that a whitish material found in samples from the seabed was evidence of the presence of a material that was continuously giving rise to life forms. This idea was very controversial. Buchanan discovered that the slime, which had been observed in collected samples of seawater, was merely an artifact of preserving seawater with alcohol. The chemical reaction of seawater with alcohol resulted in a white precipitate, which Buchanan claimed was previously mistaken by Huxley to be the mysterious *Bathybius*. He therefore falsified a major claim about the continuing origin of life on the seafloor. This discovery fit well with Louis Pasteur's earlier conclusion from lab experiments that life did not just spring from inanimate substances. The explanation for the white slime named *Bathybius* still remains an open question.

Toward the end of the nineteenth century, marine stations began to spring up over the world, starting in 1875 with the Stazione Zoologica in Naples, Italy. This station set a pattern of international participation by the scientific community. In the 1880s, marine stations were established in England and Scotland. During the same years, Prince Albert I of Monaco outfitted several yachts and larger ships that sampled the ocean, and in 1906, he eventually founded an oceanography institute and museum in Monaco. This facility came to be directed by famous inventor-oceanographer Jacques-Yves Cousteau, who died in 1998. In America, zoologist Alexander Agassiz led oceanographic expeditions, was the first to use piano wire instead of rope to lower samplers, and studied the embryology of starfish and their relatives. The now-famous Marine Biological Laboratory was founded on Cape Cod in 1886, and a number of marine stations were founded in Europe toward the end of the century. By the turn of the twentieth century, marine stations existed in many European countries. Marine laboratories such as the Friday Harbor Laboratories in Washington State made their appearance in the United States soon thereafter (**Figure 1.4**). Marine biology was now a full-fledged science with a proud history of exploration and theorization.

- **Advances in modern marine biology included the development of major research institutions, faster ships, better navigation, and greatly improved diving technology.**

The early part of the twentieth century witnessed the founding of great oceangoing institutes and a new technological ability to explore the ocean to its greatest depths. In America, the founding of the Scripps Institution of Oceanography in southern California (1903) and the Woods Hole Oceanographic Institution on Cape Cod (1930) gave the United States a unique ability to study the open sea. A large number of open-sea expeditions expanded our knowledge of marine life. The voyage of the Danish *Galathea* (1950–1952) was the last great deep-sea expedition of this era. As had happened in Europe toward the end of the nineteenth century, marine stations were opened in every coastal state in America. Marine biology also flourished in many universities. Our knowledge of the ocean expanded during World

FIG. 1.4 Friday Harbor Laboratories, located in the San Juan Islands of Washington State, are a major site for marine biological research and education in rocky-shore ecology, biomechanics, larval biology, neurobiology, and many other areas of study. (Photograph by Jeffrey Levinton)

War II owing to the need for more navigational information. Advances in navigation, deep-sea bottom drilling, remote sensing, and other techniques led to a great expansion of our knowledge of the sea. A rich diversity of open-ocean and shore biological research has since flourished to the point that scores of journals now record the activities of a community of thousands of scientists. The number of such scientists in 1850 could have fit comfortably within a single room.

- **Technology in both the laboratory and the open sea has played an important role in the development of marine biology.**

Before the nineteenth century, poor navigation, inadequate sailing vessels, and generally crude bottom dredges and plankton nets prevented researchers from sampling the ocean systematically or completely. By the late 1800s, however, steam vessels allowed for the rapid lowering and raising of samplers, and navigation was better. In the twentieth century, modern diesel-driven ships such as the R.V. *Knorr*, ported in Woods Hole, Massachusetts, could navigate accurately by means of satellite navigation (**Figure 1.5**).

Before the mid-twentieth century, the deep-sea bottom could not be seen unless a piece of it was dredged and

FIG. 1.5   The R.V. *Knorr*, one of the U.S. oceanographic research fleet, has its home base at the Woods Hole Oceanographic Institution on Cape Cod, Massachusetts. (Photograph courtesy of Woods Hole Oceanographic Institution)

FIG. 1.6   The *Alvin*, a submarine capable of diving to 4,500 m, is equipped with accurate navigation and photography equipment and underwater manipulators. The *Alvin* is the great workhorse of the world research submarine fleet and is scheduled to be replaced in the coming years. (Photograph courtesy of Richard Lutz)

brought to the surface. This has changed dramatically owing to the development of manned submarines, remotely operated vehicles, and scuba diving. William Beebe pioneered deep diving when he descended in a metal sphere, the bathysphere, to a record depth of 923 m in 1934 off the Bermuda coast. In 1960, the spherical steel bathyscaph *Trieste* made a spectacular descent into the deepest oceanic trench off the Marianas Islands in the western Pacific Ocean. By the 1970s, a number of submarines routinely dived to depths of 2,000 m and more, and scientists were able to film and collect marine life (**Figure 1.6**). Mechanical arms made it possible to perform experiments, and accurate navigation systems permitted returns to remote sites in the ocean.

A number of smaller submarines allowed longer-term observation of depths of 300 m and less. One of the more

whimsical submersibles was used in a marine station near Nice, France. The steel hull was connected to the surface by an air hose, and the investigator sat inside on a bicycle seat in a very cramped space. The first recorded observations in the Bay of Villefranche included one of a soup can on the murky bottom. Recently, researchers have used more modern submarines in the same area to observe spectacular bioluminescent planktonic jellyfish. To expand greatly the efficiency of deep-sea observation, **remotely operated vehicles (ROVs)** have been developed. These vehicles are unmanned but can make precise surveys and even take samples (**Figure 1.7**). Remotely operating vehicles are tethered to a ship by a cable, but a great deal of data are now collected by **autonomous underwater vehicles (AUVs)**, which are robots not connected to the ship. An interesting variant

FIG. 1.7 The *Ventana*, an ROV operated by the Monterey Bay Aquarium Research Institute in central California. The vehicle is connected to the mother vessel by a cable and is equipped with high-definition video, two grabber arms, and a variety of samplers, including a sample box that can be seen in the front. Newer versions have been launched. (Photograph by Jeffrey Levinton)

of AUVs are **gliders**, which use simple balancing devices to allow the vehicle to rise and fall through the water column or be moved by vanes in a constant direction by wave action. Ensembles of gliders now are being used by various shore-based laboratories and are deployed from ships because they are much cheaper than ship-based sampling.

Nothing in shallow water, however, has matched the importance of scuba diving, developed in the 1940s. This form of underwater exploration was not used often or effectively until the late 1950s, when biologist Thomas Goreau pioneered the study of coral reefs. Today, direct observations and experiments can be done on rich shallow-water marine biota.

Although many advances have been made in diving and other technologies, the coming decades will see enormous strides toward **remote sensing** of the sea by satellite imaging. In the 1970s and 1980s, an American satellite known as the *Coastal Zone Color Scanner* provided images and conducted sophisticated light-based estimates of water temperature, chlorophyll, and other parameters. Now, new satellites are investigating with far more resolution. In conjunction with the new detectors, marine biologists are trying to use "ground-truthing" to produce equations that relate color information received by satellites to measurements taken at sea. In the long run, this will allow us to process worldwide data sets, a capability that is crucial in our current studies of global climate change.

The most recent advances in **ocean observatories** have taken advantage of Global Positioning System (GPS) located fiber-optic cable systems, with ports for remote video observation, sensing of physical variables such as temperature and current speed, and chemical measurements. This exciting new area is only now being developed and will allow a series of permanent and continuous observation posts to be established within shore and estuary locales but also on the deep-sea floor and in midwater locations. Most exciting is the *Monterey Accelerated Research System* in Monterey Bay, California, where a submarine canyon (see Chapter 2) cuts the continental shelf and extends to the deep-sea floor (**Figure 1.8**). This cable has data-collection ports, including video, and allows scientists to continuously monitor and observe remote localities with Internet communication for research and education. The U.S. National Science Foundation has initiated a large-scale **Ocean Observatories Initiative**, which combines observations from moorings, autonomous vehicles, and underwater cabled observatories.

Observation and Hypothesis Testing

■ **Marine biologists, like all scientists, use the scientific method, which is a systematic means of reasoning and observation.**

Marine biology, like all science, depends on a generalized system of observation and inference of the natural world known as the **scientific method**. This may sound unduly stiff and distant, but the scientific method is merely a systematic way to reason about and observe our world and universe. It depends on observations, deduction, and prediction. We are constantly making observations about the natural world, and many of these are repeatable. For example, we might find that all fish we observe live only in water (most do!). This would lead to a conclusion about the biology of fishes: Fishes live in water.

The accumulation of specific observations to make a generalization is called **induction**. By contrast, we might take

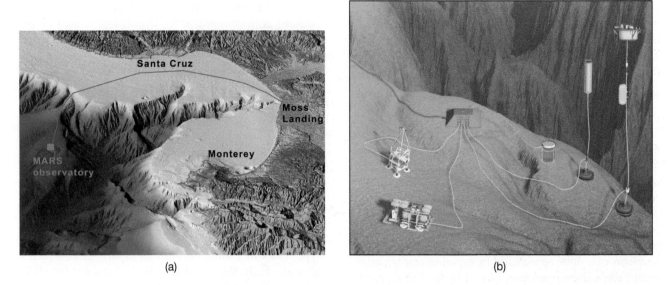

(a)

(b)

FIG. 1.8 Fiber-optic networks are being installed at many sites to create ocean observatories, which allow continuous monitoring of video, physical, and chemical variables. Here, we see the cable installation in Monterey Bay. (a) Map of cable installation in Monterey Bay, constructed by combining computer-generated topographic and bathymetric data to show the Monterey canyon; (b) deployment of instruments at the end of a 52-km-long fiber-optic cable. (Courtesy of David Fierstein and Monterey Aquarium Research Institute)

some premise and use logic to make a **prediction**. Such an inference, predicated on logical associations of conclusions with facts and premises, is a **deduction**. If you counted all the spectators in a football stadium drinking a beer, you might come to the conclusion that at 2 p.m. during the game, 10 percent of the spectators drink beer. That is an induction. Instead, you might reason that most spectators like beer, but all could not be drinking all the time because the lines at the beer concession are very long and only one beer is sold at a time. Therefore, you might deduce that only a fraction of the crowd will be drinking. If you knew the length of the game, how long it took to buy and drink a beer, how much blood alcohol it took to get drunk, and how fast alcohol is metabolized to non-inebriating products, you might be able to deduce how many spectators were holding a beer at any one time and how many are drunk. This line of reasoning is far more valuable because it has led you to develop a prediction that could be applied to other stadiums. Deduction has the beautiful property of prediction.

Here's a more biological example. We might find that there is genetic variation in a population and, knowing that the environment may change, we might predict that some variants will perform better and become more frequent in the population as the environment changes. If we know how genetic transmission works and the relative survival and reproduction rate of those variants, we can predict the rate at which they will increase in frequency in the population. This is Darwin's theory of natural selection, which uses the method of deduction. This form of inference always depends on general premises and a logical pattern of reasoning to draw some specific conclusion. Most scientists strive to develop generalizations and theories from which predictions follow by deduction. Perhaps it is worthwhile to count all the days of the year that are cloudy and then conclude that most of the year is cloudy, but it would be

much better to have a set of premises, a predictive relationship, and a theory to understand why it is cloudy most of the year. Induction, however, is a necessary part of science and can even be an inspiration for deduction.

■ **Most marine biological research requires extensive observations and correlations, but experimentation is usually the most efficient way to answer a question.**

Marine biological research involves a great deal of observation. In some cases, the observation is general and not directed toward any specific research problem. It is essential, for example, to know the distribution of temperature, salt content, water depth, and other properties of seawater because such information is required to solve a diverse array of specific problems. In other cases, observations are targeted toward more specific questions. To understand the migration route of a species of fish, it may be necessary to sample the ocean to detect tagged fish by remote signal or by catching fish directly at various times of the year and at various geographic locations and water depths.

In many instances, observations by themselves cannot solve a marine biological problem. One of the most common types of observation is a **correlation**, which is an observed relationship between one factor and another. You might discover an increase of fish abundance with increasing water depth. This would be a **positive correlation** between abundance and water depth because both variables change in the same direction. On the other hand, you might discover a **negative correlation**, which in this case would mean that fish abundance decreases with increasing depth: As one variable decreases, the other increases. In either case, however, finding such a correlation does *not* prove that depth specifically is the cause of changes in fish abundance. The negative correlation might be coincidental. A decrease of fish abundance with increasing depth might be due to an increase of predator abundance with increasing depth. The next year, the number of predators might be in a different relationship (correlation) with depth. This underscores a familiar saying among scientists: "Correlation does not prove causality."

■ Experimentation is a much sharper and more powerful way of establishing cause.

Suppose that, after finding a negative correlation between fish abundance and depth, you could perform an experiment and remove all the predators that are living in deeper water. If the prey fish then spread equally to all depths, you could reasonably conclude that the presence of predators, and not water depth itself, was the cause of the negative correlation. **Experimentation** is an important tool for both laboratory and field studies. Unfortunately, many marine problems cannot be approached by experimentation; often, organisms and environments cannot be studied except by observation. This is especially true when the spatial scale is so great that it is impractical to perform experiments. Try to imagine changing the circulation of an ocean experimentally to study nutrient transfer, and you'll get the idea. It is possible, however, to formulate hypotheses that employ tests using distributional data.

■ Marine biological research involves the testing of hypotheses and may involve experimentation or sampling.

When solving problems in marine biology, additional observations beyond a certain point are not necessarily helpful. You could count all the fish in the ocean and still not know why they are abundant in some places but absent in others. To solve a scientific problem in a satisfying way, a **hypothesis** must first be stated. A hypothesis is a *statement that can be tested*.

The following are examples of hypotheses:

- Predatory snails reduce the population size of mussels on the intertidal rocks on the coast of Monterey, California.
- Increasing temperature increases the rate of oxygen consumption of crabs.

The following is not a hypothesis:

- Mermaids can never be observed, but they exist.

I hope you can see the difference easily. One can *test* a hypothesis. To test a hypothesis, it must be possible to produce an outcome that shows the hypothesis is false. One makes a prediction, which must follow from the hypothesis. We therefore formulate an **experiment**, whose outcome will be consistent or not consistent with the hypothesis. If one has hypothesized that predators control a population, it is appropriate to remove the predator population and observe whether the prey population increases, as would be expected from the hypothesis.

It is also possible that the *premises* of the hypothesis are inappropriate. Take the following hypothesis:

- Sea stars cannot attack mussels, and they therefore have no effect on mussel populations.

Because the premise of the first clause is incorrect, the hypothesis is inappropriate. All hypotheses should be internally consistent and testable, and they should be based on correct premises.

Although some hypotheses are best tested by experiments, many cannot be. Sometimes the predictions will then be stated in terms of relationships even if the relationships could, on occasion, conceivably have more than one explanation. For example, we might state the following hypothesis:

- When circulation of a very large water body deeper than 100 m has a current speed of less than 2 cm s^{-1}, the oxygen there will decrease faster than it is replenished by circulation from shallow water, and the deep-water body will lack oxygen.

We obviously cannot perform an experiment on such a deep-water body. We might then look at current speeds in all water bodies and classify on the basis of current speed those that lack oxygen. If the results of the classification are consistent with the hypothesis, we might look for any alternative hypotheses that could explain the same information. If none are obvious, then we might lean toward the correlation study as a correlation-based test of the hypothesis.

■ Hypothesis testing is most powerful when specific predictions for an experimental treatment can be contrasted with difference from a control.

The most difficult aspect of hypothesis testing is to formulate a hypothesis that lends itself to a specific program of experimentation or data collection. **Figure 1.9** shows a technique that captures the best-known way to think deductively and formulate hypotheses. All science usually derives from initial observations that arouse curiosity,

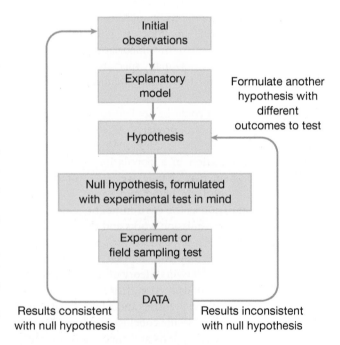

FIG. 1.9 A flowchart for the formulation and testing of hypotheses. (After Underwood and Chapman, 1995)

FIG. 1.10   A rocky shore near Bamfield, British Columbia, with abundant starfish at the base of a mussel bed. (Photograph by Jeffrey Levinton)

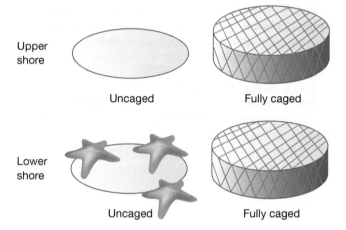

FIG. 1.11   A caging system to test whether predation affects the abundance of a rocky-shore community at different tidal levels. Results from fully caged areas, which exclude predators at high and low levels, are compared with results from uncaged areas.

such as finding that a barnacle species is abundant only on the high shore. Then an explanatory *model* is formulated, which uses general principles to attempt to explain why that barnacle would be associated with a high shore location. One might argue that predation, which occurs only on the lower parts of the shore, prevents barnacles from surviving there, but the absence of predation on the high shore allows barnacles to settle as planktonic larvae and accumulate in this upper microhabitat.

Now the crucial point arrives. One must formulate a hypothesis that is **testable**. The specific explanatory hypothesis here is that predation is more intense on the lower shore. **Figure 1.10** shows a rocky shore on the outer coast of British Columbia, and you should note the great abundance of starfish below the mussel bed. Starfish can kill and digest mussels and barnacles, but they need to move about on very delicate tube feet, which work only in a moist environment. They are therefore able to rise onto the mussel bed only as the tide rises. Hence, we might expect that predatory starfish will have only the time to seize and eat prey on the lower shore because of the sluggish movement of the starfish.

But how does one test that there is an effect of shore height on predation? We could place a wire-mesh cage over the rocks that keeps predators out but allows the rocky-shore animals kept within the cage to function normally. A cage is placed on both the lower and upper shore and compared with open-uncaged areas at the two levels. So the caged areas exclude predation, and the open areas are **controls** (**Figure 1.11**). The working hypothesis states that starfish would be able to kill mussels only on the lower shore, so we expect prey abundances to be much lower in the uncaged area on the lower shore but higher in the open area of the upper shore and also higher within cages at both lower and upper shore levels.

The **null hypothesis** in this case would state: After the cages have been in place for a set period of time, there would be no differences in mussel population density between the open and caged rocky shore regardless of shore level.

One must always remember that even an experimental result is a correlation of outcome with experimental treatment. The treatment effect, however, may have nothing to do with the hypothetical effect being studied. For example, the open low-shore area might show a decline because of full exposure to the sun, while higher-shore mussels have adjusted to this exposure. This may seem contrived, but it is consistent with the caging comparisons.

In any test of a hypothesis, one must be aware of *variation*. *Statistics* is the field that deals with the calculation of trends and differences from repeated collections of information (e.g., measuring the height of all barnacles individually in the caging experiment and calculating the mean height per treatment) and assessments of variation. The difference in barnacle abundance between treatment and control may differ, but is the difference important? Two issues must be settled. First, a test of statistical significance must be established to determine whether the average barnacle density is greater within cages than outside cages. We need an estimate of variation and therefore need replicates of each treatment. If the variation among replicates is relatively low and the magnitude of mean difference high, then the difference may be statistically significant (Sokal and Rohlf, 2011). Usually, a test is devised that can estimate the probability that the data are distributed non-randomly and to estimate the **effect size** of the factor you are studying (i.e., the presence or absence of predation). It is crucial to realize that the probability levels used in a statistical test (e.g., 0.05) are a reflection of nothing more than the probability of the data you collected and analyzed for this particular experiment, not a general truth of a scientific relationship.

Habitats and Life Habits: Some Definitions

■ **Some terms are necessary to describe life habits of marine organisms: neuston, plankton, nekton, benthos.**

It is useful to classify marine organisms by their general habitat (**Figure 1.12**). **Plankton** are organisms that live suspended in the water. They may have some locomotory power but not enough to counteract major ocean currents or turbulence. They include protists, animals, plants, and bacteria that are at most a few centimeters long. **Neuston** are organisms associated with the sea surface and include microorganisms that are bound to the surface slick of the sea. **Nekton** are usually larger animals that swim in the water column, but they can move against a current or through turbulent water. They range from small shrimp, crabs, and fish to the largest of whales. **Benthos** include animals and plants associated with the seafloor. Some animals are **infaunal**, which means they can burrow within the soft seabed, whereas others live on the seabed surface, or are **epifaunal**. Most clams are infaunal, whereas oysters and barnacles are epifaunal. Mobile organisms associated with the seabed that can swim (e.g., bottom fish) are said to be **demersal**.

Figure 1.13 gives a general classification for marine habitats based on water depth. The **intertidal zone** is the range of depths between the highest and lowest extent of the tides. In some parts of the world, there is little or no tide, and wind mainly determines the vertical range of this fringing environment (see Chapters 2 and 17). The **subtidal zone** is the entire remainder of the seabed from the low-water tidemark to the greatest depth of the ocean. **Continental shelf** (or neritic) habitats include all seafloor and open-water habitats between the high-water mark and the edge of the continental shelf. Seaward of the shelf is a series of oceanic or pelagic habitats: the **epipelagic zone** includes the upper 200 m of water, the **mesopelagic zone** ranges from 200 to 1,000 m depth, the **bathypelagic zone** ranges from 1,000 to 4,000 m depth, and the **abyssopelagic zone** ranges from 4,000 to 6,000 m depth; **bathyal** benthic bottoms range from 1,000 to 4,000 m depth, and **abyssobenthic** bottoms range from 4,000 to 6,000 m depth. **Hadal** environments include those of the seabed and the waters at the bottoms of the trenches, often far deeper than 6,000 m depth. For example, the Marianas Trench reaches about 11,000 m depth.

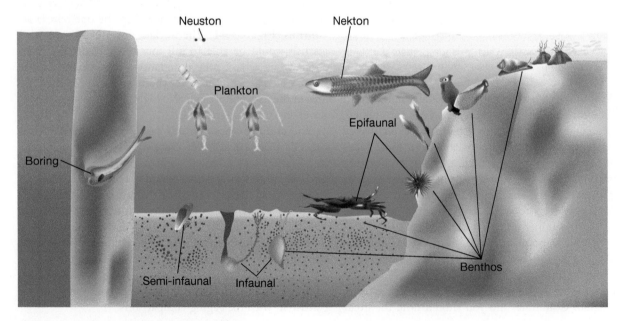

FIG. 1.12 General habitats of marine organisms.

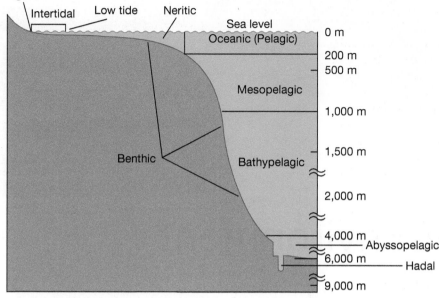

FIG. 1.13 A cross section of the ocean from the shoreline to the deep sea, showing the location of major marine habitats.

■ CHAPTER SUMMARY

- Marine biology combines functional biology, ecology, and the study of biodiversity.

- Marine biology began with simple observations of the distribution and variety of marine life. In the nineteenth century, marine biology developed into a science involving

hypothesis testing. The voyage of the H.M.S. *Challenger* gave us the first global view of marine biology. The twentieth century brought major research institutions, faster ships, better navigation, and greatly improved diving technology. Technology in both the laboratory and the open sea has

played an increasingly important role thanks to precise navigation, mapping of the seabed, and the development of submarine vehicles.

- Marine biologists use the scientific method—or systematic reasoning, observation, and experiment—to frame and test hypotheses.

■ REVIEW QUESTIONS

1. What was the azoic theory, and why could it be considered a testable hypothesis?

2. What might be the difference in potential contributions to marine biology by research done on the great oceanographic expeditions as opposed to research done at zoological stations on the coastline?

3. What was *Bathybius*, and why was its supposed existence

of importance to the basic understanding of biology?

4. Why was the use of submarines so important in the development of marine science? Why was the use of scuba important in this development?

5. Distinguish between correlation and experimentation in the understanding of scientific relationships.

6. Devise a testable hypothesis about something in the room in which you are located now. How would you test this hypothesis?

7. Explain why the following is a poor hypothesis: Because whales are very small, they must be vulnerable to predation by snails.

Visit the companion website for *Marine Biology* at www.oup.com/us/levinton where you can find Cited References (under Student Resources/Cited References), Key Concepts, Marine Biology Explorations, and the Marine Biology Web Page with many additional resources.

CHAPTER 2

The Oceanic Environment

The Open Oceans and Marginal Seas

■ **The world's oceans can be divided into open oceans and marginal seas.**

The Ocean covers about 71 percent of the earth's surface, and the Southern Hemisphere is more dominated by ocean than the Northern Hemisphere. The **Pacific Ocean** is the largest ocean and is less affected than the Atlantic by regional differences in climate or by river input. Island chains are most numerous in the Pacific, and volcanic activity is pronounced around its margins. Think of Mount St. Helens in Washington, Mount Fuji in Japan, and Krakatoa in the southwest Pacific and you'll soon understand why the rim of the Pacific is often called a **ring of fire**. The **Atlantic Ocean** is relatively narrow and is bordered by large marginal seas (e.g., Gulf of Mexico, Mediterranean Sea, Baltic Sea, and North Sea). It drains many of the world's largest rivers (e.g., the Mississippi, the Amazon, the Nile, and the Congo). The Atlantic Ocean is affected to a larger degree by terrestrial climate and river-borne inputs of dissolved and particulate substances. The **Antarctic Ocean** (also known as the Southern Ocean) has a solely water border with other oceans. Its northern boundary is the Subtropical Convergence, where colder, more saline water descends northward. A general pattern of west winds in the range of 40°–60° S generates a surface current circling eastward around Antarctica. This and other important surface currents are illustrated in **Figure 2.1**.

Most **marginal seas** have unique oceanographic characteristics owing to their restricted connections with the open ocean and their usually shallow water depths. The restrictions allow local climate to strongly influence the marginal sea. For example, a shallow-water barrier, or **sill**, restricts exchange between the Mediterranean and the Atlantic Ocean, and a local excess of evaporation relative to precipitation increases the salinity within the Mediterranean. Five to six million years ago, a global lowering of sea level severed the connection with the Atlantic, and extensive evaporation led to the formation of large-scale salt deposits. Most other marginal seas have had histories of strongly changing conditions.

Topography and Structure of the Ocean Floor

■ **The oceans share four main topographic features: the continental shelf, continental slope, the deep-sea floor, and oceanic ridge systems.**

The **continental shelf**, a low-sloping (1:500, about 1°) platform, extends from the shoreline to roughly 10 km to over 300 km out to sea (see **Figure 2.2**). Seaward of the **shelf-slope break** (usually a depth of 100–200 m), the **continental slope** increases in grade to about 1:20 (about 2.9°). The continental slope is usually dissected by **submarine canyons** that act as channels for downslope transport of sediment. The foot of the slope merges with the more gently sloping **continental rise**, which descends 2–4 km to the **abyssal plain**, which averages 4,000 m depth. In some parts of the ocean (e.g., the west coast of South America), **trenches** in the seafloor occur just seaward of the base of the continental slope and may be more than 10,000 m deep. The trenches are long and narrow in map view and may run parallel to the shoreline. Isolated

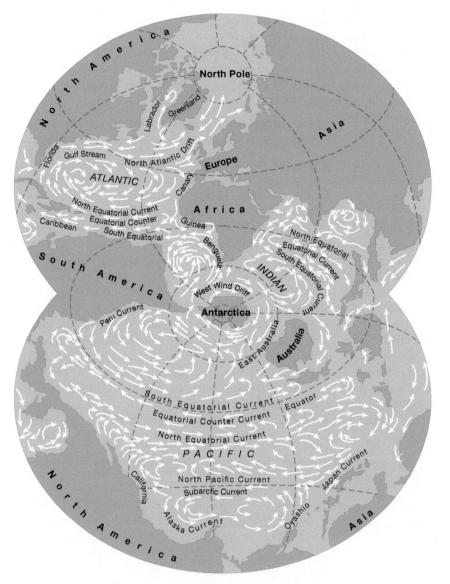

FIG. 2.1 Surface currents of the world's oceans. (From *The Circulation of the Oceans* by Walter Munk. All rights reserved.)

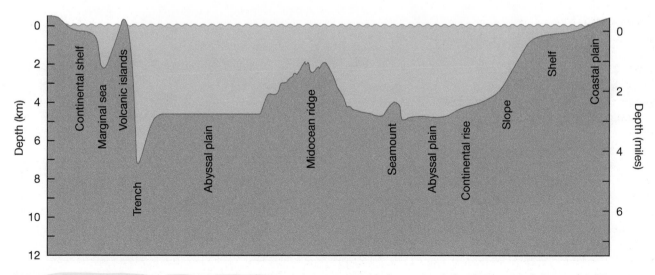
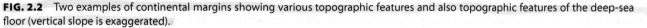

FIG. 2.2 Two examples of continental margins showing various topographic features and also topographic features of the deep-sea floor (vertical slope is exaggerated).

midoceanic islands may rise from the deep-sea floor to the surface (e.g., the Hawaiian Islands). In contrast, **continental islands** are located on the continental shelf.

The **oceanic ridges are linear features that** rise 2,000–4,000 m from the ocean floor and may reach the sea surface, forming emergent islands (e.g., Iceland on the Mid-Atlantic Ridge). The longest of the ridge systems runs the length of the middle of the North and South Atlantic Oceans, snakes around the southern tip of Africa, and runs northeastward into the Indian Ocean (**Figure 2.3**). Ridges are volcanic in origin and are often cut by **transverse faults**, or breaks in the earth's crust. Rift valleys, at a ridge system's center, are parallel to the line of the ridge.

The deep-ocean basin floor consists of volcanic rock blanketed by soft **deep-sea sediment**. The sediment contains varying combinations of mineralized plankton skeletons, clay, and other minerals deposited from continental sources, volcanic rocks, and precipitates, such as manganese nodules, which are scattered on the seabed in certain areas. The skeletons of many organisms are composed of calcium carbonate, the material that makes up chalk. Many open-ocean sediments are formed from a drizzle of calcium carbonate skeletons originating near the surface and make up, for example, **foraminiferan ooze** (made of protistans known as Foraminifera). Calcium carbonate dissolves at great depth, however, and deep-ocean sediments become progressively dominated by clays and, at certain sites, a **radiolarian ooze** of silica skeletons (made of protistans known as Radiolaria). Beneath the soft-sediment covering and throughout the world lies the **oceanic crust**, made of sedimentary rocks and an underlying layer of volcanic rock.

■ The earth's crust is in continual motion, and the main topographic features of the ocean reflect this motion.

The ocean floor can be divided into large sections, called **plates**, whose boundaries are ridges, trenches, or large transverse faults. Many geologists believed that some large-scale process caused the overall features of the oceanic crust.

In the 1960s, scientists discovered that several features of rocks and sediments occurred in bands that were parallel to the oceanic ridges. As rocks crystallize, iron-rich magnetic minerals align to the earth's magnetic field at the time of formation. Therefore, minerals that are forming now or in the very recent past have the magnetic polarity of the present-day earth. Nearest a given ridge, the magnetic polarity indicated by the oceanic crust matches that of earth's present magnetic field. In the next band on either side of the ridge, however, the magnetic polarity is reversed. The next bands are reversed again, and so on (**Figure 2.4**). The pattern exists for two important reasons. First, volcanic oceanic crustal rock is forming at the ridges and the rock

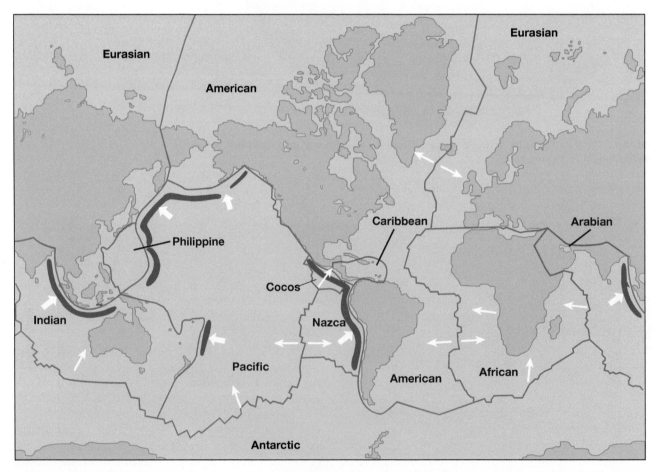

FIG. 2.3 The ocean floor showing plate boundaries, oceanic ridges, where new oceanic crust is created by volcanism (red lines with thin arrows), fault and fracture zones (red lines without thin arrows), and trench zones (thick, dark blue bands).

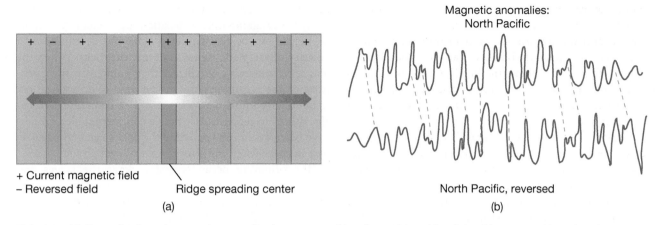

+ Current magnetic field
− Reversed field

Ridge spreading center

(a)

Magnetic anomalies:
North Pacific

North Pacific, reversed

(b)

FIG. 2.4   (a) Generalization of magnetic anomalies into a map of bands on either side of the ridge center. Note that the current magnetic field of the earth (recorded at spreading center) is positive. (b) Trace of magnetic anomalies across a spreading ridge center with transect reversed to show match of anomalies on either side of spreading center.

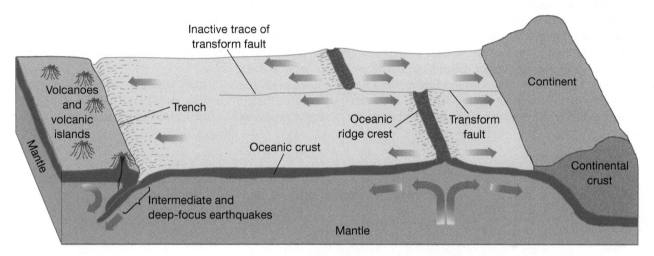

FIG. 2.5   Schematic diagram of the oceanic crust showing the formation of crust at ridges and the downward transport and destruction of crust below trenches.

is moving away from the ridges. But the earth's magnetic field has reversed periodically over time, with the south magnetic pole flipping to north or the reverse. For example, before 780,000 years ago, the earth's magnetic field was reversed relative to the present day. Thus, rocks formed at that time have magnetic minerals that crystallized with a magnetic orientation that is the reverse of minerals that are forming today or since 780,000 years ago.

The mirror image of bands of rocks of alternating magnetic polarity on either side of a ridge was the key to understanding a major process called **seafloor spreading** (see Cox and Hart, 1986). The oceanic crust, formed by volcanism at the ridge, was subsequently **displaced from the ridge in both directions**. This could happen only if new crust was formed through volcanic activity at the oceanic ridges and carried horizontally, as if on a conveyor belt, away from the ridges. Because the bands could be dated by means of radioactive minerals and the width of the bands was known, it was possible to calculate a likely range of speed of seafloor spreading: 2–25 cm y^{-1}. It was later discovered that crustal material is dragged downward at trenches and melted into the upper mantle, the layer of the earth beneath the

crust (**Figure 2.5**). These sites of downward crustal motion are known as **subduction zones**. The great depth of the trenches reflects the downward dragging of the crust.

The mechanism of seafloor spreading is a convective process occurring within the mantle. Heat is generated within the mantle, probably mainly by radioactive decay. The heat melts rock in the upper mantle, and volcanism in the midoceanic ridges is the result. Earthquakes are a direct cause of this movement, which often involves moving plates locked against each other for a time, or jerky lockup and movement along giant faults in the earth's rock crust, with catastrophic release. Many of the world's most dangerous earthquake zones are at boundaries such as the San Andreas fault in California, whose movement has caused many serious earthquakes. Earthquakes are also generated in areas of subduction, which nearly ring the Pacific Ocean. As an example, a very strong subduction zone exists in a region of the west coast of North America from northern California to southern British Columbia, Canada. Very strong evidence supports a history of very strong earthquakes about every 250 years in the Pacific Northwest of the United States. The last known earthquake that occurred in this northwest subduction zone, called the **Cascadia subduction zone**, in the year 1700, sent a **tidal wave**, or **tsunami**, toward the coast of the United States probably in 15 minutes, but also sent an enormous wave in the opposite direction across the Pacific Ocean to Japan in about 10 hours. Historical records of this tsunami allowed the accurate dating of the powerful earthquake in the Pacific Northwest.

Seafloor spreading explains many of the earth's topographic features. The continents consist of relatively low-density crustal material lying above a denser oceanic crustal layer. Continents move along with the spreading crust. The continental shelf is merely an extension of the continent below water. If you consider time spans of millions of years, the rate of seafloor spreading is significant. For example, 100 million years before the present day, the Atlantic Ocean was probably less than 2,000 km (1,000 miles) wide, and South America and Africa were fairly close together. If the continental shelves are included, those two continents are found to fit quite well together, as pieces of a giant jigsaw puzzle (**Figure 2.6**). These two landmasses were probably once part of a supercontinent that divided, eventually producing today's continents. Seafloor spreading explains such **continental drift**. Throughout geological time, continents have had radically different arrangements, and even the

major surface current systems have differed strongly from today's arrangement. The position of ancient continental blocks can be reconstructed using the orientations of polarity of magnetic minerals of ancient rocks. When the magnetic minerals crystallized, they pointed in the directions of the magnetic poles at the time of crystallization. It is therefore possible with the aid of computer-assisted algorithms to reconstruct where a continental block might have been, assuming you know the rock's age and orientation in space when the mineral crystallized.

The origin of ocean crustal rock at the ridges and destruction of the trenches, combined with seafloor spreading, creates a large-scale seascape that strongly affects marine organisms. For example, the east coast of North America has been steadily moving westward for tens of millions of years, leaving a coastline that is generally a location of sediment formation and soft shores. By contrast, the west coast of North America is along a major fault line, the San Andreas fault (see **Figure 2.3**), and geological activity and volcanism have produced a very long coastline dominated by rocky shores as opposed to the sandy and muddy shores of the east coast.

Sea level has also been strongly affected throughout geological time by seafloor spreading because mantle convective processes have vertical as well as horizontal motion.

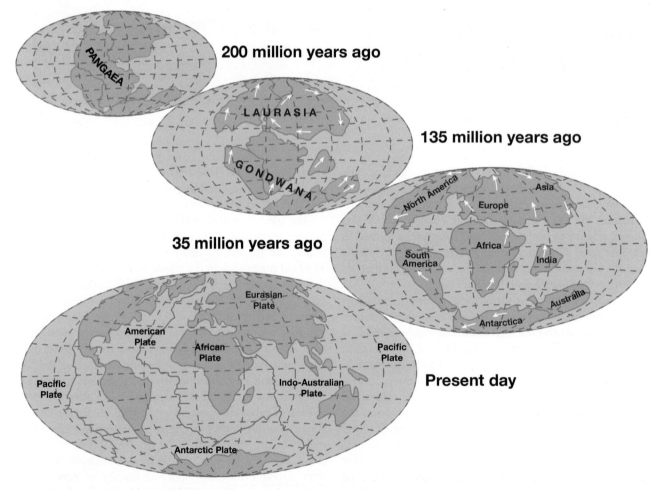

FIG. 2.6 Reconstruction of the relative positions of the continents over the past few hundred million years. Note the steady opening of the Atlantic Ocean over time toward the present.

Large parts of the seafloor flanking midoceanic ridges moved downward during the process of seafloor spreading. Using a wide range of magnetic and geological evidence, it is possible to locate ancient seafloor spreading centers and the vertical range that oceanic crust has been drawn down as it spreads from ridge centers. Seafloor spreading can lead to long-term changes in sea level in various parts of the ocean. For example, because of North America's long-term passage over a sinking plate, sea level has steadily declined from a high in the Cretaceous period, 65 million years ago, to an estimate of 170 m lower at present (see Muller et al., 2008). This long-term decline is comparable in magnitude to the fluctuations of sea level of about 120 m during the most recent Pleistocene glacial age because of expansions and retreats of continental glaciers.

The Ocean Above the Seabed

Seawater

■ **Many of the ocean's major features are due to the unique set of chemical and physical properties of water, including its dissolving power, high specific heat, transparency, and heat of vaporization.**

GENERAL PROPERTIES OF SEAWATER The formula for water, H_2O, indicates the component atoms of hydrogen and oxygen. Water molecules are **asymmetric in charge**, with a net positive charge on the hydrogen sides of the molecule and a net negative charge on the oxygen side (**Figure 2.7**). This overall polarity causes fairly strong attractions, called **hydrogen bonds**, between molecules. These attractions allow water to be liquid at atmospheric pressure and typical temperatures on the earth's surface. Charge asymmetry also enhances the ability of water molecules to combine with ions, or particles. As a result, water is an effective and versatile **solvent**, and the ocean thus contains a diversity of dissolved substances.

Water, including typical open-ocean seawater, also has a very **high specific heat**; it takes 1 calorie to raise the temperature of a gram of pure water by 1°C (at 15°C). Relative to most other liquids (and solids, e.g., soils), it takes a great deal of heat to change water temperature. Seawater can store large amounts of heat, and moving currents can transport large amounts of heat along with the water itself. Water has a high **heat of vaporization,**[1] and large amounts of heat are absorbed when water vapor is formed by evaporation. As a result, the ocean has a very strong effect, through the atmosphere, on both oceanic and adjacent continental climates. It takes a great deal of heat to change the temperature of the ocean. The ocean thus tends to smooth out temperature variation, an effect that is especially noticeable on coastlines that receive oceanic breezes.

Temperature and Salinity

■ **Seawater temperature is regulated primarily by solar energy input and the mixing of other water.**

Seawater temperature is measured by means of mercury thermometers, thermistors, and satellite detectors. **Thermometers** have been used since their invention in the eighteenth century. Since the 1980s, semiconductor devices called **thermistors** have been used to directly record temperature and have an accuracy of 0.001°C; these are often installed on ocean buoys and can either transmit data by radio or record data that can be retrieved periodically. Finally, **satellite radiometers** are sensitive to infrared radiation and can measure sea-surface temperature (SST) throughout the world ocean.

At low latitudes, there is a net capture by the earth of heat from solar energy, but at higher latitudes, the earth tends to lose heat. As a result, there is a latitudinal temperature gradient of surface seawater (**Figure 2.8**). Surface transport and deep transport of heat energy by means of water currents are extensive, however, and the high heat capacity of water permits large transfers of heat among latitudes (see the following). Nevertheless, polar surface water temperatures hover at about 0°C year-round, whereas tropical surface water temperatures are almost constantly above 25°C. Local restriction of circulation causes further increases of temperature, sometimes approaching 35°C and higher in restricted lagoons and pools.

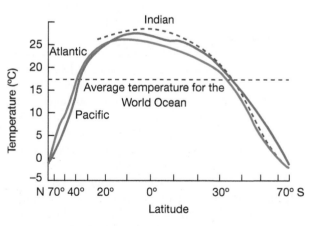

FIG. 2.8 Latitudinal variation in sea-surface temperatures in the Atlantic, Pacific, and Indian Oceans. (After Anikouchine and Sternberg, 1973)

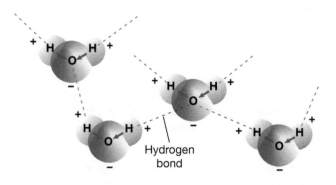

FIG. 2.7 The water molecule. A net negative charge on one side and a net positive on the other side are the basis of water's ability to form bonds with ions, making water an excellent solvent.

[1] The amount of heat required to convert a unit mass of a liquid at its boiling point into vapor without an increase in temperature.

Seasonal temperature changes are maximal in the mid-latitudes. Although temperature in high latitudes is low, the seasonal variation is minimal. Warm and seasonally constant temperatures are typical of the tropics. Seasonal change in the ocean is far more pronounced in the North Atlantic than in other oceans owing to the strong influence of continentally derived weather systems. In coastal North Carolina, temperature ranges over the year from roughly 3° to 30°C. By contrast, the ocean-moderated Pacific climate has only an annual range of less than 5°C along the coast of California, and summer maximum temperatures are far lower than at comparable latitudes in the Atlantic. Bodies of water with restricted circulation and shallow depth, such as the Mediterranean, have greater annual temperature ranges than does the adjacent open ocean.

■ Salinity is a measure of the dissolved inorganic solids in seawater.

The **salinity** of seawater is the number of grams of dissolved inorganic solids per thousand grams of seawater; it is expressed as parts per thousand (ppt). Many of the elements in seawater (e.g., sodium, chlorine, and strontium) are found in nearly constant ratios throughout the world ocean even if the total salinity varies from place to place. The constancy becomes more pronounced with the increasing **residence time** of most elements in seawater. Residence time is the average time that a unit weight of a substance spends in the ocean before it is lost to sediments or to the continents. In the ocean, the elements found to be in constant proportions have a residence time on the order of 1 million years or more, as opposed to the mixing time of seawater, which is on the order of thousands of years. The relatively short mixing time homogenizes the various elements relative to each other throughout the ocean relative to the very long time they spend in the ocean before leaving. Other elements such as nitrogen have very short residence times, which are rapidly taken up and released by biological processes and vary greatly in proportion from place to place.

Seawater is a complex solution. The solutes dissolved in it consist of both dissolved inorganic and organic matter, including dissolved gases. In addition, particles are suspended in seawater. The particles may be inorganic mineral grains, aggregates of organic particles, or living plankton. Dissolved inorganic matter enters the ocean mainly through river flow but also through atmospheric precipitation. The **major elements** are defined as those present in concentrations greater than 100 parts per million (ppm). These include chlorine, sodium, magnesium, sulfur, calcium, and potassium. By far, the dominant elements are chlorine and sodium, and indeed, when seawater evaporates, the salt sodium chloride is a very prominent residue. **Minor elements** are present at concentrations between 1 and 100 ppm. These include bromine, carbon, strontium, boron, silicon, and fluorine. **Trace elements**, those present at less than 1 ppm, include nitrogen, phosphorus, and iron. A large number of elements occur in minute quantities, in concentrations of parts per billion (ppb).

The ratio of chlorine to other elements in seawater is essentially constant despite overall changes of salinity. This fact enables you to estimate total salinity by measuring the chlorine content of seawater. **Chlorinity** is the total concentration of chloride per thousand milliliters of seawater (i.e., expressed in parts per thousand). Salinity is approximately 1.81 times the chlorinity. A simple chemical titration of chloride ions from seawater can therefore be used to estimate total salinity. Because a saline solution conducts electricity, **conductivity** (the ability to carry an electric current) can also estimate salinity, but corrections must be made for temperature. Although it is accurate only to about 2 ppt, a simple optical refractometer can also measure salinity because increased salt content increases light refraction, the degree to which water bends light.

In most oceanographic applications, you will find that salinity is not measured in terms of dissolved inorganic solids. Although the ratios of major elements with long residence times are similar throughout the ocean, the slight differences make measures based on total dissolved solids inconsistent. Oceanographers therefore generally express salinity using the **practical salinity unit (psu)**, which is a measure of the electrical conductivity of the seawater relative to a standard potassium chloride solution at one atmosphere pressure and 15°C temperature. Practically speaking, the numerical units between psu and solute-based salinity measures are nearly identical. Today, most marine biologists measure salinity by means of a relatively inexpensive device that measures conductivity. In terms of practical salinity units, salinity is expressed with no qualifier. Therefore, a salinity of 30 on the Practical Salinity Scale is described as 30.

In the open ocean, salinity ranges from roughly 32 to 37. Locally, salinity is regulated by a balance of dilution (river input, rainfall, and underground springs) and concentration (evaporation and sea-ice formation) processes. Latitudinal variation in the balance of precipitation and evaporation causes slight salinity maxima at 30° N and S latitude, with a minimum at the equator and declining salinities at higher latitudes (**Figure 2.9**). Most marginal seas differ from the adjacent open ocean owing to their restricted circulation and local factors such as excesses in rainfall, river input, or evaporation. The Baltic Sea, located in northern Europe between Sweden, Finland, and continental Europe, has an excess of river runoff and restricted circulation with the Atlantic (via the North Sea). As a result, the salinity of much of the inner Baltic is 5 or less. By contrast, the Mediterranean has an excess of evaporation relative to precipitation and runoff, and its salinity is usually greater than 36, which influences the adjacent North Atlantic Ocean.

Oxygen in the Sea

■ Oxygen is added to seawater by mixing with the atmosphere and by photosynthesis; it is lost by respiration and by chemical oxidation of various compounds.

Oxygen from the atmosphere dissolves in seawater at the sea surface. When the wind is strong, larger amounts of atmospheric oxygen are mixed into the surface waters, which may then be mixed with deeper waters. The amount of

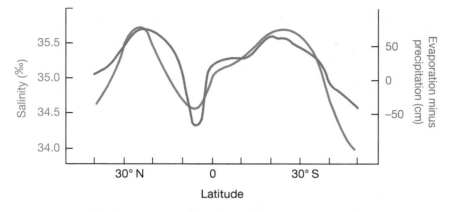

FIG. 2.9 Latitudinal variation in surface salinity of the open oceans (blue curve). Balance of evaporation and precipitation is also shown (green curve). (After Sverdrup et al., 1942)

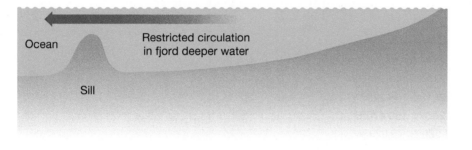

FIG. 2.10 The circulation between a fjord and the open sea. The lack of exchange of the fjord's bottom waters, combined with respiration, keeps the water low in oxygen, or anoxic.

oxygen that can be dissolved into seawater decreases with increasing temperature and decreases slightly with increasing salinity.

The balance of organism respiration and photosynthesis also affects the amount of dissolved oxygen in seawater. Photosynthesis is the light-driven process that plants use to manufacture carbohydrates; oxygen is a by-product of the process. During respiration, the opposite process occurs: Carbohydrates are oxidized to provide energy for the organism, and oxygen is consumed. Respiration losses can usually be accounted for in terms of bacterial decomposition of organic matter and by animal and plant respiration. An idealized equation for respiration is

$$C_6H_{12}O_6 + 6O_2 \rightarrow 6CO_2 + 6H_2O + energy$$

In photosynthesis, this reaction goes to the left, although the biochemical details are very different (see Going Deeper 11.1). Chemical oxidation of organic matter also may reduce dissolved oxygen in seawater.

■ Restriction of vertical and horizontal water circulation often tips the balance toward low oxygen concentration.

In some major water bodies, the supply of organic matter and its role in oxygen depletion outstrip the role played by the mixing of bottom waters of the basin with other oxygenated waters. For example, the bottom waters of many Norwegian fjords have restricted circulation with the open

sea owing to the presence of a **sill**, or shallowing, at the entrance, which permits only surface water to leave the fjord (**Figure 2.10**). Decomposition of organic matter tends to consume the oxygen in the stagnant bottom waters.

The vertical structure of seawater, with low-density seawater above and higher-density seawater below, may stabilize the water column and strongly influence local oxygen concentration. For example, particulate organic matter may accumulate in the **thermocline**, a zone of great increase of water density with depth, resulting in bacterial decomposition of the organic matter and consumption of oxygen. This results in the local absence of oxygen or **oxygen minimum layers** at depths of 200–1,200 m, which are especially common in the eastern North Pacific Ocean off the coast of California (**Figure 2.11**) but are also found in the tropics of both the Atlantic and the Pacific and throughout the eastern Pacific. In estuaries, organic matter may accumulate in deeper quiet waters, resulting in lowered oxygen and sometimes mass mortality of fish and bottom organisms (see the discussion of hypoxia in Chapter 22).

Light

■ Environmental light originates mainly from the sun and is therefore strongest in surface waters.

Light energy that reaches the earth emanates mainly from the sun, and therefore, most light enters the ocean from above at the surface. Light energy ranges from shorter-wavelength

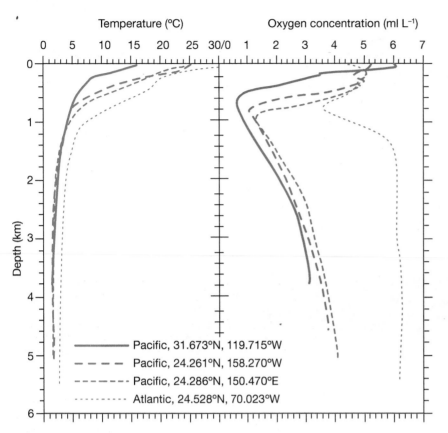

FIG. 2.11 Oxygen minimum layers: depth profiles of dissolved oxygen as a function of depth. Note the strong presence of oxygen minimum layers in the eastern Pacific, off the coast of California, compared with their absence in the North Atlantic. (From Childress and Seibel, 1998, reproduced with permission of the Company of Biologists)

ultraviolet light to light within the visible spectrum to light in the infrared and longer-wavelength spectrum. At high angles near midday, very little sunlight is reflected from the sea surface. Much of the light is either absorbed by seawater and particles or scattered by particles in the water. Light intensity declines exponentially with increasing depth, as shown in **Figure 2.12**. The ultraviolet and infrared parts of the spectrum are very strongly attenuated with depth, so only the visible part of the spectrum reaches any appreciable depth. Blue light penetrates better than red light.

Because it is the energy source for photosynthesis, light is crucial for life. Most marine animals also depend on light, so they can move about, detect prey, and spot predators. Light intensity near the surface can be sufficient to depress biological activity through the deactivation of protein and DNA. Ultraviolet light is damaging, especially in the clear waters of the tropics, where significant amounts will penetrate to depths of 10 m. Some intertidal plants use calcium carbonate to reduce damage by absorbing the ultraviolet light. Some corals have pigments that absorb ultraviolet light. In more turbid waters, ultraviolet light is no real problem, and well over 90 percent of it is absorbed in the top meter of water.

The effect of sunlight diminishes greatly with depth. In turbid shelf waters, light is strongly limiting to photosynthesis

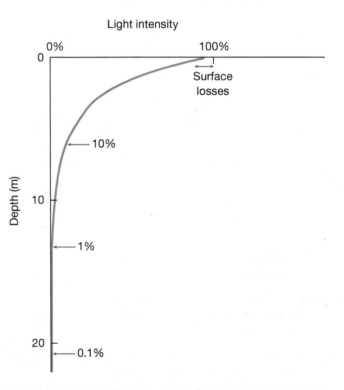

FIG. 2.12 An example of the pattern of exponential decline in light intensity with increasing depth in a coastal marine water column.

even at depths of 10 m. At depths of 30–50 m, visibility is greatly reduced. This is especially true for color discrimination, and animals living in deep water do not have especially good color vision. In the open sea, at depths much greater than 1,000 m, there is no detectable sunlight. The animals in the deep oceanic world see only a faint glimmer from above, if anything at all.

Circulation in the Open Sea: Patterns and Causes

Surface Circulation

■ **Surface oceanic currents are controlled by the interaction of the planetary wind system and the earth's rotation.**

Two factors interact to produce oceanic surface currents: the planetary wind system and the earth's rotation. More sunlight is captured near the equator than near the poles, and the heated air tends to rise and move to higher latitudes and then sink, with more such vertical cells successively created at higher latitudes. This overall air circulation system drives the **planetary winds**, which drag over water and generate surface currents. The earth rotates from west to east, which deflects the air as it rises and sinks. This modulates the direction of the winds and surface currents. Let's consider the factors that interact to produce surface water current circulation patterns.

The Coriolis Effect and Surface Circulation Patterns

■ **The earth's rotation causes a deflection in surface current direction called the Coriolis effect, which affects water flow on many geographic scales.**

The earth rotates once a day on its axis. A particle attached to the earth's surface at the equator must travel from west to east more rapidly than one attached near the pole because the particle at the equator must traverse a longer distance in each day's rotation. If a parcel of water not attached to the earth moves toward the north in the Northern Hemisphere, it is moving to a location of lower eastward velocity. Because the northward-moving parcel of water has greater eastward momentum than the water into which it is moving, it will have a relative deflection toward the east—that deflection is toward the right. If the parcel is moving south, it moves from an area of lower eastward velocity into a region of increased eastward velocity. Relative to the local regime, the water will lag behind and will deflect toward the west—once again, that deflection is toward the right. The effect of the earth's rotation on moving air and water is known as the **Coriolis effect**. It causes a deflection to the right for water traveling in the Northern Hemisphere (**Figure 2.13**). It causes a deflection to the left for water traveling in the Southern Hemisphere (you should explain to yourself why this is the opposite of the Northern Hemisphere).

The Coriolis deflection works in the same direction (rightward in the Northern Hemisphere, leftward in the

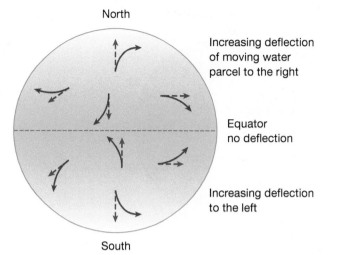

FIG. 2.13 The Coriolis effect deflects moving objects, such as the water parcels in this diagram. The dashed arrows show the paths the parcels would take if they moved undeflected in their original direction of motion. The solid arrows show their actual deflected paths. No deflection occurs at the equator, and deflection increases as the distance of the moving water parcel from the equator increases.

Southern Hemisphere) if you are traveling east or west. Let us consider the Northern Hemisphere. If you are traveling toward the east, you will be moving in the same direction as the eastward rotation of the earth. The centrifugal force tends to make you fly away perpendicular to the earth's axis, but there is a strong component of motion directed toward the equator, which makes the object turn to the right. As an object is moved toward the west, its motion is moving westward perpendicular to the earth's axis, but its tangential velocity is decreased because the earth beneath is moving in the opposite direction. Gravity tends to make the particle move to the pole, or to the north in the Northern Hemisphere. So rightward movement also occurs in the Northern Hemisphere for eastern and western motion, and the Coriolis movement is again toward the left for eastern or western motion in the Southern Hemisphere.

■ **The major oceanic surface currents are determined by the planetary wind systems modified by the Coriolis effect.**

The most conspicuous feature of the large-scale ocean surface is the presence of **surface currents**, which are driven by the **planetary wind system** (**Figure 2.14**). The winds are caused by the rise of air heated by the sun and the sinking of the air as it cools. These movements are influenced by the earth's rotation. In the latitudinal band of 30°–60° N and S, the **westerlies** move air toward the northeast (Northern Hemisphere) or the southeast (Southern Hemisphere). Between the equator and 30° N or between the equator and 30° S latitude, the **trade winds** blow toward the southwest (Northern Hemisphere) or the northwest (Southern Hemisphere). These two wind systems (westerlies and trade winds), together with the direct action of the Coriolis effect, move tremendous volumes of surface water in large circular patterns known as

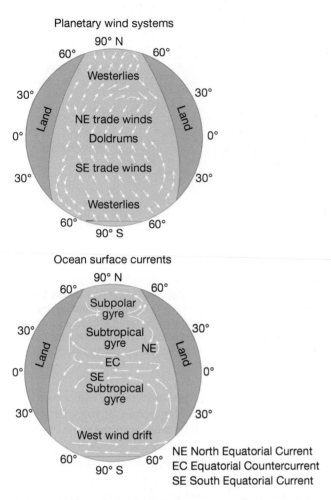

FIG. 2.14 The relationship of the ocean surface currents to the planetary wind system. (After Fleming, 1957, courtesy of the Geological Society of America)

gyres centered around 30° latitude N or S. These gyres move clockwise in the Northern Hemisphere and counterclockwise in the Southern Hemisphere. In the North Atlantic, the westerlies push water toward the northeast, but the Coriolis effect results in eastward flow. Farther south, the trade winds drive water toward the southwest, but the Coriolis effect produces westward flow. The constraining shapes of the continents in the North Atlantic contribute to the clockwise flow of the North Atlantic gyre. In summer, persistent winds tend to reinforce these patterns. On the east coast, the clockwise flow brings warm water from the south to the New England region. In contrast, the clockwise Pacific flow brings cold water from higher latitudes to the coast of central California.

The Coriolis effect deflects water moving under the dragging force of the wind. The wind causes the water to move in surface sheets. Because of friction, these sheets drag on layers beneath. In the Northern Hemisphere, the Coriolis effect causes each successive sheet to deflect to the right relative to the sheet above. Theoretically, this should produce a spiral circulation with increasing depth. In practice, the net direction of surface water is approximately 90° to the right of the wind in the Northern Hemisphere and 90° to the left of the wind in the Southern Hemisphere. This motion is known as **Ekman transport**.

■ **Western ocean boundary currents are caused by winds combined with the earth's rotation and result in large-scale transport of surface water from the tropics to higher latitudes.**

The **oceanic boundary currents** on the western sides of the major Atlantic and Pacific gyres are famous for their speed and ability to transport large amounts of water. These currents are well defined with strong thermal boundaries between the current and the surrounding oceanic water. The **Gulf Stream**, for example, originates from within the Gulf of Mexico, flows northward along the eastern coast of the United States, and moves eastward in a more diffuse current across the Atlantic toward Ireland and Great Britain (**Figure 2.15**). Off the coast of the southeastern United States, current speeds can be as high as 2 m s^{-1}, down to depths of over 450 m. More than 50 million m^3 s^{-1} of warm water may be transported northward along the North American Atlantic coast. Large amounts of heat are stored in the current at its source, and this heat is transported northward, veers to the northeast, and is eventually lost when the waters reach the mid-North Atlantic, when the waters are now cold and sink (see the next section, Seawater Density and Vertical Ocean Circulation). Westerly winds blowing across North Atlantic waters help ameliorate the climate of the British Isles, which would otherwise be colder. As a result, subtropical vegetation grows well in Ireland and southern Scotland. A similar situation exists in the western North Pacific Ocean. The Kuroshio Current brings warm water toward the north that then moves eastward and loses its heat in the North Pacific. The climate in the Pacific Northwest of the United States is strongly affected by westerly winds blowing from the Pacific toward the west coast of Oregon, Washington, and British Columbia.

■ **Eddies of meandering parts of boundary currents may form cold-core and warm-core rings.**

Western ocean boundary currents are often very well defined, but at some point downstream, they become sluggish and may meander. This is especially true in summer when there is less difference between the temperature of the oceanic boundary current and the usually surrounding colder water. Occasional meanders of the Gulf Stream form loops, which enclose pockets of cold water; these are known as **cold-core rings**. They are usually formed between the Gulf Stream and the continental slope area mainly off the coast of Nova Scotia. They then tend to travel southward, only to dissipate a few months later. These rings often have unique assemblages of plankton and fish. Occasionally, a meander of the Gulf Stream will enclose a still-warmer summer blob of water on the northeastern U.S. continental shelf and drift on the shelf toward the shore forming **warm-core rings**. Commonly, large numbers of tropical fish, such as butterflies, can be found in coastal waters of Long Island, New York. These fish originated in Florida waters, drifted northward in the Gulf Stream, and then drifted shoreward, usually in a warm-core ring.

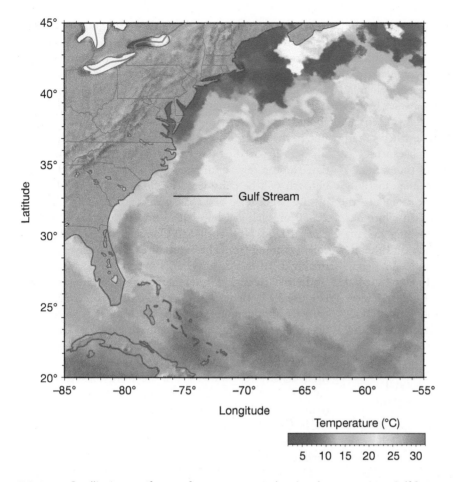

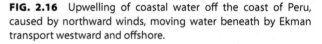

FIG. 2.15   Satellite image of sea-surface temperature showing the warm-water Gulf Stream moving northward in the western North Atlantic Ocean. Geographic variation in temperature is calculated from a 6-day average of temperatures in March 2006. (Courtesy of NOAA)

■ Winds and the Coriolis effect combine to cause upwelling, which brings nutrient-rich deeper water to the surface.

The Coriolis effect strongly influences water movement on the continental shelves of the east sides of oceans. In the Northern Hemisphere (e.g., the coast of California), a wind blowing directly from the north will deflect surface waters to the right, or to the west, owing to Ekman transport. In the Southern Hemisphere, a wind from the south (e.g., off the coast of Peru) causes a leftward deflection of surface water, also to the west. The upward movement of deeper waters near the coast compensates for these offshore surface water movements. This phenomenon, known as **upwelling**, brings nutrient-rich deeper waters to the sunlit surface and fuels large-scale phytoplankton blooms, which in turn support major fisheries such as the anchovy fishery off Peru (**Figure 2.16**). Periodic shutdowns in this upwelling system occur causing major climate oscillations in the region around Peru, but also worldwide. We discuss these oscillations, known as ENSO, in Chapter 3.

■ Extraordinarily strong surface water motion is generated by storms and earthquakes.

Cyclonic storms are rather common in both the Atlantic and the Pacific and may last for several weeks. The overall storm

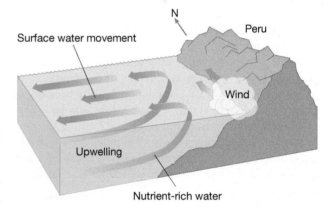

FIG. 2.16   Upwelling of coastal water off the coast of Peru, caused by northward winds, moving water beneath by Ekman transport westward and offshore.

systems may move at speeds of a few kilometers per hour and can cover tens of thousands of square kilometers. Wind velocities of 160 km or more per hour are common, and wave surges of 10 m over coastal areas are possible. **Hurricanes** can completely destroy a coral reef and cause other terrible effects. In 2005, for example, Hurricane Katrina caused tremendous damage to the people of New Orleans when the levee system broke, and ocean surge and saltwater intrusions from Hurricane Rita in the same year caused major damage to low-lying coastal forests. Earthquakes or slumping of submarine sediment can generate waves known as **tsunamis** (often called tidal waves—a misnomer, given that they have nothing to do with the tide), which may travel at speeds in excess of 600 km h^{-1}. Their wave height may be very small in the open sea, but if a shallowing coast-line causes the water to pile up, tremendous destruction can be caused by the resulting wave surge, which can exceed 5 m. In the late nineteenth century, tens of thousands of people on the coast of Japan were killed by one especially destructive tsunami. A December 2004 earthquake of magnitude 9.1–9.3 near Sumatra generated an enormous tsunami throughout the Indian Ocean, reaching heights of 30 m. More than 200,000 people perished, and the ar-riving wave caused a great deal of damage to some coral reefs in the eastern Indian Ocean. The 2011 magnitude 9.0 Tohoku earthquake off the northeast coast of Japan gener-ated waves on shore up to 40.5 m, probably killing about 20,000 people and destroying much of the large Fukushima nuclear power plant complex.

- **Small-scale changes in current systems, combined with latitudinal temperature change and coastal irregularities, may isolate adjacent coastal regions.**

Previously, we discussed the role of seafloor spreading in continental drift. Because of these processes in the past 50–100 million years, much of the world's coastlines con-sist of north–south coastlines. For example, the Pacific coast of North America stretches all the way from north-ern Alaska to the very extreme Southern Hemispheric tip of South America. These coastlines exist along a large lati-tudinal gradient of temperature and variation of current systems. Also along these very long coastlines, we see a large number of projections of land, such as Cape Cod, Cape Hatteras, and the tip of Florida as we move south-ward along the Atlantic coast of eastern North Amer-ica. These projections are also breaks of sea temperature change and isolation of coastal populations because of separated sea current systems. These breaks may be isolat-ing barriers to migration and might be limiting barriers because the temperature changes too much for species to extend past a given barrier, such as living north of Cape Cod where temperatures drop substantially relative to southerly waters.

Seawater Density and Vertical Ocean Circulation

- **Seawater density, an important property affecting the vertical movement of water, is controlled mainly by salinity and temperature.**

The **density of seawater** (symbolized by ρ, the Greek letter rho) is expressed as the number of grams per cubic centi-meter. Density increases with decreasing temperature and increasing salinity (because of its salt content, seawater is denser than fresh water). Seawater density is only a few

percent greater than that of distilled water, usually ranging between 1.02 and 1.07 g cm^{-3}.

Strong depth gradients in water density occur at many scales in the ocean. Vertical gradients in seawater density are known as **pycnoclines**. Because water density is affected significantly by variation in temperature, solar warming should play an important role in the vertical thermal struc-ture of the ocean. This can be seen even at the scale of cen-timeters if you dip your hands into a tidal pool on a hot day. The top few centimeters will be warm, but the water beneath will be considerably colder. This principle works on the larger scale of a coastal water body with a depth of 20–50 m, such as Long Island Sound, which lies between Long Island, New York, and Connecticut. In the summer, solar warming from above produces a warm layer approxi-mately 3–10 m thick, which is mixed by the wind and is isothermal (identical in temperature) throughout its depth (**Figure 2.17**). Beneath this mixed layer, a zone of rapid de-crease in temperature, the **thermocline**, ends in a colder, deeper layer of relatively constant temperature. Due to the effect of temperature on density, this arrangement is very stable because the low-density warm water rides above the high-density cool water. Wind therefore does not mix the layers very effectively. As autumn sets in, this stratification of density begins to break down, and winds more easily mix the layers until the winter decrease of temperature results in a well-mixed and cold isothermal water column from the surface to the seafloor.

On the largest spatial scale, the tropical ocean contains a surface warm layer, a deeper thermocline, and a still deeper cold layer. At the surface, the sun provides heat year round, and the deeper waters are kept cold by deep circulation. The precipitation–evaporation balance in the tropics tips toward precipitation, which further reduces the surface water density (see **Figure 2.17**).

- **Vertical circulation and deep-water circulation are regulated primarily by differences in water density, generated at the surface usually at high latitudes, followed by sinking and travel to lower latitudes.**

In contrast to the major surface current features, deep oceanic circulation is characterized by movement of large water masses whose unique temperature and salinity char-acteristics are acquired at the sea surface at high latitudes. The high-latitude water masses sink owing to their high density and move to lower latitudes. Beneath the surface of the ocean, the water masses reach different levels, which are determined by the densities of the different water masses. These water masses are stacked like a layer cake in descend-ing order of increasing water density. **Figure 2.18** shows an idealized cross section of the Atlantic Ocean. Note the continuity of deep-water masses from the surface at high latitude to lower latitudes and greater depths.

The vertical structure of the Atlantic (**Figure 2.18**) is explained as follows. In the Weddell Sea of Antarctica, the higher density of colder water is enhanced by slightly in-creased salinity generated by sea-ice formation. Thus, the very dense Antarctic bottom water (AABW) is formed

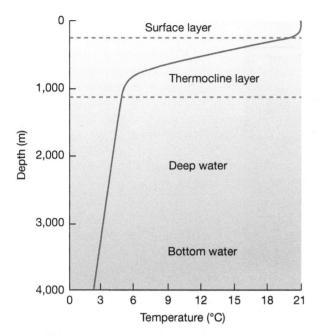

FIG. 2.17 An idealized profile of seawater temperature as a function of depth.

of Antarctica opposite the Weddell Sea, sink and eventually form cold and saline deep ocean bottom waters at lower latitudes of the Pacific Ocean. Of course, there is considerable mixing at water mass boundaries, but the masses nevertheless can be traced by characteristic temperature-salinity combinations. The time of movement is on the order of hundreds of years.

The origin of water masses is crucial for the maintenance of deep-water life because the water originates at the surface, where it is saturated with dissolved oxygen. Consumption of oxygen during transport is insufficient to make the water masses anoxic, and the deep sea bottom waters are therefore largely oxygenated. This allows deep-sea organisms to have an ample supply of oxygen, usually. Water exchange is limited in some deep-sea basins, and these are anoxic. For example, the Cariaco Trench is a 1,400-m-deep zone off a broad shelf region north of Venezuela. Oxygen is absent within this trench, and all chemical processes are affected by the absence of oxygen (Hastings and Emerson, 1988).

The Coriolis effect modifies deep circulation, and some large-scale sedimentary features show deposition that follows the expected deflection. One might intuitively expect deep-ocean currents to be sluggish, and this can be true. There are, however, some notable exceptions. Earthquakes whose foci are beneath the continental slope often set off rapid downslope movements of a water-sediment mixture known as a **turbidity current**. These currents, which are of sufficient force to transport coarse sediment, were discovered by virtue of their ability to snap submarine communication cables. Even on the abyssal seabed, currents can be fast, on the order of 50–60 cm s^{-1}. The deep sea is thus not necessarily a monotonous and static environment. In some areas, large volumes of water are funneled through narrow openings in submarine ridge systems and through narrow openings between marginal seas and the

near the surface, sinks, and moves along the bottom toward the Northern Hemisphere. The North Atlantic deep water (NADW) forms in the Norwegian Sea, but it is not as dense as the AABW. It sinks and flows southward, moving above the AABW. The Antarctic intermediate water (AAIW) is formed near the Antarctic Circle but is not quite as cold and saline as the NADW. It sinks but comes to rest just below a tropical surface layer of warm low-density water and above the NADW. These movements form a series of layers of progressively lower-density water toward the sea surface. Surface waters from the Ross Sea, on the side

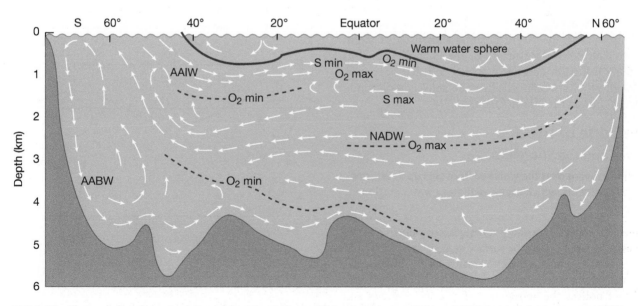

FIG. 2.18 Thermohaline deep circulation of the Atlantic Ocean. Water masses are as follows: AABW, Antarctic bottom water; AAIW, Antarctic intermediate water; NADW, North Atlantic deep water. (From Gerhard Neumann and Willard J. Pierson, Jr., *Principles of Physical Oceanography*, copyright © 1966. Reprinted by permission of Prentice-Hall, Inc.)

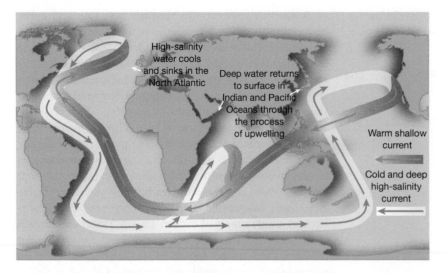

FIG. 2.19   The global conveyor belt system connects surface and deep currents.

open ocean. There, current velocities are much higher than those on the rest of the deep seafloor.

■ **The surface and deep ocean are connected by means of a global conveyor belt circulation.**

The surface currents, driven by the planetary winds, and the deeper currents, regulated by density, are coupled through the **global conveyor belt system (Figure 2.19)**. We can see this by examining the surface origin in the North Atlantic of cold seawater as the Gulf Stream moves northeastward at high latitudes. Here, the surface water, being cold and dense, sinks and moves at depth toward the south, eventually connecting with some of the AABW and later moving northward into the Pacific and warming and rising toward the surface, mixing with surface waters. This surface flow moves through the Indian Ocean, around the southern tip of Africa, and back into the Atlantic. The flow patterns are slow and not well understood. It has been recently argued that global warming might shut down the origin of the sinking water in the North Atlantic and that this might in turn shut down the strength of the Gulf Stream. Thus far, there is little evidence that such a shift has happened, although physical oceanographers are actively monitoring water motion in the North Atlantic.

The Edge of the Sea

Waves and the Shoreline

■ **Wind moving over the water surface sets up a series of wave patterns, which cause oscillatory water motion beneath.**

Wind that moves over the water surface produces a set of **waves**, which appear on the surface as a series of crests and troughs moving in the direction of the wind. This occurs even though no large current is in motion. During a storm, one first notices the appearance of a series of waves and then an increase in the height of the waves. Wave height is proportional to the wind velocity, the duration of the wind, and the fetch, or distance over which the wind acts.

■ **Waves are defined by their period, wavelength, wave height, and velocity as well as by the water depth.**

Although waves are not always symmetrical and evenly spaced, a few dimensions adequately define their geometry (**Figure 2.20**). Wave height H is the vertical distance of crests from troughs. Wavelength L is the distance between successive crests, troughs, or other specified points. Period T is the time of passage of successive crests past a reference point. Finally, velocity V is the speed at which a crest or other specified point travels. The variables V, L, and T are related by the equation:

$$V = L/T$$

When waves are symmetrical, water particles move in orbits. The diameters of these orbits decrease with increasing water depth and become insignificant at depths greater than $L/2$.

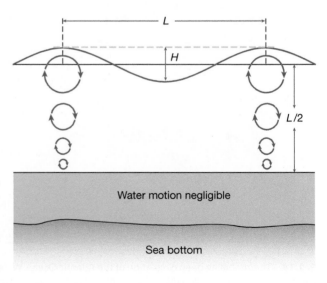

FIG. 2.20   Dimensions of ocean waves.

■ **As a wave approaches the shore, it becomes affected by the presence of the bottom when the depth is less than *L*/2; then the wave eventually breaks.**

Surface waves set the water into orbital motion at depths of less than *L*/2. When the water becomes shallow enough, the orbital motion occurs at the bottom, and the waves at the surface begin to "feel the bottom," or be affected by its presence. The geometry of the wave begins to change, and the wave height increases. These effects are due to the insufficient space available for complete orbits to occur; the water is effectively pushed upward as the wave approaches the shore. The orbital motion becomes more and more elliptical, and water and mobile bottom sediment are moved shoreward, sometimes with great force. When *H*/*L* exceeds 1/7, the wave becomes unable to sustain its own weight and collapses, or breaks.

■ **Shorelines greatly alter the pattern of waves and currents.**

Waves often arrive with their crestline at an angle to the shore. This produces currents parallel to the shore known as **longshore currents** (**Figure 2.21**). Occasionally, longshore currents encounter an irregularity in the shoreline and suddenly run offshore in concentrated and quite dangerous **rip currents**. Longshore currents are responsible for extensive erosion and transport of beach sands along outer coast beaches. The effect often makes the dredging of channels into outer beach bars a hopeless task. Often, misguided beach engineers place groins on beaches to prevent erosion. This has the effect of choking off the sand supply farther downcurrent, along the beach. As a result of these practices, many resort areas, most notably Miami Beach, Florida, have lost their bathing sand beaches in the past.

Irregularly shaped coastlines can strongly affect the direction and speed of incoming waves. As a wave front approaches the shore, horizontal water velocity diminishes as the water becomes shallower. Because the parts of a wave arriving on either side of a headland are still in somewhat deeper water, they will travel faster than the part of the wave striking the peninsula. This results in the wave refracting about the peninsula (**Figure 2.22**). The degree of refraction may focus wave energy on the headland, and this accelerates the erosion of the headland relative to the straighter coastline adjacent to the peninsula.

■ **The coast may be soft sediment or outcrops of rock under active erosion.**

In quiet water, soft sediment accumulates as sand or mudflats, where sediment movement is minimal and the slope of the flats changes little throughout the year. These flats can be hundreds to thousands of meters wide, as on the northern coast of France or on the flats of the Bay of Fundy. On more exposed beaches, strong wave action causes extensive sediment transport and a seasonal change of the beach profile owing to winter storms.

Figure 2.23 shows some features of an exposed beach. A series of windblown **dunes**, sometimes stabilized by vegetation, lies behind the beach. A relatively horizontal platform, the **berm**, extends to a break in slope. Owing to storms, the winter beach profile then increases strongly in slope to the low-tide mark. Seaward, a complex of troughs and offshore **bars** develops. Often, emergent **barrier islands** develop that protect large lagoonal complexes.

Rocky coasts develop where outcrops of rock occur in geologically youthful terrains. The topography of such coasts depends on the local rock type and wave action.

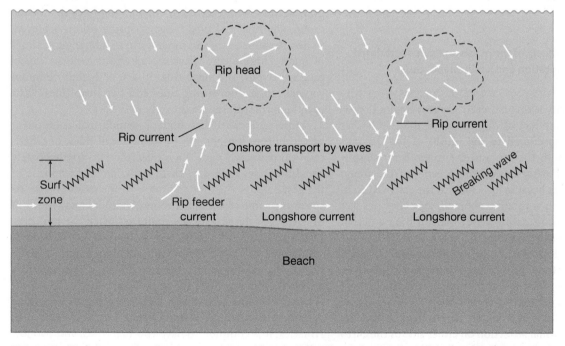

FIG. 2.21   Water transport adjacent to an exposed beach. (After Shepard, 1963. From *Submarine Geology*, 2nd ed. by Francis P. Shepard. Copyright © 1949, 1963 by Francis P. Shepard. Reprinted by permission of HarperCollins, Inc.)

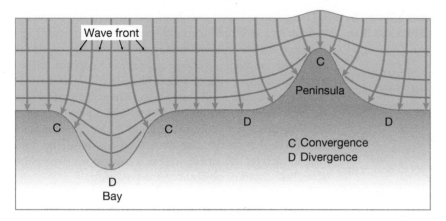

FIG. 2.22 Wave refraction. Note that wave action is concentrated at the headland, where the waves converge. (After Shepard, 1963. From *Submarine Geology*, 2nd ed. by Francis P. Shepard. Copyright © 1949, 1963 by Francis P. Shepard. Reprinted by permission of HarperCollins, Inc.)

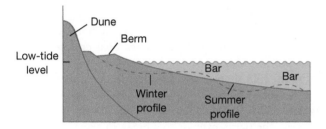

FIG. 2.23 Seasonal differences in the profile of a beach.

Sandstones, for example, are weathered and eroded into sand particles, which may produce a sandy beach at the base of a rocky outcrop (as at Santa Barbara, California). By contrast, highly cemented sedimentary deposits and crystalline rocks maintain their hard surface, and hard rock shores develop, as seen often in coastal areas of Maine and northern California.

Tides

■ **Gravitational effects of the moon and sun, modulated by the earth's rotation and basin shape, cause periodic tides.**

The force of gravity acting between any two bodies is proportional to the product of the masses of the two bodies and inversely proportional to the square of the distance between them. Both the sun and the moon exert significant gravitational attraction on the ocean, but the moon dominates the tides because it is closer to the earth. Keep in mind that the earth is rotating about an axis once per day, but the earth is also rotating about the sun with a period of about 365.26 earth days. The moon is rotating about the earth counterclockwise with a period of about 27.32 earth days. Tidal motion can be measured throughout the ocean, but it is especially noticeable at the shoreline in the form of tidal currents and vertical water motion.

The extent of the tide is largely determined by the difference in gravitational attraction on either side of the earth relative to the moon and sun. On the side closer to the moon, which has more gravitational effect than the sun, the gravitational attraction pulls water toward the moon. On the opposite side of the earth, a minimum of gravitational attraction combines with the earth's spin to produce a net excess of centrifugal force; this creates a tidal bulge away from the earth. Corresponding depressions (low tide) will exist on parts of the earth between the bulges, where there is no net excess of gravitational pull relative to centrifugal force. Because the moon "passes over" any point on the earth's surface every 24 hr and 50 min or once each tidal day, ideally there should be two low and two high tides per day. Because the moon's position relative to the earth's equator shifts from 28.5° N to 28.5° S, the relative heights of high and low water differ geographically owing to changing vectors of gravitational attraction.

The sun interacts with the moon to cause variation in the vertical range of the tide. During times of full or new moon, when the sun, earth, and moon are in line (**Figure 2.24**), the gravitational force exerted by the sun amplifies that of the moon, and maximal tidal range at **spring tide** is achieved. When the sun, earth, and moon form a right angle at times of quarter moon, the gravitational effects tend to cancel each other out, and **neap tide** occurs, with the minimum vertical range. Two spring tides and two neap tides occur each lunar month (approximately 29.5 days).

Tidal periodicity has both a semidiurnal component and a diurnal component. In **semidiurnal tides**, one finds two approximately equal high tides and two approximately equal low tides each day. These cycles are characteristic of the east coast of the United States. At the other end of the spectrum, some coasts under some conditions (e.g., the Gulf of Mexico at the time of the equinoxes when the sun's declination is zero and the moon's declination is nearly zero) are dominated by the **diurnal tide** component with a single high tide and a single low tide each day. Finally, **mixed tides** are also common, where the semidiurnal and diurnal components combine to give two high and two low tides each day but with unequal ranges. On the west coast of North America, it is common to have two high tides and two low tides of different heights each day. **Figure 2.25** shows some tidal patterns over several days under different tidal regimes.

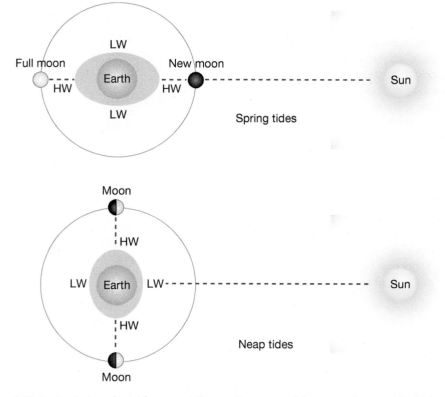

FIG. 2.24 Action of tidal forces at different alignments of the sun and moon. HW, high water; LW, low water.

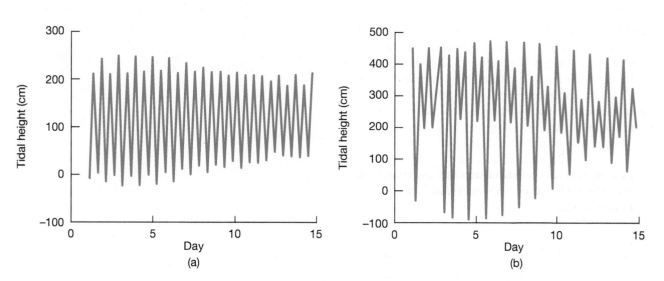

FIG. 2.25 Tidal periodicity from two different sites. (a) Bridgeport, Connecticut, which has a regular semidiurnal tide. (b) Port Townsend, Washington, which has a strong diurnal component and strongly uneven low tides.

The presence of irregularly shaped basins, the tilt of the earth's axis, and the Coriolis effect also cause significant deviations from the idealized expectation of two highs and two lows each day. Each basin develops standing oscillations that resonate with tidal forces. If the travel times of these oscillations are short relative to the period of the tide-generating force, the timing of the tide will be like that expected for an uninterrupted sphere covered with water. If travel time is long relative to the tidal period, the timing

of the tide will be out of step with that expected for such a surface.

The Coriolis effect has a great influence on tidal heights and currents. When affected by the earth's rotation, tidal currents produce in the center of seas and oceans rotating systems around **amphidromic points**, or points where no change in tidal height occurs over time.

In the open sea, both vertical tidal range and the strength of tidal currents tend to be modest. In coastal bays, however, volumes of water tend to be forced into ever-smaller blind bays, and as a result, currents and vertical tidal ranges are greater than on the adjacent open coasts. In areas with salt marshes, creeks tend to have higher tidal ranges than the adjacent open coasts. The most dramatic examples, however, are funnel-shaped basins such as the Gulf of California and the Bay of Fundy, where vertical tidal range exceeds 10 m. Tidal currents are also strong there, and a pronounced incoming tidal wave, or bore, can be seen, for example, on the tidal flats of the Bay of Fundy or in certain rivers such as the Yangtze (Chang Jiang) in China. Usually, there is a great difference of timing of the tides of bays relative to the adjacent open coast. Tides can also be extremely slight given the proper conditions. Because of the various interference effects of landmasses, areas such as eastern Denmark, the Gulf coast of the United States, and the Caribbean coast of Panama have almost no vertical tidal movement. Water levels in these places tend to be dominated by wind action.

Estuaries

■ **Estuaries are coastal bodies of water where water of the open sea mixes with fresh water from a river.**

An estuary is a partially enclosed coastal body of water that has a free connection with the open sea but whose water is diluted by fresh water derived from a source of land drainage, such as a river. On the large end of the spectrum, Chesapeake Bay (**Figure 2.26**) drains five major rivers and includes a complex set of creeks, bays, and main channels. On the small end of the spectrum are individual rivers whose salinity changes with every tide. Nearly all large estuaries have been strongly influenced by the rise in sea level since the last glacial maximum. Chesapeake Bay's present extent stems from the drowning of the Susquehanna and adjacent rivers. Many estuaries, especially those on the east and Gulf coasts of the United States, owe their general form to the rise of sea level and the formation of barrier bars, which enclose lagoons of relatively lower and more variable salinity than the open ocean. The Outer Banks of North Carolina are barrier bars that protect an inland water system of great biological richness.

Water movement in estuaries depends on the amount of river discharge, tidal action, and basin shape. Some estuaries have remarkably predictable salinity gradients and circulation. Others are far less predictable, and salinity and circulation depend on strong variations in freshwater flow from the watershed and variable connections with the open sea. Overall, there is a basic **estuarine flow**: Relatively low-density river water flows downstream and eventually comes into contact with saline water mixing in from the coastal ocean. Owing to its lower density, the river water tends to rise above the denser saline water mass moving upstream along the bottom from the adjacent ocean. In a **highly stratified estuary**, the layers are quite separate. With moderate wind action and tidal motion, mixing occurs at all depths and vertical exchange occurs, causing the salinity of both the upper and lower layers to increase seaward.

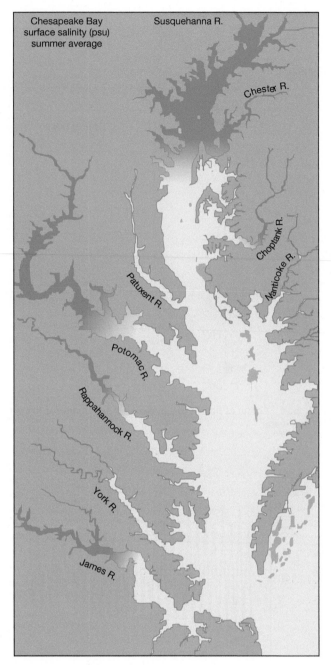

FIG. 2.26 The Chesapeake Bay estuarine system. Waters with summer salinity less than 10 are darkened. (Modified from Carter and Pritchard, 1988)

In such a **moderately stratified estuary**, a cross section would show that the isohalines, or lines of equal salinity, are inclined upward toward the sea (**Figure 2.27**). In these estuaries, water tends to flow up the estuary at depth and down the estuary at the surface. Such a flow pattern is found in the lower part of the Hudson River and in Chesapeake Bay.

The river slope and total water flow are important determining factors of salinity structure and tidal action. In fjords, a previous history of glacial scouring results in a long stretch of river where the bottom is nearly horizontal. In these cases, tidal effects tend to cause water flow far up the estuary. In the Hudson River estuary, which has such

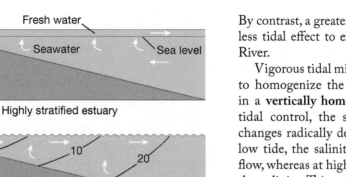

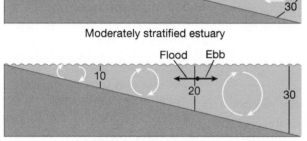

FIG. 2.27 *Salinity variation in different types of estuary. Lines of equal salinity (isohalines) are indicated for moderately stratified and vertically mixed estuaries.*

a glacial history, water is elevated only about 1 m over a stretch of about 250 km from New York City to a dam near Troy, New York. As a result, tidal action extends far upriver.

By contrast, a greater slope and higher water flow cause far less tidal effect to extend northward into the Mississippi River.

Vigorous tidal mixing in relatively small estuaries tends to homogenize the vertical salinity gradient and results in a **vertically homogeneous estuary**. Because of strong tidal control, the salinity at any point in the estuary changes radically depending on the state of the tide. At low tide, the salinity is dominated by downstream river flow, whereas at high tide, the inrush of seawater increases the salinity. This mixing can occur only in very shallow estuaries, such as the small freshwater rivers entering salt marshes.

Most estuaries experience strong seasonal effects and receive water, dissolved substances, and particulate matter from tributaries and runoff in the larger watershed drained by the estuaries and freshwater rivers. In the late winter and early spring, storms and snowmelt can cause an increase in water flow known as the **freshet**. This increase often brings into the estuary nutrients that sustain the productivity of the estuarine food web. If spring flow is small relative to the total volume of water in the estuary, as is true in Chesapeake Bay, then lines of equal salinity (isohalines) move down with the spring freshet but only slightly. In the Hudson River, however, discharge varies a great deal, and salinity in the lower part of the estuary can change greatly seasonally and between years.

■ CHAPTER SUMMARY

The world's oceans can be divided into oceans and marginal seas. Only the Southern (Antarctic) Ocean has a continuous oceanic border with other oceans. Marginal seas differ according to recent geological history and exchange with the open ocean, local precipitation, evaporation, and river input.

- The ocean floor can be divided into plates by oceanic ridges, trenches, and other features. The ocean floor is in continual motion, driven by production and movement of oceanic crust outward from the ridges, known as seafloor spreading. Crust is consumed in the trench areas, known as subduction. Continents are embedded in the oceanic crust and move on the scale of hundreds of kilometers over millions of years.

- Water has unique physical properties and is a good solvent owing to the asymmetric charge of the water molecule. Seawater temperature varies because of differential solar input at the surface with changing latitude and current mixing. Salinity is a measure of the dissolved inorganic solids in seawater; it increases with evaporation and sea-ice formation and decreases with precipitation and river input. Oxygen is added to seawater by

mixing with the atmosphere and by photosynthesis; it is lost by respiration and by chemical oxidation of various compounds. Both ultraviolet light and infrared light disappear in very shallow waters; blue light penetrates the farthest in clear ocean water.

- Surface currents are controlled by the interaction of winds and the earth's rotation. Overall, surface circulation takes the form of gyres in the north and south Atlantic and Pacific. Western ocean boundary currents result in large-scale transport of surface water from the tropics to higher latitudes.

- The Coriolis effect increases with increasing latitude; it deflects currents to the right of the wind in the Northern Hemisphere and to the left of the wind in the Southern Hemisphere. Winds and the Coriolis effect bring nutrient-rich deeper water to the surface. Every few years, persistent reversals of local current patterns occur in the eastern Pacific and cause a major shutdown of upwelling known as El Niño.

- Seawater density is controlled mainly by salinity and temperature. Cold and saline high-density water forms in high latitudes and then sinks as it moves

toward lower latitudes. The surface and deep ocean are connected by means of a global conveyor belt.

- Wind moving over the water's surface near shore sets up wave patterns, which cause oscillatory motion beneath. As a wave approaches shore, it becomes affected by the presence of the bottom when the depth is less than $L/2$; then the wave eventually breaks. Shorelines also strongly affect wave interactions and along-shore currents.

- The gravitational force of the moon and sun, modulated by the earth's rotation and basin shape, causes tides. During the full or new moon, when the sun, earth, and moon are in line, the force exerted by the sun amplifies that of the moon, and maximal tidal range at spring tide is achieved. The minimal range of neap tides occurs when the moon, earth, and sun form a right angle.

- In coastal estuaries, water of the open sea mixes with fresh water from a river. Here circulation is controlled by a freshwater flow, wind mixing, and tidal motion. In a mixing zone, lower-salinity water near the surface moves seaward, and deeper, higher-salinity water moves up the estuary.

■ REVIEW QUESTIONS

1. Why is the ocean crust about the same age on opposite sides of a midoceanic ridge?

2. What process causes the formation of trenches?

3. Why is the ocean salty?

4. Why are the relative concentrations of some elements constant throughout the open ocean?

5. How is salinity measured, and how is one method of measurement related to the use of practical salinity units as a measure of seawater salt content?

6. Why are ocean surface currents stronger and narrower on the west sides of oceans relative to the east sides?

7. Why does wind cause erosion of bottom sediments?

8. What is the main driving force of the ocean's surface currents? Of the deep circulation?

9. What processes cause the addition of oxygen to the ocean? Subtraction?

10. In what type of oceanic environment is ultraviolet light a problem for organisms?

11. What is the cause of upwelling along the western coasts of continents in the Atlantic and the Pacific Oceans?

12. What two factors play the largest role in determining seawater density?

13. Why is wave erosive action usually concentrated on headlands on the shore?

14. Suppose an oceanic wave is 50 m long and the time between the passing of two wave crests is 25 seconds. What is the wave velocity?

15. Suppose the wavelength of ocean waves in a strong Pacific storm off the coast of California is 1 km. How deep will you expect bottom erosion to occur?

16. Why is the tidal range maximal two times each lunar month?

17. What is the principal cause of the gravitational pull that causes tides?

18. What is the major cause of vertical water stratification in an estuary?

19. Suppose surface circulation of the ocean ceased to take place. How would climate and near-shore oceanic life be affected?

20. How might recent ocean history, especially the rise and fall of sea level caused by glaciation, have affected the current biogeography of the ocean?

21. It has been argued that estuaries are not very old geologically and that the resident species therefore immigrated and adapted to them recently. Does this make estuaries unworthy of protection? Why or why not?

22. Seafloor spreading resulted in India moving northward toward the Asian continent and merging with it, producing a mountain chain. If India began moving 71 million years ago and moved a distance of 6,100 km before crashing into Asia 10 million years ago, what was the average rate of seafloor spreading?

Visit the companion website for *Marine Biology* at www.oup.com/us/levinton where you can find Cited References (under Student Resources/Cited References), Key Concepts, Marine Biology Explorations, and the Marine Biology Web Page with many additional resources.

CHAPTER 3

Climate Oscillations and Climate Change

Climate Change: Oscillations and Long-Term Trends

■ **Oscillations and directional change in climate occur at large oceanic spatial scales. They interact to result in interannual climate variability.**

Short-term weather systems such as hurricanes occur on the time scale of days to weeks, but many other fluctuations in weather occur on longer time scales, the ones we associate with properties of **climate**, which refers to weather properties that characterize a region over longer time scales such as several years to decades to centuries. This leads us to a discussion of major oscillations and their interaction with longer-period trends in climate that might occur over decades or centuries.

■ **Multidecadal oscillations in climate are a major pattern observed throughout the world's oceans. World climate patterns in recent decades are usually a combination of oscillations combined with a long-term warming trend.**

A **multidecadal oscillation** in climate is a fluctuation in air pressure characteristics, wind systems, sea-surface temperature (SST), or other weather features that occur on the geographic scale of an ocean, such as throughout the North Atlantic Ocean or perhaps even the world ocean (Table 3.1). Such fluctuations occur on the scale of a few years to decades, and need not have precisely equal fluctuations. These oscillations occur in combination with longer-term trends. As we emphasize in this chapter, our planet is experiencing a long-term warming trend that began at the end of the nineteenth century. It is important to realize that oscillations can cause temporary highs or lows in temperature, for

example, that can partially obscure a longer-term trend. For instance, an El Niño in 1997 temporarily caused a major rise in planetary temperature. Several years after El Niño conditions subsided, one might think that the world was getting colder because we were on a downward slide of the El Niño cycle. Yes, this was true on a short time scale, the scale of the oscillation, but on a longer time scale of decades, world climate continued to warm toward the present day: The year 2016 is worldwide the warmest on record. The year 2015 is the next-warmest year.

Climate oscillations, or periodic fluctuations, in features such as SST or atmospheric pressure usually are captured and analyzed after long-term monotonic (steadily progressing) climate changes are **detrended** from the overall data. This means that a long-term increase in sea-surface temperature, for example, is subtracted from the fluctuations so that residual oscillations in SST can be identified. **Figure 3.1** shows a hypothetical example in which temperature is oscillating but a long-term increase also is occurring. The detrended curve shows the multidecadal fluctuation minus the long-term increasing trend. Along with detrending, complex changes over time are often **time-averaged**, so small-scale fluctuations over a few years are diminished in emphasis so that larger-scale trends can be observed. Thus, we might not plot the temperature every few minutes, but instead each data point might represent the temperature averaged over the last 10 days so that we can see broader trends and distinguish them from very short-term fluctuations. Oscillations and long-term trends need not be independent. Some recent evidence suggests that ENSO oscillations have increased in the past 100 years along with overall ocean warming.

TABLE 3.1 Some Oscillatory Climate Changes Involving Oceanic Regions

NAME OF OSCILLATION	PERIOD OF OSCILLATION	GEOGRAPHIC LOCATION
El Niño Southern Oscillation (ENSO)	2–7 years	Pacific Ocean, with effects worldwide
Pacific Decadal Oscillation	10 years with much variation	Middle latitudes of North Pacific Basin, alternating warm and cool waters
North Atlantic Oscillation	No known periodicity	North Atlantic, shift in pressure between region of Iceland and the Azores
Arctic Oscillation	No known periodicity	Shift in pressure between Arctic and lower latitudes 37–45° N, affecting distribution of cold Arctic air over varying latitudes
Interdecadal Pacific Oscillation	Several shifts in the twentieth century	Shifts in pressure in the North Pacific Ocean, affecting shifts in temperature in the North Pacific
Antarctic Oscillation	No known periodicity	Shifts in pressure at two loci, one in Antarctic Ocean and another at South latitude 40–50°
Intertropical Convergence Zone	Semiannual cycle	Northern summer development of monsoon over southern Asia with strong cyclones; Southern Hemisphere summer monsoon with weaker development of cyclones
Indian Ocean Dipole	Irregular	Eastern Indian Ocean becomes alternately warmer, then cooler, than the western Indian Ocean

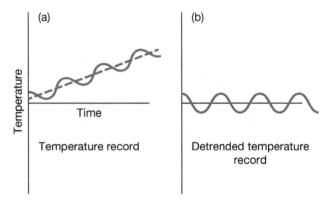

FIG. 3.1 An example of a climate variability record. (a) Temperature over time shows a steady increase of temperature with a fitted trend line, but there also is an oscillation of temperature change that is superposed on the overall increasing trend. (b) The left diagram is detrended, showing the residual, which is a temperature oscillation.

Climate Oscillations

■ **Climate oscillations are irregular alternations between coupled atmospheric-oceanic states that result in changes of wind systems, oceanic currents, and regional climate.**

World climate oscillation is partially controlled by complex shifts of coupled atmosphere-oceanic systems. Usually, these oscillations can be identified by shifts of atmospheric pressure between major regions of an ocean. For example the El Niño-Southern Oscillation comprises a major alternation of atmospheric pressure conditions between the western and central-eastern tropical Pacific, which helps to change wind systems significantly. Because wind is a major driver of current systems there is also a coupling with surface oceanic currents, which may also be coupled with vertical movements such as upwelling. The oscillations in atmospheric conditions also exert major climate fluctuations at sea and on the continents.

These fluctuations are very complex and typically the alternations of state are not evenly spaced in time. There are also interactions between different oscillation systems (e.g., the North Atlantic Oscillation and the Arctic Oscillation) that make changes very complex and often not well understood. Nevertheless, the fluctuations are real and observable, and the consequences of shifts in atmospheric pressure are predictable. Unfortunately, at this time there are few good explanations of how these oscillations are driven. One model of the well-known El Niño suggests that as warming occurs in the eastern Pacific, physical conditions are established that dampen the warming, setting climate into a cycle of cooling in the eastern Pacific.

■ **El Niño is a global-scale climate oscillation dominated by a major warming event that spreads from west to east in the tropical Pacific Ocean.**

As introduced in Chapter 2, northward winds along the coast of Peru are influenced by the Coriolis effect, and Ekman transport moves surface waters offshore to the left (and to the west). Surface waters blown offshore are replaced by the rise of deeper nutrient-rich waters, which fuels high rates of productivity in the surface waters. El Niño, named for the Christ child, was first recognized as a Christmastime weakening of these winds. On a much larger geographic scale, the trade winds (see Chapter 2) drive surface waters in the tropical Pacific toward the west. This "normal" condition results in deeper waters rising, which contributes to the upwelling seen in the central and eastern Pacific, bringing nutrients to the surface and helping to drive the very high productivity seen off the west coasts of tropical North and South America.

El Niño develops when weakening of the trade winds persists for many months or even years. Such a persistent reversal causes a cessation of upwelling and rapid warming of surface waters in the eastern Pacific from South to North America. Not only are nutrient-rich waters no longer brought from deep water to the surface, but there are mass mortalities in fishes from the increased temperature of surface coastal waters. Warm waters often cause severe mortality of normally cold-water-adapted shellfish in the intertidal region of Peru and Chile. Peter Glynn has documented extensive coral mortality in Panama during the 1982–1983 El Niño event owing to the increase of sea-surface temperatures (Glynn, 1988). The El Niño of 2015 caused extensive damage to coral reefs throughout the Pacific Ocean (see Chapter 17).

El Niños usually involve strong coastal damage from the eastward movement of storm fronts. During the El Niño of 1982–1983, about 600 people died from intense storms, and Peru suffered fishing losses and about $2 billion in storm damage. The effects were also felt along the California coastline, and many homes slid into the sea. In a disastrous flood in Peru in 1728, the Piura River became raging rapids as a result of storms triggered by El Niño. Many people were swept away, and the survivors were mainly women who were wearing hoop skirts, which served as flotation devices by trapping air below the women's waists. In 1997–1998, El Niño was extremely damaging throughout the Pacific. During the height of this period, surface production in the equatorial Pacific was very low. But upwelling commenced again at the end of the cycle, and production of the plankton increased rapidly. Warming in the 2015 El Niño has caused coral bleaching throughout the Pacific Ocean.

It used to be thought that El Niño was just a local effect along the Pacific Coast of tropical America. However, a similar cycle of fluctuation of atmospheric pressure had also been recognized in the western Pacific and Indian

Oceans, named the **Southern Oscillation**. We now know that these two pieces operate in synchrony (**Figure 3.2**), hence the name **El Niño-Southern Oscillation**, or **ENSO**. ENSO is driven by climatic oscillations that seem to be worldwide in extent but certainly involve an oscillation of the ocean-atmosphere system across the tropical Pacific and Indian Oceans (for an excellent discussion, see Enfield, 1988). Since the trade winds during El Niño are very weak, or even reversed, the westward-flowing equatorial surface current is reduced in intensity, allowing a deeper eastward-flowing countercurrent to dominate. Warm water spreads eastward across the tropical Pacific and eventually reaches the eastern Pacific continents of mainly tropical North and South America (**Figure 3.3**). Every 2 to 7 years there is a reversal of atmospheric pressure systems across the Pacific and Indian Oceans.

In the so-called normal state, the trade winds are strong and upwelling in the eastern Pacific is well developed. A persistent version of this state is known as **La Niña**, when there is a regional low-pressure system in the Indian Ocean and a high-pressure system over the eastern Pacific.

During ENSO, rain usually increases greatly on the western coasts of the Americas, but strong droughts dominate western Pacific areas such as Australia. The effects are broader in scope than just the Pacific–Indian Ocean regions; they occur worldwide. For example, when the ENSO part of the cycle occurs in the Pacific, warmer conditions may prevail in northeast North America. It has been argued that El Niño reduces wind shear in the eastern Pacific and sets up conditions for strong hurricanes but weakens such conditions in the tropical Atlantic. The effects are even farther

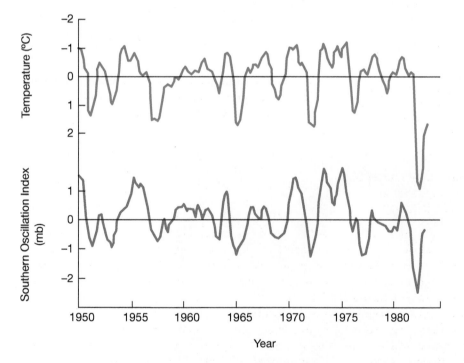

FIG. 3.2 The correspondence, over about 30 years, between departures from average sea-surface temperatures (top curve) and the Southern Oscillation Index (bottom curve), which is the difference in atmospheric pressure anomalies between Tahiti (relatively eastern Pacific) and Darwin, Australia (western Pacific). Note that because the temperature scale is inverted, positive pressure in the eastern Pacific corresponds to low sea-surface temperature. (After Shen et al., 1987)

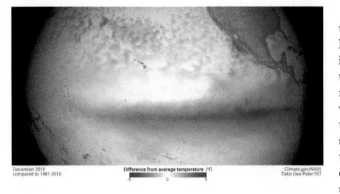

FIG. 3.3 The movement of warm water eastward across the Pacific during El Niño in November 1997 and July 2015, as shown by a satellite false color image of temperature anomalies. (Source: NOAA)

ranging. In the 2015 El Niño, strong droughts were affecting sub-Saharan Africa.

■ In the Atlantic, the North Atlantic Oscillation causes strong fluctuations in wind systems and oceanic climate.

The **North Atlantic Oscillation** (NAO) involves a periodic shift of atmospheric pressure in the North Atlantic Ocean. Normally, there is a low atmospheric pressure over Iceland at about 64° N latitude and a higher atmospheric pressure over the Azores island chain at approximately 38° N latitude. The difference in pressure between these two locations fluctuates; when the difference is strong (considered a positive NAO index, **Figure 3.4**), westerly winds are enhanced, and the eastern Atlantic and Europe have cool summers and mild winters. A large difference in pressure also tends to increase winter temperatures along the east coast of North America. In addition, the NAO may affect the position of the Gulf Stream, and large differences in pressure between the two locations may push the Gulf Stream more toward the north. Strong storms track eastward across the North Atlantic. During a negative NAO, the difference between pressure centers is weak, the Mediterranean region experiences more precipitation, and storms across the North Atlantic are very weak (**Figure 3.4**).

Such large-scale climate fluctuations have been found to strongly affect local coastal conditions, leading to a revolution in our thinking of coastal fluctuations. A surprising relationship was discovered between the NAO and the western Swedish Gullmar Fjord. When the NAO is negative, winds from the north and east tend to draw dense water from the depths of the fjord, causing exchange with the North Sea. But in the past two decades an increasingly strong positive NAO index has caused winds to come from the south and west, and there has been only a weak exchange of deep fjord waters with the open sea (Erlandsson et al., 2006). This allows organic matter, much of which comes from human nutrient additions, to settle in the deep part of the fjord. This in turn causes low oxygen events, which are very stressful for the local deep biota, causing large-scale mortalities of fish and larger crustaceans. Winters in southern Sweden are also warmer. Increased water column stratification during positive NAO index values has also been associated with toxic algal blooms in the 1980s and 1990s (Belgrano et al., 1999).

So we can again see a direct connection between an ocean-scale climate fluctuation and local environmental shifts. Before the connection was made, no one had a clue as to why these major shifts occurred in the fjord.

As mentioned, the NAO "winter" index has shown a trend of increase over the past decades (Nordberg et al. 2000), which can readily be seen by fitting a trend over time (**Figure 3.5a**). The oscillations behind this multidecadal trend can be seen in the detrended plot of the NAO index in **Figure 3.5b**. The NAO positive index (e.g., positive phase starting in 1990) is associated with warm ocean waters in the northeastern coastal waters such as the Gulf of Maine, along with the North Sea. This warming is likely partially responsible for the failure for recovery of the cod fishery from overfishing (Hurrell et al., 2003). Cod larvae in the Gulf of Maine are at their upper thermal limits, and warming causes stratification and lower productivity, which also limits cod production. In contrast, the positive NAO index is associated with very cold temperatures in the Labrador Sea, where cod larvae are at their lower temperature

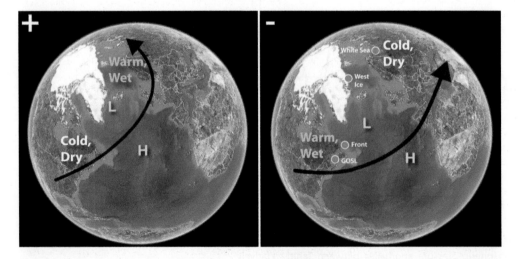

FIG. 3.4 Difference between the positive and negative phases of the North Atlantic Oscillation. (Image/photo courtesy of the National Snow and Ice Data Center, University of Colorado, Boulder)

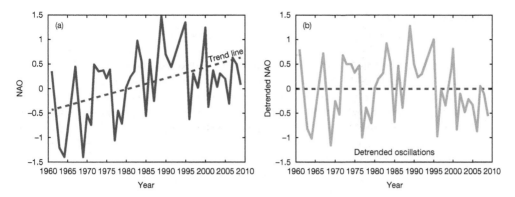

FIG. 3.5 Analysis of the North Atlantic Oscillation (NAO) winter index from 1960 to 2010. (a) Oscillations in the NAO index are superposed over an overall trend of increase of the index (blue dashed line). (b) The detrended data, showing the oscillations of the NAO index alone. (Analysis by Haikun Xu, Stony Brook University. Source of data from NOAA Climate Prediction Center)

GOING DEEPER 3.1

The Triple Key to Ocean Climate Understanding

This chapter hopefully makes it clear that climate trends and cycles are complex and our observational platforms are crucial in extending our understanding of climate cycles, long-term trends, and short time-scale events such as major storms. In order to understand the environment we need an integrated approach to observation, which depends on three important components (Box Fig. 3.1).

High-frequency (HF) radar is crucial in analyzing weather patterns that must be coupled to ocean processes. Such systems involve multiple sensors spread out over hundreds to thousands of kilometers. Current installations have approximately 1 km resolution, and shore-based detectors can measure current speeds as much as 70 km from shore at 15-minute intervals. With more than one detector and with sequential inputs to a sensor over time, directions and intensities of winds can be mapped, giving a vector field of wind over an area. Most important, these data can be used to calculate lateral water transport and wave size (e.g., wave length). HF radar inferences of water motion can be verified and calibrated by drifters in the water that measure water motion more directly. Generally there is good agreement between the two.

Satellites with sensors tuned to air and water temperature and detecting sea height can also be used to map out fields of water motion and spatial and temporal trends in ocean climate. For example, high resolution infrared and visible images can be used to track clouds, which are very useful to track storms such as hurricanes. Visible sensors can also be used to assess ice cover. Stationary satellites can be signaled by ground sensors to aim their sensors to specific regions to gather data that coincide with data being collected by high frequency radar. Satellite sensors can measure wind speed, sea-surface height, and sea-surface temperature, and time integration can give spatial movement of important ocean parameters. If cloud cover is low, sensors can estimate sea-surface temperature, chlorophyll in the surface waters, and other parameters.

Ocean glider vehicles transmit information with very little human assistance. As mentioned in Chapter 1, autonomous underwater vehicles can move in the ocean without tether to a ship. They therefore can operate efficiently in rough seas and often over great distances. Specially designed gliders can take advantage of ballast shifts, bulk density changes, and fins to allow ocean turbulence and seawater density to

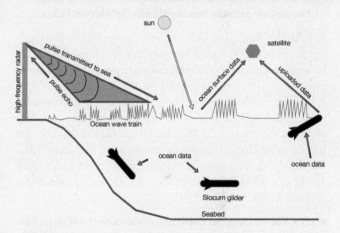

BOX FIG. 3.1 Ocean climate can be assessed from high frequency radar, satellite detectors, and sea-going gliders that collect data in the water column and upload to satellites when at the sea surface.

guide the direction of the gliders. For example, the Slocum glider can use changes in bulk density and forward-aft movement of balance weights to sink or rise in the ocean while drawing very low power, giving the glider the ability to do surveys for many months over a wide range of depths. Data are uploaded to satellites when the glider reaches the sea surface.

We can see how these three components contribute to a major new ability to make observations at sea that are relatively inexpensive and enable us to gather data under extreme conditions in which ships cannot operate. An excellent example is the approach of a hurricane toward shore, which creates conditions too dangerous for even large ships. As Hurricane Irene approached the New York–New Jersey region in 2011, continuous observations could be made by underwater gliders, high-frequency radar, and satellite estimates of sea-surface temperature. The data from wind speeds as the hurricane approached shore suggested that the intensity and coastal impact of the hurricane was significantly less than expected. The estimates of sea-surface temperature from the glider showed that the thermocline shallowed,

continues

causing surface waters to be much colder than expected, which likely diminished the hurricane's intensity. In contrast, 2012's Hurricane Sandy arrived when waters were already cool and water-depth profiles did not change and slow the storm, so its onshore progress was not diminished very much. Hence its impact was much more intense than Hurricane Irene's (Glenn et al., 2013).

On the west coast of North America, arrays of gliders are being used to study the motion and biology of the southern portion of the California Current, especially off southern and central California. Glider vehicles can report temperature, water motion, chlorophyll concentrations, among other parameters (Davis et al., 2008). This approach eliminates the need of ships and provides a dense spatial array of data reports. Such arrays of gliders are now common along the coastlines of the United States and much of the world's coastlines. The next challenge will be to develop a common data platform so we can develop a worldwide understanding of ocean climate.

limits. In this region the very low temperatures may have contributed to the collapse of the Newfoundland cod fishery. Some recent evidence suggests that there is also a more immediate connection between availability of an important planktonic food of cod larvae and juveniles, the copepod *Calanus finmarchicus* (Beaugrand and Kirby, 2010).

- **The Arctic Oscillation Index causes major alternations of climate in the North Atlantic. Its timing is strongly correlated with the North Atlantic Oscillation Index.**

The **Arctic Oscillation Index** reflects pressure shifts throughout much of the Northern Hemisphere, including the Arctic. In a positive phase, polar region surface pressure is relatively low and a jet stream confined to the north keeps frigid Arctic air confined to very high latitudes. In the negative phase, polar surface pressure is higher, which causes a shift of frigid air southward to middle northern latitudes. The timing of oscillations is strongly correlated with the NAO, but the AO has far broader latitudinal influence. In Chapter 19 we discuss the impact of the AO on shelf productivity patterns in the northwest Atlantic.

- **The Pacific Decadal Oscillation causes decadal switches in climate and may strongly affect coastal marine food webs and facilitates large movements of mobile large predators north or south along the Pacific coast of North America.**

Among the many other climatic fluctuations now known, the **Pacific Decadal Oscillation (PDO)** stands out because of its strong effects on ocean climate and ocean current systems. This oscillation shifts phases every 20 to 30 years with warm waters found above 20° N latitude in either the eastern or western Pacific Ocean. The oscillation is believed to be a complex combination of climate interactions, but the oscillations themselves drive strong biological fluctuations.

During a cool phase in the eastern Pacific, northerly species such as Dungeness crab *Metacarcinus magister* and English sole *Parophrys vetulus* move southward along the Pacific coast of California, and the northerly range limits of southerly species such as those living on rocky shores are truncated. Some of the strongest biological effects were first noticed in San Francisco Bay when the density of phytoplankton was found to increase. Under former conditions, filter-feeding clams and mussels were abundant in southern San Francisco Bay and these bivalves cropped down the phytoplankton to low densities in the water column. But as regional waters cooled after 1999, *Crangon* shrimp, juvenile Dungeness crab, and English sole increased to record levels in San Francisco Bay and decimated the clam population (Cloern et al., 2007), releasing the phytoplankton to increase in abundance in the water column. Thus an ocean-wide climate oscillation strongly changed a local bay food web by facilitating the introduction of top predators (**Figure 3.6**).

On a larger regional scale the PDO in the eastern North Pacific is strongly correlated with the state of salmon fisheries, with an especially tight correlation in Alaskan waters. We have data on fisheries only since the 1930s but evidence for PDO climate fluctuations go back further, even a few hundred years, based on tree-ring data demonstrating climate oscillations. The Alaskan salmon fisheries catch has been positively correlated with a PDO index, which is related to increased SST. For example, in the late 1970s Alaskan catches for all five species increased greatly as the PDO changed strongly to a "positive" phase. It is believed that this transition is due to oceanic changes in the entire food web, starting with phytoplankton (Hare et al., 1999).

The Greenhouse Effect and Changing Ocean Climate

The Greenhouse Effect

- **Earth and oceanic climate is moderated because of the greenhouse effect.**

The short-wavelength radiation of sunlight heats the earth and is converted to long-wavelength radiation, which is absorbed by the atmosphere (**Figure 3.7**). Atmospheric circulation distributes this heat to space but much is radiated back to the planetary surface, which makes the earth's surface warmer than it otherwise would be. This is known as the **greenhouse effect** because of the analogous role of glass greenhouses as heat traps.[1] **Greenhouse gases** absorb the infrared radiation even more and prevent it from returning to space by radiating part of the heat to the earth's surface. Greenhouse gases, in order of their effect, include: water vapor, carbon dioxide, methane, nitrous oxide, and ozone. Of other greenhouse-trapping substances, only

---

[1] Real greenhouses do not have convection and therefore do not operate like the earth's "greenhouse effect," which includes atmospheric circulation.

PDO—Warm water phase   PDO—Cold water phase

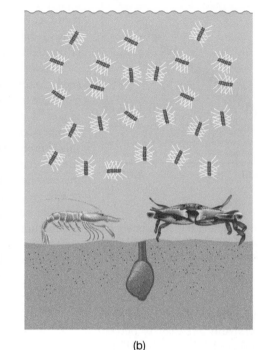

(a)   (b)

FIG. 3.6 Two alternate states of San Francisco Bay food webs. (a) The eastern temperate Pacific is warm, burrowing bivalve *Corbula amurensis* and mussel *Musculista senhousia* are abundant, and phytoplankton are relatively low in abundance. (b) The eastern temperate Pacific is cool; Dungeness crab and crangon shrimp move southward from Washington and British Columbia and invade San Francisco Bay, killing off filter-feeding bivalves and allowing phytoplankton to increase rapidly.

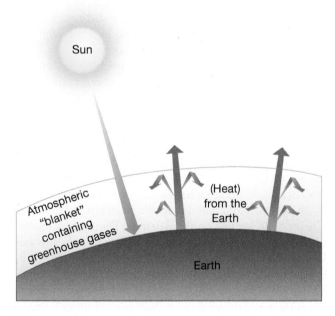

FIG. 3.7 The greenhouse effect.

the United States (NOAA) shows a steady increase in total greenhouse gases with no reduction in sight (see http://www.esrl.noaa.gov/gmd/aggi/).

■ **Atmospheric quantities of carbon dioxide have increased strongly because of the burning of fossil fuels and less so owing to deforestation; these factors are warming the earth's surface and ocean waters.**

Human activities have reached a point at which they are altering the earth's global climate. Since the Industrial Revolution in the nineteenth century, industrial activity has greatly accelerated the burning of fuels, particularly fossil fuels such as coal and petroleum products. Approximately 30 percent has been added to the storehouse of carbon dioxide in the earth's atmosphere. Measurements from Mauna Loa, Hawaii, demonstrate an increase from 315 ppm in 1958 to 390 ppm in 2011 (**Figure 3.8**), and it surpassed 400 ppm in 2015. It is believed that the atmospheric concentration was 280 ppm as the Industrial Revolution began in the nineteenth century. Most scientists believe that current trends could cause a doubling of the carbon dioxide in the atmosphere over the next 100 years.

Deforestation is a secondary source of carbon release. Trees are net absorbers of carbon owing to photosynthesis, which absorbs far more carbon dioxide than is released during respiration (yes, trees respire). Many of the tropical forests in Southeast Asia, Africa, and South America were cut down to clear land for agriculture and to provide wood for sale to the developed world. Worse than that, much of the forest is burned, which releases yet

chlorofluorocarbons, constituents of cooling fluids and some solid materials, have been nearly eliminated in production over the past few decades owing to their negative effects on the UV-protecting ozone layer. Carbon dioxide is increasing steadily, and methane is also increasing. An overall index of greenhouse gases developed by the National Oceanographic and Atmospheric Administration of

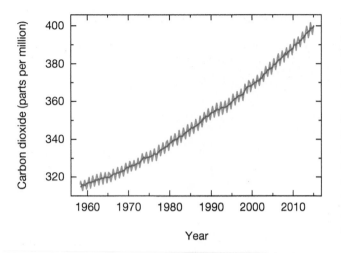

FIG. 3.8 The change in carbon dioxide in the atmosphere between 1958 and 2016 as monitored at an observatory at Mauna Loa, Hawaii. (From the Carbon Dioxide Information Analysis Center)

more carbon dioxide into the atmosphere. Nevertheless, the impact of this release is probably less than 10 percent of the total carbon dioxide released owing to burning of carbon-based fuels.

The problem with these releases is the heat-absorbing capacity of carbon dioxide, which radiates more and more heat to the earth's surface. Many have predicted that continued emissions of greenhouse gases will continue global warming. There is convincing evidence from overall world temperature records that the earth's air temperature has increased over the past 100 years by about 0.74°C. Because there were some decade-long periods of no increase in temperature, even though industrial burning remained steady, it is clear that global warming induced by industrial activity is not a linear process. Models evaluating human effects strongly support the hypothesis that human addition of greenhouse gases are the principal contributors to warming in the past 100 years (Intergovernmental Panel on Climate Change, 2014). Warming in the coastal ocean has increased greatly, especially in the past few decades. Even a couple of degrees Centigrade on average may translate into major local changes. Temperature increases may shift the latitudinal distributions of important fishery species and may even cause severe physiological stress. The magnitude and even the direction of climate change varies geographically owing to spatial differences in the effects of global warming, many of which can be predicted by climate models (Intergovernmental Panel on Climate Change, 2007).

Cloud cover is a major factor in climate, but it varies strongly both geographically and seasonally. Clouds consist of an aggregation of tiny water droplets or ice crystals that condense from the rising air as water vapor combines with atmospheric dust or other substances, known as **aerosols**. Clouds reflect short-wavelength radiation, reducing atmospheric heating, but prevent the back-radiation of long-wavelength radiation, which traps heat. On balance, clouds tend to cool the earth's surface. Current climate-predictive models suggest that global warming will reduce cloud cover,

further heating the earth. But we still do not have an adequate source of data on the geographic distribution of cloud cover, which will require more advanced remote sensing data, which will become available in coming years (Rosenfeld et al., 2014).

Aerosols consist of a variety of substances that may seed clouds and therefore increase cloud cover. Aerosols can have quite variable effects on cloud formation, which makes it difficult to predict their net effect on climate change. Combustion processes on the planet produce aerosols composed of carbon particles (Myhre et al., 2013). On the one hand, particles of soot can settle on glacial ice and increase heat absorption of glacier surfaces that would otherwise reflect heat back into the atmosphere. But carbon aerosols can also seed cloud formation. Phytoplankton production may also affect cloud cover since certain species produce the volatile substance dimethyl sulfide, which can be released into the atmosphere and seed clouds. Tiny aerosol particles containing sulfate are also major sources of cloud nuclei, and industrial emissions of sulfur may thus enhance cloud cover, mostly in the coastal zone (Falkowski et al., 1992).

■ Seawater temperatures have been increasing over the past century.

There is good evidence that seawater temperature has fluctuated at time scales ranging from millions of years to within centuries. We are now living only about 8,000 years since massive glaciers covered vast areas in upper latitudes. Glacial retreat has resulted in warming of midlatitude climate and a rise of sea level in the past few thousand years. However, in the past century, a distinct acceleration of global warming has developed both on land and in the global ocean. Plotting anomalies, which are deviations from the mean temperature since 1880, can illustrate the trends. **Figure 3.9** shows these anomalies throughout the globe for land and sea. Sea temperature has increased steadily, although not in a straight line. Most noticeably, the temperature has greatly increased in northern high latitudes and in the western Pacific. In the Arctic Ocean, sea ice has thinned, and it is anticipated that large parts of the Arctic Ocean may be ice-free throughout the year in a few decades. Melting ice has already caused reorganizations of open water communities in the northern Atlantic (see Chapter 19). Ice loss on the Antarctic Peninsula is reorganizing the local marine food web leading to penguins (see Chapter 20). However, warming has not occurred in the bulk of the Antarctic Ocean. This is apparently due to wind-driven upwelling, which draws deep and very cold water to the surface and drives surface water toward lower latitudes (Armour et al., 2016). Warming has occurred along the northern limit of the Antarctic Circumpolar Current. But it may take centuries before warming affects the bulk of the poleward Antarctic Ocean.

Throughout the coastal ocean, there are now records of temperature increase over the past century or more, and all show increases of SST over time (**Figure 3.10**). Such local increases have had a variety of biological effects.

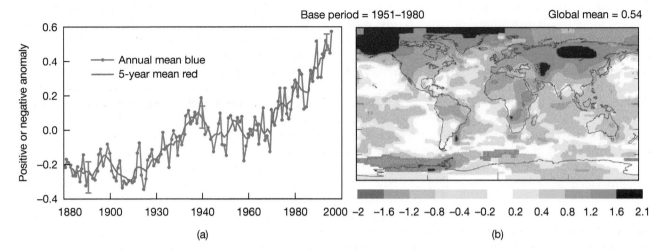

FIG. 3.9 (a) Changes in world average temperature anomalies since 1880, which are deviations from the mean temperature of a base period 1951–1980. (b) Global map of mean surface temperature anomalies for the period 2001–2005. (After Hansen and others 2006, copyright National Academy of Sciences)

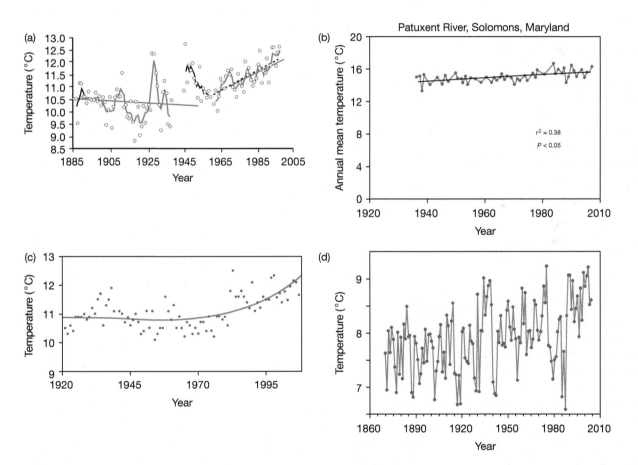

FIG. 3.10 Changes in sea-surface temperature at four near-shore localities: (a) Woods Hole, Cape Cod, Massachusetts, NW Atlantic (from Nixon et al., 2004); (b) Patuxent River, Chesapeake Bay, Maryland (from Kaushal et al., 2010); (c) Race Rocks, southern Vancouver Island, Canada (data from www.racerocks.com); (d) Baltic Sea, northern Europe, measured from ships since the 1860s (after MacKenzie et al., 2007).

In Long Island Sound, times of warming have coincided with biological invasions of warm-adapted species (Stachowicz et al., 2002) and major die-offs of cold-adapted species, such as lobsters in southern New England. In the North Sea, temperatures increased an average of 0.6°C between 1962 and 2001, and the northern extent of geographic ranges of several species has moved farther north. In the Baltic Sea, which has very low salinity, there is a record of irregular but ever-increasing sea-surface temperature from 1860 to 2007 (MacKenzie et al., 2007).

The interaction of climate oscillations and increases in planetary temperatures over the past 150 years have already produced major effects on marine ecosystems. As discussed in Chapter 17, ENSO periods of high temperature often surpass the ability of reef corals to function, resulting in the phenomenon of coral bleaching. The occurrence of bleaching has greatly increased overall since the 1970s, but bleaching frequency becomes punctuated during El Niño events. The 2015 El Niño has resulted in coral bleaching throughout the Pacific. Over 60 percent of the reefs in the northern part of the Great Barrier Reef have been bleached, and some may never recover (see Chapter 17). Droughts precipitated by global warming also have had catastrophic effects on salt marshes, increasing evaporation and salt stress on root systems. These physiological effects often result in strong interactions with the many species that depend on foundation species (e.g., corals and marsh plants) upon whom many other species depend.

Effects of Climate Change on Sea Level, Circulation, Ocean Chemistry, and Biological Factors

■ **Increased temperature has caused sea-level rise and major changes in oceanic circulation, sea-ice coverage, and water column structure.**

Global climate change has resulted in an increase of ocean temperature about 0.85°C over the period of 1880–2012 (Intergovernmental Panel on Climate Change Report, 2014). During the twentieth century, sea level rose around the world an average of 15 cm. A number of regional factors make the rate of sea level rise uneven geographically. For example, sea-level rise in the northwest Atlantic region has been greater than the world average. Sea-level rise can be explained by two factors. First, the increase of sea temperature alone increases the overall volume of the ocean. But some glacial melting has also contributed to sea-level rise by a direct addition of liquid water.

Further global warming could partially melt the large volumes of glacial ice in Greenland and Antarctica, adding a significant amount of water to the ocean, and increasing temperature will continue to expand the volume of the already liquid water (Pritchard et al., 2009). Although the great glaciers are in retreat presently in geological history, further melting could raise sea levels substantially, and the best models suggest a rise of about 1 m in the next 50–100 years. One meter of sea-level rise combined with storms would result in the piercing of many of the barrier bar systems in the world. Low-lying cities and estuaries would also suffer. Consider those who get water from the freshwater parts of estuaries; a sea-level rise might make the water too salty. Coastal biological environments such as salt marshes also can be inundated by sea-level rise. The worst-case scenarios point toward instabilities in parts of the Antarctic ice sheet, which really is a series of glacial units.

There is widespread disagreement over the exact rate and magnitude of continued sea-level rise, but the worst-case scenarios are very worrisome. The Totten Glacier in eastern Antarctica has been demonstrated to have several retreats in the recent geologic past, especially the Pliocene epoch (5.3–2.6 million years ago). Recent geophysical evidence suggests collapse and retreat of eastern Antarctic glaciers, which could be responsible for a sea-level rise of 0.9 m and perhaps 2 m if the glacier retreats even more (Aitken et al., 2016). The West Antarctic Ice Sheet is similarly unstable and may be in the process of collapse, which would produce an increase of sea-level rise of >1 mm per year (Joughin et al., 2014).

Coral reefs seem particularly vulnerable because they typically grow upward at rates of 10 mm y^{-1} at best, which is disturbingly close to some estimates of sea-level rise. Grigg and Epp (1989) suggested that sea-level rise owing to global warming might outstrip coral growth, which would trigger a worldwide catastrophe. Even if the reefs were able to grow, increasing sea temperatures might cause extensive coral bleaching and contraction of reef systems as discussed earlier.

It is also possible that global warming will cause major changes in oceanic circulation. Warming may cause significant changes in oceanic winds, and this might lead to changes in the intensity of ocean currents. Some nearshore regions are very productive because of upwelling, but these systems and their fisheries might be strongly affected if currents are reduced or changed. A study of the North Atlantic has shown that changes in winds and currents sometimes arrive very rapidly, and large local changes in temperature have occurred in the past during climate shifts. Some models suggest that global warming may reduce thermohaline circulation in the North Atlantic to the point that sinking might be reduced and Gulf Stream circulation will be inhibited. Melting of Arctic Ocean ice might reduce salinity in the sub-Arctic Atlantic and also inhibit sinking, which would inhibit Gulf Stream circulation. The outcome locally would be drastically colder regional climates in Europe and in the northwestern Atlantic. But thus far, there is no evidence of such a shift (Church, 2007).

Freshwater inputs because of warming in the Arctic Ocean have had some major impacts on ocean structure, marine plankton and food webs, and biogeographic ranges of marine species in the North Atlantic. Since 1980, summer minimum sea ice coverage has declined from ca. 8×10^6 km^2 to less than 4×10^6 km^2 today. As Arctic Ocean ice melts more and more, freshwater inputs into the northwest Atlantic have caused stratification of the shallow ocean, with low-salinity waters capping deeper high-salinity waters. These changes must be having enormous impact on the structure and composition of plankton communities on the continental shelves of the coasts of Labrador, Newfoundland, and Nova Scotia (see Chapter 19). Also, the reduction of sea ice in the Arctic Ocean in the summer has resulted in expansions of Pacific and Atlantic plankton populations northward into the Arctic Ocean. Trans-Arctic Ocean exchange is likely to be a major source of mixing of high latitude species between Atlantic and Pacific (Greene, 2008).

■ **Increased temperature may cause increases of ocean stratification, reduction of dissolved oxygen, and intensification of mid-water oxygen minimum layers.**

As mentioned in the previous section, increased warming is causing ice melting at very high latitudes and increases of influxes of fresh water into the high latitude ocean. Within the Arctic Ocean, the decline of sea ice is reducing hunting platforms for polar bears and increasing swimming distances required for Pacific walrus to reach prey such as subtidal clams.

Combined with warming from the surface, the ocean may become more and more stratified, with warm low-density water at the surface mixing less and less with deeper cool water. Within the surface waters oxygen concentration may decline (Bopp et al., 2002). Reduced oceanic circulation may also enhance already existing midwater oxygen minimum layers. Combined with reduced oxygen in coastal waters owing to human input of nitrogen, plankton blooms, and bacterial breakdown of organic matter, the oxygen minimum layers could combine with coast surface water to produce stressful low-oxygen oceanic waters over large depth ranges, which would strongly affect most marine organisms within these regions. Oxygen minimum zones at depths of 300–700 m in the eastern tropical Atlantic and tropical Pacific have been expanding over the past 50 years (Stramma et al., 2008). The expansion of oxygen minimum zones may greatly reduce the habitat available for deep-water fishes (Stramma et al., 2012).

■ **Increased temperature may cause physiological stress on marine populations, reorganizations of food webs, and changes in geographic ranges of species.**

We will address the effects of climate change on biological processes throughout the text, but several general effects of temperature rise will be at work in marine systems. In many cases, species will be in conditions of steadily increasing temperature, which may surpass the physiological limits of the species. Many of our continental coasts are approximately north–south in orientation, so we might expect that low-latitude ends of a coastal species range might be truncated owing to high temperature stress. On the high-latitude end, we would expect species ranges to extend to higher latitudes. For example, as coastal temperatures have increased in the northwest Atlantic, some shellfish species such as the snail *Crepidula fornicata* and the oyster *Crassostrea virginica* have extended their range northward along the Canadian Atlantic coast. As mentioned in a previous section, increase of surface water temperatures could cause major reorganizations of water current structure and vertical temperature distributions, which might change the regime of water exchange and nutrient supply to planktonic systems (see discussion in Chapter 12). In all these cases, the interaction of smaller-scale oscillations might produce years in which sea temperature has a high excursion, superposed on a longer-term trend of temperature increase. So, in decades in the future, an El Niño will have far greater thermal stress effects on tropical organisms than in decades past when sea-surface temperature was on average cooler.

Carbon Dioxide and Ocean Acidification

■ **Carbon dioxide additions to the atmosphere result in increased dissolved CO_2, decreased pH, and increased problems for some calcifying organisms in the sea.**

Most of the focus of global climate change has been on ocean warming. But what might be the direct effects of carbon dioxide itself? Carbon dioxide increase in the earth's atmosphere results in an increased addition of CO_2 to seawater. Warmer water can hold less dissolved CO_2 than cooler water. When carbon dioxide is added to water, the following reactions occur, which add hydronium ion, H_3O^+, to seawater:

$$CO_2 + H_2O \rightarrow H_2CO_3$$
$$H_2CO_3 + H_2O \rightarrow HCO_3^- + H_3O^+$$

You can see that seawater contains forms of carbon other than carbon dioxide and water, and these reactions involve those other major ions. Most important, the precipitation of calcium carbonate is involved in a biological process such as making a clamshell or a coral skeleton. Seawater contains three major forms of dissolved inorganic carbon (CO_2, HCO_3^-, and CO_3^{2-}), which act as a buffer system such that additions of CO_2, which result in the production of H_3O^+ acid formation, react with the other components and pH (acidity) does not change very much. But the continued addition of CO_2 and an increase in H_3O^+ cause some CO_3^{2-} (called carbonate ion) to react with H_3O^+ and to become HCO_3^- (bicarbonate ion). The reduction in carbonate ion reduces $CaCO_3$ precipitation.

$$\leftarrow \text{precipitation}$$
$$CaCO_3 \rightarrow Ca^{2+} + CO_3^{2-}$$
$$\text{solution} \rightarrow$$

Can carbon dioxide affect skeleton formation of marine organisms? Many planktonic organisms, including coccolithophorids and foraminiferans (see Chapter 8), have skeletons of calcium carbonate. Many benthic organisms, including corals and mollusks, also have such skeletons. The important questions, therefore, are whether carbon dioxide increases from global climate change and whether human-derived CO_2 additions will affect calcification. Increasing dissolved CO_2 increases water acidity, which drives chemical reactions toward undersaturation of seawater with respect to calcium carbonate.

How can we know whether ocean acidification will be a problem to marine organisms? We must know the degree of **saturation** of calcium carbonate. The value of Ω is a measure of saturation. If $\Omega < 1$, the solution is undersaturated with respect to a solid state of a compound (e.g., sodium chloride), and if $\Omega > 1$, the solution is supersaturated. If $\Omega < 1$, then calcifying organisms will have trouble precipitating a shell. **Going Deeper Box 3.2** shows how Ω is calculated.

There are two forms of calcium carbonate: **aragonite** and **calcite**. Aragonite has a different crystal structure than calcite and is found in corals and in many mollusks. At earth surface conditions, aragonite is not as stable as calcite

Solubility

Solubility is the ability of a substance to be dissolved in a liquid. For most solids, solubility increases with increasing fluid temperature and decreases with increasing pressure. The solubility of sodium chloride is 357 g L^{-1} at 25°C and 1 atmosphere pressure. A solution with this concentration is saturated with respect to sodium chloride.

If we consider the following chemical equation:

$$Na^+ + Cl^- \rightarrow NaCl$$

we can determine the equilibrium between the solid on the right side and the ions on the left side of the equation. This allows us to make the following calculation:

$$K_c = \frac{[Na^+][Cl^-]}{[NaCl]}$$

[Na$^+$] and [Cl$^-$] represent concentrations, and [NaCl] represents the amount of solid salt of NaCl.

We then have

$$K_c \times [NaCl] = [Na^+][Cl^-]$$

which means that at equilibrium (saturation), the product of Na$^+$ and Cl$^-$ concentrations gives a constant known as the **solubility product constant**:

$$K_{SP} = [Na^+][Cl^-]$$

It is now possible to answer the following question: Is the solution at saturation, supersaturated, or undersaturated with respect to a substance?

$$\Omega = \frac{[Na^+][Cl^-]}{[K_{sp}]}$$

If $\Omega = 1$, then the solution is saturated, and if $\Omega > 1$, then the solution is supersaturated. As mentioned in the text, the required value of Ω for the aragonite variant of CaCO$_3$ is even higher for the organism to be able to precipitate aragonite.

---

and more difficult to precipitate. Because of this instability, the value of Ω must be greater than 3 for aragonite to be precipitated. The value of Ω is currently >3 in most of the world ocean, but the value of Ω is declining because of increased CO$_2$ entering seawater from the atmosphere. Coccolithophores are composed of calcite; planktonic foraminiferans are mostly calcite, but some have aragonite. Pteropods, holoplanktonic snails, are one common group with aragonite shells. See Chapter 12 for discussion of ocean acidification effects on coccolithophorids and pteropods. Reef corals precipitate aragonite. Their future is very worrisome (see Chapter 17).

Atmospheric carbon dioxide increase should increase dissolved carbon dioxide and therefore reduce pH of the ocean. Current models predict undersaturation of calcium carbonate well within the next century. Effects on calcifying organisms are likely. Models from the Intergovernmental Panel on Climate Change predict a decrease of the carbonate ion concentration by almost 60 percent by the end of 2100. But there is an important interaction with ocean temperature. Warming of seawater reduces the solubility of carbon dioxide and reduces acidification. These two trends may have a canceling effect in some locations, but not all parts of the ocean will increase in temperature very much and the acidification effect is expected in higher latitudes especially but also in the tropics. In addition, a small net decline in pH might fall most heavily on aragonite-precipitating organisms, such as corals in the benthos and pteropods in the plankton. Recently discovered deep-water coral communities (see Chapter 17) may also be affected. The solubility of CaCO$_3$ increases with increasing depth, and a threshold depth of about 4,000–5,000 m where dissolution occurs will become shallower as carbon dioxide additions to the ocean continue.

■ **Ocean acidification is being accelerated in certain environments such as eastern Pacific upwelling centers where dissolved oxygen is low and waters rise toward shore environments but also in coastal environments where human-influenced degraded coastal waters result in lower dissolved oxygen.**

While carbon dioxide is being added from the atmosphere throughout the global ocean, certain environments are now experiencing sufficient acidification that effects on biological systems have already been observed. In the eastern Pacific margin of North America, persistent upwelling often occurs as a response to regional winds affected by Ekman circulation (see Chapter 2). Such upwelling often brings oxygen-depleted, organic-rich waters to the surface. As respiration exhausts the oxygen in these waters, carbon dioxide increases and combines to create acidic conditions. Along the coast of Washington to northern California, this process has created low-pH conditions, which have inhibited calcification in planktonic pteropods that often dominate the zooplankton and are part of the food of juvenile fishes such as salmon. In a few coastal areas, the localized acidification from upwelling has greatly endangered the shellfish industry, and local bivalve mollusk hatcheries have had major failures owing to poor calcification of larval shells. Failures of larval colonization have been noted from waters of Alaska all the way to Oregon. At Tatoosh Island, Washington, monitoring using a field sensor demonstrates a continuous decline of mean pH (8.3–7.7) over the period of 2000–2011, although there is a great deal of variability within each year (Wooton and Pfister, 2012). Shell thickness of large shells of the California sea mussel *Mytilus californianus* has also declined since the 1970s (Pfister et al., 2016). The mechanisms that affect dissolved carbon in this region are not well resolved, but the changes are consistent with effects on shellfish observed along a much longer stretch of coastline.

In many coastal water bodies, release of human sewage has added nutrients to coastal waters, which in turn stimulates very high productivity of phytoplankton (see Chapter 22). Much of the phytoplankton dies, organic matter is consumed by bacteria, and oxygen is therefore depleted from the water column. Again, increases in dissolved carbon dioxide occur, which can reduce the pH of coastal seawater. As low-oxygen conditions spread owing to such nutrient pollution, another form of marine acidification will threaten the survival of coastal organisms, especially shellfish.

■ CHAPTER SUMMARY

- Climate changes in the ocean on spatial scales from the size of an entire ocean to that of the world ocean. Some changes consist of oscillations over a few decades or less. But owing to human impacts on the planet mainly involving release of greenhouse gases such as carbon dioxide, climate change from the early nineteenth century to the present has also been trending toward a warmer ocean.

- Climate oscillations have been observed in all of the world's oceans, and fluctuations are correlated with a variety of climatic variables and ocean conditions, especially wind systems, atmospheric pressure, and ocean temperatures.

- El Niño Southern Oscillation (ENSO) is a change of atmospheric conditions and sea-surface temperature that has greatest impact on the tropical Pacific, but effects are worldwide. In the Pacific, warm water moves eastward across the tropical belt and creates warm conditions in the eastern Pacific, a

deeper thermocline, and a cessation of upwelling along the west coasts of North and South America. This condition alternates with La Niña, or cooler conditions in the Pacific with strong upwelling along the Pacific coasts of the Americas. This oscillation has major biological impacts.

- The North Atlantic Oscillation (NAO) involves a major shift of climate centers in the North Atlantic, causing large swings in storm systems and rainfall patterns in the Mediterranean Sea and Europe. In one part of the oscillation, a reduction of wind causes anoxia and collapses of fisheries in a northern European fjord.

- The Pacific Decadal Oscillation (PDO) has prominent effects on shifts of atmospheric pressure and SST in the North Pacific Ocean. Oscillations are associated with major shifts of cold-adapted and warm-adapted mobile species and reorganizations of food webs that have strong impacts on fisheries.

- In the greenhouse effect, greenhouse gases trap heat, increasing lower atmosphere temperature and strongly affecting ocean climate. The increase in greenhouse gases since the nineteenth century has led to warmer ocean temperatures, sea-level rise, and major changes in oceanic circulation, including increased stratification and decline in dissolved oxygen.

- The solubility of carbon dioxide in seawater is causing a trend toward lower pH of seawater known as ocean acidification. This process greatly endangers marine organisms, especially those that require skeletons of calcium carbonate. Ocean acidification has become a current problem in coastal upwelling centers off the west coast of North America from northern California to Alaska. Coastal waters with strong nutrient pollution have lowered dissolved oxygen, higher dissolved carbon dioxide, and lower pH.

■ REVIEW QUESTIONS

1. Distinguish between climate change oscillations and climate change trends.

2. What aspects of climate and ocean change are typically involved in climate oscillations?

3. In what general geographic direction do the effects of the North Atlantic Oscillation move in the Atlantic Ocean?

4. How do fluctuations in the NAO have the potential to affect coastal climate and coastal current patterns?

5. What is the main water motion process that is strongly affected by ENSO fluctuations in the eastern Pacific?

6. During El Niño, what are the major changes that occur in the ocean along the tropical belt between the central Pacific and Peru?

7. What major fisheries are known to be affected by changes in the Pacific Decadal Oscillation?

8. Give a scenario where a change in a climate oscillation can exert effects on a regional food web?

9. What are two main greenhouse gases?

10. What is the evidence that greenhouse gases have been increasing the past 50 years or so?

11. What is the principal evidence used to show that temperatures on the earth have been increasing over much of the planet?

12. What aspect of climate change might affect the pH of seawater? Why are species with aragonite shells more vulnerable to acidification than those with calcite shells?

13. What would be the effect of increasing seawater temperature on the degree of ocean acidification?

14. What processes are at work in western North American upwellng centers to cause localized ocean acidification?

Visit the companion website for *Marine Biology* at www.oup.com/us/levinton where you can find Cited References (under Student Resources/Cited References), Key Concepts, Marine Biology Explorations, and the Marine Biology Web Page with many additional resources.

CHAPTER 4

Ecological and Evolutionary Principles of Marine Biology

Ecological Interactions

- **Ecology is the study of interactions between organisms and their environment and the effects of these interactions on the distribution and abundance of organisms.**

Ecology is the study of interactions between organisms and their environment and how these interactions determine their distribution and abundance. **Biological interactions** occur between organisms, whereas **abiotic interactions** are effects of non-biological factors, such as seawater chemistry, on the functioning of organisms. In practice, the two kinds of interaction cannot be easily separated. For example, low temperature might prevent a cold-blooded creature from moving very rapidly (an abiotic interaction), and this limitation might in turn reduce its chances of escaping a predator (a biotic interaction).

- **Resources are materials whose availability or abundance may limit population growth.**

A resource is any material whose availability or abundance in the natural environment can limit survival, growth, or reproduction. Food, space, and dissolved inorganic nutrients are all potentially limiting resources. Resources that can be depleted and are no longer available are **nonrenewable.** Resources that will continue to become available are **renewable.** The issue of renewability can be resolved by scaling against the life span of the organism that is exploiting the resource. Over the lifetime of some sessile organism, the space to which it attaches is a nonrenewable resource, but it will of course be renewed once the individual dies. Microorganisms such as bacteria as a food resource

are renewable because they can often recover in population size even when grazed by larger and much longer-lived organisms such as sponges.

The Ecological Hierarchy

- **Ecology is studied at many interacting hierarchical levels, including individual, population, species, community, and ecosystem.**

Ecological processes should be studied at many levels of a hierarchy, or a nested series of sets.

The Levels Defined

INDIVIDUAL LEVEL An individual is an organism that is physiologically independent from other individuals. Examples include a single snail and an interconnected colony of coral polyps.

POPULATION LEVEL A **population** is a group of individuals *of the same species* that respond to the same environmental factors and freely mix with one another—for example, in mating.

SPECIES LEVEL A **species** is a single population, or a group of populations, that is genetically isolated from other species; that is, it does not interbreed and reproduce with other species. An example of an appropriate question at this level of the hierarchy is: Will a change in sea temperature cause a species to become extinct? Although we do not include species per se in the ecological hierarchy, they are crucial in understanding the long-term evolutionary directions of ecosystems.

COMMUNITY LEVEL A **community** is a group of potentially interacting populations, each belonging to different species and all living in the same place—for example, all the barnacles, snails, seaweeds, starfish, and other species on a rocky shore that live together and interact, for instance, as predators and prey.

ECOSYSTEM LEVEL An **ecosystem** is an entire habitat, including all the abiotic features of the landscape or seascape and all the living species within it that interact—for example, an estuary and its inhabitants. The definition of the boundaries of an ecosystem can be somewhat arbitrary. For example, we can define a coral reef ecosystem, but we sometimes might want to define a coral-reef—open-ocean ecosystem if we want to understand the processes affecting the many species that broadcast larvae into the open sea. A salt-marsh ecosystem might be protected from wave damage by a nearby oyster reef system.

BIOSPHERE LEVEL The **biosphere** is the entire set of living things on the earth and the environment with which they interact. Interactions at the biosphere level may be crucial to human welfare—for example, the carbon budget and climate change.

INTERACTIONS AMONG THE LEVELS The various levels of the hierarchy cannot always be studied separately; hierarchical levels do interact. For example, changes in climate at the biosphere level may affect an individual snail's ability to escape a predator. Here, an upper level of the hierarchy has direct effects on a lower level. As another example, the efficiency of photosynthesis of individual phytoplankton may sum up to a major change in the nutrient cycling of a marine ecosystem. Such an effect involves the impact of changes at a lower level of the hierarchy on upper hierarchical levels.

Interactions on the Scale of Individuals

At the scale of individuals, both abiotic and biotic interactions are quite important. We can define **ecological niche** as the range of environments over which a species is found. The range of environments has both biological and physicochemical dimensions, such as interacting species, water depth range, and salinity range.

■ **Many ecological interactions occur between individuals and may be classified on a plus-minus-zero system depending on whether an individual benefits, suffers because of, or is not particularly affected by the interaction.**

A plus-minus-zero system may be used to characterize ecological interactions. Plus (+) interactions benefit a species. Minus (−) interactions harm it. Zero (0) interactions do not affect it in important ways. Table 4.1 summarizes the basic interactions and the plus-minus-zero classification for the organisms involved in each. Note that an interaction is generally classified using two symbols (e.g., + and −) to represent the effect on both kinds of organism involved in it.

Territoriality

■ **Territoriality is the maintenance of a home range that is defended.**

Territoriality is the maintenance of a home range and its defense against intruders. An individual may maintain a territory to protect (a) a feeding area, (b) a breeding area, or (c) a specific nest site. In most cases, territoriality is intraspecific. An example is the maintenance of nesting territories by many species of seabirds.

Predation

■ **Mobile and stationary predators search for prey using chemical, mechanical, and visual stimuli; some lure prey by using various "deceptions."**

Predators may be either stationary or mobile. **Stationary predators** include sea anemones (**Figure 4.1**) and other cnidarians; **mobile predators** include fishes, starfish (**Figure 4.2**), gastropods, birds, and crabs. Most swimming and crawling predators can move large distances to locate prey. Although fish are obvious, crabs and even starfish may move on the order of kilometers to locate new patches of prey. Within the sediment, mobile predators, such as polychaete annelids, sipunculids, and burrowing gastropods, move over smaller spatial scales. Despite the diversity, there are some organizing principles relating to prey handling and capture and to interactions between predators and prey.

TABLE 4.1 Types of Individual Interaction

TYPE	NATURE OF INTERACTION	PLUS-MINUS-ZERO CLASSIFICATION
Territoriality	Beneficial to one and detrimental to another or detrimental to both	+ − or − −
Competition	Beneficial to one and detrimental to another or detrimental to both	+ − or − −
Predation	Beneficial to one and detrimental to another	+ −
Commensalism	Beneficial to one but no effect on the other	+ 0
Mutualism	Beneficial to both individuals	+ +
Parasitism	Beneficial to one and detrimental to another	+ −

FIG. 4.1 A stationary predator, the anemone *Anthopleura xanthogrammica*, consuming a mussel. (Photograph by Jeffrey Levinton)

FIG. 4.2 A mobile predator, the starfish *Pisaster ochraceus*, consuming a cockle. Note the extended tube feet that are attached to the bivalve shell of the prey via suction (see Chapter 14). (Photograph by Paulette Brunner, with permission from Friday Harbor Laboratories)

■ **Mobile predators may adjust their hunting behavior to optimize the rate of ingestion of prey.**

Successful predators will consume more prey, which in turn will increase growth and reproductive output. Therefore, one might expect natural selection to optimize the organism's efficiency either in maximizing the amount of energy gained per unit time or in minimizing the time spent feeding so that there is more time to carry out other vital functions such as reproduction. **Optimal foraging theory** establishes the decision rules used by predators to optimize their food intake.

Many predators are able to consume a variety of prey items. Many drilling snails, for example, can consume a variety of barnacle and mussel species. Sea otters dive and retrieve urchins, abalones, and other large benthic invertebrates. Some species are of greater nutritional value than others, and the question is whether to specialize on the nutritionally valuable items or resort to feeding on the poorer items. When a predator encounters a prey item that is not very rewarding, should the predator feed on the item or pass it up to find something better? The **diet-breadth model** predicts that when overall food density is high, it pays to

specialize on the good items and to ignore the choices of lower food quality. As overall food density decreases, it pays to become less choosy and broaden the range of prey. This conclusion can be altered if there is some cost in learning to switch from one prey item to another. For example, a snail might develop olfactory imprinting on a given prey type. It might cost more to change this imprinting than to continue to hunt for the original prey item. Satiation, or the limits of digestive activities, may also be important. A predatory animal might pass up a prey item if the predator's gut is full and it can digest no more for the moment.

The time spent in a food patch is also an important area of decision that affects the predator's total intake of prey. The **time-in-patch model** predicts that *the time spent in a patch of prey should increase with an increase of travel time between patches.* This makes intuitive sense because an increase of travel time reduces the overall opportunity to gain food. It is not worth finding a new patch unless the food in it justifies the travel time. This adjustment has been found in blue whales, which feed on large zooplankton and small fish by trapping them on huge baleen plates (see Chapter 9). A study using sophisticated tracking methods showed that the whales fed for longer periods when they dived to great depths to patches of prey and required longer transit times to these depths and recovery at the surface. Foraging time periods were much shorter when the whales dived to relatively shallow depths (Doniol-Valcrose et al., 2011).

The choice of a best-sized prey is a good example of the optimal foraging approach. Prey organisms that are too large might take an inordinate amount of time to consume. Imagine a starfish spending 2 days to open and consume a large mussel. That time might be more profitably spent on somewhat smaller mussels, whose relative ease of opening would compensate for the reduced reward per mussel. It might also be relatively unprofitable to select very small mussels because too much time would be invested in handling and opening prey with little reward. **Figure 4.3**, which illustrates this argument graphically, shows the results of a study of crabs feeding on mussels. A large mussel provides a big meal for the crab, but the length of time required to crack such a mussel open makes it more profitable to select smaller mussels. Mussels that are too small are not worth the handling time. As a result, the crab selects intermediate-sized mussels.

Predator Avoidance

■ **Resistance to predators increases individual fitness and is therefore enhanced by natural selection.**

Many marine species have evolved a large variety of traits to deter predators. For example, the large majority of tropical sponges are highly poisonous. This might be expected of a sessile group with no ability to move to hide from predators.

Like other adaptations, antipredator defenses originate as variations in natural populations of prey. The presence of any deterrent behavior, morphology, or poison would enhance survival, and the survivors that possess such traits would contribute their genes to later generations.

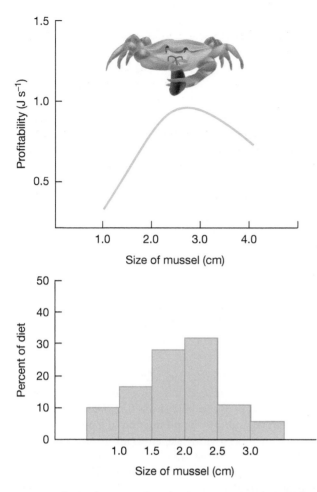

FIG. 4.3 Optimal strategy for selecting prey mussels. Top: theoretical cost-benefit analysis for the reward of a mussel prey as a function of prey size (in terms of energetic return in joules obtained per second). As expected, the shore crab *Carcinus maenas* selects intermediate-sized mussels. Bottom: actual sizes selected by the crab.

■ Marine organisms avoid predators by means of crypsis, deceit, escape responses, and mimicry.

A most obvious strategy to avoid predation is **crypsis**, or blending with the background. A variety of animals, including many fishes, crustaceans, and cephalopods, employ chromatophores, which are cells that can rapidly alter their color. Flounders, for example, can change the color pattern of their dorsal surface in a few seconds to match clean sand or a mottled bottom (**Figure 4.4**). Frogfish of the family Antennariidae have many representatives that are completely cryptic, such as species that resemble coral heads and lumps of sand.

Most of these cryptic species are drab and blend in well with the background. For example, periwinkles usually blend with the background of their rocky substrata; I have even seen a population of orange snails in an area of orange granite in Scotland. Many species blend with the background by means of camouflage coverings. To camouflage its dorsal surface, a decorator crab (spider crab) attaches fragments of seaweed, sponges, bryozoa, compound ascidians, and whole anemones (**Figure 4.5**). At night, when it is safe, it moves openly and feeds.

A number of species exhibit deceptive coloration and behavior. Many smaller reef fishes have large posterior spots. Predators are fooled into attacking the posterior of the fish as it is swimming to escape in the opposite direction. Squid, cuttlefish, and the sea hare *Aplysia* squirt dark ink, which conceals their escape.

Many species respond to predators by means of very specific **escape responses**. One of the simplest escape responses is a refuge in time. Activity at night by many mobile invertebrates allows them to avoid predators that depend on vision to find prey. But what if the prey also depends on vision? Fishes and crabs detect predators visually and can move away rapidly. Many sluggish benthic invertebrates have stereotyped escape responses. For example, when in contact with starfish, scallops escape by clapping their valves rapidly and expelling water through jet holes on either side of the hinge. Some anemones react to starfish by lifting off from the substratum and swimming into the water column. To escape from the large starfish *Pycnopodia helianthoides*, the large Pacific sea cucumber *Parastichopus californicus* violently contracts its body wall muscles and springs up from the bottom.

Resemblance of a background environment is a common strategy, but many species resemble other species to avoid predation. Although species such as frogfish may resemble corals to effectively disappear into the background, other species resemble model species. Mimicry is an evolved morphology or behavior that allows an organism to resemble another species, which serves the function of reducing attacks by predators (Randall, 2005). **Batesian mimics** may be harmless and yet resemble a model species that is dangerous and avoided by typical predators on the mimic. For example, in the southwest Pacific, species of relatively harmless snake eel have striping patterns that strongly resemble highly venomous sea snake species. Snake eels move freely in open water during the day, whereas other snake eel species are very cryptic. A spectacular example is the mimic octopus *Thaumoctopus mimicus*, which can dynamically

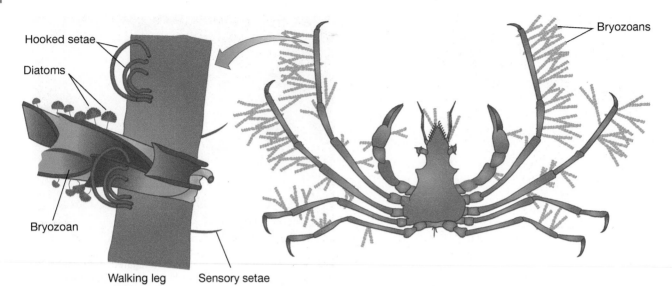

FIG. 4.5   Spider crabs have hooked setae on the sides and rostrum (anterior pointed section) and on the sides of the carapace. The California crab *Podochela hemphilli* may carry fragments of bryozoan colonies, which are ensnared in the hooked setae and camouflage the crabs from the view of predators. (After Wicksten, 1982)

change its shape and color pattern to mimic more than 15 species of nearby dangerous predators, including venomous sea snakes, jellyfish, and stingrays (Norman et al., 2001). To mimic a venomous sea snake, it buries in the sediment all of its body but two arms and uses photophores to adopt the color pattern and movement of a venomous sea snake by undulating above the bottom. Alternatively, **Müllerian mimics** may themselves be harmful and resemble other species that are also harmful. This type of mimicry is not well known in marine organisms.

- ◼ **Many marine organisms can produce various morphologic features to discourage predator attacks (e.g., spines, strengthened skeletons, and other devices).**

Mechanical defense is one of the most common adaptations in defending against predators. In some cases, simple toughening of the body wall or stiffening by means of internal structures proves very effective. In many tropical seaweeds (e.g., *Halimeda* and *Pennicillus*), the thallus is strengthened with calcium carbonate. Many gastropods have a thickened shell that deters predatory fishes. A large number of fish species have spines, some of which are poisonous. For example, members of the family Scorpaenidae, including scorpion fishes and stonefishes, are armed with poisonous spines. In the case of the Pacific coral reef stonefish, the poison is quite virulent and can kill an adult human. The Caribbean urchin *Diadema antillarum* has long sharp spines with reversed barbs. These spines deter many predators, but some fishes bite at the urchin and apparently survive piercing by its spines.

An alternative strategy to fixed defense structures is to produce the defensive structure only when predators are present. Such **inducible defenses** provide an advantage to the prey, which might otherwise waste resources for growth or reproduction when predators are absent. Inducible defenses are especially useful for sessile species, which can deploy them in the direction of predator attack. For

example, the sessile bryozoan *Membranipora membranacea*, which lives as a sheet of individuals on hard surfaces, is often attacked by a specialized sea slug, *Doridella steinbergae*. An attack induces the production of a peripheral zone of colony members whose skeletons are armed with spines (**Figure 4.6**). The spines reduce predation by about 40 percent, but the whole colony grows more slowly than colonies not exposed to predators.

The trade-off for using an induced defense is illustrated well in the acorn barnacle *Chthamalus anisopoma*, which lives in the Gulf of California. This barnacle occurs in two forms: conical and bent (**Figure 4.7**). The conical form is typical of most acorn barnacles. In the case of the bent form, the barnacle grows with the rim of the aperture oriented perpendicular (rather than parallel) to its base. This protective growth form is induced by the presence of the carnivorous snail *Acanthina angelica*, which is spatially variable in abundance. One might ask why the barnacle does not always produce the bent form. Apparently, the bent form feeds less efficiently, with a reduction of somatic growth and fecundity. Thus, there is an advantage to being conical if predation is low. This situation stabilizes the coexistence of the two-form strategy.

An interesting change has occurred fairly recently in some Maine populations of the rocky-shore periwinkle *Littorina obtusata*. In locales where the snail has been exposed to the predatory crab *Carcinus maenas*, shells are lower spired and thicker, but shells are thinner where the crabs are rare or absent. Thus, the snails do not have to pay the price of making thicker shells when predators are absent (Trussell, 1996).

- ◼ **Many marine organisms are defended chemically by toxic organic compounds, acid secretions, and toxic metals.**

Production of toxic compounds includes the secretion of acid by seaweeds and tunicates and the manufacture of toxic

(a)   (b)

FIG. 4.6 Inducible defenses. (a) Spines induced by the predatory sea slug *Doridella steinbergae* on a colony of the bryozoan *Membranipora membranacea*. Scale is 1 mm. (Courtesy of Drew Harvell) (b) Stolons armed with nematocysts (light band) induced when unrelated colonies of the hydroid Hydractinia come into contact. (Photograph by Richard Grosberg)

FIG. 4.7 The (right) conical and (left) bent forms of the acorn barnacle *Chthamalus anisopoma*. The animal develops the bent form if predatory snails are present. (Courtesy of Curtis Lively)

organic compounds by many species of marine higher plants, seaweeds, and animals. The organism usually synthesizes these substances, although some animals can eat toxic plants and store the toxic substance. For example, the sea hare *Aplysia* can graze on the alga *Laurencia* and sequester this organism's highly toxic halogen-bearing terpenes. The sea hare is thus also toxic. See the discussion in Chapter 21 of how marine toxic substances may be useful in medical applications.

If a conspicuous color can be associated behaviorally with an unpleasant dining experience, a predator might avoid the prey on other encounters. Many of the most poisonous marine organisms are conspicuous rather than cryptic. Many free-living flatworms are highly toxic but brightly colored. One often sees bite marks on individuals, which suggests that the learning process that connects color and distastefulness is continuously reinforced. The poisonous black tunicate *Phallusia nigra* is conspicuous against its usual background: white coral reef or sand. Its

tunic can contain as much as 1 percent vanadium, a highly toxic metal, and the tunicate can also produce vacuoles of sulfuric acid.

Natural selection would increase genetic variants that have conspicuous coloration, but only if predation attempts largely failed and allowed the prey to escape. Otherwise, the conspicuous prey could not live to reproduce and spread the conspicuous color trait in the population. Such an association between a bad stimulus and conspicuous color is known as **aposematic coloration**.

The work of Joseph Pawlik (2011) demonstrated that many species of coral reef sponges have a wide variety of toxic compounds that can be detected by predatory fish with very generalized physiological sensors. Many of these compounds are not only toxic to predators but also poison fouling organisms that might smother the sponges. On the other hand, about 30 percent of the Caribbean sponges are not well defended at all. These species have adopted instead a strategy of rapid growth to allow survival against predation by fish, which rarely destroys the whole sponge.

■ Mechanical and chemical defenses against predation change in frequency with latitude, habitat, and oceanic basin.

Frequencies of species with toxic defenses trend with geography. The proportion of sponges and sea cucumbers that are toxic increases toward the tropics and can reach 100 percent on tropical coral reefs. Mechanical adaptations of snails to resist crushing by crabs also increase toward the tropics. These trends reflect greater predation pressure in the tropics, which enhances natural selection for increased defense. Although predation is also often intense at some high latitudes, the high diversity of predators in low latitudes may impose the greater selective force.

■ Microhabitat can strongly affect a creature's vulnerability to predators.

Marine animals may be able to avoid predators by simply retreating to inaccessible habitats. In some cases, the organism lives in a **spatial refuge** that is inaccessible to predators. Marine animals may also alternate between a microhabitat that provides a refuge from predators and one used for feeding or reproduction. Rocky intertidal shores have strong gradients of desiccation and temperature. This is a special problem for mobile predators such as asteroid sea stars and drilling gastropods, which require long periods of time to subdue and consume their prey. As a result, predation intensity is far less intense in the highest part of the shore, and prey such as mussels can escape predation in this zone. Mark Hay (1991) has noted that many small herbivorous invertebrates, such as amphipods, feed on seaweeds that are otherwise toxic to larger herbivores such as mobile fishes. They also nestle in the seaweed. The smaller herbivores may have evolved a preference for toxic algae, which escape removal by larger herbivores and thus provide a refuge and food source for the smaller animal species.

Commensalism

■ Commensal relationships benefit one species only. The benefit usually relates to food, substratum, or burrow space.

Commensal species acquire a benefit from another species but return no benefit or harm. Commensal relationships may be facultative or obligatory. A **facultative commensal** species does not completely depend on a certain single species but may live on one of a variety of species. Barnacles, for example, may settle and live on a variety of species of mussel or on other barnacles, seaweeds, or even rock. On the other hand, **obligatory commensals** can live only with certain other species. The western North Atlantic parchment worm *Chaetopterus* often contains a commensal crab *Pinnixa chaetopterans*, which settles and invades the worm tube as a larva. The crab eventually grows too large to leave the tube and eats material swept in by currents generated by the worm's parapodia. Burrows of the eastern Pacific echiurid worm *Urechis caupo* often contain a gobiid fish, a polynoid polychaete, and a pinnotherid crab (**Figure 4.8**). The polychaete feeds on some of the mucus bag constructed by the proboscis of the host *Urechis*, which the latter uses to capture organic particles for food. The fish and polychaete probably derive protection from predators and also probably feed on detritus and prey in the burrow.

Mutualism

■ Mutualism is an evolved association among two or more species that benefits all participants.

Mutualisms involve pairs of species that exchange crucial resources. Such relationships probably began as facultative interactions, but genetic variation allowed the evolution of interdependence, which might increase reproductive

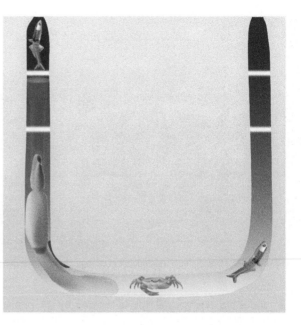

FIG. 4.8 Part of the burrow of the echiurid worm *Urechis caupo* showing the following commensals: the goby *Clevelandia* and the pinnotherid crab *Scleroplax*. (After Fisher and MacGinitie, 1928)

output of both partners in the mutualism. The obvious disadvantage of an obligatory relationship is the danger that one of the species will become locally extinct.

■ Mutualism often reduces the risk of predation or disease or provides food for one member of the species pair.

Many mutualisms are a trade-off between protection against predation on the part of one species and some other benefit on the part of the other participant. An association between species of the coelenterate genus *Hydractinia* and species of the hermit crab *Pagurus* is a good example. The coelenterate lives as a colony on hermit crab shells, and the relationship is species specific. The hermit crab is protected against predators and fouling by the *Hydractinia*. The hermit crab occupies a relatively fresh shell that serves as a substratum for the coelenterate. A number of crab species carry anemones on their claws, and some species have clearly defined rows of teeth on the claws, upon which the anemone holds on. The Chilean actiniarian anemone *Actiniloba reticulata* will move actively toward the legs of the crab *Hepatus chilensis*; on reaching its destination, it creeps along the crab's body and eventually comes to rest on the claws. In other cases, the crab collects the anemones. When disturbed, these anemone crabs wave their claws and threaten intruders with the stinging tentacles of the attached anemone.

One of the most remarkable mutualisms in coral reefs occurs between cleaner shrimp or cleaner fishes and a large number of fish species. Cleaner shrimp and fishes feed by picking ectoparasites off fishes, which approach them regularly (**Figure 4.9**). The Pacific cleaning fish *Labroides dimidiatus* maintains cleaning stations that are visited by about 50 species of fishes each day. "Customers" are attracted to the undulating movements of the

FIG. 4.9 Goatfish, *Mulloidichthys flavolineatus* at Kona, Hawaii, is being cleaned by two cleaner Wrasses, *Labroides phthirophagus*. (Brocken Inaglory)

cleaning fish. Interactions with cleaner fish result in re-duced predatory attacks, and cleaner fish that compete tend to do a better job than when cleaner fish are less dense. The fish *Aspidonotus taeniatus* mimics the cleaner fish undulation, but instead of picking parasites, it attacks the approaching fish and takes a bite out of its fins. This "cheating" is likely a part of the cost and benefit that goes behind the development of the cleaner fish mutualism and whether a cleaner fish should pick parasites only or bite its client!

Parasitism

■ **Parasitism occurs when members of one species live at the expense of individuals of another species, without consuming the hosts totally as food and thereby killing them.**

Parasites live at the expense of other species and may get nutrients or shelter by damaging their hosts. **Ectoparasites** live attached to or embedded within gills, body walls, and other surfaces. **Endoparasites** live within the body and may occupy circulatory vessels or ramify within certain organs or tissues. If parasites are ineffective in utilizing their host, other parasites may enter and displace them by competi-tion. If they are too effective, they may kill their host or even drive the host population to extinction. Because of this, parasitic species probably evolve through cycles of varying virulence. It is often difficult to draw an exact distinction between commensals and ectoparasites. Barnacles on fishes are probably harmless when sparse in density. In great num-bers, however, they create sufficient projections to increase drag and thus impede the host's swimming.

Endoparasites have **highly modified morphologies** that adapted them to life within body cavities and to food uptake and absorption of fluids. Organs needed for free life, such as sensory structures and locomotory appendages, are usually lost. The life stage of the parasite that resides in the host can seem barely related to its actual relatives, which may be typical free-living forms. In contrast to the overall degen-eration, the reproductive organs of such parasites are usually hypertrophied and acquire a central importance.

■ **Parasites of invertebrates often affect the reproduction of the host.**

Some parasites seem to affect the reproductive organs of their hosts more than they affect any other organ. As a result, the hosts often survive but are sterile. The para-sitic rhizocephalan barnacles, for example, have a typical crustacean-looking planktonic larva whose female settles, penetrates the body, and invades the fatty tissues of the reproductive organs of its crab hosts (**Figure 4.10**). The parasite uses the fat reserves for its own reproduction at the expense of the host, which may not have functional gonads as a result. Eventually, the barnacle tissue erupts through the crab's abdomen, allowing male barnacle larvae to settle, penetrate, and produce gametes that fertilize the female. The parasite, when invading a male crab, may cause its entire morphology to resemble a female and even induce mating behaviors characteristic of females.

Many animals are often in a race to grow and mature before the parasite load becomes too high for reproduction or even survival. This is a special problem for the eastern American mud snail *Tritia obsoleta*, which reproduces in its third year. By this time, females in many populations are densely parasitized by several species of trematodes and may not be able to reproduce.

■ **Parasites often have complex life cycles that depend on more than one host species.**

Because the host dies eventually or because its death may be accelerated by the presence of parasites, the parasites must have a means of dispersing to other hosts. As a result, parasites often have **complex life cycles**, with very differ-ent morphologies adapted to function in widely differing

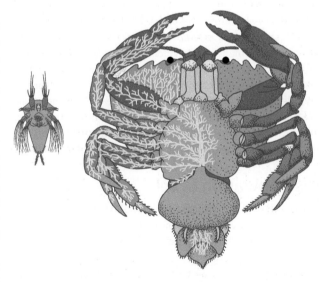

FIG. 4.10 Diagram of the extent of invasion of a rhizocephalan barnacle into the body of a crab. Swimming larvae (left, not to scale) invade a crab host and inject cells that reproduce and propagate a nutrient-absorbing tissue within the crab's body (right). (After Nicol, 1967)

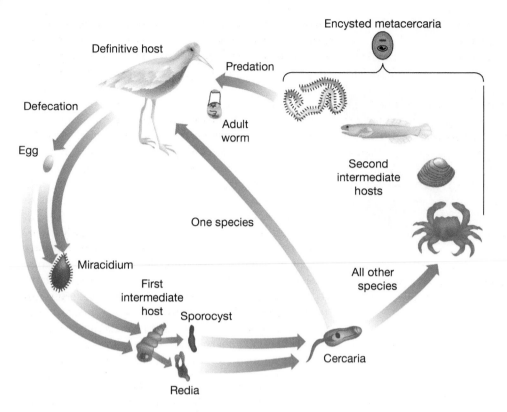

FIG. 4.11 Marine parasitic trematodes have complex life cycles with several intermediate hosts. (From Sousa, 1993)

microenvironments. There is a danger in depending on multiple hosts because one of the hosts might be absent or difficult to locate.

Many parasites have life stages suitable for specialized parasitic existence, for dispersal, and for location of hosts (**Figure 4.11**). Species of the crustacean isopod group Epicaridea may have two hosts. For example, the isopod parasite *Portunion maenadis* has a larval stage that attaches to the copepod *Acartia*; but it also has a free-swimming stage and a second parasitic stage, which lives in the visceral cavity of any of a number of crabs. The isopod parasite of the shore crab *Carcinus maenas* becomes a saclike sheath and bears no resemblance to a typical free-living isopod. In the phylum Platyhelminthes, or flatworms, a number of parasitic trematodes also have complex life cycles. Many species have a stage that inhabits mollusks, a free-swimming stage, and a terminal stage that invades fishes or birds. The fishes and birds often pick up the parasites while preying on the bodies of mollusks or even the siphons of clams.

The Population Level

■ **A population is a group of individuals that are affected by the same overall environment and are relatively unconnected with other populations of the same species.**

A species can be divided into a series of geographically localized populations. The individuals in a population share the same general influence of the physical and biological environment. Within the population, it is much more likely that individuals of a given species will breed with each other as opposed to members of other populations. Geographic barriers such as peninsulas or sudden breaks

in the environment might divide the species into a series of populations. For example, Cape Cod, Massachusetts, is a major barrier along the coast of the eastern United States, and many species do not have extensive dispersal across this barrier. Water temperature increases greatly from the north to the south of Cape Cod, and the same applies to Point Conception in southern California. The geographic ranges of many species end at such barriers.

Population size refers to the number of organisms in a defined area. **Population density** refers to the number of individuals per unit area (e.g., number of animals m^{-2}) or per unit volume (e.g., number of animals m^{-3}) and gives an idea of the degree of crowding or the degree of individual access to scarce resources, such as food or space.

■ **Population change stems from survival, birth, death, immigration, and emigration.**

Most marine populations are dynamic, and extensive change is the rule. Survival of adults is a major factor in population change. If survival is high, then the current population size plays a major role in explaining the population size in future time periods. **Generation time** is the mean time between birth and the age of first reproduction. The existence of more generations per unit time will produce more offspring and a greater potential rate of population increase. Many marine species are capable of producing hundreds of thousands of eggs per female. This is testimony to the low survival rate of adults and the extremely low survival typical of juveniles. Juveniles are often planktonic larvae, and the variability of ocean currents often dooms them to failure in that they never find the proper habitat in which to settle (see Chapter 7). Food limitation may also limit reproduction. Immigration and emigration of adults can affect the change of population abundance.

We can chart the probability of survival of different-age classes by using a graph known as a **survivorship curve**. **Figure 4.12** shows an expected survivorship curve for a species (e.g., a crab) with a planktonic larval stage and a postsettling adult stage. We begin with a starting cohort and follow the mortality of these animals with increasing age. The survivors are plotted on a logarithmic abundance scale, and in this plot, the slope of the line gives the rate of mortality. As can be seen, the rate of mortality for the planktonic stage is far greater than that for the postsettling stage. Survival can be estimated by sampling a population repeatedly as long as immigration is slight and one can distinguish newly born individuals.

Reproduction is usually seasonal and corresponds to increases of food for reproducing adults and to environmental factors such as temperature and salinity. Because of this, birth is also seasonal. Different **year classes**, or sets of individuals born in the same year, can usually be identified by distinct sizes because animals of one year class have an entire year's head start on growth relative to the next year class. It is sometimes possible to determine the age of marine organisms whose date of birth is unknown. Growth rings can be found in the otoliths ("ear bones") of fishes, in the skeletons of corals, and in the shells of clams and snails.

Population size, fluctuation, and extinction are closely related. Most populations fluctuate greatly because of changing environmental conditions that affect reproduction and mortality. When population size is very small, relatively minor random changes may cause population extinction; low population density may prevent an individual from finding a mate, which is known as an **Allee effect**, named for a famous ecologist. For example, many marine species spawn eggs and sperm into the water, and if population density is very low, sperm from a male might not encounter eggs from a female. It is an important consideration in studying the conservation of rare and endangered species (see Chapter 20).

■ Limiting resources may affect population growth.

If resources were limitless and if there were no natural catastrophes, then a population could continue to increase indefinitely. In the real world, food or space will eventually run out. As the resource becomes scarce, resource limitation of survival, growth, and reproduction will occur. **Figure 4.13** shows types of population change. In **exponential growth**, the population increases by the same proportion with the passing of a given amount of time, which might continue indefinitely if resources were limitless. In **resource-limited growth**, there is a limit, or **carrying capacity**, to the maximum population size that the environment's limited resources can sustain. As the population size approaches the carrying capacity, the rate of population growth decreases. When above carrying capacity, the population is too great for the available resources, and it declines. These situations involve intraspecific (within-species) competition for resources. In many cases, population change appears to be **random**. In this case, the factors regulating population size are too complex to show any simple pattern.

Many species occur together and require the same resources or at least overlap strongly in their resource use. This leads to **interspecific competition for resources**, and the carrying capacity of any one species is reduced owing to the similar resource requirements of other species. Competition is discussed later at the hierarchical level of the community.

■ Populations are often metapopulations, which are a series of interconnected subpopulations, some of which may contribute disproportionately large numbers of individuals to the metapopulation as a whole.

A group of populations that are living in discrete habitats but are nevertheless connected by dispersal are known as a **metapopulation** (Figure 4.14). A **source** is a subpopulation

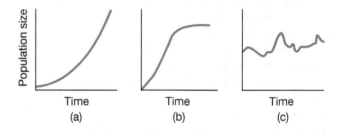

FIG. 4.13 Examples of population change. (a) Exponential growth, a continuing proportional increase. (b) Resource-limited growth, where a population's increase decelerates as carrying capacity K is approached. (c) Random change, where population-controlling factors are too complex to form any pattern.

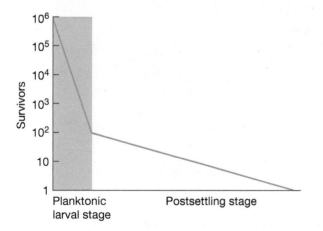

FIG. 4.12 Expected survivorship curve for a marine invertebrate species with planktonic larvae.

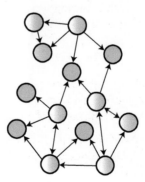

FIG. 4.14 A metapopulation is a series of subequally sized subpopulations with locations that are interconnected by means of dispersal. The light-colored areas are source populations; the dark-colored areas are sinks.

that contributes individuals to the other subpopulations of the overall metapopulation. This may occur when reproduction and dispersal are unusually high in the local subpopulation. A **sink** is a location where a subpopulation may receive immigrants from other populations but does not contribute individuals to the metapopulation and cannot sustain a population on its own. An obvious example is when all immigrants in the sink subpopulation die upon arrival and local individuals fail to reproduce.

High dispersal rates among subpopulations of the metapopulation might work against adaptation to local environments. In the opposite extreme, dispersal might be very restricted between subpopulations, which would allow selection to cause local population differentiation. You can imagine, for example, an isolated population of snails that is exposed to a visual predator. In such a place, there might be strong selection for shell color that matches the background environment. In other sites, predators might be absent, and there would be no selection. If dispersal is high and homogeneous throughout the populations, the product of selection in the subpopulations with predators would be exported randomly to those in which no natural selection has occurred, and local differences in shell color would be absent.

Metapopulation thinking is very appropriate for marine systems, where dispersal of larvae between relatively isolated subpopulations is common. A fascinating example was discovered in a study of gastropods on mangroves of central Queensland, Australia. The intertidal arboreal snail *Littoraria filosa* (**Figure 4.15a**) occurs commonly on mangrove leaves and is obviously very resistant to desiccation. The snail has separate sexes, which copulate, and the planktotrophic dispersing larvae swim in the water for about a month. Stephen and Ruth McKillup (2000) followed populations for a number of years and found that the snail appeared to be an annual: Adults died during the period of larval dispersal and settling, and there was little temporal overlap between successive generations. They were surprised when they began to investigate individual mangrove trees (see Chapter 16 for a description of mangrove forests)

and found that snails on isolated trees only 10–15 m from the main mangrove forest were not annuals at all and continued to live after reproduction. It was apparent that the snails in the main mangrove forest were not "programmed to die." But why were they dying at all? As it turned out, a previously unknown species of flesh-eating fly was the main cause. After a female fly laid an egg near a snail shell, the hatching larva would crawl into the shell and consume the snail's body. In continuous mangrove forests, this process was so effective that no snails survived into a second year or even reproduced. The main population distributed within large patches of mangrove forest is actually a large sink. The settlement of larvae had to come from some other source.

The isolated mangrove trees, often out on the beach, apparently are the source. The fly rarely parasitizes snails there, perhaps because wind prevents flies from reaching the isolated trees. The snails there reproduce well. Larvae that arrive there will not reproduce. It is instead the series of "island" isolated trees (**Figure 4.16**) that constitute sources and may be responsible for supplying the entire population. Snails found on mudflats may also contribute to the larger metapopulation.

Along coasts, populations of a single marine species may extend for many thousands of kilometers. For example, the rocky-shore sea star *Pisaster ochraceus* extends from Baja California to Alaska. Planktonic larvae can potentially disperse over many kilometers, which creates a series of interconnected subpopulations over the entire range of the species. In Chapter 7, we discuss the role of such planktonic larvae and dispersal in creating **connectivity** between marine populations.

Spatial Variation

- **Spatial distribution is a measure of the spacing among individuals in a given area.**

Spatial pattern is a useful feature of natural populations. The spatial distribution is the measure of the type of spacing among individuals. Consider a square meter of rock on a shoreline that has a population of barnacles. If a barnacle

(a)   (b)

FIG. 4.15   (a) *Littoraria filosa*, shell height approximately 2 cm. (b) An isolated mangrove island in Queensland, Australia, where the snail was collected.

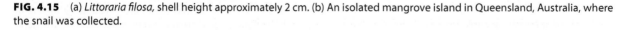

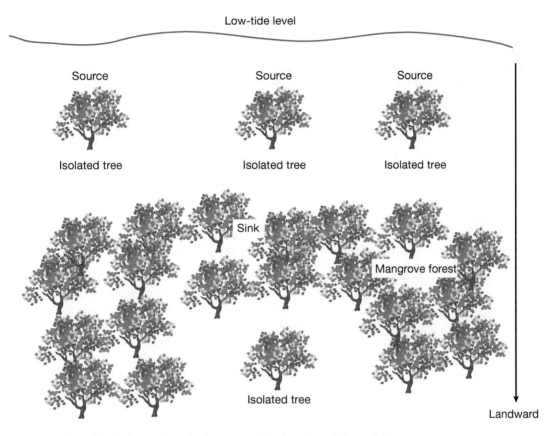

FIG. 4.16 Metapopulation structure in the mangrove leaf gastropod *Littoraria filosa*.

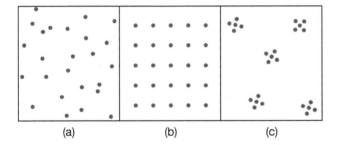

FIG. 4.17 Patterns of spatial distribution: (a) random, (b) uniform, and (c) aggregated.

has the same chance of being located in one spot as in any other spot, then the population has a **random spatial distribution**. Figure 4.17a shows such a distribution, which has the appearance of randomly sprinkled grains of salt. If every equal subarea contains a constant number of individuals, or at least a more uniform number than is expected by chance, then there is a **uniform spatial distribution** (**Figure 4.17b**). If more barnacles occur in a given subarea than are expected by chance, then other areas will be depleted of animals, giving an **aggregated** or **patchy** distribution of clusters and empty space (**Figure 4.17c**).

Spatial distributions are useful because they suggest hypotheses about the mechanisms affecting natural populations. It is rare for populations to have a random distribution. This usually occurs when larval settlement from the water column is random or animals are moving about randomly, as when mud snails move on a mudflat. Uniform distributions occur when animals are maximizing the distance between neighbors, implying the establishment of territories or interference. For example, when larvae of the tube worm *Spirorbis borealis* settle on seaweed from the water column, they usually crawl away from other settled larvae before metamorphosing into adult worms. As a result, one eventually may see a uniform array of tiny spiral tubes on the seaweed frond. Aggregations usually imply some sort of patchiness about the environment, but organisms might be socially attracted to each other for mating or to form fish schools or other aggregations to protect against predators.

■ A population may show a regular change in density along a sampling line.

If sampling is done along a transect line, many populations change in a definable pattern. A population of plant-eating snails, for example, might decrease regularly with increasing water depth because the food source also declines with depth. If the density of a population at one site can be predicted from the density at neighboring sites, we speak of the population as having **spatial autocorrelation**. A spatial autocorrelation might occur if (a) there is a change in the environment that affects survival or causes differential subhabitat selection; (b) the population is moving in a defined direction (the change in density might reflect, e.g., the tail end of a migrating population of fish); or (c) a random process occurs, which occasionally can cause a nonrandom spatial pattern.

The Community Level: Structure and Interspecies Interactions

■ **Many communities are organized around important structural aspects of the habitat or around foundation species that determine a great deal of the habitat structure.**

Ecological **communities** consist of a series of coexisting species. In many cases, the first level of explanation for the types of species that exist in a community is the **structural habitat**. The hard substratum of rocky shores usually precludes many species, especially those that depend on feeding on and living in soft sediment. Thus, the structural habitat is a major, if obvious, organizing force for marine communities. Equally important in many cases is the presence of **foundation species**, which contribute substantially to the structure of a local habitat and determine many of its physical and chemical properties. For example, the grasses we observe to dominate salt marshes are foundation species. They slow down water currents and increase the deposition of soft sediment, which creates the meadow-like soft-sediment environment that permits many species to dwell and burrow in the sediment among the grass blades. Mangrove trees have a similar effect on protected tropical shores and provide numerous habitats for animal and plant species on mangrove branches. These habitat-determining species are also called **ecosystem engineers**, as they alter substantially, sometimes even create, the structural habitat on which other species depend.

■ **Distribution and abundance of species populations in a community are determined by the combined effects of the following processes: (a) dispersal of larvae, spores, and adults; (b) competition; (c) predation and herbivory; (d) parasitism; (e) disturbance; and (f) facilitation.**

Physical features of the environment, such as temperature, salinity, dissolved oxygen, and nature of substratum, may determine the maximum environmental range of a species. However, a series of dynamic community-level processes strongly affects distribution and abundance:

Larval Access

■ **Larval recruitment patterns strongly affect the species composition of marine communities.**

Many marine fishes and invertebrates have planktonic larvae, which can disperse great distances. Interannual variation in larval settlement (recruitment) can determine the composition of marine communities. Some of this variation may result from the effects of local ocean currents, which may sweep larvae out to sea or keep them near the shoreline with strong variations among different species, which leads to different assemblages of species.

Interspecific Competition

■ **Competition within and between species derives from the limiting resources of space and food.**

Competition occurs when two individuals of the same (intraspecific) or different (interspecific) species exploit a common

FIG. 4.18 Competition affects abundance where resources are limiting. In this example, the mussel *Mytilus californianus* competes for space with the barnacle *Balanus cariosus* on an intertidal rock near Bamfield, British Columbia. (Photograph by Jeffrey Levinton)

limiting resource (**Figure 4.18**). *The two prime limiting resources are space and food.* The study of competition must focus on these limiting resources. A **guild** is a group of species that exploit the same resource. Guilds need not include closely related species. In a typical rocky-shore site on the Pacific American coast, several hundred sessile species, including stalked barnacles, acorn barnacles, mussels, brown seaweeds, green seaweeds, and crustose coralline algae, share the same space resource. In a study of competition for space, it would make no sense to study mussels without also considering competing seaweeds.

■ **Competition between species may involve direct displacement, preemption, or differential efficiency in the use of resources.**

Competition between species can proceed in several different ways. Often, when space is the limiting resource, one species may succeed through direct **displacement** of another. In such a case, we must assume that all encounters between species A and species B have the same outcome (e.g., A displacing B). By contrast, a species that holds space by colonizing a bare spot may then **preempt** invasion by another species, but only because it arrived first. Competition based on direct displacement of one individual by another is known as **interference competition**. Competitive success might involve one species overgrowing another, overtopping and shading another species if it depends on light, or a variety of other mechanisms. But if two species compete by virtue of requiring and exploiting the same resources, there might not be any direct behavioral interaction. In such **scramble** (or **exploitation**) **competition**, the more efficient species might gain more food and gradually increase in population size at the expense of the other. For example, one copepod species might be more efficient at grazing diatoms, a renewable resource, than a second copepod species and win in competition by producing more offspring, which increases the population of the first species at the expense of the second.

■ **Competition has been demonstrated in marine communities by experimental removals of abundant species followed by expansions of competitors.**

One can be overwhelmed by the variety and complexity of nature. We often cannot explain patterns in marine communities very easily. If a species is absent, are predators the cause of its absence? Has the species lost out in interspecific competition? A similar problem arises when we examine zonation, the most common feature of rocky shores, where dominant species may occur in a series of horizontal bands.

One commonly observes, especially in quiet waters, a series of horizontal bands. Classically, they consist of, in order from high to low, intertidal, lichen, barnacles, limpets, and mussels. Why such dominance by single species? Marine ecologists, inspired principally by the pioneering works of Joseph Connell and Robert Paine, have approached the problem through systematic **manipulative field experiments**, which we have discussed in Chapter 1 as a means to test hypotheses. The experiments involve removals of hypothesized predators or competitors or by caging areas against predators with careful observation of the consequences. For example, for many years, Robert Paine removed the voracious starfish *Pisaster ochraceus* from a rocky shore off Cape Flattery, Washington. It was more than 10 years before a significant change took place in the distribution of beds of the mussel *Mytilus californianus*, which extended downward and overgrew several species of seaweed (see Chapter 16). The dominance of mussels resulted in a reduction of diversity of competing rocky-shore sessile species. The displacement of species by a superior competitor is known as **competitive exclusion**.

Field experiments may be prohibitively difficult because the organisms are microscopic or because the manipulation is difficult to interpret. Some field caging experiments, furthermore, change the experimental microenvironment in unacceptable ways. Cages built to protect soft bottoms from mobile predators also alter the depositional environment, and fine sediment settles within the cage. The experimenter is then altering two factors at once. This kind of situation may preclude field experiments, but **laboratory competition experiments** may be quite informative as long as some element of realism permits one to relate the laboratory results to field conditions.

Occasionally, so-called **natural experiments** are encountered. For example, we may discover that in most sites two species are found together. However, we might find that in some locations, one species is naturally absent and the other species has expanded in abundance. We might provisionally conclude that the first species normally affects the other's abundance. Although that is a fairly safe conclusion, we should remember that this is not a controlled experiment. The factor that removed one species may also have enhanced the other's abundance. For example, many fish species decline in estuaries, but mullets are often very abundant. One might be tempted to explain this set of circumstances on the basis of relaxed competition between mullets and other fish species. However, mullets are detritivores, and estuaries often have increased supplies of detritus. Reduced salinity may have independently eliminated the other fish while detrital supply increased the mullets independently.

■ **Competition combined with differential success in different microhabitats results in niche structure.**

Niche structure is any predictable partitioning by coexisting species of a habitat into subhabitats or differential exploitation of resources. Ecologists have long believed that no two species can coexist on the same limiting resource. Although this is not always true theoretically, the presence of coexisting species using different resources has been used as evidence of the action of interspecific competition. Many of these studies are observational only. For example, species of the carnivorous snail genus *Conus* live associated with coral reefs throughout the Pacific. Alan J. Kohn (1967) found that species of *Conus* in subtidal coral reef habitats with high species numbers were highly specialized and tended to eat different foods. By contrast, the

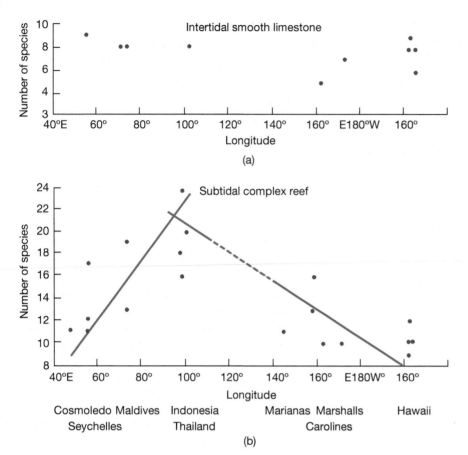

FIG. 4.19 Environmental heterogeneity promotes coexistence of many species by providing several distinct environments within which species may exploit unique resources. Diversity increases toward the southwest Pacific for the carnivorous gastropod *Conus* in complex subtidal hard substrata (b) but does not change in topographically simpler intertidal platforms (a). (After Kohn, 1967)

single species of *Conus* found on the coast of California has much more general food preferences. In the much more homogeneous intertidal smooth platform habitat, species diversity (number of species) does not increase toward the Indo-Pacific region (**Figure 4.19**). Although the evidence is circumstantial, it suggests that some niche structure exists and results from historical competition and the evolution of specialization among *Conus* species or prey.

■ Some assemblages of natural species show extensive coexistence of presumed competitors despite apparent resource limitation.

Unless there is an opportunity for niche displacement, one would expect a competitive dominant species to displace all other species. However, this **"law of competitive exclusion"** often does not seem to apply to all natural communities. For example, many species of phytoplankton coexist despite apparent resource limitation. In the open-ocean tropics, scores of species of phytoplankton coexist even though dissolved nutrients such as nitrogen are undetectable in the water column. Why have all inferior species not been outcompeted by species superior at taking up nutrients from the water? The great ecologist G. Evelyn Hutchinson termed this coexistence "the paradox of the plankton."

A number of processes can explain such a lack of competitive exclusion. These include the following:

1. *Competitive networks.* There may be complex competitive interactions combining multiple means of competitive superiority with no clear competitive dominant. It is possible that species A is competitively superior to species B but inferior to species C, leading to different outcomes of dominance depending on which species come into contact (**Figure 4.20**).

   In a simple hierarchy, one species might always win in competition. For example, intertidal mussels often smother all other competing species and win out in competition. But coral reefs appear to have network interactions. Sponge species A might be able to overgrow sponge species B, and B might be able to overgrow sponge C, but C might produce a poison that affects species A, which has poor chemical defense because it devotes it resources mainly to growth. Such complexities of competitive mechanisms delay competitive dominance by any one species.

2. *Lottery colonization.* Adult sites may be limited, but colonization is from a random larval pool in the water column. If an animal dies, a larva might settle from the water column and establish a territory, but currents

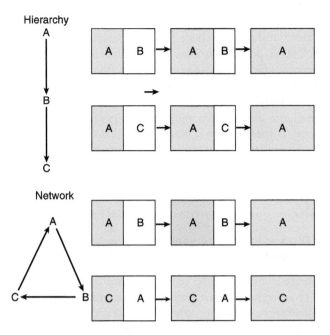

FIG. 4.20 In a simple competition hierarchy (top left), one species is superior to all others, and any given species is superior to another at a lower hierarchical level. In a network (lower left), however, species A may outcompete B, and B may outcompete C, but C may outcompete A. This can happen only if a distinctly different mechanism of competitive superiority for the C–A interaction exists to delay the eventual dominance of any particular species. The lower figure shows that networks create a variety of outcomes when the species are combined, perhaps by larval settlement of pairs of species.

and chance would determine the particular species of the colonist. The species composition of the community would be determined more by colonization than by interactions of the resident adults. Peter Sale (1977) explained the coexistence of several territorial fish species by means of the **lottery hypothesis**, which emphasizes random mortality and colonization by planktonic larvae. The fishes occupying any given site are the result of a random drawing from among the pool of planktonic fish larvae that happen to be in the water column when a benthic territory becomes open. Since gains made by a competitor are often lost because of random mortality, such random processes of adult extinction and recolonization delay or completely prevent the competitive displacement of one species by another. The **neutral theory of community ecology** states that such random interactions, combined with occasional extinctions and speciation events, result in indefinite coexistence of many species.

3. *Disturbance.* Complex patterns of disturbance may preclude the rise of any one species to dominance. Disturbance may be so common (as discussed shortly) that competitive dominance is prevented.

4. *Habitat complexity.* There may be habitat complexity, which permits the coexistence of many species. As discussed earlier, habitat differences may allow species to coexist by specializing on slightly different microhabitats.

Predation and Herbivory

■ **Predation may prevent domination by a superior competitor and may strongly affect species composition.**

The experimental removal of *Pisaster ochraceus* that resulted in the dominance of the mussel *Mytilus californianus* suggests a common effect—namely, that predation delays the competitive displacement of competitively inferior species by the competitive dominant. Herbivory can be considered a form of predation and often has the same effect on competing species of seaweeds. Experimental removal of sea urchins usually results in dominance by one rapidly growing seaweed species over the others.

■ **Seasonal influxes of predators in shallow water and in the intertidal zone may devastate local communities.**

Many habitats have a local permanent population of predators, but the spring and summer often bring on invasions of large populations of migratory predators with devastating consequences for prey populations. In the intertidal zone, the most prominent example of such predators consists of shorebirds, whose migrations may extend for thousands of kilometers. These birds often have favored feeding grounds on muddy or sandy beaches, which they visit successively during their migration. Fish often come inshore in summer and devastate local invertebrate populations.

Disturbance

■ **Disturbance opens up space in the community. Its frequency may regulate long-term aspects of species composition in a habitat.**

Marine populations suffer extensively from storms, continuous wave action, and unstable sediments. Intertidal populations often crash owing to large swings in temperature and humidity. Ice crushing (in high latitudes) and the bashing of floating logs are also major problems. Even in subtidal habitats, large swings in temperature may occur, as in the great increases in temperature during El Niño events. Any of these general physical effects is known as **disturbance**. Mobile animals may also cause mortality unrelated to predation. Such effects are known as biological disturbance. For example, while moving along rocky surfaces, limpets often bulldoze newly settled barnacle larvae from the rocks.

The effects of disturbance resemble those of predation because competitors are reduced in number. However, predation is usually a process that removes one individual at a time, although some predators come in devastating waves. Disturbance, by contrast, usually operates on larger spatial scales, removing patches of the community. Disturbance often initiates an orderly sequence of dominance by different species over time known as succession (discussed later).

■ **Species diversity may be maximized at intermediate levels of predation or disturbance.**

Let's consider first a gradient from very low to very high predation rates; we can apply the same set of causes and

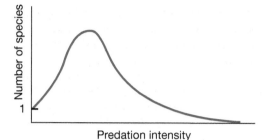

FIG. 4.21 When there are competing species and no predation, one superior species may take over. With predation that is random or targeted toward the competitively superior species, more species can coexist. With very high predation pressure, all individuals and species may decline. Thus, we might expect an intermediate graphical hump in the number of coexisting species as a function of increasing predation pressure. A similar effect is found for levels of disturbance.

effects to disturbance intensity. If there are no predators, we might expect a competitive dominant, if present, to displace all competitively inferior species. But as predation increases, resource space opens up, and more competing species may be allowed to coexist. As predation intensity increases further, however, nearly all individuals of all species will be removed, and the species diversity will decline relative to the intermediate disturbance levels. Thus, species diversity tends to be maximized at intermediate levels of predation.

Disturbance is likely to work in a similar way because it will most likely pare down the abundance of the competitive dominant. But if disturbance is very low, the competitively dominant species will displace all other competitors. If disturbance is extremely strong, all species, including the competitive dominant, will be eliminated. These combined effects are known as the **intermediate disturbance-predation effect** (Figure 4.21).

It is important to realize that this relationship is fueled by a large recruitment rate of new individuals into the area we are considering. With recruitment of all species low, coexistence might occur even at very low levels of disturbance because there would not be a sufficient influx of larvae of a competitive dominant to recruit to displace other species (Dial and Roughgarden, 1998).

Although the intermediate disturbance-predation hypothesis seems logical, it is not clear how often it works in nature, and some studies show that there are many departures. In some habitats, as we shall discuss, it is rare for a single competitive dominant to exist. Furthermore, inevitably, many species will not be eliminated by a competitively superior species. Finally, as disturbance increases, mortality may also increase, eliminating species faster than they might accumulate as competition is relaxed.

Parasites and Disease

■ **Parasites are common and can affect their hosts by reduction in growth and reproduction or by enfeeblement.**

As discussed earlier, parasites affect individual hosts by taking resources, interfering with reproductive output and

behavior, and enfeebling the host. On the population level, parasite load should reduce reproductive rate and cause the decline of population size. To examine population-level effects, we must know the percent infection (or prevalence) of the host population and the severity of parasite effects (or intensity). Lafferty (1993) was able to manipulate the prevalence and overall population density of trematode parasites on the California marsh snail *Cerithidea californica*. Snail populations with higher parasite prevalence had lower reproductive output and higher mortality rates. These results suggest that abundant and ecologically significant species may be greatly diminished by parasites in the strength of ecological interactions with other species in a community.

■ **Diseases in marine organisms are poorly understood, but they can cause swift population declines.**

Disease is a major cause of massive and widespread mortality. Infectious disease usually involves the invasion of an individual by a microorganism, such as a virus, bacterium, or protist. Unfortunately, our understanding of marine pathogens is very limited. For example, a marine protist is known to be the main cause of a disease that devastates populations of the eastern oyster *Crassostrea virginica*. This parasite caused the collapse of the very profitable oyster industry in Delaware Bay and has recently been a major cause of decline in Chesapeake Bay. Despite its obvious commercial importance, however, we still have no complete idea of the parasite's life cycle.

One of the most interesting issues of marine disease is the mechanism and rate of spread. In Chapter 18, we discuss coral reef and other tropical diseases, some of which can be interpreted as spreading by water currents. Such spread might cause a disease to invade an entire ocean basin in the time the surface currents spread the disease, which will often be less than a year. In other cases, disease vectors require direct contact. Disease organisms can be transported in hosts that are present in ballast water of ships. This means that the current use of untreated ballast water in many ships might be facilitating the spread of marine diseases throughout the ocean. Aquaculture facilities are also highly vulnerable to the spread of disease, and a virus causing white spot syndrome is responsible for enormous losses in shrimp farms throughout the world.

■ **The role of disease must be verified by rigorous use of Koch's principles, which involve identification of the pathogen, isolation, and successful experimental infection of the target organism.**

The role of a pathogen in a disease must be established by rigorous use of a set of principles named after disease biologist Robert Koch. First, the pathogen must be identified. This alone is a difficult process because a disease may be accompanied by the rapid increase of a number of microorganisms, but only one is likely the cause of the disease. Second, the pathogen must be isolated and raised in the laboratory. Finally, the pathogen in culture must be reintroduced into the target organism to confirm the cause of the pathogen of the disease. Given the difficulties of culture, let alone reintroduction into wild species, these steps are difficult to achieve.

■ Disease interacts strongly with changing environmental conditions and the increase of stressful physiological conditions of the host.

Environmental factors may also be important in the spread of disease. First, an environmental shift might facilitate the geographic spread of a disease to new areas. The oyster disease Dermo, caused by an amoeboid protist species, has been increasing in occurrence toward higher latitudes in southern New England waters. This may be related to ocean warming, which seems to promote spread of the disease organism (Powell et al., 1999). Global warming of the past few decades may be facilitating the invasion of other disease organisms toward higher latitudes.

If the environment changes rapidly, physiological stress might make marine organisms more susceptible to disease. This was discovered in humans by physiologist Hans Selye, who demonstrated that those under psychological stress were more susceptible to disease. It is an interesting question whether global warming is sufficiently stressful that many marine organisms are becoming prone to disease. This may be the explanation behind the rapid spread of coral diseases in recent decades, as stressful warming events become more prevalent (see Chapter 17).

■ Target organisms may evolve resistance to disease, resulting in cycles of virulence in marine populations, which are poorly understood.

Disease is well known to undergo cycles of virulence and spread. In the worst case, a disease can drive an entire species to extinction or near extinction. Obviously, the rate of spread will then decline because the disease has no place to spread and perhaps no hosts to infect. But an alternative possibility is that the host population will have genetic variability for resisting the disease. When a new strain of the disease occurs, resistant genotypes will increase in frequency and reduce the impact of the disease. This can be demonstrated in the laboratory. Ford and Haskin (1987) were able to select populations of the eastern oyster *Crassostrea virginica* for resistance to the disease organism MSX. After a major mortality in Delaware Bay, surviving oysters were more resistant to infection. It is not known whether there is a physiological cost that is encumbered by the evolution of resistance.

Facilitation

The study of species interactions often involves negative interactions, as in predation and competition, but ecologists have found that positive interactions between species are often important in determining the species composition of marine communities (Bulleri, 2009). In some cases, species facilitate each other's presence. In stressful environments such as the rocky intertidal, associations of sessile species may result in retention of water at low tide, which enhances survival for all species. Some species, such as intertidal plants and seaweeds, increase moisture of the substratum, which attracts burrowing animals. But in turn, such burrowers may aerate

the mud and increase the growth of plants. In soft sediments, several species might burrow and oxygenate the sediment, thus making it more hospitable for all burrowing species to live within the sediment (see Chapter 16 for more on the effects of burrowing species on sediment properties). **Foundation species** often alter the structural environment, which facilitates the presence of many other species. Sea grasses are good examples of this effect because they often make both the sediment and water column within the grass a suitable habitat for many quiet-water species.

Succession

■ Succession is a predictable ordering of arrival and dominance of species, usually following a disturbance.

Many people are familiar with the fate of small ponds in forests. The ponds fill in with sediment and are colonized by vegetation. Eventually, the soil and biota come to resemble those of the surrounding area. Succession comprises all the processes that are involved in such a progression. **Succession** is a predictable ordering of appearance and dominance of species, usually following an initial **disturbance**. A predictable final state, or **climax community**, may eventually develop. Succession is explained as either (a) a trend toward a stabler assemblage of species or (b) the simple sum of the colonization and persistence potentials of the species. Succession is not necessarily inevitable, and the rate of change is not predetermined. Much research on succession suggests that it is often more like a net trend than a closely integrated sequence of biological events.

Several factors are at work in varying degrees to determine the pattern of succession, even if the sequence is more or less predictable:

1. Differential rates of colonization might result in the early arrival of certain disturbance-dependent species. These species have high reproductive rates and short generation times. They are adapted to locate in newly disturbed environments, but such "weedy" species often are poor at holding on for very long.

2. Conditioning of the environment by resident species might facilitate the appearance of other species or prevent others from colonizing.

3. There may be monopolization of the habitat until some event (e.g., grazing) eliminates the dominant species and permits further colonization.

4. There may be irregularity in the time course of succession depending on events such as the arrival of predators or variation in recruitment to the site. On rocky shores, for example, filmy green algae often arrive first. They are frequently replaced by species with tougher holdfasts and compounds that deter herbivores. These species in turn often cannot colonize unless herbivores such as snails and urchins have eaten the green algae.

5. There may be an eventual dominance of species that are relatively resistant to predation and competitively superior to early succession species, at least under the conditions found late in succession.

■ **Succession may bring a community from one condition to another; however, other forces may also change community composition in a profound way, and local feedbacks may preserve the change.**

We usually think of the final state of succession as having a series of properties that deter a reversal to earlier stages of succession. Often in marine communities, however, major external disturbances or even differences in the time of year may cause major shifts from one community condition to another (**Figure 4.22**). Some habitats are continually disturbed with major habitat shifts. In Texas shallow bays, periods of rain and drought may drastically alter the salinity and favor very different groups of organisms adapted to differing salinity regimes. Continuing disturbance might guide the appearance of community types we see in such shifting regimes.

Dominance in marine communities might be determined by very local circumstances of disturbance and colonization. John Sutherland (1974) studied benthic colonization of ceramic plates and showed that the community composition of sessile animals depends strongly on the time of year the plates were placed in the water. For example, the colonial hydroid *Hydractinia*, would colonize and resist overgrowth by species that might settle later in the year. On the other hand, the tunicate *Styela* would colonize predictably, usurping space from colonial bryozoans. Therefore, communities do not fit the neat mold of succession as a predictable process of community condition *A* going to community condition *B* and so on. Sutherland termed the locally persistent assemblages of organisms multiple stable points, but we shall call them **alternative stable states**. The important requirement for alternative stable states is that the different community compositions arise as accidents of particular historical circumstances and manage to persist over time. If particular environmental conditions lead to a particular outcome, as in succession, we refer to the community so obtained as a new **phase shift**. For example, if the water becomes turbid and sea grasses disappear, leading to dominance of bare bottoms, we would call this a phase shift because it is predictably driven to a new state by a predictable environmental change. Alternative stable states are often preserved by feedback processes that preserve a given assemblage of species, once established. However, transitions between one stable state and another are often complex and the path of transition may differ forward and backward in transitions between states (see Petraitis, 2013, for more on this subject).

■ **The resilience of a community's ecological structure should increase with the diversity of species that have important ecological roles, such as grazers or top predators.**

Disturbance and large-scale population decline are common features of natural communities. In some cases, disturbance has large-scale random effects and removes many species. But in other cases, removals of top predators may occur simply because they are less abundant and more vulnerable to population reduction. Human effects such as overfishing might also focus disproportionally on some of the community, such as the top predators. It is therefore of great interest whether a community might return to its original state after a major disturbance. The capacity for such a return by a community is known as **resilience**.

Increased diversity appears to influence the resilience of a community. Why? First, some functional effects such as predation by top predators or grazing by herbivores might be retained if a diverse community exists. If one grazer is removed, another grazer might increase in abundance to replace it. Thus, community resilience is likely to decline as diversity declines. This is important in habitats where frequent disturbance allows only a few species to exist. A local extinction might have greater consequences than in another habitat where disturbance is less frequent and diversity is greater. It also means that a community will be more resilient when a massive disturbance or disease causes the local extinction of what might be the only top predator or grazer in a system, such as a coral reef. A removal under such circumstances might cause a major shift to an ecologically different regime with new patterns of dominance.

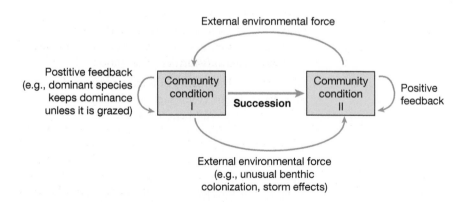

FIG. 4.22 Traditional models of succession would suggest that environments move through a series of community states. However, external changes or conditions, such as incursions of grazers, seasonal changes, or storms, may cause sudden shifts in community states. Positive feedbacks may keep the condition from changing.

Direct and Indirect Effects

■ Ecological interactions may be direct or indirect.

So far, we have discussed ecological interactions as direct effects between species. If a predator captures and consumes an individual of a prey species, the prey species is reduced in population size by one. By doing this, we are making a hidden assumption: The effect on a community is the sum of all instances of a direct effect between individuals of two species in a predation or other interaction event. But such effects may also be propagated through communities by **indirect effects** on other species. Here is a simple example. Consider a chain of interaction: Sea otters capture urchins, and urchins eat seaweed. If sea otters capture urchins, the urchin population will decline; that is a **direct effect**. But the decline of urchins would also result in the **indirect effect** of seaweeds being relaxed from grazing. Thus, the indirect effect would be that the removal of urchins causes an increase in seaweeds. From this we learn that removal of a species in an ecological system with strong species interactions will propagate throughout the community. One might imagine that if the seaweed population increases, there might be a decrease of dissolved nutrients required by the seaweed. The indirect effects will be very widespread.

■ Indirect effects in communities can involve density-mediated indirect interactions (DMII) or trait-mediated indirect interactions (TMII).

The indirect effect of predation on sea otters to seaweed abundance should be related to the density of sea otters mediated by sea urchins: The more otters, the more seaweed. This would be a **density-mediated indirect interaction**. But consider this possibility: Otters increase, and as a consequence, urchins change their behavior and start to hide in crevices, which keeps them from being spotted by the otters. This *behavioral trait* would not result in a decrease of urchins, but it would still result in less feeding by urchins on seaweeds. Therefore, *the presence of the trait of predator avoidance* would result in increased seaweed growth. As another example, dugongs are southwestern Pacific marine mammals that graze in sea grass meadows and can have major effects on grass diversity and abundance in Shark Bay, northwestern Australia. They prefer shallow sites where grass is abundant, but move to less productive deeper sites when tiger sharks are present. The dugongs are therefore modulating their use of foraging sites in proportion to the probability of attacks by sharks (Wirsing et al., 2007). These responses are known as **trait-mediated indirect interactions**. Because prey can often detect their predators by smell, feeling currents generated by the predators, or other means, indirect effects can propagate through communities without changes in density or direct consumption rates of one species by another. If an effect such as predator avoidance by reduction of activity in a species occurs, and the prey of this species is not reduced in population size or even grows as a result, this would also be known as a **nonconsumptive effect**.

The Ecosystem Level

■ An ecosystem is a group of interdependent biological communities and abiotic factors in a single geographic area that are strongly interactive.

An ecosystem consists of a group of communities that interact with the physical-chemical environment within a specific geographic area. Within the ecosystem, nutrients recycle between organisms and the environment, some of the species manufacture organic molecules using only solar energy and inorganic chemical sources (e.g., algae), and the interactions among species within the system are very strong. Under this definition, a large lake would comprise an ecosystem because the organisms, nutrients, and other environmental features interact within the lake. A coral reef and its immediate surrounding water also qualify as an ecosystem. An ecosystem is not necessarily independent of other ecosystems, but we can define the boundaries of an ecosystem. In reality, all ecosystems exchange nutrients and organisms with other ecosystems. It is crucial, therefore, to determine the boundaries of an ecosystem and the places where losses and gains may occur.

■ Nearly all ecosystems have primary producers (mainly photosynthetic), which are consumed by herbivores, which in turn are eaten by carnivores. Material escaping this cycle passes through the saprophyte cycle.

The manufacture of organic molecules is accomplished by **primary producers**. Phytoplankton, seaweeds, and sea grasses are the most familiar of these organisms, but many bacteria also manufacture organic substances with the aid of light or energy derived from inorganic chemicals. Plants are consumed by herbivores, and carnivores in turn consume these. In many marine ecosystems, most of the plant material produced is never consumed by herbivores; rather, much of it falls to the seafloor and is decomposed by bacteria and fungi, producing dissolved nutrients. The dissolved nutrients are then available for primary producers. This pathway is known as the **saprophyte (or detritus) cycle**.

Biomass, Productivity, Primary Productivity, and Secondary Productivity

Biomass (standing crop) is the mass of organisms present in a defined area or volume (expressed in units such as grams per square meter: $g\ m^{-2}$). Biomass is distinguished from **productivity**, *which is a rate*: the amount of living material or carbon produced per unit area per unit time (e.g., $g\ m^{-2}\ y^{-1}$).

In a natural environment, all organisms depend on primary producers, which use light energy, usually in the process of photosynthesis, to convert carbon dioxide and water into sugars and other essential compounds. A primary producer is also known as an **autotroph**, and consumers of autotrophs are known as **heterotrophs**. **Primary productivity** is the amount of living material produced in photosynthesis per unit area per unit time. In contrast, **secondary productivity** refers to the production of primary consumers, or herbivores, per unit area per unit time. The productivity of carnivores, or consumers of herbivores, is **tertiary productivity**. A **food chain** is a set of connected feeding levels of primary, secondary, and tertiary (and so on) sources of productivity. An example of a simple food chain is

seaweed → snail → shorebird

Each organism (primary producer seaweed, secondary producer snail, tertiary producer shorebird) occupies a **trophic** (or food) **level**. In more complicated systems, a simple chain cannot be constructed, and a more complex **food web** is a better description. We discuss transfer through food webs in Chapter 12.

In general, primary production is greater than secondary production, which in turn is greater than tertiary production. Secondary production depends on consumption of primary producers, but this process is not perfectly efficient. Some material is never eaten, and even the eaten fraction may not be digested completely. Finally, not all the food that is digested is used for growth (i.e., production). In the case of carbohydrates, for example, a large fraction of the carbon content is respired in the form of carbon dioxide. As a result of such processes, material is lost between successive trophic levels (see Chapter 12 for further discussion).

■ **Some predatory species at the apex of food webs exert strong effects on the overall ecosystem.**

Predators at the top of food webs (often called top predators or **apex predators**) may exert strong effects not only on competitive interactions but also on entire ecosystems if there are strong interactions between the trophic levels. A predator at the top of a food web exerting such strong effects is known as a **keystone species**, a distinction first recognized by rocky-shore ecologist Robert T. Paine. When linkages among trophic levels are strong, changes in abundance of the top predator causes a **trophic cascade** through the trophic levels. In Pacific U.S. kelp forests, sea otter consumption of urchins has a cascading effect on kelps.

■ **Food webs may be controlled by top-down processes usually affected by prey on lower trophic levels or by bottom-up processes.**

The strong effect of fluctuations of apex predators, especially as indirect effects at lower trophic levels, is an example of **top-down effects** in ecological systems. These effects have often been noted when humans hunt an apex predator nearly to extinction, which initiates a trophic cascade down the food chain. Top-down effects clearly occur when predators consume their prey.

Changes in lower trophic levels may also exert strong effects on ecosystems. Such **bottom-up effects** might include a large-scale increase in phytoplankton productivity, which results in greater food input at the lower levels of a food chain and indirectly allows larger populations of apex predators to exist. There are even food webs where middle-level species are crucial determinants of food web dynamics. In Chapter 19 we discuss the central role of krill in Antarctic food webs.

■ **Ecosystem studies usually account for the processes that affect movement of materials and energy through food webs and through the nonliving part of the ecosystem.**

Studies at the ecosystem level attempt to account for the processes that control the system's **throughputs of materials or energy**. For example, an ecosystem study might focus on the control of movement of nitrogen through a marine planktonic ecosystem. Clearly, this is a biologically complex problem involving anything from microbial control of conversion among different forms of nitrogen to movement of nitrogenous materials from the water to plants, herbivores, and carnivores. Ecosystem studies involve a search for general features of material flow, and species are usually treated only as **functional groups** (e.g., herbivores).

Species, Genetic Variation, Evolution, and Biogeography

Genetic Basis of Organismal Traits

■ **Organism features can be explained based on a combination of genetic and nongenetic components.**

Marine organisms are universally variable in DNA sequence, form, color, and biochemistry. A **polymorphism** is any variation that can be identified in terms of a series of discretely different forms. In other cases, variation can be measured only as continuous variation, such as differences in body size or the proportional size of a fin. We must distinguish between genotype and phenotype. The **genotype** refers to the genes that characterize an individual or to those that control a particular trait, such as eye color. The **phenotype**, by contrast, is the form the organism takes. For example, it is possible that all the brown-haired individuals in a human population will not have the same genotype. By contrast, people with gray or brown hair may be of identical genotype. The gray-hair phenotype might be associated primarily with age.

Phenotypic variation in a population can be explained with a simple equation:

Phenotypic variation = variation explained by genetic factors + variation explained by environmental factors + an interaction between genetic and environmental factors

Much of the phenotypic variation we observe has nothing at all to do with genetic variation. Environmental effects such as nutritional status and microclimate alter the course of growth and development of animals with identical genotypes. *Both genetic and nongenetic components contribute to determine a trait.* Body size is a useful example. It is almost always controlled partially by genes, but the environment also exerts a large effect. Much of the variation we see in natural populations, however, exists because of the inheritance of different genes. Shell color is a conspicuous example of this in many mollusks.

It is extremely important to realize that having a given gene does not guarantee that the form of an organism will always be the same. The same genotype may have a different phenotype when raised in different environments. This is known as a **genotype-by-environment interaction**.

In a few cases, variation is due to differences at a **genetic locus**, or single location on the genetic material, or DNA (see bonus chapter, "Molecular Tools for Marine Biology," online). In such a case, an individual has two genes for the trait, one inherited from the father and one from the mother.[1] The genes might be identical, or they may be different variants, or **alleles**. If there are two alleles *a* and *b*, then there are three possible genotypes: *aa*, *ab*, and *bb*. All three variants may look different, or one allele may be **dominant**. For example, if the *a* allele is dominant, then *aa* and *ab* genotypes may have identical phenotypes.

The mussel *Mytilus edulis* can be blue-black or brown owing to the control of a single genetic locus with two alleles (**Figure 4.23**). The brown allele is dominant over the blue-black, and the heterozygote, which inherits one brown allele and one blue-black allele, is therefore colored brown. In most traits, several or many genetic loci are in control. Body size, for example, is usually controlled by many loci. In such cases, the genetic component can be found by studying the degree of resemblance among relatives. The correlation of a trait between parents and offspring, for example, can give evidence of a genetic component. **Figure 4.24** shows the correspondence between number of vertebrae in mothers and in offspring of the eelpout *Zoarces viviparus*. The correlation is high, and we conclude that the variation in the trait is therefore controlled largely by genetic variation.

There are several types of common variation observed in marine populations. **Chromosome number** can be variable in natural populations. The Atlantic drilling snail *Nucella lapillus* is variable in chromosome number when found in different degrees of wave exposure. Many species have **color polymorphisms**. These polymorphisms may be explained mainly by genetic variation. Many **morphological characters** (e.g., body size and number of fin rays on a fish) are variable and are controlled, at least in part, by genetic variation. Variation in **biochemical and physiological traits** (e.g., presence of specific proteins and different levels of oxygen consumption) is common, and **enzyme polymorphisms** occur widely. **DNA sequences** also differ at genetic loci and, of course, give the most direct evidence of genetic differences among individuals.

What maintains genetic variability in populations? **Natural selection** is the process whereby individuals with certain genes survive and reproduce more successfully than others; this leads to dominance in the population by certain genetic variants. The relative survival and reproduction of a given genotype constitute its **fitness**. **Adaptation** occurs when natural selection causes evolutionary change in a population, which results in an increase in the ability of a typical member of the population to perform in that environment. We usually judge performance with respect to a given function, such as resisting heat shock.

FIG. 4.23   External shell color of the marine mussel *Mytilus edulis* is genetically controlled by two alleles for *blue* or *brown*, as shown by these light brown and dark blue juveniles (shell length is about 1 cm). The allele for light brown is dominant, so the light-colored mussel may be a *brown-brown* homozygote or a *brown-blue* heterozygote. The dark blue mussel is a *blue-blue* homozygote. (Courtesy of David Innes)

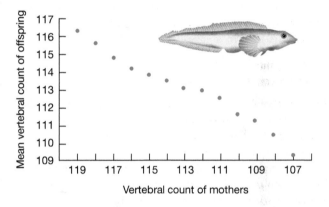

FIG. 4.24   The mean vertebral count of mothers and their offspring in the eelpout *Zoarces viviparus*. (Data from Schmidt, 1920)

Variation can be maintained by shifting of environments, which may favor one genetic variant, then another. Alternatively, a complex environment can favor several genetic variants but in different microhabitats. In some cases, a gene is favored simply because it is rare, which would cause a dynamic shifting back and forth of gene frequencies. This process, known as **frequency-dependent selection**, should work commonly when rare morphs are favored in mating. Finally, gene-level variation can be maintained if heterozygotes are favored in natural selection. This tends to keep alleles in the population because selection for a homozygote, or organism with identical alleles for a trait at a genetic locus, would favor one allele at the expense of others. Finally, immigration of different genetic variants from adjacent populations can increase variability in a local population.

Although natural selection is ubiquitous in natural populations, **random events** can also influence the genetics of marine populations. If the population is very small, chance events may cause the loss of certain variants from the population. This outcome, which is more likely when

---

[1] With the interesting exception of sex-determining chromosomes.

there is little difference in fitness among genotypes, has been claimed especially for some biochemical variation in proteins.

■ Single genotypes may have the capacity to develop into distinctly different morphologies.

A given genotype may take different forms under different circumstances controlled strictly by environmental variation. This phenomenon is known as **phenotypic plasticity**, which is the ability of a single genotype to develop into different forms, usually as a response to environmental circumstances. One can rightly say that a population has evolved individuals that are plastic and capable of responding to individual circumstances. Every individual has the capacity to respond to local circumstances. We encountered such phenotypic plasticity in our discussion of inducible defenses. Many marine organisms can grow spines, increase shell thickness, or change morphology completely in response to predators. On the west coast of the United States, the intertidal snail *Nucella lamellosa* develops teeth in its shell aperture when predatory crabs are present, which helps deter the crab from attacking at the shell aperture. *N. lamellosa* will develop these teeth even if the crab is held in a cage upstream of the snail, allowing the snail to smell the crab's nearby presence (Appleton and Palmer, 1988).

It is of great interest to ask why some species show plasticity and can adapt to all circumstances whereas, in other cases, genetically distinct morphs coexist in natural populations, with each morph better suited to function under different circumstances. In both cases, you can imagine that there might be a great cost. If you were phenotypically plastic, you might be able to generate a range of morphologies with none of them quite right. That is, you would be a jack-of-all-trades but master of none (DeWitt et al., 1998). If the environment is very unpredictable and it is not clear that you will or will not encounter a given situation (e.g., predators), then phenotypic plasticity might be selected for. In a stable set of microhabitats, on the other hand, a genetic polymorphism for specialized individuals might be selected, assuming that the specialized morphs had greater efficiency than could be achieved by the phenotypically plastic form. Plasticity is to be expected when environmental change occurs within the life span of an individual. For example, some barnacles seasonally regenerate their penis in time for mating season but resorb it as winter approaches. Some species are capable of producing a longer penis if another individual is not within easy reach (barnacles are simultaneous hermaphrodites) or can strengthen the penis if there is high turbulence and it is difficult to extend the penis to a nearby barnacle.

■ The geographic change in the frequency of genetic variants is called a cline.

It is common for members of a species to differ from place to place in morphology, color, or size. For example, the color polymorphism in mussels, described earlier, shows

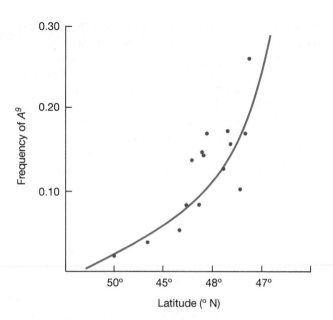

FIG. 4.25 Example of a cline: latitudinal variation of an allele, A^9, at a genetic locus coding for the enzyme lactate dehydrogenase in the crested blenny *Anoplarchus purpurescens* in Puget Sound, Washington. (After Johnson, 1971)

an increase in the brown form as one moves southward, which is related to the latitudinal gradient in solar input. Such geographic variation is also common in biochemical polymorphisms. Any directional change in frequency with geography is known as a **cline**. Figure 4.25 shows the change in frequency of an allele at an enzyme locus as one samples the crested blenny *Anoplarchus purpurescens* at different latitudes in the Puget Sound, Washington State, region.

■ New species usually originate after a species is divided by a geographic barrier.

For new species to originate, it is usually necessary for a geographic barrier to isolate a species into two or more populations. If the barrier is short lived, the populations will reconnect. If the barrier is longer lived, and especially if the populations diverge genetically, they may be relatively incompatible when reconnected. Offspring of population crosses between populations will be less fertile than crosses within populations. This would cause selection for mating with one's own kind and lead to further genetic differences between the populations, whereupon separate species would evolve. Such speciation involving geographic isolation is known as **allopatric speciation**. In some cases, we can see the recent effect of such barriers. Many pairs of closely allied species are found on either side of the Isthmus of Panama, which is only about 3 million years old. In many cases, newly evolved species are so similar that they are identical, or nearly indistinguishable, morphologically. Such species, known as **sibling species**, are very common among marine species (Knowlton, 1993). Although sibling species may have separate geographic ranges, many cases of co-occurrence have been discovered. For example, the

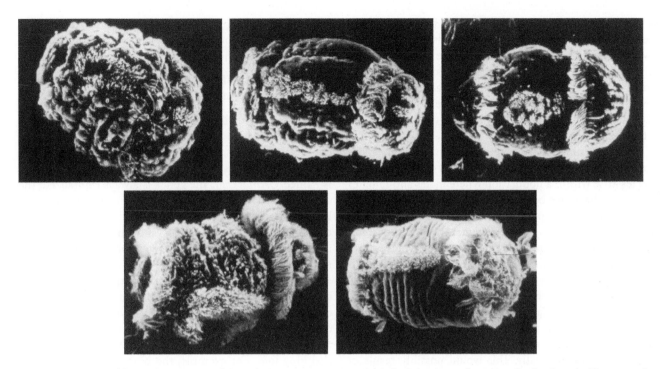

FIG. 4.26 Larvae of five sibling species of the polychaete worm genus *Capitella*, living in southern New England muds. (Courtesy of K. J. Eckelbarger and J. P. Grassle)

mud-dwelling polychaete annelid worm *Capitella capitata* is now known to consist of several closely related species that cannot be distinguished easily as adults but are quite different in the larval stage and also in chromosome number (**Figure 4.26**).

The **parapatric** model of speciation emphasizes the possibility of the origin of differentiation despite contact, as long as there are different natural selection pressures in the respective semi-isolated populations, allowing divergence even in the face of some gene flow from other populations. In effect, according to this model, natural selection is sufficiently strong to balance the influx of genes from another population, and reproductive incompatibility eventually develops between the two semi-isolated populations.

Identification of closely related species is accomplished primarily through unique DNA genetic markers that clearly identify one species from another.[2] Occasionally, some unique morphological trait may be found, as just mentioned for larvae of sibling species of the worm *Capitella*. **Figure 4.27** shows a DNA marker used to discriminate between two species of marine mussel, *Mytilus edulis* and *M. trossulus*. *M. trossulus* tends to occur in colder waters and has been found to occur in Newfoundland, eastern Canada, with *Mytilus edulis*, which is usually found farther south. Using a technique known as PCR to amplify DNA variants and separating them and visualizing unique bands (**Figure 4.27**) by a technique known

ITS DNA marker

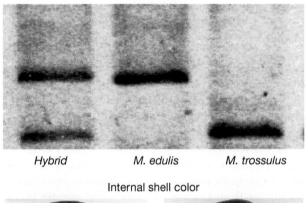

Internal shell color

FIG. 4.27 Closely related or sibling species are often difficult to identify. Top: The *ITS* gene is used to diagnose sibling species of the mussel genus *Mytilus*, especially in locations like Newfoundland where hybrids are known to occur. DNA genes are isolated and run on an agarose gel for diagnosis. Bottom: Using the *ITS* gene for diagnosis, it was learned that internal shell color could be used also to tell the species apart. (Courtesy of David Innes; top photograph by J. Toro)

---

[2] See discussion of barcode species identification in bonus chapter online.

as electrophoresis can readily pick up hybrids. As can be seen, the hybrids show up as double bands because the individual has inherited a copy of the gene that produces a different variant in each parent species. "Pure" species show single bands. In this case, Toro and colleagues (2002) discovered that all individuals with a unique DNA indicator band also had a specific shell color that differed between the two parent species.

Although most marine species probably arise from isolation across some sort of barrier, it is likely that some species arise within a population where males and females potentially have free contact. This could happen if there is a polymorphism for specificity for different microhabitats, which might lead to genetic variants mating and being located in the two different microhabitats, where selection reduces movement between the microhabitats. Such a mechanism is known as **sympatric speciation**, and has not been widely observed in the sea. Second, assortative mating might accelerate the division of two populations. If a polymorphism for color exists, it is possible that individuals of similar color and using visual mating cues might mate preferentially. This would possibly result in two differentiated species being formed if the assortative mating were continuous and without any crossing between morphs. For example, Barreto and McCartney (2007) investigated the blue hamlet and butter hamlet (genus *Hypoplectrus*), which are taken to be different species. But they used genetic markers known as amplified fragment length polymorphisms to show that these were not really different species at all but merely mated assortatively by strict color preference. No genetic variant they investigated could diagnose between these two forms. Rare matings between the color morphs seem sufficiently frequent to keep these two populations from becoming reproductively isolated true species.

As species are formed and genetic differences accumulate between closely related species, a series of isolating mechanisms develops between species. **Premating isolation** involves mechanisms that prevent members of two species from mating and producing zygotes in the first place. This is especially an interesting problem in the case of species with planktonic gametes. Even when gametes of different species are present together in the water column, the presence of different egg-sperm attractants, barriers to sperm penetration, and other mechanisms may have evolved to prevent crossing of species. In species that copulate, a series of behavioral mechanisms—including time of mating, location of mating, and mate recognition signals—may allow individuals to avoid mating with those of other species, even when very closely related. It is believed that newly formed species will evolve accentuated premating isolation mechanisms after they have reencountered each other, following a period of isolation. Even if two species cross and produce zygotes, **postmating isolation** might prevent successful production of offspring, as when gametes are incompatible. As two species are isolated and genetic differentiation occurs, a series of genetic incompatibilities might

evolve that results in improper embryonic development or early death of embryos that arise from hybridization of newly formed species.

■ Taxonomic classification involves successively nested grouping of species.

Biologists universally accept the naming system devised by Linnaeus, which gives a species a binomial (two-name) description. Every species is described by its **genus** (plural: genera) plus **species** names. For example, the killer whale is named *Orcinus orca*. The genus is always capitalized; the species name never is. Note that species names are published in italics or written and then underlined. Some species are divided into subspecies, and three names are then used. One can abbreviate the genus portion of a species name (e.g., *O. orca*).

Organisms are classified into groups larger than the species level. Each high-ranking group is made up of a cluster of groups of the next lower level. The major taxa, or classification ranks, are as follows (from the lowest to the highest): species, genus, family, order, class, phylum, and kingdom. For example, the blue mussel on the east coast of the United States has the species name *Mytilus edulis*, belongs to the genus *Mytilus*, is a member of the family Mytilidae, the order Fillibranchia, the class Bivalvia, the phylum Mollusca, and the kingdom Animalia. Note that only the genus and species names are italicized.

Species are grouped by their overall evolutionary relationships. Species in the same genus are hypothesized to be more closely related by descent to each other than to species belonging to other genera. Genera within one family are usually believed to be more closely related to each other than to those in another family.

■ Characters can be used to construct trees of relationship. Taxa are grouped by means of shared evolutionary derived characters.

All members of a given group have shared evolutionary characters, which unite them by descent and distinguish them from other groups. Thus, mollusks have an external calcium carbonate shell, differing in that respect from members of other phyla. Arthropods, such as insects, horseshoe crabs, and shrimp, all have an external cuticle, a distinct segmentation, and jointed appendages. We argue that the more distinctive characters members of a group may share, the more likely it is that the group evolved from a common ancestor. This notion allows us to construct evolutionary trees of relationship, or **cladograms**, such as **Figure 4.28**, which shows the relationships of some purely hypothetical organisms. Note that we cannot be sure about the exact history, such as who the ancestors might be. In the figure, we take the simplest creature with fewest acquisitions to be the most ancestral, but this is based only on an assumption about the simplest being the most ancestral. From the characters, we can say only who is more closely related to whom. More direct evidence, such as a fossil record, might help determine ancestry.

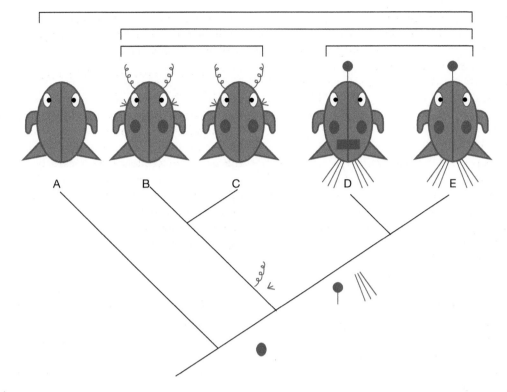

DNA nucleotide sequences

Species	n1	n2	n3	n4	n5	n6	n7	n8	n9	n10	n11	n12	n13	n14	n15
A	A	A	C	A	G	T	A	A	C	A	C	A	A	C	C
B	A	A	T	T	T	A	T	C	C	T	A	C	A	A	C
C	A	T	T	T	T	A	T	C	C	T	A	C	A	A	A
D	C	C	T	A	A	T	A	C	C	A	A	A	A	C	T
E	C	C	A	A	A	T	A	C	C	A	A	C	A	A	T

FIG. 4.28 Organisms that share a greater number of distinctive characters are likely to derive from common ancestors. Brackets on top indicate hierarchy of groups; morphological characters used to unite groups are shown along branches of the tree. Beneath the tree is a table of hypothetical DNA nucleotide sequences for the five groups. As can be seen by counting, groups B–C and D–E have the most nucleotides in common, which makes these data concordant with the morphological data. Evidence for grouping B–C and D–E groups is shown by colors of identical nucleotides.

An analysis of real critters may lead to a good deal of complexity (**Figure 4.29**) but also to an impartial analysis of evolutionary relationships. Groups are united by sets of uniquely shared evolved characters. Thus, as you go "upstream" to the base of the tree, you are encountering nodes in the tree with traits that unite all the downstream groups.

■ **DNA sequences are now used commonly to construct evolutionary trees.**

Although morphological characters may be very useful in understanding evolutionary relationships of species, DNA sequences are now commonly used for this purpose. Why? Because morphological characters often look alike only because natural selection has caused a form to evolve to converge on a single shape, which is directed by evolutionary adaptation to the local environment. For example, the presence of a predator might cause **convergent selection** in two distantly related snail species for the narrowing of a snail's aperture, which is a good adaptation to deter predators. Therefore, we cannot rely on that character to give evolutionary information about relationships. This could be true of some DNA sequences, but most are believed to be far less related to such simple cases of **convergence**. Using more genes and longer sequences of DNA will increase confidence. While it is not straightforward to quantify the difference, DNA sequences effectively give access to thousands of variable characters, many more than we can get from morphological characters. Of course, natural selection operates on DNA, and multiple mutations over time can mask the evolutionary relationships among species at any given nucleotide site, or exact location on the DNA that can vary as one of several genetic variants. But the large number of sites that are not controlled by natural selection and the large number of nucleotide sites that evolve slowly (and therefore, do not have the record of evolutionary relationships erased) should make up for these problems.

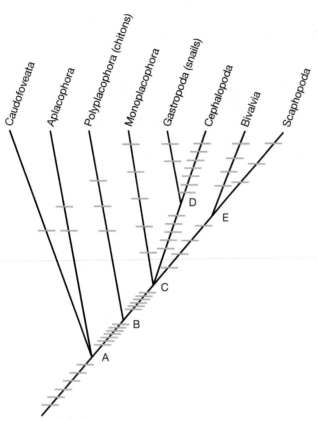

Ancestral characters

FIG. 4.29 A tree of evolutionary relationships for the phylum Mollusca. Traits (horizontal lines) between nodes unite all taxa downstream of the node higher on the page. On the stem of the tree, a number of characters unite the entire phylum, including reduction of the coelom, presence of an open hemocoelic circulatory system, and production of spicules or shell by a mantle shell gland. "Downstream" of every node (locations A–E) are traits that uniquely identify individual groups. Thus, the Bivalvia–Scaphopoda line is downstream of node E and the two groups are united by head reduction and decentralization of the nervous system, among other traits. (Modified from Brusca and Brusca, 1990)

The process of constructing phylogenies from DNA involves the following basic steps. First, DNA must be extracted from the individuals, and appropriate methods must be used to locate the specific genes to be sequenced and then to sequence those genes. This might lead to the sequences in **Figure 4.28**. Remember that each amino acid is coded by three nucleotides. We are using, for example, a DNA sequence that codes for amino acids in a protein of five amino acids. Note that one must choose the genes to sequence with care because some DNA may evolve far too slowly to obtain enough variation to analyze. In other cases, the rate of DNA evolution may be so fast that the record of evolutionary history is erased, making it impossible to relate one sequence to another. Following the sequencing step, two more steps are necessary:

1. Alignment of the DNA sequences so that individual nucleotide sites are evolutionarily related to each other

2. Construction of a tree based on the differences between evolutionarily related, or **homologous**, sites

A tree can be constructed by many methods, but the simplest is to join pairs of individuals whose sequences have the fewest differences. Then, in another round, these groups can be joined into larger groups on the same basis. Eventually, this will give a nested set of groups whose relationships arise from evolutionary transformation. The species that are most distant on the tree are most distantly related. The interested student will soon find that this is a very oversimplified representation, and many problems arise in calculating a tree (Felsenstein, 2004). In **Figure 4.28**, the DNA table is concordant with the morphology. As you can see, the groupings of B–C and D–E have the most nucleotide types in common and A is more distant, meaning that it has fewer nucleotides in common with the other two groupings.

■ CHAPTER SUMMARY

- Ecology is the study of biotic and abiotic interactions between organisms and their environment as they affect distribution and abundance. Biotic interactions such as competition are often affected by available resources.

- The ecological hierarchy consists of the individual, population or species, community, ecosystem, and biosphere. A population is a group of individuals that are affected by the same overall environment and are relatively unconnected with other populations of the same species.

- A species' ecological niche is its range of habitats. Interactions between individuals include territoriality, predation, commensalism, mutualism, and parasitism. Most populations are

dynamic, and limiting resources may affect population growth. Populations may become so rare that they have difficulty finding mates and will no longer increase.

- A population may be distributed randomly, more evenly, or in patches. A *metapopulation* is a series of interconnected subpopulations.

- Community ecology, or the interaction among species, helps determine their distributions. Often, one or more foundation species alter the habitat to allow others to live there. Other important processes are (a) dispersal of larvae, spores, and adults; (b) competition; (c) predation and herbivory; (d) parasitism; (e) disturbance; and (f) facilitation (or

positive interactions between species). The competition for resources involves space and food.

- Prey can survive by moving into refuges or escaping, evolving fixed or plastically responsive defenses, or outgrowing the predator's ability to subdue them. Still, seasonal influxes of predators in shallow water and in the intertidal zone may devastate prey populations. Disturbances such as storm damage can also greatly alter species abundances.

- Common parasites can reduce growth and reproduction as well as cause enfeeblement. Diseases can cause swift declines in marine populations.

- Succession is the ordering of species appearance, usually following a disturbance. Often, species may not be able to colonize unless another species dies off either from disturbance or predators.

- An ecosystem is a group of interdependent communities in a single geographic area. Nearly all ecosystems have primary producers, which are consumed by herbivores and in turn are eaten by carnivores. The material that escapes passes through the saprophyte cycle. While primary productivity has a bottom-up effect on ecosystems, some predatory species at the apex of food webs exert top-down controls.

- Traits vary as a function of genetic polymorphism and plasticity in responding to environmental variation. Natural selection can lead to changes in the frequency of alleles, and a regular spatial change of allele frequency within a geographic area is called a cline. New species may arise when a barrier isolates two populations and they no longer interbreed.

- Every organism is described by its genus (plural: genera) and species. Evolutionary relationships can be used to construct trees of relationships. Taxa may be grouped by morphology or DNA sequences.

■ REVIEW QUESTIONS

1. Describe the ecological hierarchy.
2. Distinguish between a population and a community.
3. What is the difference between renewable and nonrenewable resources?
4. If the distance between exploitable patches increases, should the time spent by a forager in a patch increase or decrease? Explain your answer.
5. Under what conditions might a marine creature have color that matches the background? When might it have strongly visible coloration?
6. What is the advantage of an inducible defense as opposed to a fixed trait that is always available for defense?
7. Distinguish between commensalism and mutualism.
8. Why are parasites likely to have complex life cycles?
9. How might a resource limit population growth?
10. How might several genetic variants be maintained over time in a single population?
11. What is the main determinant of biogeographic provinces in coastal marine environments?
12. What is the major effect of predation in communities of competing prey species?
13. What are the major processes that contribute to determining the relative abundance of species in a community?
14. Define ecological succession.
15. Distinguish between biomass and productivity.
16. Some species consist of genetically identical individuals, all of which are very flexible in their ability to live in different subhabitats, whereas other species consist of individuals, each of which is distinctly different and specialized for a given subhabitat but inflexible. Under what conditions might each species be favored?

Visit the companion website for *Marine Biology* at www.oup.com/us/levinton where you can find Cited References (under Student Resources/Cited References), Key Concepts, Marine Biology Explorations, and the Marine Biology Web Page with many additional resources.

CHAPTER 5

The Chemical and Physical Environment

The purpose of this chapter is to show how marine organisms respond to changes in the chemical and physical aspects of their environment. Most coastal habitats change rapidly, which presents a challenge to marine organisms. Nearly all sites out of the tropics show strong seasonal changes to which a marine organism must adjust, sometimes including strong drying from the sun all the way to coverage by ice. But very short-term changes are also common, especially in the intertidal zone. Some changes are cyclic, relating to tidal height and day–night light and temperature cycles. In a small tidal estuary, salinity can change greatly with every rise and fall of the tide, as the water is dominated alternatively by incoming seawater and outgoing fresh water. But other changes are less predictable. With a class, I once found that a strong rainstorm brought the salinity of a Florida tide pool from normal marine salinity to a freshwater state in about an hour. Interacting with all of these changes are decadal-scale climate oscillations and anthropogenic climate change, which are tipping the balance toward increased physiological stress and even extensions of geographic ranges.

Measures of Physiological Performance

■ **Measures of organismal response include whole-organism, behavioral, physiological, and biochemical factors.**

During its lifetime, any organism is subjected to a range of environmental variation. An organism's ability to survive environmental change is ultimately determined by its genes and is modified by environmental effects that may affect or even cause nongenetic transitions in physiological condition. When an environmental change occurs, such as a rise in temperature, the individual must first have receptors to sense the change. This information must then be conveyed and translated into an adaptive response,[1] such as a change of cellular function or crawling into a cool, wet burrow when conditions are hot and dry.

An adaptive response need not be confined to behavior. Many crucial adjustments are biochemical and physiological. If an animal is exposed to some toxic metals, chances are that the concentration of a metal-binding protein called metallothionein will increase within hours. Various enzymes, hormones, and other vital molecules are maintained at concentrations that maximize performance. These are examples of **biochemical changes**. Responses involving **physiological change** include such processes as change in ciliary beating rate and transport across membranes. All these responses require energy, and the metabolic rate is the total rate of energy use by the organism. Metabolic rate is often used to characterize an overall physiological response to an environmental change because it can be related to a cost of response, such as energetic need.

■ **Organisms respond to environmental change by reaching a new equilibrium through a process known as acclimation.**

Organisms rarely have fixed responses to a given environmental condition. For instance, if mussels are collected from Massachusetts in winter in near-freezing seawater, they will die quickly if placed in seawater at 20°C. No mortality

---

[1] An adaptive response is an often reversible plastic response to an environmental change that has evolved to increase fitness.

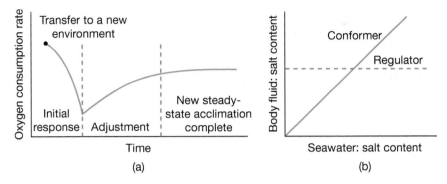

FIG. 5.1 (a) Acclimation response of oxygen consumption following a change of temperature. (b) Response of body fluid salt content to environmental salinity variation of a regulator and of a conformer.

occurs, however, if mussels collected in summer are transferred into water at this same temperature. This is because the mussel has adjusted physiologically between winter and summer conditions and has developed a capacity to function in a warmer environment. A change of function and tolerance that results in an equilibration with new physicochemical conditions is known as **acclimation**. We assume that an organism improves its functioning in a new environment by shifting an array of physiological and biochemical processes. Acclimation can be studied by changing the environment of a laboratory animal. After an environmental change, an **immediate response** is followed by an adjustment period, which is then followed by a new **steady state** (**Figure 5.1a**), which is the accomplishment of acclimation.

If the external environment changes, the individual may have a battery of adaptations to maintain constancy of a parameter such as temperature within the body. Such organisms are said to be **regulators**. Organisms whose body temperature or cellular salt concentration changes in direct conformance with environmental change are **conformers**. Some species will regulate their response to variation in some environmental factors (e.g., salt content of cellular fluids of fishes) but conform to others (e.g., temperature). **Figure 5.1b** shows the responses of a complete regulator and a complete conformer to changes in salinity.

■ **Scope for growth is a measure of the food intake that is available for growth and reproduction beyond the cost of metabolism.**

As an animal consumes and digests food, the greater the energetic cost of overall cellular reactions, or **cost of metabolism**, the less energy will be available for growth and reproduction. **Scope for growth** is the difference between the amount of energy assimilated from the animal's food and the cost of metabolism. This implies there is a zero point of **maintenance metabolism** for any given animal. At this point, an animal will be gaining just enough energy to balance its energetic needs. If scope for growth is positive, there is energy available for growth and reproduction. If scope for growth is negative, however, the animal has a cost of metabolism that is greater than can be matched by energetic intake of food. Such an animal will burn off reserve fats, lose weight, and if the process is prolonged, eventually die because of it.

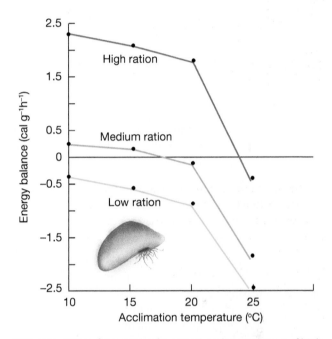

FIG. 5.2 Scope for growth of a mussel under conditions of high, medium, and low food rations and varying temperature. Note that, as temperature increases or food decreases beyond a certain point, scope for growth, in terms of energy balance, falls below zero.

To estimate scope for growth, we need to measure the feeding rate, the percentage of food assimilated (i.e., digested, transported, and incorporated into cells), and the metabolic rate, in units of energy (e.g., calories). Metabolic rate is usually estimated by respiration rate (rate of oxygen consumption), which is an estimate of the aerobic portion of the cost of metabolism. If food is abundant, scope for growth will increase. Owing to increased chemical reaction rates, increasing temperature often increases the energy need of cells. Increased temperature may therefore decrease scope for growth (**Figure 5.2**).

■ **Physiological condition can also be estimated by the scope an organism has for activity, such as swimming.**

Organisms must not only have enough excess reserves to grow and reproduce, but they also may need immediate delivery of energy for activity, such as swimming. A fish may be at rest, but if a predator is near, it needs reserves to escape in a burst of swimming. Such reserves require not only available carbohydrates to burn in metabolism but also a delivery system to bring oxygen to contractile muscle tissues and an aerobic system to oxidize carbohydrates with the muscle tissues themselves. The range of oxidative metabolism that can be performed for activity beyond maintenance metabolism is known as **scope for activity** or **aerobic scope**, when metabolism is entirely aerobic.

- **Mortality rate can also be used as a measure of the effect of environmental change.**

We can assess the effect of extreme environmental conditions by measuring mortality rate. Experimental populations are usually kept at a standard laboratory condition for a period. Then, lethal temperature, for example, can be determined (a) by inducing a slow decline or rise of water temperature or (b) rapidly transferring the laboratory-acclimated individuals to a constant new temperature. The lethal dose required to kill 50 percent of the experimental population after a specified time (24 hours is a common period) is the LD_{50}. To find the LD_{50}, one experimentally varies a factor (e.g., temperature) and observes a series of mortality rates. This produces a series of points relating the factor to percentage mortality. The LD_{50} can then be interpolated (e.g., **Figure 5.3**).

- **Changes in gene expression and protein composition of cells can be measured as a response to environmental change.**

Transcriptomic and proteomic approaches (see bonus online chapter on molecular methods) are becoming powerful tools for the study of response to environmental change, such as increasing temperature. Classes of genes are expected to respond to an environmental change through changes in gene expression, producing changes in protein composition within the cell that improves cell function under the changed conditions. Alternatively, one can expect changes in the abundance and distribution of proteins as environmental change occurs. Therefore, we can make the hypothesis that certain classes of genes or proteins would change in degree of expression or concentration as the environment changes. See, for example, the following discussion on responses of expression of heat shock proteins to thermal stress.

Temperature

- **Temperature affects the latitudinal distribution of most marine species.**

Continental coastlines of the world ocean are mostly oriented in an approximately north–south direction, and temperature tends to increase toward the tropics. This latitudinal temperature gradient is especially steep on the east coast of North America because of the continentally influenced climate. Latitudinal change on the west coast of the United States is not as strong and there is much less seasonal change in sea-surface temperature.

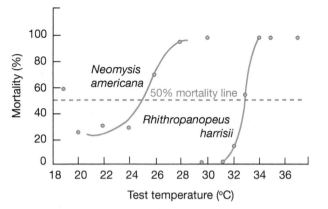

FIG 5.3 One can measure differences in temperature tolerance by exposing groups of animals to different temperatures and measuring mortality after 24 hours. The LD_{50} is the temperature at which 50 percent cumulative mortality occurs. In this example comparing eastern U.S. arthropods, a mysid shrimp, *Neomysis americana*, was taken from near the southern end of its geographic range. It proves to be less tolerant of high temperature than the mud crab *Rhithropanopeus harrisii*. (Modified from Mihursky and Kennedy, 1967)

Homeotherms and Poikilotherms

- **Homeotherms regulate their body temperature, whereas the body temperature of poikilotherms conforms to the external environmental temperature.**

There are two extremes of temperature regulation. **Homeotherms** are organisms that regulate body temperature to a constant level, usually above that of ambient seawater. A constant and relatively high body temperature enables biochemical reactions to occur in a relatively constant internal environment and at a relatively high rate. Most seabirds have a body temperature of about 39–40°C, whereas the temperature of most marine mammals is about 37–38°C. Because such temperatures are much higher than that of most seawater, marine homeotherms lose heat rapidly to the surrounding environment.

There is another completely different style of living. **Poikilotherms** are organisms whose body temperature conforms to that of the ambient environment. All subtidal marine invertebrates and most fishes fit into this category. There is an interesting intermediate status that we discuss in the next section, in which body temperature is usually somewhat higher than ambient temperature. Some intertidal animals are not true poikilotherms because they respond to hot sunny days by maintaining themselves at a temperature lower than an inanimate object of the same size, using both evaporation and circulation of body fluids. Darker-colored forms can absorb more heat than can light-colored forms; therefore, variation in color can reflect differences in adaptation to the capture of solar energy at different latitudes.

- **Homeotherms, some fish, and a species of sea turtle use restriction of circulation, insulating materials, and a countercurrent mechanism to reduce heat loss to the environment.**

As we discussed in Chapter 2, ocean temperatures are usually less than 27°C and may be less than 0°C in some

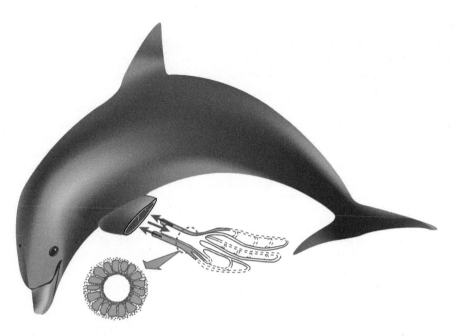

FIG. 5.4 The pattern of veins (blue) wrapped around arteries in the flipper of a dolphin. This arrangement permits the venous blood to be warmed by heat transfer from arterial blood before it reenters the body core. (From Schmidt-Nielsen, 1975)

locations and during some seasons. Therefore, most homeo-thermic mammals and birds must lose heat continuously to the environment. The skin is the main surface of heat loss, especially by direct conductance of heat from the skin to the contacting colder water. Because animals have a circulatory system, **convective heat loss** from the body surface also occurs as warm interior blood is transferred and moves into contact with the periphery of the body. As air breathers such as mammals exhale, the resulting evaporation of water involves a considerable loss of heat.

The first line of defense against heat loss is a well-insulated body surface. Marine birds deal with this problem by means of specially adapted feathers. A series of inter-locking **contour feathers** encloses a thick layer of **down feathers** that traps stationary air, which in turn acts as an insulating layer (see Figure 22.12). Whales, porpoises, and seals are insulated against the lower sea temperatures by a thick layer of subcutaneous fat. Sea otters lack such a layer, but they constantly preen and fluff up a relatively thick layer of fur. Such mechanisms are only partly successful, however, and to generate more body heat to maintain a constant temperature, marine mammals usually must have a higher metabolic rate than similarly sized terrestrial animals.

In marine mammals, the limbs are the principal sources of heat loss because they expose a relatively greater amount of body surface area per unit volume to cold water. However, warm arterial blood must be supplied to limbs, such as the flipper of a porpoise. Heat loss in porpoises is mini-mized by a **countercurrent heat exchanger**. The arteries are surrounded by veins, within which blood is returning to the core of the animal (**Figure 5.4**). At any contact point, the artery is warmer than surrounding vein, so heat is lost to the returning venous blood flow. Heat is thus reabsorbed

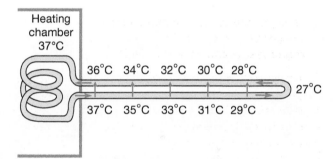

FIG. 5.5 A model of a countercurrent heat exchanger. Warm fluid leaves the hot water bath through the lower tube and loses heat to the external environment, thus reducing the temperature toward the right. However, because the return flow gains heat by exchange with the adjacent outflow tube, some of the heat is recovered and returned to the heating chamber.

and returned to the porpoise's body core. This spatial re-lationship of circulatory vessels minimizes heat loss to the flipper and thence to the seawater in the external environ-ment. A model for this is shown in **Figure 5.5**. As the fluid leaves a warmed heating chamber through a tube, lost heat is reabsorbed by fluid flowing in an adjacent tube that flows in the opposite direction back toward the heating chamber. It is the opposite and adjacent flows through circulatory vessels that minimize heat loss from the heating chamber. Although the anatomical details are quite different, fishes such as skipjack tuna have a circulatory anatomy based on the same overall design. Arteries and veins in the near-surface musculature are in contact, and in arteries and veins, respectively, blood flows in opposite directions (see Chapter 9). Leatherback turtles (*Dermochelys coriacea*) in cold waters of Nova Scotia have body temperatures 8°C higher than

Keeping Your Legs Warm: A Sea Turtle Tale   5.1

Consider the leatherback turtle *Dermochelys coriacea*, the only living species of the family Dermochelyidae, which has approximately a 50-million-year history, feeding mainly in cold ocean habitats. These creatures are truly enormous, with masses of 250–700 kg, body lengths reaching about 2 m in length, and flippers reaching over 2.5 m long. They are teardrop-shaped and glide through the water (**Box Figure 5.1**), using the front flippers for propulsion and the rear ones as rudders. They have a thick leathery skin, into which are embedded small skeletal back elements. Like other sea turtles, females come to beaches to nest (in the Caribbean, examples are Culebra in Puerto Rico and the islands of Trinidad and Nevis), mostly at night (**Box Figure 5.2**). Otherwise, adults are widespread in higher latitudes as far as the Arctic Circle, and some have been found stranded on beaches on Long Island, New York.

In cold weather, heat loss is an obvious issue. These elegant creatures swim in the water column and feed on jellyfish, often following them to great depths during the day, when jellyfish undergo diurnal vertical migrations. Their activity is correlated with a surprising high body temperature, often greater than 10°C above that of the ambient water. Paladino and colleagues (1990) recorded a body temperature of 25.5°C

in seawater of 7.5°C, and leatherbacks are known to dive into near-freezing water. They live in the Atlantic, Pacific, and Indian Oceans, and individuals that have been tagged with GPS tracking devices have been followed for thousands of kilometers across the oceans. For example, a turtle named Calypso Blue was traced from waters off of Costa Rica along the coast of North America to New Jersey and then out to the middle of the North Atlantic Ocean. It is interesting that their swimming speeds remain constant (2.5–3 km h⁻¹) whether they are swimming in tropical or high-latitude waters.

Body temperatures of leatherbacks clearly are warmer than the seawater in which they are swimming, so they are in some way gaining and retaining heat, which is probably generated during metabolism associated with activity and muscular action. Their large size, peripheral tissues, and circulatory structures all aid in preventing heat loss.

For several decades it has been well known that there is a plexus of intertwined arteries and veins at the roots of all four limbs. Is this a countercurrent exchange adaptation to retain heat in the leatherback's body core, as has been observed in the dolphin? This has been presumed to be so until John Davenport and colleagues (2015) took a more careful

BOX FIG. 5.1 A leatherback sea turtle swimming in waters of Trinidad. (Photo by John Eckert)

HOT TOPICS IN MARINE BIOLOGY

Keeping Your Legs Warm: A Sea Turtle Tale *continued* 5.1

look. They did careful dissections of the peripheral body and found that the plexuses of intertwined arteries and veins do not really supply blood to the limb blades: Instead, they supply or drain the area of the hips. The venous blood drains active locomotory muscles that are surrounded by thick blubber when the turtles are in very cold water. *The plexus therefore is designed to heat these particular peripheral muscles and not the body core.* In fact, this mechanism actively reduces loss of heat from the limb muscles to the core to increase the efficiency of swimming and to prevent overheating when females are on shore digging nests and laying eggs in air. The adaptation is to keep the turtle hot-footed.

BOX FIG. 5.2 A leatherback sea turtle nesting on a Trinidad beach. (Photo by John Eckert)

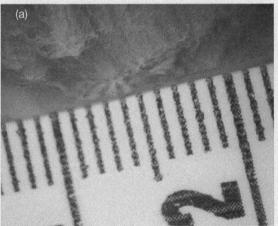

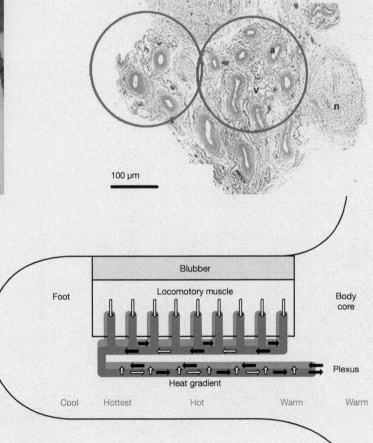

BOX FIG. 5.3 (a) Photograph of cut end of arterial-venous plexus; individual bundles are about 1 mm in diameter. (b) Cross section of plexus; a = artery, v = vein, n = nerve bundle. Circles enclose areas of mixed arteries and veins. (c) Diagram of function of hind limb vascular plexus; black arrows = blood flow, red = arterial supply, blue = venous flow. White arrows indicate net heat transfer. (From Davenport et al., 2015)

ambient temperatures, which are maintained by a combination of a large heat-retaining mass and a layer of subcutaneous fat. We discuss the controversial subject of the leatherback's countercurrent exchange mechanism in the flippers in **Hot Topics Box 5.1** in this chapter.

Temperature, Metabolic Rate, and Performance

■ **In poikilotherms, metabolic rate increases with increasing temperature.**

In poikilotherms, an increase in temperature usually increases metabolic rate and behavioral activity. Oxygen consumption is a convenient expression for overall metabolic rate because most poikilotherms consume oxygen as they burn carbohydrates for energy. With an increase of 10°C, the corresponding change in metabolic rate as measured by oxygen consumption is called the Q_{10}. For most poikilotherms, Q_{10} ranges from a factor of 2 to 3. Thus, there is a doubling to tripling of oxygen consumption with a 10°C rise in temperature.

For a poikilotherm, an increase in temperature causes an increase in energy requirement. This cost will reduce the animal's energy reserves and will therefore reduce the scope for growth. Refer again to **Figure 5.2**, which shows the scope for growth in the mussel *Mytilus edulis* at different temperatures and at low, medium, and high rations. At low ration, scope for growth is negative at all temperatures, whereas it is also negative at 25°C for all food levels. At such high temperatures, the animal cannot compensate for the high metabolic cost of living at high temperature. Over a wide intermediate temperature range, however, physiological acclimation results in a stable scope for growth (see the following).

■ **Poikilotherms respond to increasing temperature with a standard performance curve of increased performance, a maximum, and then declining performance as temperature increases further.**

Many poikilotherms experience short-term changes in temperature that are too rapid for any compensation to occur. For example, after leaving their burrows, intertidal crabs in the tropics heat up rapidly when standing on hot sand and exposed to the sun. The incoming tide may bring cool water, but the water may heat up rapidly while the crab is motionless over the sediment and exposed to the sun. This change also exposes intertidal infaunal organisms such as clams and worms to steadily changing temperature. These animals may not be the same temperature as a nonliving solid of the same size and shape because organisms use circulation of body fluids, evaporation, and other responses to avoid passive heating. However, because of the response to increasing temperature, metabolic rate and activity will increase as temperature increases to a maximum followed by a decline in performance with further temperature increase (**Figure 5.6**). Usually, the decline in performance is very strong at temperatures higher than the optimum.

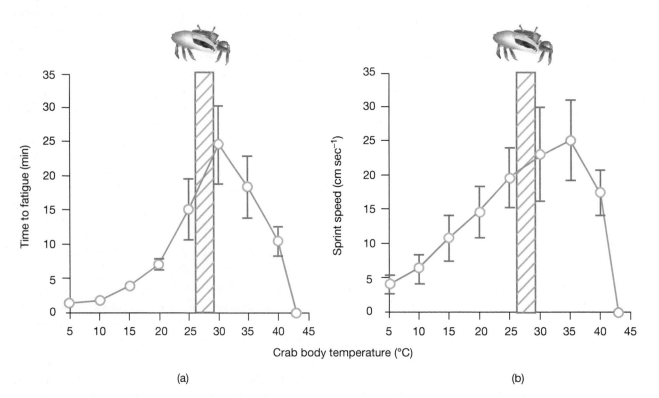

FIG. 5.6   A typical temperature-performance curve for a marine invertebrate. This diagram shows, for males of the fiddler crab *Uca pugilator*, (a) endurance time for a fiddler crab run on a treadmill at 4 m min⁻¹ velocity over a range of temperatures and (b) crab sprint speed. Both reach a peak at approximately the same temperature. Shaded area indicates range of temperatures preferred by the crabs as demonstrated in a laboratory temperature gradient on moist sand. Note that crabs prefer to be at temperatures where their performance is maximal. (After Allen et al., 2012)

■ Poikilotherms can compensate for changes of temperature by means of an acclimation process.

If temperature changes rapidly (e.g., on the scale of minutes), most poikilotherms increase or decrease their metabolic rate accordingly. If you place a tropical crab in a kitchen refrigerator, it will move very slowly, if at all. Such a relationship of temperature to metabolic rate would pose problems for poikilotherms living in seasonal environments. To avoid this restriction, many poikilotherms are able to **acclimate** to seasonal changes in temperature.

Figure 5.7 shows a common temperature response. A poikilothermic animal is capable of regulating its metabolic rate to a constant level over a broad range of temperatures as the thermal seasons progress. However, below a certain temperature, activity will drop off and the animal will be unable to function. Above a certain threshold temperature, the animal will not be able to regulate metabolic rate, and it will rapidly increase before breaking down entirely, making it impossible for the animal to function indefinitely.

Now consider a graph of metabolic rate as a function of temperature for an animal living in high summer temperatures. If temperature suddenly decreases, then metabolic rate will be very low, and the animal will not be able to generate the energy needed for activity. To get around this problem in response to seasonal reduction of temperature, the animal acclimates by shifting its metabolism–temperature curve to the winter acclimation form shown in **Figure 5.7**. This allows the animal to function in winter at much lower temperatures. But this shift will then have a cost: Metabolic rate may increase at a lower high-threshold temperature than in the case of the summer-acclimated situation. If an animal acclimated to winter temperatures experiences a sudden temperature increase, metabolic rate may rise to such a point that not enough energy would be available. Given sufficient time, however, the animal will respond by adjusting its metabolism to the summer acclimation curve.

Acclimation to changing seasonal temperature can be shown by collecting animals that have been acclimating to the temperature in one season and then shifting them rapidly to a new temperature. For example, exposure of winter-acclimated animals to a high temperature should result in metabolic rates greater than those expected for summer-acclimated animals. At 3°C, winter-collected burrowing mole crabs (*Emerita talpoida*) consume oxygen at a rate 4 times greater than do summer-collected animals tested at the same temperature. A seasonal change in tolerance usually accompanies acclimation of metabolic rate; the upper lethal temperature is greater in summer-collected animals than in members of the same species collected in winter. This adjustment can result in constant metabolic rates for acclimated individuals over a wide range of temperatures.

Similar types of compensatory mechanism dampen the range of metabolic activity expected within a species living in a latitudinal thermal gradient. At low temperature, oxygen consumption tends to be greater in animals living at high latitudes than in members of the same species living at low latitudes. Similarly, geographically separate populations of oysters (*Crassostrea virginica*) and sea squirts (*Ciona intestinalis*) have varying temperature optima for breeding depending on latitude. Populations with compensatory responses that enable them to function in different parts of the latitudinal temperature gradient have been termed **physiological races**. These races may be genetically different, or the individuals in them may merely become acclimated to different temperatures.

Tolerance to temperature is an important factor regulating the distribution of marine organisms. Because intertidal environments tend to have much greater daily and seasonal temperature ranges, *intertidal organisms tend to tolerate a broader temperature range than do subtidal marine species.* The geographic ranges of marine species indicate that natural selection has shifted the optimum response of species to that of their native temperature regime. The Antarctic fish genus *Trematomus*, for instance, has representatives that live in water temperatures close to −1.9°C throughout the year. These fish will die at an upper limit of only 6°C. Many Arctic species cannot tolerate the "high" temperature of 10°C! In the mitochondria of the Antarctic *Trematomus bernacchii*, synthesis of adenosine triphosphate, or **ATP**, the general source of energy for living cells, can be carried out at the lowest temperature of any marine animal species thus far measured (Weinstein and Somero, 1998). At any one geographic location, the marine biota consists of an assemblage of species whose optimum temperature ranges are different. Thus, after the severe winter of 1962–1963 in Great Britain, more tropically adapted species showed great mortality, whereas Arctic-adapted elements suffered no ill effects.

■ Species living over different latitudinal ranges have evolved different performances over a range of temperature. These differences have consequences as sea surface temperature changes.

An important question is the degree of evolutionary change with regard to adaptation to temperature among coexisting species and along the latitudinal range within a species. Levinton (1983) studied two sibling species of polychaete from different latitudes that could be bred in the lab. Females lay egg cases that hatch after 4 days, which allows complete studies of growth. One species, *Ophyrotrocha costlowi*, lives in

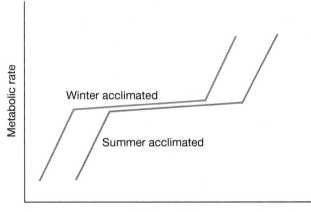

FIG. 5.7 Expected response curve of metabolic rate versus the temperature for an animal acclimated to winter low temperature and the same animal acclimated to summer high temperature.

North Carolina, and another very closely related species, *O. macrovifera*, lives in Florida waters. Highest temperatures are about the same for both latitudes, but North Carolina winter temperatures are much lower than Florida. The northerly *O. costlowi* grew faster over a range of low temperatures (15, 20, 25°C) but grew at the same rate as the southerly *O. labronica* at 30°C. This result shows that the northerly sibling species has evolved to grow more efficiently at lower temperatures than the southern sibling species. While the growth rate for species from both latitudes was the same at 30°C, the survival rate was quite different. Nine percent of North Carolina broods had less than a 20 percent survival rate, whereas 91 percent of Florida broods had the same survival rate at the same temperature. The North Carolina sibling species clearly had evolved a different temperature tolerance. A similar study demonstrated that a northern European subspecies of *Ophryotrocha puerilis* grew and survived more efficiently at low temperatures, whereas a southern European subspecies grew and survived better at high temperatures (**Figure 5.8a**; Levinton and Monahan, 1983). Similar results have been found on the west coast of North America for high-latitude and low-latitude species of the porcelain crab, genus *Petrolisthes* (Stillman, 2002). Northerly species had much higher mortality rates at high temperatures than the southerly species (**Figure 5.8b**). Species living at higher levels of the intertidal also survive better at higher temperatures. These results show clearly that closely related species and subspecies have evolved different performances that correspond to their current native temperatures, which differ as a function of latitude and tidal height. Data also support differential performance for populations of the same species living at different latitudes. It is clear that evolution has caused these differences, but we still lack specific explanations based on gene-level evidence of performance.

Because of the differences in thermal adaptation, anthropogenic climate change should change the distribution of species as ocean temperatures warm. Warm- and lower-latitude species should shift toward higher latitudes and coexist and even displace higher-latitude and cold-adapted species, subspecies, and even genotypes. Such a shift is already underway in some intertidal species, whose southerly ranges are being truncated by increased sea temperatures, moving sites beyond the upper thermal limits of the species. In Monterey Bay, California, northerly species are giving way to southerly species as ocean temperatures warm in that region. A 1990s survey of rocky shore invertebrates in Monterey Bay was compared with a similar 1930s survey. Nearly all southerly distributed species had increased in abundance, whereas nearly all northerly distributed species had declined (Sabarin et al., 1999). Ocean warming is therefore rearranging genotypes (we need more data on this) and subspecies and restructuring communities that consist of mixed northerly and southerly species.

The Meaning of Temperature Limits

■ **Temperature stress can affect organisms in three ways: by limiting growth, by limiting activity via oxygen delivery, and by limiting enzyme and cell function.**

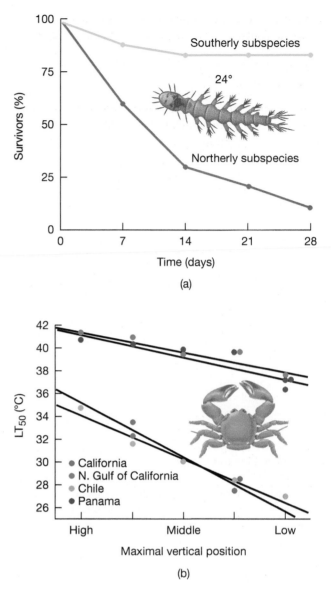

FIG. 5.8 Differential adaptation of closely related species whose main distribution is from northerly (cold) or southerly (warm) waters. (a) Survival over time of a northern European subspecies of *Ophryotrocha puerilis* versus a northern subspecies at a relatively high environmental temperature. Note that the northern subspecies has much higher mortality. (b) Survival time, until 50 percent survival, of different eastern Pacific species of the crab *Petrolisthes* from high latitude (California, Chile) and lower latitudes (Panama, Gulf of California), living at different levels of the intertidal. (After Stillman, 2002)

Because of concerns about global climate change and the steady increases that have been observed in sea-surface temperature (see Chapter 3), there has been increased attention paid to the factors that contribute to temperature stress, especially at the high end. We have discussed how the need for carbohydrate reserves interacts with temperature to reduce the scope for growth. Thus, high temperature is important in growth limitation because carbohydrates are burned more quickly. Second, at high temperatures, tissues require delivery of greater amounts of oxygen to fuel metabolism, which imposes a cost on the organism.

Finally, extreme high temperatures cause dysfunction of molecules, such as enzymes, and aspects of cell function. At extreme temperatures, enzymes lose their ability to efficiently perform catalysis, and organisms must respond to the dysfunction.

■ Large temperature changes can reduce physiological performance by affecting physiological integration.

At the biochemical level, heat death must be related to protein function, particularly the catalytic efficiency of enzymes. To function, enzymes must bind to substrates efficiently, catalyze reactions, and then release substrates and products that allow further reactions to occur. The binding sites of enzymes can be likened to furrows in a complex three-dimensional molecule, within which the substrate must fit for a reaction to be catalyzed properly. At temperatures that are too high, the binding sites are, in effect, too wide open, and enzyme function is inefficient.

High-temperature shock not only causes dysfunctional increases in binding sites, but it can also cause total unfolding of the protein, producing irreversible damage. Animals respond by producing **heat-shock proteins**, which forestall the unfolding of a protein's three-dimensional structure and in some cases reestablish folding to make a protein functional after temperature damage. If damage is extreme, the low-molecular-weight protein **ubiquitin** binds to degraded proteins, which are then digested by cellular proteolytic enzymes. Individuals can acclimate to varying temperature by changing the production of heat-shock proteins and ubiquitin. In the short term, animals such as mussels produce sharply increased levels of heat-shock proteins in response to the temperature increases observed typically in the rocky intertidal zone. The same trend should be expected between species living in different thermal regimes: Low-latitude species should be adapted to higher temperature and should have protein structures that do not fail as easily at higher temperatures. Hoffman and Somero (1996) found that a high-latitude Pacific U.S. coast mussel species, *Mytilus trossulus*, responded to high temperature exposure by producing higher levels of ubiquitin and heat-shock proteins. The lower latitude *M. galloprovincialis* responded less, which suggests that its soluble proteins function better at high temperatures. In a more extensive proteomic study of response of the two species, Tomanek and Zuzow (2010) found that expression patterns of proteins involved in molecular chaperoning, proteolysis, energy metabolism, oxidative damage, cytoskeleton, and deacetylation demonstrated similar overall responses in both mussel species, but the cool-adapted *Mytilus trossulus* showed greater sensitivity to heat stress at lower temperatures where *M. galloprovincialis* showed little evidence of stress. This difference also fits with the more low-latitude occurrence of *M. galloprovincialis* and its spread northward on the west coast as coastal temperatures increase.[2] A proteomic study of thermal stress of the

mussel *Mytilus edulis* in Normandy, France, also shows a switch to anaerobic metabolic pathways to cope with the metabolic stress of higher temperature (Péden et al., 2016).

Similarly, intertidal and subtropical snail species of the genus *Tegula* are more responsive to heat stress by producing heat-shock proteins and produce them at higher temperatures than *Tegula* species living typically below mean low water (Tomanek, 2002).

Most likely, high temperature leads to failure of a series of interdependent metabolic reactions. This is known as a loss of **physiological integration**, or a decrease in the degree of coordination among interdependent biochemical reactions. If various metabolic processes are influenced differently near the upper temperature limit, then coordination among them will be unbalanced.

The disruption of physiological integration can sometimes be observed at the cellular level by the investigation of coordinated ciliary activity of epithelial cells that can be examined over a range of temperature. A failure of ciliary activity is a special case of a more general problem of maintenance of cell membrane function. If a poikilotherm is placed in a lower temperature, the packing of the structural phospholipids in its membranes will increase and cause more ordering of the membrane structure, also known as increased **membrane order**. Conversely, an increase in temperature will reduce packing and therefore membrane order. If phospholipids are packed too poorly or too well, functions such as transport across the cell membrane will be harmed.

Intertidal invertebrates often experience both seasonal changes and tidal changes in temperature, sometimes exceeding a range of 20°C. The California sea mussel *Mytilus californianus* maintains membrane order as temperature increases (Williams and Somero, 1996). In summer, high intertidal mussels can alter membrane order rapidly, which allows them to respond to the rapid changes in temperature experienced on the upper shore.

In high-latitude marine environments and in winter in some midlatitudes, **freezing** presents a severe environmental problem. Larvae of many fishes and Foraminifera can be found encased in pack ice in Antarctic waters. Many fleshy algae and barnacles survive the winter under freezing conditions in the intertidal zone. The formation of intracellular and within-tissue ice can shear and distort delicate structures and may increase the cellular salt content of the remaining fluids. In most cases investigated, tidal animals show varying degrees of freezing under subzero temperatures. At progressively lower temperatures, increasing percentages of the body fluids are frozen. Nevertheless, intertidal fleshy algae can survive extended periods of freezing.

Dissolved salts lower the freezing point of seawater and of cellular fluids. Water with a salinity of 35 will depress the freezing point of water by about 1.9°C. It is possible that the salt content allows cellular fluids to become supercooled—that is, to remain in the liquid state below the freezing point of water. In winter, temperatures in polar waters reach the freezing points of seawater and of

---

[2] See discussion of transcriptomics in the bonus chapter, "Molecular Tools for Marine Biology," online.

the cellular fluids of many invertebrates and fishes. The problem is especially acute for bony fishes, whose cellular fluids have a lower salt content than that of seawater. The surrounding seawater may actually be in a liquid state at a temperature at which water would normally freeze within the fish. Deep-water fishes out of contact with ice can remain supercooled. The presence of ice crystals in shallow water, however, would initiate cellular freezing, especially because fishes drink seawater laden with tiny ice crystals. Shallow-water Antarctic fish of the suborder Notothenioidei (**Figure 5.9**) can synthesize special **glycoproteins**, which behave much like automobile antifreeze and depress the freezing point of cellular fluids.

Temperature also affects growth and reproduction in marine organisms. Most marine species grow and reproduce over a narrower range of temperature than the range that permits individual survival. Within an intermediate range, growth is usually faster at higher temperatures. In bivalves, members of the same species have been found to grow more slowly, but they survive to older age and reach larger size in high latitudes than in low latitudes. Growth in seasonal habitats is greater in warmer times of the year, but the increased growth may also reflect the greater availability of food.

Temperature often sets the timing and can determine the style of reproduction. Many invertebrate species will spawn only when a given temperature is reached. Some species may switch from asexual to sexual reproduction as temperature increases. Even sex may be determined by environmental temperature in some marine species. For example, in the Atlantic silverside *Menidia menidia*, embryos develop into females at low temperature but into males at higher temperature. Marine turtles have no obvious sex chromosomes, and embryos develop into males at lower temperature and into females at higher temperature (Standora and Spotila, 1985). Recent concerns about climate change and turtle conservation have led to worries that the sex ratio might become skewed. In the green turtle *Chelonia mydas*, eggs develop into males at temperatures cooler than 28°C, but warm nests greater than 29.5°C develop

into nearly all females. As beach habitats warm up, sex ratios may be skewed toward females. Changes of only 2°C or less might have drastic effects. This is especially a problem because the green turtle homes to the beach where it was born. Global warming might cause an overabundance of females. Moreale et al. (1982) warned that hatcheries paid too little attention to the temperature at which eggs were reared, which might lead to skewed sex ratios.

In general, seasonal changes in the timing and amount of egg and sperm production and release are highly correlated with temperature (as well as food and photoperiod). **Figure 5.10** shows seasonal gonad changes in the sea star *Pisaster ochraceus*.

■ **The seasonal extremes of temperature have different effects depending on the location of an individual within the latitudinal range of the species.**

Although there are exceptions, subtidal animals at low latitude generally live in warmer waters and are adapted to life in higher water temperatures than members of the same species living at higher latitudes, which are adapted to life at low temperature. This can make a very large difference in the thermal environment over a species range. The summer is the most stressful time of the year at a species' lowest latitudinal limit; high temperatures in some years probably go beyond many individuals' physiological capacity (**Figure 5.11**). At the lower latitudinal limit (high-temperature limit), winter may be the only time that reproduction is possible, because the species is generally adapted to low temperatures. For example, the tomcod is a northern Atlantic estuarine fish that reaches its southerly latitudinal extreme in the Hudson River; there, it reproduces only in the winter because only then is the temperature appropriate

FIG. 5.9 The Antarctic "ice fish" *Pagothenia borchgrevinski* lives in Antarctic waters just beneath the ice. It has antifreeze glycoproteins, which help prevent freezing of tissue fluids. (Photograph by Ian McDonald)

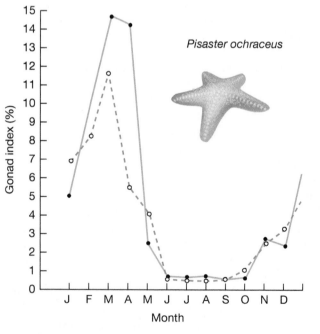

FIG. 5.10 Seasonal changes in two successive years of the gonad development of the intertidal sea star *Pisaster ochraceus* on the coast of California. (After Boolootian, 1966)

for such a cold-water species to spawn. The tomcod can survive and grow for the rest of the year at higher temperatures. **Figure 5.12** shows the case of a bryozoan near the southern end of its range. In this case, individuals of colonies can increase in number when the environment is thermally more favorable. Colonies are most abundant in winter and are rare in summer because high temperature is stressful. At the highest latitudinal limit, the winter would be stressful, whereas reproduction may be possible only in summer.

While water temperatures usually change consistently from colder temperatures at high latitudes to warmer temperatures at low latitudes, important details complicate the latitudinal gradient more than one might think. For example, on the east coast of the United States, summer water temperatures increase from Maine to Cape Hatteras. But at Cape Hatteras, the continental shelf is quite broad, and shallow water becomes quite warm in summer in the shallow coastal depths, often reaching 29–30°C. Farther south, in Florida, the shelf is narrower, and summer coastal water temperatures are slightly cooler.

Intertidal species may also have complex patterns of temperature with changing latitude that are not as simple as warm–low latitude and cool–high latitude. Intertidal invertebrates are exposed to air and may obtain much higher temperatures in summer than the local seawater or much colder temperatures in winter air. In summer, high-intertidal mobile animals such as fiddler crabs and snails often become stressed by high air temperatures. They respond by retreating to burrows or wet cracks in rocks. But others, such as barnacles and mussels, may bear the full brunt of the sun's heat. In the intertidal zone, high air temperatures can even denature proteins and shut down cellular function.

One might expect that physiological stress from temperature would increase toward lower latitudes, but this expectation might be strongly modulated by the timing of the tides. On the west coast of North America, the latitudinal thermal gradient of water temperature is not strong, and waters of southern California are just a few degrees warmer than those of Oregon and Washington. Helmuth et al. (2002) found that more northerly intertidal sites such as Friday Harbor, Washington, had low tide in midday in some years, when high temperature stress was likely, but in the same years, southerly locales such as Monterey, California, had more low tides at night, which reduced total temperature stress. These differences can change from year to year, so the location of maximum temperature stress is variable. In some years, therefore, global warming could exert maximum stress and cause mass mortality in different latitudes. Therefore, there is no simple north-south gradient in thermal stress in the intertidal zone because of the daily timing of low tide.

Latitude \ Season	Highest latitude limit	Lowest latitude limit
Winter	Survival limiting	Reproduction limiting
Summer	Reproduction limiting	Survival limiting

FIG. 5.11 Marine species often live along coasts that extend over a large latitudinal gradient, so seasonal change in temperature interacts with where the species is located along its latitudinal range. The effects of the highest and lowest latitudinal limits of a species range act to limit reproduction and survival depending on the season.

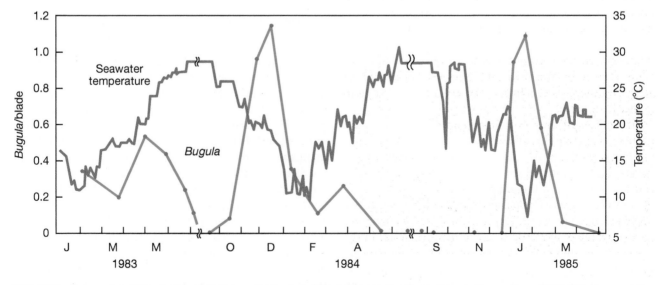

FIG. 5.12 Seasonal variation in the abundance of colonies of the bryozoan *Bugula neritinea* in the northern Gulf of Mexico near the lowest latitudinal limit of its range. Because the organism is near the warm limit of its distribution, its abundance is greater in the colder part of the year. (Courtesy of M. Keough)

Planktonic larval stages may drift to thermally unfavorable habitats. In spring in the 1980s, larvae of the mussel *Mytilus edulis* were often carried from the north to rocks and dock pilings south of Cape Hatteras, North Carolina. They settled, metamorphosed, and then grew quite rapidly. In summer, however, the temperature exceeded the thermal limit for the species and the mussels died. The maximum summer temperature clearly limits the low-latitude geographic range or extent of this species, whereas the high-latitude extent is likely limited by minimum winter temperature. Climate change, causing recent increases of shoreline temperature, has resulted in a northward migration of the southern limit of dense populations of this mussel species, and mussels are now very sparse near their (former) southern limit (Jones et al., 2009).

As we discuss in Chapter 17, mussels often dominate rocky shores in midlatitudes and usually occur in multiple layers, with mussels beneath attached directly to the rocky surface with byssal threads. But often mussels at the surface may be attached to a layer of dead shell, and therefore are isolated from moisture and vulnerable to heat stress during a summer heat wave when the tide is low and air temperatures are very high. The mussel layers beneath may be moist, but isolated mussel clumps may become very dry and will succumb to heat and desiccation stress (see Mislan and Wethey, 2015).

■ **Climate change is exerting thermal stress at the low-latitude end of species ranges and permitting northward extension of ranges as warm temperatures extend to higher latitudes.**

In recent decades, warmer temperatures have characterized the coastal zones of many regions. We have discussed the steady increase of SST in coastal waters throughout much of the world. As mentioned earlier, thermal stress is maximized in summer at the low-latitude end of a species range. This is a major effect, since so many of our coastlines are oriented north-south. On the other hand, as coastal waters increase in temperature, the high-latitude end of a coastal species' range becomes less stressful in winter. These two range-end changes suggest a simple hypothesis: Thermal stress should be foreshortening the low-latitude range of coastal species and allowing the expansion of range to higher latitudes.

The effects at the low-latitude end of the range have been observed in a change of the southern end of the range of the blue mussel *Mytilus edulis*, as mentioned earlier. Similar signs of thermal stress have been found by Pörtner and Knust (2007) in the eelpout *Zoarces viviparus*, which lives in nonmigratory populations in southern North Sea populations in north coastal Germany. Water temperatures have increased in recent decades and population size of the eelpout was found to be inversely related to water temperature in the previous summer, showing that temperature variations at the upper extreme are having direct effects on population dynamics. Beyond a critical upper and lower threshold, the eelpout's circulatory system cannot deliver enough oxygen by means of its cardiovascular system to satisfy demand. Past the upper temperature threshold, the fish relies on anaerobic sources of metabolism, which in the long run are insufficient long-term for functions such as swimming, growth, and reproduction. Eelpouts and other fish species in this region have to extend their range to the north in order to survive the increasing heat stress in waters adjacent to the German North Sea coast. Nearly two-thirds of fishes in the North Sea have extended their range northward in response to climate change increases of sea-surface temperatures (Perry et al., 2005).

Salinity

■ **Salinity can change rapidly, which may have a detrimental effect on marine organisms.**

In near-shore habitats, salinity may change rapidly over very short spatial scales. This poses a challenge for marine organisms. To operate efficiently, using a specific set of biochemical reactions, marine organisms must maintain rather constant chemical conditions within cells. Any process that causes significant changes in cellular chemistry will harm marine organisms. Significant changes in dissolved cellular inorganic constituents such as potassium and sodium will strongly affect the function of important proteins. The quantity of salts affects catalytic rate and the interaction of substrates with enzymes. Membrane transport depends on the precise regulation of inorganic constituents such as potassium. For example, sodium channels are proteins that conduct sodium ions through the cell membrane, especially in nerve and muscle cells. Exact regulation of sodium is crucial. Similarly, potassium must be regulated exactly because it is a crucial element in the membrane function of excitable nerve and muscle cells. Salinity change may pose a major challenge to cell function unless the organism can respond to dampen the change.

■ **Changes in salinity affect marine organisms through the processes of osmosis and diffusion.**

Why is external salinity change a problem to the internal cells of marine organisms? When salinity changes, marine organisms may face difficulties owing to **osmosis**, which is the movement of pure water across a membrane that is permeable to water but not to solute (material dissolved in the water). If the solute differs on the two sides of the membrane, pure water will move across the membrane in the direction of higher solute content. In an enclosed container, this creates **osmotic pressure**, which can be counteracted by pushing a plunger, as shown in **Figure 5.13a**. A solution is said to be **hyperosmotic** if water flows into it across a semipermeable membrane and **hypoosmotic** if water leaves that solution. The osmotic effect generated by dissolved substances in body fluids can be estimated by the amount of depression of the freezing point below 0°C. Greater solute concentration, which produces greater osmotic strength, causes a greater freezing point depression.

TABLE 5.1 Ionic Compositions of Seawater and of Fluids of Marine Animals (mmol kg⁻¹ of water)

SOURCE	Na	Mg	Ca	K	Cl	SO₄
Seawater	478.3	54.5	10.5	10.1	558.4	28.8
Jellyfish (*Aurelia*)	474	53	10	10.7	580	15.8
Polychaete (*Aphrodite*)	476	54.6	10.5	10.5	557	26.5
Sea urchin (*Echinus*)	474	53.5	10.6	10.1	557	28.7
Mussel (*Mytilus*)	474	52.6	11.9	12	553	28.9
Squid (*Loligo*)	456	55.4	10.6	22.2	578	8.1
Isopod (*Ligia*)	566	20.2	34.9	13.3	629	4
Crab (*Maia*)	488	44.1	13.6	12.4	554	14.5
Shore crab (*Carcinus*)	531	19.5	13.3	12.3	557	16.5
Norwegian lobster (*Nephrops*)	541	9.3	11.9	7.8	552	19.8
Hagfish (*Myxine*)	537	18	5.9	9.1	542	6.3

Source: Modified after Potts and Parry, 1964, with permission from Pergamon Press, Ltd.

Osmotic pressure develops when an animal is exposed to a change in salinity. Let's consider a worm that does not differ osmotically from the external seawater environment and whose body is permeable to pure water. If we expose the worm to a lower salinity and no regulation occurs, then pure water will move from the external environment, across the body wall, into the worm. Inevitably, the worm's cell volume and total body volume will increase. If the external salinity is low enough, one might suppose that the body volume will increase rapidly and cause the animal to explode. (After all, the osmotic pressure effect will continue until the salt content is equal on both sides of the body wall!) **Figure 5.13b** shows what really occurs in such an experiment performed on a sipunculid worm, which is placed in seawater with a salinity of 15. At first, the body volume does increase, but then it gradually decreases to approximately the same volume the worm had before it was plunged into the lower-salinity water. At first, osmosis results in pure water entering the body across the permeable body wall. However, the animal regulates its body volume by excretion of salts through its nephridiopores. As salts are lost, water moves osmotically across the body wall into the seawater medium, and the body volume returns to normal.

Marine organisms must also counteract the process of **diffusion**, which is the random movement of dissolved substances across a permeable membrane. Diffusion occurs until the concentration equalizes on either side of the membrane. If the salinity decreases suddenly, salts leave the body for the external seawater.[3] This is bad for the organism because, for cells and biochemical reactions to proceed efficiently, the overall cellular concentration of salts, or the concentration of particular constituents, must be regulated.

Table 5.1 shows the concentrations of common ions in seawater and in some marine animals. For the most part, there is a strong similarity between cellular concentrations and those of the open marine environment. However, the hagfish and the Norwegian lobster maintain magnesium concentrations far different from that of seawater. Jellyfish actively eliminate sulfate. Diffusion would tend to equalize these concentrations to the detriment of the organisms, which must therefore regulate their cellular concentrations to counteract this effect.

Most marine organisms are not completely permeable to the external seawater medium. For example, the outer skin of fishes and the thicker part of the chitinous body wall of arthropods are relatively impermeable to dissolved substances and to pure water. These organisms are able to localize exchange of dissolved salts at special sites, such as gills and excretory organs.

■ Marine organisms regulate both inorganic and organic cellular constituents to adjust to changing salinity.

As mentioned, most organisms must precisely regulate the cellular concentrations of inorganic constituents such as potassium and sodium. As a result, most marine organisms

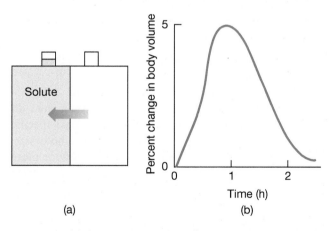

FIG. 5.13 (a) Osmosis: movement of pure water occurs across a membrane in the direction of higher solute concentration. (b) Change in the original body volume of the sipunculid *Golfingia gouldii* when transferred into diluted seawater at time = 0 hours.

---

[3] It is important to remember that osmosis is occurring at the same time as diffusion.

do not vary concentrations of these elements to adjust to osmotic stress. Instead, they regulate the concentrations of a variety of **organic osmolytes**, which are small carbon-based molecules generated by cellular reactions that act to reduce osmotic stress. Many unicellular algae, seaweeds, and salt-tolerant plants employ glycerol, mannitol, and sucrose as osmolytes. These substances have little effect on protein function despite wide changes in cellular concentration. **Free amino acids** are also used by several phyla of marine invertebrates (including mollusks and polychaetes), bacteria, and hagfishes. These organisms employ only those amino acids (e.g., glycine, alanine, taurine, proline, and b-alanine) that have little effect on protein function, for example, by being uncharged. Other amino acids, such as arginine, are strongly reactive and are not employed to respond to osmotic stress. **Urea** is used by cartilaginous marine fishes (e.g., sharks) and by coelacanths to counteract osmotic stress. Urea affects protein function, but this effect is counteracted by the presence of various methylamines, which usually occur in a 1:2 ratio with urea. Adjustments to new salinity can be slow, and as much as several weeks may be needed for complete acclimation.

■ **The body fluids of bony fish have low osmotic strength; therefore, fish must regulate salt content in full-strength seawater.**

Bony fishes maintain their body fluids at concentrations of only one-third to one-fourth that of normal seawater. The reason is unclear, but it may be related either to optimal functioning of fish enzymes or to the early origin and evolution of fishes in seawater that was less than full concentration (e.g., in an estuary). In any event, in the absence of a bony fish's ability to regulate the composition of its body fluids, the process of osmosis would cause a continual loss of fresh water to the external environment (**Figure 5.14**). Salts are usually taken up as teleosts drink to maintain water balance; they then are actively eliminated to maintain a lower (hypoosmotic) salt content. The gills help maintain the salt balance by actively excreting salts through chloride cells (for more information, see Evans et al., 2005). In elasmobranch fishes, such as sharks and rays, a high concentration of urea is used to maintain osmotic balance, as in the case of free amino acids for bivalve mollusks. Sharks and rays also actively eliminate inorganic ions, such as sodium.

Many fishes migrate between water bodies of widely differing salinity. The Atlantic eels *Anguilla rostrata* and *A. anguilla* reproduce in the Sargasso Sea, and juveniles return to salt marshes and other inshore water habitats. They mature and can then live in fresh water. Salmon hatch in freshwater rivers, migrate from fresh waters to the ocean in a few months, stay at sea for 1–3 years, and then return to fresh water to spawn. The eastern American killifish *Fundulus heteroclitus* can live in fresh water and in seawater. These species have a great capacity for regulation of cellular ionic content; this capacity is coordinated through a complex interaction of different hormones mediated by the pituitary gland.

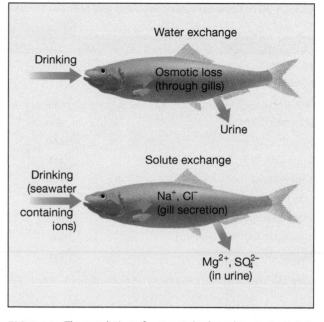

FIG. 5.14 The regulation of water and solutes by a typical marine teleost fish. Because the body fluids of the teleost are osmotically more dilute than the surrounding seawater, the fish must drink substantial amounts of seawater and excrete the excess salts across the gills. (After Schmidt-Nielsen 1975)

Oxygen

■ **Oxygen is required by most organisms to oxidize energy-yielding compounds.**

All organisms require energy to fuel chemical reactions within the cell. Although some organisms (e.g., some bacteria and protists) can live in the complete absence of oxygen, most require oxygen for the manufacture of necessary reserves of ATP. As a result, the availability of oxygen and mechanisms for its uptake are of great importance in understanding the function of most marine organisms.

Consumption Rate of Oxygen

■ **Although total oxygen consumption increases with increasing overall body mass, the mass-specific consumption of oxygen decreases with increasing overall mass.**

Oxygen consumption is usually expressed as milliliters consumed per unit time per unit body mass ($mL\ O_2\ h^{-1}\ mg^{-1}$). This is the mass-specific oxygen consumption rate. Although large individuals consume more oxygen than do small members of the same species, the mass-specific oxygen consumption rate is inversely related to body size. Thus, on a per-unit-mass basis, small snails consume more oxygen than do large snails. We can put this another way: If oxygen consumption of the whole animal is graphed as a function of body size, the curve will rise, but the rate of increase will decelerate (**Figure 5.15**). This change can be quantified (see **Going Deeper Box 5.1**).

There are several possible reasons that metabolic rate does not increase proportionately with body mass. In protists, the decrease in surface area/volume with increasing size is a

Quantifying the Relationship Between Body Size and Oxygen Consumption Rate

The relationship of oxygen consumption rate to body mass can be expressed as follows:

$$\text{mL O}_2 \text{ consumed} = kw^b$$

where b is a fitted exponent, W is body mass, k is a constant, and oxygen consumption rate is expressed in milliliters of oxygen per hour per gram: mL O_2 h^{-1} g^{-1}). Most marine poikilotherms have b values of less than 1.0 (typically, 0.66–1.0, with a central tendency to 0.75), indicating that their metabolic rate does not increase linearly with increasing body mass. The rise of the curve therefore decelerates. The rate of increase of oxygen consumption (value of b) is similar in poikilotherms and homeotherms, but the curve for homeotherms is displaced upward, indicating that homeotherms of a given mass always have a higher oxygen consumption rate than poikilotherms.

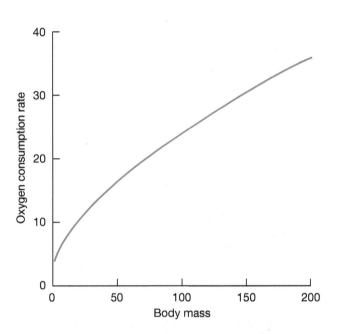

FIG. 5.15 Graph showing general relationship of oxygen consumption rate to body mass in a typical animal. Line is concave down, indicating that the rate of oxygen consumption decelerates with increasing body mass.

probable limiting mechanism. In organisms equipped with respiratory and circulatory systems, other mechanisms may be important. An increase in the proportion of nonrespiring mass (nonliving parts of skeletons, fat, and connective tissue), changing activity and growth patterns, and limitations of respiratory organs such as gills may be important.

Oxygen consumption rate increases in animals with increased activity.

Active species require more oxygen for energy and therefore consume more oxygen, assuming other factors (e.g., body mass) are kept constant. **Figure 5.16** depicts such a difference among benthic crustacea of differing swimming ability. Sponges, sea squirts, and most bivalve mollusks consume much less oxygen than do decapod crustaceans, cephalopods, and fishes. Activity may change within the life span of a single individual. Thus, species moving and feeding during the day require more oxygen during that time.

Some organisms are obligate anaerobes, but most organisms require oxygen. Aerobic animals, however, may rely on a varying mix of metabolic pathways with or without the need for oxygen.

Aerobic environments contain oxygen, whereas **anaerobic** (or **anoxic**) **environments** lack oxygen. Organisms that require oxygen are known as **aerobes**, whereas those that live in the absence of oxygen are **anaerobes**.

Nearly all eukaryotic organisms (those having a nucleus, endomembrane system, cell organelles, and often, true cilia or flagella) require oxygen for life. There are a few exceptions to this general rule. In 1970, Fenchel and Riedl described a biota of microscopic benthic animals living in the anoxic interstitial water of sediments (see Chapter 15 for a description of the sediment habitat). Tom Fenchel (see Fenchel, 1993; Fenchel and Bernard, 1993) found ciliate groups that can live only in the absence of oxygen. Typically, these ciliates have endosymbiotic bacteria that metabolize substrates anaerobically and provide nutrition for the ciliate. These forms are quite small—less than 1 mm long. Energy acquisition is very inefficient in the absence of oxygen, so such a lifestyle is restricted to small and relatively slow-moving forms.

Although all larger eukaryotic organisms require oxygen, many use a mix of sources to obtain energy to manufacture ATP. If oxygen is available, it is used to oxidize carbohydrate, and the energy is stored in the form of energy-rich ATP, which can later be used as a source of energy for muscle contraction, ciliary movement, and so on. During intense muscular action in vertebrates, energy demands can exceed the available oxygen. At these times, glucose or glycogen is broken down anaerobically. This process, known as **glycolysis**, has less than 10 percent of the efficiency of aerobic breakdown. The end product of glycolysis is lactic acid, which accumulates in muscle tissues. Whales and dolphins that are in the process of diving provide an extreme case of oxygen in short supply. See Chapter 9 for a description of diving by mammals.

At low tide, intertidal animals requiring submersion in water for oxygen uptake are subjected to a protracted period of oxygen depletion. In high latitudes during the winter, respiration is also depressed, and low temperature causes lowered transport rates of oxygen to cells. During these

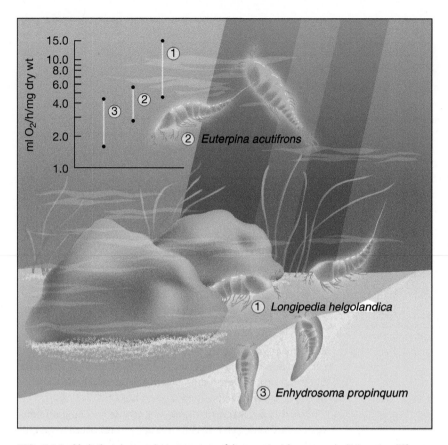

FIG. 5.16 Variation in respiratory rates of harpacticoid copepods living in different habitats. *Euterpina* is a swimming species; *Longipedia* is an active benthic form; *Enhydrosoma* is a sluggish benthic form. (After Coull and Vernberg, 1970)

times, many marine invertebrates sidestep the need for oxygen by using metabolic pathways involving the anaerobic breakdown of organic materials. As in the vertebrates, anaerobic breakdown is less efficient than breakdown in the presence of oxygen. At the time of low tide, end products of anaerobic metabolism (alanine and succinic acid) build up in tissues. In mollusks, a portion of the succinic acid is neutralized by dissolution of part of the calcium carbonate shell. In winter, the inner layer of the shell of the marsh mussel *Geukensia demissa* becomes pitted because of this sort of dissolution process.

■ **Animals only a few millimeters thick rely on diffusion for oxygen uptake; larger animals have specialized organs, such as gills, for this purpose.**

Mechanisms of oxygen uptake vary with body size, phyletic origin, habitat, age, and activity. Many small organisms—such as protists, nematodes, embryos, and larvae—rely on diffusion of oxygen across the body wall. This mechanism of uptake cannot be employed for bodies thicker than a few millimeters. Larger polychaetes, mollusks, and most crustacea use gills and circulatory systems for respiratory exchange. All gills have a large respiratory surface for oxygen exchange. Burrowing species, such as polychaetes, always create a water current to irrigate burrows with oxygenated water. Fishes have respiratory gills on gill arches. In many cases, gills play a role in feeding (e.g., in mollusks) and in ion exchange (e.g., in crustaceans and fishes) as well as in

respiration. Mammals have lungs with an enormous surface area to enable them to acquire oxygen rapidly. A whale has a lung surface area on the order of 1,000 m².

■ **Animals respond to lowered oxygen by regulating oxygen consumption, but their eventual response to very low oxygen levels is to leave the habitat if possible or to reduce activity levels.**

In many marine habitats, marine organisms are exposed to quite varied oxygen conditions, sometimes on very short time scales. For example, a fish that swims in oxygenated surface waters may then hunt prey in low-oxygen bottom waters. When confronting low-oxygen conditions, the simplest response for mobile organisms is to "vote with their fins" and leave the area. In many cases, such as in estuaries in summer, this option is not available because whole basins may be low in oxygen. The response of a typical aerobic animal is shown in **Figure 5.17**. There is usually a broad zone of environmental oxygen concentration within which oxygen consumption is relatively constant; within that zone, the animal can apparently regulate its oxygen consumption. However, at some minimum environmental oxygen concentration, the rate of respiration will decline with decreasing oxygen. Beyond that point, the animal cannot regulate its oxygen consumption efficiently and may respond by reducing its activity levels, which in turn reduces its requirements for oxygen to fuel aerobic muscular metabolism.

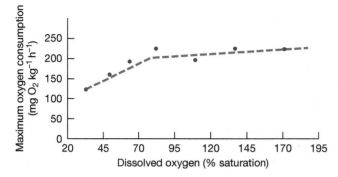

FIG. 5.17 Relationship between the rate of oxygen consumption of the turbot *Scophthalmus maximus* and the concentration of oxygen in the water. Over a range of relatively high oxygen concentrations, the animal regulates respiration rate, but the rate declines past a lower threshold and declines steadily with decreasing environmental oxygen. (After Mallekh and LaGardére, 2002)

Regulation of oxygen consumption at low oxygen levels can occur only if more oxygen can be transported into an uptake structure such as a gill or if more oxygen, once taken up, can be retained and delivered to the needy tissues. At low oxygen concentrations, crabs may increase ventilation rates and bivalves may increase heartbeat rate, which are both good enhancers of transport rates of water across the gills. Changes of blood chemistry and certain blood pigments, which are discussed in the following section, accomplish an increase of retention efficiency of oxygen within the circulatory system.

Oxygen-Binding Pigments

■ **Animals requiring great amounts of oxygen or living in environments where oxygen is difficult to acquire may have blood pigments that greatly increase the blood's capacity for oxygen transport.**

Many animals contain oxygen-binding "pigment" compounds that greatly increase the capacity for oxygen in blood and other tissue fluids. They are often colored—hence, the name blood pigment. One of the most common of these compounds is **hemoglobin**, which is a combination of a protein unit (globin) and a heme unit, which contains iron. Such pigments increase the carrying capacity for oxygen by a hundredfold. They are widespread; hemoglobin, for example, is found in many phyla. The main pigments found in marine animals are the following:

- **Hemocyanin.** Copper-containing protein, carried in solution having a molecular weight (MW) of 300,000–9,000,000 Da. Present in mollusks and arthropods.
- **Hemerythrin.** Iron-containing protein, always in cells, nonporphyrin structure, MW 108,000 Da. Present in all sipunculids and some polychaetes, priapulids, and brachiopods.
- **Chlorocruorin.** Iron-containing porphyrin protein, carried in solution, MW 2,750,000 Da. Present in some polychaetes.
- **Hemoglobin.** Iron-porphyrin protein, found in solution or in cells, MW 17,000–3,000,000 Da. Present in echinoderms, arthropods, annelids, mollusks, nematodes,

flatworms, protists, plants, bacteria, and in all vertebrates except some Antarctic fishes.
- **Myoglobin.** Like hemoglobin (but a much smaller molecule), an iron-porphyrin protein, binding to oxygen within skeletal or cardiac muscle tissues of active vertebrates and serving other oxygen-carrying functions in other organisms, including mollusks.

The Caribbean clam *Codakia orbicularis* contains hemoglobin and is a resident of shallow soft sediments rich in decaying turtle grass and other organic debris. Hemoglobin is found in other invertebrates living in anoxic sediments but is found universally in animals requiring large oxygen supplies (e.g., fishes, birds, and mammals).

The binding and release of oxygen to and from a blood pigment such as hemoglobin (Hb) can be described by the following equation:

$$Hb + O_2 \longrightarrow HbO_2$$

As the blood oxygen concentration increases, more and more oxygen binds to hemoglobin until the hemoglobin is saturated. At a given concentration of oxygen, a definite proportion of the hemoglobin present is bound to O_2. We can draw an **oxygen dissociation curve** (**Figure 5.18a**) that portrays the percentage of the hemoglobin bound with oxygen as a function of dissolved oxygen concentration. If the oxygen concentration is lowered, the hemoglobin releases oxygen.

The pH of the blood and coelomic fluids also affects hemoglobin-binding characteristics. In nearly all species studied, lowering the pH can shift the oxygen dissociation curve to the right—the **Bohr effect** (**Figure 5.18a**). Oxygenated hemoglobin tends to release oxygen into the blood when CO_2 is abundant (thus lowering the pH) and when O_2 is at lower concentration. This effect is adaptively significant: As oxygenated blood reaches tissues that are poor in oxygen and rich in carbon dioxide, the hemoglobin releases oxygen to the blood, thus making oxygen available for the cells.

We would expect active animals requiring large supplies of oxygen to have their dissociation curves to the right of the curves for less active forms (**Figure 5.18b**). This appears to be the case for invertebrates containing the copper pigment hemocyanin. Cephalopods, for example, have high levels of hemocyanin and rely greatly on the substance for oxygen transport. Inactive animals living in environments low in oxygen concentration should have dissociation curves shifted to the left, which allows the hemoglobin to bind to oxygen at relatively low blood oxygen concentrations. Blood pigments may serve as oxygen reservoirs for burrowing animals living in sediments that contain little oxygen. Bivalves living in tropical low-oxygen soft sediments, for instance, tend to have blood pigments, whereas pigments are not present in species living in environments that are more highly oxygenated.

■ **Animals with blood pigments can respond to low oxygen levels by changes in the character of the oxygen-carrying molecule.**

Marine animals with blood pigments not only can carry more oxygen in the blood, but they can also use changes in

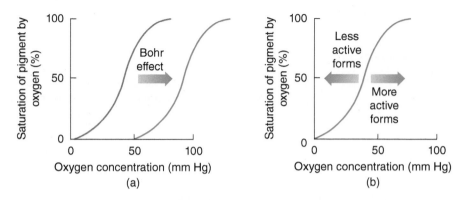

FIG. 5.18   (a) Oxygen association–dissociation curve (red) for a typical respiratory pigment, such as hemoglobin. Lowered pH (increased acidity) due to an increase of carbon dioxide released into capillaries by respiring cells causes the Bohr effect (blue curve), which reduces the hemoglobin's capacity to hold oxygen. Then oxygen is released and absorbed by needy cells. (b) Relationship of oxygen dissociation curves for animals of varying activity.

their oxygen-carrying molecules to respond to low-oxygen conditions. Under such conditions, bony fish hemoglobin and crustacean hemocyanin are responsive to changes in blood chemistry and become capable of binding more efficiently to oxygen. During shortages of oxygen, both calcium ion and lactate tend to build up in the blood, which, in crustacea, has a direct effect on the three-dimensional structure of hemocyanin, raising its affinity for oxygen. Hemocyanin can occur in hexamers (6 joined repeats of variants of the protein unit) or in dodecamers (12 joined repeats). Under low-oxygen conditions, the hexamers have a higher affinity for oxygen and are produced by crabs in greater proportions. In the horseshoe crab *Limulus polyphemus*, increased lactate concentration usually correlates with exposure to low environmental oxygen, which causes a **reverse Bohr effect**—a higher binding of hemocyanin to oxygen.

■ **The highest intertidal zone and oxygen minimum layers pose special problems for aerobic animals.**

High-intertidal animals may spend more time exposed to air than immersed in seawater. Several bivalve species open their valves when exposed to air. In the marsh mussel, such gaping allows direct access to the air for a relatively large surface area of water trapped in the mantle cavity. This consumption can occur to a meaningful extent. The eastern Pacific coast mussel *Mytilus californianus* consumes oxygen in air at rates comparable to its rates of respiration in water.

Organic matter may accumulate at the base of thermoclines in the open ocean; breakdown of this organic matter by oxygen-consuming bacteria causes a zone of reduced oxygen—an **oxygen minimum layer**. The oxygen minimum layers in the tropical oceans cause low-oxygen stress (see Figure 2.11). Some zooplankton pass through this zone daily during vertical migrations (see Chapter 10). Some of the zooplankton spend much of their time at this depth. The deep-water mysid *Gnathophausia ingens* moves water rapidly across its gills and can extract 80 percent of the oxygen from the water in which it is located. The animals have larger heart and arteries than do similar-sized animals from more highly oxygen-enriched habitats.

Light

■ **Light in the ocean comes mainly from the sun: Sunlight is the source of photosynthesis in the surface layers and adds heat at the surface. It also allows animals to use vision to help them function.**

Without the sun's light, there would be no marine biology. The sun emits 2×10^{45} photons each second, but the ocean captures 1/10,000 millionth of the sun's energy. Only a thin surface layer, some 50–100 m thick, receives sufficient light for photosynthesis, yet this drives nearly the entire biological engine of the entire ocean. Nearly all oceanic environments depend on this surface production.

Phytoplankton that live at the surface die and sink to the deepest part of the ocean, where their remains are eaten by deep-sea creatures. The sun is also the ocean's main source of heat. As discussed in Chapter 2, the great heat capacity of seawater also makes the ocean a storehouse for solar heat, which can be transferred thousands of kilometers horizontally when the water is driven by the winds.

Vision is impossible without light, which declines in intensity with increasing depth (see Chapter 2). In the surface layers, nearly all animals have well-developed eyes or at least some ability to sense the presence of light, and they depend on vision to detect prey, avoid predators, and find mates. In the deepest parts of the ocean where surface light is essentially extinguished, many organisms make their own light, and eyes are everywhere.

■ **Marine animals may detect images with the aid of a simple layer of sensory cells, lenses, concave mirrors, or even a structure functioning as a pinhole camera.**

To see, an animal must have a means of gathering light and a degree of nervous integration to form and interpret an image. In the air, the human cornea and air combine to bend light and help focus images on the retina with the aid of a lens. Light bends when it moves between two substances of differing density. It bends between air and the human cornea, which allows focusing of light, but it is pretty useless as a focusing mechanism in water because the

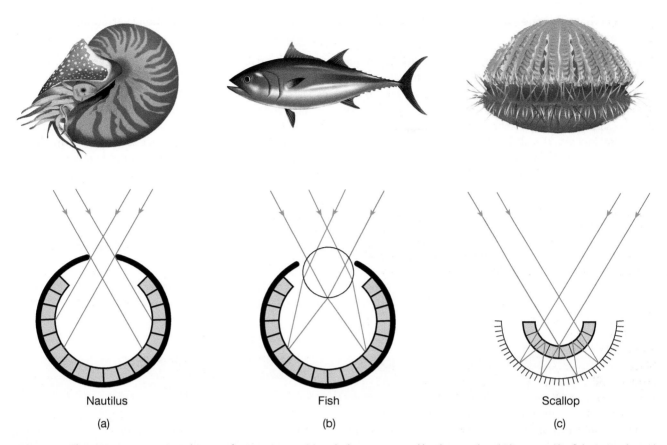

FIG. 5.19 Three ways marine animals use to focus an image: (a) pinhole camera, used by the nautilus; (b) lens, used by fishes, squids, and lobsters; and (c) curved reflector, used by scallops.

density of seawater is close to that of the cornea. As a consequence, marine organisms must use a lens made of dense material (transparent proteins) almost exclusively to bend and focus light if they are to form a sharp image.

The simplest kind of eye forms no image at all. Many animals, including marine larvae and flatworms, simply have a small pit lined with light receptor cells. All the animal receives is a signal: light on or off. It may also be able to monitor the light intensity and use the information to move away from or toward the stimulus depending on how the animal processes the signal. Shadow detection is important for predator detection. On an even simpler level, light captured by plants stimulates hormone production, which in turn causes the plant to grow in the direction of the light.

Marine animals use any of three basic ways of forming an image: lens, pinhole camera, or reflector. A cleverly constructed marine eye belongs to the chambered nautilus, a relative of squids and cuttlefish that secretes a spiral external shell and uses its arms to capture animal prey. Light enters through a tiny hole, and rays directly hit the retina. The hole acts as a pinhole camera, permitting an inverted image to form on the back of the eye (**Figure 5.19a**). The nautilus uses muscles to adjust the pinhole opening, but if the opening is too large, image quality decreases. The major disadvantage of a pinhole camera (and of a nautilus eye) is that the design allows vision only in high light. Somehow, over approximately 400 million years, the nautilus and its ancestors never evolved a lens.

Fishes, squids, some annelids, and some arthropods (e.g., lobsters and crabs) use a lens or a series of lenses much like a telescope to focus light gathered by the eye (**Figure 5.19b**). A fish lens differs from a glass lens and consists of material of continually increasing density in the direction of the center of the lens. This variation allows the lens to bend light and focus an image in a very short distance; it also reduces distortion. Fishes have muscles that move the lens forward and back to adjust for focus (as opposed to terrestrial vertebrates including mammals, which alter the shape of the lens itself). Lobsters and their allies, a few clams, and even some annelids have a large number of adjacent lenses, which together form a compound eye. A brain or some other part of the nervous system must integrate all the individual images into a mosaic image.

A small number of marine animals form an image much like that formed by an astronomical reflecting telescope. If you look at a live scallop, you will probably notice hundreds of shiny bluish spots around the periphery of the shell. These are eyes, and the silvery blue is a concave surface. Light passes first through a crude lens, then through a nearly transparent retina (layer of light receptors), and reflects from the shiny layer, to be detected on return by the retina (**Figure 5.19c**). A deep-sea arthropod species has a similar eye structure, which, in the description of Sir Alistair Hardy, resembles the reflectors of automobile headlights. See **Hot Topics Box 5.2** for a spectacular example of eyes with lens in a chiton.

■ Vision relies on a series of pigments that filter and absorb light.

Marine animals have a series of **photoreceptor pigments (rhodopsins)** located in a retina. Typical vertebrate retinas contain rods and cones. Rods are more sensitive but do not discriminate wavelengths; cones are less sensitive and vary in their sensitivity to different wavelengths allowing color vision. Invertebrate eyes differ but also have a series of light-sensitive pigments that vary in peak wavelength sensitivity. They use a rhabdomeric system, which is biochemically distinct from the vertebrate system (see Arendt, 2003). An eye structure focuses light on the retina, and the pigment molecule is excited when absorbing the light energy. Sensitivity of the pigments may range from ultraviolet to the red end of the spectrum. A series of photochemical steps ultimately causes phototransduction, which produces a neural signal that is interpreted for the organism by a brain or ganglion (for more information, see Warrant and Nilsson, 2007).

■ Color vision is widespread among vertebrates and invertebrates.

Although many animals are incapable of discriminating colors, a large number can identify color, which must at least involve discriminating between two wavelengths of light independent of their respective intensities. Individuals use color to discriminate among objects, such as predators, mates, and the general seascape. As might be expected, marine organisms are adapted best for vision in the wavelengths most common in seawater. Many crustacea can discriminate between two basic color types (usually blue and green-yellow), but scallops and others are able to discriminate among three types.

■ Light is an important cue in behavioral adaptation of marine organisms.

Many mobile intertidal animals use positive and negative responses to light to adjust their optimum position relative to

HOT TOPICS IN MARINE BIOLOGY

Crystal Eyes: Past and Present   5.2

Vision can require astounding degrees of image and color resolution to be effective. Predatory diving seabirds must be able to resolve images rapidly as they approach their prey. Cormorants perform less efficiently in turbid water, because they must resolve the images of swimming prey fish as they scan the bottom before they dive. But in many other cases, shading is a good enough indicator of the approach of a predator, allowing a prey fish to escape without resolving the image of the approaching diving bird. Similar strong variations in image perception occur among the invertebrates. Scallops have a large array of cup-shaped light detectors located on the mantle edges whose perception of approaching shade might be good enough to detect the approach of a predatory crab. But fiddler crabs have a highly concentrated line of ommatidia (arthropod light-capturing units, each with a neuroreceptor) whose main purpose is to detect vertical motion above the horizon, which warns of the approach of a predatory bird. These crabs can also resolve images to detect mates of differing size and color.

How are images resolved with great detail? Humans have an elegant adjustable lens that focuses images on the cornea. It stands to reason that other organisms must have structures that focus an image on a retina or retina-like structure. A challenged human could go to the optometry shop and buy a glass or plastic lens to help increase the focus of an image on the retina when things go a bit wrong. Do animals have such evolved capacities?

Some of the fossil group Trilobita have eyes that usually look superficially like the compound eyes of insects, with a honeycomb pattern of individual units with some sort of clear facet on the surface that gathers light and focuses it at the base of the unit, where a nerve-embedded structure translates a light stimulus into a nervous transmission to the brain or an organized group of nerves known as a ganglion. The orientation of the array of units, or ommatidia (**Box Figure 5.4**), can be used to infer the visual field, even of an extinct trilobite.

Two genera of trilobites have schizochroal eyes, where the lens is composed of two units, one above the other, of calcite, a crystalline

mineral form of calcium carbonate. The upper units have been found in two forms (**Box Figure 5.5**). What is amazing about the two different upper crystalline lens types is that they strongly resemble one of two lenses that were designed by Descartes and Huygens, two major eighteenth-century pioneers in philosophy and the design of telescopes and microscopes. The top units were combined with bowl-shaped lower units that completed the focusing of the image in seawater (see Clarkson and Levi-Setti, 1975).

The use of natural materials and structures to help in the design of engineering schemes for people is known as Biologically Inspired

BOX FIG. 5.4   Side view of a rolled-up specimen of the Devonian trilobite *Phacops rana* showing the large eye with surface parts of individual units, or ommatidia. (From www.fossilmuseum.org)

HOT TOPICS IN MARINE BIOLOGY

Crystal Eyes: Past and Present *continued* 5.2

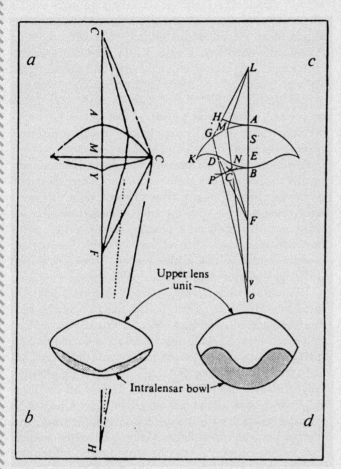

could roll into a ball, and eyes would allow them also to perceive approaching predators before they could be bitten.

Chiton eyes have lenses made of aragonite, another form of calcium carbonate, like the trilobites we mentioned earlier (Speiser et al., 2011). The lenses are translucent and are surrounded by a dark pigment, which makes the eyes show up as dark dots (**Box Figure 5.6**). The eyes are embedded in larger structures known as aesthetes. The lens is either a single crystal or many microcrystals with similar alignment. A thin (~5-mm thick) concavo-convex corneal layer covers the lens and is continuous with the surrounding eye microstructure.

Li and colleagues (2015) showed through experiments that the lenses can form clear images of objects such as fishes. In a previous study, Speiser et al. (2011) demonstrated that chitons could perceive dark objects and appear to have an angular resolution of the objects that is consistent with the perception angle that could be afforded by the aragonite crystal in a single eye lens. This could allow the chiton to distinguish between approaching objects and perhaps distinguish predators. Given the geometry of the lens, a fish at a distance of ca. 2m could be resolved through one lens unit on the chiton's dorsal surface, according to the measurements of Li et al. (2015).

The eyes are lenses embedded in a soft tissue structure, which may be great for detecting approaching mobile predators, but the soft tissue interrupts the otherwise continuous armored dorsal surface of the chiton. The chiton is therefore paying a price in defense for the ability to have windows on the outside world to detect the approach of prey. This overall structure helps us understand that this biologically inspired design suggests that there may be a trade-off when designing armor. In the twelfth century, medieval knights and infantry used a cylindrical helmet, known as a great helm, which had slits for vision. This provided great protection but greatly reduced the field of vision and had poor air circulation. A helmet with more openings for vision weakened the helmet but might have provided quicker response to attackers. Such compromises are the nature of industrial and biological design.

BOX FIG. 5.5 (a) Aplanatic lens design in air, using two oval lens, designed by Descartes. (b) Lens of the trilobite *Dalmanitina socialis*. (c) Aplanatic lens in air, making use of spherical first surface and a Cartesian second surface, designed by Huygens. (From Clarkson and Levi-Setti, 1975)

Design. Our example of trilobite lenses, however, puts the whole process backwards. Lenses were invented over 300 years ago by Descartes and Huygens, but the trilobites evolved these lenses hundreds of millions of years ago only to be discovered and compared with those of Descartes and Huygens in 1975. But many living structures today give us insight in the design of new structures and materials for human purposes. For example, in Chapter 6 we mention the bumpy skin of some sharks, which reduces skin friction as the sharks swim through the water. This principle was employed in the design for swimming skin suits for Olympic swimmers, although the outcome was not as beneficial as expected.

So let us consider a remarkable armor found within the shell of the West Indian fuzzy chiton *Acanthopleura granulata* (**Box Figure 5.6**). Many chiton species have hard shells, dotted with tiny eyes. Maybe it's no accident that the chitons superficially resemble trilobites. They both are (were) small epibenthic creatures, which may be very vulnerable to crabs and fish predators. The chitons can clamp to the substratum so the predator cannot pry it off and get to the soft foot beneath. Many trilobites

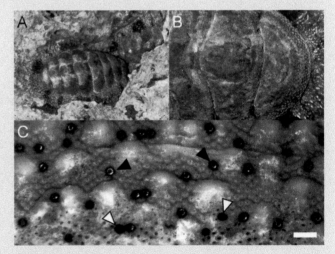

BOX FIG. 5.6 (a) *Acanthopleura granulata* on rocks near Tavernier, Florida. (b) Closeup with anteriormost valve to right; ocelli are small black dots. (c) Chiton eyes: newer, less eroded eyes (black arrowhead); more eroded eyes (white arrowheads). Scale is 200 µm. (From Speiser et al., 2011)

tidal height (see Chapter 16). Diel vertical migrations in response to daily light changes are well known in zooplankton and are discussed in Chapter 10. Many intertidal fishes, such as blennies, use well-developed vision to navigate excursions and returns to preferred shelters. Some fishes and invertebrates are believed to use the sun as a compass to accomplish migrations on tidal flats or to and from feeding grounds. Parrot fishes living in Bermudian coral reefs migrate from offshore caves to feeding grounds near shore by monitoring the orientation between their direction of movement and the sun's location. This behavior is depressed at night and on cloudy days. On a smaller scale, many invertebrates use images, light intensity, and color to detect approaching predators and to discriminate among mates. Fiddler crabs, for example, are very brightly colored, and females use color to help choose among males with which to mate.

Many salmon species and sea turtles accomplish spectacular migrations over thousands of kilometers. The role of solar navigation in these migrations is questionable, however. Salmon accomplish their long journey from freshwater spawning grounds to the open sea and back under all conditions of cloud cover. Chemoreception and geomagnetic orientation are more likely mechanisms for the navigation (see Chapter 7). Sea turtles are extremely myopic in air and can see stars from under water only in perfectly calm seas. But Kenneth Lohmann and colleagues (Lohmann, 1991; Lohmann et al., 2004) discovered that loggerhead and green turtles can detect the strength and inclination of the earth's magnetic field and can use these cues to leave the beach as hatchlings and to locate themselves at sea, as if on a map.

■ **Bioluminescence, a widely distributed property among marine organisms, is produced by specialized structures in animals or by (symbiotic) bacteria.**

The sun is not the only source of light in the ocean. Many organisms are **bioluminescent**; that is, they can produce their own light. In the deep ocean, this is probably the only light encountered by marine organisms. Bioluminescence occurs in bacteria, algae, protozoans, sponges, comb jellies, coelenterates, annelids, crustaceans, cephalopods, echinoderms, and fishes. Many animals exhibit *intracellular luminescence*. Squids and cuttlefish, for example, have elaborate photophore cells with focusing lenses and reflectors. Comb jellies have luminescent structures along the eight radial canals and the tentacles. At night, one can see them by the thousands emitting a pale greenish light. Some animals emit a luminous mucous secretion. Some squid and fish species have special sacs within which lie populations of symbiotic luminescent bacteria.

The biochemical mechanisms of light emission usually involve the reaction of a protein with some other substance. Most widespread is the reaction of **luciferin**, which emits light when it reacts with the enzyme **luciferase**. The jellyfish *Aequorea* contains a protein, aequorin, that lights up when in contact with calcium ion. This luminescent protein, discovered at a marine biology laboratory, is used widely in laboratory biochemical applications.

Bioluminescence is very common in species living in midwaters of the open ocean. The function of the bioluminescence is not always clear, but it has been related to a mechanism of confusing predators and even finding mates. A group of phytoplankton known as dinoflagellates can startle and confuse potential predators by glowing. Jellyfish and comb jellies can glow on and off rhythmically, which may also confuse predators (see Chapter 10 for more discussion). Luminescence may sometimes serve the function of camouflage. Many deep-water animals have luminescent organs on the ventral surface. The color and intensity of this ventral light match the pale sunlight from above, making the animal invisible to predators below.

■ CHAPTER SUMMARY

• Measures of organismal condition include whole-organismal, behavioral, physiological, and biochemical components. The food available beyond the cost of metabolism determines an organism's scope for growth. Mortality rate can also be used to measure the effects of environmental change. Organisms respond to environmental change by reaching a new equilibrium, a process known as acclimation.

• Poikilotherms have the same body temperature as water. Their metabolic rate increases with increasing temperature, although many species can maintain a near-constant metabolic rate. Poikilotherms can also

compensate for seasonal changes of temperature by readjusting metabolic rate. Homeotherms regulate body temperature by maintaining a high metabolic rate, by restricting circulation, by using insulating materials, or by a countercurrent mechanism to reduce heat loss to the environment.

• Marine species evolve to function at highest efficiency in a given thermal regime. This is important because of common latitudinal thermal gradients. Coexisting species may have differing thermal adaptations because their principal thermal adaptation may be to different thermal regimes in different latitudinal zones.

• Responses to extreme temperature include the production of heat-shock proteins, which help stabilize the catalytic activity of enzymes. Membrane properties are also disrupted during heat shock. Because most species have upper and lower thermal limits, temperature limits the geographic range of marine poikilotherm species.

• Changing salinity presents challenges to marine species because of diffusion and osmosis, which occur when dissolved materials within the organism differ from the external environment. Marine organisms regulate both inorganic constituents within cells but also produce and regulate organic cellular constituents, known as organic osmolytes, to maintain a constant cell volume. The body fluids of bony fish have low osmotic strength, and therefore fish must regulate salt content in full-strength seawater by drinking seawater and continually excreting salts.

- Aerobic environments contain oxygen, whereas anaerobic (or anoxic) environments lack oxygen. Organisms that require oxygen are aerobes, whereas those that live in the absence of oxygen are anaerobes. Animals only a few millimeters thick rely on diffusion for oxygen uptake; larger animals have specialized organs, such as gills and lungs, for this purpose.

- Total oxygen consumption increases with body mass, but weight-specific consumption actually decreases with increasing weight. Some animals require great amounts of oxygen or live in environments where oxygen is difficult to acquire, and they respond by regulating oxygen consumption. For example, blood pigments such as hemoglobin greatly increase the blood's capacity for oxygen transport. Eventually, however, they must leave the habitat or reduce activity levels.

- Sunlight is the source of photosynthesis in the surface layers and adds heat at

the surface. It also allows animals to use vision.

- Marine animals may detect images with the aid of a simple layer of sensory cells, lenses, concave mirrors, or a structure functioning like a pinhole camera. Vision relies on a series of photoreceptor pigments that filter and absorb light. Color vision is widespread. Light is also an important cue to behavior, such as migrations.

- Bioluminescence is produced by specialized structures in animals or by symbiotic bacteria.

■ REVIEW QUESTIONS

1. Distinguish between conformance and regulation.
2. What is the advantage of homeothermy? What is the cost?
3. What in principle underlies the operation of a countercurrent heat exchanger? What is its function in dolphins?
4. Why do some animals that maintain activity in winter temperatures cease to be active in summer if moved suddenly to the same low temperature?
5. What are the advantages and disadvantages of exposing a poikilotherm to higher temperatures in the spring?
6. At what time of year is a species liable to reproduce at the low-latitude extreme of the species range?
7. Distinguish between the processes of osmosis and diffusion.
8. What do most invertebrates do to acclimate to changing salinity?
9. What special osmotic problem do marine bony fishes have?
10. Why do some migrating fish have special salinity problems?
11. What are the major ways organisms take up oxygen from the environment?
12. What is the Bohr effect, and why is it important?
13. What factors principally explain variation in oxygen consumption rates among species?
14. How do marine animals deal with lowered oxygen in the environment?
15. What are the three main ways marine animals focus light to create an image?
16. Why may a bony fish in a pool of water on sea ice experience problems as more and more ice forms in the pool?
17. Why are many marine animals less sensitive in the red part of the spectrum and more sensitive in the blue and yellow-green parts?

Visit the companion website for *Marine Biology* at www.oup.com/us/levinton where you can find Cited References (under Student Resources/Cited References), Key Concepts, Marine Biology Explorations, and the Marine Biology Web Page with many additional resources.

CHAPTER 6

Life in a Fluid Medium

Introduction

This chapter will introduce you to the important properties of water that strongly affect the functioning of marine organisms. Seawater is a fluid in motion that surrounds marine organisms. Its motion can be stressful because the push, or drag, of moving water can displace an organism and even snap off parts of the body of a sessile organism such as a coral or a seaweed. But moving water also delivers signals to marine organisms, including odors and even information about the water motion itself, if the organism has mechanical sensors to detect water speed. The very density and other physical properties of water have profound effects on marine organisms, depending on their size and shape and the water velocity conditions. It is therefore essential to understand the basic physical properties of fluids and water motion, in order to understand how marine organisms function and how fluids affect them and deliver signals, to which organisms can react.

Density, Viscosity, and Reynolds Number

■ Water is relatively dense and viscous.

Water has physical properties far different from those of air. Seawater bathes marine organisms and protects soft moist tissues from drying. Oxygen can be obtained from solution. Water is a more supportive medium than air because it is denser. This eliminates the need for the strong supportive skeleton required by a large terrestrial organism. Water is also a viscous medium, and this poses some unique challenges. Imagine a frigate bird diving for fish through 10 m of water instead of through the same distance in air. Viscosity makes movement through the water far more difficult.

The important properties of density and viscosity must be defined more clearly if you are to understand how flow affects marine organisms. **Density**, or ρ (Greek letter rho), is the mass per unit volume and is expressed as grams per cubic centimeter (g cm^{-3}). The salt content makes seawater somewhat denser than fresh water. The density of seawater decreases with increasing temperature. **Dynamic viscosity**, or μ (Greek letter mu), is a measure of the molecular "stickiness" between layers of a fluid. The greater the dynamic viscosity, the more energy is needed to move one part of the fluid past another part of the same fluid. For instance, stirring a pot of honey is more difficult than stirring a pot of water. Honey, therefore, has greater dynamic viscosity than water, and water is more dynamically viscous than air. Dynamic viscosity decreases with increasing temperature.

■ The Reynolds number is an estimate of the relative importance of viscous and inertial forces in a fluid.

In any fluid, there are two basic competing forces: viscous forces and inertial forces. Viscous forces are "sticky." They keep the fluid together and flowing in smooth streamlines. In a fluid, particles affected by viscosity move, or stay still, depending upon the movement of the fluid. As viscosity increases, molecular stickiness keeps different parts of a fluid from separating easily, and any object in the fluid will be less able to move unless the surrounding fluid is also moving. Inertial forces are those that relate to inertia, the tendency of an object to continue moving after a force is applied to it. If you throw a steel ball through air, for example, it continues to move; that is an example of inertia. Inertial forces make a fluid break up into uneven streamlines, or allow an object to "drop" through a fluid like a stone in water—in other words, to not go with the flow.

The **Reynolds number (*Re*)** is a measure of the relative importance of inertial and viscous effects of a fluid and on objects in a fluid. *As the Reynolds number (Re) increases, the inertial forces come to dominate.* Under high *Re*, objects in a fluid are dominated by inertia; that is, they tend to keep on moving when a force is applied to them. Under low *Re*, objects do not move unless a force is applied because viscous forces dominate.

Re is simply the product of the velocity *V*, size *l*, and density *ρ* divided by the dynamic viscosity *μ*.[1]

$$Re = \frac{Vl\rho}{\mu}$$

 *Reynold's number*

sea water < Re <

As long as we are dealing with seawater at a certain constant temperature, we can take the density and the dynamic viscosity to be constants. So we don't have to worry about them for now. *Re*, therefore, **increases with an increase of either velocity or size.** That is what you really have to know! Therefore, we must measure the remaining two variables, velocity *V* and size of the object *l*.

There are two different approaches that we can take to measure the velocity (**Figure 6.1**). First, if we place an object in a moving fluid and keep it fixed to the bottom, we can measure the fluid's velocity past the object (e.g., water flowing past a coral). As an alternative method, we can take a nonmoving fluid and measure the velocity of motion of an object through this stationary fluid (e.g., the swimming speed of a fish). At the scale that we are examining, the size *l* can be measured fairly generally. **Figure 6.1** shows some examples.

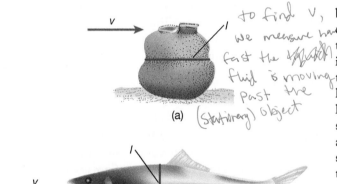

V

l

to find V, we measure how fast the (water) fluid is moving past the (stationary) object

(a)

l

V

(b) *or we can measure how fast an object is moving through stationary fluid*

FIG. 6.1 Two different situations in which one can find the size *l* and velocity *V*, parameters necessary to determine the Reynolds number. (a) An object (a sea squirt) is stationary in moving water. (b) An object (a fish) is moving through the water.

TABLE 6.1 Reynolds Number (*Re*) for a Range of Swimming Organisms of Different Sizes and Swimming Velocities

ANIMAL AND VELOCITY	*Re*
Large whale swimming at 10 m s⁻¹	300,000,000
Tuna swimming at 10 m s⁻¹	30,000,000
Copepod swimming at 20 cm s⁻¹	30,000
Sea urchin sperm swimming at 0.2 mm s⁻¹	0.03

Source: Data from Vogel, 1994.

Table 6.1 gives *Re* for a range of organism sizes and swimming velocities. Empirical research has shown that **when *Re* is less than about 1.0, viscous forces predominate. When *Re* is much greater than 1,000, inertial forces predominate.** The threshold (actually a broad band of transition) makes all the difference in terms of how organisms move in and react to their environment. Viscous forces dominate a small organism traveling at a low velocity; in other words, it is living in what amounts to a sticky medium. That is why a paramecium swimming through still water can stop swimming, seemingly instantaneously. The instant the protozoan ceases to move its cilia, it is entrained in the sticky still water and stops. By contrast, a supertanker may require several miles to come to rest from full speed after a pilot cuts the throttle. The supertanker, which operates at high *Re*, is only very slightly affected by viscosity, and considerable inertia must be overcome to bring the vessel to a stop. The protozoan, by virtue of its minuscule size and low velocity, lives in a world completely different from that of a ship or a fast and large fish, even though all move in the same fluid.[2]

If a protozoan ceases movement, it stops instantaneously because, at low *Re*, inertia is unimportant relative to viscous forces. To swim, a protozoan must, therefore, continuously exert a force against the surrounding medium; it cannot depend on inertia for movement. Measured in terms of body lengths per unit time, paramecium should have a better reputation for being an excellent swimmer, at least in relation to its body size. A protozoan can sustain swimming speeds of 100 body lengths per second, whereas a tuna cannot swim much faster than 10 body lengths per second. See **Going Deeper Box 6.1** for an important exception to these conclusions, the case where *Re* is in the gray zone between 1 and 1,000.

---

[1] The units, such as length and time, cancel out, and *Re* is therefore a number with no dimensions.

[2] The dominance of viscosity at low *Re* allows you to perform a parlor trick, if you are delicate. You need corn syrup, black artist's ink, an eyedropper, and a chopstick. Place corn syrup to a depth of about 1 cm in a broad cylindrical glass container. Then very carefully place a drop of ink at one spot on the surface of the corn syrup. Put the chopstick in the drop and very slowly rotate it, but do not remove the chopstick. You will create a black swirl. Now, retrace the path of the chopstick to the starting point. If you did this carefully, you will recreate the black dot of the ink! This cannot be done in water, where the density is lower and the *Re* is, therefore, higher than for corn syrup.

Is Seawater Always Seawater? A Tale from the Regions of Intermediate Reynolds Number

We have discussed the role of Reynolds number (Re) in the life of marine organisms. That role is a profound one. If $Re < 1$, viscosity dominates the organism's lifestyle. Effectively, a small organism swimming at low velocity, such as a tiny zooplankter or even a tiny fish larva, is living in a very different world from a tuna. The tiny larvae are working against viscosity and can never rest if they are to move. They have a large boundary layer about them and must move from one place to another to expose their body surface to fresh nutrients and even oxygen. By comparison, the tuna flexes its body, spurts forward, and then continues to glide.

In calculating Re, we mentioned that we could more or less forget about viscosity because we are talking about seawater, whose viscosity is always the same. Or is it? Well, on the gross scale, where we compare $Re = 0.05$ with $Re = 20,000$ that generalization is true. We can pretty much assume that seawater is seawater, as viscosity cannot change enough to matter. But when we are in the gray zone of $1 < Re < 1,000$, other factors come into play: most importantly, temperature. At low temperatures, kinematic viscosity changes a great deal—for example, from 1.6 to 1.1 m² s⁻¹ × 10⁶, or about 45%, over a temperature range of 5–15°C (von Herbing, 2002). This effect is very important for smaller fish larvae. At low temperature, they are strongly affected by viscosity. This can be proven by taking a fluid at higher temperature and placing neutral substances that increase the viscosity to the same degree that cold water does. As a response, the smaller larvae of the herring *Clupea harengus* change their behavior entirely and reduce the amplitude of their tail beating. The stride length coming from a tail beat is much less at the higher viscosity, as would be expected if inertia was not as important but viscosity was (Fuiman and Batty, 1997). Small larvae moving at low Re and, therefore, higher viscosity encounter additional resistance to tail movement, but they lack the muscle power to achieve sufficient transverse tail speed. To maintain tail beat frequency, they reduce tail beat amplitude. Overall, this effect is not as important when temperatures are higher as in the tropics. Therefore, only higher-latitude fish species live in this low Re trap.

An interesting complication is the metabolic cost of swimming. In Chapter 5, we concluded that a poikilotherm should have increased metabolic rate if temperature increases. But what about small fish larvae? Oddly, it can be shown that at higher temperature (10°C) the metabolic rate of cod larvae is actually less than at lower temperature (5°C). The effect of lowering viscosity more than compensates for the temperature effect, so larval fish pay a lower metabolic price to swim at the higher temperature at lower viscosity.

Moving Water

■ **Whatever is part of the flow will not cross streamlines in a flowing fluid.**

The movement of fluids can be readily diagrammed as a series of approximately parallel streamlines as in **Figure 6.2**—but keep in mind that all water movement is not usually so parallel. **Whatever is part of the flow will not cross the streamlines.** One effect of this simple rule is that fluids and entrained particles (e.g., protozoans and dye) move in the same direction. This helps greatly in tracing fluid motion because particles can be used to characterize flow; it's easier to watch dye, for example, than to try to look at the movement of transparent water! Thus, an important question is this: What is the character of this flow? That is the subject of the next few sections.

■ **Laminar flow is regular, whereas turbulent flow is irregular.**

We can distinguish between two main types of flow. **Laminar flow** is regular, and lines describing movement of water molecules characterized by such flow are parallel (**Figure 6.3**). By contrast, **turbulent flow** is characterized by lines that are very irregular, and the overall direction of flow can be determined only as an aggregate of individual irregular motions. The famous hydrodynamicist Osborne Reynolds (for whom Re is named) discovered that flow in a pipe becomes irregular (i.e., turbulent) if velocity, pipe diameter, or fluid density increases beyond a certain point. These factors contribute to increasing the Re. The same principles can also apply in open water. **If Re is high, flow is turbulent; if Re is low, flow tends to be laminar.** When Re is high, a fluid encountering an object may change velocity rapidly and inertia may cause the fluid to break up into complex vortices and wakes behind the object.

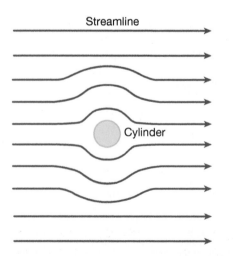

Streamline

Cylinder

FIG. 6.2 Water flow can be visualized as streamlines, which indicate the path that individual particles would take. Particles move along streamlines, not across them. In this illustration, water is flowing around a fixed cylinder, which is viewed in cross section.

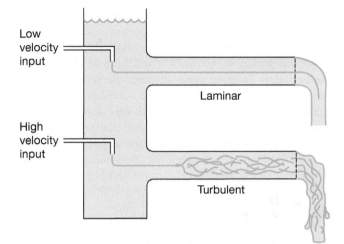

FIG. 6.3 Laminar and turbulent flow. Blue lines represent paths of flow. (After Vogel, 1994)

■ Turbulence has different effects at different spatial scales.

Turbulence operates at many spatial scales, ranging from many meters to the scale of millimeters. In all cases, turbulence involves multidirectional movement of water. The sum of flow directions does not have to be random but is usually too complex for an organism to be able to perceive the next direction of water movement and respond with the aid of its sensory organs. Turbulence has quite different meanings at different spatial scales. On the scale of meters, turbulence implies movement of large parcels of water operating at relatively large *Re*. In these cases, the organism will either be carried passively in these parcels of water or may be able to swim against the complex flow, as in the case of a seal or a large shark. Large-scale turbulence also may involve large-scale overturn of the water column, which might sweep many organisms from a good environment to a bad one, or the reverse.

On the scale of millimeters or less, however, many processes occur at low *Re*. A single bacterial cell might benefit from very-small-scale turbulence because the cell might be transported to a locally new parcel of water with new nutrient supplies. This local process, happening with mainly laminar flow conditions, is called **shear**. If turbulence did not exist at the small spatial scale, the low *Re* conditions would require that uptake of dissolved nutrients occur only by diffusion at one location, and the bacterial cell would deplete the surrounding water with replenishment coming only by diffusion from greater distances, which is a slow process. Microturbulent tumbling of cells and even small feeding animals would, therefore, be beneficial, but at a very different scale than larger-scale turbulence. It has been argued that turbulence on the microscale might cause local concentrations of cells and nutrients, but Fenchel (1988) calculated that average concentrations of nutrients in nutrient-poor conditions in the ocean are still sufficient to allow for the growth of planktonic microorganisms by means of diffusion. Therefore, just microturbulent motion

of cells might be sufficient to move them around to have continuing access to nutrients in the water as new parcels replace depleted ones and as cells are mixed to new sites for nutrient uptake.

Turbulence and diffusion at the microscale are also very important when we consider **chemical signaling** by organisms. In most cases, organisms have a variety of sensors to detect chemical concentrations. For example, in Chapter 8, we discuss copepods, which are small planktonic animals with chemical sensors on their antennae. They operate at low *Re*, which means that they will detect chemical signals from food either by diffusion or by flicking the antenna, which creates local shear and moves water around them. Another very different example is a planktonic egg, which is produced by many invertebrates and fishes. How does a sperm know the egg is there? Many eggs are known to produce chemical attractants, which diffuse from the egg. This creates a cloud of attractant around the egg, which increases the target of the egg for the moving sperm to detect. If there is a bit of water movement around the egg, shear might create a plume of sperm attractant, which might increase the size of the chemical signal for sperm.

Water Moving over Surfaces and Obstructions Such as Organisms

■ As a fluid moves over a solid surface, very near the surface, velocity steadily decreases with depth, the water reaching a standstill at the solid surface.

Consider water moving over a perfectly smooth-surfaced bottom. It is a hydrodynamic necessity that **water velocity will decrease to zero at the bottom surface (Figure 6.4)**. This is called the **no-slip condition**. In effect, the grab of the bottom is perfect at the surface but rapidly loses hold as you go into a mainstream, where the current velocity is not affected by the solid surface. The exact decline of velocity is

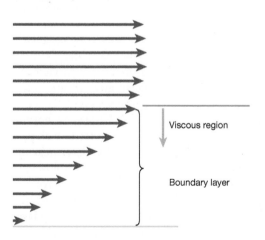

FIG. 6.4 Flow over the bottom at slow velocity. Water velocity maintains an average mainstream laminar velocity well above the bottom, but velocity decreases to zero at the bottom surface. Near the bottom is a thin boundary layer, where velocity decreases linearly down to the bottom surface. (Not to scale: boundary layer is often less than 1 cm thick.)

complex but is determined in large measure by the velocity of the fluid and by the dynamic viscosity.

If the flow is regular and the bottom is level, very close to the bottom there is always a **boundary layer**, where velocity declines approximately linearly with decreasing distance to the bottom surface (**Figure 6.4**). The boundary layer is thin, ranging from the scale of millimeters to centimeters, depending on the flow speed in the mainstream above. Outside this boundary layer, velocity increases asymptotically toward the **mainstream flow**. The border of the boundary layer is better described as a transition zone than as having a precise depth. **Within the boundary layer, viscosity must be factored into studies of flow and signal detection by organisms.** The thickness of the boundary layer decreases with increasing *Re*. When *Re* is small—that is, when the mainstream current is very slow—the boundary layer is relatively thick, making it something of a barrier to the exchange of materials and mechanical energy from the mainstream to the bottom surface. But with faster flow, there will be a very thin viscous layer at the bottom and a turbulent transitional boundary layer above. In the latter case, it is easier for materials to be mixed toward the bottom, although the problem of viscosity becomes important at the thin viscous layer.

Boundary layers affect two major processes in marine biology. First, the **boundary layer affects water motion near the bottom**. Within the boundary layer, viscosity dominates, but outside, water motion is dominated by inertia. Particles literally slow up as they approach and enter the boundary layer. The difference in transport creates a boundary layer of physical transport, or **momentum boundary layer**. Equally important, the **boundary layer affects concentration** of substances, such as dissolved ammonia released as an excretory product by many animals within the seabed. Within the boundary layer, organisms may locally deplete materials such as nutrients. It takes some motion to bring new nutrients into the boundary layer, by events such as the stirring of a tentacle or small-scale turbulence that is generated by irregular water motion or bottom irregularities that disrupt flow along the bottom. There also is an important effect on materials released by organisms into this boundary layer. As soon as the materials move from the boundary layer, which is dominated by viscosity, they will be swiftly diluted and taken away in the waters above. In the reverse direction, particles mixed from the water column above could be affected by feeding bottom organisms within the boundary layer. Such feeding on particles can create a **concentration boundary layer** near the bottom. A bed of mussels living in a bottom with slow current speeds will deplete the boundary layer of their favorite food, phytoplankton cells. Unless the current speeds up, the mussel feeding will create a concentration boundary layer where the phytoplankton concentration is low right near the bottom.

Objects in the middle of the water column also have a boundary layer surrounding them. They too have a no-slip condition at their surface (e.g., the surface of a fish or the surface of the cell membrane of a planktonic protozoan).

A microscopic organism, such as a protozoan, lives enmeshed in a relatively viscous environment. Because the *Re* is very small, the boundary layer is relatively thick. If the protozoan wishes to get food, it will have to reach out of the boundary layer or move through the water to get to new sources of food. Otherwise, the boundary layer will be a prison where food is depleted with very slow resupply. It used to be believed, for example, that cilia on the gills of bivalves or setae on the feeding appendages of copepods trapped food particles between the fibers, much like a sieve, as the particles drifted by on currents and impacted on the cilia or setae. The boundary layer prevents this, however. As the particles approach the feeding micro-structures, they may even be deflected from the boundary layer. If they are trapped within the boundary layer, the cilia must reach out and touch or grab the particles or the particles must collide directly with the cilia or other collecting fibers. The new understanding of this process has changed the way we look at suspension feeders, or animals that collect particles as food from seawater usually with tiny structures such as cilia.

The boundary layer also affects detection of odors. If a fish's nostril is flush with its body, odors will arrive slowly through the boundary layer. This creates a lag time between the arrival of the odor to the fish and its transport through the boundary layer to the nostril. Some fishes use projections upon which sensory structures are located, simply to place them out of the boundary layer so that water can approach the structures more quickly in the mainstream flow. A copepod can flick its antenna to disrupt the boundary layer and induce flow past an olfactory sensory hair.

■ Water flow over a surface can be turbulent.

As water flows over a surface, it may at first move regularly. But at some point downstream, the water even in part of the boundary layer will not move in a laminar fashion. It will break up into a series of irregular vortices, although there will remain a very thin viscous layer at the surface. The roughness of the bottom can greatly alter the nature of flow. On a smooth bottom there is a regular decline of horizontal water velocity through the boundary layer to the bottom, as we have discussed. But consider a bottom covered with pebbles or small mounds of sediment. The bottom current will not be smooth but will be broken up into an irregular series of vortices. For example, the blue crab *Callinectes sapidus* is a voracious predator along the east and Gulf coasts of North America. On a smooth bottom it can detect the regular plume of odor emitted from a burrowing clam's exhalent siphon. The crab moves back and forth and follows the odor plume up the odor concentration gradient by flicking its antennules, which arises from the regular flow from the bivalve siphon, which bears indicator molecules such as ammonia and urea. The crab locates the bivalve much like a heat-seeking missile detects a heat plume from a smokestack. But if the bottom is rough and lined with pebbles (**Figure 6.5**), moving water will be broken up into irregular flow, which makes it much more difficult for the crab to detect the odor plume (Weissburg and Zimmer-Faust, 1993).

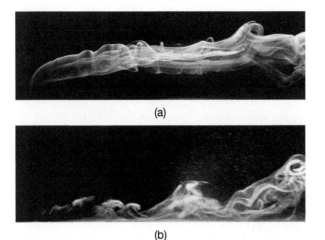

(a)

(b)

FIG. 6.5 Dye traces of odor plumes. The blue crab *Callinectes sapidus* must locate the odor plume emanating from the exhalant siphon of the clam *Mercenaria mercenaria*. (a) If the current is relatively weak, the flow is laminar and the odor plume maintains its integrity. (b) If the current is strong, or if the bottom is rough, the boundary layer becomes turbulent and the crab will have trouble following the odor to its source. (Courtesy of Marc Weissburg)

Using Water Motion for Biological Advantage

■ The principle of continuity allows one to calculate flow velocity in a biological circulatory system.

Consider a unit volume of fluid flowing through a rigid pipe. Assume that the fluid is incompressible; therefore, if a certain volume enters at one end, an equal volume will leave at the other. **The product of the velocity and the cross-sectional area always remains constant** (neglecting friction). If the fluid is then forced to flow through a pipe of half the cross-sectional area, its velocity will be doubled (**Figure 6.6**). This **principle of continuity** applies equally to changes in cross section of a single pipe and to the case in which a pipe splits into several smaller pipes. If, for example, a pipe splits into several equal subsections, the product of the velocity and cross-sectional area of the main pipe will equal the sum of the products of the velocity and the sum of the cross-sectional areas of the smaller pipes.

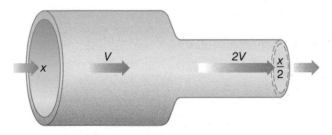

FIG. 6.6 Flow velocity through a pipe and the principle of continuity. The product of cross-sectional area and velocity is constant. Therefore, if cross-sectional area x decreases by half, the velocity doubles.

The principle of continuity permits organisms to regulate water velocity. Consider a sponge, which consists of networks of chambers, each of which is lined by flagellated cells known as choanocytes (**Figure 6.7**). The chambers are all connected to one or several main water expulsion channels, which guide wastewater from the sponge. If you dive, an application of food coloring to a sponge will quickly show you that the velocity of these excurrent openings is on the order of 1 cm s^{-1} (= 0.01 m s^{-1}). How can a collection of choanocytes, each able to produce a water velocity of only 50 μm s^{-1} (= 0.00005 m s^{-1}), manage to produce such a rapid exit speed together? The total cross-sectional area of choanocyte chambers adds up to several thousands of times the cross-sectional area of the excurrent canal. The velocity through the excurrent canal must therefore increase proportionally. Suspension-feeding polychaetes and mollusks employ a similar principle to drive water at fairly high speeds through interfilamental openings or siphons. In a bivalve gill, tens of thousands of cilia may operate at low individual velocity, but the total cross-sectional area is great, relative to, for example, that of the exhalent siphon. Many small-velocity motions add up to generate a larger velocity at the exit.

■ Bernoulli's principle states that pressure varies inversely with fluid velocity.

Bernoulli's principle applies the principle of conservation of energy to pressure changes in pipes and burrows or along surfaces. If total energy must be constant, then pressure will vary inversely with the velocity of the fluid. The simplest case is represented by a pipe whose diameter changes somewhere along its length. If the diameter decreases, then velocity in that section of the pipe will increase, as just discussed. Pressure, however, will decrease in this section. If you punctured a pipe of this sort filled with flowing gas and then lit a match, the flame would be higher in the wide section of the pipe, owing to the higher pressure (please don't do this at home!).

This principle has broad biological applications. The design of a cross section of a wing is based directly on an application of Bernoulli's principle. The lower surface is flat, whereas the upper surface is curved. As air encounters the wing, it moves more rapidly along the upper surface and the pressure is thus lower than along the lower surface. The pressure difference creates lift. The same principle applies to a flatfish (**Figure 6.8a**). As the fish pushes through the water, greater pressure is exerted on its flatter bottom than on the curved top, and lift develops.

Pressure differences on either side of a tube can also be used by an organism to create a current. Consider a U-shaped tube in the mud (**Figure 6.8b**), with one entrance to the tube opening to a small rise, higher than the adjacent entrance. Because the entrance in the rise is probably exposed to a slightly higher current speed, the pressure will be lower than at the other entrance. As a result, water will flow passively through the tube. This principle also reduces the work needed to drive water through worm tubes or through sponges because a moving current above the sponge reduces the pressure at the exit.

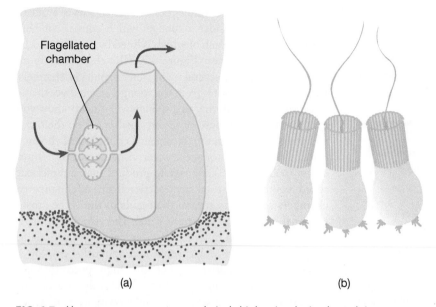

(a)   (b)

FIG. 6.7   How a sponge generates a relatively high exit velocity through its excurrent channels. (a) The low velocity of the water from flagellated cells in flagellated chambers is compensated by the far greater total cross-sectional area of the flagellated chambers relative to the excurrent opening of the sponge. (b) Diagram of flagellated sponge cells. (After Vogel, 1994)

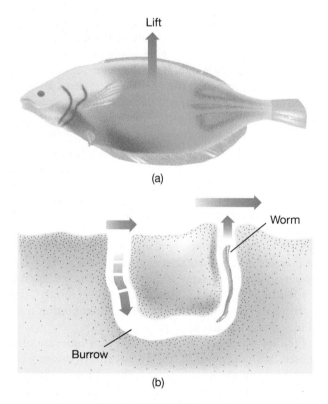

(a)

(b)

FIG. 6.8   Bernoulli's principle. (a) Differences in pressure above and below a moving flatfish create lift. (After Vogel, 1994) (b) A raised mound on one end of a buried U-shaped tube places it in a slightly higher current velocity relative to the other opening, which is flush with the sediment. Water moving past the two holes creates a pressure difference, with lower pressure on the raised area, and this drives water through the tube. Length of arrows is proportional to water velocity.

Drag

■ **Water moving past an object creates drag, a force that operates differently at different Reynolds numbers.**

Consider a cylinder sticking up into the water from the bottom. The flow will be disrupted and will be irregular on the downstream side of the cylinder. This means that there will be a pressure difference between the upstream and downstream surfaces of the cylinder. Pressure will be lower on the downstream side than the upstream side. This change in pressure is a force known as **pressure drag** and the difference tends to push the object downstream (**Figure 6.9**). Drag, however, can be dissected into two components, either of which may dominate depending upon the *Re*. At low *Re* (more sticky, or viscous, situations), **skin friction**

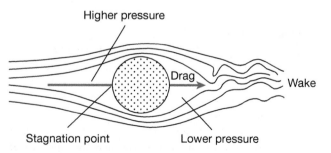

FIG. 6.9   Flow patterns around a cylinder (view is down the axis of the cylinder, looking at the cross section). Note the irregular flow in the wake of the cylinder.

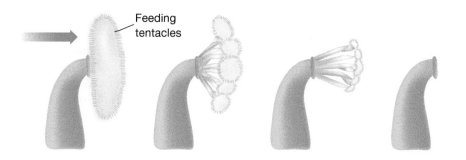

FIG. 6.10 Changes (left to right) of the sea anemone *Metridium senile* as the water velocity increases. Withdrawal of the tentacle crown reduces drag.

dominates. It is a force resulting from the interlayer stickiness (dynamic viscosity) of the fluid, and it acts parallel to the surface of the object and in the local direction of fluid flow. The more surface is exposed to the flow, the more skin friction there is. Although skin friction never disappears, it becomes far less important at higher *Re*.

Pressure drag occurs because pressure exerted on the upstream part of the object (e.g., a stationary coral in a current) is not exactly counterbalanced by an equal pressure on the downstream side. The stream effectively pushes along the object. **Pressure drag increases proportionally to the cross-sectional area exposed to the current and to the square of the current velocity.** For example, a flat plate oriented perpendicular to a current exerts a maximum amount of pressure drag; the pressure drag is minimized when the flat surface of the plate is parallel to the current. This principle also applies to objects moving through a fluid, such as air or water. Anyone who has driven an old van or a big truck knows that its flat front creates sufficient drag to increase fuel consumption. Modern vans have much more streamlined shapes to reduce drag, but it's still more efficient (and fun) to drive a highly streamlined vehicle, such as a Ferrari.

The best way for an engineer to minimize pressure drag is to orient elongated objects with the long axis parallel to the current and give them a long, tapering tail on the downstream side. This allows the fluid, after passing over the front of the object, to decelerate gradually in the rear. The object is pushed forward by the closure of the fluid around the object toward the rear. This principle explains the streamlined teardrop shape of fast-swimming fish, such as skipjack tuna, and the shape of submarines. (See further discussion in Chapter 9.) Many organisms are fixed to the bottom, and pressure drag on them may be considerable. Seaweeds, corals, sea pens, and sea anemones all project into the flow from the bottom. The work of Miriam Koehl (1976) has contributed much to our understanding of how flexible organisms can reduce drag, both by the structure of their body wall and by alteration of their behavior. Drag can be reduced by flexibility and by bending over in strong flow, much as palm fronds conform to a strong wind, streaming downwind from the palm's trunk.

When currents and drag are too strong, some animals, such as feather stars, either contract the body or retreat to a crevice. The sea anemone *Metridium senile*, for example, normally protrudes its bushy tentacles into the flow to feed, but this creates drag. When the current increases and pressure drag is too great, the animal withdraws its tentacles, which greatly reduces drag (**Figure 6.10**). Many seaweeds simply bend over, and this allows their long axis to be nearly parallel to the current, which reduces cross section exposed to the flow and, therefore, reduces pressure drag. The kelp *Nereocystis luetkeana* has a series of fibrils in the cortical cell walls that have an average angle of 60 degrees to the axis of the stipe. This increases the extensibility of the stipe, which prevents breakage in a strong current.

Let's consider a specific example of how pressure drag may affect an organism. If we visit an exposed rocky shore, we immediately see waves crashing against the organisms attached to the rocks. In many cases, the mobile invertebrates such as drilling snails and limpets found on exposed coasts are much smaller than members of the same species inhabiting protected rocky shores. It is possible that larger animals would simply be swept away by the waves. Larger animals project above the surface to a greater degree and are more exposed to the drag effect of passing waves. Michael Judge (1988) tested the latter idea by gluing vertical copper plates to the West Coast rocky shore limpet *Lottia gigantea*. The limpets with the plates spent less time moving around than those lacking plates, which gave the plated limpets less time to feed. The loss of feeding time means fewer resources for growth, which may also explain the smaller size of the exposed shore limpets. Drag is also a factor in mobility of snails crawling on a soft sediment surface. An orientation that places the axis of coiling of the snail parallel with the current will minimize pressure drag. The mud snail *Tritia obsoleta* often finds itself in tidal creeks with flow exceeding 30 cm s^{-1} and orients with the apex pointing upstream.

The eastern Pacific stalked sea squirt *Styela montereyensis* is remarkable for its occurrence in a wide array of environments on the California coast, ranging from wave-swept outer coasts to quiet bays. In quiet water, this species resembles typical solitary sea squirts, and the siphons orient upward.

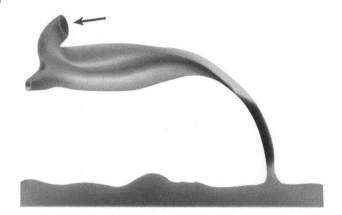

FIG. 6.11 Orientation of the outer-coast form of the sea squirt *Styela montereyensis*. Note that a current causes the individual to bend, but the incurrent siphon is bent, forcing its opening to face the current, which facilitates intake of water and particles for food. (After Young and Braithwaite, 1980)

In outer coasts, most animals are attached by relatively slender stalks and sway with the wave surge (**Figure 6.11**). The incurrent siphon is bent backward approximately 180 degrees. When the animal is bent over by the current, which reduces drag on the body, the water with food particles can still ram into the siphon, which facilitates flow of water and particles.

In Chapter 9, the form of fish is discussed. However, it is worthwhile at this point to think about how the characteristics of a fluid contribute to influencing the swimming efficiency of a fish. Most fast-swimming fishes move by rhythmic contractions, which pass through the body as a wave. At any one time, part of the fish body is pushing against the water, propelling the fish forward.

As a fish moves forward, the forces on the fish surface include pressure drag and frictional effects. As water in streamlines passes over the fish, friction causes the water to lose some kinetic energy. This loss prevents the water from penetrating the steep pressure gradient behind the fish, and the water that leaves the surface behind the fish forms a wake (**Figure 6.12**). If a fish is short and squat, there is a very steep pressure gradient from front to rear, and this leaves a large and irregular wake as the fish moves along. The difference in pressure creates drag. In effect, the fish is being pushed back as it swims. Through streamlining, the wake is diminished and the drag is reduced greatly. That is why fast-swimming and continuously swimming fish exhibit some variation of the classic shark or tuna shape, which creates the smoothest wake. Fast-swimming sharks also have adaptations to reduce skin friction, including arrangements of scales that present surface irregularities. Fish surface slimes also reduce skin friction, and some rib-bearing fish scales may be adaptations to hold onto the slime.

Conflicting Hydrodynamic Constraints

■ **Hydrodynamic forces often present conflicting constraints.**

Hydrodynamics may suggest simple rules for both behavior and morphology, but many marine organisms find themselves having different functions that might require different responses to the same environmental change. In other words, there may be a trade-off between increasing one function, such as feeding, and another, such as minimizing drag and potential mechanical stress. Consider a blade of eelgrass. It projects upward from the bottom and thereby is able to capture light and nutrients. However, this upward projection causes a significant pressure drag when a current impinges on the blade. There is a conflict of different functional requirements because projecting a blade of grass into a current is good to gain food and nutrients but also increases the chance that the current will push on the blade and perhaps tear it from the bottom if the current is strong enough.

The size and velocity scaling of hydrodynamic effects create additional conflicts. As an organism grows in length, the *Re* increases. Consider **Figure 6.13,** which shows patterns of flow downstream of a cylinder under conditions of varying *Re*. When *Re* is about 30, a pair of attached vortices reside just downstream of the cylinder. This might give a small coral the opportunity of feeding on particles that are relatively stationary. As the coral grows larger (or as the current increases in velocity), these vortices become more erratic, however, and food may not be held in a predictable pattern. Size and velocity can increase together; as an organism grows larger, it often projects into a rapid "mainstream" current.

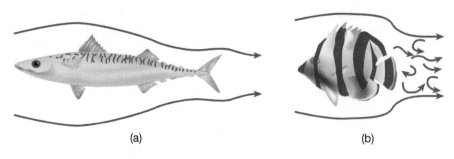

(a)   (b)

FIG. 6.12 Drag on a fish is affected greatly by streamlining. (a) If a fish is well streamlined, the wake is reduced, streamlines are maintained behind the fish, and the drag is much reduced. (b) If a fish is poorly streamlined, a wake is created at the rear, producing a pressure gradient and drag.

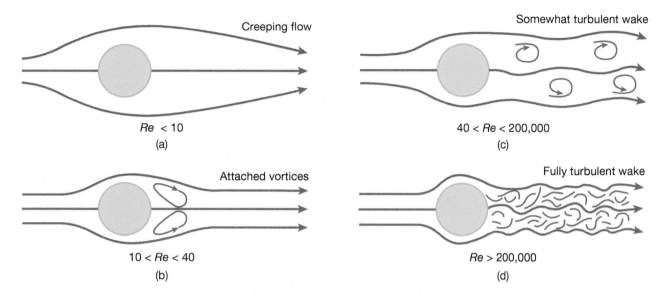

Creeping flow

$Re < 10$

(a)

Somewhat turbulent wake

$40 < Re < 200,000$

(c)

Attached vortices

$10 < Re < 40$

(b)

Fully turbulent wake

$Re > 200,000$

(d)

FIG. 6.13 Different types of wake downcurrent of a cylinder, at different Reynolds numbers. (After Vogel, 1994)

The Use of Flumes

■ Flumes are useful for studying the effects of moving fluids on organisms, although flumes must be scaled carefully.

It is usually quite tricky to study the effects of flow in the marine environment. Some clever investigators have devised field current meters and have been able to characterize the flow field around an organism. Because of inaccessibility, this becomes impractical for studying the streamlining of a tuna or the flow about a deep-sea organism. Even in accessible habitats, it is very difficult to measure flow on the smallest of scales. Electronic devices, such as flow meters based upon thermistor sensors, are usually difficult to use accurately in the field.

In still waters, microcinematography allows us to study the behavior of very small creatures at low Re. In moving waters, flumes of various types are used to study the effects of flow. A **flume** is a device that includes a source of moving water, a working area where the organism and flow field are characterized, and a drain-return system (**Figure 6.14**).

Most flume designers seek two objectives: maintenance of laminar flow and maintenance of scaling by Re (and there are also other parameters beyond the scope of this text). A long flume is desirable because it takes a while for the flow over the bottom surface to stabilize and produce a predictable boundary layer and velocity profile above the bottom. A wide flume, relative to water column height, prevents effects of the walls on flow. Scaling by Re is also essential to keep the proper ratio of inertial to viscous forces. This refinement has an advantage, however, in that you can study a very small object, such as a copepod, by making a larger model and placing it in a more viscous medium.

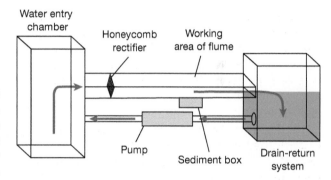

FIG. 6.14 A flume designed to study the effects of flow on small epibenthic animals. In this design, water is recirculated by means of a pump. Water enters a honeycomb-like material to rectify the flow and then flows into the working area of the flume, where organisms are placed. Water leaves the working area and drains into a sump, from which it is pumped to the water entry chamber.

Using a flume, one can study the hydrodynamic forces at work on a biological object. For example, I have studied the reaction of the siphon of a sediment-ingesting bivalve mollusk, *Macoma secta*, which feeds much like a vacuum cleaner (Levinton, 1991). At low mainstream current velocity, the siphon is protruded into the water and swirls around, picking up sand grains. If the current increases to about 15 cm s^{-1}, the pressure drag on the siphon makes the siphon difficult to control, and it is withdrawn. The animal then feeds on sediment within the burrow. At velocities above 35 cm s^{-1}, the bottom sediment is stirred up and the bottom is very unstable. At that point the animal ceases to feed. Using a flume with a video recording device allows such qualitative observations quite easily.

■ CHAPTER SUMMARY

• The Reynolds number is the product of velocity, size, and density divided by dynamic viscosity. When $Re < 1$, viscous forces dominate fluid properties; when $Re > 1,000$, inertial forces dominate. Small and slow-moving organisms operate in a very viscous world, whereas large, fast-moving organisms are dominated by inertia.

• If an object is entrained in moving water, it will move along with streamlines. Laminar flow is regular and occurs under relatively slow flow and low Re. Turbulent flow is irregular and occurs under higher velocities and high *Re*.

- Turbulence on the scale of meters occurs at high *Re*. On a smaller scale, it involves shear, which allows new water to arrive more quickly to the surface of small organisms.

- Water moves in a free stream above a surface, but its velocity is zero at the surface (the *no-slip condition*). A boundary layer of lower velocity and low *Re* exists at the surface, even if water is moving some distance away. Cells in the middle of the water column also have

no velocity at the cell surface and very low *Re*. Shear due to cell motion can move materials toward the cell surface.

- The principle of continuity states that velocity is inversely proportional to cross-sectional area. If the area doubles, velocity is cut in half.

- Organisms can propel water at low velocities with thousands of tiny structures called flagellae. Water velocity through an exit channel is increased when the total cross-sectional area over which many flagellae act is much larger than the exit.

- Bernoulli's principle states that pressure varies inversely with fluid velocity. Organisms use differential velocity to create pressure gradients, which can create lift and also drive water through burrows.

- Drag is produced when pressure differs upstream and downstream of an object. Drag can be reduced by streamlining, by reducing the presentation of flat surfaces perpendicular to a water current, or simply by bending over (as in seaweeds and soft corals). Streamlined form in fast-swimming fishes minimizes pressure drag and also minimizes skin friction.

■ REVIEW QUESTIONS

1. What is dynamic viscosity? Give an example of a fluid that is more dynamically viscous than water.

2. What information does the Reynolds number provide?

3. In seawater, what are the principal factors that vary to determine the Reynolds number?

4. Why can a ciliate stop nearly instantaneously, whereas a large swimming fish takes much more time and a higher number of body lengths to stop?

5. What is the no-slip condition?

6. What conditions are different within the boundary layer, relative to mainstream flow conditions?

7. Marine larvae try to find a place to settle and metamorphose into an adult. If the larvae are hovering a meter or so above the bottom, what must they do to find a clue that a member of their species is present in the bottom?

8. How can a sponge have an exit velocity through its exhalant siphon that is orders of magnitude greater than the flagellated sponge cell velocity?

9. How might an attached organism reduce drag, even when it protrudes into the mainstream flow?

10. How might a burrowing worm take advantage of the Bernoulli principle to enhance the flow of water through its burrow?

11. What are the advantages and disadvantages encountered by a sessile marine organism that projects its body into the mainstream flow?

12. Consider a fish swimming in moving water. What creates drag on this fish?

Visit the companion website for *Marine Biology* at www.oup.com/us/levinton where you can find Cited References (under Student Resources/Cited References), Key Concepts, Marine Biology Explorations, and the Marine Biology Web Page with many additional resources.

CHAPTER 7

Reproduction, Dispersal, and Migration

Reproduction, dispersal, and migration are the fundamental processes that allow living populations to grow and exploit new habitats. **Reproduction** is the replication of individuals and is necessary for population growth. Nearly all species have some form of **sex**, which allows exchange of genetic materials among individuals. Sex and reproduction are intimately related, but many organisms can reproduce without sex, so the two processes can be discussed separately. **Dispersal** is the spread of progeny from one location to another, usually to locations that differ from that of the parent. Dispersal is a one-way process and often is controlled by the vagaries of water currents. Because water is a supportive medium and because water circulates rapidly over great distances, many newly hatched young can disperse over large distances, often as much as hundreds of kilometers. As a result, marine species often have broad geographic distributions and the capability of rapid extension of their geographic ranges. **Migration**, a directed movement between specific areas, allows an expansion of types of habitat use and can increase a species' efficiency at exploiting the best resources at optimal feeding and spawning sites.

Ecological and Evolutionary Factors in Sex

Costs and Benefits of Sexual Reproduction

■ **Sex is a nearly universal characteristic, despite the considerable costs to organisms of maintaining it.**

Sex is a species property whereby different individuals have the capacity to exchange or combine DNA, which causes offspring to differ genetically from their parents. Sex may involve simple transfer of DNA between bacterial cells or the union of gametes to produce zygotes. Nearly all organisms have some form of sexuality, ranging from the mating types of bacteria to the separate sexes and mating dance of humpback whales.

The near universality of sex is a paradox because of its considerable cost. In a typical diploid organism, each parent contributes the same number, n, of chromosomes to the offspring, giving the proper diploid number of $2n$. However, the mother usually invests much more in her offspring than does the father, even though half the offspring's genes are his (think of the energy put into egg production and the greater amount of parental care usually contributed by the female). Why not devote all her energy to nurturing offspring that carry only her own genes? Otherwise, she invests a lot to place only half of her genes in any offspring.

There are other costs to maintaining sex. Finding a mate can be costly in terms of time and energy. In fiddler crabs (genus *Uca*), the males have two claws, one much larger than the other, which comprises over 30 percent of the body weight (**Figure 7.1**). This larger claw is used only for sexual displays and for combat with other males. Because the claw is not used for feeding, males must compensate for this handicap by feeding longer and faster. In one form of sexual display the male's major claw is stretched out to the side and the crab has exaggerated musculature and stouter legs to support the weight! Why should so much cost be attached to sex and sexual differences such as these?

The most obvious benefit of sex is the **generation of genetic diversity**. Genetic variation allows offspring to survive in a broader variety of habitats. There is also an increased potential for evolutionary change when the environment changes.

FIG. 7.1 Displaying male of the fiddler crab *Uca pugilator*. Note the large major claw, which is used for displays and male–male combat. (Photograph by Jeffrey Levinton)

Without sex, all offspring are genetically identical to the parent. No evolutionary change is possible except by means of mutations, which occur only rarely. Asexual populations are, therefore, **clones**, with little or no genetic diversity, except by mutation within the clone. In contrast, sex provides continually new combinations of genes and changes of gene arrangements due to chromosome crossover during meiosis. The process of natural selection results in the eventual dominance of the adaptively superior genetic variants.

A study of the snail *Potamopyrgus antipodarum* demonstrates how genetic variation allows adaptation to the challenge of disease (Lively, 1996; Jokela et al., 1997). This species, living in New Zealand fresh water, has populations that reproduce sexually and asexually. The asexual populations are far less able to resist invasions of parasites. Sexual populations grow more slowly than clonal populations, which accords with our assumption that sex does have a cost in population growth. The continuous evolutionary battle between snails and their parasites results in many local sexual populations that are each adapted to different suites of parasites (Lively et al., 2004).

■ **Sexual selection increases differences between the sexes and is driven by mate selection and polygynous behavior patterns.**

In most animals, males compete for mates. Females invest more energy in gametes and energy in rearing offspring. Males invest in form and tactics that maximize the number of matings they can achieve. Once sex exists with intermale competition, the differences between the sexes can be enhanced by **sexual selection**, the selection for male secondary sex features (e.g., the antlers of deer, the major claw of fiddler crabs, and bright coloration in male fishes) that increase mating success. Genetic variants that have the most successful mating abilities will increase in the population. For example, fiddler crab males with larger claws may be better at attracting mates or holding high-quality

territories where females can incubate eggs (**Figure 7.1**). This superiority would set in motion a selective process that could lead to selection for larger claw size. Success in sexually selected characters can result in occupation of high-quality habitats by means of intermale combat, where matings are gained and winners emerge in a process of **intrasexual selection**. Alternatively, conspicuous coloration in males might reflect their high fitness, which may be used as a criterion in **mate selection** by females, also known as **intersexual selection**. While such selection may result in more extreme structures, trade-offs with function may prevent a structure from becoming too extreme. In the case of fiddler crabs, too large a male major claw might increase fatigue as males run from predators. In many species, selection for bright male color might be countered by increased detectability by visual predators. See **Hot Topics Box 7.1** for more on this subject.

Intersex differences are also driven by social structure in mating patterns. In many species of fish and mammals, one male may mate with a group of females. A male elephant seal (see Figure 9.25) is about five times the mass of a female and may inseminate a group of 10–100 females during the breeding season on the beach. The male's dominance over other males first involves vocalizations toward other males to establish dominance but often escalates to bloody battles. If the ability to be dominant is correlated with fitness, then the female gains an advantage in the fitness of offspring she produces.

Types of Sexuality and Their Value

■ **Sexes may be separate (gonochoristic species), simultaneous, or sequential in the same body.**

Organisms of nearly all kinds (with the exception of bacteria and some protozoa) have two sexes. Species that have separate sexes, or **gonochoristic** species, are the most familiar. While males may mate in direct contact with females, seawater permits many species to simply shed sperm (and possibly eggs) into the water. Planktonic gamete transport often involves simultaneous spawning of males and females, which is often keyed into tidal or lunar rhythms.

A **hermaphrodite** can produce gametes of both sexes during the individual's lifetime. In **simultaneous hermaphrodites**, sex cells for both eggs and sperm are active within a single individual at the same time. Despite this, self-fertilization is rare, probably because inbreeding often produces inviable offspring, which imposes selection against such behavior. Being a simultaneous hermaphrodite means never having to say you're sorry, because an encounter between any two individuals guarantees that mating can occur. For example, acorn barnacles need to copulate to reproduce, but they are stuck with a sessile lifestyle and can reach only nearby barnacles. Simultaneous hermaphroditism is a great advantage because the nearest individual will always be of the complementary sex. An extraordinarily long penis allows a barnacle to reach relatively distant fixed mates. Acorn barnacles often have a seasonal cycle of growth and degeneration of the penis. They show phenotypic plasticity and can grow a longer penis if a potential mate is far away or can even grow

HOT TOPICS IN MARINE BIOLOGY

A Lover and a Fighter   7.1

Darwin realized that there was a strong force in nature that involved selection for mating ability. He termed this **sexual selection**, to distinguish it from natural selection. Natural selection occurs among genetic variants and is driven by selection for improvements in performance in physiological function and resistance to predators, among other changes. But sexual selection involves competition among members of the same sex to get matings. In most cases, this involves competition among males, and from this we see the most amazing array of sexual dimorphisms, including the antlers of deer, the enormous feather trains of peacocks, and the bright coloration so characteristic of males of many species. Overall, these extreme traits emerge from two factors in mating success: conspicuous display, resulting in mate selection by females; and combat, involving intrasexual competition among males for females.

Fiddler crabs are a group of species in which the males are displaying lovers *and* fighters. About 100 species occur throughout the world's tropical and subtropical regions, and in all cases males have an enormous major claw (**Box Figure 7.1**) that is used for both display and combat. Males use the claw in a wave characteristic of its species, and this wave attracts either females, on an open mudflat, or males that are guarding a hole. What is remarkable about this attraction is that it is largely visual; the crabs court while the tide is out and the animals are, therefore, communicating in air, much like terrestrial species. Females approach males because of their acute eyesight, which is trained on the horizon. When a male waves high, the wave at first stimulates an ancient response system in the female that is geared to sensing visual stimuli as coming from predators. But immediately the female recognizes the wave and sees the male as a potential mate, not a predatory bird. After a stereotyped approach, the female may accept the male and copulate with him. But why *him?* Why this male, instead of that?

In combination with this visual signaling, males use their enormous major claw in combat with other males. In some cases, other males just come too close and this stimulates a bout. Males brush up against each other and may grip each other's major claws (**Box Figure 7.1**) until one slightly damages the other or even flips the other on its back. Once such a flip occurs, the losing male is in danger of being attacked by a bird, so it quickly retreats. Some males may attempt to remove a resident male from its territory and burrow. This may involve hours of combat.

Males often autotomize their major claws when they are attacked by predatory birds. They are then unable to attract females, let alone fight with other males. Being arthropods, they cannot regenerate a claw until several molts occur. But a major claw is often an elaborate and robust structure, requiring extensive resources for growth. In some species of fiddler crabs, males regenerate claws rapidly that are essentially phonies. Patricia Backwell and colleagues (2000) investigated the South African fiddler crab *Uca annulipes* and found that males that lost their claws could regenerate deceptive major claws that were used in signaling to attract females. The deceptive claws were long but lacked ornamentation useful in combat and were weaker than the major claws of other males. Nearby males apparently failed to notice this deception and did not challenge the "dishonest" males to combat. Obviously, if most males bore these dishonest claws, the system would break down, so rarity is essential for the deception to work.

So let's return to the female. How can her selectivity for mates be a driving force in selecting for the male's display, which comes down to a bigger claw, claw wave, and brighter colors? What does a male have to do to attract her? Does bigger mean better? A few clever experiments have helped us understand this process. Maybe you will wonder about just how clever these females are when you learn that female *Uca tangeri*

BOX FIG. 7.1 Two males of the sand fiddler crab *Uca pugilator* fighting. Location is a salt marsh on the north shore of Long Island, New York. (Photograph by Bengt Allen)

continues

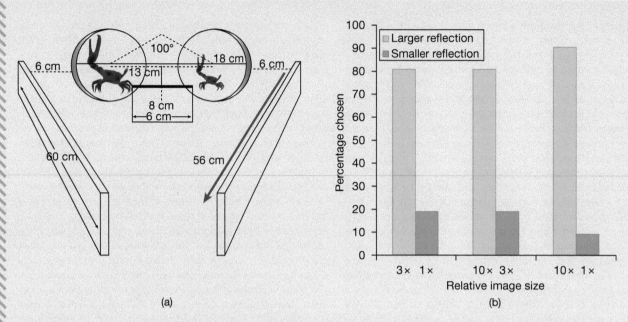

BOX FIG. 7.2   An experiment to demonstrate that females prefer larger males. (a) A male is hidden from a female's view, but she can see a reflection of the male and a magnified reflection. The investigator monitored the female's choice in her approach to one of the two mirrors. (b) Percentage chosen by females of males magnified different amounts compared to unmagnified males.

from Portugal will approach larger dead major claws mounted on a stick more frequently than smaller claws (Oliveira and Custodio, 1998). This was certainly suggestive of a size advantage in the males. A clever experiment used mirrors that can be obtained in any store. On one side these mirrors usually reflect just your image, but the other side magnifies the image. McLain and Pratt (2007) placed a male behind a barrier so that a female could not see him directly. What she could see was a mirror image, either of the normal or magnified variety. Females decidedly selected the "larger" males (**Box Figure 7.2**).

Of course, the selection is just the beginning. Why select a bigger male? What is the difference? This is especially a problem because the male's size is a bit of an ambiguous signal: A large male could be a more robust male, but it also could just be older and therefore larger. Two alternatives come to mind: Big males may be selected because

size is a reflection of fitness or the presence of genes that will give the female more robust offspring; it is also just possible that big males confer a benefit to females. In the case of the sand fiddler crab, males mate in the burrow and cover the female with sand, and the female remains in the burrow for about 2 weeks until she emerges to release the hatched eggs into the water. It may be that males high in the intertidal, usually the larger ones that win places by combat, are preferable for the reason of protection alone. But this second explanation is not so clear when asking why one male high in the intertidal would be chosen over a smaller one, since both microsites might be equally useful to the female. Bengt Allen demonstrated that males with proportionally larger major claws also had more body mass, suggesting that such males were healthier. Big claw size may therefore be an honest signal of good male physiological condition.

a more strengthened penis if the local microenvironment is more turbulent (Hoch et al.. 2016).

One of the most fascinating aspects of simultaneous hermaphroditism is the common continual trading of mating roles. In many simultaneously hermaphroditic fishes, for example, members of a mating pair change sex roles every few minutes, first producing sperm, then spawning eggs (as the other individual, who also has changed roles, now deposits sperm).

Sequential hermaphrodites start their mature sex life as one sex, and then transform into the opposite sex. If they are male first, they are said to be **protandrous**, whereas if they are female first, they are **protogynous**. Many invertebrates,

including some oysters, polychaetes, prawns, and coelenterates, are protandrous, whereas some fishes, particularly coral reef fishes, are protogynous. In some cases, such as the diminutive protandrous polychaete species of the genus *Ophryotrocha*, sex change can involve a mere switch from the manufacture of oocytes to the manufacture of spermatocytes. In protogynous reef fishes, however, the sex change also involves color, size, and morphological transformation. The snail *Crepidula fornicata* usually occurs in stacks, with larger and older females below and smaller and younger males on top (**Figure 7.2**). Top-layer members of the stacks are oriented with right anterior margin in contact with the same margin of the lower member. This allows insertion of

FIG. 7.2 The sequentially hermaphroditic snail *Crepidula fornicata*, stacked on a gravel beach. Females are on the bottom (right). Males are the smallest and topmost individuals. (Photograph by Jeffrey Levinton)

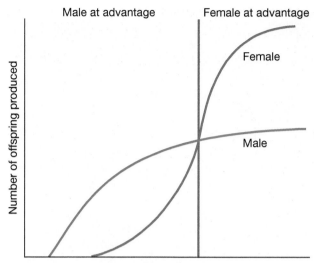

FIG. 7.3 Graph showing how to find the optimal size for a switch from male to female in a protandrous hermaphrodite. The curves show the number of offspring an individual would parent if it remained either male or female for its entire life. The curves cross once, and output as a female surpasses that for a male as the crossover is reached and passed. For sizes smaller than the crossover, output is greater as a male.

the penis of the upper snail into the female gonopore of a lower snail. The stimulation to change sex involves contact with other individuals. To change to a female, a male must be in contact with a female below in the stack.

■ **The relative contributions of different sexes to zygotes in the next generation determine the value of hermaphroditism and the relative sizes of males and females.**

Why be a simultaneous hermaphrodite? In a sense, one might turn the question around and ask: Why aren't all organisms simultaneously hermaphroditic, given the obvious advantage of finding a potential mate in every individual? The answer may lie in the limitations of being a jack of all trades and master of none versus being a specialist. Simultaneous hermaphrodites must invest not only in an apparatus to form gametes but also in the secondary sexual attributes necessary to attract another mate for both sexes. If one started with a population in which genetic variants for hermaphroditism and for separate sexes coexisted, one might find that sexual selection would favor genetic variants for "pure" males and females. The hermaphrodite would lose out in the competition because it serves both functions but each less efficiently. Hermaphroditism is favored only when the value of always being able to find a mate compensates for the disadvantage of not being a sexual specialist. The acorn barnacle may be just such a case. Simultaneous hermaphroditism is far more common in freshwater invertebrates, where it is generally more difficult to find a mate owing to the lower population densities and the uncertainties of freshwater existence.

Sequential hermaphrodites would seem to have all the advantages, given that they can be both sexes at different times. Unlike simultaneous hermaphrodites, however, they lack the advantage of always being able to find a mate in another individual. Then what is the advantage of changing sex? The answer lies in the advantage of being a male or female at different ages or sizes. Consider **protandry**, or the quality of first being a male and later a female. As a general rule, producing eggs costs more energy than producing sperm. It is also usually true that a larger and older animal has more energy at its disposal. Then it would be best to be a male while still small, when a relatively small investment in sperm could yield many offspring. Above a certain threshold size, however, females can parent more offspring because the available energy can be used to produce more eggs. Male success would reach a plateau with increasing size. This is known as the **size-advantage model**. The threshold size of switching (**Figure 7.3**) should be that size at which the number of offspring parented would be the same regardless of the sex of the animal. Below that threshold size, the animal would sire more offspring if it were male; above that threshold size, it would parent more offspring were it a female. The considerations are thus simply those of a cost-benefit game. This model was tested in a protandrous species of the polychaete *Ophryotrocha*. Larger males were less successful in gaining mates than smaller males, whereas female reproductive output in females continued to increase with increasing body size. These results favor the hypothesis that size is the advantage that selected for the switch from male to female function (Berglund, 1990).

Why then should an organism be female first, then male? Reproductive success as a male often involves intermale competition, involving combat, bright colors, and rapid swimming. Most of these traits are enhanced with increased size and experience. Many territorial coral reef fish have flashy males and are **protogynous**. For example, the cleaner wrasse *Labroides dimidiatus* is usually found in a group of 10–15 fish. All but the largest fish in the group are relatively dull in color and are females. By contrast, the largest fish is brightly colored and male. The male constantly tries to

FIG. 7.4 A terminal-phase male of the bluehead wrasse *Thalassoma bifasciatum* courting a female. The two are about to mate. (Photograph courtesy of Robert R. Warner)

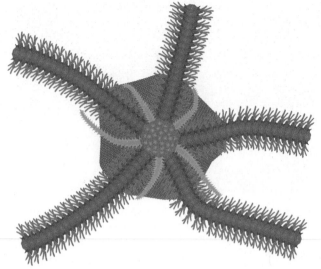

FIG. 7.5 A female brittle star, *Amphilycus androphorus*, carrying a dwarf male. The two are attached mouth to mouth. (From Hyman, 1955)

prevent the females from mating with interlopers. If this male is removed from his harem, the largest female will transform into a male. In other species of wrasse, individuals can either be terminal-phase males (**Figure 7.4**), which hold territories, initial-phase males, or females. Initial-phase males do not hold territories and either spawn in groups with females or attempt to sneak into the territories of the terminal-phase males and gain matings. Both initial-phase males and females are capable of changing into terminal-phase males. The proportion of initial-phase males varies from place to place. This suggests that there are places where being a small male results in more successful matings.

Although it is possible to classify types of sequential hermaphroditism, it is important to realize that many natural populations contain considerable geographic variation in sex change. This is especially true of coral reef fish. Certain species may be locally protogynous but gonochoristic in other sites. Hermaphroditism is not necessarily a fixed trait of a species and may be under active natural selection in different directions in different populations.

Male morphology has been found to be dimorphic in some species. Kurdziel and Knowles (2002) examined **male dimorphism** in an amphipod; some males had a large structure known as a gnathopod, which was used in aggression, but other males lacked this large structure and more resembled females. While the aggressive males guarded females until they molted and were ready to mate, the other males were "sneakers" and approached unguarded females. As it turns out, the sneakers were as successful as the large aggressive morph in getting mates, which explains why the male dimorphism is maintained over time. Male dimorphism is found in many other species, including several species of rockpool blennies, which have a nesting male morph, who attracts females to lay eggs in the male's small territory, which are cared for by the nesting male. But an alternative smaller male morph can sneak in and achieve matings within the nesting male's territory.

■ **Dwarf parasitic males are found in some species that live in situations where it is difficult to find mates.**

Where the sexes are separate and mate location is very difficult, males may be small animals that either attach to larger females or reside very close to them. Good examples are found among some barnacles, where tiny dwarf males attach within the mantle cavity of normal-sized females (e.g., species of the stalked barnacle *Scalpellum* and *Ibia* and the boring barnacle group Acrothoracica). Many deep-water fish have dwarf males. Males show varying degrees of modification and may be parasitic upon females. In some cases, the male is a miniature version of the female, except for the reproductive organs. This is common in some brittle stars (Echinoderm class Ophiuroidia), whose dwarf males cling to the much larger females (**Figure 7.5**). In some cases, dwarf males of deep-sea anglerfishes find and attach to females and even have their blood circulation systems in connection with each other.

■ **A rare case of eusociality can be found in groups of alpheid shrimp in spaces within large sponges, where one female is served by a large number of males.**

When individuals are closely related and found within an isolated space, one might expect a division of labor that serves a genetically closely related group. In ants and bees, all members of a colony are closely related and such division of labor into queen and different types of workers can be found. Why is there such altruistic behavior, in which workers serve the queen? William Hamilton noted that workers are extremely close relatives of the queen; their role in doing the work of the colony, therefore, increases not only the genes of the queen but also their own. The evolutionary force leading to the evolution of cooperative behavior among close relatives to proliferate similar genes is known as **kin selection**.

The extreme case of an entire colony of individuals behaving in cooperative fashion is known as **eusociality**. This behavior usually is found in a group of individuals that

FIG. 7.6 A mature queen of the eusocial shrimp *Synalpheus regalis*. The clutch of late-stage embryos, some with visible eyes, can be seen in the brood chamber beneath the queen's abdomen. Normally, shrimp would rarely be found on the exterior of the sponge. (Courtesy of J. Emmett Duffy)

serve a queen's reproduction. It is extremely rare beyond social ants, bees, wasps, and a few other groups of insects. Naked mole rats (*Heterocephalus glaber*) are also eusocial: There is one "queen," plus several mating males and numerous other closely related nonmating siblings or very close relatives that gather resources and take care of the tunnels in which the mole rats live. A very similar case has been found in some species of synalpheid (snapping) shrimp, where large sponges in the Caribbean have colonies each with one female (**Figure 7.6**) and 100–200 males and juveniles. When snapping shrimp from outside the colony are placed near the opening, males swarm to defend the colony, attacking the intruders and killing them. The shrimp recognize members of their own colony and can even signal to each other, warning of intruders. J. Emmett Duffy (1996a, 1996b), who investigated these shrimp, found through genetic analysis that individuals of the colony were closely related and the relatedness was consistent with the individuals being siblings. Thus the snapping shrimp fit the condition of closely related individuals in a confined space, all serving the colony by a specialization of function, with the single female being an analog of the queen bee.

Fertilization Success

- **Fertilization success is affected by the mode of sperm transfer, the volume of gamete production, the distance between males and females, water turbulence, timing, and behavior.**

Many animal species use a variety of specialized means of sperm transfer involving direct male-to-female contact. In the simplest cases, found in some polychaetes and fishes, the male applies sperm to the body surface of the females or to an egg clutch as the eggs are released by the female.

Males of many species (e.g., gastropods, crustacea, and fishes) have copulatory structures. Many species, ranging from migratory fishes to intertidal drilling gastropods, form breeding swarms at the time of mating to further ensure the finding of a mate. These modes of sexual contact increases the probability of fertilization and allow for mate selection.

Although direct contact and copulation guarantee sperm transfer, many marine species take advantage of the presence of water and shed their gametes directly into it, which is known as **free spawning**. At best, the gametes will mix and fertilization frequency will be high because the sperm are so much more numerous than the eggs. However, turbulence or large distance between spawning individuals greatly reduces fertilization success. We discuss the costs of free spawning below.

Spawning by one or a few individuals may induce mass spawning of an entire local population. Mussels will spawn in response to phytoplankton in the water; the mussel larvae then can eat the phytoplankton. Timed spawning is usually keyed into lunar tidal cycles. In some reef sponges, all male individuals spawn simultaneously, spreading a fog of sperm over the reef. Many polychaete annelids can change morphology radically to transform themselves into epitokes, individuals that are essentially swimming sacs of gametes. Operating on a lunar cycle, they may swim to the surface and perform a nuptial dance, in which males and females swim rapidly about each other while releasing gametes into the water. South Pacific peoples take advantage of this behavior by harvesting swimming polychaete Pololo worms. Many coral reef fish shed their gametes, but the fishes often locate themselves strategically on the downstream end of patch reefs to minimize turbulence and gamete dilution. As a result, fertilization success is surprisingly high, often exceeding 90 percent.

■ Free spawning has a number of costs relating to turbulence, polyspermy, timing of spawning, and avoidance of interspecific hybridization.

With free spawning, the game changes because both eggs and sperm must encounter each other, sometimes based upon hydrodynamics as opposed to sex appeal. In free spawners, the eggs suddenly are part of the gamete collision equation, and there might be strong selection for egg traits that maximize sperm encounters and even attraction.

But there are also significant costs of free spawning:

1. *Fertilization success.* Fertilization success is often low when distance between males and females is great and when turbulence is strong. At low population densities and with few spawning individuals, one might expect to see the effect of sperm limitation, and this has been demonstrated in the eastern Pacific temperate sea urchin *Strongylocentrotus franciscanus* (**Figure 7.7**). Some species, such as the Pacific sea star *Acanthaster planci*, produce such large volumes of gametes that high rates of fertilization can occur between males and females separated by tens of meters.

2. *Timing of egg and sperm release.* Gamete release of both sexes must be simultaneous for fertilization to be successful. Gametes often do not last much more than a few hours, meaning that an environmental signal to ensure simultaneous spawning, such as a response to a lunar cycle, must be evolved. We discuss mass spawning in coral reefs in Chapter 17, where many species cue into lunar cycles. Alternatively, some species of mussel are known to respond to the spawning of nearby individuals. This may mean that certain local individuals will make a disproportionate contribution to the next generation, especially when a male invertebrate spawning millions of sperm successfully fertilizes a large number of gametes of nearby females.

Because water turbulence should dilute gametes and reduce the percentage fertilization, we might expect that some species would tend to release gametes only in quiet water, when the turbulent dilution effect would be minimal. The seaweed *Fucus vesiculosus* releases gametes only at current speeds of less than 0.2 m s[1]. At these low speeds, usually occurring in the early evening, fertilization success is nearly 100 percent. In the lab, experimental increases of water speed causes the seaweed to reduce its release of gametes (Serrao et al., 1996).

3. *Avoidance of interspecific fertilization.* What if several species evolve a response to the same lunar cycle? Many species would be spawning at the same time. Gametes would likely encounter those of other species. If reproductive isolation has occurred, this would cause lowered fitness unless there was extreme specificity of gametes.

4. *Fertilization at the microscale (polyspermy).* In quiet water, many sperm may simultaneously encounter a single egg. High population density of males might select for more sperm production because male genotypes are competing with each other for matings, and increased sperm release might increase the production of successful fertilizations. So more sperm is better. Or is it? If sperm concentration is very high, then polyspermy, which entails the beginning of zygote formation by two or more sperm and one egg, may develop. When this happens, successful fertilization is usually foiled. All known animals have elegant blocks to polyspermy, some of which act within seconds to prevent another sperm from attaching to the egg. But the phenomenon is still possible and perhaps likely in a concentrated cloud of sperm, as might be produced by some invertebrates.

5. *Fertilization and small-scale turbulence.* On the scale of meters, turbulence and current flow are essential for

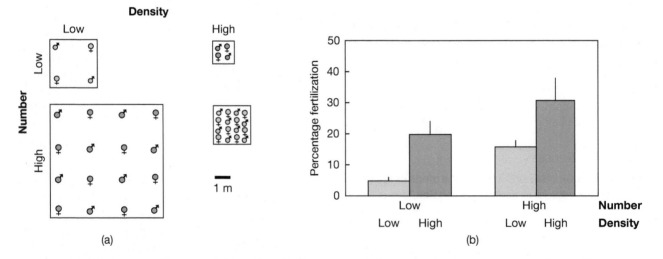

(a)   (b)

FIG. 7.7   A field experiment showing the effect of population density and abundance on fertilization success of eggs of the eastern Pacific sea urchin *Strongylocentrotus franciscanus*. Eggs were placed in cups and sperm was injected in the water from a syringe. Populations were high or low, and density (number per unit area) of egg cups was high or low. As can be seen, population density was very important in percentage fertilization success. (After Levitan et al., 1992)

gamete encounters. Otherwise, gametes would sit still. But, on the scale of millimeters and less, turbulence also is very important for fertilization. With no turbulence, gamete encounter would be reduced because of the problem of viscosity generated by low Reynolds number (see Chapter 6). So a small degree of turbulence increases gamete encounter and fertilization success (Riffell and Zimmer, 2007). But when turbulence increases further, gametes will not have the time and opportunity for sperm–egg recognition to occur and fertilization rate will be reduced.

■ **There are two main phases in small-scale gamete encounter: attraction on small scales of sperm to eggs by means of sperm attractants, followed by surface compatibility of sperm when attempting to fertilize eggs.**

Sperm are usually motile, using the flagellum to swim. But on scale of millimeters, sperm in the neighborhood must detect the presence of eggs, and, of course, eggs might signal their presence. To accomplish this, many species use **sperm attractants**. For example, eggs of the Pacific coast red abalone *Haliotis rufescens* release the amino acid 1-tryptophan (Riffell et al., 2002). When abalone sperm encounter a threshold concentration, they swim toward the egg, which is now just a couple of millimeters away. The egg is surrounded by a chemical plume of this sperm attractant, and the sperm probably encounters this plume randomly. The chemical plume increases the target size that the egg can present to recruit a sperm cell.

Sperm attraction is also known for seaweeds. In the Laminariales (kelps) pheromones are produced by a microscopic female gametophyte stage, which induces male gametophytes to release sperm. The pheromone Lamoxirene may exist in several isomers, but there does not appear to be specialized species recognition based on difference of gamete recognition between species, as we shall now discuss for some animal sperm–egg recognition systems (Maier et al., 2001).

■ **In some groups, specialized gamete recognition proteins are employed to avoid interspecies fertilization.**

Because different species will be selected to spawn when gamete contact is optimal, at a specific time when simultaneous spawning of males and females can occur and turbulence is low enough that gametes can collide, it is to be expected that different species might have gametes come into contact. So, is there a mechanism that keeps gametes of different species from attempting fertilization?

We can see several steps at fertilization that might be disrupted when gametes of different species come into contact:

1. Receptors in the egg coating induce the sperm acrosome reaction, which is an initiation of attachment.

2. Sperm and egg attach, owing to the presence of complementary binding (sperm) and receiving (egg) proteins.

3. Further reactions cause the uniting and fusing of gamete plasma membranes.

We have information on only two major groups of organisms—mollusks, especially abalones (Phylum Mollusca, Class Gastropoda, Order Archaeogastropoda, Family Haliotidae), and sea urchins (Phylum Echinodermata, Class Echinoidea, several genera)—but a surprising similarity has emerged. Both have a highly specific sperm–egg binding system. In urchins, sperm use the protein bindin, which unites with a bindin receptor in the egg membrane. In abalones and other mollusks, gamete recognition is controlled by the sperm protein lysin, which creates a hole in the egg envelope, but only for eggs of the same species as that of the sperm. In the abalone egg envelope, a receptor binds specifically to lysin. Hybrids are thus rare, even though seven species of California abalone free-spawn gametes and overlap in geographic range and breeding season (Swanson and Vacquier, 1998).

Different species of urchin rarely, if ever, have gametes that will fertilize each other. Bindins from different urchin species are different proteins and, therefore, have different amino acid sequences, and this makes the crucial difference in the sperm–egg recognition process. The same is true for lysins in abalones. Proteins are typically folded in a complex manner. Some of the amino acids are involved in regulating the three-dimensional structure, while others are involved in binding with other molecules. Thus, only a portion of the amino acid chain is involved in gamete recognition.

One of the fascinating aspects of these gamete recognition proteins is the apparent high rate of evolution of amino acid sequences compared to other proteins. The accelerated rate might be due to a response to evolution in the recognition protein found on the complementary gamete. For example, the bindin protein found on an urchin sperm binds with a protein receptor known as VERL. There may be a constant tugging in evolution between the two proteins. When one evolves, then the other must evolve as well. Stephen Palumbi (1999) found that there is genetic variability of bindin sequences within populations of species of the urchin genus *Echinometra* and that eggs may be selective for sperm with some bindins relative to others. This is a form of sexual selection that may drive the evolution of bindin and VERL proteins. For example, when sperm are abundant, rare bindin genetic variants apparently are selected and increase in the population; when sperm are rare, it is the common genotype that is likely to have success in fertilization, simply by virtue of its high frequency (Levitan and Ferrell, 2006). However, not all urchin species have this sort of polymorphism.

Ultimately, if two populations are separated, the evolution of bindins can lead to complete incompatibility between species. Complete gamete incompatibility in sea urchins can be accomplished with as little as 10 differences in amino acids, and this amount of change can occur in as little as 1.5 million years (Zigler et al., 2005). So, the mechanism of speciation in free-spawning species may be very much tied into the dynamics of protein evolution of these binding proteins in many unrelated species groups. Sympatric (coexisting) groups of closely related species of sea urchin, for example, have been found to have comparatively rapid rates of evolution of bindin when compared to species that are geographically separated from their closest relatives (Palumbi and Lessios, 2005). This suggests that bindin evolutionary rate may be setting the pace for the evolution of reproductive isolation between species.

Parental Care

■ **Parental care is nonexistent in many marine animal species, but in some cases females or males care for eggs, embryos, and young.**

Marine species often exhibit absolutely no parental care. Many of these species spawn eggs and sperm, which unite to form zygotes, which, in turn, develop into free-swimming larvae that are completely at the mercy of the seas. However, in some groups, such as the fiddler crab *Uca*, the female incubates the eggs for a couple of weeks and then releases swimming larvae into the water. Many species care for their young to an even greater degree. In some species, the young are reared to a juvenile stage within the mother. This is, of course, true for all marine mammals, but it is also true in invertebrate groups and fishes. The Atlantic and Pacific intertidal clam *Gemma gemma*, which is only a few millimeters long, nurtures its young within the mantle cavity and releases them as shelled juveniles. The Chilean oyster *Ostrea chilensis* broods bivalve larvae within the mantle cavity and can tell the difference between its own young and other particles of about the same size, which they reject (Chapparo et al., 1993).

In some species, courtship is strongly related to male parental care. Conceptually, male care should be linked to the assurance that the offspring are its own, and the investment in care therefore increases the probability of survival of the male's offspring. Sea horse mating pairs engage in a complex mating dance, often intertwined by their tails. Eventually they face each other, and the female releases eggs as the male opens up a pouch; the eggs are then fertilized and deposited within the male pouch, where they are then nurtured for 2–4 weeks, until they hatch and are released. The male of the three-spined stickleback *Gasterosteus aculeatus* has a vivid red belly during courtship. This anatomical feature serves to lure females to lay eggs in its nest, usually hidden among rocks on the bottom. Many aquarium enthusiasts have observed the male dance in a zigzag motion, to lure the female to the nest. The great behaviorist Niko Tinbergen showed that the reproductive females have a stereotypical response: They are strongly attracted even to silver disks painted red on the bottom. Behaviorist Susan Foster (1990) has observed that in some locations, the female participates, perhaps to a dominant degree, in mate selection and often swims up rapidly to a male and jumps onto his back! During the courtship process, a female enters the nest and lays her eggs, which are fertilized externally by the male. Subsequently, the male aerates the eggs until they hatch. It is in the male's interest to rear the young of all females that have laid eggs in his nest because they are his progeny. When a female enters, however, the eggs that have already been laid there by other females are, well, just food. Commonly, an intruding female will gobble up the eggs already present and bolt.

Nonsexual Reproduction

■ **Nonsexual reproduction permits the same genetic type to increase rapidly in an open environment.**

Although sexuality is nearly universal in marine organisms, many of them are capable of reproducing without the formation of a zygote. Asexual reproduction lacks the cost of sexual reproduction and permits the spread of a genotype that has successfully colonized a given habitat. A population of genetically identical individuals, all deriving from one founder, is known as a **clone**.

The exact style of asexual reproduction varies, depending upon the biology of the individual group. Diatoms, a major group of marine phytoplankton, consist of cells or cell chains and can reproduce asexually by fission.

In multicellular organisms, **fragmentation** can serve the same function as fission. Some annelids fragment, while others divide by forming head segments midway down the body and then splitting into two new individuals. Many seaweeds (e.g., *Gracilaria* and *Polysiphonia*) and many corals (e.g., the Caribbean staghorn coral *Acropora cervicornis*) may reproduce mainly by fragmentation. Tropical storms may benefit a staghorn coral population by dispersing newly created fragmented individuals.

Colonial invertebrates may reproduce asexually by fission of whole individuals in the colony. As opposed to the production of new independent individuals, this process produces genetically identical and morphologically similar **modules**. Such colonial animals as encrusting sponges, coelenterates, and bryozoans utilize this mode. Marine algae and angiosperms include many vegetatively reproducing species. The marsh grass *Spartina alterniflora* usually consists of a large number of plants connected by a rhizome system. It can be demonstrated by biochemical marker techniques that genetically identical plants, all of which have derived from one progenitor, may cover tens of square meters. A marsh meadow is thus typically a series of clones.

The value of vegetative spread varies with the biology of the individual group. In some corals and sponges, the overall form of the colony is important in the collection of food. In certain cases, the strength of the colony is helpful in resisting strong currents. Large colony size may also be an advantage. Larger colonies will have a greater ratio of living surface to periphery exposed to moving sediment. Therefore, they will be superior to small colonies in their ability to survive the movement of sediment along the bottom. For the same sort of reason, colonial forms may survive the attacks of predators, at least those attacking from the side.

Colonies of the colonial ascidian *Botryllus schlosseri* are usually founded by a single sexually produced larva, which attaches to a hard surface, metamorphoses (transforms into an immature adult), and then reproduces asexually. However, sometimes the larvae settle in aggregations. Using a unique genetic marker at an enzyme locus, Richard Grosberg and J. F. Quinn (1986) demonstrated that such aggregations consist of larvae that are genetically similar in

(a)

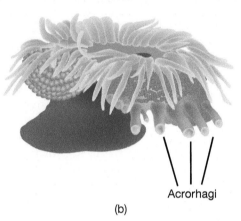

Acrorhagi

(b)

FIG. 7.8 The rocky-shore intertidal anemone *Anthopleura elegantissima* (a) often occurs as clones of a few hundred, but different clones have a bare zone between them (yellow line, upper left). (b) When an individual at the edge of one clone encounters members of another, they rear up and expose acrorhagi, which sting the individuals of the other clone.

tissue compatibility. These larvae can fuse and form larger initial colonies, which are more resistant to predation and, because the initiation of reproduction is size dependent, may reproduce sexually at an earlier age.

The phenomenon of immunological recognition of genetically related individuals seems to be widespread among groups that may benefit from group living. On eastern Pacific shores, the anemone *Anthopleura elegantissima* (**Figure 7.8**) lives in large clones of several hundred to several thousand individuals. All derive from a single colonizing larva that repeatedly divides asexually. At any time, it is usually possible to catch some individuals in the middle of fission. Lisbeth Francis (1973) found that contacts between individuals of different clones result in a stereotyped aggressive response. The affected anemones raise their tentacles and expose acrorhagi, which sting individuals from the other clone. No such aggression occurs between individuals within the same aggregation. This behavior can be interpreted as the defense of a communal territory, but it is not known what benefit is obtained from the defense.

Reproduction, Demography, and Life Cycles

Natural Selection on Reproduction and Life Cycle

■ **Age of first reproduction, reproductive effort (resources devoted to reproduction), and longevity may evolve in response to different age-specific patterns of mortality and predictability of reproductive success in the population.**

Not all marine organisms spawn for the first time at the same age, nor do all devote the same proportion of available resources to reproduction. Consider the five species of Pacific salmon. All spend 3 years or so at sea while feeding as adults. They then make the long return trip to their spawning grounds: the tributary from which they originally came. During the trip upriver, they change morphology significantly, and by the time they spawn, they have no ability to return to the sea. They are so weak after spawning that they lie on the spawning beds and soon die. By contrast, Atlantic salmon spawns but then may go through at least one more migration–spawning cycle.

Why should there be such a difference among species? Additional differences among species can be found in the age of first reproduction, proportion of energy devoted to reproduction, or **reproductive effort**, and whether the animal spawns only once (**semelparity**) or more than once (**iteroparity**). The theory explaining these differences is based partially on the premise that there is a cost of reproduction. We would expect evolution to maximize the total reproductive output over the lifetime of an organism. *If there is a cost to reproduction, it then matters how reserves are allocated to reproduction versus somatic growth. If more reserves are devoted to reproduction, there will be fewer that can be devoted to growth.* If mortality during reproduction is heavy, then any investment of resources into growth may be wasted because the organism will probably die before getting a chance to reproduce again. High adult mortality, therefore, selects for earlier reproduction. If adult mortality is low, it may pay to invest in additional growth to ensure repeated reproduction because larger adults can often produce more gametes and are often more experienced at winning mates than are smaller adults. Low adult mortality, therefore, selects for later age of first reproduction and for repeated reproduction.

Environmental uncertainty is also an important factor and can be related to the potential success of a new year class. If environmental uncertainty is high, a new group of juveniles may not survive. This imposes a selective force for repeated reproduction to ensure that some year class of juveniles will carry on the population. In effect, a female engaged in such repeated reproduction is hedging bets against a future bad year. **Table 7.1** compares age of first reproduction and reproductive span with variation in spawning success for five species. Fishes with relatively high spawning variability tend to have a longer reproductive span than those with low variability.

The age of first reproduction, the degree of investment in gonads, and even longevity are all known to be genetically

TABLE 7.1 The Relationship Between Variation in Spawning Success (recruitment of young in the best year divided by recruitment in the worst year) and Age of First Reproduction and Total Reproductive Span

POPULATION	RESULTS		
	AGE AT FIRST MATURITY (Years)	REPRODUCTIVE SPAN (Years)	VARIATION IN SPAWNING SUCCESS*
Herring (Atlantic-Scandinavian)	5–6	18	25
Herring (North Sea)	3–5	10	9
Pacific Sardine	2–3	10	10
Herring (Baltic)	2–3	4	3
Anchoveta (Peru)	1	2	2

*Variation in spawning success seems to cause natural selection for increased age of first reproduction and prolonged total reproductive life span.
Source: Data from Murphy, 1968.

variable. It therefore is likely that natural populations are continually subjected to varying pressures of natural selection that act on this variation to change the life history traits of marine species. Some correlations we see in the ocean fit our expectations, but we would like a more direct approach to verify whether such changes are acting to cause evolutionary changes.

■ Fishing causes natural selection on life histories.

Fishing can be seen as an unplanned experiment that might alter life histories of natural populations very rapidly. Commercial fishing pressure should strongly influence the distribution of life histories of the exploited fish species. After all, species with low reproductive effort and late age of first reproduction will be the first to be seriously affected by intensive fishing. By contrast, species with early reproduction and high reproductive effort would be best suited to withstand the onslaught. It also stands to reason that life histories may evolve as a response to fishing. Fishers have inadvertently performed an experiment for us. The spiny dogfish *Squalus acanthias* is fished heavily in Europe but until recent years was hardly fished at all off the American Atlantic coast. As a result, spiny dogfish populations were far denser in American waters. The European populations have an earlier age of first reproduction and higher reproductive effort in the first spawning season. Assuming that all other things are equal, this would indicate that fishing pressure has selected for reproductive tactics that ensure more reproduction. Recent increased fishing pressure on North American populations will probably yield the same effect. In Chapter 21, we discuss an experimental approach that demonstrates strong life history responses of fish populations to simulated fishing, which may slow recovery of the populations once fishing is relaxed.

Commercial fishers inadvertently performed an experiment that tested certain aspects of the theory explaining the size at which sex switching occurs. Species of the prawn genus *Pandalus* are fished widely in both the North Atlantic and Pacific Oceans. The prawns mature first as males but, after a variable period of time, switch to being females for the rest of their lives. A prawn fishery commenced in Danish and Swedish waters about 1930. Following the

rapid increase of fishing pressure, the average body size of the catch decreased and the threshold size of switching to females also decreased. This development can be interpreted as follows: Increased mortality (from fishing) will cause females to become relatively rare if males continue to switch sex only when they have reached a large body size. If a new genetic variant appears that switches at a smaller size, it will produce more offspring. Owing to natural selection for these variants, the size of sex switch will decline over time. If fishing pressure is very severe, it might "pay" for some individuals to mature first as females rather than go through a male stage. Such populations exist.

Migration

■ Fishes, crustaceans, turtles, and marine mammals often migrate between spawning and feeding grounds.

Many marine species have a reproduction and migration cycle as shown in **Figure 7.9**. Juveniles drift or swim from a spawning area to a nursery area, and then move to an adult feeding area as they grow older. Adults then migrate back to the spawning area. The cycle of spawning, drift, and active migration back again reflects a maximization of reproductive success by feeding and spawning in different optimal sites. The movement from nursery grounds to different adult feeding grounds may reflect competition between juveniles and adults for limited food resources, or it may result from differing food requirements of the different age classes. In many cases, fish spawn for more than one reproductive season, so there is repeated migration between the adult stock area and the spawning area. Such migratory patterns are strongly timed with seasonal changes that strongly affect temperature and food availability. Spawning must be timed precisely to allow development of larvae in water conditions with favorable temperature for development and with seasonally abundant food supplies for feeding larvae.

Marine turtles are a good example of the great difference in optimal breeding sites versus feeding sites. Females of all species lay and bury eggs in sandy beaches in the subtropics and tropics. The warm temperature within the sand ensures proper development, and the specific temperature even determines the sex of hatchlings. On the other hand,

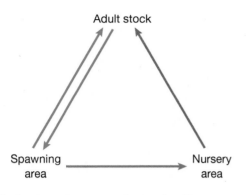

FIG. 7.9 A schematic pattern of migration. The two-way arrow between the spawning area and the adult stock area applies to cases of adults that migrate more than once to spawn. (After Harden Jones, 1968)

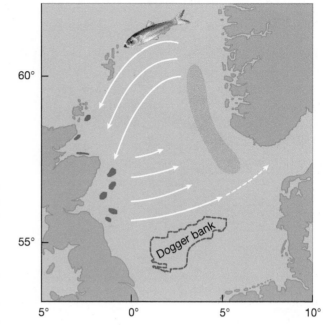

FIG. 7.10 The migration of the herring *Clupea harengus* in part of the North Sea between spawning grounds (dark ovals) and feeding grounds (lavender area). (After Harden Jones, 1968)

different turtle species migrate to feeding sites that depend on the food specialization of the turtle species. Green turtles usually feed on marine benthic plants and some attached bottom invertebrates and, therefore, migrate to sites where they are abundant. This may involve trips of hundreds or thousands of kilometers. Leatherback turtles feed on jellyfish. They, too, lay eggs on tropical beaches, and those that nest in the Caribbean migrate to cool high-latitude waters of Canada and Europe. Concentrations of leatherbacks can be found in association with continental shelf areas where jellyfish are usually abundant, which suggests that migration patterns and homing might be influenced by uneven prey distributions of jellyfish in the ocean (Houghton et al., 2006).

Migratory patterns involving estuaries are classified on the basis of the location of spawning and adult feeding grounds. **Diadromous** species divide their lives between residence within the freshwater parts of estuaries and in the open sea. **Anadromous fishes** (e.g., salmon, shad, and sea lamprey) spend most of their time in the sea but breed in fresh water. **Catadromous fishes** (e.g., eels of the genus *Anguilla*) usually spend their adult lives in fresh water or tidal creeks but move to the sea to breed. Many species (e.g., herring, cod, and plaice) feed *and* breed near the oceanic coast or in the open sea, although they migrate between different localities.

Migratory species vary in their degree of homing. Pacific salmon species are born in freshwater tributaries and migrate to the sea after a few months. They return 2–3 years later to the coastal area of their native river with the aid of detection of the earth's magnetic field and then return to their native tributary using their olfactory system, spawning in the very small rivulet where their parents spawned. The timing of the return is crucial to allow favorable conditions for spawning. By contrast, the early larval stages of the herring *Clupea harengus* (**Figure 7.10**) drift shoreward from the spawning grounds. As they grow, the herring move to deeper water and feed upon larger zooplankton. They return to a spawning ground after a year, but homing is not exact. The herring (like cod and other species) often

occur in distinct stocks or separated populations that maintain separate migratory routes.

The American eel *Anguilla rostrata* and the European eel *A. anguilla* undergo one of the world's most spectacular and enigmatic migrations (**Figure 7.11**). Both species migrate from freshwater rivers, ponds, and estuaries in eastern American or European waters to spawn in partially overlapping areas in the Sargasso Sea in the middle of the subtropical North Atlantic gyre. The trip to the spawning site is poorly known, but adults are believed to swim in deep water. After spawning and zygote formation, the larvae drift north and eastward in the Gulf Stream. American eels then move from a range of latitudes, westward across the continental shelf, toward shallow-water eastern American rivers and estuaries, whereas European eels continue to drift across the Atlantic and are carried by currents into shallow waters there. When the larvae reach an estuary, they metamorphose into the glass eel phase, which takes advantage of tidal behavior to move upstream by locating themselves up in the water during flood tide stage. The two species are distinctive in chromosome count, in various biochemical genetic markers, and even in the number of vertebrae (which is determined at birth). Other species of *Anguilla*, such as the Australian *A. reinhardti*, also undergo very long migrations that depend on currents to carry larvae back to adult habitats.

It is not clear at all why these eel species migrate so far to their spawning grounds. Could these particular areas be optimal for spawning? If so, how were they "discovered" at such a great distance? The movement of larvae must be passive, so the further passage of European eels must be

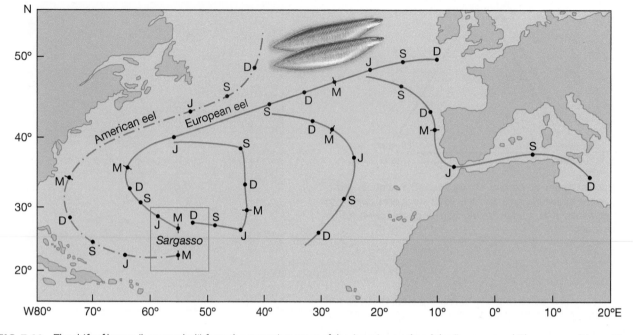

FIG. 7.11   The drift of larvae (leptocephali) from the spawning areas of the American eel and the European eel. The geographic positions of larvae hatched in March (M) in the two spawning areas are plotted at quarterly intervals: June (J), September (S), December (D), and March (M). (From Harden Jones, 1968)

determined genetically. A fascinating population has been discovered in Iceland, which appears to include an excess of eels that are intermediate in morphology and genetic markers. The Icelandic population includes hybrids of *A. rostrata* and *A. anguilla*, which hybridized thousands of kilometers away in the Sargasso Sea (Avise et al., 1990). A recent study employing single nucleotide polymorphisms from DNA sequencing that isolated species-specific alleles suggests that about 11 percent of Icelandic eels are hybrids, and some are even backcrosses between hybrids and either European or American eels (Pujolar et al., 2014). These hybrids appear to be migrating halfway across the Atlantic and winding up in Iceland.

One wonders whether the migration cycles of the North Atlantic species started millions of years ago, when the Atlantic was narrower. Then, the opening of the Atlantic might have isolated new species on either side. Alas, the data do not support this idea. Minegishi and colleagues (2005) used mitochondrial DNA sequences to estimate the time of divergence of eels, based on the notion that greater DNA distance between groups would imply greater time since evolutionary divergence. The estimate of divergence between *A. anguilla* and *A. rostrata* is 5.8 million years, which is much younger than the age of divergence of the Atlantic.

Green turtles (*Chelonia mydas*) are still something of an enigma to marine scientists. It is clear that they migrate hundreds to thousands of kilometers between feeding grounds and beaches where females lay eggs. Turtles that lay eggs at Tortuguero on the Atlantic coast of Costa Rica, for example, can be found at other times throughout the Caribbean, including Venezuela, Mexico, Cuba, and Puerto Rico. At least some tagged females return repeatedly to the same beach to lay eggs (**Figure 7.12**). It is unclear whether this behavior starts at birth, when hatchlings move to the sea, or whether females just follow others to good beaches to lay eggs. Right now, the evidence favors the hypothesis of imprinting at birth (Meylan et al., 1990). Female turtles from different beaches are genetically distinct, but turtles from the same beach are very similar. This can be shown by studies of the DNA in their mitochondria, which are inherited maternally. Populations from different beaches have probably been isolated from each other long enough to diverge genetically.

The mechanism of migration routes of sea turtles is a major question, even at first hatching, escape from the nest, and running to the sea. On small scales, sea turtles, when hatched, first respond to light and wave action in order to find the water. At larger scales, however, it appears that perception of the earth's magnetic field is a means of providing a map for sea turtles that may swim thousands of kilometers. For example, young loggerhead turtles *Caretta caretta* migrate from the waters of the southeastern United States clockwise around the North Atlantic and return back to American waters. Lohmann and Lohmann (1996) subjected turtle hatchlings to variation in magnetic fields and demonstrated that they can respond to variation in both the intensity and the direction of a magnetic field. Both of these factors vary with latitude, which means that sea turtles have the means to map their location and respond accordingly. For example, in the laboratory, loggerhead turtle hatchlings orient toward the east in a field magnetic intensity that corresponds to being along the southeastern United States coast. This would put them in the Gulf Stream. But when they are subjected to a magnetic field intensity corresponding to the eastern Atlantic, they orient

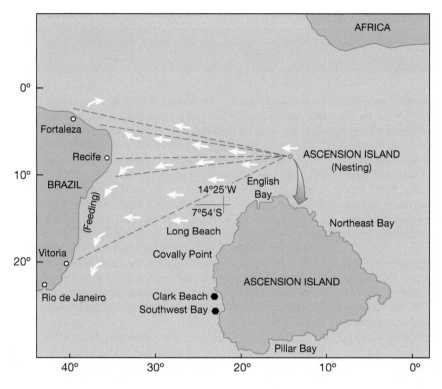

FIG. 7.12 The location of egg-laying and feeding grounds of the Ascension Island nesting populations of the migrating green turtle *Chelonia mydas*. Nesting sites are shown in tan at lower right on the map of Ascension Island, disproportionately enlarged in relation to the mainland. (After Koch et al., 1969)

toward the west, which in the Atlantic would keep them within the North Atlantic gyre and move them toward the tropics and eventually back to the tropical western Atlantic (Lohmann et al., 2001).

■ **Anadromous fish species are more common in high latitudes, and catadromous species dominate low latitudes. This pattern may be related to the availability of adult food.**

Fish migration is one of the most perplexing phenomena in marine biology. Some species breed in fresh water but migrate, sometimes thousands of miles, to feed in salt water. Others do the reverse. Why should some species move in one direction, whereas others take the opposite tack?

Migration should make sense as a way to increase the growth rate of a fish population. After all, if migration had costliness as its only characteristic, genetic variants that did not migrate would win out in the evolutionary race. And indeed, migration costs a lot. Such a journey increases the risks of starvation, predation, and simply becoming lost! It must be that migration produces increased benefits in terms of increased reproduction at a suitable spawning ground and increased growth in a good feeding area.

As it turns out, there is a systematic variation in the relative abundance of anadromous and catadromous fish species in different regions. Anadromous fishes are more

common in high latitudes, whereas catadromous fishes seem to dominate tropical habitats. Despite some glaring exceptions—the Atlantic eels are catadromous but live in temperate and boreal North Atlantic latitudes—overall the pattern holds well, at least for the Northern Hemisphere.

Zoologist Mart R. Gross and colleagues (1988) discovered an important key to success in migration. Food abundance is obviously important for fish survival and growth. As it turns out, high-latitude rivers and streams have lower overall productivity, and therefore less fish food, than the adjacent oceans. Thus, anadromous fish spend their time feeding in the ocean but come into freshwater bodies briefly to spawn. By contrast, tropical oceans are very low in productivity, but tropical rivers and streams are very productive. This suggests that the difference in migration is not really paradoxical. Fishes are feeding where the food is. An interesting experiment confirms the hypothesis. The Arctic char is anadromous, but experimental additions of food to fresh waters decreased the migrations back to the sea. McDowall (2008) has argued that the relationship holds only in the Northern Hemisphere, where anadromous fish diversity is greatest in the range of 50–60° N latitude. But there is no corresponding peak in the Southern Hemisphere. He suggests that after the continental glaciers retreated 8,000–10,000 years ago, a large number of previously anadromous species (e.g., salmon, shad) just reinvaded these northern latitude rivers as they reopened.

Sharks Find Their Way Home   **7.2**

Our discussion of migration demonstrates some striking examples of homing during migrations of fish, sea turtles, and other marine organisms. The great evolutionary biologist Ernst Mayr devised the term *philopatry* to describe species in which individuals show a propensity to return to the general area of their birth. But can organisms return again and again to the exact place where they were born? That is, do they have natal philopatry? This idea was first championed by the great sea turtle biologist Archie Carr, who identified green turtles by their specific blotches of color as they returned to what appeared to be natal beaches. But if they did this, could you dismiss the possibility that others moved to other, faraway beaches to lay eggs?

The presence of natal philopatry has a special significance for fishery management. This behavior creates a series of geographically closed subpopulations that respond to local conditions with little mixing of individuals from other sites. Intensive fishing in a bay might result in a local decline and perhaps local extinction of the species of interest. There might *not* be a constant replenishment of fish from other locations to take the place of the fish removed by fishers.

Shark biologist Robert E. Hueter (1998) argued that most coastal sharks exhibited natal philopatry. He felt the supporting evidence was weak but that the characteristics of shark brains reflected an intelligence that was up to the task. Some previous information using population genetic data supports the fidelity of sharks to local regions, but only recently have we seen strong support for very local reproductive site fidelity.

On a large scale, fish navigation is likely controlled by responses to the geomagnetic field. But how would sharks find their natal breeding ground on smaller geographic scales? To find out, Robert Hueter and colleagues (2015) tagged blacktip sharks, *Carcharhinus limbatus*, with acoustic transmitters in a bay in southwestern Florida. Then they displaced the sharks 8 km from their home area. The sharks returned within 34 hours and remained there. Hueter et al. experimentally blocked the olfaction of some sharks by inserting wads of cotton soaked with petroleum jelly into the nares. Such blocking reduced the number of sharks that returned to the natal breeding area (**Box Figure 7.3**). Other experiments using displacement show that Port Jackson sharks and lemon sharks returned to the site from which they were removed.

It is hard to get a long-term assessment of the extent of natal philopatry, but a study of lemon sharks over 2 decades provides strong support of breeding-site fidelity. Shark specialist Sam Gruber has studied lemon sharks for decades, following their behavior and physiological responses to environmental change in the region around Bimini,

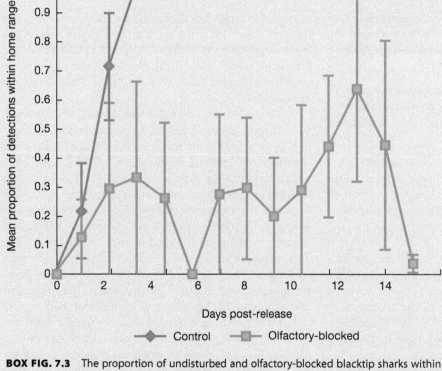

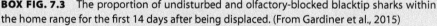

BOX FIG. 7.3 The proportion of undisturbed and olfactory-blocked blacktip sharks within the home range for the first 14 days after being displaced. (From Gardiner et al., 2015)

HOT TOPICS IN MARINE BIOLOGY

Sharks Find Their Way Home   *continued*   7.2

an island complex in the western Bahamas. Feldheim and colleagues (2014) caught sharks and used dart tags and PIT tags (**Box Figure 7.4**) to allow identification when the sharks were recaught. They also used DNA markers at 11 microsatellite loci. These DNA-based loci have the advantage of high polymorphism with strong evidence that there is no strong selection among alleles. The large number of loci and alleles makes it very improbable that any two sharks will have the same genotype (p of identity ~ 10^{-15}).

Their evidence is compelling (**Box Figure 7.5**). Some females returned to the same site repeatedly over a 20-year period. At least six females born in 1993–1997 returned to give birth 14–17 years later. Because it is difficult to spot the same sharks repeatedly, the numbers accumulated in this study are not large. A careful survey in 2008 recovered a number of sharks with either PIT tags of previously caught sharks at the site or identical microsatellite DNA genetic profiles. The data strongly support the apparent homing instinct for highly localized shark breeding sites.

BOX FIG. 7.4   A lemon shark being tagged in Bimini waters, with inset (lower right) showing insertion of tag. (Photography by Matthew Potenski)

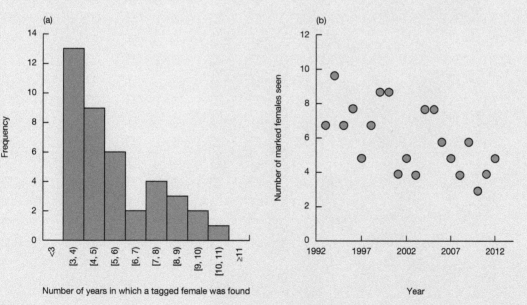

BOX FIG. 7.5   Repeated return of lemon sharks to North Island, Bimini. (a) Frequency of years in which a reproducing female was found to return to the island again and again; (b) number of originally tagged individuals that were observed at North Island over time. The decline could be due to mortality or movement of reproduction to another site (calculated from Feldheim et al. 2014)

Larval Dispersal: The Long and the Short Haul

Modes of Dispersal

■ **Marine invertebrate offspring may be (1) brooded or released as small adults; (2) dispersed usually short distances by means of relatively short-lived, yolk-dependent lecithotrophic larvae; or (3) dispersed great distances by means of longer-lived, plankton-feeding planktotrophic larvae.**

The hallmark of larval production and dispersal is a **complex life cycle**. An adult stage gives rise to a dispersing **larva**, which moves to a new site and completes the cycle by establishing itself and eventually growing to reproductive maturity. **Figure 7.13** shows an example of such a cycle. Although there is a continuum of dispersal distances, several qualitatively different types of release and spread of larvae result in modes of dispersal distances. **Direct release** of individuals next to adults is the shortest type of dispersal. Many species are **viviparous**; that is, they bear live embryos and then release juveniles as crawling miniature adults. Such young may crawl directly onto the mud or rocks. The Atlantic periwinkle *Littorina saxatilis* broods fertilized eggs in a modified oviduct, and fully shelled young snails are later released. Many Antarctic species of feather star (comatulid crinoids) develop their eggs in brood chambers located near the gonads in the arms at the pinnule bases.

Other species are **oviparous**, producers of young that hatch from egg cases. Several species of the drilling genus *Nucella* lay egg capsules that they attach to rocks. Embryos develop in the eggs, then hatch and crawl away as fully shelled juveniles. It is extremely common for meiofauna to have direct release, since any sort of current-driven dispersal will likely take them to inappropriate habitats. The small size of meiofauna also restricts them to releasing very few young, which puts a premium on maximizing survival.

Most invertebrates have larvae that swim for varying amounts of time before settlement and metamorphosis. **Lecithotrophic larvae (Figure 7.14)** are swimming larvae that depend on nourishment from the yolk provided in a relatively large egg; the larvae have no feeding or digestive structures. Capable of limited swimming, lecithotrophic larvae of most marine species spend a few hours to a day or so moving either along the bottom or up in the water column. The development time of this mode of larval development cannot permit dispersal for usually more than hundreds of meters. Lecithotrophic larval development occurs in species of many phyla, including mollusks, annelids, and ectoprocts. An interesting exception is echinoderms, whose high-latitude species commonly have lecithotrophic larvae that remain in the plankton for weeks, which is unusual for this type of dispersal.

Planktotrophic larvae feed while they are in the plankton. They usually have specialized larval feeding structures and digestive systems. They feed on planktonic bacteria, algae, and smaller zooplankton and usually drift for one to several weeks

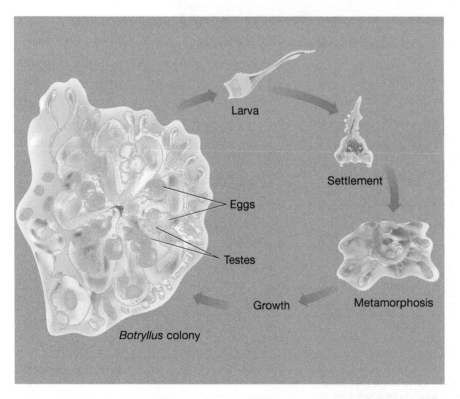

FIG. 7.13 Example of a complex life cycle of a marine invertebrate with a dispersing phase. The colonial sea squirt *Botryllus schlosseri* is a colonial hermaphroditic species. Sperm move through the water from one colony and fertilize a nearby neighbor. Larvae emerge and move only a few meters to settle, metamorphose, and establish a new colony.

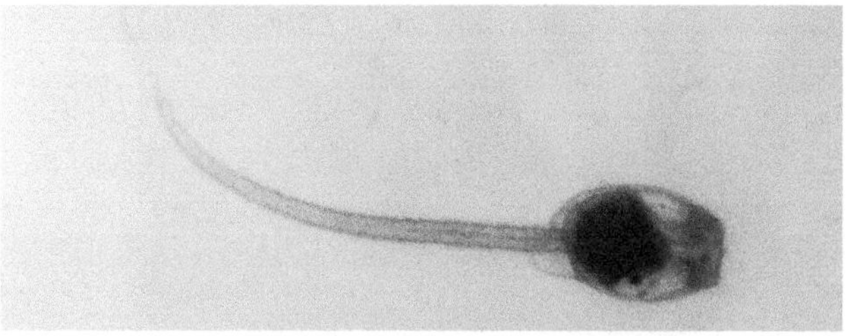

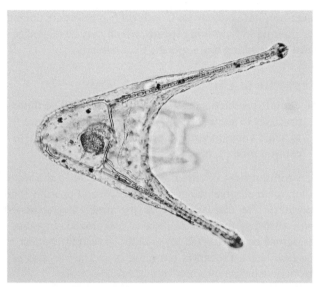

FIG. 7.14 An example of a lecithotrophic larva: a "tadpole" larva of the colonial ascidian *Botryllus schlosseri*. This larva lives for only a few hours, is a poor swimmer, and moves only a few meters from the parent colony. (Courtesy of Richard Grosberg)

in the water column (**Figures 7.15** and **7.16**). Because of the time they spend in open waters, they can disperse over long distances. Development often proceeds through a number of planktonic stages, when the larva is competent to settle and metamorphose on the bottom. If a suitable adult substratum is not found, the larva may be capable of delaying metamorphosis. Metamorphosis often involves a major reorganization of morphology from the larval to the adult form.

While closely related species with planktonic larvae may have either planktotrophic or lecithotrophic development, it is quite rare for a single species to have the ability to produce both larval types. Species with more than one larval developmental mode are called **poecilogonic** and are found in a few groups, including polychaete annelids, sea slugs, and a caenogastropod. For example, Lisa Levin (1984) found that the small mud-dwelling polychaete *Streblospio benedicti* is widespread on North American mudflats and has either planktotrophic or lecithotrophic larvae, depending on the local population. A southern Californian species of the saccoglossan sea slug *Alderia* also shows poecilogony and tends to produce lecithotrophic larvae in summer and planktotrophic larvae in winter (Ellingson and Krug, 2006). The seasonal switching may be due to the presence of planktonic food, but the explanation is still open.

Lecithotrophic larvae are often covered with **ciliary bands**, which are used for locomotion. However, planktotrophic larvae use ciliary bands or tufts to move *and* to feed upon the planktonic food (**Figure 7.17**). The small size

FIG. 7.17 Planktotrophic larva of the red sea urchin *Strongylocentrotus franciscanus*. Dark pigment spots are concentrated near the ciliary bands used for feeding and swimming. (Courtesy of Richard Strathmann)

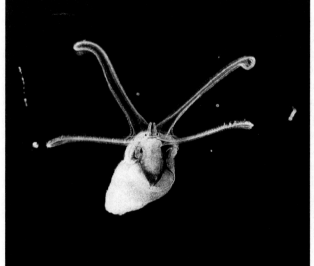

FIG. 7.15 Planktotrophic pluteus larva of the sea urchin *Lytechinus* sp., Florida. (Photograph by Will Jaeckle)

FIG. 7.16 Planktotrophic larva of the hairy triton *Cymatium parthonopetum*, a teleplanic larva found in the tropical Atlantic. (Photograph by Rudolph Scheltema)

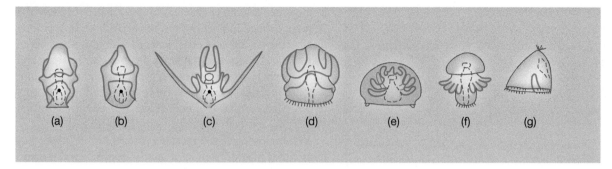

FIG. 7.18 Many different groups have evolved ciliary bands (blue) for feeding on plankton: (a) sea star, (b) sea cucumber, (c) brittle star, (d) hemichordate, (e) inarticulate brachiopod, (f) phoronid worm, and (g) bryozoan. (Courtesy of Richard Strathmann)

of the cilia and the low velocities of ciliary movement cause planktonic larval feeding to operate at low Reynolds number. This means (see Chapter 6) that viscosity dominates inertial forces in fluid behavior. The cilia must beat to propel water and food particles. Although there is a great deal of variation in morphology, a surprising number of taxonomic groups of organisms have ciliated bands (**Figure 7.18**). The cilia beat and create a current across the band. To capture a particle, it appears that a cilium suddenly reverses its rowing motion toward downstream. The reverse beat captures the particle and traps it upstream of the ciliary band. The ciliated bands are also generally used to propel the larva through the water, which exposes the larva to new micropatches of food. It is likely that although the same cilia serve for propulsion and feeding, the ciliary action for the two functions involves tradeoffs. Capture of algae involves adding small vortices by ciliary reversal. But these vortices are added at the expense of larger propulsion of the larvae. Ciliary propulsion in sea star larvae involves the production of large vortices around the swimming larvae, but feeding may involve a different regime of ciliary control (Gilpin et al. 2016).

The larva of the blue mussel *Mytilus edulis*, which lives on intertidal and shallow subtidal rocks and cobbles, is a good example of a typical planktotrophic larva. Sexes are separate, and the animals release eggs and sperm into the water. Within 9 hours after fertilization, the larvae are completely ciliated and are strong swimmers. Within 5–7 days, fully developed feeding veliger larvae develop. Normally, larval life may be 4–5 weeks. By contrast, the Atlantic mud snail *Tritia (formerly Ilyanassa) obsoleta* produces fertilized eggs laid in cases that are attached to rocks. Larvae emerge from the cases and swim away. Planktotrophic larvae may also be released directly from the mother, as in the acorn barnacles and many species of crab.

Larvae of bony fishes are obviously different from the foregoing descriptions of invertebrate larvae. Most marine fish species spawn buoyant eggs and sperm into the water. Hatched embryos first develop into larvae with a prominent yolk sac. Once the sac is absorbed, the larvae usually feed on zooplankton. Fish larvae remain in the upper water column from 2 weeks to as much as several months (eels). They then metamorphose into adults, with adult organs. The degree of morphological change during metamorphosis varies from very little to a major reorganization, as when the American and European eel leptocephalus larvae

metamorphose into elvers. Coral reef fish larvae are notable for their rapid swimming speeds, which may be related to a need for reaching more specific final habitats.

Reproductive Timing and Egg Size

■ **Gamete production and larval life must often be timed precisely to allow settlement and promote dispersal, to avoid being swept to inappropriate habitats, and to counter predation.**

The commitment of either gametes or larvae to the water column entails great dangers of extensive and possibly total mortality. We discussed earlier the problems associated with ensuring that fertilization is maximized. In estuaries, larvae in the surface low-salinity layer may be lost to sea (see later). Predation on gametes and larvae also can strongly depress reproductive success.

These dangers require precise timing in many cases to avoid gamete and larval death. For example, animals living or breeding in the highest part of the intertidal zone must have precise larval release schedules because the animals may be inundated only once every 2 weeks during spring high tides. The Atlantic marsh snail *Melampus bidentatus* lives in the highest part of the tide zone and breathes air, but it lays eggs that develop into swimming larvae. These larvae are launched at a spring high tide and return to settle during the next spring high tide. The California grunion *Leuresthes tenuis* comes onto sand flats to spawn at spring high tide at night. A female rides the swash onto the beach accompanied by several males. She twists her body, digs a pit in the sand, and lays eggs at the highest level of the tide, while males curl around the female and deposit sperm, which fertilizes the eggs as they are laid. Males and females are washed down the beach slope to the water and swim away. For about 2 weeks, the high tides fail to reach the level of the beach where the eggs were buried, but the next spring high tide washes out the embryos, which detect the cue provided by wetting and hatch into swimming larvae.

Tidal environments are most challenging because gametes and larvae may be trapped in the surface layer during the day, when visual predators are active. Therefore, one might expect strong selection for larvae to be moving in the surface waters at night. This has been found in the coral reef-flat damselfish *Pomacentrus flavicauda*, which lays eggs at dawn; its embryos hatch as larvae on the fourth sunset. Egg production is timed so that hatching occurs at the time of the month when sunset coincides with ebb tide, which at once lowers predation and sweeps the larvae offshore (Doherty, 1983).

Many intertidal crab species are exquisitely timed to release larvae at high tide. This release must be coordinated with copulation and embryo development. Larval release itself must be coordinated with tidal amplitude, tidal cycles, and light–dark cycles. Ideally, larvae should be released during highest tide and at night, to maximize larval release but also to be in the dark to avoid predators. Species are often capable of releasing larvae in different rhythms, depending on variation in tidal cycles. As mentioned in Chapter 2, tidal cycles can vary considerably in different locations. For a number of crab species, Steven Morgan and John Christy (1994) found that crabs were able to adjust to radically different tidal cycles on the Pacific and Caribbean sides of Panama. In nearly all cases, females released larvae at the time that high tide and dusk plus 1 hour coincided. It is of interest that there are some tidal regimes where this is not possible, as within an estuary. Fiddler crabs release larvae in time with the tides, but the larvae also have an endogenous rhythm that allows them to rise up in the water column in an estuary when the tide is going out. They drop to the bottom when the tide rises. This allows them to efficiently move out to sea and not get swept up-estuary when the tide is rising (Lopez-Duarte and Tankersley, 2007).

■ **Variation in egg size is considerable among marine species, and this may relate to different consequences in terms of mortality.**

Although variation in age of reproduction is considerable, equally noticeable is the great variation in egg size in marine animals, sometimes between closely related species. One expects larger eggs to have more yolk nutrients, but this is only broadly true, and there are exceptions. In large measure the variation remains unexplained, but there are some general patterns. As mentioned earlier, lecithotrophic larvae develop from eggs that are much larger than planktotrophic eggs. Eggs of directly released juveniles are largest of all. As a general rule, planktotrophic eggs produced per female are also much more numerous than lecithotrophic eggs, which are, in turn, more numerous than the eggs of directly releasing species. This suggests a strategy that, in turn, may determine egg size. Planktotrophic larvae swim in the water far longer and probably have much higher mortality rates than do lecithotrophic larvae, which are in the water for only a few hours. In turn, directly releasing species deposit their young in appropriate habitats (namely, those of the adults), thus further minimizing mortality. It seems likely then that the female, by producing a large number of planktotrophic eggs, "hedges her bet" because far fewer young will survive relative to lecithotrophic eggs. Some have suggested that the strategy of producing an intermediate-size egg with intermediate numbers may offer the worst of both worlds, producing high mortality with relatively low numbers of eggs and young. Thus, we expect to observe either the low-number/quality-care approach or the high-number/leave-them-alone approach.

The interested student will find that the subject of egg size is quite controversial (Vance, 1973; Levitan, 2000).

The Microscale: Larval Settlement, Metamorphosis, and Early Juvenile Survival

■ **Planktonic larval life is a transition from presettlement morphology and behavior to competence for settlement and metamorphosis.**

Planktonic larvae develop from the zygote into a free-living organism that develops through a series of stages adapted to remain in the water column to stages involved in settlement and metamorphosis. The stage at which a larva can touch the seabed and metamorphose into a benthic adult is the time of **competence**. A competent larva faces the important choice of settling and metamorphosing on the best adult substratum possible. But what if an acorn barnacle larva encounters only mud? In some species, delay of metamorphosis is possible, but in others the larva is programmed to settle and metamorphose at a specific time, with no scope for delay. Morphology and behavior may change during the time of planktonic larval life. For example, planktotrophic larvae of the eastern oyster *Crassostrea virginica* spend about 2 weeks in the upper water column, but move more slowly and spend time in bottom waters in the third week as the larva approaches metamorphosis. During the third week the larva develops a foot (pediveliger stage), which can probe the bottom for suitable habitat to metamorphose. In the third week the larva is competent to settle and will be especially stimulated to settle and metamorphose on the shells of adult oysters.

We have an incomplete understanding of the internal signaling that leads from larvae remaining as feeding individuals in the water column to finally transitioning to a stage of competence to settle and metamorphose. In a wide variety of distantly related phyla, high activity of the **nitric oxide (NO) pathway** interacting with **heat-shock protein 90 (HSP90)** contribute to forestall the tendency to metamorphose. NO production relies on the enzyme nitrogen oxide synthetase (NOS) and serves as a signal to regulate larval development. As larval life proceeds, the activity of NO declines. Signals that suppress NO or HSP90 lead to metamorphosis. Details have been worked out in a few species but external environmental signals must interact with the NOS-HSP90 complex and other pathways to lead to larval competency and settlement and metamorphosis (see, e.g., Castellano et al., 2014, but also see Ueda and Degnan, 2014).

■ **Planktonic larvae use light and pressure cues to maintain depth for optimal feeding and location of settling sites.**

Planktonic larvae reach a stage at which they must locate the adult habitat, where they will settle and attach before metamorphosis. At the stage of competence, larvae of some species can delay metamorphosis for some time, but other species appear to have a biological clock for settlement within a specified time and will die otherwise.

Although the vast majority of larvae die before reaching a potential adult habitat, the road ahead is still dangerous for the survivors. Natural selection and evolution favor any larval features that enhance location and settlement on the proper substratum. **Figure 7.19** provides a general outline of the stages of larval selection.

In the plankton, larvae pass through one or more stages of photopositive and photonegative behavior. These permit the larvae to remain near the sea surface to feed and then to drop to the bottom and settle on the proper habitat. Larvae of intertidal animals are photopositive during their entire larval life so that they may feed in the surface waters and settle in the intertidal zone. Richard Grosberg (1987) examined the water column distribution of intertidal barnacle species in California and reported that the highest intertidal species was to be found at the surface during larval life; species living in the mid-intertidal, however, were found a meter or two below the surface. Intertidal larvae are apparently capable of subtle adaptations to maintain the proper tidal height at the time of settlement. Subtidal species, by contrast, are first positively phototactic but then switch to negative phototaxis just before the time of settling. Larvae are also capable of responding to pressure differences that might enable them to select a specific depth during the period of feeding, dispersal, and settlement. Pressure is a good indicator of water depth.

Planktonic larvae must maintain their proper depth in the water column during the dispersal phase either to feed on phytoplankton that are abundant near the surface or simply to avoid sinking before development is completed. Most larvae have slightly negative buoyancy, but simple turbulence retards sinking. Larvae can be simple and ciliated, but planktotrophic larvae usually have a swimming organ (e.g., the velum of gastropod larvae, which is lost at the time of metamorphosis). Young postlarval mussels (e.g., *Mytilus edulis*) secrete monofilamental byssal threads more than a hundred times longer than the larval shell. The filaments exert viscous drag force sufficient to reduce greatly the sinking rate, thereby enhancing dispersal.

Even if turbulent mixing prohibits the sort of vertical stratification observed by Grosberg in larvae, there is good evidence that larval selectivity strongly influences where settlement occurs. Jenkins (2005) studied settlement of planktotrophic larvae of two species of the barnacle genus *Chthamalus* in waters of the United Kingdom. While larvae of both species were thoroughly mixed in the water column, settlement patterns of each of the two species were quite specific and corresponded to the different environmental occurrences of *Chthamalus* species, one of whose adults were found in exposed habitats and the other in more protected habitats. Larval supply may, therefore, be important in total abundance of settlers but not in determining the microhabitat where they wind up. Larval behavior is clearly important in habitat selection at the microscale.

Amazingly enough, deep-sea hot-vent environments (see Chapter 18) have species with planktotrophic larvae. Hot vents are scattered throughout the deep-sea floor and are ephemeral because the hot nutrient-rich sources give out after a time. Dispersal might thus be expected. The protoconch (larval shell) size of hot-vent bivalve mollusks is of the typical small size for planktotrophic larvae, and isotope data from the larval shell (isotope ratios of oxygen in the shell can be used as a thermometer) show that the larvae spend a time in water that is warm, like surface oceanic water and unlike the deep sea. Vent bivalve larvae must make the seemingly impossible trip to the surface waters, and then some manage to locate an incredibly rare appropriate deep-bottom environment. By contrast, archaeogastropods (a group of snails) near the vents have large yolky eggs and probably have larvae with quite limited dispersal. It may be that these eggs hatch as swimming but nonfeeding larvae, with strong dispersal abilities, relative to their nonvent relatives.

■ At the scale of meters to centimeters, larvae use chemical and biological cues for a final settlement site.

After touching the bottom, larvae must be able to determine whether the substratum is suitable for adult existence. Larvae use chemical cues and mechanical cues to detect suitable settling sites. Almost all larvae prefer surfaces coated with bacteria. Sterilized sand or rock usually inhibits larval settling. Larvae of some sand-dwelling polychaetes can select sand grains of the appropriate size, whereas rock-dwelling barnacles have larvae that move preferentially to surfaces with pits and grooves, which provide a secure attachment against wave action, predators, and competing barnacles that might overgrow or undercut them. Although barnacles are often found in cracks, it is often not clear whether this is the result of preferential settlement or differential survival after settlement (**Figure 7.20**).

An important physical cue is the local hydrodynamic environment. Many bivalves, for example, cannot suspension feed efficiently in high flow, but zero flow is also bad because no current means no phytoplankton delivery. The final settling stage (cypris) larvae of the barnacle *Balanus improvisus* encounter substrates and actively evaluate the current speeds at different microsites. If water movement exceeds 5–10 cm s⁻¹, the larvae actively reject the site and move on to sites where currents are less. This selection

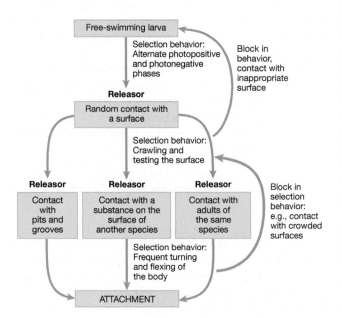

FIG. 7.19 Stages in the selection of a suitable substratum by planktonic larvae. (Modified after Newell, 1979)

FIG. 7.20 Preferential recruitment of barnacles into cracks on a rocky shore near Nahant, Massachusetts. It is not clear whether this pattern results from preferential larval settlement, from differential postsettlement mortality, or from a combination of these processes.

relates directly to the poor efficiency at which the adult barnacles will be able to feed at the higher current speeds (Larsson and Jonsson, 2006).

In **gregarious settling**, larvae settle on adults of their own species. This usually requires direct contact of larvae with adults because the chemical cue is a relatively insoluble molecule, such as arthropodin in barnacles. Zimmer-Faust and Tamburri (1994) discovered that the cue for settling of oyster larvae is a soluble peptide (which is a string of amino acids). Such a waterborne settling cue exists in effective concentrations only within a few millimeters of the bottom, but larvae sinking within the bottom boundary layer can detect this cue and settle. The greatest concentration of such a settling cue would likely emanate from the water coming from adult oysters, which would concentrate settling of oyster larvae on adults of their own species. Larvae are often entrained in currents too strong to swim against, but they frequently drop down toward the substratum and make arclike swimming patterns. On encountering the cue, they can slow down, settle, and metamorphose. Field experiments show that oyster larvae actively settle on artificial substrates that "leak" such peptides into the water (Browne and Zimmer, 2001).

Gregarious settling allows a larva to settle in a site where adults have already settled and survived. There is the disadvantage, however, of being eaten by adults of the same species. There also is the problem of how, in a gregarious species, a "pioneer" can ever manage to colonize a new site. Toonan and Pawlik (1994) discovered that the colonial tube-dwelling polychaete worm *Hydroides dianthus* has two distinct types of planktonic larva: One seeks adults of its own species, and the other is specifically a "pioneer," a larva that seeks new, bare, hard substratum. The cue for gregarious settlement is a waterborne compound emanating from benthic adults (Toonan and Pawlik, 1996). The proportion of larvae in a given spawning that adopts one of the two modes is a highly inheritable trait, which suggests that the two-strategy approach is under strong natural selection (Toonan and Pawlik, 2001).

Planktonic larvae may also use the chemical characteristics of other benthic species as cues. For example, adults of the hydroid family Proboscidactylidae live on the tubes of

members of the polychaete annelid family Sabellidae. At settling time, the larvae can detect and settle on the tentacles of the worm. After settling, they move down the tentacles and live as adults on the tube. The mechanism probably involves chemical attraction. Some bryozoan larvae are attracted to seaweeds, and organic substances can be extracted from some seaweeds that will attract larvae onto another nonliving surface. In other species, mechanical attraction is involved. Planktotrophic larvae of the mussel *Mytilus edulis* are attracted to filamentous red algae, but the larvae will also settle on fibrous rope that has approximately the same texture. This attraction is used to great advantage in mussel mariculture.

After settling, larvae may move a short distance, no more than a few centimeters, to a better site. This is very important for larvae of sessile species, whose movements at this point commit them to life at one location. Newly settled barnacle larvae can move to a small degree, locate optimal microsites, and space themselves away from other barnacles. The tubeworm *Spirorbis borealis* settles randomly on the seaweed *Fucus serratus*, but individuals then crawl and space themselves evenly. Avoidance of crowding reduces overgrowth and competition for food. In barnacles, among which settlement is often very dense, the reduction of crowding that follows a period of mortality after settlement greatly enhances the subsequent expectation of life.

When they reach the substratum, marine planktonic larvae encounter a complex physical and chemical environment. As we have discussed, larvae may select settling sites based upon chemical cues. However, there are probably a number of chemical cues that may lead to avoidance of a local site. Unfortunately, we know very little about these cues, even though they may be a major means of local site selection. Sara Ann Woodin and colleagues (1993) demonstrated that noxious chemicals released by benthic animals may reside in the sediment pore waters and can inhibit settlement by larvae. In soft-sediment intertidal flats of Washington, the infaunal terebellid polychaete *Thelepus crispus* is locally abundant and harbors toxic bromine-containing aromatic compounds. These compounds are found in the sediment pore waters adjacent to the worm. Such sediments are actively rejected by settling larvae of the nereid polychaete *Nereis vexillosa*, a common resident of sandy shores in the area. This result suggests that negative chemical interactions may also be important in determining the settling and metamorphosis of planktonic larvae.

Finally, a larva commits itself and metamorphoses. (Recall that metamorphosis is the process whereby a larva changes dramatically into the adult form.) In the case of barnacles, for example, the football-shaped cypris develops adult appendages, and a basal plate is laid down. The energetic cost of metamorphosis often is so severe that the animal must feed immediately afterward. Some sea star species die if they fail to find a prey item within a couple of days after metamorphosis. If larvae of the plaice *Pleuronectes platessa* are starved of zooplankton for more than 8 days, a point of no return is reached, and they will not have the energy to feed subsequently. Older larvae can survive at least 25 days before the point of no return is reached. It may be that early larval development in these fish is also very costly.

The Mesoscale: Transport by Currents, Loss of Larvae, and Retention Mechanisms

■ **Planktonic dispersal success of marine species is strongly controlled by currents that transport larvae and may have important ecological consequences.**

Planktonic dispersal does not guarantee the ability to survive a long trip across the ocean or to colonize new habitats successfully. The overwhelming majority of planktonic larvae die in the water column from predation or starvation, or are swept to inappropriate habitats. Larvae of rocky-shore invertebrates are released in currents and may easily be swept to inappropriate habitats. Along the southern New England coast, a longshore current would tend to sweep larvae from the rocky coasts of Rhode Island to the sandy south shore of Long Island, New York, and most larvae of the rocky-shore species would be doomed. Species with nonplanktonic larvae dominate the Galapagos Islands, off Ecuador, where currents carry surface waters offshore for a large portion of the year. As a result, species with planktonic larvae are relatively rare in the Galapagos.

Most recruits, even of planktonic larvae, probably come from nearby. **Figure 7.21** shows the general possibilities of larval transport. Depending on the local current regime, larvae might be swept offshore, or they might even be brought onshore by wave trains, internal waves that move material toward shore. Longshore drift is common, and larvae usually are moved parallel to coastlines to some degree. This may permit rapid extension of a species' biogeographic range. Finally, inhomogeneities in the coastline often create eddies, which may concentrate planktonic larvae.

Although average dispersal distance per individual planktotrophic larva is probably on the scale of 10^1–10^2 km, the potential exists for longer-distance dispersal in a significant fraction of the planktotrophic larval cohort, and this potential must increase with the planktonic larval life span and with the favorable nature of ecological opportunities at the distant site. Rudolph Scheltema (1989) discovered that many planktonic larvae of coastal invertebrate species live in the open sea (**Figures 7.16** and **7.22**), particularly in major transoceanic surface currents. He termed such larvae **teleplanic larvae** and found transoceanic similarities in species whose larvae were common in the open ocean. Another particularly interesting case is that of the larvae (**Figure 7.23**) of the coral *Pocillopora damicornis*, which is widespread and dominant throughout Pacific coral reefs. Robert Richmond (1987) has shown that these larvae can live in the plankton and are competent to settle and metamorphose for periods greater than 100 days. This may explain the coral's broad geographic distribution. The larva's symbiotic zooxanthellae (see Chapter 17) photosynthesize and provide food, which fuels the larva's journey over long distances across the open sea.

Most larval exchange probably occurs along coasts, and strong coastal currents parallel to shore may increase the interchange of populations from various locations along the coast. The degree of interchange is known as **connectivity**. Connectivity can be strong between populations of a species if larvae typically move great distances along coastlines and are periodically reversed in direction. Alternatively, as we have mentioned, some circulation patterns cause larvae to be retained within local areas, causing the local population to be self-seeding. Patterns of connectivity can vary from region to region, and even from season to season, since strong coastal currents might move larvae in one season, but not in another (Becker et al., 2007).

If some ecological opening exists, arriving larvae may flourish, reproduce, and continue to extend the species' geographic range along a coastline. Such invasions have happened several times during the last century, and the sudden expansions may have been aided by commercial shipping. In the mid-nineteenth century, the shore periwinkle *Littorina littorea* was first noticed on the shores of Nova Scotia and spread southward. It had arrived at Cape Cod, Massachusetts, by the turn of the century and has now reached the Middle Atlantic states. It is unlikely that the species will spread much farther southward, given its thermal and geographic ranges in Europe. Many invasive species also have spread rapidly along a new coastline, once having arrived (see Chapter 20).

■ **Despite the potential for dispersal, planktonic larvae often come to settle quite near their origin, owing to cyclonic or returning currents.**

Given the long lives of many planktotrophic larvae, one expects that dispersal should occur over great distances. A larva

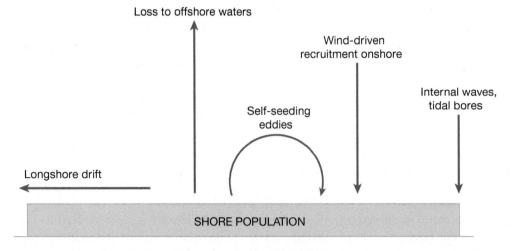

FIG. 7.21   General types of movement of larvae of benthic invertebrates with respect to the coastline.

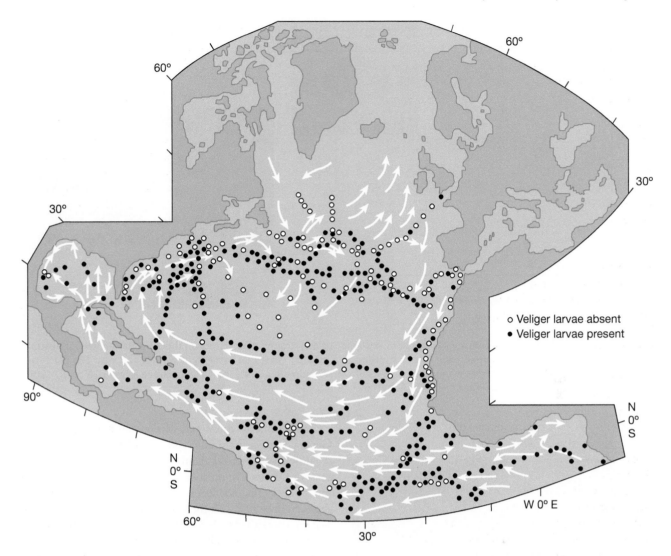

FIG. 7.22 Geographic distribution of teleplanic gastropod veliger planktonic larvae, based on samplings taken in the tropical and North Atlantic oceans. (After Scheltema, 1971)

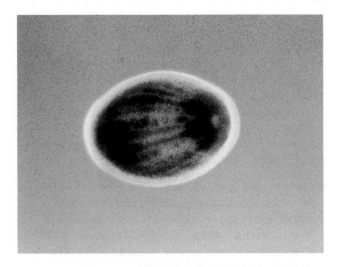

FIG. 7.23 Planula larva of the coral *Pocillopora damicornis*. Note the rows of zooxanthellae. (Photograph by Robert Richmond)

with a 30-day life span and a steady and straight current of 2 cm s⁻¹ should disperse over 50 km. This would suggest extensive mixing of marine species with planktotrophic larvae. The assumption of straight and steady currents is the problem

with such a prediction. Not all currents move steadily or in one direction. Shapes of basins and large obstructions to flow, such as islands, often create eddies that cause larvae to travel in cyclonic flow patterns, sometimes to their parent populations.

Paul Sammarco monitored larval settling in the vicinity of an isolated patch reef in the Great Barrier Reef of Australia and demonstrated surprisingly reduced dispersal of planktonic coral larvae (Sammarco, 1988). Settling plates were placed at various distances from the patch reef, and a detailed study of local water currents allowed a prediction of sites where larvae might be concentrated. Settling occurred mainly in the close vicinity of the patch reef (**Figure 7.24a**), suggesting that most larvae come from within the patch system itself. Settlement was greatest in a vortex of currents, where water was trapped and recirculated for awhile before being swept to sea.

A similar story emerges from the dispersal and current patterns associated with a population of the European lobster *Nephrops norvegicus* in the western Irish Sea. Alan Hill and colleagues (1996) discovered a cyclonic flow in spring and summer. This flow lies above a large patch of mud, which has the main population of the lobster in this region. In spring, newly hatched larvae spend approximately 50 days in the water column, and successful recruitment

FIG. 7.24 (a) Settlement of planktonic coral larvae onto settlement plates around a patch reef on the Great Barrier Reef of Australia, as estimated by counts of newly settled coral colonies (distance in meters). Note that settlement was highest near the patch reef, indicating that most settlers came from within the patch itself. (After Sammarco, 1991) (b) A coral planula larva. (c) A newly settled coral, only a few millimeters across. (Photographs by Robert Richmond)

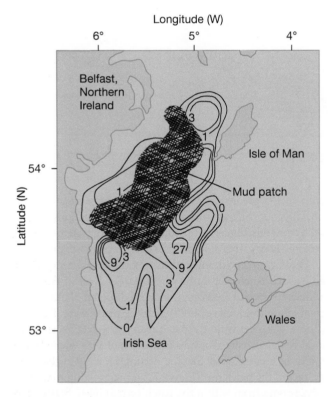

FIG. 7.25 The distribution of stage III larvae of the lobster *Nephrops norvegicus* in the western Irish Sea. Larvae are retained in this area by a cyclonic flow pattern and thus find their way back to a large patch of mud (hatched area) suitable for adult survival. (After Hill et al., 1996)

depends upon finding the mud patch. Larval distributions show a strong correspondence to a stratified water body enclosed by the small gyre system. This ensures that the larvae will settle in the vicinity of the mud patch (**Figure 7.25**).

Currents may cause net movement toward the shore. General eastward currents and winds tend to keep eastern Atlantic larvae pressed against the shorelines of the Atlantic coast of Europe. Off most open-ocean coasts, one can observe surface slicks that contain jetsam that seems to move toward shore. Surface floats off the California coast are often carried 1–2 km shoreward in 2–3 hours. Planktonic invertebrate and fish larvae are 6–40 times more concentrated in these slicks than in the surface waters between the slicks. The slicks are formed by the interaction with the sea surface of tidally driven internal waves (waves within the water column). Planktonic larvae may therefore be "trapped" against the shoreline. In recent years, episodic recruitment has been discovered to be the rule for settling megalopa larvae of the blue crab *Callinectes sapidus*, both on the east and Gulf coasts of North America. In some cases, these recruitment events last only a day and may be simultaneous over several hundred kilometers of coastline. Recruitment in the mid-Atlantic states region appears to be unrelated to internal waves but is correlated with onshore transport caused by wind stress and by spring flood tides, although other factors may also be important.

■ **Many coastal species have evolved mechanisms for retention of larvae near suitable adult sites, take advantage of current systems to return to suitable sites, or even benefit from seasonal reversals of flow that allow them to move out to sea and then return to suitable coastal adult habitats.**

Many coastal species have apparently taken advantage of persistent currents, tidal reversals, and even seasonal changes to reduce loss of planktonic larvae to unsuitable sites. Many estuarine species spawn within an estuary, but larvae and juveniles may spend some period of time in coastal waters

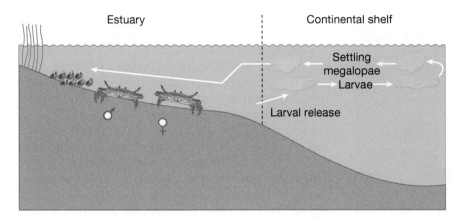

FIG. 7.26 Mating and egg hatching of many invertebrates and fishes, such as the blue crab *Callinectes sapidus*, occur within the estuary. Larvae and juveniles, however, may move into coastal continental shelf waters and later return to the estuary to spend their adult lives. (Courtesy of Steven Morgan)

(**Figure 7.26**), perhaps to avoid predation within the estuaries, which can be intense because of large predatory fish populations. Because they are entirely dependent on currents for transport, they must take advantage of the various wind- and tide-driven transport sources discussed earlier to return to the estuary from adjacent coastal waters.

As was discussed in Chapter 2, surface water flows seaward in moderately stratified estuaries, whereas the more saline bottom layer moves landward. If estuarine planktonic larvae held their position near the surface, they would be carried out to sea. This is a strategy employed by a number of species (e.g., species of fiddler crabs of the genus *Uca*, the blue crab *Callinectes sapidus*), which brings their larvae out into continental shelf waters. They may return to the estuaries after a larval life in coastal waters, and larval recruitment depends very strongly on often quite irregular onshore winds. But many other estuarine species avoid the seaward surface flow. Larvae of the eastern oyster (*Crassostrea virginica*) studied in the James River estuary, Virginia, differ in depth distribution from inert particles and are found in surface waters more frequently at the time of flood tide, which would tend to move them upstream. The larvae use the halocline as a cue for vertical motion, rising with the flood and sinking at the time of ebbing tide. Larvae of the mud crab *Rhithropanopeus harrisii* move upward on the rising tide and downward on the ebbing tide and stay near the bottom, which gradually concentrates the larvae farther up the estuary.

The evidence suggests therefore that there are two alternative strategies for planktonic larval habits of estuarine species (**Figure 7.27**). Some species have larvae with adaptations for estuarine retention, rising on the flood and moving to the bottom during ebb tide. A consequence may be increased predation by large populations of smaller planktivorous fishes in the estuary. Alternatively, larvae may rise on the ebb to move to coastal waters and then depend upon wind-driven currents to recruit back to the estuary. While such larvae would be living in waters of lower predation, the strategy is plagued by a higher probability that the larvae will never return to the estuary.

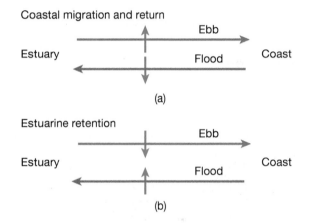

FIG. 7.27 Two alternative strategies for larval recruitment in estuaries: (a) movement of early stage larvae to offshore waters with a subsequent return and (b) retention of larvae within estuaries.

The water column is filled with predators that cause considerable planktonic larval mortality. In temperate waters, planktivorous fishes and ctenophores are important grazers in the spring and summer. Ctenophores are especially abundant in the inner shelf waters and are major predators of zooplankton. Predation is a major problem for larvae that are adapted to be retained in estuaries. Steven Morgan (1990) has shown that three species of crustacea retained in a North Carolina estuary have evolved swimming larvae with pronounced spines. The mud crab *Rhithropanopeus harrisii*, for example, has movable spines, which are erected when fishes come close (**Figure 7.28**). Other crustacean species whose larvae spend time in coastal waters experience less predation and have less completely developed spination.

The life history of the blue crab demonstrates the precarious situation of species that export their larvae from estuaries onto the continental shelf. The work of Charles Epifanio and colleagues on the biology and coastal oceanography of the blue crab *Callinectes sapidus* in Delaware Bay shows just how fragile a marine planktotrophic life

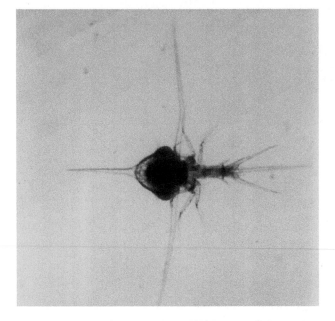

FIG. 7.28 Larvae of the mud crab *Rhithropanopeus harrisii* have erectable spines, which make it very difficult for fishes to attack. This is a necessary defense for larvae that are retained within estuaries, where predation by fishes is intense. (Courtesy of Steven Morgan)

cycle can be. As mentioned previously (**Figure 7.26**), adults copulate within the estuary. Larvae are released at the time of high tide, and the early naupliar larval stages move out to sea on the ebbing tide. The movement offshore occurs in the early summer (**Figure 7.29**), when low-density and low-salinity water is moving from northeastern United States estuaries (e.g., Delaware Bay) onto the continental shelf. Although most flow from this low-salinity water is toward the south, a northward wind combined with the Coriolis effect pushes some water eastward, which carries some of the larvae out on the shelf, just seaward of the estuary from which they came. It is this fraction of larvae that makes it back to the estuary. In the late summer and early fall, winds move southward and combine with Coriolis-driven Ekman transport to bring later-stage megalopa larvae back westward across the shelf to the coastal estuaries. This whole process probably results in a loss of over 95 percent of the larvae that were originally exported from the estuary, but apparently enough return to sustain the population of Delaware Bay. What is amazing is how complex the relationship can be between larval transport and physical oceanography (Epifanio and Garvine, 2001). Once the final megalopa stage larvae arrive in the estuary, they rise in the flood tide to be washed inward and retreat to the bottom on the falling tide to avoid being washed out to sea (Tankersley et al., 2002).

The two alternative estuarine strategies, estuarine retention (e.g., mud crab) and larval export to the continental shelf (e.g. blue crab), likely have strong consequences for connectivity among populations and evolutionary divergence. In species with strong retention within estuaries, populations along the coast are probably strongly isolated from each other. But in species where larvae are exported to the shelf, larval exchange along a coast greatly increases

connectivity. Species with the retention strategy would be predicted to show strong genetic differences among estuaries along the coast because isolation is strong and individual estuaries may have differing environments. On the other hand, local extinction of retained species within estuaries may be frequent and recolonized by sporadic immigrants from other estuaries along the coast.

Coral reef fish usually have planktonic larvae of several weeks' duration, and one might expect similarly a great deal of larval loss to sea by currents. But instead, many species show strong evidence of adaptations to take advantage of currents to return to localized reef areas of their parents. It may be that adaptations for return are crucial, since coral reefs are literally islands surrounded often by inhospitable soft sediments. Many coral reef adults are associated with very specialized microhabitats. Clown fish species, for example, often live in association with the tentacles of sea anemones. Butterfly fish feed on the polyps of corals but also must live close to sites where they can rapidly swim to crevices for shelter from predators. Still, larvae swim in the open sea, often spending several weeks in currents feeding on small zooplankton. Are they randomly mixed throughout hundreds of miles of coral reef tracts? The evidence suggests otherwise. Almany and colleagues (2007) studied a butterfly fish on a marine reserve in Papua, New Guinea, and labeled females by injecting them with an isotope of barium that is rare in nature, relative to the common ^{135}Ba. The isotope was translocated to eggs, and the larvae also had identifiable amounts of ^{137}Ba. The vagabond butterfly fish had larvae that spent about 38 days in the plankton, which might lead to the expectation that most would be lost from this small island reserve of 0.3 km^2. But instead, about 60 percent of the butterfly fish returned to their native reef. Similar results were found for a shorter-dispersing species of clown fish.

The work of Robert Cowen and colleagues provides a great deal of insight into the mechanisms and implications of the homing ability of planktonic larvae of many coral reef species. The bicolor damselfish *Stegastes partitus* is common in the Caribbean island of Barbados, and larvae enter the water column after egg hatching and remain for about 30 days until settlement. Paris and Cowen (2004) followed larvae in the ocean by means of closely spaced samples and found that larvae tended to stay in discrete patches for many days. Early larvae were found in surface waters, but later-stage larvae descended to depths where currents carried them back to the patch reefs from which they originated. These results demonstrate that active behaviors facilitate homing and are analogous to those mentioned previously for retention of oyster and mud crab larvae in northeast United States estuaries.

Larval retention must also work on the very large scale of upwelling systems, such as are found of the Pacific coasts of North and South America. Upwelling during La Niña periods tends to move planktonic larvae of benthic species in surface waters out to sea, where adult habitats will never be found. Some evidence in recent years demonstrates that some planktonic larvae can descend to great depths offshore and return back appropriate adult coastal shore habitats in

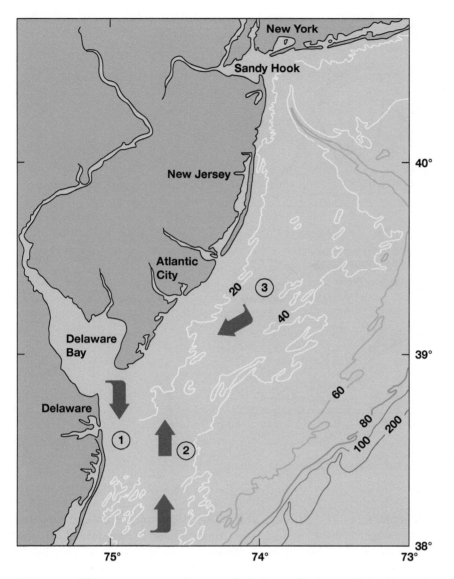

FIG. 7.29 Offshore transport and return of planktotrophic larvae of the blue crab *Callinectes sapidus* occurs in three stages in the Delaware Bay region. (1) Estuarine buoyant flow in late spring carries larvae out of the estuary and mainly southward, but (2) a northward wind combined with the Coriolis effect pushes some of the surface water eastward onto the continental shelf. Finally, (3) southward winds in fall combined with the Coriolis effect brings water and larvae back to shore, and some larvae return to Delaware Bay. (After Epifanio and Garvine, 2001)

coastward-moving countercurrents. Diel vertical migration by crab larvae often results in a net landward movement within such upwelling systems (Marta-Almeida et al. 2006).

Planktonic Larvae: Getting Through Major Obstacles to the Final Destination

- **Planktonic larval populations may fail because of a lack of planktonic larval food or because of concentrations of toxic algae.**

Although primary production is often sufficient to feed larval populations, shortages or failures of phytoplankton can devastate a planktonic feeding larval population that might be produced only once a year for a brief period. Harold

Barnes encountered a bad phytoplankton year in British waters in 1951 (**Figure 7.30**). Early larval stages of barnacles could be found at first, but these soon disappeared because of starvation. It would be best for larvae if their release could be coupled with the spring phytoplankton bloom. Such coupling has been found in sea urchins and mussels living in the St. Lawrence estuary of eastern Canada, which spawn in response to high concentrations of phytoplankton. The response is similar when urchins and mussels are exposed to only the filtered water in which the phytoplankton lived. The response is also advantageous because zooplankton predators tend to be rare at the peak of the phytoplankton bloom, so larvae can eat and not be eaten.

There may be cases in which larvae do not suffer especially from food shortages. Echinoderm larvae may survive well

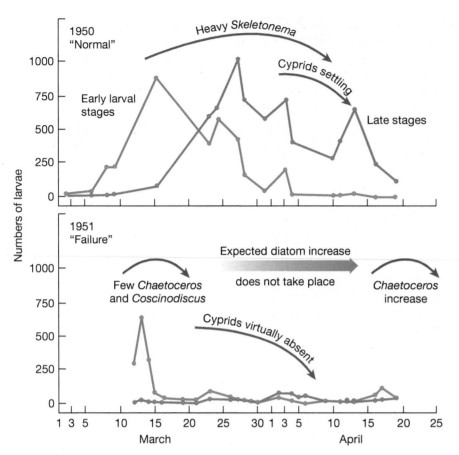

FIG. 7.30 Successful settlement and failure in the barnacle *Semibalanus balanoides* in good and bad phytoplankton seasons. (After Barnes, 1956)

despite an absence of phytoplankton, and many bivalve mollusk larvae are also resistant to the effects of food shortages. It is possible that these groups are able to survive on dissolved organic matter in seawater. (The interested student should consult Olson and Olson, 1989, for more on this subject.)

■ **Some planktotrophic larvae can increase the chances of survival by cloning or other responses to appropriate environmental conditions.**

We must continue to emphasize the high mortality rates suffered by planktotrophic larvae from loss by currents, predation, and food shortage. In some species of echinoderms, including asteroid sea stars and brittle stars, larvae have the remarkable ability to reproduce asexually while in the water column (Knott et al., 2003). The question is under what circumstances would this type of response be advantageous. A laboratory study of the common Pacific sea star *Pisaster ochraceus* demonstrated that larvae tended to reproduce asexually by budding when food was more abundant (Vickery and McClintock, 2000). It would make sense that larvae in the field would bud when food was more concentrated, which would increase the larvae population and, therefore, provide more individuals that might allow survival into the next generation. In a sand dollar species, larvae were found to reproduce clonally in response to chemical cues from fish predators. The clonally produced offspring larvae were smaller, which might have made them less visible to predators (Vaughn and Strathmann, 2008).

There are now scattered examples of invertebrates that can produce either lecithotrophic or planktotrophic larvae. One marine nudibranch (shell-less gastropod) revealed an interesting strategy: Starved adults laid eggs that eventually developed into planktonic larvae, but when adults were fed, larger eggs tended to produce some individuals that developed directly, hatched, and had no planktonic stage at all (Chester, 1996). This suggests that plasticity in development might be a response to food conditions that allows a female to disperse her young away from locally poor food conditions. Much more research is needed to see how this affects the fate of marine populations.

The Macroscale: Major Separations Lead to Biogeographic Structure

■ **The geographic range of a species with planktonic dispersal is greater than the range for species without planktonic larvae.**

Planktonic dispersal gives some species the opportunity to invade distant shores merely by being passengers on the ocean's transport system of currents. We rarely see this process in action, but several species have invaded new coasts in the past 100 years, and their spread has been documented. We mentioned earlier in this connection the shore periwinkle *Littorina littorea*, which invaded Nova Scotia waters in the late 1800s and has now spread as far south as New Jersey.

It stands to reason that current systems that bring water toward shore should also contribute toward increased larval recruitment. We have noted, for example, that the Galapagos Islands stand in a current system that is not conducive to successful larval settlement on the islands. Most coasts are likely to vary much more, owing to variations in current transport. On the west coast of North America, for example, areas of upwelling are often localized, often on a scale of tens of kilometers. When an upwelling system is close to shore, larvae may be transported mostly out to sea in surface waters, which would strongly reduce recruitment. In El Niño years on these coasts, one might expect a great increase in recruitment, since the influx of warm surface water will trap larvae at the California coast and even transport coastal water northward. During El Niño years, the usual isolation that is enforced by current systems at Point Conception, California, is broken down and coastal water is transported northward, bringing larvae to the coast of northern California, often during subtropical storms. Connolly and Roughgarden (1999) found that barnacle recruitment increased strongly over a broad region in central and northern California from 1996 to 1997, a probable result of El Niño. Thus, infrequent changes in climate provide opportunities for breakthroughs in dispersal between previously isolated biogeographic zones.

While there are some exceptions, invertebrate species with planktonic larvae appear to have far greater geographic ranges than those species with direct release. **Figure 7.31a** shows a series of biogeographic zones on the Atlantic coast of North and South America, identified for the distinctness of their respective benthic species. Note that species with nonplanktonic dispersal tend to occupy two to three zones, whereas those with planktonic larvae occupy four to five zones (**Figure 7.31b**). Although planktonic dispersal increases the geographic range, planktotrophic larvae with dispersal times of many weeks or months do not seem to have any broader range than those whose larvae spend only 2–6 weeks in the plankton.

■ Molecular genetic variation and other markers can be used to identify barriers to coastal zone dispersal.

In Chapter 20, we discuss the large-scale biogeographic structure in the ocean, where groups of species are found in similar widespread biogeographic areas, known as **provinces**. Such differences would not occur unless there were barriers to dispersal, which many different species have in common. Why would there be such barriers? Two important factors are involved. First, some physical barriers to dispersal often prevent larvae from spreading. Teleplanic larvae may occasionally traverse broad oceanic distances, as mentioned previously, but most marine planktonic larvae cannot make it across broad oceans such as the Atlantic or the Pacific. Such barriers also exist along coastlines. Separate current circulation systems appear to isolate groups of species north and south of Point Conception in southern California. South Florida isolates populations of the southeast United States from those within the Gulf of Mexico. On either side of the barrier one observes closely related species or strong genetic differences between populations of the same species, as can be measured using DNA sequence differences.

We might expect that species with long-distance planktotrophic larvae would have coastal populations with high connectivity. Many current systems tend to move parallel to coasts, and these should carry planktotrophic larvae long

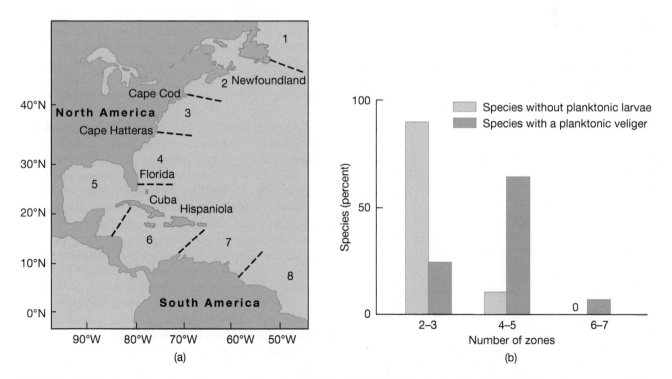

FIG. 7.31 (a) Diagram of biogeographic zones of shore benthic invertebrates in the western Atlantic. (b) Number of zones occupied by invertebrate benthic species with and without planktonic larvae. (After Scheltema, 1989)

distances. It is very difficult to study this process directly because larvae are difficult to follow at sea, although this has been done in some cases. Instead, we would seek indirect evidence, such as genetic similarity among coastal populations, as evidence of interchange. This evidence would be especially strong if it is obtained from different genetic sources, such as different genes coming from different parts of the genome. Along the Pacific coast, the barnacle *Balanus glandula* has a broad geographic distribution, which suggests a great deal of transport of larvae along the coast. It is no surprise, therefore, that there is a great deal of genetic homogeneity of this species in DNA sequences from British Columbia in Canada to northern California. However, a major difference in DNA frequencies was found between northern and southern California (**Figure 7.32**). The strong cline coincided with an apparent isolation of current systems that could be demonstrated by simply placing labeled oceanographic drifting floats in Oregon and southern California: The drifters moved hundreds of kilometers in 40 days but never overlapped each other. The major genetic difference found, therefore,

is a reflection of a broad-scale geographic isolation that is probably reinforced by currents.

A more recent study on the red abalone, *Haliotis refescens*, also shows genetic homogeneity from Monterey, California, to Oregon over an enormous number of single nucleotide polymorphisms. But still there were a number of genes whose single nucleotide polymorphisms showed significant geographic variation. The online bonus chapter, "Molecular Tools for Marine Biology," discusses procedures, but large number of sequences collected could be identified as distinct genes and those that were significantly variable geographically were involved in hypoxia, energy metabolism, calcification, and other physiological processes in which climate change might come into play as a major selective force now and in the future (De Wit and Palumbi, 2012).

Recent geological history may be very important in explaining isolation among populations within species along coasts. There is a rather complex geographical pattern of genetic differences in southern California in populations of the small harpacticoid copepod *Tigriopus californicus*. This species lives in tide pools and does not have planktonic dispersal, and we would therefore expect little dispersal between sites by currents (Burton, 1998). But a very different pattern was found toward the northern part of the Pacific coast of North America. There was very little differentiation at the same genetic locus along the coast as you go northward from Puget Sound in Washington State into Canada and Alaska. This homogeneity derives from the glacial history of the past few tens of thousands of years. It is likely that *Tigriopus* populations have only recently spread northward after the retreat of the glaciers, and this makes the populations resemble each other genetically along great distances (Edmands, 2001).

For species with calcareous shells or skeletons, the trace elemental composition can sometimes be used to trace the origins and connectivities of populations along coasts. For example, planktotrophic larvae of bivalves often have larval shells with elemental compositions (e.g., Mn, Pb, and Ba content) that are, in effect, a signature of their point of origin. Thus, larvae originating in different locations along a coastline may have specific compositions that reflect the origin of the larvae. When juveniles are collected, one can examine the elemental composition of the youngest part of the shell and see if it matches the local area or the signature formed by an origin in another area. Using this technique, the California sea mussel *Mytilus californianus* was found to originate in the northern parts of the coast of southern California but spread evenly toward the south (Becker et al., 2007). This may relate to a spawning–larval period that occurs when currents are moving strongly along the coast. Another mussel species had much more localized elemental compositions, which reflected larval spread during a time of year when longshore currents were not strong, so recruiting individuals reflected more their local point of origin.

A second barrier to dispersal is strong natural selection. If a larva disperses to a location where it cannot survive or where its adult phase cannot survive, then individuals at that point are liable to be genetically different and isolated from the source of those dispersing larvae. A good example is

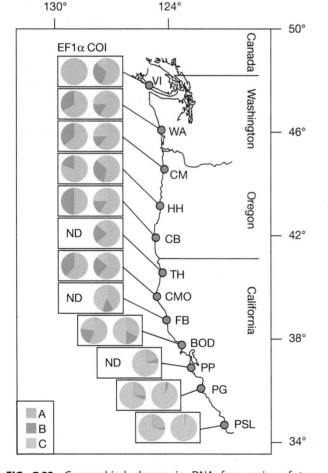

FIG. 7.32 Geographical change in DNA frequencies of two genes for the barnacle *Balanus glandula*, which has planktotrophic dispersal. There is a strong transition of frequencies at both genes from northern to southern California, which you can see in the pie charts by an increase southward in the frequency of the tan-colored variant C from site CMO to PSL. (From Sotka et al., 2004)

Cape Hatteras in North Carolina. Summer water temperature increases greatly to the south of this cape. It is not unusual for planktonic larvae of the blue mussel *Mytilus edulis* to be carried by coastal currents in spring to the south of this cape followed by settlement and metamorphosis into juvenile mussels. But by midsummer, the temperature is simply too high for this species and individuals die. This explains the southerly extent of *Mytilus edulis*. Recent experiments with planktonic larvae of the eastern United States fiddler crab *Uca pugilator* show that their larvae cannot tolerate the low temperatures found north of Cape Cod (Sanford et al., 2006). This may explain the northern limit of this species.

A particular type of intraspecific genetic change with distance along the coast can be an excellent indicator of isolation that is caused by natural selection. Consider a hypothetical coast that has a very strong spatial gradient in temperature. It might be that one allele at a genetic locus might be favored at one end and an alternative allele might be favored at the other end of the gradient. Thus, natural selection might be isolating populations on either side of this genetic transition where the frequencies of variants change over space. As discussed in Chapter 4, this spatial change of variation is known as a cline (see Figure 4.25). Natural selection favors different variants on either end of the cline, which may isolate the two populations from each other if the selection gradient is very strong. If natural selection were operating on two genetic loci, then the strength of isolation would be even greater (Sotka et al., 2004).

■ **On the macroscale, isolation between regions may result in a large group of separated populations of a species, which might have only occasional interchange.**

Because of numerous barriers to dispersal along coasts and across open seas, we would expect that widespread marine species would comprise a series of isolated populations.

These populations might be affected by very different local conditions, and population change might be independent in each local region. In the long run, if isolation were never broken, the populations might diverge genetically and separate species might be produced. But occasional exchange of far-traveling individuals might keep the chain of connection open and new species might not be formed.

Robert Cowen and colleagues have analyzed the structure of the entire Caribbean basin and have found that currents tend to isolate distinct island groups from each other, creating a series of subpopulations of coral reef fishes. Typically, larval dispersal is on the order of 10–100 km, although many of the subregions are not isolated and receive immigrants in substantial numbers. Still, the adaptations coral reef fish larvae have that cause them to be retained in their natal reefs tend to maintain isolation among the regions into a series of island groups among which there is no major population replenishment from other areas (**Figure 7.33**). There also are four major regions that are largely isolated: the eastern Caribbean, the western Caribbean, the Bahamas and the Turks and Caicos Islands, and the region at the periphery of the Colombia-Panama Gyre. This may be a common pattern in island groups and along coasts where barriers are common. Such conclusions would be understood if we had more information on genetic isolation based upon DNA sequences. Unfortunately, we are still at a primitive stage in marine biology in this area, but DNA sequencing will rapidly allow us to advance our understanding of how much isolation there is among populations in the sea. In Chapter 20, we discuss how planktotrophic larval dispersal makes for large geographic ranges for species of cone snails in the Pacific (genus *Conus*), but lecithotrophic larval dispersal is reflected in complexes of isolated species of *Conus* among islands in an eastern Atlantic oceanic archipelago.

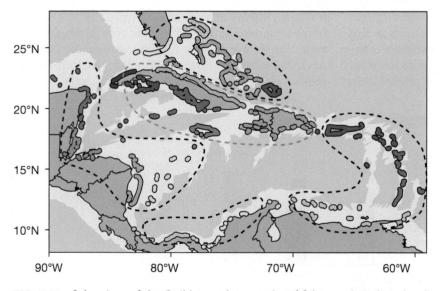

FIG. 7.33 Subregions of the Caribbean where coral reef fish are relatively isolated from other subregions. The black dashes enclose four major isolated domains, while the red dashes enclose a broad area of interchange with the peripheral areas. (Drawn after description in Cowen et al., 2006)

On a smaller scale within a species range, geographically based genetic differentiation patterns can be seen to reflect different modes of larval dispersal within a single species, for example, the sea slug *Costasiella ocellifera*, which has both planktotrophic and nonplanktonic development in different populations. These populations were studied by sequencing mitochondrial and one nuclear gene in populations of differing larval dispersal (Ellingson and Krug, 2015). The planktonic larval populations in the Caribbean constituted one relatively homogeneous set of populations with strong evidence of high connectivity. By contrast, the Floridean nonplanktonic dispersing populations (larvae of a lecithotrophic type that are retained in the mother and released at the same site) were much more genetically differentiated from each other and also showed evidence of reduced genetic variability. In this difference one can see the role of larval dispersal type in affecting connectivity and causing strong differences in patterns of local evolution according to dispersal type. These results fit a number of other studies that show that species with no planktonic dispersal show much more regional differentiation than species with planktonic dispersal. Apparently even some planktonic larval dispersal tends to swamp out local genetic divergence that might develop from local environmental influences on the direction of natural selection on gene frequencies.

Planktonic Dispersal: Why Do They Do It?

■ **Why disperse? Dispersal of planktonic larvae ensures that local habitat destruction will not lead to extinction.**

Given the dangers of planktonic dispersal, one might wonder why this mode evolved at all. Richard Strathmann (1987) has pointed out that many evolutionary lines have lost the ability to produce feeding planktonic larvae simply because it is fairly easy, in the evolutionary sense, to lose a swimming or plankton-feeding structure but difficult to reacquire complex structures such as swimming organs and ciliated bands for feeding. However, the great majority of marine invertebrates have planktonic feeding larvae. Planktotrophic dispersal is rare only in high latitudes (**Figure 7.34**), where the phytoplankton

season is very short and the water temperature is low. Gunnar Thorson argued that this may stem from the danger of failing to synchronize reproduction with the short phytoplankton production season, combined with the very long development times due to low temperature, which would increase the danger of predation and of being swept to sea.

Avoidance of crowding is one significant benefit of dispersal. Many marine habitats are severely space-limited, and dispersal to an open habitat would ensure population increase. The rate of destruction of local marine habitats may also make dispersal beneficial. All local marine habitats eventually experience major disturbances. Coral reefs, which are often represented to students as benign environments, are continually disturbed by major storms, and some are affected by temperature fluctuations. No rocky shore long escapes the force of waves, ice, hot and dry weather, or an influx of predators. If an organism's offspring could not disperse, the chances are that its genes would become extinct. Dispersal, therefore, ensures that some of the progeny will escape an eventual catastrophe. It is of interest in this regard that paleontologists have discovered that mollusk species characterized by long-distance dispersal have lower extinction rates. One should not forget, however, that many species with reduced dispersal manage to survive. They have the advantage of giving their young access to a habitat that has at least proved suitable to the parents.

A study on the southern Australian soft-sediment snail *Nassarius pauperatus* makes a convincing case for the dispersal value of planktotrophic larvae (McKillup and Butler, 1979). In some sand flats, the mud snails are clearly in a poorer nutritive state than in others. This can be estimated by a measure of how responsive snails are to food, which is in effect an estimate of hunger. As food availability decreases, *N. pauperatus* produce more eggs per capsule, suggesting a shift to investment in dispersing progeny. By contrast, snails living in flats with abundant food invest less in progeny because selective pressure is not intense enough to warrant massive devotion to early reproduction. Some reserves are instead devoted to somatic growth, which allows more reproductive seasons. Planktonic dispersal, therefore, serves the successful dispersal of progeny of food-starved populations to sites that may be better trophically.

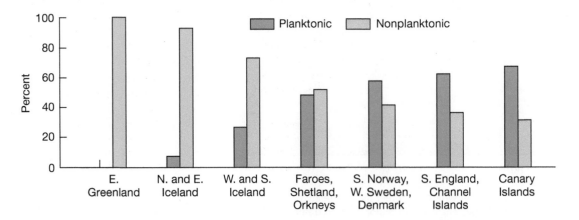

FIG. 7.34 Latitudinal variation in the abundance of prosobranch gastropod species with planktonic and nonplanktonic larvae. (After Thorson, 1950)

Long-distance dispersal also has the advantage of spreading the young over a variety of habitat types. In carrying out such dispersal, a parent could be hedging its bets on the success of its offspring. Instead of investing heavily in one site, dispersal allows settling on many sites. In any year, one site may be very poor for settling, whereas another might be good. In the long run, survival is averaged over all habitats.

The minimum time required for a planktotrophic larva to develop and the distance the larva travels in the water column during this period are puzzling. Why should a larva travel for 2 weeks or more when it could sample a diversity of new sites only a few hundred meters away? Strathmann (1985) suggests that planktotrophic larvae might not have evolved as an adaptation for long-distance dispersal but that the larvae may have evolved planktotrophy to take advantage of the phytoplankton food source, much as a juvenile fish exploits a nursery ground. The exact length of dispersal may, therefore, be something of an evolutionary accident, even if wider dispersal does have the consequence of lowered species extinction and the potential for rapid colonization.

Although long-distance dispersers are common, many colonial animals, such as bryozoans and coelenterates, are known to have short-distance-dispersing lecithotrophic larvae. Indeed, this form of dispersal has been underemphasized in the past. The bryozoan *Bugula neritinea* lives in Florida on blades of turtle grass. Areas with dense populations are self-maintaining, but nearby areas without *Bugula* also exist for several years without any significant colonization. Thus, marine populations in certain areas may be limited by dispersal potential, especially in the case of species with no planktonic larvae or with lecithotrophic larvae.

■ CHAPTER SUMMARY

- Reproduction, or the replication of individuals, can occur without sex. Dispersal is an undirected movement of an individual to a new place. Migration involves directed movement between specific places.

- Sex is a nearly universal characteristic, despite its considerable costs to organisms. Its adaptive value is the maintenance of genetic diversity, which increases adaptability to disease and other challenges.

- Sexes may be separate (in gonochoristic species), simultaneous, or sequential in the same body. The contributions of different sexes to the next generation determine the value of hermaphroditism as well as the relative sizes of males and females. For example, if the relative advantage of a sequential hermaphrodite functioning as a large female and small male have an advantage in maximizing offspring production, this will select for male-female transitions. Dwarf parasitic males may occur where it is difficult to find mates.

- Fertilization success is affected by the mode of sperm transfer, the volume of gamete production, the distance between males and females, water turbulence, the timing of spawning, and behavior. Free spawning has a number of costs, and planktonic gametes have special problems in ensuring fertilization, but specialized sperm attractants may exist widely.

- Marine species differ in parental care. Most species have free-swimming larvae, but some guard or brood eggs, brood young within body cavities, or have live birth like mammals.

- Nonsexual reproduction permits the same genetic type to increase rapidly in an open environment. A population of genetically identical individuals, all deriving from one founder, is known as a clone. In algae and corals, nonsexual reproduction involves colonial individuals (modules) that are connected to each other and exchange nutrients.

- The age of first reproduction, the resources devoted to reproduction (or reproductive effort), and longevity may evolve in response to patterns of mortality and reproductive success. High mortality selects for an earlier first age of reproduction and sometimes for reproduction only once. Uncertainty in spawning success can be countered by repeated spawning over more years. Commercial fishing imposes natural selection.

- Fishes, crustaceans, turtles, and marine mammals often move between spawning and feeding grounds. This migration allows optimal functioning of spawning, egg laying, and feeding. In some cases, as with turtles and salmon, adults return to exact egg-laying locations, while homing is less specific for others. Migration routes can be traced by tagging individuals and by genetic markers.

- Gamete production and larval life must be timed to allow settlement and promote dispersal, to avoid being swept to inappropriate habitats, and to counter predation. Egg size decreases with increasing egg numbers.

- Marine invertebrate offspring may be (1) brooded or released as small adults; (2) dispersed only to a small degree by means of short-lived, yolk-dependent *lecithotrophic larvae*; or (3) dispersed great distances by means of longer-lived plankton-feeding *planktotrophic larvae*.

- Despite the potential for dispersal, planktonic larvae often are adapted to settle quite near their origin, owing to behavior and cyclonic currents. On the microscale, larvae use a number of cues to find the final site of settlement.

- The geographic range of a species with planktonic dispersal is greater than the range for species without planktonic larvae. Genetic variation can be used to identify barriers to coastal zone dispersal. Larval dispersal, oceanographic, and geographic barriers combine to determine regional differences.

1. Why are not all sexual species hermaphroditic?

2. What is the "cost of sex," biologically speaking?

3. What might be the benefit of sex to organisms?

4. Under what circumstances does it make sense for a hermaphrodite to be protandrous? Protogynous?

5. When might you expect the presence of dwarf males attached to females?

6. What is the advantage of clonal reproduction? The disadvantage?

7. Why do fishes such as Pacific salmon species reproduce only once?

8. What might be the value, if any, of long-distance dispersal across an ocean?

9. What are the advantages of having planktotrophic larvae capable of settling and metamorphosing upon any substratum? What are the disadvantages?

10. Why are species with planktotrophic larvae more common in the tropics than at polar latitudes?

11. How might planktonic larvae be able to find adults of their own species?

12. What is the value of planktonic feeding larval development?

13. What are the potential sources of mortality for planktonic larvae?

14. What effect does planktotrophic larval dispersal tend to have on the geographic range of a coastal marine invertebrate species?

15. Do you think that the total evolutionary life spans of species with planktotrophic larvae are liable to be greater than those with lecithotrophic larvae? Explain your answer.

16. Anemones often occur in clones of large numbers that have arisen by fission from a founder individual. What experiment might be performed to determine the benefit of large numbers of adjacent anemones as opposed to smaller groups or solitary individuals?

17. Very few species are known to be poecilogonous. Why do you think species tend to have only one dispersal type?

18. What evidence might you use to demonstrate that there is little connectivity between populations of a coastal invertebrate along a long stretch of coast?

19. How might oceanic structure affect the migration routes of long-distance traveling species? What factors might determine a specific migration route across the ocean?

20. Imagine you are in charge of studying the swimming routes of blue whales. How would you follow their movements along the west coast of North America, from Mexico to Alaska?

21. We now know that many long-distance swimming fish species follow major current systems. How might climate change affect the feeding success of these migrating species?

Visit the companion website for *Marine Biology* at www.oup.com/us/levinton where you can find Cited References (under Student Resources/Cited References), Key Concepts, Marine Biology Explorations, and the Marine Biology Web Page with many additional resources.

CHAPTER 8

Plankton

Introduction and Definitions

Plankton live in the water column and are too small to be able to swim counter to typical ocean currents. **Phytoplankton** are photosynthetic planktonic protists and bacteria and usually consist of single-celled organisms or of chains of cells. Although some can move using a flagellum, phytoplankton movements in the water column are nearly completely controlled by water turbulence and currents and by the bulk density of the organisms. Some taxonomic groups include phytoplankton species, which are photosynthetic, as well as nonphotosynthetic forms that are heterotrophic and therefore gain nutrients by consuming small planktonic cells or organic matter in the water column. The **zooplankton** comprise nonphotosynthetic planktonic protists and animals, ranging from single-celled forms to smaller animals, such as eggs and early larval stages of fishes and larval invertebrates. Although they may be able to swim at small spatial scales of a few centimeters or fewer, their movement is mostly determined by major water currents, turbulence, and bulk density. Some plankton do not fit either the zooplankton or the phytoplankton mold. Some protists, for example, are photosynthetic but also can ingest other plankton and retain the chloroplasts of these prey organisms. Such plankton are called **mixotrophic plankton**. **Meroplankton** are zooplankton that spend only part of their lifetime in the plankton but are benthic or nektonic as juveniles and adults. They include the planktonic larval stages of many fishes and benthic invertebrates. **Holoplankton,** by contrast, are those planktonic organisms that spend their entire life cycle in the water column. **Neuston** are those organisms associated with the air-water interface, such as bacterial films, whereas **pleuston** are plankton that live at the surface but protrude into the air. The latter include animals that have floats, such as the Portuguese man-of-war.

Planktonic organisms are also classified by size classes, as shown in **Table 8.1**. The sizes correspond fairly well to different groups of plankton. Plankton nets (**Figure 8.1**) can be used successfully to sample patches of water at specific times and depths and can be used to capture without damage microplankton and even nanoplankton so long as they have fairly rigid skeletons (e.g., diatoms, copepods). But water bottle samplers that are submerged and automatically closed at depth must be used to capture nanoplankton and smaller organisms that are fragile, such as ciliates.

Marine Viruses

Viruses are not complete living organisms as we typically define them. They are particles in the femtoplankton range (0.02–0.20 μm) and consist of strands of DNA or RNA enveloped in a protein coat. They cannot be seen by ordinary light microscopy but usually can be visualized by electron microscopy (**Figure 8.2**). They may be spherical or have elegant polygonal shapes, sometimes with a tail-like structure. While viruses have long been known to be in the plankton, their importance in the recycling of nutrients and pathogenic infection of phytoplankton has been understood only very recently. Their abundance can be of the order of 10^9 viruses per liter.

TABLE 8.1 Size Classification of Plankton

SIZE CLASS	SIZE RANGE	EXAMPLES
Megaplankton	> 20 cm	Jellyfish
Macroplankton	2–20 cm	Pteropods, krill
Mesoplankton	0.2 mm–2 cm	Copepods, Foraminifera
Microplankton	20–200 μm	Ciliates, coccolithophores
Nanoplankton	2–20 μm	Diatoms, dinoflagellates
Picoplankton	0.2–2 μm	Smaller protists, bacteria
Femtoplankton	0.02–0.20 μm	Viruses

FIG. 8.1   A MOCNESS (Multiple Opening/Closing Net and Environmental Sensing System) plankton sampler being retrieved after a tow in Saanich Inlet, Vancouver Island, Canada. Multiple nets are attached to a series of stacked bars and released at specific locations and depths by command from a computer-controlled program. (Courtesy of Peter Wiebe)

Viruses have no metabolic activity as such. Instead, they invade other living organisms and take advantage of those organisms' genetic machinery to reproduce. Viruses may invade and attach to the host's DNA and rapidly initiate synthesis of new viruses, which eventually reach very large numbers, bursting the host's cell in a process known as **lysis**, which facilitates dispersal and infection of other cells. Alternatively, they may become a permanent part of the host's genome and replicate for long periods. Some trigger mechanism may eventually cause these viruses to replicate rapidly and initiate cell lysis. Viruses are important in affecting the abundance of marine bacteria (see discussion in Chapter 11 on the microbial loop) and may also be a major factor in controlling populations of some phytoplankton species. For example, coccolithoviruses are double-stranded DNA viruses that infect and cause cell lysis of large fractions of populations of the abundant photosynthetic species *Emiliana huxleyi* (see later section on coccolithophores).

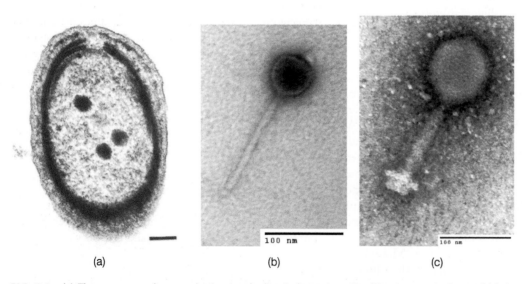

(a)	(b)	(c)

FIG. 8.2   (a) The common photosynthetic picoplanktonic bacterium *Prochlorococcus marinus*, which is infected by a number of types of virus of which we show two examples: (b) P-SSP7, a bacteriophage (a virus that infects bacteria) belonging to the family of the Podoviridae (icosahedral capsid with a short tail), and (c) P-HM2, a bacteriophage belonging to the predominant marine viral family of the Myoviridae (icosahedral capsid with a long contractile tail). (*Prochlorococcus* photo courtesy of Claire Ting; virus photos courtesy of Simon Labrie and Penny Chisholm)

Marine Bacteria and Archaea

A large number of species of Bacteria and Archaea[1] are found in open waters of the ocean. Marine bacteria are a numerically dominant component of the picoplankton. While they occur in high densities, on the order of 10^6 cells per ml in open waters, they are especially abundant on particles of organic matter, whose decay they accelerate by decomposition processes. In typical ocean waters, bacteria are aerobic, but ocean water may be devoid of oxygen in certain places, such as in trenches with poor circulation or in quiet water columns of the low-salinity zones of estuaries, especially in summer. In such waters, a number of bacteria adapted to anoxic conditions are abundant, such as **denitrifying bacteria**, which reduce nitrates to nitrites or nitrogen gas. **Methanogenic bacteria** belong to the Archaea and sometimes live in the water column without oxygen, producing methane as a byproduct of carbon dioxide (carbon source) and hydrogen (reductant).

Cyanobacteria

The cyanobacteria, which are members of the class Cyanophyceae (often called blue-green algae, although they are not true algae), are found in and may dominate near-shore waters of restricted circulation, brackish water, and waters of the open ocean. Members of the genera *Prochlorococcus* and *Synecococcus* are widespread in the open ocean, especially central ocean areas with low nutrient concentrations. *Prochlorococcus* with oval or spherically shaped cells of a diameter of about 0.6 μm, are the smallest and the most abundant photosynthetic organisms in the world. Many cyanobacteria are vulnerable to attack by viruses.

Cyanobacteria carry out **nitrogen fixation**, in which gaseous nitrogen is converted to NH_4^+, or ammonium ion, which is then available for use in the synthesis of amino acids and proteins. The filamentous cyanobacterium *Trichodesmium* (**Figure 8.3**) is abundant in the nutrient-poor waters of the warm oceanic gyres and was first noticed in the ocean near Australia by the explorer Captain Cook. It is very abundant and important in nutrient cycling in tropical oceanic waters. *Trichodesmium* is unusual in that it simultaneously photosynthesizes and carries out nitrogen fixation. Normally, oxygen is poisonous to the process of nitrogen fixation, but in this case the oxygen is detoxified because its production is spatially and temporally segregated from nitrogenase within the cell. Along with filamentous cyanobacteria, spherical or coccoid cells are also found throughout the world ocean. In Chapter 11, we discuss the role of cyanobacteria and other bacterial groups in the ocean's nitrogen cycle.

Eukaryotic Phytoplankton
Single-Celled Protists

Here we consider those phytoplankton groups that occur as photosynthetic cells belonging to the informal taxonomic group Protista.

FIG. 8.3 The colonial planktonic blue-green cyanobacterium *Trichodesmium thiebautii*. (Photograph by Edward Carpenter)

DIATOMS
■ **Diatoms are single-celled phytoplankton that usually have a silica shell and often dominate the phytoplankton, especially at high latitudes.**

Diatoms (Phylum Chrysophyta, Class Bacillariophyceae) dominate the phytoplankton in waters from the temperate to the polar zones. They occur either as single cells or chains of cells (**Figure 8.4**) in the size range of nanoplankton and microplankton. Each cell is encased in a silica shell (actually a cell wall impregnated with silica) composed of two valves that fit together much like a pillbox (**Figure 8.5**). The shell may be covered with spines, or it may be ornamented with a complex series of pores and ridges. The pores are the only connection between the cell and the external environment. Planktonic diatoms are usually radially symmetrical (shaped like a pinwheel) and are usually centric diatoms (**Figure 8.4**), in contrast to pennate diatoms such as *Pseudo-nitzschia*, which are bilaterally symmetrical and are often dominants of the phytoplankton, frequently as harmful algal blooms (see Chapter 11).

Diatoms reproduce by binary cell division, with one half of the cell and one valve going to each daughter cell and serving as the larger valve. A smaller valve is then formed to pair with the valve from the parent cell. After several successive generations of division, cell size usually decreases to a lower threshold, when gametes are often formed and released into the water column. Alternatively, the smaller diatom forms an **auxospore**, which increases in size and casts off the valves, synthesizes new large valves, and then undergoes asexual divisions. Many diatom species can form asexual resting spores, which settle to the seabed and may be regenerated later as a planktonic form. Because of their asexual mode of reproduction, diatoms can increase in population size rapidly, with doubling rates on the order of 0.5–6 per day.

---

[1] See Chapter 13 for an introduction to the major domains of living organisms, including Bacteria and Archaea.

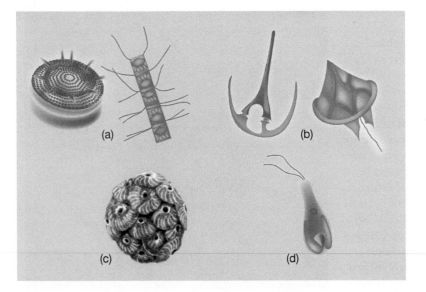

FIG. 8.4 Some members of the phytoplankton: (a) diatoms, (b) dinoflagellates, (c) coccolithophore, and (d) the microflagellate photosynthetic protist *Isochrysis*.

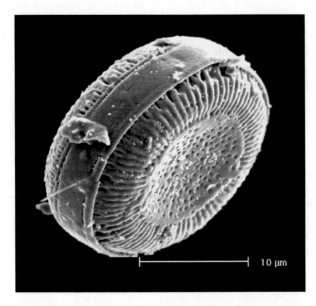

FIG. 8.5 Electron micrograph of the silica shell of a centric diatom.

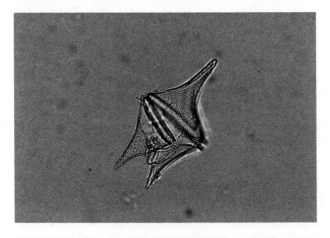

FIG. 8.6 Light micrograph of the dinoflagellate *Peridineum* sp., common in estuarine and shelf waters. About 50 μm across. (Photograph by George Rowland)

DINOFLAGELLATES

■ **Dinoflagellates usually have two flagella, are somewhat mobile, and have an organic skeleton. They are dominant members of the phytoplankton later in the summer in midlatitudes and may always be dominant in lower-latitude phytoplankton.**

The **dinoflagellates** (Phylum Pyrrophyta, Class Dinophyceae) are unicellular species that often dominate the subtropical and tropical phytoplankton and also may dominate the summer and autumn phytoplankton of the temperate and boreal zones. Zooxanthellae, which are endosymbionts of a number of invertebrate groups, are also dinoflagellates (see Chapter 17). Free-living dinoflagellates have two flagellae and are in the size range of nanoplankton and microplankton (**Figure 8.6**). The transverse flagellum is located in a groove (the girdle) that divides the cell into two subequal parts. The other flagellum is oriented perpendicularly to the transverse flagellum and extends toward the posterior end of the cell. The organism is generally covered with a series of contiguous cellulose plates, which comprise the **theca**, whose pattern is usually diagnostic of the species.

Dinoflagellates reproduce asexually by binary fission. Like diatoms, they have a capacity to reproduce up to several times per day. During cell division, the theca may be shed, or a part of the theca may be inherited by each of the two daughter cells. Sexual reproduction also occurs, and dinoflagellates can form resting stages, or **cysts**, that sink to the bottom and may be regenerated later as planktonic forms. Some dinoflagellates are bioluminescent, and the gentle flickering of shallow waters is often due to luminescence by species of *Noctiluca*. Rapid population increases of dinoflagellates often become concentrated enough to color the water, causing **red tides** (**Figure 8.7**).

Some dinoflagellates, such as the genera *Protogonyaulax*, *Gonyaulax*, and *Gymnodinium*, are known to cause **harmful algal blooms**, in which rapid increases of dinoflagellate populations are of sufficient magnitude to color the

FIG. 8.7   Red tide photographed from vessel *Ronald H. Brown*, off of South Korean coast. (Courtesy of NOAA)

seawater a dull red-brown (red-colored blooms are often not toxic). The dinoflagellate *Alexandrium* sp. produces **saxitoxin**, which depresses sodium ion transport and therefore inhibits nervous transmission and can cause death in the suspension-feeding bivalves that consume them. People can be poisoned by consuming the shellfish. You can usually tell immediately that you have eaten a saxitoxin-contaminated shellfish because your lips feel numb. Within 12 hours of ingestion of a toxic bivalve, human respiration is inhibited and cardiac arrest may follow. Another toxin, **brevetoxin**, is produced by the dinoflagellate *Karenia brevis* and is the major cause of toxic red tides in Florida and Gulf coast states. Brevetoxin also binds to sodium channels but rarely is fatal to humans. The origin of red tides, like many other sudden blooms of phytoplankton, is an incompletely explained phenomenon. Red tides are often associated with sudden influxes of nutrients or washout of nutrients from land sources into the sea. Storms may remobilize populations of cysts buried in the bottom sediment, setting the stage for red tides of some species.

Dinoflagellates have a wide range of trophic status. While many species are completely photosynthetic and autotrophic, many others are completely heterotrophic. Some species are known to attack and ingest tissues of animals and may attack and ingest marine invertebrate planktonic larvae.

COCCOLITHOPHORES

■ **Coccolithophores secrete a skeleton consisting of calcium carbonate plates and are sporadically dominant in phytoplankton throughout the world.**

Coccolithophores (Phylum Chrysophyta) are unicellular and usually nanoplanktonic. They are important in the phytoplankton of the tropical open ocean but also form enormous blooms in subpolar seas that produce a strong bluish-white reflection that may cover thousands of square kilometers. They are nearly spherical and are covered with a series of calcium carbonate plates, or **coccoliths** (**Figure 8.4**), which comprise approximately one-third of

the total calcium carbonate production in the entire ocean, blanketing the deep pelagic seabed in some locations. Also in this group are a large number of naked forms that are poorly preserved. A great deal of worry about such calcareous plankton has arisen since the recent discovery of decreased pH in the open ocean, which can inhibit calcification, but initial experiments show that additions of CO_2 can benefit coccolithophore growth (see Chapters 3 and 12 for discussion of ocean acidification and responses of coccolithophores).

SILICOFLAGELLATES Silicoflagellates (Phylum Chrysophyta) are unicellular and biflagellate; they have numerous chloroplasts and an internal skeleton of silica scales. They are usually less abundant than diatoms but are abundant in Antarctic phytoplankton and in the plankton of many other open-ocean locales.

GREEN ALGAE The true green algae (Phylum Chlorophyta, Class Chlorophyceae) are rare in marine waters but can dominate the phytoplankton of enclosed estuaries or enclosed lagoons, especially in late summer and fall. They can be flagellated or nonmotile. Several species cause nuisance phytoplankton blooms associated with coastal pollution. Green algae are much more important as components of the benthos and dominate intertidal soft-bottom seaweed assemblages.

CRYPTOMONAD FLAGELLATES Members of the class Cryptophyceae (Phylum Cryptophyta, Class Cryptophyceae) are widespread and locally abundant in estuaries. They are unique in having chlorophyll types *a* and *c* in addition to the photosynthetic light-absorbing pigment phycobilin.

Zooplankton

Crustacean Zooplankton

Crustacea (phylum Arthropoda, subphylum Crustacea) have distinctive features that include (1) an external skeleton of chitin, a flexible but stiff material, that is relatively impermeable to the external environment and (2) some degree of body segmentation, and paired, jointed appendages (e.g., legs, antennae). The crustaceans possess antennae, mandibles, and maxillae, which are modified head appendages and usually have compound eyes. They include the copepods, crabs, shrimp, lobsters, crayfish, and sow bugs (a land organism: type of isopod). They are notable for their mobility, armored exoskeleton, and generally good vision.

COPEPODS

■ **Copepods, important worldwide members of the microplankton, include dominant consumers of phytoplankton and also include species that are predators on smaller zooplankton.**

There are more copepods (crustacean order **Copepoda**), at least by mass, than any other kind of zooplankton, and these organisms dominate nearly all the oceans and marginal seas. They range in length from less than 1 mm to a few millimeters. The calanoid copepods dominate. The harpacticoid copepods usually have benthic adults, but the larvae of some species may dominate the estuarine zooplankton. Members of the Lernacopodoida parasitize marine mammals and fish. They are wormlike and can stick out from their hosts over 30 cm into the water.

Calanoid copepods (**Figure 8.8**) are usually barrel shaped, with a body composed of head, thorax, and abdomen. They swim up to 200 body lengths s^{-1} by means of rhythmic strokes of the five posterior pairs of thoracic appendages. Inertia dominates over viscous forces. Calanoid copepods have a median naupliar eye but lack the compound eyes found in many other crustaceans, and therefore have poor image-forming vision.

Calanoid copepods are large enough and active enough that they live in the transition between situations where viscous forces are important and inertia may come into play (Yen, 2000). When calanoid copepods sink without moving any appendages at all, they generate no mechanical movement of the surrounding water and streamlines are close to the body. But movements of the appendages generate mechanical motion of water, which may be detected by predators and potential mates. Predatory copepods have elaborate rows of hairs on the antennules that detect motion of smaller prey copepods.

Mating involves rapid swimming motions as the male copepod darts after the female. The male uses chemoreceptors to detect the odor trail that is left in the wake of the female's swimming motion. The male can move accurately in three-dimensional space to locate the female (Doall et al., 1998). The female lays eggs (**Figure 8.9**) in clutches of about 10–50 every 10–14 days. Larvae go through a series of naupliar and copepodite molts before the adult stage.

Calanoid copepods feed mainly on phytoplankton, some organic particles in the water column, and smaller zooplankton. Detection of food is mediated by the first antennae, which are festooned with mechanosensory and chemosensory hairs (**Figure 8.8**). The first antennae move water around, and this motion brings blobs of water across the chemosensory hairs that detect odors of food, such as phytoplankton cells. Otherwise, the copepod would have to depend on molecular diffusion to detect odors, which would be very inefficient. When the chemosensory hairs detect the presence of food, a feeding motion begins on anterior appendages specialized for feeding. The copepods trap particles on hairlike maxillary setules, located on setae, which in turn are located on the maxillipeds (**Figure 8.10**). Recall the conditions for low *Re* (see Chapter 6), where water velocity is low and the size of the objects in the fluid is small. In these circumstances, water does not flow freely past a copepod's feeding appendage. Rather, viscous flow dominates around the appendage. The animal flaps four pairs of appendages (**Figure 8.11**) to propel water past itself, which creates a slow swimming speed. As a diatom advances near the copepod, the maxilliped reaches out and grabs it, rather than sieving it through the setules (as was originally thought). Because of the viscosity of its environment, the copepod grabs the diatom with a surrounding envelope of water, which may have to be strained off the particle before it is passed to the mouthparts. In larger copepods, *Re* is about 1–6 during feeding, and there is some contribution of inertial forces. Under these circumstances, viscosity dominates but sieving by the setules may occur. The animal must detect the presence of a diatom by means of chemosensors on the feeding appendages before reaching out for the food.

FIG. 8.8 The planktonic copepod *Euchaeta norvegica* detects prey food particles by means of sensory hairs concentrated on the large first antennae. (Courtesy of Jeanette Yen)

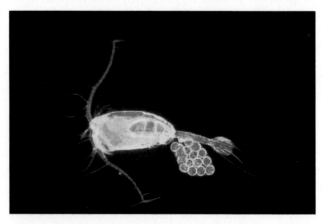

FIG. 8.9 An egg-bearing female of the copepod *Euchaeta elongata* taken from Dabob Bay, Washington. (Photograph by Steve Bollens)

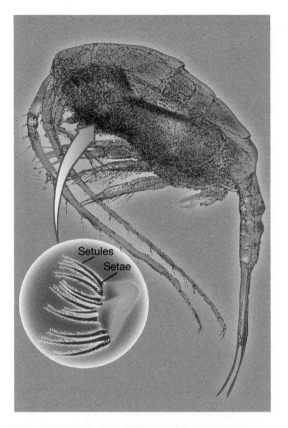

FIG. 8.10 Side view of *Temora longicornis*, a common calanoid copepod of inshore waters of New York. The second maxilliped is shown in detail. (Courtesy of Dom Ninivaggi)

One might wonder about the turbulence of the ocean. These copepods are living in a very energetic water column, especially when wind and tide interact on the continental shelf. Does this affect the small-scale currents generated in feeding and detection of water motion by mechanoreceptors on the antennules? The size scale of oceanic turbulence is usually larger than the scale over which these behavioral activities occur, although there can be some overlap in the most turbulent of waters. This is also true for the time scale: The usual turbulence encountered in water motion is of a slower time scale than the detection and feeding by copepods on phytoplankton cells. Turbulence might interfere with larger-scale copepod motions such as rapid swimming to either attack prey or avoid attacks by predators.

Many copepod species can transition into resting stages in a process known as **diapause**. In some species, this involves the production of special diapause eggs that sink to the bottom and do not hatch until later conditions are favorable, such as the advent of spring. The eastern North American coastal copepod *Labidocera scotti* has a latitudinal cline of dormancy; dormant eggs are produced with increasing frequency toward the north. In contrast to this strategy, species of the genus *Calanus* use dormancy of adult stages and arrested development to live through low food availability in winter or during other low food periods (Hirche, 1996). Late developmental stages of *Calanus finmarchicus* arrest development and sink to deeper depths, ceasing to feed and lowering metabolism. This behavior may reduce their exposure to predators. In Chapter 10, we discuss daily vertical migration in copepods. Arrested development behavior may reduce exposure to predators, but Atlantic Northern right whales, *Eubalaena glacialis*, apparently follow dense layers of inactive *C. finmarchicus* to their abundant resting zone at about 100 m depth in the Bay of Fundy, Atlantic coast of Canada (Baumgartner et al., 2003).

KRILL

■ **Krill are members of the crustacean order Euphausiacea, are larger than copepods, and are important grazers in surface waters of upwelling systems throughout the world.**

Krill (**Figure 8.12a**), which are the crustacean order Euphausiacea, are elongate creatures ranging up to 5 cm long. They dominate the zooplankton of much of the Antarctic Ocean but are also common in highly productive pelagic waters throughout the world, such as in the Benguela Current off Africa. In the Antarctic, they are associated with sea ice and are the main food of baleen whales, other marine mammals, and seabirds such as penguins. Their abundance is crucial in many open-ocean marine food webs. Recent declines of sea ice in parts of Antarctica have had a major negative impact on krill (see Chapter 19).

Krill feed on phytoplankton and smaller zooplankton using anterior thoracic appendages, which are covered in fine setae for filtering food particles. The feeding

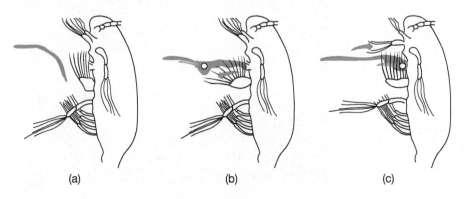

FIG. 8.11 Left-side view of a calanoid copepod, showing (a) the feeding current created by flapping of the appendages, (b) the outward stroke to reach the particle (diatom), and (c) the inward stroke to capture the particle. (After Koehl and Strickler, 1981)

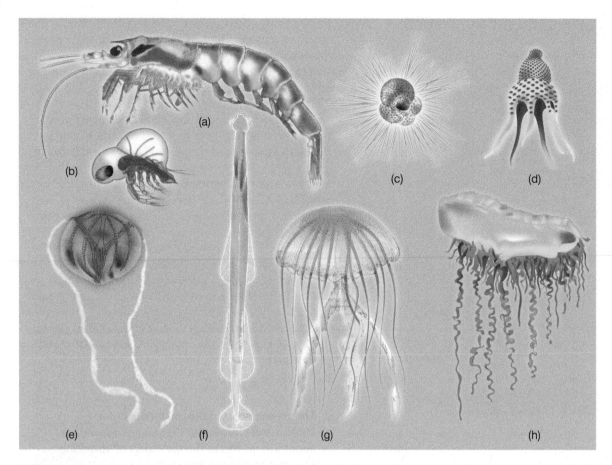

FIG. 8.12   Some zooplankton: (a) krill, (b) the cladoceran *Podon*, (c) a foraminiferan, (d) a radiolarian, (e) a comb jelly, (f) an arrow worm, (g) a scyphozoan jellyfish, and (h) a pleustonic siphonophore, the Portuguese man-of-war *Physalia physalis*.

appendages are held together to form a basket. As the basket opens, water is sucked from the anterior and then filtered by the setae of the feeding appendages as the basket compresses. Other setae comb the food off and transfer it to the mouth appendages (Hamner, 1988). Sperm are transferred in spermatophores, and eggs are carried in the thoracic basket or in ovisacs attached to the ventral thorax.

Krill are famous for swimming well and swarming. This swarming behavior has been associated with movement toward patches of phytoplankton and has been also interpreted as a response to avoid the attacks of predators. While this tactic might work against smaller predatory fish, the swarms are more vulnerable to consumption by baleen whales and flocks of predatory seabirds when the swarms are near the surface. Krill are crucial parts of Antarctic food webs leading to seabirds and some whales (see discussion in Chapter 19). All but a single genus of euphausiacean are luminescent. The luminescent material is intracellular and is located in light-producing photophores, usually on the upper end of the eyestalk.

CLADOCERA   Cladocera are more common in fresh water than in open marine waters, but they are sometimes abundant in estuaries. The genus *Podon* (**Figure 8.12b**) preys on other zooplankton.

OTHER CRUSTACEANS   A few mysids, ostracods, and cumaceans are truly planktonic but rarely dominate the zooplankton. Some mysids rise into the plankton at night but live on the bottom during the day. They are often very abundant but are not well studied. A few amphipods (**Figure 8.13**) are holoplanktonic (e.g., the genera *Euthemisto* and *Hyperia*) and are important members of the zooplankton in many parts of the world.

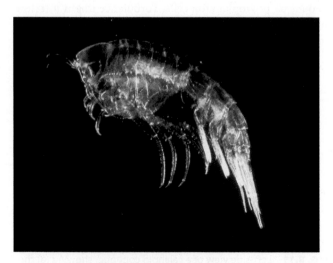

FIG. 8.13   The planktonic amphipod *Themisto compressa*, about 25 mm long. (Photo by Marsh Youngbluth)

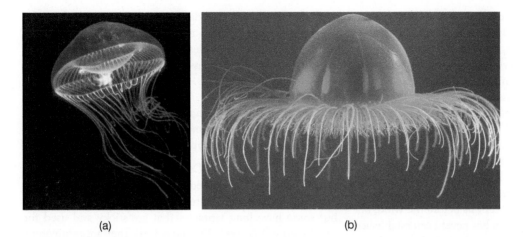

FIG. 8.14 Scyphozoan jellyfish: (a) *Aequorea victoria*, Puget Sound region; (b) the trachymedusan *Benthocodon pedunculata*, collected at 900 m depth off the shore of the Bahamas. (Photographs by Claudia Mills)

(a)   (b)

Gelatinous Zooplankton
Jellyfish and Siphonophores (Phylum Cnidaria)

■ **Scyphozoa are the true jellyfish, and siphonophores are colonial animals with polyps specialized for different functions (e.g., feeding, reproduction). Both groups are abundant in the world ocean and feed on food ranging from phytoplankton to smaller animal prey, including both invertebrates and fishes. A colonial hydrozoan, *Velella*, is sometimes found on the sea surface in great numbers.**

The class **Scyphozoa**—the **true jellyfish**—is divided into several taxonomic orders. Jellyfish swim by rhythmically contracting the bell (**Figure 8.14**), which propels them forward (**Figure 8.12g**). Overall, jellyfish include species that eat food ranging from phytoplankton to small fish, but most species feed on zooplankton. Members of most species capture live food on their tentacles, by means of stinging or sticky structures called **nematocysts**. Some nematocysts produce toxins, and some of the sea wasps (Cubomedusae) can sicken and even kill human beings.

The **siphonophores** are specialized colonial and polymorphic cnidarians (**Figure 8.15**) belonging to the class Hydrozoa. Specialized polyps with different morphologies serve the functions of feeding, reproduction, and floating in some groups. For example, in the Portuguese man-of-war *Physalia physalis*, one individual is a pneumatophore, or float, and may be as long as 10–30 cm. The tentacles dangle below and contain feeding-defensive and reproductive individuals (**Figure 8.12h**). A small fish, *Nomeus gronovii*, lives among the tentacles of *Physalia physalis* and is well protected against predators. It is probably a commensal, but some have suggested that the fish attracts predators into the tentacles of its stinging host.

Subsurface siphonophores (**Figure 8.15**), living at all depths of the world ocean, are by far the most abundant siphonophores in the sea. They consist of a series of swimming individuals that beat in synchrony, propelling the colony forward. Feeding and reproductive individuals are

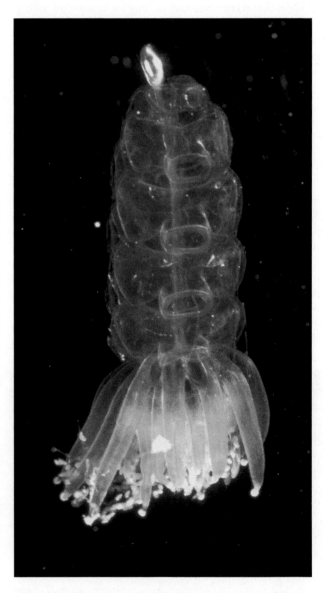

FIG. 8.15 The siphonophore *Pandea conica*, common in continental slope waters of the western Atlantic, is a predator of other gelatinous zooplankton. Up to 30 mm long. (Photo by Laurence P. Madin)

found at the posterior end of the colony. The feeding polyps bear tentacles, which in turn have smaller projections with stinging cells. Siphonophores may be just a few centimeters long or can be several meters in length.

One genus, *Velella*, is a colonial hydrozoan often found in great numbers, sometimes washed on shore. The by-the-wind sailor *Velella velella* is small but has a remarkable sail-like structure, which catches the wind (**Figure 8.16**). These organisms can be seen in the millions off the Pacific coast, often washed ashore from the outer coasts of California to Washington, and on Atlantic coasts as well. A remarkable but enigmatic feature is the presence of left-hand and right-hand sails. They should sail in different directions before the wind, but the reason for this polymorphism has never been established conclusively. Usually a population washed on shore is dominated by either left- or right-handed forms. The animal has a prominent skirt around the gas-filled float that stabilizes the animal as the wind hits the sail. The sail is a low-profile flat sheet that takes advantage of drag for downwind sailing. Unlike modern tall triangular yacht sails, *Velella's* sail does not use the Bernoulli principle to create lift. The closely related *Porpita porpita* seems also to have a stabilizer, in the form of radiating tentacles, but it lacks a sail.

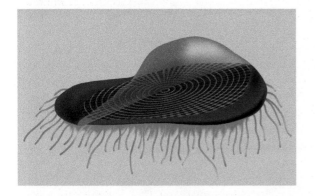

FIG. 8.16 The by-the-wind sailor *Velella velella* has a sail, which catches the wind. The skirt stabilizes the animal as the wind strikes the sail. Millions of these creatures may be found stranded on North American coasts from California to Washington.

Other Hydrozoan cnidarians are usually found in the plankton only as a small medusa a few millimeters wide, which superficially resembles a true scyphozoan jellyfish. They are the medusa stage of a usually larger benthic polyp stage.

The Comb Jellies

■ **Comb jellies are gelatinous and usually feed by means of ciliated rows and sometimes tentacles that capture zooplankton such as copepods.**

Members of the phylum **Ctenophora** are gelatinous, nearly transparent, and egg shaped (**Figure 8.17**). **Comb jellies** are distinguished by eight external rows of meridional plates, but some have long tentacles that are sticky and used for prey capture (**Figures 8.12e and 8.17b**). They are carnivorous and are especially effective at consuming copepods. *Pleurobrachia* captures copepods on its long looping tentacles and draws the tentacles toward its mouth. *Mnemiopsis* lacks the long tentacles, and cilia move the prey into grooves, where they are entangled by a row of short tentacles that pass the food to the labial trough, a structure that leads directly to the mouth. Food enters the mouth, is digested by externally secreted enzymes and is directed to a series of canals where digested products are absorbed. It has been recently discovered that sphincters at the end of the canals pass excreta to the external environment, which means that comb jellies have a complete gut with a type of anus, which was not previously believed to be so (see Maxmen, 2016). This feeding mechanism is very efficient, and comb jellies are believed to cause strong declines in shellfish larval populations. A recent invasion of *Mnemiopsis leidyi* into the Black Sea resulted in the decline of a number of species of fish, owing to predation by comb jellies on fish eggs and larvae. The *M. leidyi* population has since been greatly reduced when it was largely consumed by another invasive species, the comb jelly *Beroe ovata*. Gametes of comb jellies are usually shed into the water, and the embryo develops into a larva that resembles the adult. Comb jellies are often strongly **bioluminescent** and light up like flash bulbs when disturbed. In other species, tightly packed cilia on the beating comb rows refract light and produce a

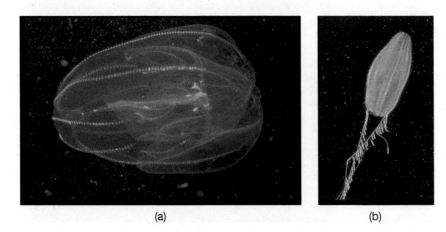

(a)   (b)

FIG. 8.17 Some comb jellies: (a) *Bolinopsis vitrea*, a lobate ctenophore, which is common in the Caribbean and other subtropical regions, up to 60 mm high, and preys on copepods (photograph by Laurence P. Madin); (b) an undescribed oceanic tentaculate ctenophore (photograph by Marsh Youngbluth).

lovely **iridescence** (Welch et al., 2006; **Figure 8.17a**). The function of these flashes is not well known.

Salps

The **Thaliacea** (subphylum Urochordata; see also Chapter 14), or **salps**, are specialized for a free-swimming planktonic existence. Unlike their relatives, the benthic sea squirts, they have incurrent and excurrent siphons at opposite ends of the body. *Salpa* is barrel shaped and solitary. *Pyrosoma*, by contrast, is colonial and shaped like a cylinder closed at one end. Colonies reach over 2 m in length, and each individual has its intake siphons oriented outside, whereas its exit siphon empties into the central cavity. Salps strain phytoplankton and fine particulate matter on a ciliary mucus net, which is ingested. Salps feed on particles in the range of 1 µm to 1 mm. Some salps are important predators on fish larvae.

Larvaceans

The **larvaceans** (Phylum Urochordata, Class Appendicularia; see also Chapter 14) are another group of specialized planktonic gelatinous zooplankton. Usually only a few millimeters in length, these organisms have some features, such as a tail, that are typical of tunicate swimming larvae. The animal constructs a "house" a few centimeters long, to which it is attached or in which it is enclosed. The beat of the tail generates a current through the house. The current is strained by a grid of fine fibers that is stretched across the anterior opening of the house and traps food. Particles are passed to the mouth and are sorted in the pharynx. The houses are periodically discarded because they are clogged with particles; they then collapse and are a major source of organic matter particles that sink to deeper waters.

Protistan Zooplankton

■ **Protists are common in the plankton and usually feed on smaller prey such as bacteria, or they may be photosynthetic. They are an important link between microorganisms and larger animals in the oceanic food web.**

Single-celled protists are common in the plankton. **Protists**[2] are of major importance in the plankton, especially because they consume very small organisms, such as bacteria, that are largely unavailable to most other zooplankton. Protists are consumed by larger zooplankton and are therefore a major link between microbial forms and the rest of the planktonic food chain (see Chapter 11).

Ciliates

Ciliates (Phylum Ciliophora) are ubiquitous in the plankton and often very abundant. They are usually elongated and are covered with rows of cilia, whose coordinated beating propels water, thus setting the protistans in motion.

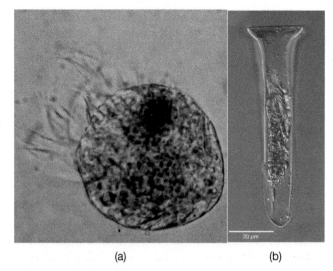

(a)   (b)

FIG. 8.18 (a) Transmitted light micrograph of the marine planktonic oligotrich ciliate *Strombidium capitatum*. Oligotrichs are common ciliates in the plankton. (Courtesy of Diane Stoecker) (b) Abundant tintinnid in the Mediterranean and the world ocean, *Steenstrupiella steenstrupii*. (Courtesy of John Dolan)

Ciliates have an oral opening (**Figure 8.18a**), and cilia around this opening move food particles into the body, where they are engulfed by food vacuoles and digested. Ciliates usually feed on bacteria and phytoplankton. Some species may ingest phytoplanktonic cells but retain the chloroplasts, which remain functional within the ciliate. Photosynthetic products are transferred into the ciliate's cytoplasm.

The **Tintinnida** (**Figure 8.18b**) are an ecologically important group of ciliates and often are a major, even dominant, grazer of the phytoplankton. They have a distinct lorica, a conical organic proteinaceous skeleton, and bristle-like cilia at the mouth region arranged in a circle, which capture phytoplankton cells. Like many ciliates, they reproduce by binary fission but also reproduce sexually through a process known as conjugation. These zooplankton have been found abundantly from tropical to high-latitude waters and are major consumers of the phytoplankton. In turn, they are consumed by copepods and other small zooplankton consumers.

FORAMINIFERA Members of the phylum **Foraminifera** (**Figure 8.12c**) are common throughout the ocean. They range from less than 1 mm to a few millimeters in size. They usually secrete an external calcium carbonate skeleton that is divided into chambers. High-latitude forms usually have simple spiral groupings of spherical chambers, whereas low-latitude species include forms with elaborately spined sculpture. Cytoplasm of this single-celled organism occupies the chambers and streams out through perforations in the shell to form contractile pseudopodia, which capture phytoplankton and bacteria.

Reproduction of foraminiferans usually involves several cycles of asexual cell division, alternating with a cycle of gamete formation. After the gametes have fused, the zygote produces a microspheric shell, which is smaller than

---

[2] The kingdom Protista includes a number of single-celled groups and also several macroalgal groups. It is not a clear evolutionary unit, and the status of the evolutionary-related taxonomy of these groups is unsettled at present.

the megalospheric form, which reproduces asexually by fission. Some tropical Foraminifera have intracellular algal symbionts.

Foraminifera are abundant in the open sea, and certain species are good indicators of the identity of different water masses. Foram calcium carbonate tests sink to the bottoms in great numbers. In depths of less than 2,000 m, where calcium carbonate does not dissolve due to a higher pH, they form deep-sea sediments known as globigerina ooze, named after a common foraminiferan genus.

RADIOLARIA The **Radiolaria** (**Figure 8.12d**) range from less than 50 μm to a few millimeters in size, and colonial forms can attain several centimeters. They are common, especially in tropical pelagic waters. A layer of pseudochitin separates the body into a central capsule and an outer layer of cytoplasm, the calymma. Straight, threadlike pseudopodia (axopods) radiate from the central capsule. The silica skeleton is usually a combination of radiating spines and spheres, producing a complex lattice of great beauty.

Radiolarians feed much like foraminiferans, and some species have symbiotic algae, known as zooxanthellae, within the calymma. The radiolarians derive some nutrition from these algae. Radiolarians reproduce asexually by binary fission, but gametes are also produced. Radiolarians are sufficiently abundant in some parts of the ocean to sink to the bottom and dominate the sediment, forming radiolarian ooze.

Other Zooplankton

ARROW WORMS **Arrow worms** (Phylum Chaetognatha) are torpedo-shaped macroplankton (length is 4–10 cm) with one or two pairs of lateral fins (**Figure 8.12f**). They swim rapidly by means of fast contractions of longitudinal trunk muscles. Armed with grasping spines, the head is adapted for grabbing prey, and arrow worms feed voraciously on other zooplankton, especially copepods. Some species produce a venom containing tetrodotoxin,[3] a deadly neurotoxin that attacks sodium channels that can be used to kill prey. The toxin is derived from bacteria that live in association with the arrow worms. Individual species are often confined to specific water masses, and some of the species can be used to distinguish pelagic from neritic waters. They are hermaphroditic, and eggs may be shed into the water or attached to floating objects. The larva develops directly into a free-living juvenile.

PTEROPODS **Pteropods**, or sea butterflies, are holoplanktonic snails that swim by means of lateral projections from the foot, which is otherwise reduced relative to the bottom-living snails (**Figure 8.19**). Pteropods are sometimes quite abundant, and the shells of one group—the thecosomes—sink to the bottom in great abundance and form sediments known as pteropod ooze.

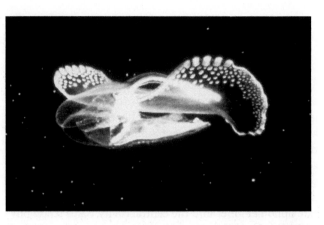

FIG. 8.19 *Gleba cordata*, a sea butterfly. (Photograph by Marsh Youngbluth)

PLANKTONIC POLYCHAETES A few families of polychaetes are holoplanktonic and have well-developed locomotory appendages (parapodia) and sense organs (e.g., the genus *Tomopteris*).

Molecular Techniques to Identify Planktonic Microorganismal Diversity

■ **Microorganismal planktonic diversity cannot be assayed by use of visual techniques alone.**

A variety of molecular techniques allow us to identify and enumerate a suite of species whose small size and diversity make it difficult to make accurate surveys based upon microscopy alone.[4] Accurate identification is necessary because many of these forms are responsible for diverse mechanisms of processing of nutrients in the sea. Identification is thus turning toward a number of biochemical and DNA-based techniques. There is particular urgency in identifying species that might cause harm to humans, such as phytoplankton responsible for harmful algal blooms.

■ **Polymerase chain reaction (PCR) can be used to develop probes to sample and enumerate major plankton groups.**

An example of a molecular approach is to use PCR (see bonus chapter, "Molecular Tools for Marine Biology," online) to probe for known genes that might occur in wholly new and yet undiscovered organisms in the sea. Fuhrman and colleagues (1992) accomplished this by using PCR primers designed to obtain sequences of 16S ribosomal RNA, which is ubiquitous in bacteria and eukaryotic organisms. One can use the same primers for all species because the primers correspond to parts of the gene that have not changed over the course of evolution. Using this approach, the investigators sampled waters in the Pacific at 100 and 500 m depths. Previous studies at the surface had found more common bacteria, but the samples at depth produced DNA sequences that could be related only to the Archaea, a very ancient group distinct

---

[3] Tetrodotoxin is found widely in the sea, including in cone snails, flatworms, blue-ringed octopus, and some puffer fishes.

[4] See bonus chapter online.

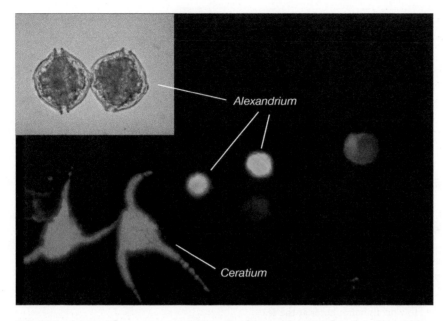

FIG. 8.20 Immunofluorescence staining: cells of the red-tide dinoflagellate *Alexandrium tamarense* stained yellow with an antibody probe to cell surface proteins, along with some *Ceratium* cells that are autofluorescing red under ultraviolet light. Light microscopic image of *A. tamerense* in upper left. (Photograph by D. M. Anderson)

from the bacteria that includes a number of species that live in extreme environments of high temperature (e.g., hot springs), acid environments such as the guts of cows, and in petroleum.

PCR-based techniques now allow us to make a relatively complete survey of the different functioning groups of microbial plankton. A complete survey of known microbial groups has been surveyed within the North Pacific Gyre (DeLong et al., 2006). Sequences associated with photosynthesis, as expected, were recovered with great frequency in shallow waters. DeLong and colleagues report many such depth-related variations, which can be interpreted because of the presence of extensive DNA libraries against which functional gene comparisons can be made. This type of approach identifies groups that are abundant and important in processing materials in the ocean.

■ Immunofluorescence is a useful immunological technique to probe for phytoplankton species.

Immunofluorescence can be used to identify the presence of species in a natural water sample and can even be used to identify expression of specific genes. Immunofluorescence involves tagging antibodies with fluorescent dyes, which fluoresce under ultraviolet light for easy identification. First, as is done in regular immunological methods, an **antibody** is produced that binds specifically to an antigen, which may be an extract of cells of a phytoplankton species in culture or even a purified protein from that species. These antibodies are usually generated against antigens from the cell surface of the species to be probed in the plankton. Then, the fluorescent label is added, which binds to the antibody synthesized upon exposure to the phytoplankton species under study (**Figure 8.20**). When the antibody reacts with an appropriate antigen in the plankton sample, the fluorescent dye is released and fluoresces under ultraviolet light. This technique can be used to identify species of nanophytoplankton that would be otherwise difficult or impossible to identify with light microscopy. Another related immunological technique, **monoclonal antibodies**, is far more specific to a single protein and has also been used successfully to identify smaller phytoplankton (Caron et al., 2003).

■ CHAPTER SUMMARY

- Phytoplankton include *diatoms* (single cells or chains of cells that reproduce by cell division, have sex, and are abundant in high latitudes); *dinoflagellates* (single biflagellated cells, often heterotrophic and common in late spring–summer midlatitude blooms and at lower latitudes, some causing harmful algal blooms); *coccolithophores* (individuals with calcareous test, dominants of oceanic blooms); and *cyanobacteria* (single cells or colonies with nitrogen fixation).

- Among zooplankton, crustaceans include *copepods* (dominant in ocean, suspension feeders or microcarnivores, and swimmers on a small scale) and *euphausiids* (larger than copepods, suspension feeders, and abundant in upwelling areas such as the Antarctic).

- Gelatinous zooplankton are often bioluminescent. They include *cnidarian scyphozoans* (true jellyfish, usually feeding with nematocysts,

microcarnivores); *cnidarian siphonophores* (colonial and polymorphic, most swimming and some with floats, such as Portuguese man-of-war, microcarnivores); *ctenophores* (gelatinous, egg-shaped individuals with ciliary feeding on bacteria, fish eggs, and smaller zooplankton); *salps* (relatives of benthic sea squirts, colonial or solitary, and ciliary suspension feeders); *larvaceans* (often colonies of small individuals, feeding within an organic house); and *chaetognaths* (arrow worms and carnivores).

- *Protists* include *ciliates* (single cells, usually suspension feeders that consume bacteria); *foraminifera* (calcareous, having pseudopods feeding on bacteria); and *radiolaria* (silica skeleton, single cells feeding on bacteria).

- Microorganism diversity cannot be assayed by visual techniques alone, such as morphology and traditional staining. Instead, marine biologists use molecular identification to assay the relative abundance of different plankton with different abilities to process nutrients.

- Immunofluorescence is a useful immunological technique for probing phytoplankton species. Microorganisms are also identified by DNA sequence probes used in conjunction with the polymerase chain reaction (PCR).

■ REVIEW QUESTIONS

1. What is the difference between holoplankton and meroplankton?

2. Why is it important for planktonic organisms to remain in the surface waters?

3. What is the relative value of the Reynolds number (high or low) for most plankton? Why does this matter for their biology?

4. What is the difference between the hard parts of diatoms and dinoflagellates?

5. Why do some dinoflagellates represent potential difficulty for other organisms?

6. How does low Reynolds number influence the feeding of marine planktonic copepods?

7. Where are krill especially dominant in the zooplankton?

8. How do siphonophores at the sea surface stabilize themselves and keep from tipping over in a heavy wind?

9. Upon what element do foraminiferans depend in the making of their skeletons?

10. Why are DNA probes particularly useful in identifying microbial organisms in the water column, relative to other techniques?

Visit the companion website for *Marine Biology* at www.oup.com/us/levinton where you can find Cited References (under Student Resources/Cited References), Key Concepts, Marine Biology Explorations, and the Marine Biology Web Page with many additional resources. You also can have access to a bonus chapter on Molecular Methods in Marine Biology for more information on plankton research.

CHAPTER 9

Marine Vertebrates and Other Nekton

■ **Nekton can swim to the degree that they can overcome many ocean currents. Nekton usually live in a world dominated by high Reynolds number (*Re*) conditions.**

Nekton include fishes, cephalopods, marine mammals, birds, and reptiles. Unlike the plankton, they can swim, often against strong currents. This allows them to move great distances within a day. Some nekton migrate over thousands of kilometers.

Most nekton live at high Reynolds numbers (see Chapter 6). Nekton have momentum while moving through the water. In other words, if a fish thrusts through the water, it will not stop immediately when it ceases to undulate or move its fins. There is no relatively thick boundary layer around it to interact viscously with the surrounding fluid. This fact allows nekton to dart about, coasting along on the strength of periodic thrusts. Contrast this with the situation for small and slow-swimming nanoplankton, which are dominated by viscosity. When a ciliate moves through the water, it effectively drags a small envelope of water along with it and stops moving as soon as the cilia stop beating.

Cephalopods

■ **Cephalopods belong to the phylum Mollusca, are nearly always carnivorous, and are characterized by complex behaviors, a well-organized nervous system, a circle of grasping arms, and a powerful beak.**

The **cephalopods** (**Figure 9.1**) belong to the phylum Mollusca (see Chapter 14 for a more complete introduction to this phylum) and include squid, cuttlefish, chambered nautilus (**Figure 9.2**), and octopus. Cephalopods are the largest of the invertebrates. The open-ocean giant squid *Architeuthis* exceeds lengths of 15 m, including the two long tentacles. The colossal squid *Mesonychoteuthis hamiltoni* is comparable in size and probably the most massive of any squid or invertebrate. But on the other end of the spectrum, the Southeast Asian pygmy squid *Idiosepius paradoxus* is a few centimeters long and the Japanese firefly squid *Watasenia scintillans* is less than 10 cm in length and lives for 1 year. Squids are commonly found in open water, but octopods, the chambered nautilus, and cuttlefish are **demersal**, or associated with the bottom, even though they are good swimmers. The chambered nautilus (**Figure 9.2**) may descend to deeper open waters during the day to avoid predators.

Cephalopods are carnivores and grasp prey by means of a circle of arms (plus tentacles in squids), which are often covered with suckers. The mouth is armed with a powerful beak. Squids feed on a wide range of prey, from large fish to crustaceans such as krill. Squids have eight arms and two longer tentacles, with which they typically capture and draw prey to the mouth. Squids can contract the mantle muscles, expelling water through a siphon, called the hypnome, which propels the squid toward the prey. The siphon must be directed so that the animal attacks with the arms first. It then bites into the prey, using its beak. Prey are seized by the arms, which in the case of squids and octopods are lined with suckers. Smaller squids can swim through the water, spread their arms suddenly, and create local turbulence, which can draw crustacean zooplankton to their arms and suckers, which transfer the prey to the mouth. Octopods capture prey by entwining their

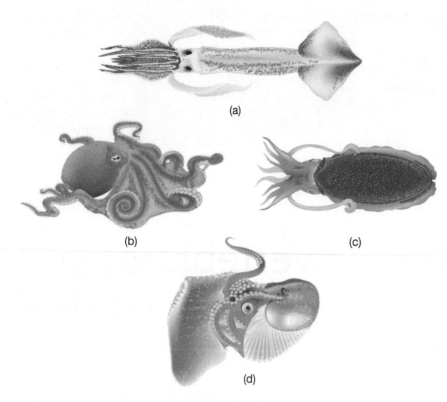

(a)

(b)   (c)

(d)

FIG. 9.1   Some cephalopods: (a) the squid *Rossia pacifica*, (b) the Pacific octopus *Octopus dofleini*, (c) the western Pacific cuttlefish *Sepiella japonica*, and (d) the paper argonaut *Argonauta argo*.

FIG. 9.2   The chambered nautilus *Nautilus pompilius* in its natural midwater habitat. (Photograph by Peter Ward)

The cephalopod nervous system is well organized, and most cephalopods have extremely well developed eyes, organized like the eyes of vertebrates. The giant squids (*Architeuthis*) have the largest eyes of any animal, with eyeballs as wide as 27 cm in diameter and pupils of 9 cm. The large pupil size allows the detection of prey such as sperm whales at distances of as much as 120 m (Nilsson et al., 2012), but absorption and scattering of light place an upper limit.

Cephalopods have great powers of color change. They use specialized pigment cells called **chromatophores**, which are controlled neurally and rapidly enlarged and contracted by attached muscles. The combination of chromatophores with different pigments allows squids, cuttlefish, and octopods to produce nearly instantaneous and spectacular color changes, which often enable them to camouflage against visual predators. The most remarkable change by some cuttlefish species is a moving pattern of ripples that pass rapidly along the body surface. The pattern is believed to resemble the ripple effect caused by light dappling on the seabed. Squids and octopods have an ink gland, which produces ink that is expelled, affects olfaction in predators, and allows escape.

Male cephalopods transfer a spermatophore to the female and internal fertilization occurs. In most species, the eggs are attached to a substratum or are free floating. The female of the paper argonaut (**Figure 9.1d**) *Argonauta argo* spreads two arms and secretes a beautiful spiral shell, within which the eggs are deposited. The paper argonaut is sexually dimorphic: Females are 10–15 cm in size, but males are much smaller.

arms around prey and attaching suckers, which transfer the victim to the beaked mouth. The chambered nautilus lacks suckers but has over 90 arms, as opposed to eight for octopods and 10 for squids (eight shorter arms and two very long tentacles). It lives and feeds near the bottom and is often a scavenger. The paper argonaut grabs prey with tentacles and then injects a poison as it bites the prey with its beak. The cuttlefish *Sepia* usually feeds on bottom crustacea by shooting a jet of water at the sand and uncovering its prey, which move into the water and are then attacked.

■ Some cephalopods can regulate buoyancy by altering the gas content within a rigid structure.

Some cephalopods are able to use gas production and absorption to regulate their bulk density and depth. The chambers of the shell of the chambered nautilus are filled with gas and water in varying proportions that are controlled by the animal to change its buoyancy in order to maintain a desired depth. There are limits because of the water pressure, and the chambered nautilus cannot live in waters much deeper than 800 m because of the danger of implosion. In squids and cuttlefish, an internal chambered and rigid structure also holds varying proportions of gas and water.

The cuttlefish uses a rigid dorsal structure, the cuttlebone, made of calcium carbonate. It consists of a series of lamellae, separated by pillars that help to form a series of chambers (**Figure 9.3**). A yellow secretory membrane surrounds the cuttlebone. Because the cuttlefish can dive between the surface and a depth of about 200 m, it must have a mechanism for gas and water exchange between the rigid chambers and the external medium. The animal uses an osmotic pump for this purpose. As discussed in Chapter 5, distilled water will diffuse across a membrane until the overall salt content is equal on both sides. To remove water from the cuttlebone, salt is actively pumped into small ampullae adjacent to the cuttlebone chambers, which causes water to diffuse from the cuttlebone. The reverse process forces water to move across a membrane into the cuttlebone chambers. The gas content of the chambers is essentially a by-product of diffusion of gas into the chambers as water is removed, or displacement, as the water is added to the cuttlebone chambers. As a result of these processes, the

cuttlefish can regulate the density of the cuttlebone to vary between 0.5 and 0.7 of the density of seawater. At the upper end of the range, the cuttlebone's density balances against the rest of the cuttlefish's body, which is denser than seawater and would otherwise sink.

The overall mechanism of buoyancy adjustment is similar in the chambered nautilus, although the details are quite different. As the nautilus grows, it lays down new calcareous septa within the shell, creating new chambers. At first, the chambers are filled with a fluid that is similar in salt content to seawater. Gradually, however, some of the fluid is removed as the animal grows, and the body and shell become heavier. A calcified tube, the siphuncle, connects all the chambers. The tube is coated with a soft tissue, which contains large numbers of small canals. These canals are filled with fluid of high salt content, creating an osmotic gradient between themselves and the shell chambers. Fluid flows from the shell chambers into the soft-tissue chambers and then empties further into the nautilus's body until neutral buoyancy is achieved.

Fish

■ The marine fishes, which occupy nearly all marine habitats, include mainly the cartilaginous fish (sharks, skates, and rays) and the bony fish.

Fishes are ubiquitous members of marine communities, and it is impossible to do justice to them in a short section. There are about 30,000 species, which can be found in coral reefs, estuaries, submarine canyons, and midwater–bathyal environments and on the deep-sea bed. It is difficult to distinguish between water-column and bottom fish owing to the great mobility of the many bottom-living fish and to occupation of multiple habitats.

All fishes are members of the phylum Chordata, subphylum Vertebrata, and are characterized by a dorsal hollow nerve cord (the spinal cord and brain), an internal skeleton, a complete digestive system, and various arrangements of fins to control movement. The **cartilaginous fishes (Chondrichthyes)** all have a skeleton of cartilage, whereas the **bony fishes (Osteichthyes)** produce a true bony skeleton. Although they are a much smaller group, the cartilaginous fishes (**Figure 9.4**) include the predatory sharks, the bottom-feeding skates and rays, and the whale shark, a huge animal that cruises through the water and eats zooplankton by filter feeding. Cartilaginous fishes usually have a mouth that is ventral, with multiple rows of replaceable teeth, whereas bony fishes tend to have fixed teeth, on the upper and lower jaws and associated with bones on the roof of the mouth, and the gill arches (pharyngeal teeth), which allows fish to chew or process their food. Far more diverse in form and habitat, bony fishes (**Figure 9.5**) include salmon, herrings, and flounder with greater maneuverability and speed, and a greater diversity of feeding adaptations. The jaws of bony fishes are generally directed forward. The upper jaw of bony fishes may protrude, which enhances their ability to suction feed. Other bony fishes may bite their prey or ram feed (filter feeders).

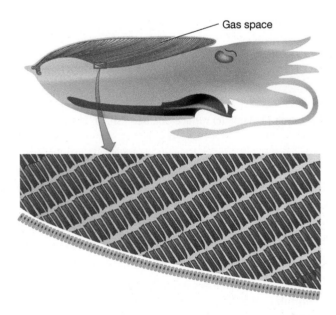

Gas space

FIG. 9.3 Top: the location of the cuttlebone of a cuttlefish. Bottom: cross section showing chambers, which may be filled with water or gas.

FIG. 9.4 Cartilaginous fishes: the shark *Squalus acanthias* and the ray *Dasyiatis akajei*.

FIG. 9.5 Variation in the form of bony fishes: (a) rover predator, (b) lie-in-wait predator, (c) surface-oriented fish, (d) bottom-feeding flatfish, (e) deep-bodied fish, and (f) eel-like fish.

■ **While bony fishes have a rigid skeleton for muscular attachment and contraction in swimming, larger sharks strengthen their cartilaginous skeletons with mineral matter within cartilage and with external mineral plates.**

Bony fishes usually have a well-ossified bony skeleton, which is inherently strong and provides support and sites for muscular attachment. Cartilaginous fishes lost such mineralization during the course of evolution, but many are quite large and also may bite large prey, which requires strengthening. Many skeletal elements are under great stress, such as the jawbones that bite large prey or crush prey, such as mollusks. The skeletal elements of such sharks

and rays are usually strengthened with mineralized plates, known as tiles. The plates can be visualized by the use of CT (computed tomography) scans (Summers et al., 2004). Cartilagenous vertebrae of sharks are usually mineralized and as strong as regular bone.

Swimming

■ **Swimming is often accomplished by means of undulatory body movements, with the aid of paired and single median fins.**

Most fishes swim by throwing the body into **undulatory waves** that pass from head to tail, producing a component of thrust tangential to the body surface and a normal component (**Figure 9.6**) that causes the fish to move forward. The side force component represents wasted energy. There are great differences among fishes in the parts of the body that are used for the undulatory movements. Elongate fish such as eels undulate the entire body in nearly equal waves when swimming at top speed, but faster-swimming and accelerating fishes are stiffer and concentrate most of their movement in the rear part, and tunas concentrate flexing nearly exclusively in the region of the tail (**Figure 9.7**). Gars and barracudas use rapid movements of their strong flattened caudal region to achieve rapid acceleration. In fish that specialize in flexing during swimming, the pectoral fins are used only in stabilization and adjustments for maneuverability. Rays use flapping wing-like projections for swimming, and a few fishes, such as the ocean sunfish *Mola mola* (**Figure 9.8**), rely on the flapping of their median dorsal and ventral fins for propulsion through the water. Many species of fishes generate forward thrust by flapping of the pectoral fins, at slower speeds, but also use adjustment of the orientation of these fins to allow for accurate short-distance maneuverability. Sharks flex their body in swimming but also create dynamic lift by means of stiff pectoral fins that direct the shark upward as it thrusts with body movements.

In Chapter 6, we discussed drag. In fish swimming, skin friction creates little drag relative to pressure drag. Pressure drag is generated as the fish undulates and distorts the normal flow around the body as it moves forward. Pressure drag is minimized if fast-swimming fish are streamlined, which reduces the distortion of

flow around the body as it undulates through the water. Another type of pressure drag is generated as fish flap their fins. The water shed behind the fish appears to be in the form of vortices, each produced by an undulation of the body or the flapping of the paired pectoral fins in some species (**Figure 9.9**). The vortices can be visualized by laser light that is trained on reflective particles, whose motion can be traced by frame-to-frame analyses of high-speed video recordings (Lauder and Drucker, 2002; Lauder, 2015).

■ There are three main functional components to swimming: accelerating, cruising, and maneuvering.

Accelerating (**Figure 9.10**) is maximized by the propulsion generated by a strong caudal fin. This allows rapid escapes or strikes at prey. A deep body allows maximum contact with the water and maximum thrust. By contrast, **cruising** involves continued undulation of the body. The skipjack tuna is a specialist at cruising. It has a stiff body and a streamlined shape, with the greatest body depth about midway down the body length. The fastest forms, such as swordfishes and tunas, have quarter-moon-shaped tails that "shed" water easily as the fish moves, thus minimizing disruptive turbulence at the posterior of the fish.

Maneuvering is best accomplished with a disk- or diamond-shaped body (**Figure 9.10**), which permits body flexure and sudden changes of direction. The optimal shape for maneuverability, therefore, is in conflict with the shape that is optimal for high-speed cruising. The presence of oscillatory fins permits further refinements in maneuverability. Butterfly fishes and triggerfishes are good examples of this general type, and they use the pectoral and other fins extensively in steering. In fish requiring short-range maneuverability, the pectoral fins tend to be broad, as in sculpins, which dart along the bottom and into crevices. Dorsal and anal fins are high on the body and may extend much of the body length (**Figure 9.11**).

Although some species of fishes are specialized for one function, most fish species have forms that are compromises to permit use of all three component functions to some degree.

Fish Reproduction

Nearly all marine fishes are oviparous and shed eggs directly into the water column (cod, herring, striped bass) or lay eggs in nests within the sediment (salmon) or on a hard substratum (sticklebacks, garibaldis). Female fishes may spawn thousands to millions of eggs into the water column, which usually float and are fertilized by spawning males. The eggs soon hatch and develop into yolk-sac larvae, which use the yolk for food, and when the yolk is exhausted they usually feed on zooplankton and then on larger prey. On a smaller scale, some coral reef fishes spawn relatively few eggs near a male in a quiet water site, such as the downcurrent side of a patch reef.

Fish that make nests usually have more elaborate mating behaviors, involving male courting and competition for

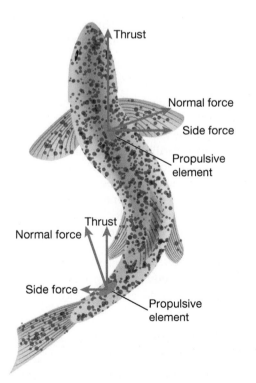

FIG. 9.6 Swimming in fishes. Components of force generated by undulation are shown.

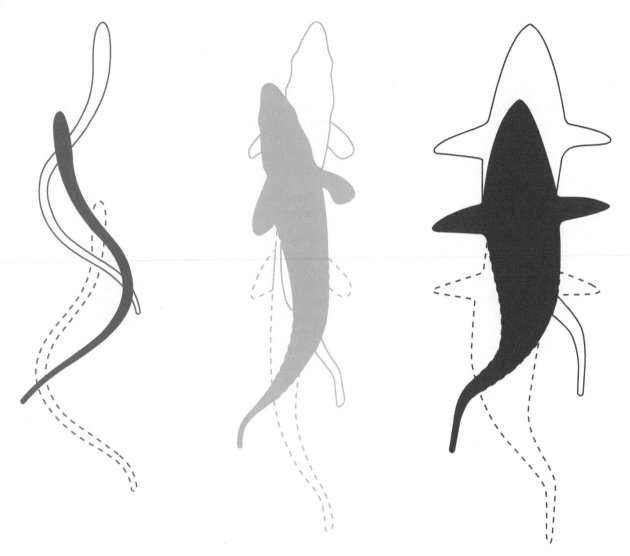

FIG. 9.7 Swimming undulations of fish range from those that are nearly equally distributed throughout the length of the body, such as in an eel (at left), to undulations that are focused in the tail region, such as in a tuna (at right). (From Sfakiotakis et al., 1999, © IEEE)

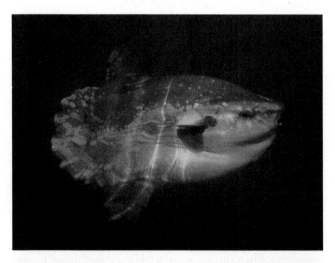

FIG. 9.8 The ocean sunfish *Mola mola*, seen here off the coast of Long Island, New York, is found in all tropical and temperate seas. It is enormous, weighing up to 2 tons. It swims slowly with the aid of large pectoral fins and feeds on jellyfish. (Photograph by Sam Sadove)

females. As sockeye salmon migrate up their natal streams, males become brightly colored, and jaw morphology changes substantially. At the spawning grounds, males fight vigorously for the ability to fertilize the eggs laid by females in gravel nests. Atlantic salmon, which may migrate more than once, have alternate male strategies: a fighting male, whose combat often reduces survival into the next migration cycle, and a nonfighting parr male, that tries to sneak fertilizations without combat and invests more in testes in order to be successful in sperm competition (Fleming, 1996). The garibaldi (**Figure 9.11**) is a damselfish that lives on subtidal rocky areas of southern California and Baja California. Garibaldi males tend groups of egg clusters; they attract females that lay eggs in the nest areas.

A minority of fish species use internal fertilization and retain the eggs until the young are more well-developed. The coelacanth, *Latimeria chalumnae*, is **ovoviviparous**; eggs are retained by the female and fertilization is internal, but developing embryos receive no nutrition from the mother during development. Live-bearing **viviparous**

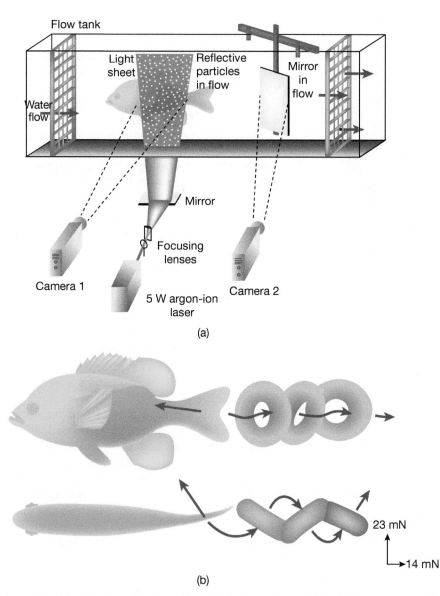

(a)

(b)

FIG. 9.9 Visualization of vortices shed behind a swimming fish. (a) Apparatus, which consists of a fish moving upstream in a flume. Neutrally buoyant particles are illuminated by a sheet of laser light, and two video cameras record the reflection from two different angles. The movement of specific particles can be followed from frame to frame with the aid of a computer-calculated position program. (b) Description of ring-shaped vortices generated by tail undulations of a sunfish. (From Lauder and Drucker, 2002)

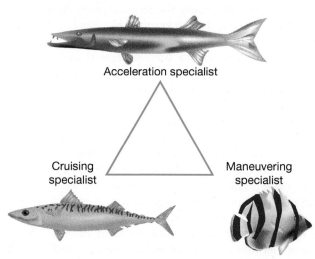

FIG. 9.10 The overall form of fishes can represent an intermediate among the end-member forms that would be ideal for the separate activities of accelerating, steady cruising, and maneuvering.

fishes are relatively rare and include some sharks (lemon sharks give birth to fewer than 20 young) and a few bony fish species.

Classification in Terms of Feeding and Habitat

■ **The body form of fishes is often a good indicator of feeding and habitat requirements.**

Fishes can be classified by their feeding ecology and habitat (**Figure 9.5**).

Cruising predators (e.g., tuna, marlin) are usually long and torpedo shaped, with the fins spaced along the body for maneuverability but with nearly all of the thrusting power in the tail. Lie-in-wait fish predators are also torpedo shaped, but the fins are often positioned near the tail, to help provide the sudden thrust necessary to capture prey. Surface-oriented fishes (e.g., flying fishes) often have the mouth oriented upward, to capture prey at the surface. Benthic

FIG. 9.11 The garibaldi *Hypsypops rubicundus* is a temperate damselfish common in kelp areas such as the shallow subtidal regions of California. Territorial males guard nests within which females lay eggs. These fish are short-range swimmers that maneuver with a laterally flattened body, large dorsal and anal fins, and broad pectoral fins. (Photograph by Carol Eunmi Lee)

fishes are quite variable in shape: The flounders, plaice, and soles are notable for their flattened shape, with one side preferentially on the sea bed and both eyes on the other side, facing upward. Some bottom-roving sharks have strongly flattened heads. Some fishes such as sculpins and gobies have modified pelvic fins that allow them to adhere to the bottom. Deep-bodied, diamond-shaped fishes are flattened laterally and are excellent at maneuvering, although they are relatively slow swimmers. They are often small and able to maneuver among grass blades (**Figure 9.11**) or among crevices in coral reefs. The fins of many are armed with spines that deter predators. Eel-like fishes are well adapted for moving in crevices, as moray eels do in coral reefs.

Oxygen Use and Buoyancy

■ **Fish gain oxygen by passing water over gills, within which blood flows counter to the external water current.**

Fish are active creatures, and nearly all have very high oxygen requirements to fuel their activity. Fish muscle is composed of red muscle fibers, which are responsible for rapid contraction, and white muscle fibers, for slow contraction. Active fishes, such as tuna and sailfish, have much higher red muscle fiber content than do fishes that rest on the bottom. As fishes swim or actively ventilate by moving their gill covers, water passes over structures called **gills**. Each of the **gill arches** has a series of gill filaments, which are in turn divided into thin secondary lamellae (**Figure 9.12**). As the water flows over the gill lamellae, oxygen diffuses into the tissues of the gills. Oxygen cannot diffuse into the gill, however, unless its concentration in the water is greater than its concentration in the blood. Blood flows in a direction opposite to that of the external water flowing over the gills. By moving the blood against the direction of the external water flow, a new supply of relatively

deoxygenated blood will be exposed to more highly oxygenated water. This is a **countercurrent exchange mechanism** (see Chapter 5). Hemoglobin in the blood allows the efficient uptake and transport of oxygen by the circulatory system to needy tissues. Gills are also used for osmoregulation (see Chapter 5) and excretion of nitrogenous wastes.

■ **Fishes differ in buoyancy owing to bulk composition, but most bony fishes use a swim bladder to alter buoyancy by changing the gas content.**

The maintenance of neutral buoyancy can save considerable energy, because a fish can remain suspended in the water column at no energy cost. Moreover, the energy expended to avoid sinking when swimming can be reduced. When a fish is swimming at one body length per second, 60 percent of the total power of movement is expended to avoid sinking. This percentage is considerably reduced at higher swimming speeds (due to the generation of dynamic lift), but a normal cruising speed of three to four body lengths per second still would involve a 20 percent cost to avoid sinking. Neutral buoyancy therefore confers a considerable reduction of energetic cost.

Alterations of **bulk chemical composition** may contribute to buoyancy in bony fishes. The relatively low salt content of cellular fluids reduces bulk density of fishes. Sharks have **high lipid content** in the liver, an organ that may account for up to 25 percent of body volume. The use of the lipid squalene in some elasmobranch fishes can reduce bulk density considerably. Bony fishes generally use a **swim bladder** to adjust bulk density. This organ is generally located mid-dorsally, sits below the vertebral column, and occupies about 5 percent of the animal's volume. When filled with gas, this volume compensates for a fish's 5 percent higher density relative to seawater. Some bony fishes (the more ancestral fishes) have a connection between the esophagus and the swim bladder and can swallow air at the surface. In more derived bony fishes, the swim bladder absorbs or secretes gas to adjust the depth at which the fish is neutrally buoyant. Fishes have no muscular control over swim bladder volume. Rather, variable amounts of gas are secreted or absorbed to keep the bladder at a constant volume as a fish changes depth. Hydrostatic pressure increases by approximately 1 atmosphere with each increase of 10 m depth. Thus, when a fish descends, the increased pressure will compress the swim bladder unless more gas is secreted into the swim bladder. Similarly, when a fish ascends, the decrease in pressure will result in the expansion of swim bladder volume unless the gas is removed. If a deep-water fish is brought rapidly to the surface, the swim bladder expands rapidly and may rupture.

A swim bladder into which gas is actively secreted is illustrated in **Figure 9.13**. The **rete mirabile** consists of two intertwined networks of capillaries, in which blood flows in opposite directions. For simplicity, the diagram shows contact between one arterial and one venous capillary. The swim bladder usually contains nitrogen, oxygen, and carbon dioxide in varying proportions, depending on the species.

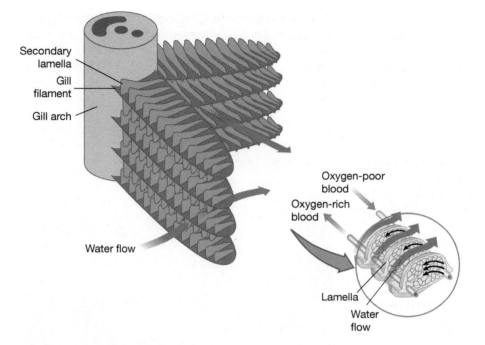

FIG. 9.12 Gill filaments of a bony fish. Within the secondary lamellae, blood flows counter to the direction of the external oxygenated water, to ensure that the external water will be higher in oxygen and that oxygen will therefore diffuse into the blood. (After Hughes and Grimstone, 1965)

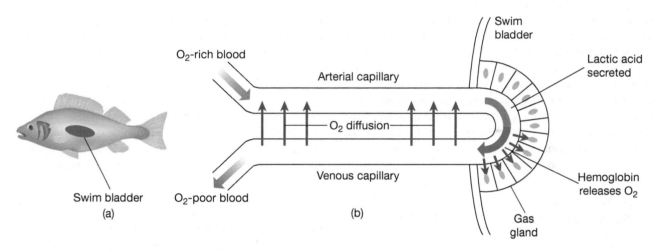

FIG. 9.13 (a) Location of the swim bladder of a teleost fish. (b) Function of the rete mirabile, showing the fate of incoming arterial blood with 10 mL of O_2 per 100 mL of blood. Countercurrent flow retards loss through venous capillary return. (After Schmidt-Nielsen, 1975)

A gas gland controls gas secretion, which occurs against a steep concentration gradient. The cells of the gas gland secrete lactic acid, which causes the hemoglobin to release its oxygen into the blood (this process is discussed shortly). As oxygen is removed from the venous blood, its pressure is greater than in the capillaries carrying the arterial blood, and oxygen passes across into the arterial capillaries. The oxygen is then carried back and deposited into the swim bladder. This **countercurrent exchange mechanism** works because the oxygen tension is always greater in the venous than in the arterial blood, thus causing diffusion of gas into the swim bladder.

The blood chemistry in the capillaries of the gas gland also facilitates the exchange of oxygen. A decrease in pH

(due to an increase in glycolysis) in the capillaries reduces the capacity of hemoglobin to bind oxygen (see discussion of the Bohr effect in Chapter 5), allowing oxygen to be released into the blood. However, the solubility of oxygen in the capillary blood is reduced, owing to an increase in the concentration of lactate, so dissolved oxygen comes out of solution. It is the sum of these two effects that forces more oxygen into the swim bladder.

Gas can also be retained in the swim bladder by the same overall mechanism. As gas in the arterial blood flows toward the swim bladder, the arterial capillaries are in intimate contact with the venous capillaries. Gas tending to diffuse out from the venous capillaries diffuses into the arterial capillaries and is available for secretion into the swim bladder. This tends to maintain the gas content of the swim bladder.

Not all fishes have swim bladders, however: Many bottom-living fish such as gobies and blennies lack them. Sharks, rays, and mackerel also lack them. This may be advantageous to them because rapid vertical locomotion would be difficult for an animal obliged to contend continually with the changing gas pressures of a swim bladder. As mentioned earlier, sharks use squalene to obtain near-neutral buoyancy. Bony fish employ other lipids for the same function. The nearly horizontal pectoral fins of sharks provide some lift by means of the Bernoulli effect. The asymmetrical tail also causes sharks to have vertical mobility while swimming. The upper lobe sheds vortices, which pushes water downward and backward (Wilga and Lauder, 2004).

Feeding by Fishes

■ Most fishes feed by means of suction and ram feeding.

Most fish use **suction** feeding to obtain their food. The rapid opening of the mouth cavity creates a flow field outside the mouth and negative pressure within the mouth, which sucks both water and prey into the mouth, which is then closed rapidly. Suction feeding can also be used to draw much smaller prey, such as zooplankton, which is captured on extensions of the gill arches known as **gill rakers** that form a sieve through which water, but not prey, can move. Fish that feed on smaller plankton often have finely arrayed and long rakers. In other fishes, larger prey may be grasped by jaw teeth and transported inside the mouth to a set of pharyngeal jaws in the back of the buccal cavity, which manipulate and process (chew) prey and help in swallowing. Moray eels have the surprising ability to thrust the pharyngeal jaws forward to seize prey and pull it into the mouth (**Figure 9.14**). This remarkable adaptation allows for rapid seizing of prey from the crevices in which morays usually live. As they say, when an eel reaches out and bites you in the snout, that's a moray!

Larger carnivorous fish may be **ram feeders**. They move forward and their open jaws encounter the prey directly, which can hurt! Some sharks, for example, protrude the upper jaw and depress the lower jaw as they attack. Protrusion of the jaw and enlargement of the mouth cavity coincide with the widening of the jaw. As the encounter with a prey animal occurs, the jaws are closed. The enlargement of the mouth cavity as it opens may also cause suction. In smaller fish, ram feeding may enhance the effectiveness of suction by focusing entering water streams toward the middle of the mouth, which will ensure that the prey enter the mouth as it opens to suck in the prey item. Ram feeding also increases the approach speed of the prey toward the opening mouth, which increases the accuracy of guiding the prey into the mouth (Wainwright et al., 2007).

■ Fishes use teeth to scrape algae from surfaces, tear prey, and crush shells. Many fishes use gill rakers to suspension feed.

Feeding in fishes includes scraping of algae from surfaces (e.g., by parrotfishes and surgeonfishes), crushing of shelled

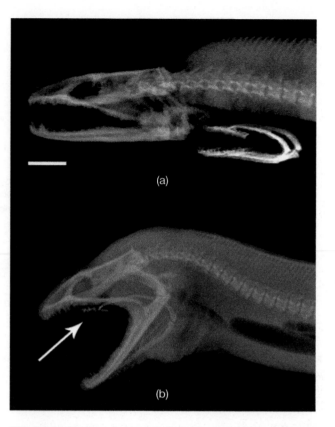

FIG. 9.14 Moray eels protract the pharyngeal jaw to seize prey in the oral cavity. (a) X-radiograph of moray with jaw closed. Pharyngeal jaw beneath. (b) X-radiograph of jaw opened with pharyngeal jaw protruded (arrow). (Courtesy of Rita Mehta and Peter Wainwright)

prey (by pufferfishes), filtering of organic detrital particles (by mullets), filtering of zooplankton (by basking sharks), and active seizure of prey and carrion (by bass, tuna, mackerel, and sharks). Mouth morphology varies extensively. Active carnivores have rows of sharp teeth. The jaw teeth of parrotfishes (Family Scaridae) are fused into plates that can nip off chunks of coral skeletons (see Figure 17.28). A spiny puffer (*Diodon*) can take a shell into its mouth and, by a series of apparent inhaling and exhaling motions, position it on crushing plates embedded in the jaws. The shell is crushed repeatedly until the soft parts can be separated and swallowed. Some sculpins have sharp structures that puncture snail shells, allowing digestive juices to penetrate the shell (**Figure 9.15**). In filterers such as the basking shark, modified gill rakers strain out zooplankton, and the jaw and dentition are relatively weak.

An exciting discovery has been made concerning the use of tools by the tuskfish, *Choerodon schoenleinii*, a resident of the Great Barrier Reef in Australia. The fish holds a bivalve in its jaws, rotates its body from side to side and slams the clam against a rock on the bottom, which is used as an anvil, to fracture the bivalve's shell (Jones et al., 2011).

Filter-feeding fishes are widespread. Some are continuous ram feeders, and water is strained as the fishes move through the water. Basking sharks (**Figure 9.16**), for

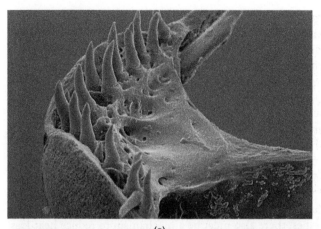

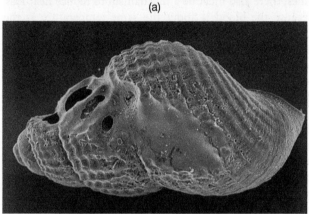

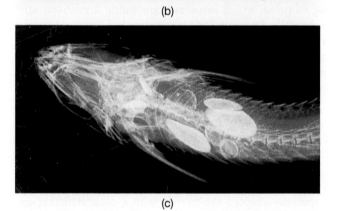

FIG. 9.15 (a) Closeup of the operative mouthpart of the shell-crushing sculpin *Asemichthys taylori*, found in the U.S. Pacific Northwest. (b) A snail shell with holes punched by sculpin teeth. (c) X-radiograph showing bivalve mollusks in a predator's gut. (Courtesy of Stephen F. Norton)

example, have a series of gill arches, on which gill rakers are located, which are in turn festooned with tiny projections that catch particles. Herrings, manta rays, and some mackerels also feed in this way. By contrast, some fishes apply suction within the mouth and strain food particles, such as zooplankton, across the gill rakers. Laurie Sanderson and colleagues (1991), who used a surgical telescope to study blackfish, found that the gill rakers guide particles to a degree, as does a dorsal fleshy organ toward the rear of the mouth.

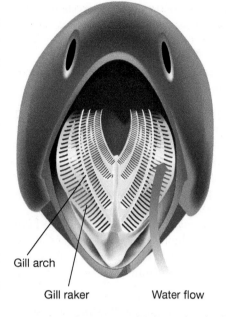

Gill arch

Gill raker   Water flow

FIG. 9.16 Suspension feeding of a basking shark. Water is directed across the gill rakers as the fish is swimming, and plankton is caught on barbs projecting from the gill rakers. (Modified after Sanderson and Wassersug, 1990)

Sensory Perception

■ **Fishes perceive other fishes, prey, predators, and the environment using many structures—the eyes, a lateral-line system, olfactory organs, otoliths for hearing, and in some groups, electroreceptors.**

Fish usually have excellent sensory perception. Elasmobranchs and bony fish have a **lateral-line system** with canals along each of the fish's sides, within which is a series of mechanoreceptors with separate nerves leading to the brain. As the fish moves, slight disturbances created by the presence of rocks or prey stimulate the mechanoreceptors, and the integral of the signals to the brain from these nerves aids the fish in navigation. Blind freshwater cavefish can maneuver around stones on the bottom by means of the lateral-line system alone. In some fishes (e.g., all elasmobranchs), there also is a system of electroreceptors that can detect extremely small voltages arising from prey. Cartilaginous fish can detect prey electrically at distances of greater than a meter. Fish have variously developed olfactory organs, whose ciliated surfaces can detect dissolved molecules as water enters one or two nasal openings.

The inner ear is used both for hearing and balance, as it is in all vertebrates, including humans. **Otoliths** (composed of calcium carbonate) are in contact with the cilia of hair cells, like those in the lateral-line system. Sound causes relative movement between the otolith and hair cells that provides the fish with information about the intensity and frequency of sound. Relative movement of fluid inside the canals and the hair cells in a small organ, called the crista, provide information about the fish's spatial orientation. Vision is well developed in most fishes. In well-lit, shallow-water habitats, many species have excellent color vision, and mating signals often involve displays of color.

Sound Production and Detection in Fishes

■ **Sound travels efficiently through water and therefore is important for detection and for communication.**

Fish propagate sound for detection of predators, but also sound can be generated by fishes for communication. In water, sound travels farther and with less attenuation than visual and chemical stimuli. Therefore, it makes sense that marine organisms such as fish would have evolved the ability to detect sound, probably before the ability to make sounds appeared in evolutionary history.

Fishes may detect sound with their body, but most have sensitive sound detection through the otolith organs of the inner ear. The swim bladder may be connected to the inner ear, which helps amplify sound from the environment and enhance reception.

■ **Sounds can be produced by muscle vibration against the swim bladder, external body part stridulation, and other techniques.**

Drumming sounds are made by a number of species, such as toadfishes, using muscles attached to the swim bladder (sonic muscles), which are able to contract faster than all other vertebrate muscles! External rubbing of body parts to make sound also occurs. For example, marine catfish can rub pectoral spines to make a squeaky sound. Some sounds are made unintentionally. For example, parrotfishes feeding on corals make a squeaky sound with their teeth that is readily heard by divers. Changes in velocity by swimming fish can cause low-frequency hydrodynamic signals, which can be detected by prey and conspecifics.

■ **Sounds are produced by some species during the reproductive season.**

Sound is used by males of a number of species, especially in turbid environments where vision is limited. Principally, sounds are produced as mating calls and signals of territories. Most notable are the toadfishes, which produce long-playing whistles to attract females and shorter grunts that may express aggression.

Body Temperature of Fishes

■ **Tunas, some billfishes, and the butterfly mackerel are warm-blooded, which aids in sustained activity and in maintaining temperature while moving into environments at which temperatures vary.**

Nearly all bony fishes are poikilotherms. Their body temperatures hover within 1–2°C of the ambient water temperature. In some species, elevated muscle temperature allows production of more power and is an advantage to fishes that must swim for extended periods at high velocity. It also allows species such as tunas to function with high metabolic rates even when they dive to deep, cold waters. However, being warm-blooded requires expenditure of a tremendous amount of energy, and only a very few species

groups have evolved this character. Endothermy occurs in the suborder Scombroidei, an assemblage of large oceanic fishes that includes the tunas and mackerels. It also occurs in the shark families Lamnidae (mackerel sharks) and Alopidae (thresher sharks).

For endothermy to evolve from an ectothermic ancestor, a mechanism must be developed to elevate metabolic activity, which generates heat. Such activity is aided by the delivery of oxygen by myoglobin to muscle cells. Myoglobins in endothermic fishes have higher oxygen affinity than in ectothermic fishes, which may allow more complete binding of oxygen to myoglobin and steady efficient release of oxygen to tissues (Marcinek et al., 2001). In endothermic fishes, there also must be a mechanism to reduce heat loss. Because the blood is about the same temperature as the surrounding seawater when passing through the gills, heat can come only from metabolic chemical reactions. In most fishes, the blood is delivered through main vessels that lie just beneath the vertebrae, with branches that deliver blood to the outer part of the body. Tunas, however, have their main blood vessels just beneath the skin, with smaller vessels delivering blood to the interior, which allows heat generated in the body to be conserved. In tunas, heat is generated by red muscle, which is located near the axial region, instead of closer to the periphery of the body (an arrangement that characterizes most other fishes). The central placement of red muscle may be related to the thrusting of tunas, which involves maintaining a relatively stiff body while flexing the tail and peduncle. Heat loss in tunas is reduced through the whole body with the aid of a **countercurrent heat exchanger** (see the Chapter 5 discussion of countercurrent flow in porpoises). The heat exchange involves a series of arteries whose flow is in the direction opposite to that of the adjacent veins. The exchanger allows tunas and sharks to maintain body temperatures of greater than 10°C over ambient temperature (**Figure 9.17**). Atlantic bluefin tuna can be as much as 21°C over ambient.

Billfishes, such as the swordfish, heat only the brain in a process that is accomplished by passing blood through a specially adapted eye muscle with a countercurrent exchange blood flow. The butterfly mackerel uses a different eye muscle to accomplish heat production, also with a countercurrent exchange blood flow mechanism.

Physiologist Barbara Block has used molecular sequences to analyze the evolutionary relationships among the tunas, billfishes, butterfly mackerel, and other scombroid groups (Block et al., 1993). Apparently, tunas, billfishes, and the butterfly mackerel have evolved endothermy independently. In all three cases, this evolution has accompanied an expansion of habitat into cool temperate oceanic habitats. This suggests that the evolution of endothermy was an adaptation for expansion of habitat exploitation, as opposed to merely an increase of potential for activity. This is most apparent in the billfishes and the butterfly mackerel, which heat only the brain. These "hot-headed" fishes control and sustain metabolic activity in the brain, abilities that afford them maximum sensory control while hunting over a wide range of seawater temperatures. For example, warming the

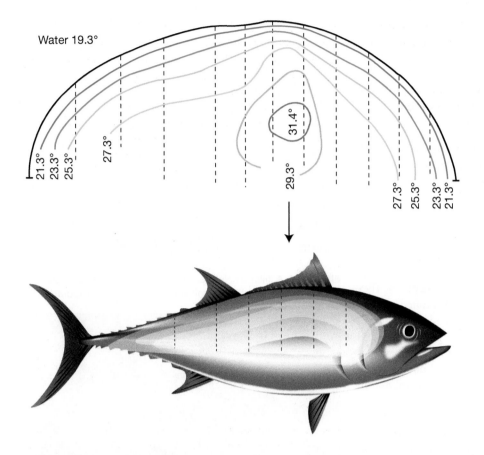

Water 19.3°

21.3° 23.3° 25.3° 27.3° 31.4° 29.3° 27.3° 25.3° 23.3° 21.3°

FIG. 9.17 Temperature distribution in a bluefin tuna. Long thermistor needles were used to measure temperature shortly after the tuna was caught. Isotherms, or lines of equal temperature, show the differential of about 10°C between the core body temperature and the environment. (After Carey et al., 1971)

brain allows swordfish to hunt squids that migrate vertically across a great range of depth and temperature.

Mammals

Whales and Porpoises

■ **The mammalian order Cetacea includes the toothed whales and dolphins, Odontoceti, and the baleen whales, the Mysticeti. All are streamlined and highly adapted to a fully marine existence.**

The mammalian order Cetacea (Phylum Chordata, Class Mammalia) includes the whales, dolphins, and porpoises. The group includes the suborder **Odontoceti**, the toothed whales, and the **Mysticeti**, the baleen whales. In both groups, the body is nearly hairless, elongated, and streamlined, properties that reduce drag as the animal swims. Unlike their terrestrial relatives, these animals have forelimbs that have been modified into stabilizing paddles. The hind limbs have been completely lost and can be identified only by vestigial bones, which do not protrude from the main trunk. The posterior of the body is strongly muscularized and ends in a pair of flukes, which resembles the horizontal tail of an airplane. A strong twisting motion causes the flukes to push against the water and propel the animal forward. This body design enables very efficient swimming.

Some large whales can travel continuously at speeds of over 16 km h⁻¹ (10 miles per hour), and baleen whales can sustain burst speeds of over 32 km h⁻¹ (20 miles per hour). (These estimates come from investigators on whaling ships, who observed whales attempting to escape.)

Cetaceans are homeothermic, and all have a thick subdermal layer of fat that retards heat loss. In extreme cold, blood circulation can be restricted to the vital body organs and brain. Heat is also conserved in the limbs by a counter-current exchange system (see Chapter 5).

Cetaceans reproduce the same way as terrestrial mammals. Some species have elaborate courtship rituals including vocalizations. Female gray whales have been commonly observed in multiple matings over a short period, and DNA studies have confirmed multiple paternity in offspring of individual humpback whales (Clapham and Palsbøll, 1997). After a gestation period of usually several months, or as much as a year for blue whales or 16–17 months for killer whales, the young are born live and under water, which is unique for marine mammals. The young suckle, and the milk has a far higher fat content than that of terrestrial mammals. It has a fishy flavor and is cheesy in consistency. Birth size is also quite large; a blue whale, for example, is about 12 m long at birth. By the time of weaning, after 7 months or so, it has already doubled in length. Cetacean mothers care for their young for as long as 2 years.

Being mammals, cetaceans are air breathers and must return to the surface for oxygen. The nasal opening is the **blowhole**, located on the back dorsal surface of the head. When a whale reaches the surface, there is often a characteristic loud sneezing sound as carbon dioxide is expelled through special flaps that seal off the nasal opening when the animal dives.

TOOTHED WHALES, DOLPHINS, AND PORPOISES

■ **Odontocetes generally prey on large animals and have sophisticated oral communication and complex social behavior.**

The suborder Odontoceti includes species such as the sperm whale, the killer whale (**Figure 9.18**), the beluga whale, and porpoises (**Figure 9.19**). All toothed whales and porpoises actively hunt large prey, such as fishes, and smaller marine mammals, such as seals and sea otters. They have typical mammalian teeth, which are modified to various degrees. There is a single blowhole, and most species are excellent divers, a characteristic that is probably related to their hunting behavior for mobile prey. Sperm whales can dive routinely to depths of 1,000 m, and one has been spotted from shipboard sonar at a depth of 2,250 m! Sperm whales are squid eaters, and they probably dive to great depths, wait there, and then attack unwary squids. Toothed whales and dolphins range in size from just a few meters in length (dolphins and porpoises) to lengths of 15–20 m (the sperm whale).

Most odontocetes are capable of sophisticated oral communication and can generate a series of sonic and ultrasonic clicking signals. Sound travels efficiently and rapidly in water with much less attenuation than in air. The auditory canal is reduced in toothed whales, so sound is likely transmitted to the middle ear through another route, such as the bones of the lower jaw, which are thin and seem to be good transmitters. Sound produced by odontocetes is further focused through the melon, which is a bulbous structure on the anterior part of the upper skull. The melon is filled with a fine oil that was once prized by whalers for use in oil lamps. The melon most likely serves as a device to focus reception and transmission of sound waves, especially those produced in the process of echolocation. Facial muscles may be able to adjust the melon's shape, which could alter sound transmission. In killer whales, individuals produce sounds specific to their pod (group of whales). The clicks are also used as a means of echolocation; the animals can accurately estimate distance on the basis of the travel and return time of the clicks.

Communication and social interaction are of major importance in odontocetes. In most of the species, traveling in small pods is the rule. Killer whales (*Orcinus orca*) travel in pods of usually less than 10, with a dominant male, several females, and a few subordinate males. The dominant male usually mates with the females, but the subordinate males are sometimes successful in mating. Pods are matrilineal and usually consist of a mother, her offspring, and sometimes members of the third generation. When the group reaches a large size, it tends to split into two pods, so a local

FIG. 9.18   A killer whale breaching the surface, San Juan Islands, Washington State. (With permission of Friday Harbor Laboratories)

FIG. 9.19   The bottlenose dolphin *Tursiops truncates*. (Photograph by Sam Sadove)

group of pods can therefore be placed on a family tree. Bottlenose dolphins travel in herds of about 15 and sometimes actively spread out and then corner schooling anchovies in a confined area, to facilitate capture. Individuals can learn to recognize one another on the basis of signature whistles, a recognition ability otherwise known only in humans (Janik et al., 2006). Some evidence exists for specialization within different populations of the same odontocete species. In the

HOT TOPICS IN MARINE BIOLOGY

Whales and Wonder Dogs   9.1

In Chapter 10, we discuss the great difficulty of following large swimming animals in the ocean to track large-scale migration routes. Now, let's move to an entirely different level. What if you want to know the activities and physiological state of large mobile animals such as cetaceans in a relatively small but dynamic region? For example, odontocetes such as killer whales and dolphins live throughout the year usually in small pods under widely varying conditions of food availability, seasonality of mating cycles, and even stress from toxic substances that are released into the environment. Some of the most important marine conservation problems involve declines in regional populations. For example, the Alaskan and Bering Sea populations of the Steller sea lion *Eumetopius jubatus* have declined dramatically in the past few decades. It is not practical to capture marine mammal individuals continuously: This imposes further stress on the creatures and alters the very cycles we want to study. What is needed is a noninvasive way to study such large and yet sensitive creatures.

The same problem arises in studies of terrestrial mammals. The capture of primates causes great stress, and many people have raised ethical issues concerning the potential damage done by shooting them with tranquilizer darts in order to collect blood samples and to make measurements. But such animals also leave a record of their physiological state and even their genes in a very unlikely form: feces. Just by picking scat samples apart, one can see fragments of the recent meals of mammals. I often show students the crushed carapaces of crabs and remnants of berries in raccoon scat. But conservation biologist Samuel K. Wasser and colleagues at the University of Washington have raised this type of study to a far different level.

Over the past 20 years, field biologists have been able to gather a number of types of data from feces of marine mammals, such as killer whales, in the wild:

Genetic identification and variation: DNA sequencing allows the determination of an individual's genotype from fragments of tissue caught in feces, with proper attention paid to contaminants from microbes and food eliminated from the gut.
Diet: By microscopic examination of feces and also DNA extraction, the individual's diet can be determined.
Stress hormones: Many hormones indicative of stress are excreted and can be measured in feces. The large range of known hormones in vertebrates allows an examination of physiological state, especially stress from low food availability.
Contaminants: While contaminants such as organic toxic substances (like hydrocarbons—see Chapter 22) may be retained by chemical bonding with fatty acids and other substances in body tissues, a considerable amount of material is excreted into feces.

A very interesting system is the series of pods of killer whales, *Orcinus orca*, that live in the inland waters of the U.S. Pacific Northwest and British Columbia. Because of long-term observation under the supervision of a number of government and private organizations, nearly all of the killer whales over 2 years of age have been identified, mainly by color patterns on the dorsal surface and fluke. Most of the population is broken up into pods, which are comprised of members of a few generations of whales, centered around a matrilineal lineage. When an orca is sighted, it would be great to collect its scat to determine its relations to others in the pod, and its physiological condition. But how can one get scat from an orca?

A very convenient fact is that, at first, feces released by killer whales rise to the surface! Yes, *it* floats, if only for a while. So with the right sharp observer, floating scat can be related to the whale that has just swum by. As it turns out, the "right" observer is a dog trained to the scent of killer whales, or more correctly to the scent of killer whale feces. Dogs have a strong sense of smell, at least a thousand times better than humans. The methods for training scent dogs derive from those used by government agencies for detecting drugs and explosives. Wasser discovered that he could entice dogs to pick up on scents using a simple reward: playing with a ball! As a result, he and his group have searched dog pounds for dogs that simply love to chase balls. They are the best candidates to train to the scent and are used on boats to sniff out killer whale scat.

Amazingly, dogs can detect the smell of floating whale feces over 1 nautical mile from the boat, so when they are collected the whale has swum away undisturbed. The dog sits in the bow and reacts strongly after a piece of scat rises to the surface and the airborne scent reaches the boat (**Box Figure 9.1**). Once the floating scat is located and brought on board, the dog is rewarded with a ball and the sample is placed in a sampling tube for transport to the laboratory. This approach has already been used by Wasser's group for many species of land mammals, but the adaptation of these methods to the orca is beginning to produce exciting results, identifying individuals by their genes, concentrations of stress hormones, and presence of toxic substances.

There is an urgency about understanding all aspects of the orca populations of the San Juan Islands of Washington and southern British Columbia. Total abundance has declined about 20 percent from 1995 to 2001, and some loss has occurred through 2016. It is not clear why these declines have occurred, but they have been related to disturbance from tour boats; a general decline in their main food, Chinook salmon (*Oncorhynchus tshawytscha*); and the presence of toxic substances such as PCBs, DDT, and others that concentrate in fat. In the Puget Sound region, concentrations of PCBs are believed to be harming the orcas. Recent studies of scat shows that analyses of toxic substances are an accurate indication of concentrations within the body (Lundin et al., 2015).

BOX FIG. 9.1 Tucker, a mixed-breed Labrador retriever who has been trained to sniff and locate floating orca scat at the sea surface, off San Juan Island, Washington. (Courtesy of Sam Wasser)

continues

One discovery is the connection between DNA-based genetic analysis and family relationships among individuals in an orca pod (Ford et al., 2011). Killer whales mate nearly exclusively within the small local group, as evidenced by the genetic markers collected from potential parents and offspring. This result counters previous ideas of frequent mating with other distant pods of orcas. But although mating tends to occur within the pod, there is an apparent avoidance of inbreeding. Mating pairs are more distantly related than you would expect from random combinations of males and females within the pod. Wasser's group has also found that the ability to sire offspring increases with age of a male in the pod, which suggests that age and experience count in a male's mating success (**Box Figure 9.2**).

Food source is also an important factor in orca ecology, and whale scat has been examined by extracting DNA matching sequences to known food sources. Wasser's group used the 16S gene in mitrochondrial sequences of fish to match DNA extracted from San Juan Island whale scat. About 98.6 percent of the sequences indicated salmonids and orcas fed mainly on chinook in early summer and coho salmon later in the summer (Ford et al.. 2016).

Finally, an important objective is to understand causes of stress, and these can be studied by analyzing concentrations of stress-related hormones that are extracted from scat samples. Some very interesting results have been obtained using scent dogs trained to locate scat of the Atlantic right whale off the coast of New England, whose numbers are also in worrisome decline. Comparative study of scat samples from a number of whales showed variation in stress hormones known as glucocorticoids, but a whale enmeshed in a fishing net showed especially high levels (Hunt et al., 2006). Another study has found the compound domoic acid, a highly neurotoxic substance that can be a major danger to human and perhaps whale health, in right whale scat (Leandro et al., 2010). Right whales consume enormous numbers of copepods, which probably feed on a diatom that makes this highly neurotoxic substance. Such studies have enormous potential to piece together several lines of evidence that can be combined to help understand the factors that regulate marine mammal populations.

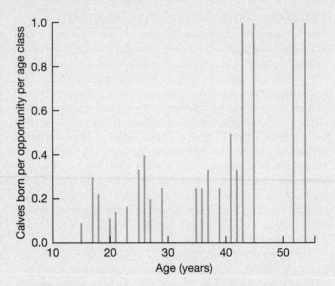

BOX FIG. 9.2 Reproductive success of male orcas as a function of their age. A mating opportunity is defined as a potential father being alive during the year prior to the birth year of a calf that can be inferred to be the offspring of that father. Data for these results were partially obtained from DNA identifications from floating scat of orcas within a pod. (After Ford et al., 2011)

A study of killer whales demonstrates the effects of reduced food, in the form of Chinook salmon, and how this stress compares with the nearby presence of tour boats that follow the whales for much of the year (Ayres et al., 2012). As it turns out, fecal glucocorticoid (GC) hormones decline when Chinook salmon are abundant as food, indicating a reduced food stress. GC concentrations respond rapidly to abundances of salmon. During this same time, tour boats are abundant, but there is no evidence that the boats are a stress factor.

coastal waters of Washington and British Columbia, resident pods of killer whales consist of fish specialists, whereas other, more mobile, pods are specialized for feeding on pinnipeds. Mobile pods are smaller, usually three, which is optimal for hunting seal prey. It would be interesting to know whether this difference has led to language differentiation and mating isolation among the two feeding groups. See **Hot Topics Box 9.1** for a new way to study the genetics and physiological state of cetaceans in the wild.

Some odontocetes have a social structure that is remarkably convergent with that of terrestrial mammals (Connor et al., 1998). For example, sperm whales, *Physeter macrocephalus*, and elephants both have complex behavior and are relatively large-brained. Female sperm whales move in groups of about 10 related females, and parental care is extended for at least 6 years, likely more. Young are vulnerable to predators when their mothers are diving for food, and there is evidence for baby care by related females while the

mother is hunting. Young male Pacific sperm whales leave their elders and spend several years as solitary individuals in highly productive, high-latitude regions such as New Zealand waters, opting for growth instead of trying to compete for mates in low-latitude breeding areas with larger and older males. They return to breeding grounds when they are large and competitive with other males (male sperm whales are highly sexually dimorphic, weighing up to three times as much as females).

BALEEN WHALES

■ **Mysticetes have keratinous baleen plates that replace the function of the more typical mammalian teeth found in odontocetes. They feed on smaller animal prey such as larger crustacean zooplankton and small fishes.**

The Mysticeti, or baleen whales, include blue whales, humpback whales (**Figure 9.20**), and gray whales and are

FIG. 9.20   A humpback whale breaching near Montauk Point, Long Island, New York. (Photograph by Sam Sadove)

distinguished by **baleen plates** (**Figure 9.21**) that are made from keratin, like hair and fingernails, and are derived from dermal tissue. They are attached to the margin of the upper jaw, and are each composed of tightly packed fibers that are frayed on the edge that faces the inside of the mouth. When water containing fish or crustacean prey enters the mouth, the baleen plates strain out the water, allowing the prey to remain in the mouth. The right whales, such as bowhead and right whales proper, are continuous ram suspension feeders (**Figure 9.22**). Their baleen plates are longer, and the whales swim relatively slowly through the water, continuously taking in water and straining finer zooplankton,

FIG. 9.21   Baleen plates (inset, showing black-and-white plates) of a beached finback whale, *Balaenoptera physalis*, ca. 20 m in length. (Photograph by Jeffrey Levinton)

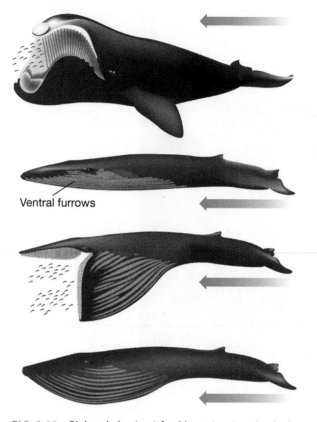

FIG. 9.22 Right whales (top) feed by swimming slowly through the water with the mouth open and baleen plates suspended. Rorqual whales, such as the blue whale, periodically open the lower jaw (bottom sequence) and extend an enormous ventral pouch. When the pouch closes, water is forced through the baleen plates, which are shorter than those of right whales. Zooplankton are trapped on the plates. (From Pivorunas, 1979)

such as copepods. By contrast, rorqual whales, which include blue whales, are intermittent ram suspension feeders. Rorqual whales are characterized by longitudinal pleats in the skin on the outer lower surface of the jaw extending down the body. As the mouth opens, the pleats expand, allowing the whale to engulf as much as 70 tons of water (**Figure 9.22**). Periodically, after engulfing a large volume of water, the blue whale closes its mouth and the tongue is raised, thus forcing water out through the baleen plates. The zooplankton trapped on the baleen are then swallowed.

Whales feeding on small fish known as capelin consume only dense schools. The apparent threshold of feeding activity (Piatt and Methuen, 1993) above a certain fish density must be a behavioral adjustment to increase foraging gains relative to the considerable costs of increased drag while a whale is swimming with its mouth open and baleen exposed to the water, combined with diving time away from air.

Baleen whales range in size from small rorquals and gray whales, which are 10–15 m long, to the blue whale, which can exceed 30 m. Baleen whales often migrate great distances through waters with little food. Gray whales spend the summer in the Bering Sea and the Arctic Ocean. They winter in breeding grounds in bays on the Pacific coast off Baja California and off Korea and Japan. There are several

independent migrating populations of humpback whales. One, for example, winters in the Hawaiian Islands and spends the summers in Alaskan waters. Another population divides its time between waters off California and Mexico.

All mysticetes have two blowholes, whereas odontocetes have one.

SEALS, WALRUSES, AND SEA LIONS

■ **Pinnipeds include seals, sea lions, and walruses and have hair but lack the fat layers of cetaceans. Sea otters are marine mammals belonging to the family Mustelidae that are coastal carnivores.**

Members of the suborder Pinnipedia, mammalian order Carnivora, include the seals, sea lions, walruses, and elephant seals. Pinnipeds have hair and lack the subdermal fat layer so characteristic of whales, but they are streamlined and expert swimmers. The rear legs are modified as flippers and can help maneuver or propel the animal through the water. The group ranges in size from the relatively small (1 m long) sea lions to the enormous southern elephant seal, which can reach 4,000 kg (males) and 6 m in length. Seals spend a great deal of time out at sea diving for fish prey. Most spectacular are the elephant seals. At sea, males spend a great deal of their time underwater, and one has been traced to a depth of over 1,500 m.

The pinnipeds (**Figure 9.23**) are divided into the true seals (Family Phocidae) and the eared seals (Family Otariidae). True seals include the Weddell seal, *Leptonychotes weddellii*, and the northern elephant seal, *Mirounga angustirostris*; they have a small external ear opening and short backward-pointing hind flippers that are used for propulsion in water. These seals move about clumsily on land, in comparison to their elegant propulsion in the water. By contrast, eared seals, which include the common American sea lion, *Phoca vitulina*, and the Australian fur seal, *Arctocephalus pusillus* (**Figure 9.24**), can fold the rear flippers forward, sit on them, and use them for propulsion on land. They have an external ear and relatively longer necks; they use the anterior flippers for propulsion during swimming. In the walrus, the upper canine teeth have become modified into large tusks. In most pinniped species, the teeth are simpler and less specialized than those of terrestrial mammals. Pinnipeds are carnivores and are usually excellent divers.

Pinnipeds are well known for their interesting mating and reproductive habits. Although most species spend the majority of time at sea, they come to beach and rocky-shore areas to mate and rear young. Usually, large males arrive first and establish territories. This is true for species such as the northern fur seal and the elephant seals. The females and subordinate males then arrive, and the dominant males maintain harems of one to several females. Subordinate males risk severe injury if they attempt to mate with a harem female. Dominant males maintain their territories by means of fierce displays and pitched battle. Male elephant seals are approximately five times the mass of females, and males of the southern elephant seal *Mirounga leonina* may mate with 100 females on the beach during

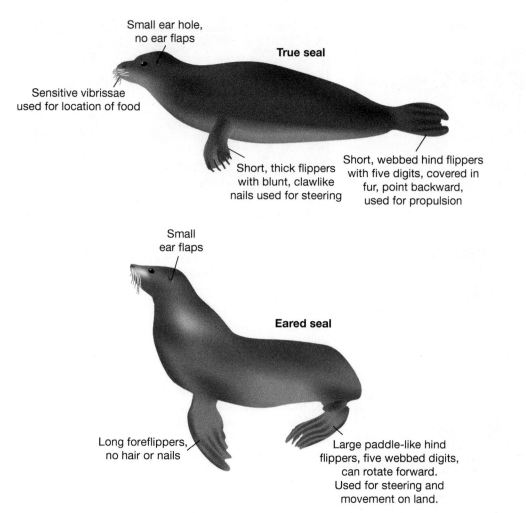

FIG. 9.23   General external differences between true seals (top) and eared seals (bottom).

FIG. 9.24   Australian fur seal, *Arctocephalus pusillus*, mother and pup, Kangaroo Island, Australia. (Photograph by Jeffrey Levinton)

the mating season. Females arrive after the males and give birth a few days later, having been inseminated the previous winter. Females bond with their pups, which grow from approximately 30 kg at birth to 150 kg a month later. Mating begins again about a month after the pups are born.

Males vocalize in order to advertize their territories, but fights often ensue and male combat can result in severe wounds with profuse bleeding (**Figure 9.25**). Subordinate males usually congregate on the fringe of the main territories of the dominant males. Some female southern elephant seals mate with males offshore, so not all mating success of males results from success of dominant males on the beach (de Bruyn et al., 2011).

Sea otters (**Figure 9.26**) are only distantly related to the pinnipeds and belong to the mammalian family Mustelidae, which includes otters and badgers. Sea otters have evolved a number of traits for life at sea, including streamlined body, modified appendages, and hair with an air layer that can be preened and piled into a thermal barrier to retard heat loss to seawater. They are carnivores, preying mainly on benthic invertebrate species such as sea urchins and mollusks and on fishes. They often dive tens of meters to pull abalones and urchins from hard bottoms. They bring these prey to the surface and often crush them with the aid of rocks. Their effectiveness as predators makes them extremely important in the structure of kelp forest ecosystems (see Chapter 17).

SIRENIA

■ **Sirenians include manatees, dugongs, and sea cows. They are hairless and usually herbivorous.**

Members of the mammalian order Sirenia are generally hairless and streamlined, and they superficially resemble

FIG. 9.25 Two bull elephant seals fighting on a California beach. Intermale combat for beach territories often involves fierce and bloody battles. (Photograph by Laura Rogers-Bennett)

FIG. 9.26 The sea otter *Enhydra lutris* is an important predator in kelp forests. (Photograph by Jeffrey Levinton)

whales. They have blunt and broad muzzles, and their bodies are rounded with a horizontal, paddle-shaped tail. They are generally sluggish and live in shallow water, where they eat aquatic plants and seaweeds. West Indian manatees, *Trichecus manatus* (**Figure 9.27**), winter in the shallows

FIG. 9.27 A West Indian manatee in the Crystal River in Florida. (Photograph by Patrick M. Rose, courtesy Florida Marine Research Institute, Department of Environmental Protection)

of the Caribbean and in Florida but may migrate northward and westward in summer. They feed on water hyacinths in fresh water and green seaweeds in saline water. Because of their slow movement and continuous grazing on vegetation in shallow water, the propellers of boats often injure them. Dugongs live in shallow waters from the Red Sea to Australia, but their oil and meat are highly prized, and hunting has severely reduced their numbers. Explorers in the North Pacific discovered Steller's sea cow, but hunters exterminated the species by the 1700s.

Diving by Marine Mammals
OXYGEN DEBT AND CIRCULATION PROBLEMS
■ **Diving causes problems of oxygen shortage, which require a number of conservation responses.**

As mentioned earlier, marine mammals can be excellent divers. Toothed whales dive regularly to depths greater than 1,000 m, but deep diving is also common in elephant seals, otters, and many other species. The Weddell seal, a resident of Antarctica, dives deeper than 500 m and can stay below for nearly an hour, and we have already noted that elephant seals spend much of their time underwater. Diving is clearly valuable because it gives access to prey living at great depths, but the animals must deal with several physiological challenges. Marine mammals save a great deal of energy by gliding during diving. Still, over such long periods under water, oxygen is consumed but carbon dioxide and lactic acid build up in the blood.

Marine mammals must get their air supply at the surface. Moreover, these animals have relatively high oxygen consumption rates, which is necessary to maintain the internal body temperature of mammalian homeotherms. One might therefore suppose that they risk running out of oxygen during extended dives. Special mechanisms, however, have evolved to accommodate these circumstances. For example, seals can carry much more oxygen than humans by making use of the following mechanisms: (1) increased volume of arteries and veins; (2) storage of oxygen attached to myoglobin in muscles and high concentrations of other globins in brain tissue, which prevents hypoxia in the brain;

(3) ability to carry more oxygen per unit volume of blood, owing to increased red blood cell concentration; (4) decreased heartbeat and oxygen consumption rate; and (5) restriction of peripheral circulation to limbs and maintenance of circulation to abdominal organs (to reserve the oxygen supply for essential functions, such as nervous transmission and operation of main circulatory arteries).

It may be surprising that seals and toothed whales do not have extraordinary lung capacity. Inhalation of large amounts of air at the surface would be dangerous to them because then toxic amounts of oxygen might be released at depth. Also, nitrogen taken up at low pressure might then be released, giving the animal the bends (see the following). The animals are, however, extremely efficient at absorbing oxygen, which is then bound to hemoglobin in the blood. During a dive, oxygen in the blood decreases and a "debt" builds up. After surfacing, diving mammals usually increase their metabolic rate and provide needed oxygen to tissues.

GAS BUBBLE PROBLEMS When a diving animal ascends to the surface, pressure decreases about 1 atmosphere for every 10 vertical meters. If the bloodstream is saturated with gas, this decrease in ambient pressure will cause the gas to bubble out of solution. In human scuba divers, bubbles of nitrogen (which comprise 70 percent of air) accumulate in the circulatory system at the joints, causing the painful and potentially fatal syndrome known as the bends. The only possible treatment is to place the victim in a recompression chamber, where pressure is first increased and then lowered very slowly so that the gas leaves the bloodstream slowly without the formation of large bubbles. This is important because bubbles that accumulate in capillaries of the nervous system can suddenly cut off blood supply to part of the brain and cause death. Seals and whales avoid this problem to a large extent. Not only is their lung capacity relatively small, but also they are able to limit blood flow between the lungs and the rest of the circulatory system and allow their lungs and rib cage to collapse to limit the retention of any extra gas (especially nitrogen) that would

be released into the blood during ascent after a dive. They also do not breathe compressed air at depths the way scuba divers do, so they have fewer problems with gas bubbles.

Marine Birds

Seabirds

Seabirds (Phylum Chordata, Class Aves) may travel great distances across the sea and typically breed on offshore islands or in isolated coastal areas. Seabirds consist of a surprisingly diverse array of groups adapted to life at sea. They have a **salt gland** that efficiently excretes salts gained from seawater and food. They range from the flightless cormorants to the frigate birds, which are completely dependent on long-term flight. They range from species feeding upon small zooplankton to those, such as pelicans, that feed on large muscular fish. Some are faithful to a relatively small feeding and breeding area, whereas others migrate for thousands of kilometers. Seabirds are long-lived; albatrosses probably often reach the age of 50 years. In most seabirds, apart from gulls, evolution has involved a strong degree of loss in the ability to walk efficiently on land. Many species breed in colonies, often comprising several thousand pairs, and some tern colonies probably number in the millions of pairs. Breeding often occurs only in isolation from predators, and large breeding groups are often found on remote islands on cliffs or beaches.

■ **Seabirds include the penguins, the petrels and their allies, the pelicans, and the gulls and their allies.**

Seabirds can be conveniently divided into four major groups (**Table 9.1** and **Figure 9.28**). **Penguins** are flightless, and their flippers are modified forewings that are derived from flying ancestors. When swimming, they appear to be flying through the water. They live in cold Antarctic and sub-Antarctic waters in colonies that vary from a few pairs to thousands. They are well insulated from cold air and water by means of a layer of blubber and a thick layer of feathers. They dive from the surface for their food, which

TABLE 9.1 Taxonomic Groups of Seabirds

ORDER	FAMILY	NUMBER OF SPECIES	COMMON NAME
Sphenisciformes	Spheniscidae	16	Penguins
Procellariiformes	Diomedeidae	13	Albatrosses
	Procellaridae	55	Fulmars, prions, petrels, shearwaters
	Hydrobatidae	20	Storm petrels
	Pelecanoididae	4	Diving petrels
Pelecaniformes	Phaethontidae	3	Tropic birds
	Pelecanidae	7	Pelicans
	Phalacrocoracidae	27	Cormorants, shags
	Fregataidae	5	Frigate birds
Charadriiformes	Stercorariidae	6	Skuas
	Laridae	87	Gulls, terns, noddins
	Rynchopidae	3	Skimmers
	Alcidae	22	Auks

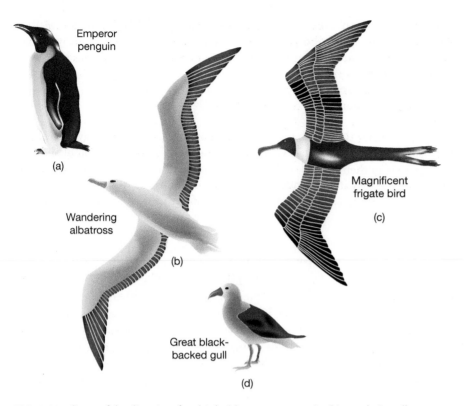

Emperor
penguin

(a)

Wandering
albatross

(b)

Magnificent
frigate bird

(c)

Great black-
backed gull

(d)

FIG. 9.28 Some of the diversity of seabirds: (a) emperor penguin, (b) wandering albatross, (c) magnificent frigate bird, and (d) black-backed gull.

usually consists of small fish. Diving depths can exceed 30 m but tend to be less than 10 m, with dives lasting about 30 seconds. Penguins employ specialized lung sacs to greatly increase aerobic capacity. Emperor penguins equipped with a Critter Cam (**Figure 9.29**), a video camera attached to their back, have been observed diving beneath the ice to depth, ascending to near the ice, and hunting for smaller fish (Ponganis et al., 2000). Other penguins are very dependent upon krill, and climate change has greatly endangered them (see Chapter 19).

Penguins range in size from the little blue (1–2 kg, about 40 cm tall), to the majestic emperor (30 kg, >1 m tall). Protection against the cold is a major factor in penguin biology. Like dolphins, penguins have a countercurrent heat exchange in blood circulation in the wings and feet. A complex nasal passage heat exchange system retains heat as the birds exhale. Penguins such as the emperor huddle in aggregations to keep warm, the individuals on the periphery often moving into the middle of the pack. Father emperor penguins incubate the mother's single egg throughout the entire Antarctic winter, balancing the egg on his feet to prevent it from touching the ice. The baby bird hatches in spring to take advantage of the increase of plankton and fish. King penguin chicks do not leave the care of their parents until about 14 months after hatching from the egg.

Petrels and their allies—the albatrosses, petrels, shearwaters, and diving petrels—have large external nostrils, which may be useful to smell prey, and a hooked bill. Gabrielle Nevitt and colleagues discovered that many species of this group can accurately smell krill-related odors

FIG. 9.29 Emperor penguin equipped with a Critter Cam, a video camera used to observe waters in the penguin's natural underwater environment. (Photograph by Greg Marshall)

in the Antarctic Ocean and therefore will aggregate over patches of krill (Nevitt et al., 1995; Nevitt, 2000). They can also detect dimethylsulfide (DMS), which is a breakdown product of a compound produced by phytoplankton that is transferred into the atmosphere. DMS is often associated with upwelling centers, which are often found in the region of seamounts. Seabirds are attracted to phytoplankton blooms in these local upwelling regions, because zooplankton and small fish accumulate around the phytoplankton. Nevitt has argued that these long-distance-flying birds may store a mental map of the distribution of patches of high DMS concentrations over broad areas of the ocean.

The albatrosses (**Figure 9.30**) can have wingspans of over 3 m and are superb gliders, taking advantage of the steady winds in the southern oceans. Seabirds are nearly all colonial and breed on open and windswept ground. They range from the giant petrel, which preys on other birds, to the small-fish-eating puffins to the zooplankton-straining prions, which have comb-like plates on each side of the mouth. They may nest in colonies from several thousand to just a few, and some species participate in long-ranging migrations.

Pelicans and their relatives—the boobies, gannets, and cormorants—include many brightly colored and ornamented species. They are mainly tropical, but some species nest in the Arctic and Antarctic. While some, such as frigate birds, fly far out to sea, most species of this group stay closer to land. They are diverse in hunting methods, from the plunging dives of gannets to the underwater pursuit of cormorants. As opposed to the broad gliding habit of albatrosses, frigate birds are capable of tight maneuvering in the air. Feeding in this group is restricted to fishes.

Gulls, **terns**, and **auks** comprise by far the most diverse group of seabirds. They can be found in the millions, although their breeding colonies are in the thousands and

not particularly larger than others, such as those of gannets. The herring gull, a complex of closely related species, extends over vast areas of the Northern Hemisphere and can be found breeding in a wide variety of shoreline and island habitats. Terns are smaller and more marine in prey hunting and habitat than gulls. They are most abundant in diversity in the tropics, although the Arctic and Antarctic terns nest in extremely high latitudes. The auks and puffins are found in high latitudes. Puffins nest in holes or dig their own nests under grasses on exposed cliffs. Auks can be found in very large breeding colonies, but some species nest as single breeding pairs. They feed mainly on small fish and zooplankton.

BREEDING BEHAVIOR, NESTING GROUNDS, AND MIGRATION

■ **Seabirds are nearly always monogamous and nearly always breed communally in large colonies.**

The mother and father almost always share care of young, and this may explain the ubiquity of monogamy in seabirds. Monogamy appears to ensure earlier breeding and increased survival of young. The male may share even egg incubation; the male emperor penguin incubates a single egg for the entire Antarctic winter. Nearly all species nest in colonies, some numbering in the thousands. Colonial breeding (**Figure 9.31**) may allow sharing of information about feeding grounds among foraging parents, but the practice more likely arose as the result of the shortage of suitable sites that are remote from predators of nesting birds and their eggs. Communal defense, such as "dive-bombing" by herring gulls, also is an effective deterrent against predators. Colonial living, however, has the disadvantage of putting birds in intense competition for limited space resources in the breeding colony.

FIG. 9.30 The yellow-nosed albatross *Thalassarche chlororhynchos* flies with minimal effort, owing to its extraordinarily long wings that permit it to soar on modest air drafts. (Photograph by Douglas Futuyma)

FIG. 9.31 Colony of northern gannets *Morus bassanus*, Newfoundland. This detached near-shore island has thousands of pairs nesting at this site. (Photograph by Jeffrey Levinton)

■ **Territoriality and combat are common in seabirds because high-quality nesting sites are in short supply.**

Dense colonies of seabirds are found on remote islands, and space on these islands is usually in short supply. It should be no surprise, therefore, that there is intense competition for limited breeding sites. Seabird pairs often spend much time selecting and defending sites, even well before nesting occurs. Every small area is defended aggressively. When a bird moves through a colony—for example, to take off from a high promontory—it inevitably moves through the protected areas of many nesting pairs. This usually involves gripping of beaks, pushing, and flailing around. Still, a male without a nesting site will continually search and attempt to find an unoccupied location.

When a bird first lands at its nest, its mate often confronts the arrival aggressively until recognition occurs. Fighting is usually more intense in species that nest on cliffs or in holes, where space is more limiting than, say, on open-ground sites. In many cases, fights are avoided by stylized displays that demonstrate the mood or size of the potential combatants (**Figure 9.32**). Gulls, for example, can adopt an upright posture, which signifies a readiness to peck rapidly and downward. Gannets bow continuously during nesting to signify site ownership. When one member of a mated pair lands, the two greet by crossing bills and calling. To see this occur by the thousands in a colony is among the most exciting sights one can ever witness.

■ **Mating pairs may be formed before or after a nesting site is chosen. Courtship often includes elaborate displays, which may involve groups of males.**

Mate selection can be intimately involved with choice of a nest site. In strongly territorial species such as gannets, the male first establishes a nesting site and displays to attract females. But female choice clearly involves both the male himself and an assessment of the quality of the nesting site he has chosen. In some cases, however, pairs are formed outside the nesting area, whereupon the pair establishes the nesting site together. Because the relationship between mates is so strong, it is not surprising that courtship is involved and

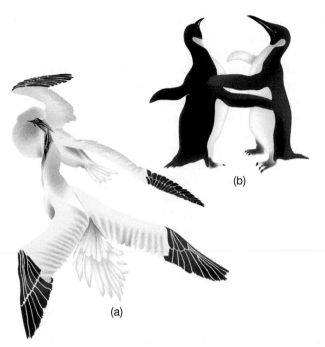

FIG. 9.32 Aggression in seabirds: (a) beak grabbing in Atlantic gannets and (b) flipper fighting between male penguins. (After Nelson, 1979)

occupies a great deal of time and energy. Male herring gulls attract females by means of a call and eventually by feeding the female. Once attracted, the male leads the female to the male's territory, and a new series of head-bobbing behaviors continues the process of mate attraction. In some terns, males catch fish and present them to prospective mates. Albatrosses have extremely elaborate displays, which involve rotating of male and female bills, flapping the lower mandible, occasional dances, and pronounced strutting about by the male. Nuptial coloration, communal displays by males, and elaborate facial adornments also aid in mate attraction by some species. Male king penguins, for example, raise their heads to expose the orange neck and bill patch.

■ **Nesting sites vary in substratum type, degree of slope, and degree of isolation. Many species may share a given area.**

Steep cliffs are often sites for nests of many species of seabird. They provide isolation from predators. A rich island may include nesting cormorants, gulls, guillemots, and many other species. Within these areas, however, individual species have more specialized requirements. Gannet colonies, for example, are best developed on the tops of cliffs, where there are broad horizontal ledges. By contrast, cormorant nests are often found on very narrow ledges in otherwise vertical cliffs. Gulls seem to occupy a broader variety of sites and may be found on cliffs but also on broader, flatter, and lower rocky areas.

■ **The breeding cycle includes periods of nest site establishment, egg laying and incubation, and fledging.**

Seabirds become sexually mature rather late, usually at least in their second year. They then progress through a series of breeding stages, which in most cases are strongly seasonal. In temperate and higher latitudes, breeding must be timed to coincide with the plankton production cycle, when plankton and planktivorous fish are most abundant. This reaches an extreme in Antarctica, where birds must arrive in the spring (September–October in the Southern Hemisphere), dig through the ice and snow, establish nests, and fledge their young by the time of Antarctic summer, when the water is rich in fishes and euphausiid shrimps. Eggs are incubated sometimes for long periods, over 60 days in the case of albatrosses, and larger eggs are produced in areas where food is relatively scarce. Single-egg clutches are very common among seabird species, although females of some species may lay several eggs at a time. After hatching comes the fledging period, lasting until the young birds are able to fly. The first flight is often a glide from a cliff, and a sudden downdraft may result in death. Fledging periods can be short, as in Antarctica, or as long as 6 months, especially if food is relatively scarce. A longer fledging period is associated with a chick's resistance to starvation, which can be crucial in birds that must travel a great distance to the first meal. An extreme case is the albatross, where parents desert the nesting grounds or at least the chicks leave the nest to fend for themselves. Chicks must swim in the water and take off on an air draft. Before this can happen, they may very well be attacked and eaten by sharks and other predators. After a safe takeoff, albatross young clearly have inborn mechanisms to guide them along predictable routes at sea.

■ Seabirds migrate to maximize use of feeding and nesting areas.

Although, as we have seen, nest sites are chosen for their suitability to avoid predators, for social interactions, and for proximity to local food, after the breeding season birds may migrate to distant areas where food is more abundant. If breeding occurs on a mid-Pacific island, for example, this migration may involve moving hundreds to thousands of kilometers in any of several directions in the pursuit of food. Albatrosses and frigate birds fit this pattern. At the other extreme are many gull species, which move no more than a few tens of kilometers from the nesting grounds. In some species, however, the migration is much more directed, especially in high latitudes when the winter season is very harsh. Arctic terns breed during the Northern Hemisphere summer and then fly to Antarctica to take advantage of the austral summer. The short-tailed shearwater migrates between southern Australia and the northern Pacific Ocean. Unlike many land birds, such long-distance migrants often touch down to feed during their journey.

Large colonies of seabirds are probably limited by food availability in the surrounding area. In general, chick-fledging weight is inversely related to breeding colony size, which suggests food limitation by the fish available in the vicinity. Philip Ashmole and Myrtle Ashmole (1967) suggested that seabird population size is regulated by food; in very large colonies, parental birds would have to travel very far from the nesting grounds to find food for their young.

This would put an upper limit on the number of young that could be fledged in a given area.

A study by Thorne et al. (2015) demonstrates how climate oscillations (see Chapter 3) change food availability to nesting seabirds by changing the distance to dense fish prey. Laysan and black-footed albatrosses nest in dense aggregations in the northwest Hawaiian Islands in the central North Pacific. Parent birds of both species must fly north over 600 km from these islands through to a broad and biologically rich front, the Transition Zone Chlorophyll Front (TZCF) in the North Pacific. Laysan albatrosses fly farther northward than the black-footed albatrosses. The TZCF moves toward the north during a full La Niña Pacific Ocean climate state. These conditions increase the distance that the birds must fly northward from the northwest Hawaiian Islands to reach the TZCF and get food for their young. The time lost in making these foraging trips is especially stressful on younger chicks.

Albatross wanderings were followed using radio telemetry tracking of birds that transmit GPS data over a period of a decade. During times when the TZCF is displaced farthest north, the number of albatross chicks fledged was much lower, relative to the number of eggs laid, and effects were most extreme on Laysan albatrosses.

FEEDING

■ Food gathering depends on the alternative strategies of efficient long-distance flying and shallow diving.

Seabirds employ two strategies for hunting food: flying and underwater swimming. To a large degree these strategies are mutually exclusive because efficient flying requires long and relatively inflexible wings, whereas underwater swimming requires short wings that are usable as flippers. **Figure 9.33** shows the methods by which seabirds implement these strategies. Underwater swimmers may dive from the surface (e.g., penguins) or from the air (some shearwaters). Penguins have flightless wings that are highly specialized for underwater swimming. Cormorants use their feet as flippers to swim below the surface. Gulls and terns are good fliers but can dive to shallow depths by plunging into the water from the air. Some seabirds simply rest on the sea surface and pluck fish from the water; others dip their bills in the water and strain out zooplankton. Gulls also feed on land, either by scavenging or by taking prey, such as mussels and urchins, and smashing them on the rocks or even the pavement of roads and parking lots.

The dichotomy of flying and underwater swimming influences the geographic location of seabirds. Efficient underwater swimmers cannot fly very far—or indeed at all, in the case of penguins. Thus, one tends to find specialized underwater swimmers in coastal areas of very high productivity, such as upwelling and polar regions. By contrast, efficient fliers can be found throughout the relatively less productive tropical waters of the open Pacific.

Shorebirds

Shorebirds differ from seabirds in that they have a greater dependency on terrestrial sites for nesting and often migrate

FIG. 9.33 Methods by which seabirds obtain prey: (a) feeding from surface (fulmar), (b) plunge diving (gannet), (c) diving from air (tern), (d) underwater pursuit diving using wings (puffin), and (e) use of feet in underwater propulsion (shag). (After Furness and Monaghan, 1987)

between two sites on continents. Like other migrating birds, they usually migrate between winter feeding grounds and spring–summer nesting grounds. Shorebirds, including the sandpipers, plovers, and oystercatchers (**Table 9.2** and **Figure 9.34**), feed principally on intertidal soft-bottom and rocky-shore organisms and may exert profound effects on these ecosystems. Many species migrate in extremely large numbers and often stop for feeding at specific sites (e.g., Plymouth Bay, Massachusetts; Jamaica Bay, New York; Delaware Bay, Pennsylvania), where they devastate the local invertebrate populations. Sadly, hunting has reduced many of these species to numbers far below those that predated European colonization of North America.

■ Shorebirds migrate great distances between nesting and feeding grounds.

Many shorebird species migrate truly great distances between summer nesting grounds and winter feeding grounds. Phalaropes, for example, nest in high northern latitudes, but they spend the winter in the Southern Hemisphere. The white-rumped sandpiper, which breeds on grassy or mossy tundra in northern Alaska and on islands in the Canadian Arctic, migrates mainly within the continental interior and

winters east of the Andes Mountains, from Paraguay to the southern tip of South America. As mentioned earlier, many stop at major migratory feeding areas, where benthic invertebrates are very abundant. Nests are often very simple, perhaps scrapes of the ground, adorned with just a few pebbles or shells. Predation on eggs by other birds and mammals is common, because of the vulnerability created by the ground-nesting of shorebirds. Some species use a deception: Parents move away from the nest and fake a broken wing to deflect the predator's attention from the nest (**Figure 9.35**).

■ Shorebird feeding is diverse and is related to strong differences in beak morphology.

Shorebirds have a diversity of foraging behaviors and beak types. Unfortunately, this diversity allows us to predict their prey types only partially. Nevertheless, there are several distinct types of feeding. Running and stabbing is done by plovers, which pursue prey (e.g., a crab) and stab it with the beak. Chiseling and hammering are employed by turnstones and oystercatchers. Ruddy turnstones use their bills to excavate clams from the sand and chisel them open. Oystercatchers may chisel a hole directly in an oyster or mussel

TABLE 9.2 General Classification of Shorebirds

ORDER	FAMILY	NUMBER OF SPECIES	COMMON NAME
Charadriiformes	Haematopodidae	ca. 6	Oystercatchers
	Charadriidae	ca. 65	Plovers, turnstones, surfbirds
	Scolopacidae	ca. 85	Sandpipers
	Recurvirostridae	7	Avocets, stilts
	Phalaropodidae	3	Phalaropes

FIG. 9.34 Some of the diversity of shorebirds: ruddy turnstone, purple sandpiper, marbled godwit, and northern phalarope, with detail of webbed foot. (After Schneider, 1983)

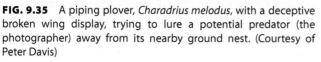

FIG. 9.35 A piping plover, *Charadrius melodus*, with a deceptive broken wing display, trying to lure a potential predator (the photographer) away from its nearby ground nest. (Courtesy of Peter Davis)

FIG. 9.36 The sooty oystercatcher *Haematopus fuliginosus* (seen here on Heron Island, Great Barrier Reef, Australia) is an effective predator on mollusks such as limpets and bivalves. Note its robust bill, used for bashing shells against rocks and severing adductor muscles. (Photograph by Jeffrey Levinton)

or bash a mollusk's shell against a rock (**Figure 9.36**). They also can plunge the beak into a gaping bivalve and sever the adductor muscle, thus creating an opportunity to consume the rest of the bivalve's flesh. Sandpipers use pecking and probing, sometimes penetrating the sand to a depth of a few centimeters, to find infaunal prey. Probing is often accompanied by a shaking of the beak, which liquefies the sand. Godwits have heavily reinforced long bills, and the skull musculature is modified for vigorous probing into the sediment for razor clams and lugworms. Sanderlings simply pick out burrowing animals such as mole crabs that have been washed out by the waves on a beach. A number of species wade in the water and capture prey at the surface or perhaps by probing into the sediment. Phalaropes swim

in the water and use a whirling motion to concentrate crustaceans, which are concentrated by a repeated tweezering motion of the beak, which moves droplets containing food items further into the mouth (Prakash et al. 2008).

Marine Reptiles
Sea Snakes

Sea snakes (Phylum Chordata, Class Reptilia) are relatives of the cobras and are found commonly throughout the Indo-Pacific region, although a single species, the black and yellow-orange colored *Pelamydrus platurus*, occurs along the Pacific American coast. None is found in the Atlantic or the Mediterranean. Sea snakes have fixed fangs and are largely venomous, although very few humans have died from sea snake bites and some species are surprisingly docile. Many snakes lay eggs on shore, but some complete their breeding cycle out at sea and young are born alive in the water.

Sea snakes usually prey on fish, but because they are air breathers, they must periodically go to the surface to breathe. The lung is large and extends to the base of the tail. Some can dive to 150 m, but dives to 5 m are common. Some snakes can stay submerged for several hours and must be tolerant of anoxia. Some species actively trap fish prey in coral crevices and then grab them with the mouth, whereas others lie quite still, like sticks, and strike fish that approach them.

Sea Turtles

Sea turtles (Phylum Chordata, Class Reptilia) were once common throughout tropical waters but are now nearly all threatened, owing to direct hunting and indirect trapping (bycatch) in gear intended for fish and shrimp. The species in United States and Caribbean waters (**Figure 9.37**) are carnivores, with the exception of green turtles, which eat seaweeds and sea grasses (**Figure 9.38**). Kemp's ridley and loggerheads eat bottom-dwelling invertebrates, leatherbacks consume jellyfish from the water column, and hawksbills feed on sponges. Green turtles swim elegantly in shallow water, chewing on sea grasses or seaweeds. By contrast, leatherbacks suck in their gelatinous zooplankton prey (Bels et al., 1998). Leatherbacks are phylogenetically

FIG. 9.38 A green turtle *Chelonia mydas* resting on the bottom at a depth of about 12 m near the coast of southwestern Maui. (Photograph by Jeffrey Levinton)

distinct from other sea turtles and maintain body temperatures much higher than ambient; they apparently conserve heat by means of their large body mass, a layer of subcutaneous fat, and a countercurrent exchange mechanism in the blood circulation of the flippers.

The estuarine diamond-backed terrapin *Malaclemys terrapin* is not a sea turtle and is closely related to freshwater turtles but can live in full-strength seawater. Terrapins may walk along the bottom and seize a mussel by means of a rapid, downward protrusion of the head. Alternatively, the turtle may swim in a burst by means of a rapid backward movement of its forelegs, thrusting itself forward to allow for an attack on a more mobile prey such as a crab.

■ **Female sea turtles lay eggs on specific beaches and migrate between the beach and seasonal feeding grounds great distances away.**

Sea turtles are well known for their beach nesting habits and extensive migrations (see Figure 7.12). Males and females mate near nesting beaches, and females come onto shore several times per breeding season. On each occasion, females lay approximately 100 eggs each, which are buried in the sand. Nighttime nesting serves to avoid predation and minimizes overheating of eggs. After about 2 months, young hatch from the eggs, crawl to the surface, and scramble to the sea. Their movements are guided by three distinct signals (Lohmann and Lohmann, 1998). After hatching, they orient and move toward the horizon, which gets them to the strandline (**Figure 9.39**). They then move opposite to the direction of arriving waves, which moves them out to sea. These movements create an orientation that can be locked in by orienting to the earth's magnetic field; loggerhead turtles can respond to both the strength and the declination of a magnetic field. The magnetic field provides information on geographic position and migration direction, but it is not known how turtles manage to migrate long distances to feeding grounds or locate their natal beaches. Beach development, especially beach use by tourists and light emanating from buildings, has hampered hatchling orientation in many tropical areas.

FIG. 9.37 Sea turtles found in U.S. coastal waters. (From the National Research Council, 1990)

FIG. 9.39 (a) A leatherback *Dermochelys coriacea* hatchling crawling toward the sea. (b) A female leatherback nesting at night on a beach on Isla Culebra, Puerto Rico. (Photographs by Sam Sadove)

Sea turtles spend some time in open water, but juveniles may move into coastal waters or estuaries, depending on the species, to feed until they reach sexual maturity, which usually takes more than 10 years. Long-distance migration is common, and recent evidence demonstrates repeated homing by green turtles to the same beach (see Chapter 7).

■ CHAPTER SUMMARY

- Nekton must overcome many ocean currents to swim. Nekton usually live in a world dominated by high Reynolds numbers.

- Cephalopods are characterized by complex behavior, a well-organized nervous system, a circle of grasping arms, and a powerful beak. Squids and the nautilus are commonly found in open water. Octopods and cuttlefish are nektonic but demersal, or associated with the bottom. Their nervous system is well organized, and most cephalopods have extremely well developed eyes and can change surface color and patterns in order to mimic other species and the background environment. All are carnivores.

- Fish, which occupy nearly all marine habitats, include cartilaginous fish (sharks, skates, and rays) and bony fish. Bony fishes use a rigid skeleton for muscular attachment and contraction in swimming, and larger sharks strengthen their cartilaginous skeletons with mineral matter and external plates.

- Swimming often involves undulatory movements with the aid of paired and single fins. Sharks flex their body in swimming, but they also swim upward by means of stiff pectoral fins. Body form arises from shape adapted to movement and the need to combine acceleration, cruising, and maneuverability to different degrees.

- Fish exhibit a range of feeding strategies, from suspension feeding to carnivory. They may combine suction, ram feeding, and food manipulation by jaws. All use gills to absorb oxygen, and most use a swim bladder to regulate buoyancy. Fish often move in schools, which may afford protection from predators.

- Fish are able to detect sound and water motion by means of a lateral-line mechanoreceptor system and with an ear system, based on vibrations of a stony otolith. Many can produce sound by vibrating a muscle against the swim bladder and rubbing external body parts.

- Cetacea are mammals and include carnivorous toothed whales (sperm whales and porpoises) and baleen whales (blue and humpback whales). They are homeotherms with sophisticated communication, sound generation and reception, complex social behavior, and migrations at many geographic scales depending on the species. Diving causes some problems of oxygen shortage.

- Pinnipeds include true seals, eared seals, and walruses. Many are known for their breeding on beaches, with male territoriality and combat. They are carnivores, generally preying on fish, and may migrate thousands of kilometers between breeding beaches and feeding areas.

- Sea otters are in the family Mustelidae and are associated with eastern northern Pacific kelp beds. They feed on larger invertebrates, such as sea urchins and abalones.

- Seabirds include penguins, petrels and their allies, pelicans, and gulls and their allies. Seabirds migrate to maximize use of feeding and nesting areas. Nearly all feed on fish, crustaceans, and squids. Seabirds travel great distances and typically breed on offshore islands or coastal areas protected from predators. They are long-lived, usually monogamous, and colonial in breeding habit. Territoriality and combat are common in seabirds because high-quality nesting sites are in short supply.

- Shorebirds differ from seabirds in that they have a greater dependency upon terrestrial sites for nesting and often migrate between two sites on continents. Shorebirds migrate great distances between nesting and feeding grounds. Feeding is diverse—chiseling mollusks, probing for burrowing invertebrates, and stalking for fish—owing to differences in beak morphology.

• Sea turtles consist of few species, most of which are threatened by fishing and habitat alteration. They may migrate thousands of kilometers between nesting and feeding grounds. Food may include sea grasses, seaweeds, jellyfish, and crabs.

■ REVIEW QUESTIONS

1. How do cephalopods move to capture prey?

2. How do cuttlefish and the nautilus regulate their vertical position in the water column?

3. There are a few species of nautilus in the ocean, but these are the only cephalopods with external shells. Why were there many more externally shelled species in the geological past?

4. How does the fish swim bladder use a countercurrent exchange mechanism to regulate its vertical position in the water?

5. What is the difference in the pattern of tooth emplacement between bony and cartilaginous fishes?

6. How does a shark maintain a rigid body while performing undulatory swimming despite the lack of a rigid bony skeleton?

7. What is the best body form for a fish that must maneuver among rocky crags in a coral reef?

8. Why is it that in any single marine habitat, one finds fish species of many shapes rather than a single general shape?

9. Fishes often swim in large schools of a single species. Why might this be so?

10. Why is there a hydrodynamic conflict between feeding and rapid movement through the water in baleen whales?

11. What are two different mechanisms that marine mammals employ to insulate against heat loss?

12. Why may language tend to diverge among different groups of killer whales?

13. What might cause a diving carnivorous mammal to dive deeper for prey than before? What are the costs and benefits that might enter into such a decision?

14. Why do you think a related female might be altruistic and babysit for a juvenile sperm whale while its mother dives to hunt?

15. Male animals are about the same size as females when the species is monogamous. Male seals that maintain harems, however, are often much larger than females. Why should there be such sexual size difference when the species is polygamous?

16. What are the advantages of colonial breeding in seabirds? Disadvantages?

Visit the companion website for *Marine Biology* at www.oup.com/us/levinton where you can find Cited References (under Student Resources/Cited References), Key Concepts, Marine Biology Explorations, and the Marine Biology Web Page with many additional resources.

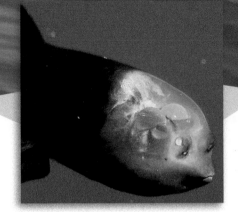

CHAPTER 10

The Open Sea
Distributions and Adaptations

Introduction

The open sea is occupied by a staggering diversity of organisms, many of which are poorly known. In this chapter we will focus on important processes that distribute plankton and some nekton in time and space, and function in a wide range of light and depth conditions in the open sea. We will focus on physical processes that disperse plankton, behavior that causes aggregations of plankton and attraction of plankton feeders to plankton, and oceanographic conditions that result in ocean-scale movements of predators at the top of food webs such as whales, tunas, and large sharks. We will discuss interactions of processes that regulate spatial differences in phytoplankton abundance and relate these on many spatial scales to consumers. We will cover the major zooplankton phenomenon of diel vertical migrations. Then we will explore transfer of material to deeper depths and adaptations of plankton to avoid predation in the vast deeper open-ocean ecosystem.

Near the surface, a diverse group of phytoplankton species are abundant and produce much of the base of the food web for the bulk of the ocean. Within the surface waters, phytoplankton rely on light for photosynthesis but they also depend on a supply of nutrients, which arise in spatially and temporally complex patterns. We will show how light, turbulence, and nutrients interact to determine primary productivity in Chapter 11.

A portion of the phytoplankton are consumed by zooplankton in the surface waters and eventually move through the food web to top carnivores. But this productive **surface system** is connected to **deeper water systems** by a cascading transfer of sinking material and swimming animals from surface waters to the deep ocean. The transfer of material to the deep sea can occur because a good deal of material is *not* consumed by zooplankton near the surface but drifts to deeper waters, to be consumed by a vast array of zooplankton and microorganisms adapted to the fall of detrital material. Many other planktonic animals in deeper waters are eventually supported by this subsidy of sinking **particulate organic matter (POM)** from the surface waters. These consumers are preyed on by a diverse group of predators, often beautifully adapted to a rare meal. As we will discuss in Chapter 18, much of the remote deep ocean floor is the final catchment for sinking material and much of the open-ocean abyssal ocean floor has very little organic input.

Spatial Processes that Create Plankton Patches in the Open Sea
Detecting and Explaining Patchiness of the Plankton

■ **Plankton are rarely distributed homogeneously throughout the ocean. Rather, we see strong spatial differences in plankton abundance, to the degree that many plankton species occur in discrete patches.**

Because light is so important in photosynthesis, **most primary production occurs in the surface waters**. But plankton are also spatially variable. Both phytoplankton and zooplankton are usually **aggregated** in space, on many spatial scales. Commonly, discrete **patches** of plankton are encountered. A plankton net will capture large numbers of a species in one location, but very few may be caught nearby.

Phytoplankton patches can be detected by extraction of chlorophyll, which is a good indicator of the presence of photosynthesis; strong spatial variations are found on the scale of meters to kilometers. Fish and larger zooplankton can often be detected as large patches by **acoustic detectors** using narrow- or wideband **acoustic signaling** whose reflections can be used to analyze the strength and spatial distribution of reflected sound pulses, which may represent krill, small fish, or other groups (see Benoit-Bird and Lawson, 2016). Such acoustic methods show that zooplankton are found more frequently together in discrete patches than one could predict from an expectation of randomness (**Figure 10.1**).

The following are the major causes of patchiness in the plankton:

1. **Horizontal spatial changes in physical and chemical conditions**, such as light, temperature, nutrients, and salinity to which plankton are adapted in differing combinations of conditions, and may reproduce at different rates in different microhabitats.

2. **Specific depth gradients in salinity and temperature**, as can be seen in many estuaries and coastal water bodies, which attract zooplankton and mobile phytoplankton to different microhabitats or depths.

3. **Water turbulence and current transport**, which transport plankton and often concentrate them in localized aggregations.

4. **Grazing** in some areas and reduced grazing in others, leading to patches of phytoplankton and zooplankton.

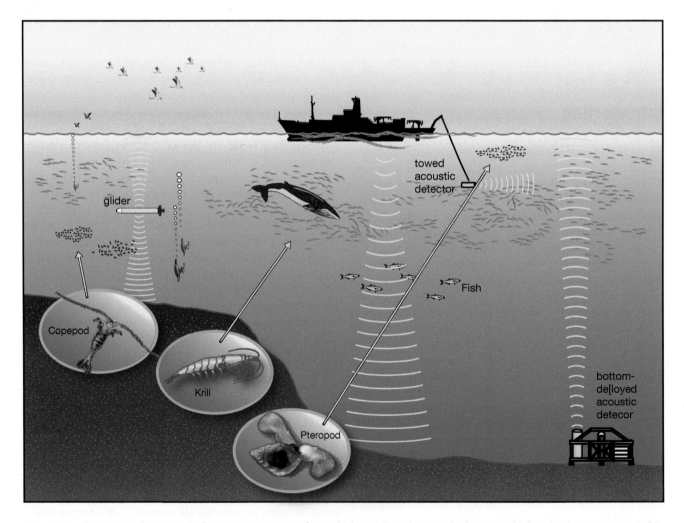

FIG. 10.1 The water column is a challenging environment from which to collect data on planktonic and other marine organisms. A ship or an underwater vehicle can deploy a range of acoustic signal sources to detect a wide size range of open-ocean creatures, from whales to zooplankton, such as krill and copepods. Complex acoustic signals can be resolved to distinguish among species in the water column (After figure supplied by Kelly Benoit Bird).

5. Localized **reproductive behavior**, causing aggregations of zooplankton.

6. Aggregating **feeding behavior** in many larger zooplankton and fish, resulting in concentrations or schools.

Phytoplankton and smaller zooplankton distributions are often determined by localized wind patterns.

Wind moving over the water surface generates **spatial structure** differing from random, over a wide range of spatial scales. **Langmuir circulation** results from the creation of vortices by wind-driven water movement. The vortices are small in scale and result in small divergences and convergences of water (**Figure 10.2**). Phytoplankton may be concentrated in the convergences, where downwelling water cells drag particles. Zooplankton may be trapped in an upward current while attempting to swim downward to avoid light originating from the surface. Langmuir circulation forms linear convergences at the surfaces and plankton may be concentrated by physical transport to these convergences (**Figure 10.2a**).

Wind-generated turbulence can generate patches of phytoplankton on a scale of 100–1000 m. Differences in abundance from place to place may reflect larger-scale turbulence but this condition might also be due to local differences in phytoplankton growth rates, perhaps owing to localized differences in nutrient overturn from deeper waters. Directional water currents can also cause persistent spatial patterns in circulation. Obstructions such as islands and narrow passes at the mouths of estuaries can strongly alter flow patterns, which may concentrate or disperse phytoplankton patches.

Phytoplankton and smaller zooplankton are often concentrated in layers at different depths below the surface. Sometimes these variations are associated with a vertical water density gradient due to variations in temperature and salinity. Even limited swimming ability may allow some phytoplankton species to establish small scale depth-related differences in population density. For example, dinoflagellates can swim vertically and adjust their depth if the water column is not too turbulent. Photosynthetic dinoflagellates may often be found closer to the surface, whereas heterotrophic nonphotosynthesizing dinoflagellates may be found deeper in the water column as they are less dependent on light (Mouritsen and Richardson, 2003). A lack of turbulence is more likely in a stratified water column, as when there is a vertical density gradient in an estuary due to lower salinity–warmer water located above higher salinity–cooler water.

Phytoplankton density results from a balance between population growth and dispersive processes caused by turbulence and wind.

Phytoplankton can reproduce rapidly and under conditions of strong light and high nutrients will grow rapidly. But a concentrated patch of phytoplankton must inevitably spread outward because of the transfer of wind and current energy into kinetic energy that spreads a phytoplankton cell away from a specific point because the cells are entrained in the moving cells of seawater. Dye that is placed in relatively quiet water will gradually disperse in all

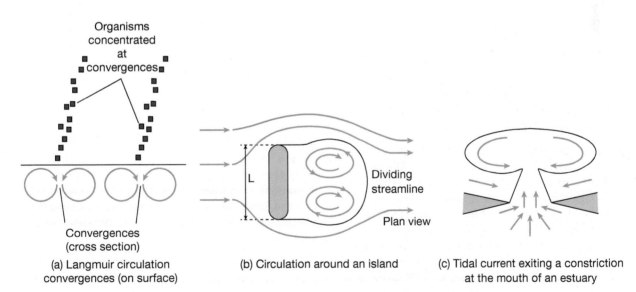

FIG. 10.2 Spatial heterogeneities generated by wind and moving water. In all these cases, plankton would be entrained in these currents and eddies. Streamlines are in red. (a) Langmuir circulation cells, which are circular in cross section. Converging currents, however, produce a series of parallel linear aggregations of plankton at the surface. (b) Vortices formed downstream of a current moving past an island. (c) Vortices formed as a tidal current move through a narrow pass, such as the mouth of an estuary.

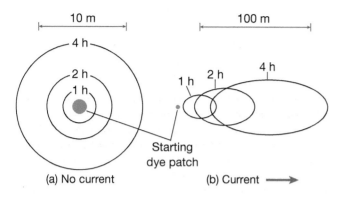

FIG. 10.3 The spread of a concentrated patch of dye when added to the water: Size and shape of patch are shown at the start and at several subsequent times. (a) Case of no current and random turbulence; dye is diffusing equally in all directions. (b) In a unidirectional current, shape of dye patches becomes larger and elongated in the direction of the current.

directions (**Figure 10.3a**). If the dye is placed in a current, the dispersion of dye will spread mainly in the direction of the current (**Figure 10.3b**). However, a phytoplankton population is obviously not a jar of dye; phytoplankton reproduce, often with a few generations per day. Therefore, **if zooplankton grazing is absent, the persistence of a phytoplankton patch is a balance between the growth of the phytoplankton population and the rate at which turbulence disperses the population.** As the rate of turbulence decreases and the rate of phytoplankton reproduction increases, there is an increased likelihood of appearance of a concentrated patch of phytoplankton. Thus, the introduction of a rapidly growing phytoplankton population into a sluggish body of water with adequate nutrients is the probable condition necessary for a rapid growth of phytoplankton into a concentrated patch. Predation by dense associated zooplankton can also cause the failure of the development of a bloom. So the development of a phytoplankton patch depends on nutrient supply, turbulence, and grazing rate.

■ **Zooplankton often occur in patches as a result of behavioral aggregations and periodic movements. Such aggregations may attract predators.**

Zooplankton swimming behavior, combined with attraction between individuals and other forms of adjustments of water depth, are major causes of patchiness (Folt and Burns, 1999). Zooplankters such as protists and copepods are small and undoubtedly subject to redistribution by currents, turbulence, and localized cells of water movement, as shown in **Figure 10.2**. But swimming behavior, even at slow swimming speed, allows zooplankton to move toward local resource patches but also to behaviorally associate with one another and to form dense patches that are not caused directly by water motion alone. Mysid shrimp and krill are well known for their **swarming behavior**, resulting in great concentrations of individuals. Copepods are also known to swarm. One interesting question is the nature of biological signaling that causes such attractions.

Concentrations of phytoplankton can cause zooplankton to swim to a food source, but detection of signals between zooplankters might also be important and aggregation may arise, for example, from attraction between mates. For example, male copepods can detect females at some distance when females release pheromones. Turbulence at small scales can spread such **odor trails** and increase the distance over which the copepods can detect each other. Copepods, for example, have chemoreceptors on the large first antennae, which can be flicked to create small-scale turbulence, which increases the chance of detecting the odor trail of a nearby copepod.

Aggregations of zooplankton may also arise from similar behavioral responses to environmental variation. In Chapter 19 we will discuss krill in the Antarctic, which aggregate in the vicinity of floating sea ice. As seasonal changes of ice cover occur, krill move to respond to changes in algal food resources. In coastal waters and estuaries, mysid shrimp are often associated with the bottom, but substantial parts of the population rise and fall in accordance with seasonal, lunar, diel, and tidal cycles, leading to strong aggregations that can be detected by acoustic reflections (Sato and Jumars, 2008). Such mysid aggregations are important in the foraging behavior of gray whales, which seek out these aggregations (Kim and Oliver, 1989). Gray whales will feed only when mysids are above a threshold density, because otherwise it does not pay to expend the energy to dive for food (Feyrer and Duffus, 2015). See **Hot Topics Box 10.1** for a discussion of foraging strategies of blue whales on threshold aggregations of krill.

When sonar was first employed, **deep-scattering layers** were discovered at depths of 5–1,500 m. These consisted of dense populations of vertically migrating fish and cephalopods that reflected sonar. These layers reflect the attraction of zooplankton and small nekton to aggregations of particulate organic matter that accumulates at certain depths in parts of the ocean at 500–1,000 m depth. Because of regular daily migrations (discussed below), many zooplankton groups are aggregated as they move up and down.

Diel Vertical Migration of the Zooplankton

THE PHENOMENON

■ **Many planktonic organisms undergo diel vertical migrations: They move toward the surface at night and descend during the day. The behavior is controlled by a biological clock that is reinforced by environmental light cues.**

In the 1950s, the Norwegian explorer Thor Heyerdahl set off across the Pacific in an outrigger canoe, one he reckoned was like those used by ancient Polynesian explorers. At night in the open ocean, he noticed a fantastic abundance of jellyfish and other zooplankton near the sea surface but far fewer at midday. Zooplankton were migrating to the surface waters at night and spent the daytime at some depth beneath the surface.

Blue Whale Diving: Balancing Food and Energy Needs   10.1

It is not clear which is more unlikely: a creature 30 m long that is warm-blooded and must feed on millions of tiny prey, or our ability to successfully understand how such a creature evolved! Blue whales are the largest animals in the ocean—indeed, in all of the history of life—with credible sightings over 30 m long and weighing about 40 tons, about the same as 25 large elephants. They are likely the largest individual creatures ever to have existed. We know of whale sharks and manta rays, which are attracted to great concentrations of larger zooplankton and slowly move through these crowds of (usually) larger crustacea, filtering out the prey using large, specially adapted gill rakers. But these creatures are poikilotherms, usually sluggish for much of the day, and generally feeding in shallow waters. Rorqual whales, in contrast, are homeotherms with high body temperatures and therefore large metabolic needs. They regularly dive, seeking swarms of krill.

Much of the blue whale's anterior body consists of a mouth, compressed in large folds when the mouth is closed. But periodically the mouth expands and the folds are stretched. The whale lunges forward, engulfs a large volume of water, and closes its mouth, aided by the tongue in expelling water, which is filtered through large baleen plates that capture large zooplankton, such as krill, and smaller fish (see Figure 9.22).

On the large spatial scale of hundreds to thousands of kilometers, whales, like other megaplanktivores, must find where the great concentrations of zooplankton are located. This often involves migrations from upwelling center to upwelling center, where the great concentrations of krill are to be found. But what happens when they get to these sites? At smaller spatial scales, krill are very patchy, yet the whales have to dive, lunge with the mouth open, close it, and finally strain the zooplankton. This energy expenditure occurs during a period when they cannot get oxygen at the surface, oxygen they need for the high metabolic rate required for hunting by lunges and maintenance regulation of high body temperatures.

This unlikely lifestyle raises many questions: How does the whale manage to maximize its efficiency at finding and capturing prey? Because the whale is warm-blooded, isn't this type of feeding chancy, given that zooplankton in the ocean are highly patchy? In addition, when the whale encounters a prey patch, it must open its mouth, which greatly enlarges its cross-sectional area and increases drag relative to the streamlined shape of the swimming whale when its mouth is closed and folded.

Lunging has a large metabolic cost, so one expects that there must be a compensating payoff. Equally perplexing is the searching behavior for dense patches of prey. How do they adjust their behavior to the tremendous spatial patchiness of their prey in the water, which varies by orders of magnitude in density? Isn't the metabolic cost of maintaining body temperature and swimming too great to pull off this type of lifestyle? Of course they do manage this, so our goal is to figure out how to even observe a whale that typically dives tens to hundreds of meters beneath the surface, out in the open ocean, with likely complex moves below the surface before the whale must eventually return to the surface to breathe (remember it is not a fish but an air-breathing mammal!).

Before measuring anything, we must think about the constraints of an organism with high metabolic rates and energy needs in maximizing encounters with patches of prey. We need to apply a model that optimizes efficiency, much like the optimal foraging models discussed in Chapter 4. We need rules to predict how much time and effort a whale should apply to a patch of zooplankton, given that there are patches of

many different sizes and densities (zooplankton per volume of seawater). Because most of the ocean probably has zooplankton densities too low to support the high metabolic needs of the whales, we must find out how they locate patches above the benefit-cost threshold of food energy intake relative to the cost of searching, lunging, and feeding.

Elliot Hazen and colleagues of the University of California at Santa Cruz (2015) have argued that oxygen uptake at the surface is a driving force in decision making on the time and depth of dives, and even the decision to feed actively once a whale encounters a patch of krill. If a dive is very long and deep, the whale will have to spend a great deal of time at the surface breathing the oxygen needed to match the aerobic metabolic needs encumbered during the dive. Thus, diving pattern and length must be strongly linked to the occurrence of patches of food and especially to the prey density that will allow for a lunge in which to gather enough food to compensate for the time needed to recoup oxygen at the surface. It would be very inefficient to just search and lunge during long, deep dives if the density of prey is too low to make up for the time needed to gain oxygen when the whale returns to the surface to breathe.

Hazen and colleagues conceived of two energy-feeding strategies, **energy maximization** and **oxygen conservation**, with the following features:

STRATEGY	DIVE DURATION	FEEDING RATE	OXYGEN CONSUMPTION
Energy maximization	Extended	Increased	Increased
Oxygen conservation	Shortened	Decreased	Decreased

Underlying these strategies is the increased oxygen consumption debt encumbered with time, which can be calculated in terms of energy consumption (**Box Figure 10.1**). In a high-prey-density patch, more prey will be consumed and eventually as krill density increases and increases the feeding-gain function will cross the oxygen debt function and the whale will make a profit (lower left of figure). But a whale would have to spend far more time in a low-prey-density patch to make a profit (lower right of figure).

This leads to the need for some measurements:

- Time at the surface
- Time to dive to a krill patch
- Time feeding within the krill patch
- Time to return from the krill patch to the surface
- Energy costs of all these components of hunting

As you can see, there is more to the decision than just the amount of krill in a patch and the energy expended while feeding on a patch. Diving time to the patch at depth, plus the return to the surface, also correspond to periods of the whale's oxygen consumption and therefore its energy need.

Until recently, it was not possible to make proper measurements of blue whale foraging activities in terms of time, depth, and prey patch density encountered. But now whale location and activity can be monitored simultaneously with the measurement of prey patches using

continues

HOT TOPICS IN MARINE BIOLOGY

Blue Whale Diving: Balancing Food and Energy Needs   *continued*   **10.1**

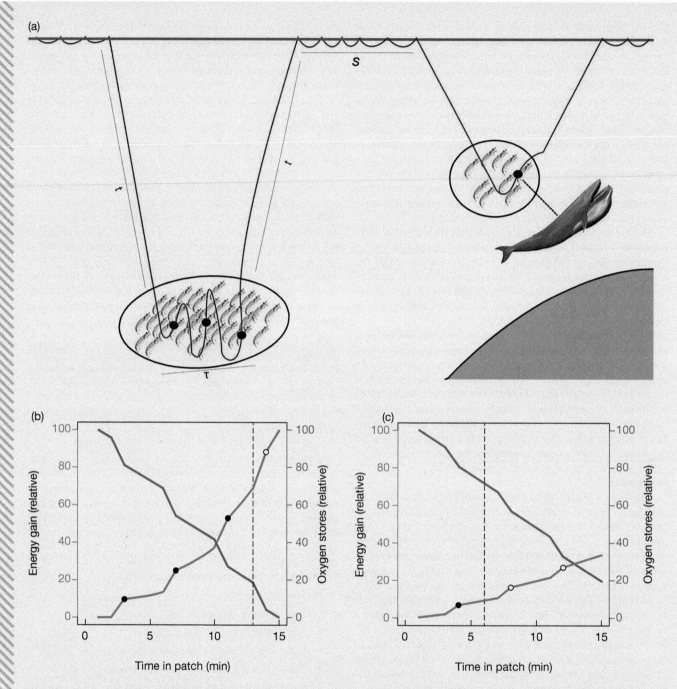

BOX FIG. 10.1   (a) Blue whales dive to patches of krill, which may have widely varying densities, found to range from 30 to 550 krill m⁻³. The gain by performing several lunges within a patch must be considered against the cost of oxygen over a certain dive time and the cost while the whale is lunging within the patch. (b) A dense patch. You can see the gain over time (in blue) in terms of energy, which exceeds the cumulative cost (in red) as the whale feeds within the patch. In this case, the whale might adopt an energy maximization mode. (c) Here the krill density is much lower and the cumulative gain (in blue) never equals the cumulative loss (in red). In this case, a whale should shift to an oxygen conservation mode. (From Hazen et al., 2015, with permission from The American Association for Advancement of Science)

remote sensing technology. Hazen and colleagues used noninvasive digital recording acoustic tags, attached to the whale when it surfaces, to record location below the surface; the tag eventually detaches from the whale and reaches the surface so that data can be collected from

the sensors on the tagging device. Lunging and closure of the mouth is recorded by an accelerometer, located in the sensor suite of the recorder. Echo sounders are deployed from boats at the surface simultaneously with the whale's dive. These devices collect data on densities of krill. The

Blue Whale Diving: Balancing Food and Energy Needs   *continued*   10.1

echo sounder can estimate the size of the patch and the relative density of krill within the patch.

What did they find? In low-krill-density patches, maximum lunge rate occurred much less frequently than in high-krill-density patches. This suggests that in the low-krill-density patches the whales were adopting an oxygen conservation strategy by avoiding lunging. There was a threshold density of about 100–200 krill m⁻³, above which the whales switched to an energy maximization strategy. This nicely fits a simple predictive model based on energetic gain versus cost. Hazen et al. used an equation for gain of energy in a patch of krill, minus the energetic cost (calculated from oxygen consumption) of lunge feeding under conditions of varying depth, maximum lunge rate, and varying krill density.[1] As can be seen in **Box Figure 10.2**, positive values occur as krill densities surpass about 100 krill m⁻³, which corresponded well to the data threshold of 100–200 krill m⁻³ at which whales switch to the energy maximization strategy of maximum lunge frequency within a krill patch.

These results have larger implications beyond the behavior of blue whales in a given region. They explain why the whales might travel long distances to new areas once a seasonal peak of krill within patches has declined in any geographic location. We now know that bony fish, sharks, and whales migrate many thousands of kilometers seasonally. Similarly to wildebeest following the appearance of grass in different subregions of the Serengeti in Africa, whales must keep moving to stay alive. The greatest beasts in the world are shackled to the cycles and vagaries of ocean-scale change—and perhaps the challenges of human influence via climate change.

---

[1] See the original paper for details on the equations of feeding gain and energetic cost within a patch.

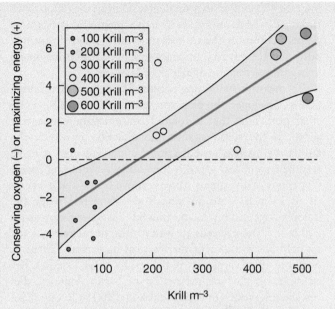

BOX FIG. 10.2 Prediction of a shift from the oxygen conservation strategy, in which whales do not feed at high lunge frequency, to the energy maximization strategy. The y axis shows the difference between the energy gain function and the energy loss function: A negative value predicts that whales should not be actively feeding, and a positive value predicts that whale should be actively lunging within a patch. Data points in color show whales actively feeding, which appears to occur at krill densities that exceed about 100–200 krill m⁻³. This fits the model line prediction quite well.

The phenomenon of **diel vertical migrations**, which is widespread throughout the ocean, has been noticed and quantified by many marine biologists (**Figure 10.4**). The first detailed scientific observations were made during the famous *Challenger* expedition of 1872–1875. Copepods were found only in surface samples taken at night. In 1912, Esterly found that copepods off the Pacific coast of California could migrate as deep as 400 m during the day and return to the surface at night. Since the early twentieth century, diel migrations have been observed in a wide variety of zooplankton groups, including copepods, euphausiids, jellyfish, ctenophores, and arrow worms, and also in fishes. The great diversity of groups that undergo diel migrations would suggest that there must be a common underlying selective force for this behavior.

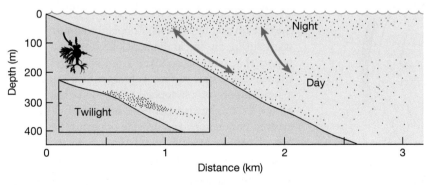

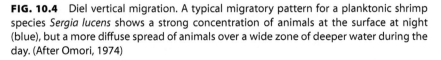

FIG. 10.4   Diel vertical migration. A typical migratory pattern for a planktonic shrimp species *Sergia lucens* shows a strong concentration of animals at the surface at night (blue), but a more diffuse spread of animals over a wide zone of deeper water during the day. (After Omori, 1974)

Diel vertical migrations are common in coastal shallow waters, but they also have been observed to depths of greater than 1,000 m, where light is not detectable from above. Some animals, such as copepods and jellyfish, can migrate 400–800 m in a single day. During the day, passive sinking can be as important as active downward swimming. At night, however, the animals must swim upward continuously and actively to counteract sinking, which is inevitable in organisms that are denser than seawater.

To measure swimming speeds, Hardy and Bainbridge (1954) developed the plankton wheel (**Figure 10.5**), a circular glass tube that is rotated continuously at constant speed. If the speed against the direction in which the animal swims is adjusted, a zooplankter will appear to be stationary, and this speed is the animal's swimming speed. Copepods of the genus *Calanus* move upward at a velocity of 15 m h^{-1} and downward at 100 m h^{-1}, while the holoplanktonic polychaete annelid *Tomopteris* can swim 200 m h^{-1}. These speeds are well within the range predicted by field observations of diel vertical movement. There is, however, much variability in the results of field studies. For example, on the northeast coast of North America, diel vertical migrations of copepods can be 100 m in one place and virtually absent in others. Vertical migrations are often strong in spring and early summer, but copepods frequently move to greater depths in the autumn, and vertical migration is dampened somewhat.

Diel variation in light seems to be the major cue for the timing of the migrations. Onset of dawn and dusk would seem to be the only stimuli that can set the clock. There are zooplankton populations, however, for which downward migration begins before dawn and whose surface movements commence before dusk. Thus a direct response to variation in light intensity is not sufficient to explain the diel pattern of movement. An internal biological clock is probably reset continually by the day–night cycle. A number of studies confirm this by bringing copepods into the laboratory and keeping them in a vertical tube under constant conditions of light and temperature. For several days, the copepods continue to rise at night and fall in the daytime, although the cycle eventually breaks down. A series of clock genes are known that regulate diel cycles, but we know relatively little about this in zooplankton because our understanding of the genome is in its infancy. Clock genes have been found that vary in daily expression in the freshwater flea *Daphnia pulex*, whose genome has been completely sequenced (see Bernatowicz et al., 2016).

ADAPTIVE EXPLANATIONS FOR DIEL VERTICAL MIGRATIONS

■ **Convincing explanations of this behavior include the effects of predation and the potential of energy savings of poikilotherms by diving to cooler waters.**

The widespread occurrence of diel migrations is matched only by the great mystery of why such migrations occur at all! Why should zooplankton move several hundred meters down, only to rise to their original depths again a few hours later? Even if it is known that diel light change is required to continually reset a biological clock, this information does not reveal the adaptive value of this behavioral cycle. Many adaptive explanations have been proposed. They fall into the following overall categories:

1. **Strong light hypothesis.** Zooplankton are adversely affected by UV radiation and strong light and, therefore, leave surface waters during the day.

2. **Phytoplankton recovery hypothesis.** Zooplankton exploit the phytoplankton but dive to allow the phytoplankton to photosynthesize and to recover during the day so that they can be exploited again the following night.

3. **Predation hypothesis.** Predators (e.g., fish, diving birds) use vision to capture prey, so zooplankton leave the surface waters during the day to avoid being seen. They return to the surface waters at night to feed on the phytoplankton when they are invisible to visual predators.

4. **Energy conservation hypothesis.** It is energetically advantageous to spend the day in colder, deeper water, where metabolic rate and energy needs are lower. The animals come up at night to feed on phytoplankton or on other zooplankton that rise to the surface.

5. **Surface mixing hypothesis.** Zooplankton move downward during the day, in the hope of returning to surface waters that have been driven from another locale by the wind. The new surface waters are likely to contain a new supply of phytoplanktonic food. The mixing of waters might also promote mixing of differently evolved populations and provide opportunities for the use of new genetically based variants to evolve and adapt to changing environments.

Let us now consider the merits of these hypotheses. Light intensity probably has nothing to do with the evolution of diel vertical migrations in most species, because the animals usually migrate to depths far greater than those at

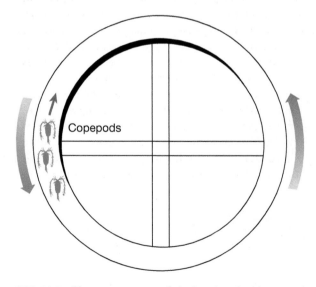

FIG. 10.5 The measurement of planktonic swimming speed in a Hardy–Bainbridge plankton wheel.

which surface light intensity can be damaging. The phytoplankton recovery hypothesis is also unlikely, because cooperation among zooplankton organisms would be required for all individuals to swim downward to allow a complete recovery of phytoplankton to occur. However, natural selection favors individuals, and those that would "cheat" by remaining in the surface waters could take advantage of the remaining phytoplankton. The surface mixing hypothesis has the same logical problem.

The **predation hypothesis** is consistent with the idea of avoidance of visual predators, because predators would be unable to detect prey at night. In Gatun Lake (freshwater), Panama, the calanoid copepod *Diaptomus gatunensis* avoids fish predation through a diel vertical migration. The lake has no vertical gradient in temperature, which eliminates the energy conservation hypothesis from consideration, at least in that lake. A small copepod, *Pseudocalanus* sp., in Dabob Bay, Washington, actually performs a reverse vertical migration, spending daylight hours in the surface waters and nighttime at depth. This seems to be a response to an arrow worm predator *Sagitta elegans* and a larger carnivorous copepod, *Euchaeta elongata* (**Figure 8.9**), both visual predators located in that bay, which themselves carry out a typical vertical migration (upward at night, downward during the day). At a time when arrow worms were rare, the copepods reverted to a typical night-up and day-down vertical migration (Ohman, 1990).

In many cases, diel vertical migrations are much stronger when planktivorous fish are locally abundant. Species that are more pigmented and more visible tend to have stronger diel vertical migration patterns. Also, smaller zooplankton arrive at the surface earlier in the evening and leave later than larger forms. This suggests that predators are avoiding the smaller size classes because they provide less reward in feeding. Finally, lipid-rich copepods, which are in a good nutritional state, tend to stay in deeper water and come to the surface less frequently than copepods with smaller lipid reserves. Staying in cooler waters with more lipid reserves is also consistent with the energy conservation hypothesis. The interested student should see Hays, 2003.

There are some shortcomings to the predation hypothesis, however. Many zooplankton migrate far deeper than is needed to avoid predators. Also, many vertically migrating species bioluminesce at night, which would seem only to attract predators. A third objection is that members of species that are relatively invisible (e.g., transparent gelatinous zooplankton) nevertheless participate in vertical migrations.

The **energy conservation hypothesis** is based on the decrease of metabolic rate with decreasing temperature in poikilotherms. By spending some time in the cooler deep waters, an animal can save energy. This gain would have to be balanced against the energy required to migrate and against the time lost from feeding at the surface. Also, being in deeper cold waters might slow development and reduce reproductive rates. So the value of being at the surface would have to be balanced against a cost of being at depth. Apparently, the energy expended in downward and upward movements is quite modest and not a consideration.

Dawidowicz and Loose (1992) measured reproduction and growth rate in moving and in stationary water fleas and found no differences between the two groups. McLaren (1963) calculated that copepods should have an energetic advantage in spending time in deeper waters, where temperatures are cooler and metabolic costs lower. Unfortunately, there is not a great deal known definitively about the cost of swimming in vertical migrations, although it is hypothesized that the cost is low. An energetic advantage of living in cold water would be realized in the production of additional eggs and greater potential population growth. Copepods often retire to deeper waters after the spring peak of phytoplankton production. It may be that this is an adaptation to lower energy expenditure that comes into play once the surface food source is no longer abundant. As mentioned above, lipid-rich copepods tend to stay in deeper water, often throughout the season. On balance, it appears that both the predation and energy conservation hypotheses must contribute in varying degrees to selecting for the adaptation of diel vertical migrations.

Some predatory species can dive rather rapidly and seek out dense layers of zooplankton that might be hiding at depth. The North Atlantic right whale, for example, appears to be associated with depths in which the high-latitude copepod *Calanus finmarchicus* lurks in an arrested developmental stage (Baumgartner et al., 2003).

Movement of Nekton on Local and Ocean-Level Scales

Fish Group School Formation and Motion on Small Spatial Scales

■ **Schooling is coordinated aggregation and movement and common among fish and some cephalopods.**

Schooling (**Figure 10.6**), a coordinated directional movement, is a common phenomenon in fishes and squids. It involves aggregation and movement by mutual attraction and interindividual signaling. A related phenomenon, **shoaling**, involves attraction of individuals as an aggregation to an attractive microhabitat, usually in shallow water where food is abundant. Some fish species, such as the Atlantic herring *Clupea harengus*, are obligate schoolers, but many fish species form schools only occasionally or not at all.

Schools are notable for the elegant maintenance of precise distances between individuals. Schooling is probably coordinated by movements in response to visual cues and information received by the fish's lateral-line system. Schools of even obligate schooling fish species break down in the evening. In nearly all cases, schooling is confined to a

(a)   (b)

FIG. 10.6 Schooling in fishes: (a) traveling school and (b) feeding school.

single species. But some feeding schools may consist of one or many species, especially in coral reef environments. It is not clear, however, whether these multispecies groups have the coordinated directional movement characteristic of single-species schools. Schooling is associated commonly with migration and movement to feeding sites.

Schooling may benefit fish by providing protection against predation. Schools can confuse predators, and fishes may either form a ring around the predator or swim in an undulating line. As the number of fish in the school increases, the predator's apparent confusion increases as well. Most tightly schooling fishes have silvery sides, which serve to reflect sunlight and confuse predators. Fishes wandering from the school are usually more prone to predation. Schools approaching predators may confuse the predator and buy enough time to reduce the predator's efficiency in making a decision to attack a specific fish before the entire group has escaped. In some cases, fish such as herring aggregate into approximately spherical schools known as **bait balls**, which minimize the peripheral area of fish exposed to predators. In the Caribbean, one can see bait balls of thousands of fish with predators such as tarpon remaining at the periphery. While the tarpon does not dive into the bait ball, it can attack fish that swim out alone.

Schooling also provides the benefit of **drag reduction** for individual fish, which would be a benefit in general movements, migration, or predator avoidance. When an entire school swims in a tight pack, the fishes within the pack or at the rear experience less drag, perhaps by a form of slipstreaming fish in front. A member of the school usually drops behind and slightly to the side of the fish before it, which locates it in a vortex where drag is minimal. This may permit the aggregation to move with less expended energy per individual than if all individuals swam alone. Experiments show that fish located in the rear of schools tend to have lower aerobic capacity for bursts of activity and appear to be able to keep up with the school with fewer tailbeats (Killen et al., 2011). By contrast, fish at the head of schools have higher aerobic capacity. As a fish tires, it can retreat within the school to regain capacity in the same way that a bicycle rider can retreat within a pack of riders to reduce drag and build up oxygen available for muscle action.

Migration by Nekton over Large Distances: Control by Shifting Productivity Hotspots

■ **New archival satellite tag technologies allow long-distance tagging of large migratory fish. The results are startling: Large oceanic fish typically move across and between oceans.**

Many large oceanic fish move over great distances, which has become apparent from a large-scale tagging program of fishes. Satellite GPS tags have been implanted in marine mammals and sea turtles that surface frequently to breathe, which allows direct communication of the tag with a satellite GPS receiver. But many large fish such as tuna and sharks swim at depth and cannot transmit underwater. So **pop-up satellite archival tags** have been developed that are attached to the dorsal skin. A microchip embedded in the tag continually records water temperature, which can be used to estimate not only the depth to which the fish has swum over time but also longitude and latitude by means of temperatures inferred at the surface over time. The tag's connector is designed to eventually rust, break away, and pop up to the surface, and its position is recorded through communication with a GPS-recording satellite.

The results so far have been surprising and very important in our understanding of migration and conservation of large open-ocean fishes. For example, bluefin tuna are of great interest because they are major open-ocean predators and have been severely overfished. Satellite tracking demonstrates their remarkable journeys. Atlantic bluefins reach lengths of 3 m and weights of 750 kg and have been routinely tracked across the Atlantic. One fish was tagged off the coast of Cape Hatteras, spent some weeks there, moved to New England waters, then to the Caribbean, and then across the North Atlantic, where it was finally spotted in waters of just west of North Africa (Block et al., 2001). Similar evidence shows migration of the Pacific bluefin tuna across the entire Pacific. In August 2011, Pacific bluefin tuna caught off the coast of southern California were found to have radioactive cesium-134, which has a half life of 2 years. This means that they had picked up contaminated material from the region of the Fukushima, Japan, nuclear power plant meltdown (following the March 2011 tsunami and peak of leakage in April) and had swum across the entire Pacific Ocean to California in only 4 months (Madigan et al., 2012). Luckily the tuna have cesium at levels that are not dangerous for human consumption, but the detection shows how extensive and rapid the migration can be.

Perhaps even more amazing is a study (Bonfil et al., 2005) of the great white shark, *Carcharodon carcharias*, which is found worldwide in temperate waters but is especially abundant off South Africa. Previously, it was well known that male great whites could travel long distances. But some molecular population genetic data suggested that females might be much more faithful to small areas.[2] The data were based on a comparison of differences in mitochondrial and nuclear genes. Mitochondrial genes are inherited maternally, so differences between areas may reflect the movements of females. In contrast, nuclear genes are inherited from both parents. Genetic variants in great white shark mitochondrial genes showed more geographic differences than nuclear genes as predicted by the hypothesis that females tended to stay put. The satellite tracking, however, suggested that great white travel was not so simple. Both male and female great whites traveled great distances. Tracking was done by satellite, but identification of sharks also could be done by characteristic indentations on the dorsal fin that were photographed when a fish was caught and released. Most amazing was a recorded trip by a female from South Africa to northwest Australia and back, a round trip of 20,000 km, including dives to depths of nearly 1,000 m at water temperatures as low as 3.4°C. These long-distance connections have completely changed

---

[2] See bonus chapter, "Molecular Tools for Marine Biology," online.

our conception of conservation of these apex predator fish stocks. What affects one of these majestic creatures in one part of the world may affect the species everywhere.

■ **Large-geographic scale movements of nekton and seabirds in the ocean are selected to maximize efficient use of habitats for feeding and breeding, which may shift seasonally in availability.**

The large-scale migration of many marine species, including fishes, mammals, birds, and turtles, reinforces the idea that we have to look at the large ocean scale to understand how many marine species and communities function. We discussed in Chapter 7 that an ideal place for breeding may therefore be thousands of kilometers from an ideal feeding site. This large migration scale can also be regulated because suitable feeding sites shift location as seasons and weather favor high productivity in widely different places.

The following factors tend to determine or allow large-scale movements and migrations of mobile species such as bony fish, sharks, and whales.

- *Migration routes.* Optimal breeding sites and feeding sites are very distant from one another and may alternate in quality with seasonal change.
- *Large current systems with favorable conditions.* Open-ocean current systems may be favorable for feeding of mobile species over very large spatial scales, allowing for long distance movement within the current system.
- *Shifting favorable conditions over large spatial scales.* As current systems and sites of upwelling change, often under seasonal cycles, favorable sites for food exploitation may move great distances. For example, seasonal shifts of upwelling centers would enhance primary productivity and the stimulation of local zooplankton and small planktivorous fishes that attract migrating larger exploiters such as baleen whales.
- *Climate change.* As climate oscillations change regional ocean conditions, optimal feeding sites for long-distance traveling species may shift over great geographic distances. Global warming trends should also rearrange the location of preferred feeding sites.

■ **Movements of whales and large fishes demonstrate that large nekton can adjust large-scale movements to maximize local and geographically shifting food supplies, especially at sites of upwelling.**

A well-known migration cycle demonstrates the obvious role of shifting food abundance and the well-known movement of large marine mobile nekton between feeding areas and reproductive sites. The blue whale *Balaenoptera musculus* has large-scale migration routes in all oceans, but we will focus on the eastern north Pacific region. Individual whales have been identified by specific pigment patterns, satellite tags, and even vocal calls, which can be detected by arrays of acoustic buoys. The big picture is as follows: In late summer and fall, blue whale adults move along the northeastern Pacific coast, tracking coastal upwelling that moves northward from the southern California Bight in late summer to coastal waters of Washington and Vancouver Island in fall

(Burkenshaw et al., 2004). While breeding and calving, blue whales prefer warm waters and in winter swim southward to warmer coasts of western Mexico and Central America.

Satellite tagging of the eastern Pacific blue whales tells us much about the ecological factors that shape the details of the migration. For instance, whales often travel in similar tracks. Even more interesting, whales are often clumped in space and their tracks often zigzag, which indicates that the whales are tracking local patches of food (usually krill). A popular area to feed on the way south is the California Bight, approximately offshore of San Diego, California (see Mate et al., 1999). For example, whales congregate in the region of the Channel Islands, south of Point Conception, where krill are abundant (see **Hot Topics Box 10.1** for a discussion of their diving feeding strategy). They then arrive in the calving regions in Central America but can feed through the winter calving season in a major upwelling region off the Pacific coast of Costa Rica. The situation is quite different for gray whales and humpback whales, which experience very low food supply during their respective calving seasons.

The movement of the great ocean sunfish *Mola mola* (see Figure 9.8) in this same region reinforces the idea of shifting foci of food abundance stimulated by changing upwelling conditions. Ocean sunfish are found in tropical and temperate regions of the Atlantic and Pacific Oceans, but we will focus on those in the eastern Pacific. They feed mainly on jellyfish and other gelatinous zooplankton, but they can feed on a wide variety of other zooplankton. They reach lengths of 3 m, can weigh as much as 2,000 kg, and are the most massive fish in the world ocean. They are known to dive to deep water during the day to follow their zooplankton prey, sometimes deeper than 500 m.

One individual tagged with a permanent satellite tag moved over a distance of 800 km, usually keeping within 150 km of shore. It followed shifting upwelling fronts, and large numbers of salps were found on the warm coastward sides of these fronts. Ocean sunfish migrate seasonally between the northern California Bight (where the above-mentioned blue whales were found) and other sites off Baja, California. Overall, this fish, with different food preferences and very different reproductive biology, also was found to track upwelling fronts, where its preferred gelatinous zooplankton food was found in abundance. In 2013, waters of the northeast Pacific warmed significantly and ocean sunfish were found in Alaskan waters, along with other more typically low latitude species. As the El Niño of 2015–1016 develops, this trend will only continue.

■ **A multispecies tagging program throughout the North Pacific Ocean has demonstrated the presence of very large scale current systems and long-distance swimming of many large-bodied nekton, including fish, turtles, and mammals.**

The foregoing examples from the eastern Pacific suggest that we need a far more sophisticated multispecies approach to tracking, and this has led to the formation of the Tagging of Pacific Predators (TOPP) group of the Census of Marine Life (www.topp.org), which has followed for over a decade

FIG. 10.7 This Pacific white shark, photographed at the Farallon Islands off Northern California, has been tagged with an acoustic tag (front) and a pop-up satellite tag (rear) as part of the TOPP research program.

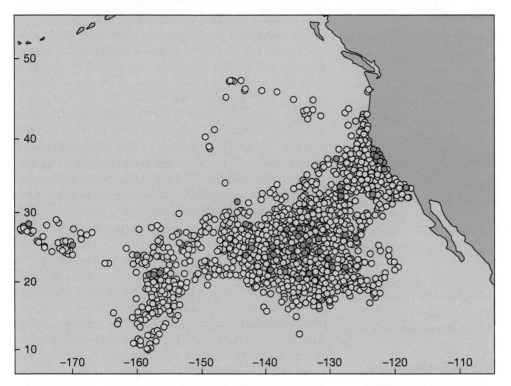

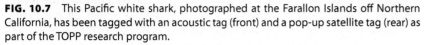

FIG. 10.8 Tagging locations together identify the migration pattern of the Pacific white shark *Carcharodon carcharias*, which migrates between coastal California waters and the Hawaiian Islands.

the movements of 23 predatory species in the North Pacific Ocean.[3] Implanted pop-up archival satellite tags record the data—including temperature, water pressure (indicating depth), and GPS location—that disattach and float to the surface after a few years in order to transmit data to satellites.

The results inferred from thousands of tags have produced a map of major North Pacific migration routes (Block et al., 2011). Taken one by one, many species of predators have somewhat different patterns. For example, the Pacific white shark (**Figures 10.7** and **10.8**) shows a consistent homing migration between the Hawaiian Islands

and California (Jorgensen et al., 2010). But salmon sharks move between the waters of the Pacific subtropical ocean and Prince William Sound, Alaska. When combined with other species, we see that this great system is composed of more than one major migration route.

The great marine predators are associated with highly localized and often spatially changing huge sources of food driven from the base of the oceanic food web. This association is quite evident with tuna, which are usually concentrated in upwelling regions where nutrients and plankton are abundant. Plankton-eating baleen whales show this within a very short time frame: As mentioned earlier, a group of whales may be found in a

---

[3] This program has recently been expanded to the global ocean scale.

very localized upwelling zone with abundant zooplankton and then move hundreds of kilometers away after the resource is exhausted. These movements demonstrate that roving predators have been able to map the locations of a large number of areas with predictable but shifting food abundance. Even seabirds that rove large expanses of the ocean for prey can smell zooplankton or detect compounds released from phytoplankton blooms, which allows them to find upwelling centers where zooplankton and fish are both abundant (Nevitt et al., 1995; Nevitt, 2000).

On the grand scale, there are two great belts of predatory activity and migration in the North Pacific Ocean. First, the **California Current Large Marine Ecosystem**, along the west coast of North America (**Figure 10.9**), includes a large number of tuna, whale, seabird, and other large fish species that move over tracks stretching hundreds to thousands of kilometers, usually in annual migrations with high site fidelity between zones of high productivity, often related to upwelling.

Another large group of species move across the Pacific through the **North Pacific Transition Zone (NPTZ)**, another region of high productivity that extends as an approximate east–west belt across the North Pacific Ocean between approximate latitudes of 32° N and 42° N (**Figure 10.9**). This zone is bounded by strong oceanic fronts where primary production is very high, which attracts prey of large predators. Prominent species included the white shark, Pacific bluefin tuna, elephant seal, sea turtles, and sea birds. The NPTZ is very vulnerable to climate change (Hazen et al., 2012).

Descending to the Depths: Organic Matter Transfer and Adaptations to Infrequent Meals and Low Light

Transfer of Organic Matter Downward from Surface Waters

■ **Surface waters are the ultimate source of material that sinks to the deep, but many plankton are adapted to reduce sinking.**

We can envision the surface layer of the ocean as a zone of strong light and high plankton density. But two immediate considerations arise: First, organisms that are denser than seawater will sink from their preferred depth zone unless they respond in some way or physical factors prevent sinking. Second, much material must slip away to greater depths. For example, dying phytoplankton if not consumed will sink to greater depths. Zooplankton fecal pellets sink rather rapidly (**Table 10.1**). This might provide a source of food for deeper parts of the ocean. Thus, there must be a range of adaptations to prevent sinking of both phytoplankton and zooplankton, while other material will sink to great depths.

■ **Small zooplankton and phytoplankton live in a world dominated by low Reynolds numbers.**

If a cell or a body is denser than seawater and the water is perfectly still, then the cell or body will sink to the bottom. In Chapter 6, we discussed the conditions of a body that is very small and moving at low velocity. Under such conditions of low Reynolds number ($Re < 0.5$), there is a **boundary layer** around the body that causes it to be dragged by the

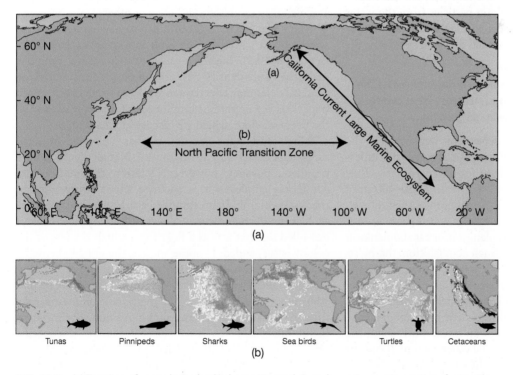

FIG. 10.9 (a) Tagging of many large-bodied oceanic predators shows two major routes of migration, one along the eastern Pacific within the California Current Large Marine Ecosystem and the other through the North Pacific Transition Zone. (b) Migration routes of individual migrating species groups, as indicated by tagging locations. (After Block et al., 2011)

TABLE 10.1 Sinking Rate of Various Particles and Living Organisms in the Plankton

GROUP	SINKING RATE (m day⁻¹)	TIME TO BOTTOM OF 4,000 m (days)
Phytoplankton		
Living	0–30	≥ 133
Dead, intact cells	< 1–510	≥ 8
Fragments	1,500–26,000	0.15–3
Protistans		
Foraminifera	30–4,800	1–133
Radiolaria	350	11
Other zooplankton		
Chaetognaths	435	9
Copepods	36–720	5–111
Pteropods	760–2,270	2–5
Salps	165–253	16–24
Fecal pellets	36–376	11–111
Fish eggs	215–400	10–19

Source: Modified from Parsons, T. R., and M. Takahashi, 1973, Biological Oceanographic Processes (Oxford: Pergamon Press). Used with permission of Pergamon Press, Ltd.

surrounding water. If a particle is small and the velocity slow, its sinking velocity will increase with increasing weight of the particle. These organisms will reach a constant terminal velocity. This phenomenon is part of **Stokes' law**. Larger organisms will sink faster than smaller organisms of the same density. Remember, though, that this applies only to nanoplankton and smaller, relatively slow-swimming microplankton. Also, irregularly shaped plankton tend to fall at lower velocities than predicted by Stokes' law. Still, this law allows one to predict sinking of plankton particles from the water column, perhaps bringing them to depths that are deeper than the penetration of light needed for photosynthesis.

■ **The vertical position of plankton is determined by the bulk density of the organism, structures that create drag, water motion, and swimming ability.**

Planktonic organisms often depend on remaining in the surface waters for survival. Phytoplankton will die unless they are near a source of sunlight for photosynthesis. They will be inviable if they sink below a depth of 50–100 m in the open sea but will be unable to photosynthesize even in much shallower depths in estuaries and inshore waters. Zooplankton usually eat phytoplankton, or eat zooplankton that consume phytoplankton. Such zooplankton, therefore, also must remain in the surface waters, although there is a surprisingly large and diverse community of deep-water zooplankton and vertically migrating species, as mentioned earlier. How do plankton maintain their preferred depth?

Observation of a pot of soup can reveal much about the distribution of plankton in a natural water column. If you stir up the soup, many pieces will be lifted up from the bottom into the fluid but will eventually settle again as the broth calms down. Some pieces are less dense than the broth, and they will float. The only other way for a particle to remain above the bottom in still water is if it can swim or is less dense than the fluid. In the case of seawater this means that

a particle will remain suspended if (1) it is less dense than seawater or of equal density so it is neutral, (2) it has a shape that increases drag and reduces settling velocity, (3) it can swim, or (4) water turbulence keeps the particle suspended in the water column. In the sea, turbulence is usually generated by wind, which mixes the water column and forestalls the sinking of plankton denser than seawater. All four of these mechanisms combine to keep plankton from sinking rapidly to the bottom or from sinking lower than the depth at which they can photosynthesize or otherwise survive.

Many planktonic organisms are somewhat denser than seawater and will, therefore, sink in a quiet water column (see **Table 10.1**). The silica valves of diatoms, for example, make their bulk density greater than that of seawater. Many dead particles may take days to weeks to reach the deep-sea floor. This has important implications for supply of food to the deep-sea floor, which we discuss in Chapter 18.

Most crustaceans are denser than seawater and will surely sink unless they swim upward. Not all plankton have negative buoyancy, however. Flotation mechanisms are commonly employed to keep these animals suspended in the water column. In an extreme case, some siphonophores such as the Portuguese man-of-war *Physalia physalis* have a gas-filled sac that acts as a float from which the rest of the colony is suspended. In some species, the gas can be withdrawn to permit the colony to sink below the surface during storms, thereby avoiding the surface turbulence. A second adjustment mechanism is the *regulation of bulk density* through variations in chemical composition. For instance, the dinoflagellate *Noctiluca* accumulates ions of low specific gravity, which reduces their density, while some cyanobacteria are believed to have vacuole-like structures that contain low-density gaseous nitrogen. Many zooplanktonic organisms become more neutrally buoyant by replacing dense magnesium, calcium, and sulfate ions with lower-density ammonium, sodium, and chloride. Diatoms use oil droplets to increase their buoyancy. Many copepods have high lipid contents, which aid in buoyancy (Pond, 2012). Copepods acquire lipids by feeding on diatoms. Copepods can change their lipid profile to adjust bulk density during the life cycle in order to adjust to needed changes in depth. For example, some copepods spend the winter months in deeper water and change their lipids to more condensed phases.

Gas secretion is used by many species of smaller nekton as a means of maintaining neutral buoyancy. The nektonic cephalopods are able to regulate the gas content of their bodies and thereby achieve neutral buoyancy under different conditions of depth and feeding. In the chambered nautilus, the gas content of the inner chambers of its shell can be filled or emptied to suit individual conditions. In the cuttlefish *Sepia*, the inner spaces of the cuttlebone, an internal skeleton, can be regulated similarly. In fish, the **swim bladder** is used to adjust bulk density (see Chapter 9). In all these cases, gas adjustments help reduce the energy required for movement, and especially the energy required to counteract the sinking expected of otherwise strongly negatively buoyant bodies. Many deeper-water fish species employ lipids rather than gas bladders, and fish associated with the bottom often lack swim bladders because they do not change depth. Neutrally buoyant deep-sea squaloid sharks use squalene combined with other lipids to maintain

level. They have relatively small pectoral fins that are used for maneuvering but not much for lift generation. Adults of several species of the fish family Myctophidae undergo vertical migrations of as much as 500 m. Although all juveniles have swim bladders, adults of many species have reduced gas bladders and use lipids instead to maintain neutral buoyancy. Still, there is a great diversity of fish that have swim bladders, especially at depths less than 1,000 m (**Figure 10.10**). Many undergo vertical migrations and therefore must be able to monitor their depth and be able to resorb gas on ascent and release gas on descent.

Many planktonic animals have shapes that retard sinking. The bell shape of jellyfish is an obvious example. A jellyfish's flat bottom and tentacles combine to create a drag that slows sinking when compared to a sphere of the same shape and bulk composition. Some diatoms, such as *Chaetoceros*, form rather twisted chains that spiral as they sink slowly. Many tropical zooplankton have elaborate projections, which also retard sinking but may in fact serve primarily as a defense mechanism against predators.

Swimming is the final major means of avoiding sinking. Flagellated phytoplankton can swim to a degree, and dinoflagellates such as *Noctiluca* use flagellae to aid in keeping near the surface. Swimming is best developed in

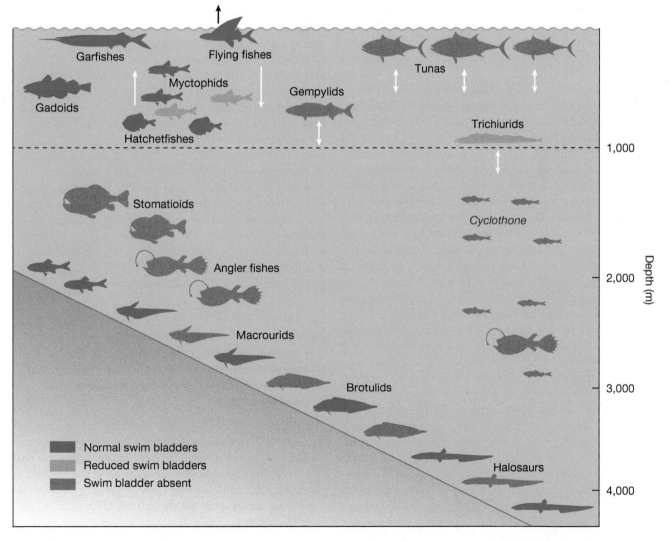

FIG. 10.10 Fishes have a diversity of adaptations in the water column, ranging from full use of swim bladders to complete lack of swim bladders (After Bone et al. 1994).

zooplankton, especially holoplanktonic forms. In pteropods (holoplanktonic snails), for example, the foot has two lateral wing-like projections, and the snails flap through the water. Jellyfish move in pulses, through rapid compressions of circular muscles, which compress the bell and force water backward. Crustaceans such as copepods use their appendages to push against the water to move forward, while arrow worms swim by undulation, as do many fish. All these swimming methods are effective means of reducing sinking rate, allowing organisms to stay in surface waters.

Dealing at Depth with Reduced Food Input

■ **Organisms living at greater and greater depths experience decreasing food input and rapidly extinguishing light.**

While many planktonic organisms are adapted to reduce or even eliminate sinking, dead particulate organic matter sinks to the deeper ocean. On the continental shelf this deposition of organic material hits the sea floor, where it is decomposed and then returned to the surface waters by seasonal mixing (see Chapter 11). But seaward of the shelf-slope break, particulate organic matter (POM) tends to keep sinking to much greater depths. Much of the economy of deeper waters depends upon the sinking of POM and often feces of zooplankton from surface waters. By the time you reach mesopelagic and bathypelagic depths, this input has been worked over by numerous animal consumers and microorganisms and is very small, and population sizes of consumers in deeper waters (>1,000 m) may be very low. Particulate organic matter accumulates in **deep-scattering layers**, common in the eastern North Pacific and eastern tropical Atlantic Oceans, which attract decomposing organisms and even zooplankton adapted to feed on the localized decomposing particles and decomposing bacteria. But the total food input to greater depths is more scarce.

At bathyal depths, there are no actively growing phytoplankton populations, just a few dead and live cells that drift downward. Usually, the dead particles are aggregated, and the aggregates, called **marine snow**, often have embedded microorganisms and smaller zooplankton. This may be the base of the deep-water food web, and these particles may be consumed by smaller zooplankton such as copepods, but also larger zooplankton and even nekton. Larger gelatinous invertebrates and fishes may consume the smaller zooplankton and provide food for other larger fishes such as anglerfishes. Studies in the past 20 years have resulted in the seemingly endless discovery of new species in mesopelagic and bathypelagic waters. But the other amazing outcome is the very diversity of these deep-water forms spread over an enormous volume of the deep ocean. Considering the volume of the deep-sea environment, probably the largest marine community type on the planet, it is amazing that we know so little about it. Gelatinous zooplankton are amazingly diverse in their morphology, ecology, and adaptations.

Observations from submersibles, usually equipped with high-definition video, have captured deep-sea animal behavior and to some extent their distribution in the water column. Work by Bruce Robison (2004) of the Monterey Bay Aquarium Research Institute and colleagues has been

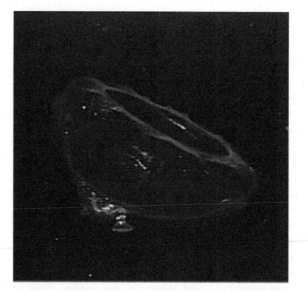

FIG. 10.11 View from a submersible of the salp *Pseudusa bostigrinus,* which is modified for more infrequent meals of zooplankton in the deep sea. (From Robison et al., 2005)

especially important in bringing us a sense of the diversity and abundance of deep-water organisms and the dynamics of population change.

Early work showed that bathypelagic animals were rather sluggish and had reduced metabolic rates when compared to similar animals living in the upper parts of the water column (Childress, 1995). Cold temperature explains part of this, but also there probably is less food available and also less need to escape fast-swimming visual predators. Many deep-water plankton and nekton are poorly muscularized and seemingly adapted for low food supply. An excellent example is the bathypelagic doliolid salp *Pseudusa bostigrinus,* which is found in the eastern Pacific in depths of about 1,100–1,900 m. As mentioned in Chapter 8, salps are usually continually beating a tail that moves food particles through a "house." But deep-sea particle density is much lower, and this group has lost in evolution its mucus-filtering house, found in shallow-water relatives. The deep-sea salp *P. bostigrinus* has evolved a form that more resembles a jellyfish that faces its mouth opening upward (**Figure 10.11**). By doing this, it can propel itself upward to capture zooplankton prey.

Predation and Defense into the Depths

DEFENSE MECHANISMS OF PLANKTON

■ **Predation is intense in the plankton at all depths. Both phytoplankton and zooplankton rely on evolved armature, chemical defenses, and transparency to avoid predation.**

The composition of planktonic communities must be affected by differential susceptibility of planktonic species to predation. However, few studies have demonstrated this process to be important in the ocean. Among the freshwater plankton, differential susceptibility is probably a common underlying explanation for the relative abundance of species. The evolution of body spines and armature in many phytoplankton and zooplankton is probably an adaptation to increase the difficulty of capture and ingestion. Common diatoms, such as those of the genus *Chaetoceros*, have large projecting spines that increase the effective body size (and may also aid in flotation). Spines of some *Chaetoceros* species can damage the gills of fish and likely have evolved as a defense against predation. Many planktonic crustaceans are armed with elaborate spines. The presence of spines may place the prey out of the size range of predators or may make the prey difficult to handle and seize. On the other hand, we cannot exclude the possibility that the spines on many phytoplankton species are designed to disrupt flow around the phytoplankton in order to encourage mixing and nutrient replenishment of cells at the scale of individual cells.

Many planktonic organisms are nearly transparent. Jellyfish, comb jellies, arrow worms, and many other groups are abundant throughout the deeper water column and are difficult to spot in turbulent water. The animals become inconspicuous under these conditions and tend to be overlooked by predators. It is likely that this feature has evolved independently in many groups. In deeper waters, many other zooplankton can bend over in order to resemble jellyfish, which are often avoided by fish as prey items.

■ Many phytoplankton species are defended by toxic substances.

Many phytoplankton species, notably the cyanobacteria and dinoflagellates, are often toxic to their grazers, suggesting an evolution to resist consumption. This toxicity undoubtedly influences the distribution of both species of phytoplankton and their potential grazers. Species of the haptophyte alga *Phaeocystis*, which are common in the Antarctic Ocean, produce large amounts of acrylic acid, an effective antibiotic capable of sterilizing the guts of consumers. It may inhibit microbe-mediated digestion and could induce future avoidance of the alga. Bioluminescence in some dinoflagellates may serve as a "warning" system that has coevolved with the production of toxins and the threat of grazers. In some species, great variability in toxicity is known. The small phytoplankton organism *Aureococcus anophagefferens* is the cause of so-called brown tides that strongly affect cellular function and inhibit feeding in bivalves such as scallops in mid-Atlantic North American waters. Different strains of this alga have widely different toxicity.

BIOLUMINESCENCE: A MAJOR ADAPTATION IN THE DEEP

■ As depths increase toward 1,000 m, light diminishes and organisms must use special adaptations to deal with diminishing and even absent sunlight originating from above.

While chemical and mechanical defenses of plankton can be effective at all depths, a special change occurs as waters become deeper than about 500 m and pass 1,000 m in depth. Sunlight from above steadily diminishes in this depth transition zone. If you are a fish looking downward you will see darkness, but upward glances will still show a dappling of light. The color spectrum also is strongly altered since wavelengths at the red end of the spectrum are essentially eliminated and therefore a red flash will appear to be black. Most flashing by bioluminescent creatures is in the blue or green end of the color spectrum.

The eyes of fish will still function in the strongly diminished sunlight even at depths of 1,000 m. As a consequence, a predator can look upward and see a prey as it moves and breaks the dappling of sunlight from above. At these deep depths and even in shallower waters, many fish are lightly colored on the ventral surface and dark on the dorsal surface. In either case, an upward- or downward-looking predator may not be able to see the prey fish against the light-dappled upward view or the dark downward view.

One of the most peculiar animals living in this transition zone is the barreleye fish (**Figure 10.12**) *Macropinna microstoma*, found at depths of 600–800 m off of Monterey Bay (Robison and Reisenbichler, 2008). It has barrel-shaped eyes located within a transparent dorsal shield and the eyes can rotate from facing vertically upward to forward, as in a more typical fish. It uses large fins to stabilize its position while pointing the eyes upward, scanning for food. Then the eyes can be rotated to orient forward, which puts prey within sight and also smell of the nasal organs. In another barreleye

FIG. 10.12   The barreleye fish *Macropinna microstoma*. (Courtesy of Monterey Bay Aquarium Research Institute)

species, each eye has a reflective element that can gather light ventrally and laterally, which focuses an image and strongly expands its field of view (Partridge et al., 2014). Its prey is not known for certain but may be small nekton that are captured by siphonophores that live at the same depth and are then stolen by the barreleye.

- **At bathypelagic and deeper depths, animals use bioluminescence based on the luciferin-luciferase system or the photoprotein-calcium system in a variety of adaptations to avoid predation.**

Bioluminescence is the biological production of light. It is used widely by many marine animal groups and microbes living at all depths, but it is especially common in organisms living in deeper waters at mesopelagic and bathypelagic depths. It has evolved many times in the history of life. In most cases light is emitted by the organisms themselves, but sometimes bacterial symbionts may be the source of light. Despite the widespread occurrence, light emission is based on one reaction type: the oxidation by oxygen of an organic molecule, **luciferin**, which can occur in a number of distinct classes in different organisms. The reaction is catalyzed by the enzyme **luciferase**, sometimes with the help of cofactors, resulting in release of a photon by the luciferin. In another case, light is emitted by a molecule but only after it absorbs a photon itself, thereafter releasing light at a longer wavelength. This type of light reradiation is known as **fluorescence** and occurs, for example, in green fluorescent protein, a photoprotein found in a species of jellyfish. The distinction between bioluminescence and fluorescence is not so clear-cut, and the interested student should consult Haddock et al. (2010).

- **Bioluminescence has often involved the elaborate evolution of interaction and cooperation between a host and symbiotic bioluminescent bacteria.**

In many animal species, certain organs have cells or groups of cells that are bioluminescent. But in a number of species elaborate symbioses have evolved between an animal host and bioluminescent bacteria, which live in a specialized organ in the host. Shallower-water bobtail squids (*Euprymna scolopes*, order Sepiolida), relatives of the cuttlefish, have special light organs in the mantle that contain bioluminescent bacteria. The symbiosis is very highly developed and gives us insight of the degree to which interactions between a host and bioluminescent bacteria can evolve. Colonization by the bacterium *Vibrio fischeri* from the outside environment into the squid initiates development of the bioluminescent organ in the squid. A number of studies using transcriptomics and microarray studies have demonstrated the intimate interactions between the host squid and bacterial symbiont, especially identifying the genes employed by both squid and bacterium in recognizing changes in the microenvironment of the luminescent organ (Rader and Nyholm, 2012). These recognition factors allow responses by both the squid and bacterium. The squid feeds sugars and amino acids into the organ, which provides nutrition for the luminescent bacteria, *Vibrio fischeri*. In some species, the bacteria are expelled each day and taken up by the squid from the external environment. The bacteria grow rapidly within the luminescent organ. They use small signaling molecules to sense when there are enough bacteria in the light organ to stimulate gene expression to produce a light signal. Their glow enables the squid to blend into the dappling and glistening waters while foraging at night, which makes them difficult to discern when viewed by predators (Jones and Nisiguchi, 2004).

- **Low densities of prey in meso- and bathypelagic depths have resulted in a set of offense and defense adaptations to detect or attract prey and to thwart predators.**

Deep-sea consumers live in a world of low food supply and are adapted to rapidly consume a rare meal. This has resulted in a large number of meso- and bathypelagic predators with remarkable capabilities to capture prey, and a matching set of capabilities to prevent being captured by other predators. A pervasive feature of deeper-water creatures is the frequency at which bioluminescence plays such a role in defense and attack (**Figure 10.13**).

Many deep-sea animals use bioluminescence to play defense against the many predators that lurk in midwater. A bright flash may startle, whereas exuding a milky luminescent fluid may act as a smokescreen, allowing the potential prey to escape. The vampire squid *Vampyroteuthis* (**Figure 10.14**) lives to depths of 1,200 m, where light is nearly absent Although *Vampyroteuthis* can swim by flapping its fins, it is very much a "sitting duck" for predatory fish and diving mammals, especially those with good vision in low light. When approached, however, it uses a battery of flashes to momentarily deceive and startle an approaching predator. It has luminescent organs at the tips of its eight arms and in more extreme situations will release a fluid loaded with luminescent particles. When touched, the animal flares its arms, which glow and dim in synchrony. With more intense stimuli, the arm tips are raised above and then release the fluid, which envelops the animal and serves as a disguising shroud, allowing the vampire squid to escape (Robison et al., 2003).

An even more bizarre case is the benthopelagic sea cucumber *Enypniastes eximia*. Sea cucumbers normally live on the ocean bottom, and this species does spend time feeding on soft sediment. But it spends a great deal of its time in the water column as a zooplankter. When a potential predator contacts this species, the sea cucumber sloughs off some of its body surface onto its attacker, which is sticky and bioluminescent. The material slimes the predator with a glowing "paint," which now forces the predator to take evasive action from its own predators!

The use of luminescent flashes is in itself a good defense since it reveals the position of a predator, even after the prey has been swallowed. Lie-in-wait predators, such as many anglerfishes, often have opaque, black digestive tracts that may block off the light produced by bioluminescent prey that have been swallowed. This would prevent an obvious cue for yet another predator, who might otherwise obtain a double meal!

- **Midwater fishes employ color-shading and counterillumination generated by ventral bioluminescent organs, to hide their position in low-sunlight conditions.**

In the low-sunlight conditions of mesopelagic waters of 150–1,000 m depth, midwater fishes, cephalopods and zooplankton,

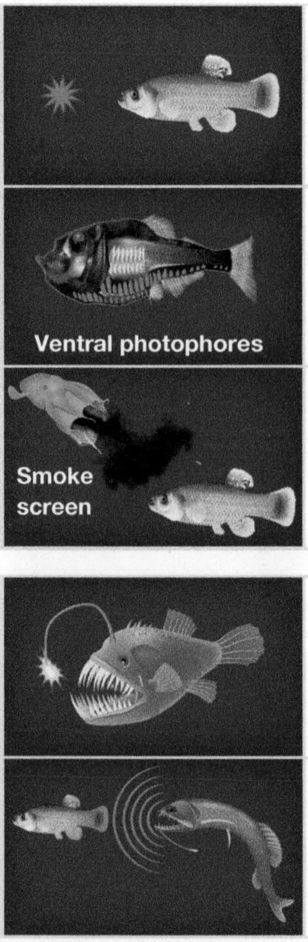

Defensive		Startle - flash to confuse predators	Dinoflagellate flash, squid, stern chaser myctophid fish
	Ventral photophores	Counterillumination	Many fish, crustaceans, squids
	Smoke screen	Lit smoke screen, release of luminescent slime	Many fish, crustaceans, squids, ctenophores, larvaceans, and others
Offensive		Lure prey or attract host (bacteria)	Angler fishes, siphonophores. Cookie cutter shark with photophores around gill slits that attract prey
		Illuminate prey	Dragonfish, flashlight fish

FIG. 10.13 Defensive and offensive adaptations of animals using bioluminescence in deep water. (After Haddock et al., 2010)

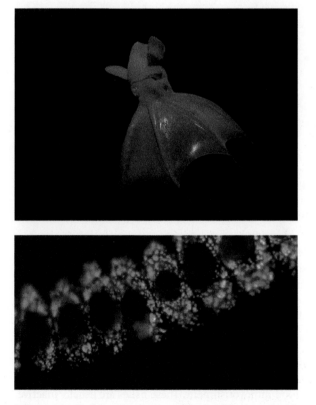

FIG. 10.14 Top: the deep-sea vampire squid *Vampyroteuthis* as seen from a submersible. Bottom: closeup of arm tips, showing bioluminescence. (Photos by Steve Haddock)

are defenseless against predators unless they can reduce their obvious visibility as they cast shade from sunlight above. To be less visible to predators from below, they have to blend in with the sunlight. Many midwater fishes possess **photophores**, or light organs, probably for several functions, including attracting mates and luring and illuminating prey. But the photophores of many bioluminescent mesopelagic fishes are concentrated on the ventral surface (**Figure 10.15**). The light from this array provides **counterillumination**, which eliminates a shadow when predators beneath the fish look upward into the downwelling light from the sun and moon. Many other mesopelagic fish also have a silvery color that blends with the downwelling light.

■ **Bioluminescence can be used for offense by projecting a light signal that is invisible to prey or by using luminescent lures to attract prey.**

Bioluminescence for offense is used to great effect by some fishes. As mentioned earlier, nearly all bioluminescence is in the blue and blue-green part of the light spectrum. Light at these wavelengths travels a greater distance through the water column, and nearly all marine creatures detect light well in the blue end but are very poor at detecting red light, which is not prevalent in deep water. Deep-sea fish in the family Malacosteidae, or loosejaws, have formidable teeth but also have distinct bioluminescent organs on the head beneath the eye that emit both green and red light. The red light is essentially invisible to other fish until they are just a few millimeters away. Loosejaws have a source of red

Squalidae · Sternoptychidae · Chuasnidibtidae

FIG. 10.15 Ventral counterillumination areas in members of three families of mesopelagic fish. (After McAllister, 1967)

FIG. 10.16 The deep-sea dragon fish *Malacosteus niger*. (From commons.wikimedia.org)

and infrared light, which projects toward unsuspecting prey. But to have "night vision" you have to be able to detect the reflected light that returns to you. Calculations suggest that the retina of the loosejaw *Aristostomias tittmanni* can detect such light about 1 m away. One species of *Aristostomias* uses a retinal pigment that is very sensitive in the red end of the spectrum. Even more amazing is the detection system of the predatory deep-sea dragon fish *Malacosteus niger* (**Figure 10.16**). It uses a derivative of chlorophyll, which acts like an antenna, receiving the red light and then transferring the energy to more ordinary blue-green sensitive photopigments in the retina (Douglas et al., 1998). Imagine being a prey fish. You can be 1 m from your death, have eyes, and yet never see the final moment coming.

Some meso- and bathypelagic fish take offense with lures. The spectacular anglerfishes (**Figure 10.17**) use a lure protruding upward from between the eyes. This lure is a fin ray modified by evolution into a structure that attracts other fish. The lures also have bioluminescent organs, which contain luminescent bacteria that produce a long glow that attracts prey fish. The anglerfish opens its wide mouth and rapidly consumes the prey.

BIG-MOUTHED PREDATORS WIN AT DEPTH
■ Mesopelagic fishes are adapted for dealing with consumption of rare prey.

Mesopelagic fishes live in depths between 150 and 1,000 m. Trawls from these depths usually bring up an astounding assortment of fantastic fish, some with mouths that are enormous in comparison to the fish's overall size. The spectacular anglerfishes seem to be all mouth, and they have a lure protruding upward from between the eyes (**Figure 10.17**). The deep-water viperfish *Chauliodus* has a specialized backbone, which accommodates the enormous mouth as it is opened (**Figure 10.18**). Fishes of this zone may have well-developed musculature and feeding apparatus and are commonly excellent hunters. Many species have tubular eyes that point upward to spot zooplankton prey. Such fishes have relatively short and small mouths, used to rapidly ingest small prey.

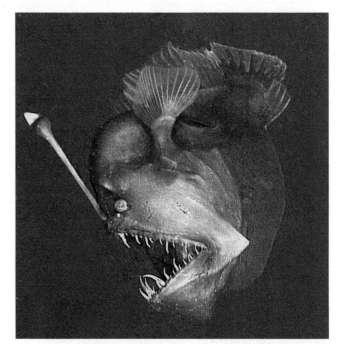

FIG. 10.17 A black sea devil anglerfish, *Melanocetus niger*, photographed from a submersible in the deep sea of the California coast, showing a bioluminescent organ containing symbiotic bioluminescent bacteria, which acts as a lure. (Courtesy of Research Photographers of Monterey Aquarium Research Institute)

Because of the difficulty of finding a mate, many anglerfishes have dwarf males, which attach to females and produce sperm that is released in the immediate vicinity of the eggs released by the female. Many other species are simultaneous hermaphrodites.

■ Bathypelagic and abyssopelagic fishes are not very active.

Bathypelagic fishes (depths of 1,000–4,000 m) and abyssopelagic fishes (depths of 4,000–6,000 m) inhabit a world of great food impoverishment. These fishes are usually inactive, feeding only occasionally. There is therefore little reason for the development of a strong skeleton and musculature as observed in fishes in shallower water. Indeed, muscle contraction consumes oxygen and is thus a disadvantage when food is so scarce. Consequently, it is not surprising that fishes in this zone have poor musculature and incompletely ossified skeletons. In addition, swim bladders are usually absent since these fishes do not swim over great ranges of depths to shallow water, which would be facilitated by changes in bulk density or buoyancy. Abyssal benthic fishes are somewhat more active than those in the water column at abyssal depths, which may reflect a greater food supply. Elongated fish species dominate the fish fauna of deep-sea bottoms. This shape may be related to the evolution of a particularly effective lateral-line system, which is important for detection of prey and other fishes in a dark environment.

The Big Picture: Scales of Processes in Time and Space

The adaptations discussed in this chapter involve a large range of spatial and time scales, and we should try to summarize this variation in order to appreciate the context of the physical factors that influence spatial distributions of plankton, large-scale movements of large nekton, and the enormous depth gradient in the ocean, leading to transfer of organic matter from the shallows to deep water and the great transitional decline of light with increasing depth.

Figure 10.19 shows the enormous range of spatial-temporal scales associated with the great range of processes

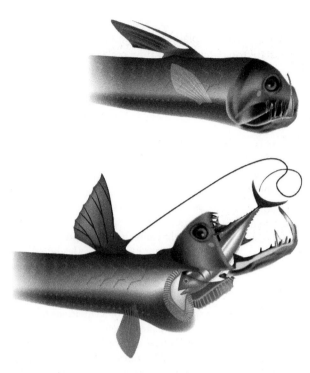

FIG. 10.18 The deep-water viperfish *Chauliodus* uses its specialized backbone to enable the opening of its enormous mouth in order to consume large prey.

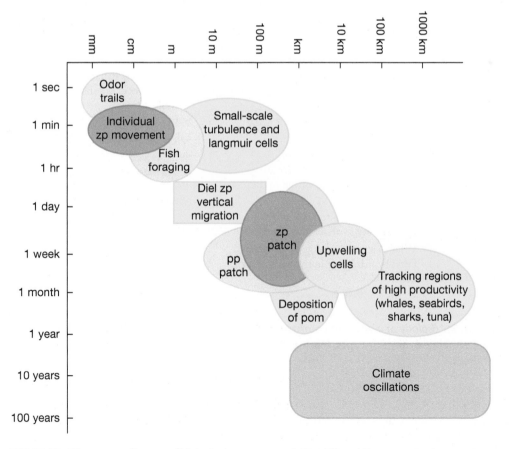

FIG. 10.19 Time-space diagram of biological processes scaled to different time-space scales. zp = zooplankton; pp = phytoplankton; pom = particulate organic matter.

involving biological functions, turbulence, and climate. As can be seen, biologically facilitated movements and detections range over the mm–m scale over a time range of sec–hr, ranging from odors produced by zooplankton to behavioral movements, resulting in zooplankton swarms to diel vertical migration. A larger spatial scale of m–km over times of 1 day–week explains the formation and dissipation of phytoplankton patches, over time scales of days–months. Movements of large predatory nekton occur over 10–1,000 km over time scales of 1 week to 1 year.

■ CHAPTER SUMMARY

- Plankton are rarely distributed homogeneously throughout the ocean. Rather, we see strong spatial differences in plankton abundance, to the degree that many plankton species occur in discrete patches.

- Phytoplankton and smaller zooplankton distributions are often determined by localized wind patterns and a balance between population growth and dispersive processes caused by turbulence and wind.

- Many planktonic organisms undergo diel vertical migrations: They move toward the surface at night and descend during the day. The behavior is controlled by a biological clock that is reinforced by environmental light cues. Convincing explanations of this behavior include the effects of predation and the potential of energy savings of poikilotherms by diving to cooler waters.

- Organisms living at greater and greater depths experience decreasing food input and rapidly extinguishing light.

- The vertical position of plankton is determined by the bulk density of the organism, structures that create drag, water motion, and swimming ability.

- Predation is intense in the plankton. Both phytoplankton and zooplankton rely on evolved armature, chemical defenses, and transparency to avoid predation.

- At bathypelagic and deeper depths, animals use bioluminescence based on the luciferin-luciferase system or the photoprotein-calcium system in a variety of adaptations to avoid predation and communicate.

- Schooling is coordinated aggregation and movement. Schooling by fish and other nekton is strongly involved with foraging and defense against predators.

- On the oceanic scale, species of large-bodied nekton such as whales, tuna, and sharks follow specified large-scale migration routes, stopping at foci of food abundance such as upwelling centers.

- Bioluminescence is a major feature of deep-water species, used for producing ventral illumination in fishes to disguise the fish against the dappled sunlight from above, deceptive cloaks of luminescent material that lead predators astray, and even as startle devices to slow predator advances.

- At bathypelagic depths fish have a spectacular array of adaptations to hunt rare prey, often in the dark. Many species have extraordinary large mouths with bioluminescent adaptations to lure prey, including lures suspended over the head filled with luminescent bacteria.

- Biological and physical processes over the depth-time-space gradients of the deep sea operate over a very large range of time and space scales.

■ REVIEW QUESTIONS

1. What major physical process in the ocean retards sinking of planktonic organisms?

2. What adaptations of planktonic organisms may retard sinking from the surface waters?

3. How can we efficiently and rapidly detect, count, and distinguish among different planktonic and nektonic animals in the water column without direct sampling and identification of every animal?

4. Describe how two processes in the ocean interact to determine the abundance of diatoms at a site at the ocean surface.

5. Why do large predators such as whales and sharks travel distances of many km over an annual cycle?

6. Describe two adaptations of zooplankton that allow them to adjust their vertical position in the water column.

7. How do zooplankton reduce their vulnerability to attack by predators?

8. What is a major ecological problem faced by fish living in the bathypelagic zone?

9. Why do many fish have bioluminescent organs on their ventral surfaces?

10. What is an advantage of living in a fish school? A disadvantage?

11. What is a convincing reason that zooplankton do not undergo diel vertical migrations to avoid ultraviolet light at the surface?

12. What is a disadvantage of being bioluminescent at bathypelagic depths?

Visit the companion website for *Marine Biology* at www.oup.com/us/levinton where you can find Cited References (under Student Resources/Cited References), Key Concepts, Marine Biology Explorations, and the Marine Biology Web Page with many additional resources.

CHAPTER 11

Processes in the Water Column

Critical Factors in Plankton Abundance

Plankton are extremely variable in abundance, both spatially and temporally. We have discussed the spatial patchiness of the plankton in Chapter 10. But temporal changes, driven by seasonal cycles of nutrient supply, light, and turbulence, are also major drivers of plankton abundance. This is especially true of the phytoplankton. Although many factors such as light and nutrients contribute to variation in abundance, water motion is a major driving factor. Water motion affects the exchange of nutrients from deeper waters and also drives the mixing of waters and entrained phytoplankton downward and, therefore, away from access to sunlight. Water motion is important for phytoplankton growth in all parts of the ocean, but its influence is best introduced in a discussion of the **spring phytoplankton increase**, which occurs in nearly all mid- and high-latitude areas where there is a seasonal change in day length. Animal grazing also affects phytoplankton abundance. It is the purpose of this chapter to discuss the importance and interaction of these various factors on plankton dynamics. This material will be important for our understanding of the next chapter on productivity, food webs, and global climate change.

The Seasonal Pattern of Plankton Abundance

■ **In midlatitudes, phytoplankton increase in the winter-spring, decline in summer, and may increase to a lesser extent in fall.**

There is a predictable seasonal pattern to plankton abundance in temperate and boreal waters of depths of an approximate range of 10–100 m[1]. **Figure 11.1** traces the seasonal changes in phytoplankton, zooplankton, light, and nutrients near the surface during the year in a temperate–boreal coastal zone, where water depth ranges from about 10 m to the depths of the continental shelf. Usually in the early spring, but sometimes in later winter, phytoplankton populations increase dramatically and are dominated by a few diatom species. This is known as the **spring phytoplankton increase**. Although the exact time varies with latitude and year, phytoplankton abundance in waters of southern New England usually reaches a peak in February–March. During this period, the surface waters are dense with phytoplankton. The diatom-dominated phytoplankton begin to decline as late spring approaches, when the early dominants give way to other phytoplankton species. In northeastern U.S. coastal waters and estuaries, microflagellates and dinoflagellates become abundant, but in higher latitudes, diatoms dominate and bloom also later in the year. In some locations, the phytoplankton increase again to a smaller maximum peak in the fall, followed by a decline to very low abundance again in the winter. While temperate and boreal inshore spring phytoplankton blooms are dominated by diatoms, other phytoplankton groups, such as flagellates, may dominate spring blooms in other areas such as the open waters of the continental shelf.

Phytoplankton blooms are not restricted to the coastal zone but can also be found in surface waters of the open sea. Diatom blooms occur in later winter–early spring on the continental shelf. In subarctic latitudes such as the Bering

---

[1] The **temperate zone** is the range of latitudes between the tropics and the Arctic Circle; the **boreal zone** refers to the north temperate zone in the Northern Hemisphere.

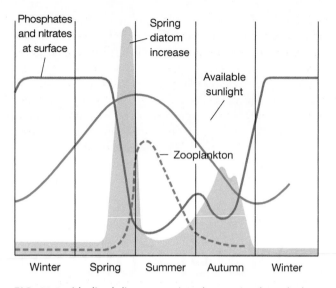

FIG. 11.1 Idealized diagram tracing changes in phytoplankton, zooplankton, light, and nutrients during the year in a temperate–boreal inshore body of water. (Modified after Russell-Hunter, 1970)

Sea and northernmost Atlantic, enormous blooms of the coccolithophorid *Emiliania huxleyi* occur, often spanning hundreds to thousands of kilometers. While such blooms are found in other parts of the ocean, most appear in sub-polar latitudes. The living and dead calcareous plates of these organisms reflect a lovely bluish light that allows the bloom to be readily followed by satellites (**Figure 11.2**). Coccolithophores have increased in the North Atlantic over the past 50 years, which has led to speculation that they are benefiting from the addition of carbon dioxide to the ocean from anthropogenic climate change (Rivero-Calle et al., 2015). Experimental studies demonstrate that additions of carbon dioxide may reduce pH but still have a net positive effect on the growth of coccolithophores, per-haps by the positive impact of carbon additions to water, which help to drive photosynthesis. Thus coccolithophores may be winners in a carbon dioxide–charged ocean of the future.

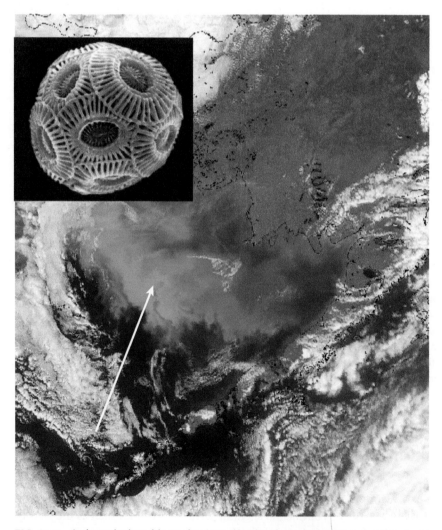

FIG. 11.2 A phytoplankton bloom dominated by the phytoplanktonic coccolithophorid *Emiliania huxleyi* (inset upper left) in the Bering Sea, September 2000. The blue-white waters seen here may represent living or dead cells, as their reflective properties do not change significantly when the cells die. (Courtesy of NASA)

■ **Zooplankton start to increase as the phytoplankton bloom reaches its peak, attaining a maximum following the phytoplankton peak in the late spring or early summer.**

In temperature-boreal shelf waters, zooplankton reach their yearly maximum after the spring phytoplankton increase begins to decline. **Calanoid copepods** dominate this burst of abundance and are the principal **grazers** of the diatoms, but they also feed on ciliates and even larger bacteria in the plankton. At any one time, several different developmental stages (nauplii) of copepods coexist in the plankton. **Ciliates** are also important grazers of phytoplankton and bacteria, and may dominate the herbivorous zooplankton, especially in some estuaries. Planktonic larvae of benthic invertebrates are common in the late spring and in early summer. Following this burst, zooplanktivorous fish and invertebrates become abundant. Comb jellies and jellyfish become especially abundant and are effective predators on copepods and planktonic larvae. These gelatinous creatures clog plankton nets in summer, when arrow worms and tunicates may also become abundant. Zooplanktivore fishes, such as menhaden in the eastern United States, also are important predators in the plankton but they have been overfished in many places.

■ **The spring phytoplankton peak and the later zooplankton peak are shortest and sharpest in high latitudes, becoming indistinct in the tropics.**

The temperate–boreal pattern is a good starting point for discussion, but this pattern is not universal. The strength of the spring phytoplankton bloom varies with latitude. In the Arctic, a single summer peak of phytoplankton abundance is followed by a zooplankton maximum (**Figure 11.3a**). The phytoplankton production lasts as little as 2–4 weeks. In temperate–boreal waters, as we have discussed, a spring phytoplankton increase is followed by a decrease, coinciding with a zooplankton increase (**Figure 11.3b**). In late spring and summer, the zooplankton decline and a smaller peak of phytoplankton may follow in the fall—for example, in the Gulf of Maine and Nova Scotia shelf region. In the tropics there is no clear alternating pattern of phytoplankton and zooplankton abundance (**Figure 11.3c**). This correlates with the relative lack of seasonality in tropical waters.

Water Column Parameters and the Spring Phytoplankton Increase
Water Column Stability and Light

■ **Because light intensity decreases exponentially with increasing depth, there is a compensation depth below which respiration for a given phytoplankton cell exceeds photosynthetic output.**

Light intensity decreases exponentially with increasing depth (**Figure 11.4**) and becomes a limiting factor to photosynthesis. The **compensation depth** is that depth at which the amount of oxygen produced by a phytoplankton cell in photosynthesis equals the oxygen consumed in respiration. We can estimate the compensation depth by placing phytoplankton cells in a clear bottle. At depths shallower than the compensation depth, there is a net increase of oxygen over time, whereas at depths greater than the compensation depth, there is a net decrease of oxygen over time. The compensation depth is thus an indicator of the potential of a photosynthesizing cell to be a net producer. The light intensity corresponding to the compensation depth is the **compensation light intensity**. Remember, we are just considering the potential of a cell that is held at a given depth and are not considering the important additional effect of turbulent water or changing light conditions.

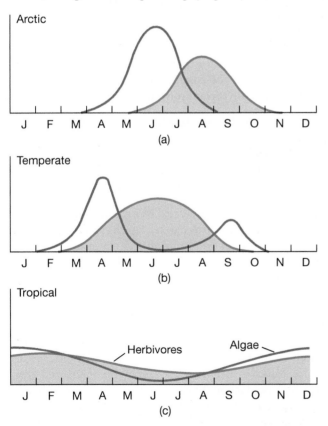

FIG. 11.3 Phytoplankton and zooplankton in a seasonal cycle: (a) Arctic, (b) temperate–boreal, and (c) tropical. (After Cushing, 1975)

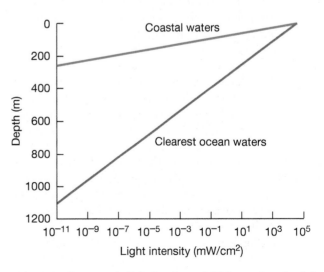

FIG. 11.4 Decrease in light irradiance with increasing depth in coastal water and clear ocean water. Note that the light intensity is plotted on a logarithmic scale; light is decreasing exponentially with depth.

The compensation depth is controlled by season, latitude, and transparency of the water column. As the temperate–boreal spring progresses, the increasing photo-period (day length) tends to increase the compensation depth to an eventual maximum. The Arctic winter photoperiod is zero, and therefore there is no light for photosynthesis. Suspended matter in inner shelf waters intercepts light and reduces the compensation depth, relative to the clearer open sea. Similarly, as a phytoplankton bloom develops and as suspended matter (seston) becomes trapped in the water column, the compensation depth decreases due to light absorption and shading by particles. A yellow pigment originating from rivers and other terrestrial sources is also important in the extinction of light with depth.

The Sverdrup Model Explaining the Spring Phytoplankton Increase

■ **During winter, the water density is about the same throughout the water column, and phytoplankton cells are stirred on average to depths that prevent average photosynthesis from permitting the phytoplankton population to become dense.**

The Norwegian oceanographer Harald Sverdrup proposed a model that has guided our thinking about how a spring phytoplankton bloom develops. Consider the state of the plankton and of the water column during the winter before the spring phytoplankton bloom occurs. At this time the water column can be isothermal to great depth, or even to the ocean bottom in deep bays and estuaries. The term *isothermal* implies that there is little or no density variation with depth given that salinity is invariate with depth. Near the shore, temperate–boreal winters in bays of 10–30 m depth (e.g., Long Island Sound, New York, continental shelf waters of the Mid-Atlantic Bight) are times of high wind stress, resulting in extensive **wind mixing** of the water column. Because there are no density differences, the water column is unstable, and winds cause extensive vertical mixing. In the open ocean away from land, wintertime mixing is even deeper and is driven by wind mixing combined with the process of convection. **Convective mixing** occurs when the air temperature is colder than the surface ocean temperature. When this happens, water near the air–sea boundary cools and becomes denser, causing it to sink and be replaced by the warmer water below. So wind mixing turns over the water column in the inner shelf and bays, and convection adds to mixing of the water column offshore on the shelf in deeper waters of approximately 30–100 m. In all cases, vertical mixing is the source of high nutrients brought to the surface. In bays such as Long Island Sound, late fall and winter winds mix nutrients from bottom waters to the surface. On the continental shelf, vertical mixing in late fall and winter also brings deep nutrient-rich water to the surface.

The **mixing depth** is the depth above which all water is thoroughly mixed by wind or convection. Because the winter mixing depth is very deep, phytoplankton cells are exposed to a wide range of light levels from the surface to great depths where photosynthesis is light limited. Over the course of a winter day, the growth rate of phytoplankton in the mixed layer is light limited and slow. The Sverdrup model suggests that phytoplankton cannot bloom—indeed, that their concentrations must be decreasing—during this period because their total growth from the mixing depth to the surface, measured in photosynthetic oxygen output, is slower than the rate of loss resulting from phytoplankton cellular respiration. In effect, wind mixing is trading off with light input to determine whether a phytoplankton bloom can take off.

The Sverdrup model predicts that the winter phase of declining phytoplankton continues until springtime increases in sunlight, longer days, and shallower wind mixing act together to increase phytoplankton growth rates past a critical threshold. This critical point is where the total rate of phytoplankton division equals the rate of phytoplankton loss above the mixing depth. Once this threshold is crossed, the model suggests that phytoplankton will begin blooming and will continue to bloom until growth rates once again fall below the critical value. The Sverdrup model argues this: Along with increases in solar elevation and photoperiod, shallower water column mixing by weaker winds is an essential part of the development of the phytoplankton bloom.

But when exactly will a phytoplankton bloom begin? A phytoplankton bloom should develop only when the volume of water in which photosynthesis (phytoplankton oxygen production) occurs has a net excess of production over consumption (phytoplankton cellular respiration). The depth above which total production of oxygen from photosynthesis in the water column *equals* total consumption from respiration is known as the **critical depth**. The Sverdrup model argues: If the mixing depth is less than the critical depth, phytoplankton should increase, but if the mixing depth is deeper than the critical depth, phytoplankton growth rates will be less than loss rates and their biomass will decrease (Figure 11.5). As the spring progresses, the mixing depth shallows above the critical depth; this is the condition required by the Sverdrup model for the phytoplankton bloom to begin.

■ **As the spring temperature increases, the surface waters warm up, and the water column stabilizes and allows the phytoplankton bloom to take off. But eventually nutrients are lost to deeper waters, and the phytoplankton bloom is cut off.**

A spring thermocline generated by solar heating reduces the water density in the surface water, which stabilizes the water column and retards wind mixing and convection overturn. As the water column stabilizes, the mixing depth becomes shallower than the critical depth. The onset of summer conditions with lower wind speed enhances the decrease of mixing depth. In inshore waters, the water column may also be stabilized by lowered salinity due to freshwater influxes from terrestrial sources. The late spring and early summer runoff from the Fraser River in British Columbia, for instance, causes a shallow-water salinity minimum and typically high production throughout the northern Puget Sound and the Strait of Georgia region. The earlier timing of inshore Norwegian fjord phytoplankton blooms, which appear earlier than open-water offshore phytoplankton blooms, may be related to the greater stability of the water column inshore.

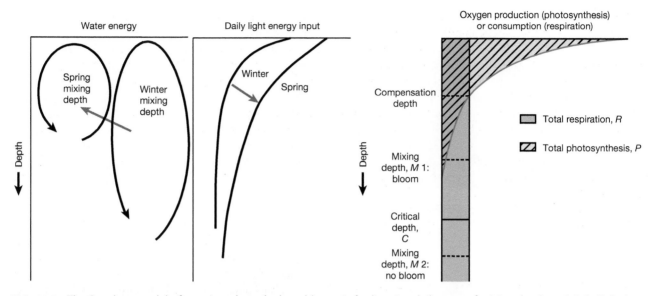

FIG. 11.5 The Sverdrup model of a spring phytoplankton bloom. Left: directional changes of mixing depth and daily light input from winter to spring, setting up conditions for the spring phytoplankton bloom. Right: the quantitative relation of critical depth (C) and mixing depth (M). In the absence of vertical mixing, $P = R$ at the critical depth. If the mixing depth is less than the critical depth C (e.g., at mixing depth 1), then $P > R$ and a bloom will develop. If $M > C$ (e.g., at mixing depth 2), a bloom fails because some phytoplankton cells are swept below to waters of light intensity low enough to yield the condition $P < R$.

The stabilization of the water column allows the birth but also causes the eventual demise of the spring diatom bloom. The spring stabilization of the water column maintains phytoplankton in the upper layer, thus reducing the rate removal from the zone of active photosynthesis. As the water column stabilizes, however, and the thermocline is established, phytoplankton die and sink, or are ingested and egested by zooplankton and sink to the sea floor. Because of the stabilization of the water column, these materials and other nutrients are not returned immediately to the surface from greater depths and from the bottom. In a shallow area, like Long Island Sound, New York, there is extensive exchange between the bottom and the overlying water in terms of resuspension of detritus and dissolved nutrients in the winter and early spring. Once the thermocline has been established in the spring and summer, however, this exchange is greatly diminished. Toward the end of summer, with the advent of fall storms, the thermocline may be disrupted, bringing nutrients toward the surface from the bottom in shallow water. This may result in a fall increase of phytoplankton. Otherwise, decomposition continues, and dissolved nutrients are not returned to the surface of the water column until later in the fall and early winter, when vertical mixing is enhanced but light is low.

As the water column stabilizes in late spring and summer, phytoplankton organisms denser than seawater, such as diatoms, start to sink from the water column. Such considerations do not hold, however, for phytoplankton (e.g., dinoflagellates) whose swimming abilities circumvent this tendency to sink. The difference must contribute to the increasing abundance of dinoflagellates and other flagellated phytoplankton species from spring to summer.

The hydrological conditions tied to seasonal variation play the primary role in the birth, development, and demise of the spring phytoplankton increase. The stabilization of the water column in spring initially permits the development of the spring increase. In the Gulf of Maine, this occurs in March (**Figure 11.6**). The stability of the water column, however, prevents nutrients lost from the surface waters from returning to the surface where light is available for photosynthesis. Furthermore, dense phytoplankton organisms sink out of the water column when late spring–summer stability sets in and grazing further reduces phytoplankton in the surface waters. Zooplankton excretion of fecal pellets results in export of nutrients to the bottom, which further depletes the surface waters of nutrients in late spring and summer. The poor nutrient situation prevails until the fall and winter overturn of the water column. Close to the shoreline, however, natural drainage of nutrients and human-influenced inputs cause primary production to be high for much of the year, even in midlatitudes.

Critique of the Sverdrup Model of the Spring Diatom Increase

■ **The mixing depth–critical depth hypothesis has been criticized because phytoplankton blooms often develop in late winter when waters are still strongly vertically mixed.**

The Sverdrup model of the initiation of a spring phytoplankton bloom just discussed depends on the stabilization of the water column and the increasing light penetration as spring progresses. The combination of these changes traps the phytoplankton in the surface waters under favorable light conditions, fueled by the high nutrients brought to the surface during winter mixing.

The Sverdrup model for phytoplankton blooms has been criticized (Behrenfeld, 2010) because many phytoplankton populations in near-shore and open-ocean regions begin blooming under winter conditions when sunlight levels are lowest, temperatures are coldest, and mixing depths are deepest. The timing of such blooms has been recognized

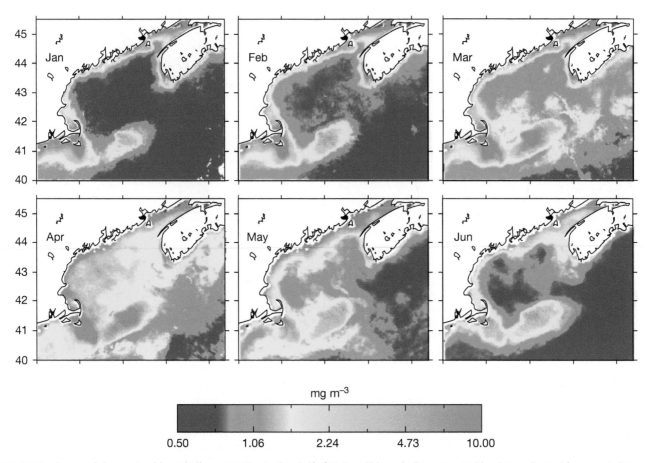

FIG. 11.6 Seasonal change in chlorophyll composition in the Gulf of Maine. Chlorophyll is estimated by data collected from a satellite-based photometer sensitive to chlorophyll. Note widespread increase in spring, followed by reduction in summer. Also note high chlorophyll values through the year very near the shoreline. (From Thomas et al., 2003)

in open waters of the North Atlantic, based mainly on our new developing database from satellite observations.

Beherenfeld argues that the phytoplankton in late winter *are* mixed to deeper water but are also decoupled for a time from the effect of zooplankton grazing—for example, copepods, who in winter are very diluted in the water column and fail to keep the phytoplankton from increasing. The phytoplankton density (cells per liter) might be low at this time in winter, but the total phytoplankton population in the total water column is high. As the water column stabilizes and light and nutrients are high in surface waters, the phytoplankton continue to increase faster than any possible losses from grazing, respiration, or downward loss. Thus, Behrenfeld is arguing that phytoplankton start to increase in winter while downward mixing is high, nutrients are high, but zooplankton are just too dilute to prevent the phytoplankton from increasing in abundance overall. It is therefore only a reinforcement of this initial phytoplankton growth that the bloom growth continues to accelerate in later spring as stratification develops and the mixing depth shallows. Conditions of water column stability, high nutrients, and high light just continue the process initiated in late winter and the phytoplankton grow faster than they can be grazed. What matters is growth rate increases as light increases in late winter, grazing is very low, and eventually stratification

begins, allowing further phytoplankton growth in the well-lit warm surface layer. Finally, in spring, the phytoplankton *are* growing most rapidly in the surface stratified layer. The bloom will end in later spring–early summer, owing to sinking of nutrients from the surface layer or by zooplankton grazing. This part of the cycle resembles the Sverdrup stratification model discussed earlier.

The difference between the Sverdrup and Behrenfeld models is illustrated in **Figure 11.7**. The two models show important differences. The Sverdrup model sees winter as a time when the mixing depth is greater than the critical depth and phytoplankton concentrations are decreasing so a bloom cannot initiate. This decrease continues until the mixing depth is shallower than the critical depth, and then the bloom increases rapidly in direct proportion to growth rate. The Behrenfeld model suggests that phytoplankton blooms can begin in early winter if mixing is sufficient to decouple phytoplankton growth and grazing. Under all conditions, it argues that the phytoplankton will bloom whenever division rates are on the increase, even if those rates are very slow (such as late winter and early spring). It further predicts that the spring bloom will end when low nutrient levels in surface waters or grazing reduces the rate of phytoplankton increase to zero or negative.

A pattern of bloom development in winter during isothermal conditions has also been found in Long Island

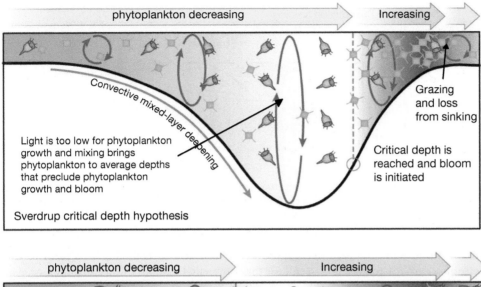

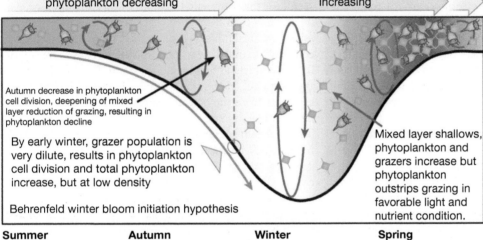

FIG. 11.7 Comparison of the traditional Sverdrup model of vertical mixing control of phytoplankton blooms (top) compared with new Behrenfeld model. In the Sverdrup model, grazing is not important, and the phytoplankton bloom is set off when the mixing depth shallows to be less than the critical depth. In the Behrenfeld model (bottom), phytoplankton are low in density but start to increase in low numbers when mixing is quite deep but grazing is at a low rate and lags the slow phytoplankton increase. As the mixed layer shallows the phytoplankton increase rapidly, still outstripping zooplankton grazing, which allows a bloom in the surface waters (after Behrenfeld and Boss 2015) (modified from Behrenfeld and Boss, 2014).

Sound, New York. It is well known that the so-called spring phytoplankton increase often starts in winter, often in February when the water column is cold and nearly isothermal (Rice and Stewart, 2013). George et al. (2015) performed experiments on Long Island Sound water in tanks and found that an increase in temperature resulted in rapid increase of zooplankton, whose grazing effect exceeded phytoplankton growth. Thus we conclude that the bloom initiation was controlled by phytoplankton growth that exceeded grazing potential, but at cold winter temperatures in the absence of the development of stratification. Much of the grazing in Long Island Sound is due to feeding by copepods and tintinnids. Earlier work by John Ryther and others demonstrated that much of the phytoplankton production in Long Island Sound was not grazed and was deposited on the bottom in early spring, which suggests that later loss from a stratified water column occurs in conjunction with grazing, and may help to end the spring diatom increase. Much further work needs to be done to understand the extent of these phenomena, but so far Behrenfeld's ideas

on the initiation of a bloom seem to fit the data we have, so a modified model may now be evolving to explain the classic spring diatom increase-demise cycle so well known in marine waters in temperate latitudes. It is not yet clear that the initiation of the "spring diatom increase" is always so early in winter in all coastal areas.

Water Column Exchange in Very Shallow Waters and in Estuaries

■ **In very shallow water estuaries, nutrient exchange, or benthic–pelagic coupling, occurs between the bottom and the water column, fueling more phytoplankton growth.**

Many water columns exchange nutrients between the bottom and the overlying water column. This process is known as **benthic–pelagic coupling**. In shallow temperate estuaries such as Long Island Sound, between New York and Connecticut, the spring phytoplankton bloom ends when the water column is strongly stratified because

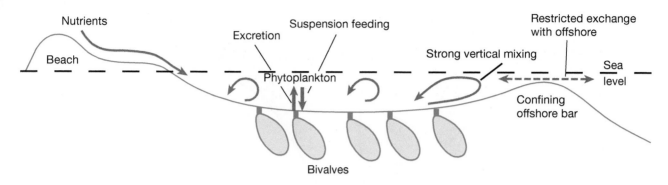

FIG. 11.8 A benthic–planktonic coupling system in coastal bottoms with restricted exchange with shelf waters. Bivalve excretion provides nutrients for phytoplankton uptake, as do inputs from the beach. Bivalves in the bottom feed on the phytoplankton. A large confined circulation landward of an offshore bar recirculates the water. Water exchange with offshore waters is relatively restricted.

nutrients are not regenerated from the bottom to the surface. Regeneration from the bottom occurs the following winter, as the water column homogenizes in density and is turned over by winter storms. Benthic–pelagic coupling therefore lags on a seasonal time scale. In shallower waters, however, the nutrient exchanges between bottom and surface waters are on shorter spatial and time scales and are not shut off by stratification. Many bays and estuaries are very shallow, and even modest winds turn over the water column for most of the year. Examples in the United States include Narragansett Bay, Rhode Island, and the south shore bays of Long Island, New York. In late spring and summer, decomposition of organic matter in shallow bottoms, combined with vertical mixing, brings nutrients back to the surface, and phytoplankton continue to bloom into the summer. Erosion of the bottom also may bring benthic algae and bacteria directly into the overlying water column.

Benthic–pelagic coupling figures importantly in a high-energy near-shore phenomenon known as **beach blooms**. These are phytoplankton blooms of a few species of very large diatoms and are especially well developed off the sandy coasts of Oregon and South Africa. Even though the coast appears to be very energetic with strong waves, an offshore submerged bar confines the vertically recirculating water to the near-shore region (**Figure 11.8**). Dense populations of benthic animals excrete nutrients into the water column that are thoroughly mixed and fuel enormous phytoplankton growth, enough to color the water a deep brown. The diatoms sink to the bottom and are eaten by the invertebrates, although some planktonic consumption occurs as well, even by filter-feeding fish, in South Africa. Input from waters offshore and from the exposed beaches adds nutrients to the system.

In deeper water, on the continental shelf, benthic–pelagic coupling is reduced or absent on seasonal or smaller time scales. But there is still extensive vertical mixing in late fall and winter, caused by increasing wind stress and convective overturn, because surface waters become cold and therefore higher in density. Combined with wind, these waters sink, which results in vertical mixing of the water column. This brings nutrient-rich water from greater depths toward the surface, because deep water is low in photosynthesis and therefore not depleted of nutrients.

Figure 11.9 illustrates three modes of vertical mixing seen near the continental margin. In very shallow bays with depths less than 5 m or so (e.g., Narragansett Bay, Rhode Island) vertical mixing is continuous, so benthic–planktonic coupling occurs all year long. In deeper bays and estuaries (e.g., Long Island Sound and the main stem of Chesapeake Bay), late fall and winter is a time that wind mixing brings nutrients from the bottom water, that have been produced by microbial decomposition all summer and early fall, which is the nutrient source for the late winter–spring phytoplankton bloom. Thus these deeper bays show benthic–planktonic coupling on a seasonal cycle. Later spring–summer stratification allows sinking of particulate (sinking diatoms and zooplankton fecal pellets) nutrients back to the seabed. In deeper continental shelf waters, wind mixing and convective mixing brings nutrient-rich deeper water toward the surface, which helps fuel the winter–spring phytoplankton increase. Later spring–summer stratification allows sinking of nutrients back to the deeper nutrient rich layer below.

■ **In estuaries, the spring freshet combines with net water flow to the sea and water mixing to determine the nutrient regime.**

Nutrients recycle extensively in the ocean, but the extraordinarily high primary production of estuaries owes itself to large inputs of nutrients from the tributaries of the watershed, with a special seasonal peak during the **spring freshet**, which is the increased flow following spring snow melt and rains. To appreciate the pattern of input and loss in an estuary, consider **Figure 11.10**, which shows a budget of carbon input and loss in the Hudson River estuary. Carbon comes principally from tributaries and runoff from the land, although sewage input also occurs to a minor degree. Carbon then enters the main part of the estuary and is taken up by bacteria and phytoplankton, to be later passed on to consumers. Some of the carbon, however, is respired as carbon dioxide and leaves the estuary as gas. Another fraction is lost to the sediment. There is general downstream movement of the remainder, and finally about half the original carbon reaches the saline portion of the estuary. Similar budgets could be made for nitrogen and phosphorus, but the proportions lost, and even the processes of loss and gain, would differ. For example, nitrogen can be lost as

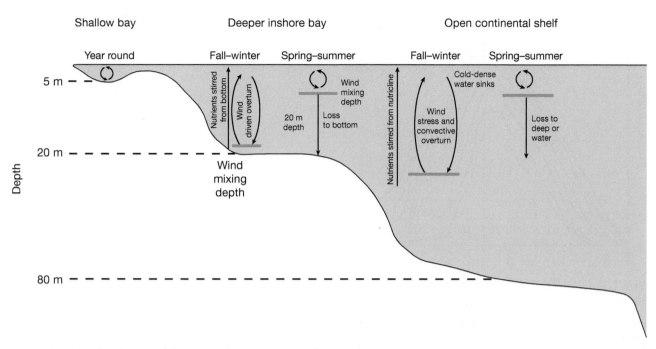

FIG. 11.9 Vertical exchange in fall-winter and spring-summer of nutrients in shallow bays, deeper inshore bays and estuaries, and open continental shelf water columns.

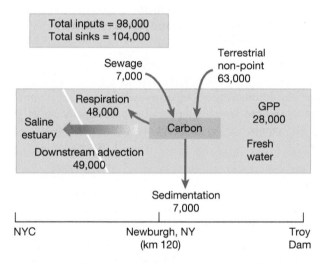

FIG. 11.10 A carbon budget for the Hudson River estuary, showing the inputs and losses (in tons) of carbon into the freshwater tidal part of the estuary and the degree of downriver loss of carbon to the saltwater part of the estuary: GPP, gross primary productivity; point sources, carbon from sewage treatment plants. (The total input does not match the total losses and sinks of carbon because these were measured independently and there is some error.) (Courtesy of Robert Howarth)

nitrogen gas to the atmosphere via denitrification, a bacterially mediated process.

Nutrient input to estuaries occurs as a pulse during the spring freshet, but some estuaries have rather short residence times for water, which is rapidly lost to the sea. The Hudson estuary loses water over its main channel in a few tens of days, whereas the Chesapeake Bay estuarine system has a water residence time of hundreds of days. Carbon input into the Hudson system is about 30 times that into the Chesapeake Bay estuary. The longer residence time in the Chesapeake allows for more extensive recycling of

nutrients within the estuary. The shorter water retention time in the Hudson River estuary results in more export of nutrients downriver into the saline part of the estuary and into the open sea. In most estuaries, primary productivity increases as the nutrient input of the freshet increases and decreases as the degree of flow from the estuary increases in late spring and summer. In North Carolina estuaries, strong spring precipitation and nearly windless summers combine to trap larger concentrations of nutrients within the estuaries, which fuel phytoplankton blooms.

Light

■ **Light may be inhibitory to photosynthesis near the surface, but a series of photosynthetic pigments captures light over much of the visible spectrum.**

Energy from solar sources is expressed in terms of energy units hitting the sea surface (e.g., watts per square meter). The angle of the sun at different times of day, the latitude, and other factors contribute to the spectral distribution of light that strikes the sea surface and the amount of back-scattering. The light striking the sea surface includes a large part of the ultraviolet–infrared spectrum; however, only visible parts of the spectrum penetrate to great depths. At temperate latitudes in clear weather during the summer, the maximum energy striking the sea surface is about 120 kw h m^{-2}. About one-half the total radiant energy is in the infrared region of the spectrum and so is not available to marine photosynthetic organisms.

Light is attenuated in the water column through **absorption** (interaction of photons with atoms in water) and **scattering** (change of direction of photons by collision). Water molecules, dissolved organic matter, particulate organic and inorganic particles, and living plankton can

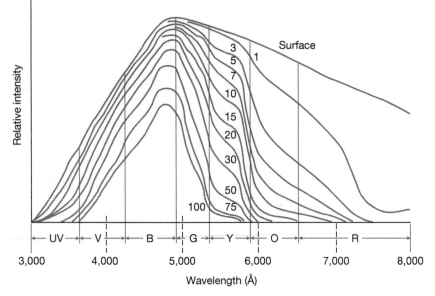

FIG. 11.11 Attenuation of different wavelengths of light with increasing depth (each curve is labeled by depth in meters) below the sea surface. Wavelength is in angstroms (1 Å × 0.1 = 1 nm). (From Clarke, 1939)

accomplish scattering of light. **Figure 11.11** shows attenuation values for different wavelengths of light. In the clear open ocean, the attenuation spectrum of light transmission maximizes transmission at about 480 nm. In turbid inshore waters, however, a more pronounced maximum occurs at longer wavelengths, 500–550 nm. Because ultraviolet light has detrimental effects on DNA, its penetration is of great interest. In moderately turbid coastal waters, UV incident light with a wavelength of 380 nm or less is almost all attenuated at a depth of 1–2 m, but in very clear parts of the ocean, 20 m may be required to remove 90 percent of the radiation entering the surface.

Incident light near the surface is intense enough to inhibit photosynthesis through bleaching of photosynthetic pigments, such as chlorophyll *a*, or the arresting of pigment production. With increasing depth, light energy is absorbed to the extent that the inhibitory effect disappears.

Photosynthetic phytoplankton use chlorophyll *a*, chlorophyll *c*, and a group of "accessory" pigments, such as protein-bonded fucoxanthin and peridinin, to utilize light energy from most of the visible spectrum (for a review of photosynthesis, see **Going Deeper Box 11.1**). The **action spectrum** is the extent of utilization of different wavelengths of light and can be determined by using different monochromatic sources of light and measuring the amount of photosynthesis. Within the usable wavelengths of 400–700 nm, the light absorbed by phytoplankton pigments can be divided into (1) light of 400–450 nm, which is mainly absorbed by chlorophyll, and (2) light of 450–700 nm, which is mainly absorbed by accessory pigments. The combined absorption of chlorophyll and accessory pigments allows the yield of photosynthesis to be constant over a large portion of the visible light spectrum. The total range of light wavelengths used in photosynthesis is known as the **photosynthetically active radiation (PAR)**.

■ **Photosynthesis increases with increasing light intensity, up to a plateau, and then is inhibited by high light intensity.**

Figure 11.12 illustrates a theoretical photosynthesis–light curve showing photosynthetic rate as a function of light intensity. Photosynthesis increases with increasing light intensity and then reaches a plateau at a maximal value P_{max}. At the compensation light intensity, the **photosynthetic rate** (in this case, measured in terms of oxygen evolution) equals the amount of oxygen consumed in respiration. Because the amount of light reaching a phytoplankton cell varies over a day, the compensation light intensity is usually expressed on a 24-hour basis. It is assumed here, for simplicity, that respiration occurs at the same rate in the light and the dark (light-enhanced respiration, known as **photorespiration**, could affect this under some circumstances). An average 24-hour compensation light intensity is in the range of 3–13 langleys per day in temperate seas. **Figure 11.12** shows **photoinhibition** at high light intensities.

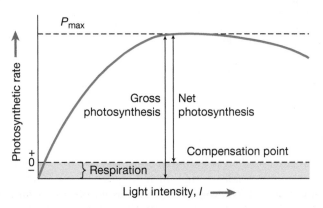

FIG. 11.12 The relationship between photosynthesis and light intensity.

GOING DEEPER 11.1

The Basics of Photosynthesis

Photosynthesis is the transformation of light energy into chemical energy for use in living cells. The overall process can be described by means of the following relation:

$$CO_2 + 2H_2O \longrightarrow (CH_2O)_n + H_2O + O_2$$

With the aid of photosynthetic organisms and light, water is split into hydrogen and oxygen, and carbohydrates (CH_2O), such as sugars and starches, are produced. In the equation, $n = 6$ in the case of glucose. In eukaryotic photosynthesizers such as plants and algae, photosynthesis occurs in the chloroplasts, which are cell organelles bounded by a double membrane. Within the chloroplast is a membrane system, the thylakoids, where light is captured and transformed into chemical energy, through the reaction of carbon dioxide and water. The membranes are surrounded by spaces where this energy is used to produce the carbohydrates.

Light energy is captured by photosynthetic pigments. All photosynthesizers have chlorophyll *a*, which absorbs light energy most efficiently in the blue and green regions of the spectrum. Other pigments, known as accessory pigments, also absorb light, but in the parts of the light spectrum where chlorophyll *a* is relatively inefficient.

The overall process of photosynthesis can be divided into the light reactions and the Calvin cycle (**Box Figure 11.1**). As one would expect from the name, the light reactions involve the capture of light energy. They involve two processes. First, light energy is used in conjunction with a complex series of reactions to oxidize water and to split it to produce molecular oxygen, protons (H^+), and electrons (e^-). This is accomplished in two distinct biochemical photosystems, also with a chain of reactions to transport electrons. Photosystem II is most important for our purposes. It is a complex of chlorophyll and proteins. A **light-harvesting complex** surrounds a **reaction center** (there are many in any cell). When the photosystem absorbs photons, electrons are raised to a higher energy level and are trapped by primary electron acceptors. Water is split in an oxygen-evolving complex, which produces oxygen, protons, and electrons. The electrons are continually taken to the chlorophyll, which continues to move electrons. The electrons then enable ADP to be converted to ATP, which is the major source of energy for the Calvin cycle. Also, NADP is reduced by the electrons to NADPH, which is a major source of chemical reducing power in the Calvin cycle.

Carbon is "fixed" to carbohydrates in the Calvin cycle. The reaction is shown in **Box Figure 11.2**. The enzyme ribulose biphosphate carboxylase (RuBP carboxylase, or Rubisco) enables the reaction of carbon dioxide, water, and the five-carbon sugar ribulose biphosphate (RuBP) to produce a three-carbon molecule that is a precursor of the six-carbon sugars. Energy from ATP and the chemical reducing power of NADPH are both required to make the reaction go. Most of the reaction series (five out of every six three-carbon molecules) involves the regeneration of RuBP.

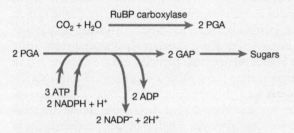
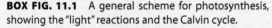

BOX FIG. 11.1   A general scheme for photosynthesis, showing the "light" reactions and the Calvin cycle.

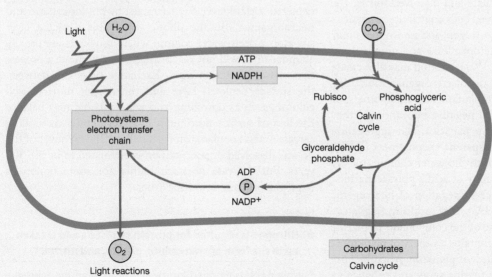

BOX FIG. 11.2   The Calvin cycle of photosynthesis: PGA, a three-carbon molecule, 3-phosphoglyceric acid; GAP, another three-carbon molecule, 3-phosphoglyceraldehyde; RuBP carboxylase, an enzyme, catalyzing the first reaction.

The physiological adjustment to changing light conditions involves changes of some of the following morphological and biochemical factors: total photosynthetic pigment content, pigment proportions, morphology of the chloroplast, chloroplast arrangement, and availability of Calvin cycle enzymes. For example, under strong light conditions, diatom chloroplasts may shrink and aggregate. On the other hand, a phytoplankton cell may increase the number of active photosynthetic centers when favorable conditions for photosynthesis arise. Under nutrient limitation, chlorophyll *a* increases relative to carbon as a response to increasing nutrients. Adaptations to changes in light intensity usually occur within 1 day. Deep-water phytoplankton can respond partially to low light intensities by increasing the total photosynthetic pigment content.

Nutrients Required by Phytoplankton

- **Nutrients are required by photosynthetic phytoplankton. They may occur in dissolved and particulate form.**

Photosynthetic phytoplankton require nutrients. We can speak of required nutrient elements (e.g., nitrogen) that occur in dissolved inorganic form (e.g., ammonium, nitrate) or in organic form (e.g., amino acids). **Major nutrient elements** are required in great amounts and include carbon, nitrogen, phosphorus, oxygen, silicon, magnesium, potassium, and calcium. **Trace nutrient elements** are required in far smaller amounts and include iron, copper, and vanadium. **Organic nutrients** are those nutrients that are synthesized by other organisms and include vitamins. **Autotrophic uptake** and **auxotrophic uptake** are the uptake of inorganic and organic nutrients, respectively, in association with photosynthesis. **Heterotrophic uptake** refers to uptake of organic substances for nutrition in the absence of photosynthesis. Many phytoplankton can absorb peptides and even engulf particles.

Many elements essential for phytoplankton nutrition can be found in both particulate and dissolved forms. The influxes from rivers and seaweed beds contribute to particulate organic matter. Carbon, nitrogen, and phosphorus may be found in carcasses of phytoplankton and zooplankton sinking in the water column. Water motion may also create particles from collision of dissolved organic molecules. Toward the end of rich phytoplankton blooms, some species become sticky and exhibit negative buoyancy. As they sink, they aggregate and form particles. The aggregation is greatly accelerated by **transparent exopolymers (TEP)**, which is a product released into the water column, mostly by diatoms. TEP consists mainly of acidic polysaccharides, which are sticky and enhance aggregation of fine organic particles (Alldredge et al., 1993). This resulting fragile organic aggregate material is one of the components of **marine snow** (Figure 11.13), which contains a variety of planktonic organisms and detrital products of plankton. Microbial organisms are very common in marine snow and enhance its decomposition as it sinks through the water column.

The work of Mary Silver of the University of California at Santa Cruz and Alice Alldredge of the University of California at Santa Barbara and their colleagues has greatly

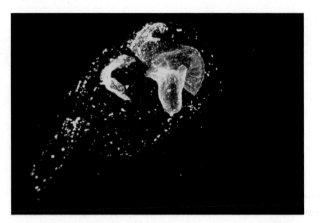

FIG. 11.13   Marine snow in open-ocean surface waters off the coast of California. Fragments of gelatinous zooplankton have been colonized by bacteria, and other fragments have adhered to the original larger particle. Marine snow derives from collisions of macromolecules and from the degradation of dead plankton. (Courtesy of Alice Alldredge)

illuminated our understanding of the global extent and importance of marine snow in transport of organic material in the ocean (Silver and Alldredge, 1988). Marine snow persists in relatively quiet water and can be a major mechanism of transport of material to deeper waters. Marine snow aggregates are rich in organic matter, derived from various sources including decaying phytoplankton cells, mucus from various gelatinous zooplankton, and marine bacteria. In the eastern Pacific, gelatinous houses are used by larvacean tunicates to filter material; but the houses are shed several times daily. These houses often entrap particles and are a major part of the larger marine snow particle aggregates. Sometimes the aggregates are rich enough in organic matter and bacterial activity to cause the material to become anoxic. Bacterial decomposition often results in low oxygen concentrations in the water in the vicinity of marine snow aggregates.

The ocean is a solution of nutrients, with a nutrient-depleted surface layer due to uptake by phytoplankton and microorganisms in the photic zone. This depleted layer sits atop a deep-water column where nutrients are usually abundant but will not reach the surface without a process that causes vertical mixing. Exchange processes between the surface depleted layer and the deeper nutrient-rich column, such as upwelling and wind mixing, may balance the loss of surface nutrients to greater depths via sinking carcasses and zooplankton fecal pellets. The turnover of the ocean's dissolved organic carbon is estimated to be 30–300 years, but the rate is much higher for some molecules (e.g., glucose) and lower for others.

Nitrogen

- **Nitrogen is required for protein synthesis and is taken up in the form of ammonium, nitrate, and nitrite.**

Nitrogen (N) is required for the synthesis of amino acids, the building blocks of proteins, and occurs in three principal inorganic dissolved forms: **ammonium** (NH_{4+}), **nitrate** (NO_3), and **nitrite** (NO_2). Nitrogen also occurs in dissolved organic forms, such as urea, amino acids, and peptides.

Ammonium is usually the preferred form of nitrogen from a kinetic perspective because no chemical reduction is required to be used in protein synthesis, which makes ammonium uptake and processing faster than for other forms of nitrogen. Phytoplankton must use the enzymes nitrate reductase to reduce nitrate and nitrite reductase to reduce nitrite, making their uptake a chemically slower process. Dissolved ammonium can inhibit the uptake of nitrate. Ammonium can also be taken up more efficiently at lower light levels than is the case for nitrate.

The highest concentration of dissolved nitrogen in the ocean and usually in eutrophic coastal waters is nitrate (roughly 100 micromolar: 100 μM). Upwelling and storm-induced turbulence carry nitrate to the euphotic zone. Under certain circumstances, however, ammonium can surpass nitrate in abundance (usually when the nitrate is used up and near sewage treatment plants). Nitrite is generally the least abundant of the three nitrogen-bearing nutrients and behaves similarly to nitrate in phytoplankton nutrient uptake. The dissolved concentrations of all three forms of nitrogen increase in the temperate–boreal winter and decrease in the spring and summer when phytoplankton populations build up.

▪ Nitrogen supplied to phytoplankton can be divided between that provided from new production and that provided from regenerated production.

Our earlier discussion demonstrated that nutrient supply in coastal and shelf waters depends strongly upon mixing of water from a nutrient-rich pool of deeper water. In the open sea, wind can mix deeper and nutrient-rich water to the surface and supplies the phytoplankton with nutrients. As long as phytoplankton cells are not mixed downward too far, this supply, combined with light, allows phytoplankton to grow. The amount of primary production attributable to nutrient supply from deeper waters is known as **new production**.

By contrast, the surface waters also contain zooplankton and bacteria, which excrete nitrogen, usually in the form of ammonia. This, too, may be used in primary production and is known as **regeneration production**. Because regeneration production comes from a recycling process that occurs within the system, the new production is often of greater interest in regional estimates of primary productivity because it represents the nutrient supply coming from outside the system.

▪ Nitrogen recycles between phytoplankton and the bottom in shallow-water environments. Zooplankton excretion is another major source of recycling.

We discussed benthic–pelagic coupling earlier in this chapter. In shallow coastal bays and estuaries, coupling of nitrogen with the benthic system may influence phytoplankton nutrient dynamics. During the process of decomposition, dissolved forms of nitrogen are released from the bottom into the overlying water. In very shallow bays, much of the return of nitrogen from the bottom to the water column is probably in the form of organic nitrogen, such as urea. Some phytoplankton species are capable of taking up urea, uric acid, and amino acids.

Recycling of different forms of nitrogen depends on the habitat and the nature of the nutrient regeneration cycle. In coastal areas of high upwelling (e.g., off the coast of Peru), nitrate is the main form of nitrogen regenerated from the bottom and from deeper waters. Tracer studies employing the ^{15}N isotope as a tracer of nitrogen show that over half the nitrogen uptake in upwelling areas is in the form of nitrate. The remainder is in the form of ammonia that recycles from zooplankton excretion and decomposition, and back to the phytoplankton. Excretion of the anchovy may be the principal source of regenerated nitrogen in the Peru upwelling region. In contrast, in the nutrient-poor gyres, less than 10 percent of the measured nitrogen uptake is in the form of nitrate. Most nitrogen uptake must, therefore, involve efficient recycling of ammonia and organic nitrogen between the zooplankton and phytoplankton.

▪ Nitrogen cycling is intimately involved with microbial transformations.

Figure 11.14 shows the pathways of nitrogen exchange in the sea. Several distinct groups of bacteria transform

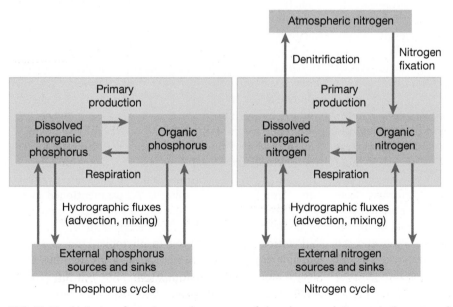

FIG. 11.14   Major transformations and movement of phosphorus and nitrogen in the water column.

nitrogen from one form to another. Nitrogen cycling involves a large pool of gaseous nitrogen in the atmosphere. **Nitrogen-fixing bacteria** convert nitrogen gas to ammonia, NH_3, by means of the enzyme nitrogenase. **Nitrogen fixation** occurs in the absence of oxygen, although this may often be in a microzone within the phytoplankton cell. The nitrogen:phosphorus ratio of nitrogen-fixing bacteria is 16:1, which has an important influence on oceanic ratios. One group of **nitrifying bacteria** convert ammonium ion to nitrite, whereas other species of nitrifying bacteria oxidize nitrite to nitrate. Together, these processes are known as **nitrification**. Both processes require the presence of oxygen. Under anaerobic conditions, **denitrifying bacteria** reduce nitrate to ammonium ion, accomplishing the process of **denitrification**. These nitrate-reducing bacteria can return nitrate to the atmosphere, in the form of nitrogen gas. They occur only in anaerobic conditions, such as near-shore sediments, anoxic waters in estuaries, and oxygen minimum layers.

Nitrogen constitutes about 79 percent of the atmosphere, but the N_2 molecule is chemically nearly inert and can be altered naturally only by lightning, unless a crucial biological process occurs. Nitrogen is incorporated into marine food chains mainly through the process of nitrogen fixation (accomplished by some bacteria, cyanobacteria, and yeasts). Gaseous nitrogen is converted eventually to nitrogen in proteins. Small cyanobacteria cells are found ubiquitously in the ocean. Species of *Prochlorococcus* are the most common cyanobacteria in the ocean and also the smallest of the phytoplankton, existing at cell sizes of less than 1 μm. *Prochlorococcus* dominates low-nutrient and low-latitude regions (Partensky et al., 1999). A filamentous nitrogen-fixing cyanobacterial species, *Trichodesmium thiebautii* (Figure 8.3), is found in tropical gyre centers, and surface waters may be thick with filamentous colonies. The Red Sea is reputed to owe its name to a filamentous cyanobacterium that was dominated by reddish photosynthetic pigments. In some seas, nitrogen fixation is responsible for as much as 20 percent of the input of nitrogen into phytoplankton, but elsewhere the role of nitrogen fixation is trivial. Nitrogen fixation is an anaerobic process, and *Trichodesmium* and other water column cyanobacteria have local anoxic zones within groups of nitrogen-fixing cells.

In the global ocean, nitrogen is gained by nitrogen fixation. It is also lost, however, because of the process of denitrification, which is most common in anaerobic environments such as inshore water columns and sediments. Denitrification and nitrogen fixation are not coupled processes, so there is a possibility that one will dominate the other. Current evidence suggests that there is a slight excess of denitrification, which implies that nitrogen is a more limiting nutrient element than phosphorus, at least as a world average. Also, a historical increase of nitrogen fixation might therefore increase global primary productivity.

Because of large uncertainties over the role of microbial transformations of nitrogen, the global cycle and the limitation of nitrogen are difficult to determine. We will return to this theme shortly when we discuss nitrogen and phosphorus limitation.

Phosphorus

■ **Phosphorus occurs in seawater mainly as inorganic phosphate, is required for the synthesis of ATP, and is a crucial energy source in enzymatic reactions.**

The biochemical role of phosphorus (P) is different from that of nitrogen because phosphorus is used primarily in the energy cycle of the cell. Adenosine triphosphate (ATP) is a crucial energy source in all enzymatic reactions.

Phosphorus occurs in the ocean as inorganic phosphate PO_4, dissolved organic phosphorus, and particulate phosphorus. Phosphate is the form preferred by phytoplankton and exchanges rapidly between phytoplankton and seawater (**Figure 11.14**). Phosphate is taken up very rapidly by phytoplankton, and the concentration in surface waters is usually quite low. Phosphorus is recycled rapidly between the water and phytoplankton and is often a rate-limiting step in primary productivity as a result. Grazing and excretion by the zooplankton allow rapid regeneration in the plankton. As phytoplankton detritus settles from the water column, the sediments accumulate phosphorus. Benthic decomposition results in the diffusion of phosphorus from the bottom. The remixing of the water column in fall and winter returns phosphorus to the surface waters. Within food webs during the spring diatom bloom, phosphorus may be locked up in parts of the food web such as fish, creating ephemeral strong phosphorus limitation (Trommer et al., 2013).

Nitrogen/Phosphorus Ratio in the Sea

■ **The nitrogen-to-phosphorus ratio in the sea is generally about 14.7:1 and is regulated by uptake and decomposition of phytoplankton, whose N:P ratio is about 16:1.**

The growth of phytoplankton results in the simultaneous depletion of both nitrogen and phosphorus. Those elements are available in deeper ocean water (the majority of the ocean) (N:P ~ 14.7:1) in very nearly the proportions usually required by phytoplankton. Phytoplankton particles are more enriched in nitrogen (N:P ~ 16:1), which suggests that inasmuch as N and P are taken up in photosynthesis, nitrogen is a limiting factor in primary production on the scale of the entire ocean. These proportions are known as **Redfield ratios**, after their discoverer. The near coincidence of nitrogen-to-phosphorus ratios in the sea and in phytoplankton requirements has led to the conclusion that growth of phytoplankton cells, followed by their sinking and decomposition, controls the N:P ratio in both the phytoplankton and seawater. A balance of nitrogen fixation and denitrification would fix the overall value of nitrogen relative to phosphorus. As we mentioned in the discussion of nitrogen, there appears to be a slight excess of denitrification, leading to the apparent difference in N:P ratios in seawater and in phytoplankton. It is also possible that phosphorus is lost at a greater rate from deeper ocean water than nitrogen, although the mechanism is not known. Despite the probable correctness of the phytoplankton nitrogen control hypothesis, numerous examples exist of phytoplankton species that deviate from the "typical" N:P uptake ratio of 16:1.

It appears that nitrogen is usually a short-term controller of primary production, with some important exceptions that we will discuss shortly. We can draw this conclusion from studies of change of nitrogen and phosphorus during phytoplankton growth. Open-ocean, nutrient-poor waters are often found to have undetectable nitrogen but a small amount of phosphorus, suggesting that nitrogen is limiting to phytoplankton growth. Near-shore surface waters are relatively enriched in phosphorus. In a famous experiment, Ryther and Dunstan (1971) added phosphorus to near-shore mid-Atlantic water samples and found no increase in primary production, but nitrogen additions greatly stimulated primary production. Most data on phytoplankton cultures similarly suggest that nitrogen limits marine production in much of the ocean. We do have to remember, however, that these experiments are done in small containers and there are no nitrogen fixers present, which might have made nitrogen less limiting. On the ocean scale, as we have discussed, nitrogen cycles through a number of bacterial groups, through the atmosphere and through the ocean. In the tropical ocean, nitrogen fixation is probably a major driver of primary productivity and changes in the activity of nitrogen fixers are crucial in determining the amount of photosynthesis (Capone et al., 2008). Increases of nitrogen fixation could result in increased photosynthesis in the tropical ocean with a greater potential for organic matter to be created and lost to the seafloor. This could be a mechanism that affects the world's carbon budget and therefore climate change. We will discuss the attempts to fertilize ocean surfaces with iron, which can stimulate nitrogen-fixing photosynthetic phytoplankton.

By contrast, phosphorus ultimately comes from weathering of mineral matter, entering the ocean through rivers and leaving the ocean by formation of phosphorus-rich minerals. In the long run, it is therefore possible that fluctuation of phosphorus input is also a controller of primary production on geological time scales (the student with mathematical interests should see Tyrrell, 1999). In the coming century, however, climate change may increase the relative importance of denitrification, because increases in water column stratification may increase the frequency of oxygen-minimum zones throughout the tropical ocean. This would increase denitrification and make nitrogen more limiting over time. While these arguments work on a global scale, we must remember that in many cases, P may be diverted into species in some part of the food web, placing it in short supply.

Iron

Iron is a crucial limiting nutrient element needed in the synthesis of cytochromes, ferredoxin, and Fe-S proteins.

Metals, such as iron, manganese, and zinc, have important functions in oxidase systems (iron is the cofactor in the oxygen evolution step of photosynthesis) and serve as cofactors for enzymes essential for growth of organisms (e.g., molybdenum, zinc, cobalt, copper, and vanadium also play such a role). All marine photosynthesizing phytoplankton use **ferredoxin** to catalyze electron-accepting reactions in the light reactions of photosynthesis (see the scheme for photosynthesis shown earlier in **Box Figure 11.1**) and need iron for the protein structure. Iron is especially limiting for nitrogen-fixing cyanobacteria such as *Trichodesmium*. Nitrogenase, the important enzyme in nitrogen fixation, has two subunits that require iron and molybdenum for function in catalyzing electron transfer reactions. The reactions are inefficient with regard to iron and require much more iron than in nitrogen metabolism, where phytoplankton cells use nitrate or nitrite.

Iron normally is insoluble in typical ocean water, and must be kept in a chemically complexed form to be taken up by phytoplankton.

Normally, iron occurs chemically as Fe(II) and Fe(III). In typical ocean water with oxygen, Fe(III) is the stable form and is very insoluble and therefore not bioavailable to phytoplankton. Fe(III) can be bioavailable if it is complexed with dissolved organic molecules. Phytoplankton release such organic molecules in order to complex and solubilize iron, allowing uptake by the cells. Viruses can lyse natural cells, which accelerates the release of iron from the cells and may allow rapid uptake of iron by other phytoplankton cells.

Iron is the important limiting nutrient in high nitrogen–low productivity areas of the ocean, which are remote from terrestrial windborne sources of iron.

We have discussed the importance of nitrogen as a limiting nutrient element to phytoplankton, but iron is a major exception in large parts of the ocean. Oceanographer John Martin (1992) first showed experimentally that iron may be a strongly limiting nutrient in parts of the North Pacific Ocean. These conclusions derived from experimental findings that phytoplankton growth was strongly increased in water in bottles to which iron had been added. The key to the geographic distribution of regions where iron appears not to be limiting is the source of iron, which is largely in aerosols from land that are deposited on the sea surface.

Experiments at sea with iron addition demonstrate rapid mesoscale increases of primary production.

Larger-scale experiments on iron addition to stimulate phytoplankton growth were conducted in 1994 in the equatorial Pacific, and phytoplankton production was found to increase strongly (Kolber et al., 1994). Since that time, shipboard mesoscale iron-addition experiments have been performed in many parts of the world ocean: in particular, in the so-called **high nitrogen–low productivity (HNLP) sites** where nitrogen seems to be abundant yet primary productivity is low. These are mainly the northern Pacific, the equatorial eastern Pacific, and parts of the Antarctic Ocean. The effect on primary production is strongest when wind mixing is minimized (for a good summary, see de Baar et al., 2005). Iron additions have stimulated phytoplankton growth in experimental sites in the North Pacific and in the Antarctic Ocean, although the rate of phytoplankton growth was lower than in the equatorial experimental sites. Temperature, therefore, is likely also a major factor in iron stimulation of primary productivity. The results on iron additions show that phytoplankton in HNLP regions is controlled by a bottom-up process such as nutrient control, as opposed to a top-down ecosystem process such as grazing by zooplankton. The surprise is that about one-third of the global ocean is limited by iron (Boyd et al., 2007).

Wind from dry land sources is the crucial factor in supplying iron to the ocean. Areas remote from windborne iron-bearing dust, such as the Southern Ocean, are more likely to be limited by iron than areas nearer to such supplies (e.g., as the Atlantic is to the deserts and winds of northern Africa—**Figure 11.15**). Even in the very Fe-limited Southern Ocean, local peaks in primary productivity are correlated with windborne dust supplied from areas in southern South America, Africa, Australia, and New Zealand (Cassar et al., 2007). It appears that during glacial retreats in the Pleistocene epoch, iron was eroded from the land and supplied at a much higher rate to the ocean. Because iron is so intimately related to nitrogen fixation, we can expect that iron limitation in oceanic areas remote from windborne iron dust sources will affect the rate of nitrogen fixation.

Silicon

Silicic acid, which contains the nutrient element silicon, is a constituent of seawater and is an essential nutrient for the secretion of diatom skeletons. Depletion of silicon inhibits cell division and eventually suppresses the metabolic activity of the diatom cell. In natural waters, depletion of silicon can limit diatom populations and may direct the course of subtropical succession toward phytoplankton lacking a siliceous test. As diatoms sink from surface waters, they remove silicon, which may partially cause the demise of the spring diatom increase found in so many temperate and higher-latitude waters. In some regions, such as parts of the North Pacific, waters are doubly depleted of silica and iron during diatom blooms, and this may exert an additional limitation on primary production.

Silica is delivered to the ocean by wind and river transport. Thus silica usually also is transported when iron is delivered to the ocean by wind, which should favor diatom growth. During glacial advances in the Pleistocene, more silica probably entered the ocean through glacial erosion and river input. This might have favored diatoms, which would sink and remove carbon from surface waters to the seabed for hundreds to thousands of years. This has been suggested as a major cause of reduction of carbon dioxide in the atmosphere and, therefore, a cause of global cooling because of a relaxation of the greenhouse effect. During glacial retreats leading to interglacial times, which are accompanied by the reduction of input of silica, other plankton, such as coccolithophores, might have been favored. Because production of calcium carbonate results in a release of carbon dioxide into the atmosphere, reduction of silica input would

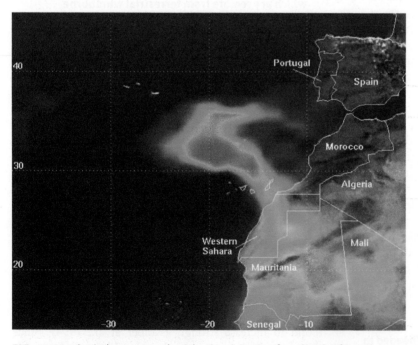

FIG. 11.15 A windstorm over the Atlantic originating from North Africa. The green to red false colors in the dust plume image represent increasing amounts of aerosol, with the densest portion over the ocean. Under the densest portions of the dust plume (red), the amount of ultraviolet sunlight is reduced to half its normal value. Such storms from dry terrestrial origins are the principal source of iron nutrients in the open sea. (Data from Total Ozone Mapping Spectrometer NASA Goddard Space Flight Center Ozone Processing Team)

have had the indirect effect of warming the earth's surface through the greenhouse effect. While this large-scale trade-off makes sense, it has been difficult to demonstrate a correlation between glaciation, silica input, and phytoplankton dominance by diatoms in areas such as the eastern equatorial Pacific (Matsumoto and Sarmiento, 2008).

Trace Substances

Organic trace nutrients, particularly vitamins, may also be of great significance in the sea. Almost all marine phytoplankton species are auxotrophic and require cobalamine, thiamine, or biotin. In mixed cultures, vitamin production and release by one species may stimulate the growth of another species, although most vitamin production is probably by bacteria. In the Sargasso Sea, small diatoms requiring vitamin B_{12} increase in abundance relative to coccolithophores (which need only thiamine) at the time of year when the vitamin B_{12} concentration is great.

Rate of Nutrient Uptake

■ Nutrient uptake varies with taxonomic group of phytoplankton, cell size, and conditions of microturbulence.

Because nutrient concentration varies so much in the ocean, we expect that different types of phytoplankton may take up nutrients under different conditions. The external environment of phytoplankton cells is simultaneously a source of dissolved carbon dioxide, nitrogen, phosphorus, iron, and other nutrients. Cell membrane transport is therefore an important control of nutrient uptake. But internal cell processes such as photosynthesis, other metabolic processes, and storage within the cell may determine the rate of uptake of nutrients from the outside water. These cellular processes are complex and not completely understood.

Cell size and surface area are important determinants of nutrient uptake. If a cell is spherical, the surface area/volume ratio increases as cell size decreases.[2] Smaller cells, therefore, should be able to take nutrients up proportionally at a higher rate than larger cells. This would be an advantage in oligotrophic parts of the ocean where nutrients are in low concentration, such as in nutrient-poor gyre centers. On the other hand, larger cells can store more nutrients than smaller cells, and this would be an advantage where nutrient concentrations are high. These predictions are verified in oceanic environments. In oligotrophic environments the phytoplankton are dominated by picoplankton and nanoplankton, except when nutrient flux increases from upwelling, and then larger cells predominate. In temperate latitudes, early season high-nutrient conditions are dominated by larger-celled diatoms, whereas at later times of the summer, when mineral nutrients are scarce, the phytoplankton is dominated by tiny phytoflagellates and cyanobacteria (Fenchel, 1988).

Turbulence at a small spatial scale may be an important influence on nutrient uptake. Turbulence on the large spatial scale moves large parcels of water, and processes such as upwelling operate on this scale and provide nutrients in the well-lit surface waters of the ocean. But small-scale turbulence is also important because it moves cells through the water and exposes the cells to new micropatches, where nutrients may be higher. If water was completely still, then cells might take up nutrients by diffusion and the immediate volume around the cell would be depleted. The positive effect of cell motion through the water column is important for larger phytoplankton, such as larger swimming dinoflagellates and perhaps larger diatoms as they sink through the water. Ramon Margalef (1978) suggested that the spikes commonly found on larger phytoplankton may have evolved to increase shear around the cells to increase nutrient delivery.

Different groups of plankton have different abilities to take up nutrients, at least as judged by cell division rate. Diatoms, for example, have higher chlorophyll concentrations than dinoflagellates and also have higher cell division rates.

■ Nutrient uptake increases with increasing nutrient concentration, eventually leveling off to a plateau.

Given that nutrients vary in concentration, it is desirable to determine the relationship between nutrient concentration and uptake rate by phytoplankton. Uptake rate may be measured directly (e.g., in terms of nitrate taken up per cell per unit time) or indirectly (in terms of the cell doubling rate). We assume here, for simplicity, that the faster the nutrient uptake, the faster the cell doubling rate. (This is not always a good measure on the time scale of a short laboratory experiment because some phytoplankton are known to take up nutrients at night and use them in photosynthesis the next day.)

Nutrient uptake usually follows the general pattern illustrated in **Figure 11.16**, which shows the relationship of cell doublings per day, D (which is an estimate of nutrient uptake, as mentioned earlier), as a function of nutrient concentration, C. The cell doubling rate increases with increasing nutrient concentration but then reaches a plateau, at a value of D_{max}. The nutrient concentration at which half the maximum cell doubling rate occurs is known as the half-saturation concentration, or K.[3] This is a useful measure of nutrient uptake, which we shall use later. Originally, this measure was developed as an analog to measures of enzyme kinetics, but it is not clear that the values of D_{max} and K can be directly compared to enzyme reactions.

■ Inshore phytoplankton live at higher nutrient concentrations and would be expected to be able to take up nutrients at higher nutrient concentrations in the environment relative to open-ocean phytoplankton, which would be expected to be more efficient at low concentrations.

Open-ocean environments usually have lower nutrient concentrations than do inner-shelf waters. As a result, we may expect phytoplankton living in these two different habitats to evolve differing patterns of nutrient uptake.

---

[2] Divide the surface area ($4\pi r^2$) of a sphere by its volume ($4/3\pi r^3$) and plot a graph of this ratio as a function of r and you will see this.

[3] K, formally estimated by plotting C/D as a function of C, is the negative value of the point at which this line intersects the abscissa.

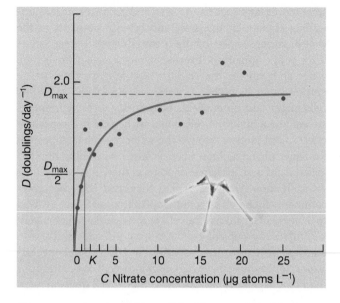

FIG. 11.16 Growth rate (D = cell doublings per day) of the diatom *Asterionella japonica* (circles) as a function of C (which, in this case, is the nitrate concentration in μg atoms L^{-1}). The half-saturation constant K is estimated at 1.5 by finding the value of C on the blue curve that corresponds to $D_{max}/2$. (Modified from Eppley and Thomas, 1969)

Open-ocean phytoplankton should be able to take up nutrients efficiently from low environmental concentrations. They would be expected, however, to have low maximum uptake rate because they never encounter high nutrient concentrations. By contrast, inner-shelf phytoplankton should be relatively inefficient at low nutrient concentrations but able to take up nutrients at far higher concentrations than the open-ocean phytoplankton. Expected curves for both types are shown in **Figure 11.17**.

The difference leads to a prediction that the half-saturation concentration K should be greater for inner-shelf phytoplankton relative to oceanic phytoplankton. For coastal phytoplankton, K is usually greater than 1 μM for nitrate uptake, whereas oceanic phytoplankton have much lower values of about 0.1–0.2 μM. Clones of the same diatom species show high and low K values, depending on whether

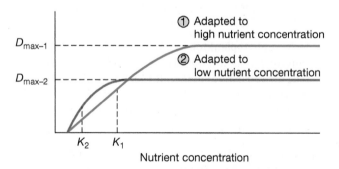

FIG. 11.17 A theoretical representation of the differences in nutrient uptake expected for a phytoplankton clone (1) adapted to high nutrient levels near-shore versus another clone (2) adapted to open-ocean low nutrient levels. D_{max-1} is the maximum doubling rate for clone 1, and K_1 is the estimated corresponding value of K (same procedure for clone 2).

the clones are isolated from near-shore or oceanic waters, respectively. Oceanic phytoplankton take up nutrients more efficiently at low nutrient concentration. Thus, they may be competitively superior to coastal forms in the low nutrient concentrations of the open sea but unable to take up high concentrations. This type of analysis probably does not tell the whole story. It is not clear that phytoplankton are always exposed to the same average nutrient concentration; short-term temporal and spatial variability might occur when patches of zooplankton excrete nitrogen in one place. The general significance of this problem in nature is poorly understood.

Harmful Algal Blooms

■ **A stable water column, input of nutrients, and sometimes an initial input of resting stages all combine to promote dense harmful phytoplankton blooms.**

In the book of Exodus we read, "All the waters that were in the river turned to blood. And the fish that was in the river died; and the river stank, and the Egyptians could not drink of the water in the river." Today, many interpret this passage as a description of a **red tide**, where intensely colored seawater is caused by dense populations of some species of dinoflagellates (**Figure 11.18**). Both dinoflagellates and cyanobacteria (only a few species of each) are some of the organisms responsible for **harmful algal blooms (HABs)**, which are found mainly in inshore waters and estuaries. Usually, the blooms are dominated by one species of phytoplankton. The blooms often result in subsequent population crashes, and bacterial decomposition may reduce the oxygen in the water, helping to reduce water quality and conditions for marine animals.

Red tides and other harmful algal blooms are often accompanied by release of toxins, which may poison organisms in higher levels of the food chain. Most notably, members of the genera *Alexandrium* (**Figure 8.20**), *Karenia*, and *Pyrodinium* cause **paralytic shellfish poisoning (PSP)** by producing any one of 18 neurotoxins that generally bind to nerve cell membrane–associated sodium channels. Some bivalves consume red tide dinoflagellates and sequester the toxin class known as saxitoxins. These toxins affect bivalve mollusks such as clams and mussels and cause reductions in metabolic rate and feeding rate, and may even cause death. They are also virulently neurotoxic and strongly affect predatory marine vertebrates such as seabirds and sea otters, serving as an antipredation mechanism. Humans are also endangered by eating shellfish that have sequestered saxitoxin. Such toxins also suppress other phytoplankton species, thus allowing the nuisance species to dominate the phytoplankton completely. The dinoflagellate *Karenia brevis*, found abundantly in waters of Florida and the Gulf of Mexico, produce **brevetoxin**, cyclic polyether neurotoxins that bind to sodium channels in nerve cells. Brevetoxin has caused high mortality in manatees and can cause respiratory problems in humans when seawater with *K. brevis* is blown onshore. The toxin also may have

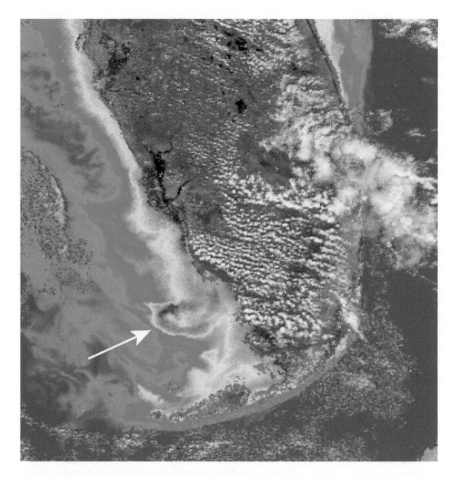

FIG. 11.18   Red tide on November 21, 2004, in the Gulf of Mexico, near Cape Coral, Florida, detected by satellite imagery. Blue to red represents increasing concentrations of chlorophyll. (Courtesy of NASA)

severe effects on the gametogenesis of adults and survival and size of larvae of the eastern oyster *Crassostrea virginica* (Rolton et al., 2016).

Common causative species of harmful algal blooms—dinoflagellates and cyanobacteria—are both motile and can migrate through the water column. They are often able to move several meters per day and to cross a strong thermocline. This gives them a tremendous advantage over nonmotile phytoplankton, since they can spend the daylight hours near the well-lit surface but can then dive at night to more nutrient-rich waters below. Dinoflagellates accomplish this by flagellar movement, whereas cyanobacteria adjust their depth by changing cell buoyancy. This advantage can be enjoyed only if the water column is stable. Harmful algal blooms, therefore, do not usually develop unless the surface part of the water column is stable or the phytoplankton population is trapped near a discontinuity of water density. One type of harmful bloom known as a **brown tide** (*Aureococcus anophagefferens*) tends to develop in response to the presence of organic sources of nitrogen and carbon. Its use of organic and inorganic nutrients gives it a competitive advantage over other phytoplankton. Viruses in the water column apparently lyse cells and accelerate the supply of organic nutrients, which helps the brown tide organism to grow (Gobler et al., 2004). Toxins produced by this alga are

cytotoxic to suspension-feeding bivalves, especially the scallop *Argopecten irradians*, which has resulted in mass mortalities in Long Island, New York; and Narragansett Bay.

Although a stable water column is usually necessary for the generation of a harmful algal bloom, a couple of other factors are also necessary. First, the blooms are usually associated with some major input of nutrients, including those related to sewage. Dinoflagellates often require phosphorus more than other phytoplankton, and pulse inputs of phosphates can help set off dinoflagellate blooms. In cyanobacterial blooms, the nitrogen-fixing species do better under conditions of low oxygen, and organic matter that promotes bacterial reduction of oxygen in the water column will promote cyanobacterial growth. In some cases, nutrient inputs from runoff include metals such as iron, which may be limiting to phytoplankton growth. Runoff may also include necessary organic nutrients, such as vitamins. Large-scale nutrient and blue-green algae releases from Lake Okeechobee in Florida have recently caused major blue-green algae blooms in Florida coastal waters.

Second, phytoplankton species in these blooms often appear to have resting stages (e.g., cysts of some dinoflagellates, cyanobacterial resting cells in estuaries), which may be abundant in the sediment and in the water column. The resting stages are often produced under unfavorable environmental conditions but may be mobilized by, for example, a storm that erodes cysts from the sediment into the water column and thus ends their dormancy. If, following this erosional event, the water column is stable and nutrient input is strong, then favorable conditions exist for a bloom. This is the typical means by which *Alexandrium* dinoflagellate blooms develop each spring in the Gulf of Maine (Stock et al., 2005).

The diatom genus *Pseudo-nitzchia* has toxic strains that produce **domoic acid**, inducing **amnesic shellfish poisoning (ASP)**, which can cause amnesia, neurological damage, and even death when humans consume ASP-contaminated shellfish. Toxic strains were first identified in Prince Edward Island in eastern Canada (Subba Rao et al., 1988) but have since been found worldwide and are a major danger to coastal populations of marine mammals and birds of the coast of California (see **Hot Topics Box 11.1**). *Pseudo-nitzchia* blooms in California are associated with El Niños and are eaten by planktivorous fish, which is the vector that delivers the toxin to seals and dolphins. Another important group in this ignominious list is **neurotoxic shellfish poisoning**, caused by brevetoxin, which is produced by the dinoflagellate *Gymnodinium breve*. Brevetoxin concentrates in shellfish, but it also can be breathed by humans in aerosols near the coastline, causing gastrointestinal and neurological problems as well as asthma-like symptoms.

The dinoflagellate *Pfiesteria piscicida* is one of the most fascinating and troubling harmful algal bloom organisms, and it appears to have been increasing in extent and abundance in recent years. It was first observed in North Carolina estuaries but has also been the source of fishery closures in Chesapeake Bay. As in other dinoflagellates, encysted stages are in the bottom, but they can apparently rapidly develop into biflagellate swimming forms that emerge from the bottom and attack fish. They release an organic-metal complex, which releases metal-mediated free radicals that are apparently highly cytotoxic (Moeller et al., 2007). Using a well-developed feeding organ known as a peduncle, they attack the skin of fish and leave bloody lesions, especially on the thin-skinned menhaden *Brevoortia tyrannus*. *Pfiesteria* has been implicated in at least half of the large-scale fish kills observed in North Carolina inshore waters (Burkholder et al., 1999). Worst of all, the toxin is very dangerous to humans, causing a number of symptoms including dementia and loss of vigor.

Harmful algal blooms are becoming more and more frequent in coastal waters (Shumway, 1990; Berdalet et al., 2015). This may be related to increased nutrient input, owing to increasing coastal populations of humans and increased sewage input. The blooms often have very negative effects because toxic substances are often produced. The nuisance phytoplankton choke or poison suspension-feeding benthic invertebrates, and the phytoplankton population crashes are followed by death and decomposition, which in turn may reduce the oxygen content of the water. The surface scums of phytoplankton are also unappealing aesthetically.

Phytoplankton Succession and the Paradox of Phytoplankton Coexistence

■ **During the production season, there is a successional sequence of phytoplankton species whose general properties correspond to the seasonal trend of nutrient availability. Differential dependence on substances excreted by phytoplankton and production of toxic substances may influence the succession of phytoplankton species during the production season.**

Predictable seasonal changes in dominance are common in the phytoplankton. These changes are known as **phytoplankton succession**, and the causative mechanisms are not well understood. Groups such as dinoflagellates may depend on exudates and nutrients produced in the excretion and decomposition of species earlier in the successional sequence. Diatoms early in succession are autotrophic, requiring only inorganic nutrients for their survival, whereas species later in succession may be auxotrophic, requiring nutrients such as vitamins that they cannot produce themselves. Earlier diatom species often have large cell sizes, which allow them to store nutrients. Later species often are smaller in cell size, and thus the surface area (relative to cell volume) that is exposed to lower nutrient concentration is increased. This adds up to a shift of advantage for different species as the season progresses. But the auxotrophy of later species suggests that later species cannot reach great abundance until the flowering of earlier species. Dinoflagellates are known typically to require more nutrients that they cannot manufacture themselves than is the case for diatoms, which perhaps explains part of the seasonal order of species appearance in the plankton. Many dinoflagellates are largely heterotrophic, which further increases the dependency on other plankton. The demise of the diatoms in late spring may also be explained in terms of sinking from the surface waters of silica, which is a crucial nutrient for diatom growth. The depletion of silica at this time may favor phytoplankton groups that do not have a siliceous test, like dinoflagellates.

Allelopathy—the production of toxic compounds by one organism to inhibit another—may play a role in succession. Cyanobacteria may dominate eutrophic lakes and can inhibit the development of diatom populations. Cell-free filtrates of cyanobacteria cultures inhibit the growth of diatoms isolated from the same lake. Cyanobacteria blooms may thus alternate with diatom outbursts in lakes and some polluted estuaries.

During succession, changes in relative abundance of phytoplankton species occur, but it is important to remember that all species are present at all times of the year. Otherwise, there would be no seed population from which a population increase of a given species could develop. In some cases, a population of cysts in the bottom sediment may help initiate blooms of some species, especially dinoflagellates, but diatom resting cysts are also common in sediments (Lewis et al. 1999). Many dinoflagellates survive well as heterotrophs.

HOT TOPICS IN MARINE BIOLOGY

Angry Birds and Lost Seals: Solution of the Effects of a Mysterious Poison   11.1

An angry seagull swoops down and cuts a gash in a woman's head as she walks along Bodega Bay, a sleepy town in northern California. Soon thousands of angry birds attack the town. And so unfolds *The Birds*, a great horror movie by Alfred Hitchcock. The story was inspired by a real flock of seabirds whose erratic behavior had become a major cause of concern and led to some exciting research connecting blooms of some diatoms to a major threat to marine mammals and seabirds of California.

The story begins far from California, on Prince Edward Island (PEI) in the Atlantic Canadian province of New Brunswick. PEI is a lovely place with an enormous aquaculture industry (see Chapter 21) where mussels and oysters are grown. In 1987, three deaths and 129 other cases of apparent poisoning—causing symptoms such as cramps, diarrhea, memory loss, and even coma—were related to cultured marine mussels, *Mytilus edulis* (Perl et al., 1990). At first, researchers tried to find the toxin in a dinoflagellate, which is one of the usual suspects in shellfish poisoning. But it took a rebooting of research perspective to realize that the poison was to be found instead in a very abundant diatom, a strain of *Pseudo-nitzchia pungens*. This diatom, along with a number of related species, secretes domoic acid, which is related to the amino acid glutamate and binds strongly to neuroreceptors in the brain, severely affecting neurotransmission and memory and even sometimes causing death. Domoic acid binds and causes opening of the ion channels, leading to degeneration of neuronal cells. Other cases affecting humans popped up—for example, four deaths in coastal waters of British Columbia, Canada, in the Pacific. Several domoic acid incidents have resulted in closures of shellfish grounds in Washington State in the past 20 years.

Now, back to the angry birds. Although there was a great concern for human health after domoic acid was identified as the poison on Prince Edward Island and other locations, a great concern was raised over a series of peculiar phenomena surrounding marine bird and mammal populations on the California coast. First, those angry birds we mentioned: In 1961, a large number of sooty shearwaters, *Puffinus griseus*, began diving seemingly without control into various buildings in the seaside town of Capitola, California, on the north coast of Monterey Bay. The film director Alfred Hitchcock, who vacationed nearby, became fascinated with newspaper accounts of the event, and it inspired his scary movie about angry birds and their attack on a more northerly coastal town.* Looking back, this was the beginning of a disturbing series of poisonings that continue to this day.

Mary Silver, a phytoplankton researcher at the University of California at Santa Cruz (just down the coast from Capitola), helped establish the connection between phytoplankton blooms of *Pseudo-nitzchia* and poisonings of marine vertebrate populations. Silver and colleagues, especially Sibel Bargu, found domoic acid in a wide variety of coastal marine vertebrate and invertebrate animal species in many habitats (Kvitek et al., 2008). For example, domoic acid has been found in squid (Bargu et al., 2008), which provides a possible link between phytoplankton and higher levels of the food web. The squid did not feed directly on diatoms, but probably consumed

*It was also inspired by a story written by Daphne du Maurier.

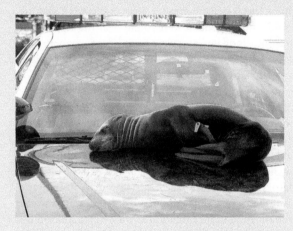

BOX FIG. 11.3   A California sea lion disoriented by domoic acid hopped on the hood of this car and stayed there. (Courtesy of Marine Mammal Center)

diatom-feeding species such as planktonic crustacea. Subsequently, important vectors of domoic acid in *Pseudo-nitzchia* have been identified as planktivorous fish and krill, which are consumed by seals, dolphins, and seabirds.

In recent decades, strandings of a number of marine mammals, especially the sea lion *Zalophus californianus*, have occurred on the central and southern California coast. Strandings have multiple causes. During strong El Niños, sea lions must swim far out to sea to get food, become weakened, and often suffer from an infectious bacterial disease known as leptospirosis. But in recent years, since the late 1990s, a prolonged multidecadal period of strong upwelling has developed that has brought nutrients to the surface and stimulated *Pseudo-nitzchia* blooms and greater abundances of anchovies. Strandings of marine mammals caused by domoic acid poisoning have become common, and stranded sea lions are often found disoriented, suffering from seizures and even falling into coma (Bejarano et al., 2008). Disoriented sea lions have even been found on paved roads (**Box Figure 11.3**).

Now, finally back to those angry shearwaters in Capitola. Can we find a smoking gun to connect domoic acid to the incident that led to the great Hitchcock horror film? It seems likely that the shearwaters were exposed to domoic acid by feeding on anchovies, which in turn would have fed far offshore on toxic *Pseudo-nitzchia*. Silver's group realized that a long-term set of survey-collections of plankton off the California coast known as CALCOFI (California Cooperative Oceanic Fisheries Investigations) might provide important clues. Since 1949, collections of oceanographic data, phytoplankton, and zooplankton have been made to document conditions relating to fisheries. Silver's group went to the archives and found domoic acid in preserved salps and in *Pseudo-nitzchia* cells from offshore. This at least provides a possible connection with the mysterious attack of the angry birds on Capitola.

Combined with an ability to swim by means of flagellae, both photosynthetic and heterotrophic dinoflagellates have an advantage in summer when dissolved nutrients are less abundant in surface waters and the water column is stable. A study of a Danish bay showed these effects nicely. The bay was well stratified in summer. Diatoms were distributed throughout the water column because many were sinking. Autotrophic dinoflagellates were found near the surface and heterotrophic dinoflagellates were found deeper, where light was less available but other food sources were present (Mouritsen and Richardson, 2003).

The diversity of phytoplankton also tends to increase as succession progresses. This may reflect an increasing diversity of nutrient sources, especially an increase in organic substances in the water. Later phytoplankton successional dominants also are often ornate and adorned with spines. This may be a response to a temporal increase in the presence of predators or to an increase in surface area/volume ratio, which would increase nutrient uptake efficiency. It may also be a mechanism to induce rotation as the phytoplankton cell is sinking, which may break up the boundary layer around the phytoplankton cells and increase nutrient access.

The pattern of seasonal succession seems generally to reflect the geographic distribution of phytoplankton. Phytoplankton species that bloom early in succession are typical of nutrient-rich (eutrophic) coastal waters, whereas those that occur later dominate nutrient-poor (oligotrophic) offshore environments. Successional stages of phytoplankton dominance are most pronounced in mid- to high latitudes and are indistinct in tropical locations.

■ Coexistence of many photosynthetic and heterotrophic microbial groups under nutrient limitation presents an ecological paradox of coexistence.

In Chapter 4, we discussed the principle of competitive exclusion, which states that when resources are limiting, one species will tend to dominate. This clearly rarely applies to the phytoplankton, as many species tend to coexist, to the point that molecular techniques have become necessary to investigate the diversity. Yet we know that nutrients are typically limiting to phytoplankton growth. This contradiction was named the **paradox of the plankton** by the great theoretical ecologist G. Evelyn Hutchinson. On the sea bottom in habitats such as coral reefs, complex habitats can be invoked to explain the coexistence of a number of similar species, but the water column is clearly homogeneous and too well mixed on a small scale to permit many species to coexist in micropatches. It is true that in the higher-diversity tropics the water column tends to be more thermally stratified, which might create a set of microhabitats at different depths. Even in higher latitudes, stratification may allow species with different lifestyles to coexist, as mentioned earlier. But many phytoplankton species coexist even within small volumes of water in the tropics and throughout the water column in higher latitudes. Stratification is, therefore, an insufficient explanation for coexistence.

The paradox of coexistence has eluded marine biologists for many decades and remains to be solved. Essentially, one would expect that if nutrients were limiting, equilibrium would be reached and one competitively superior species would come to dominate, but this is rarely the case except in extremely dense nuisance phytoplankton blooms. It might be that conditions even on a very local scale are changing constantly, which might shift the advantage from one species to another. This might prevent competitive dominance. Differences in nutrient limitation among competing species may also contribute to coexistence. For example, species A may be superior at phosphate uptake, whereas species B may be superior at silica uptake. Thus, under conditions of varying nutrient concentration (spatial patchiness), different species may be favored. Such coexistence has been shown for lake phytoplankton. DNA sequencing allowed the discovery of a number of coexisting and very closely related species of the cyanobacterium *Prochlorococcus* (Moore et al., 1998). These *Prochlorococcus* species functioned very differently at different light intensities, which demonstrates that some of the diversity present reflects diversity in physiological responses to different conditions. Such localized differentiation may allow one type to increase when conditions are not so favorable for others.

Models of competition with many species produce very complex outcomes, sometimes including cycling behavior that involves alternating shifts of dominance among the different competitive species (Scheffer et al., 2003). This would be true if the species' competitive abilities were nearly equal. Still, even in these cases, random events would eventually cause one or more species to become extinct when they were rare. It might be that the reproductive capacity of phytoplankton is so great, usually involving cell division, that competitive exclusion is never possible. After all, a density of one cell per cubic meter would be undetectable to an investigator, but that one cell could reproduce when times are good and a population could bounce back from an unobservably low density.

Some recent work on spatial structure and dominance of phytoplankton types may lead to important advances in solving this paradox. We are still in a primitive phase, but researchers are starting to use satellite-based color detection and statistical analysis of color (see Chapter 12 for further discussion) to identify distribution of some basic phytoplankton types in the ocean, such as diatoms, coccoliths, nannoplanktonic phytoplankton, and even smaller cyanobacteria. This is far from the total diversity of phytoplankton, but d'Ovidio and colleagues (2010) used this information to analyze spatial variation of dominance of various basic phytoplankton groups that could be identified by satellite-based color detection. They connected maps of phytoplankton groups to physical maps of sea-surface height, which is a good measure of current structure. They found a remarkable match between local dominance of specific phytoplankton groups and patches on the scale of tens to hundreds of kilometers wide, which corresponded strongly to current structure. They made a speculative but important conclusion: Within a patch with specific physical features, one phytoplankton type—for example, diatoms—will be favored, and another patch type will favor another phytoplankton group. But the dynamic change that occurs in the ocean continually rearranges these physical patches, and therefore the advantages shift around and phytoplankton are being mixed constantly. It is also possible that major phytoplankton blooms, as they are developing, cause the release of decomposition products that create even more localized different microhabitats that favor the presence of different phytoplankton species (Teeling et al., 2012). Thus, the solution to the "paradox of the plankton" may lie in continually rearranging microhabitats that favor different phytoplankton types, ensuring coexistence of all species over long periods of time.

- ■ **The roles of nutrient concentration and turbulence can be integrated into a general model of phytoplankton dominance.**

Throughout this chapter, it has become clear that turbulence and nutrient concentrations are key factors in determining conditions that might determine dominance of particular phytoplankton groups. We see these factors plotted together in **Figure 11.19**, an integrated model that was suggested by the great phytoplankton ecologist Ramon Margalef. As discussed earlier, the spring diatom increase requires both high nutrients and an overall environment of high turbulence. Turbulence in the open sea brings nutrient-rich water from depth, and in shallow regions turbulence occurs in late winter when conditions are favorable for the spring diatom increase. Hence, diatoms are predicted to increase in abundance when nutrients and turbulence are both high (see upper right end of **Figure 11.19**). At the other end of the diagonal, low nutrients and low turbulence favor many dinoflagellates and cyanobacteria in extremely low nutrient concentrations, which occur in tropical waters of low turbulence and therefore low upwelling. When nutrients are extraordinarily high but turbulence is low, conditions are favorable for dominance by nuisance phytoplankton blooms, such as dinoflagellate-dominated red tides.

Margalef's integrative model can also be applied to succession. As the spring transitions into summer, nutrient concentrations and turbulence both decline to very low turbulence and lower nutrient concentrations, which both favor a transition from diatoms to dinoflagellates.

The Microbial Loop: Nutrient Cycling by Viruses, Heterotrophs, and Chemoautotrophs

- ■ **Phytoplankton may take up organic molecules, but bacteria are the major heterotrophic consumers in the water column.**

Phytoplankton take up the majority of dissolved inorganic nutrients in well-lit surface waters. Phytoplankton

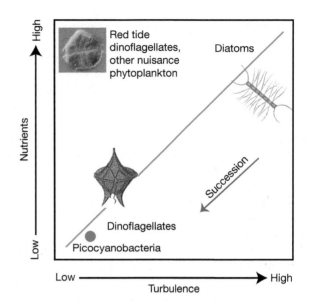

FIG. 11.19 Graphical model of dominance of different phytoplankton groups as a function of turbulence and nutrient concentration.

such as dinoflagellates are also responsible, especially near shore, for heterotrophic consumption of dissolved organic nutrients. Bacteria, however, are probably the principal heterotrophic consumers in the water column; they efficiently utilize both dissolved and particulate organic material, including chitin and cellulose. In some areas, bacterial consumption of dissolved nutrients can even exceed the uptake by phytoplankton. Jonathan Cole and others (1988) have shown that bacterial heterotrophic production overall averages about 20 percent of primary production.

Although ubiquitous, bacteria are most abundant in association with suspended particulate matter. In shallow basins during spring and summer, they often coincide with concentrations of particulates at the base of the thermocline. In the open ocean, they are similarly associated with the oxygen minimum layer below the thermocline, which is itself associated with microbial consumption of organic matter.

Because bacteria in the water column can rapidly take up organic substrates, these dissolved substances can be efficiently converted to a particulate form available to small-bodied consumers, such as protistans, and even smaller copepods. This part of the planktonic food web is known as the **microbial loop (Figure 11.20)**.

Marine viruses invade a wide variety of phytoplankton and bacteria (Fuhrman, 1999). Typically, electron microscopy surveys show that a small percentage of marine bacteria and cyanobacteria cells are infected by viruses; the actual rate of infection is probably much greater. Specialized viruses invade abundant phytoplankton groups such as coccolithophorids and the common cyanobacteria *Prochlorococcus* spp. The invasion of viruses causes the lysis of cells, which releases molecules such as peptides, proteins, and pieces of DNA and RNA into the water. Lysis actively

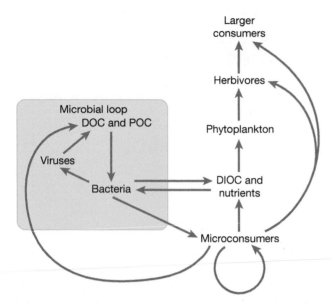

FIG. 11.20 Diagram of the cycling of organic material and nutrients through the phytoplankton and through the microbial loop (shaded box): DOC, dissolved organic carbon; POC, particulate organic carbon; DIOC, dissolved inorganic carbon. Microconsumers include protistans and smaller copepods. Herbivores include protistans and copepods.

prevents the invaded cells from being consumed by organisms such as protozoa. Viral activity therefore short-circuits the entry of material into the protozoan or smaller zooplankton part of the food web and causes dissolved organic molecules to be taken up by bacteria in the plankton. Viral lysis can increase bacterial abundance greatly but at the same time causes a decrease of transfer of carbon to higher trophic levels, since much of the bacteria will not be transferred up the food web.

A wide variety of bacteria are specialized to utilize specific inorganic dissolved substrates to generate the energy (ATP) required for synthesis. Such chemoautotrophs oxidize substances such as ammonium (by nitrifying bacteria) and H_2S (by means of sulfur bacteria). These bacterial groups tend to be inviable or very inefficient in oxygenated waters. Consequently, chemosynthesis is largely restricted to anoxic or poorly oxygenated waters, and some microbial transformations occur only at the border of oxygenated and anoxic water.

■ **Protistans are the major consumers of water column bacteria and are themselves a major component of the food of zooplankton.**

Planktonic protistans are diverse and include such groups as heterotrophic nannoflagellates, heterotrophic dinoflagellates, ciliates, naked amoebae, foraminiferans, and radiolarians. Experimental studies show that ciliates promote growth of many suspension-feeding invertebrate zooplankton and fish larvae. Protistans turn out to be a major component of the lower part of planktonic food webs. In the past, most oceanographers believed that the bacterial-feeding protistans were not consumed and, therefore, constituted a sort of trophic dead end in the plankton. It is now

known that bacterial production is efficiently transferred to the zooplankton via these protistan intermediate trophic levels. Smaller microzooplankton such as small copepods also have been found to feed on smaller protistans and even bacteria. The microbial loop is therefore of considerable importance in planktonic food webs. In Chapter 12, we will discuss food webs and their importance in understanding productivity in the sea.

Zooplankton Grazing in the Sea

■ **Zooplankton abundance usually follows phytoplankton abundance in coastal and shelf planktonic systems. Although zooplankton growth depends on phytoplankton growth, zooplankton grazing can, but often does not, control phytoplankton abundance.**

In temperate-zone and higher-latitude near-shore systems, the zooplankton usually increase after the peak of the phytoplankton bloom has passed, owing to the time lag in reproduction of the zooplankton and the grazing down of the phytoplankton standing crop as the zooplankton population increases (see **Going Deeper Box 11.2**). Zooplankton such as copepods have a much longer generation time (weeks) than phytoplankton (hours to days), and a large population of zooplankton could graze a phytoplankton population to near extinction. This would cause a collapse of the zooplankton until the phytoplankton recovered. Such overgrazing would thus result in strong oscillations in the zooplankton population. When the phytoplankton organisms reach very low densities, however, the zooplankton may not be able to find them. A low-density refuge for phytoplankton allows a subsequent increase when zooplankton decrease in abundance, and the severity of zooplankton oscillations may diminish.

The impact of zooplankton grazing is quite variable, and zooplankton are not necessarily a major cause of the demise of the spring phytoplankton bloom. In the coastal waters of California, Nova Scotia, and Great Britain, zooplankton grazing can exceed 50 percent of the phytoplankton growth per day. However, in near-shore and estuarine waters of high production, much of the early spring phytoplankton production is not grazed by the zooplankton, if only because of a lag time for buildup of the zooplankton. By the time the copepods become abundant, nutrient depletion from the surface waters may have already caused the phytoplankton to decline. For a brief time, the advent of zooplankton may even stimulate phytoplankton growth, owing to animal excretion. Eventually, zooplankton may cause significant reductions in the phytoplankton, but this does not seem to occur in estuaries and inshore bays. Rather, the stabilization of the water column and loss of nutrients to deeper water in summer seems to halt phytoplankton production there. Calbet (2001) analyzed a large number of cases of grazing and found that the dominant mode of grazing rates was about 6 percent per day. The amount of grazing did increase with increasing phytoplankton density but reached

GOING DEEPER 11.2

Quantification of the Effect of Grazing

Phytoplankton (e.g., diatom) population growth can be described in terms of the equation

$$C_2 = C_1 e^{rt}$$

where cell concentration C_2 is produced by a previous concentration C_1 growing after t time units with a rate r. With a grazing rate g, this equation becomes

$$C_2 = C_1 e^{r-g}$$

If g exceeds r, the zooplankton will graze the phytoplankton population to extinction.

---

a plateau at fairly high phytoplankton densities. These results suggest that on the global scale, grazing is not very high, and the possible controlling top-down effect of grazing becomes very unlikely in productive near-shore systems. It is possible that smaller zooplankton such as protistans have a greater role in grazing in the open sea. These protistans may be a major food for grazers such as copepods and krill.

In some open-ocean conditions, phytoplankton production may be smaller and not so confined to a very brief peak. Under these conditions zooplankton may have stronger grazing effects on the phytoplankton. In the subarctic North Pacific, copepod grazing is strong enough to exert control over the phytoplankton, although iron may also contribute to limiting phytoplankton growth (Frost, 1991). Spatial alternations of abundance of phytoplankton and zooplankton occur in some regions, especially in surface waters of mid- and high-latitude open oceans. This would suggest that an area currently rich in zooplankton but poor in phytoplankton had been rich in phytoplankton until the zooplankton grazed them down. Spatial patterns of alternation between phytoplankton and zooplankton are, therefore, indications of temporal patterns of phytoplankton growth and subsequent zooplankton population growth and strong grazing. For example, in the waters near Plymouth, England, there is a sudden drop in the phytoplankton population as the zooplankton first appear. This finding suggests that the zooplankton strongly influence phytoplankton decline. In the North Sea, alternating areas of phytoplankton and zooplankton dominance occur, suggesting cycles of dominance by phytoplankton production and then by grazing (**Figure 11.21**). Phytoplankton in the Antarctic are grazed down by the zooplankton, resulting in large numbers of broken diatom tests being deposited on the seafloor.

Zooplankton Feeding

- **Feeding behavior varies with phytoplankton cell size and cell concentration (cells per unit volume).**

Figure 11.22 shows the relationship between phytoplankton cell concentration and ingestion rate, I, which

is the cells ingested per animal per hour. I increases linearly with cell concentration until a maximum is reached, whereupon ingestion increases no further. Below the maximum, the feeding response of the copepod is indicative of an animal that searches, encounters, and feeds on particles in direct proportion to their concentration. At higher phytoplankton concentrations there appears to be a saturation level, and ingestion increases no further. Feeding may cease because the feeding appendages are saturated with food, or because digestion efficiency would decrease if material were moved through the gut any faster. There is also some evidence for a lower threshold, below which the copepod does not expend energy to search for food and to feed. Studies of spatial variation in the water column show that there are areas where phytoplankton concentrations are too low to support copepod growth. These areas may be within meters of dense patches of phytoplankton. Copepods are probably able to regulate their energy expenditure to the degree that feeding efforts are reduced in water parcels with very low phytoplankton concentrations.

Cell size and species type also influence feeding in copepods and other zooplankton. Copepods feed more rapidly on two-celled chains of diatoms than on single cells and also feed at higher rates on diatom species with larger cell size, relative to smaller cells. Many zooplankton avoid toxic phytoplankton, but tintinnids actually feed preferentially on these forms.

- **Zooplankton can select particles by size. Owing to the low Reynolds number for copepods, this involves direct plucking of preferred particles rather than straining of particles on a feeding sieve.**

In natural waters, copepods are presented with a great variety of particle sizes that continually change in relative abundance with season. Adaptation to the particle size spectrum would maximize the efficiency of particle uptake. There is some evidence that copepods can shift the preferred particle size, depending on the abundance of various food particles in the environment. As mentioned earlier, larger particles are usually preferred. As mentioned in Chapter 9, copepod feeding involves movement of feeding

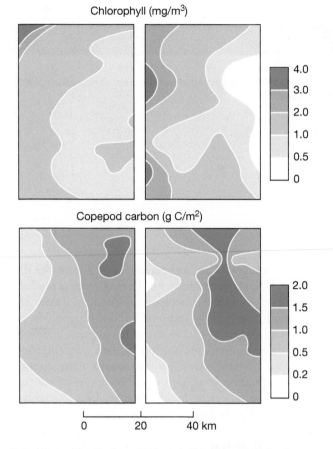

Chlorophyll (mg/m³)

Copepod carbon (g C/m²)

0   20   40 km

FIG. 11.21 Distribution of chlorophyll *a* and copepod carbon on a survey in the North Sea, showing an inverse relationship between phytoplankton and zooplankton standing stock. Phytoplankton are most abundant toward the left, whereas zooplankton are most abundant toward the right. (Modified from Steele, 1974)

appendages and particles at low Reynolds numbers. Under these conditions, the water is viscous and it is not possible to strain phytoplankton cells out of the water on a sieve. Instead, feeding appendages actively select different phytoplankton cells, but the exact mechanisms of search and sensory detection are poorly understood. In copepods, selection of particles involves chemical detection, and the animals tend to avoid inert particles relative to live phytoplankton cells.

If copepods and other zooplankton can track changes in the size spectrum of the phytoplankton and graze the most abundant size classes, then selective grazing could favor the less abundant species and the less common size classes. Copepod grazing thus could affect the species composition of the phytoplankton and prevent one species from competitively excluding all others. If there is a threshold below which feeding is suppressed, then the phytoplankton always have a refuge, thereby preventing grazing to extinction. Recovery is possible because zooplankton populations will decrease and the lag time for zooplankton reproduction will allow the phytoplankton to recover if nutrients are available. Because the phytoplankton generation time is generally shorter than that of the zooplankton, the phytoplankton can recover quickly. This factor suppresses strong oscillations in both phytoplankton and zooplankton abundance. Shifts of grazing may thus offer an explanation for the so-called paradox of the plankton.

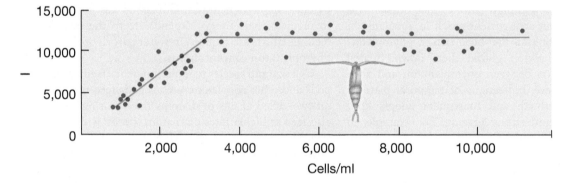

FIG. 11.22 Ingestion rate, *I* (cells copepod⁻¹ h⁻¹), for the copepod *Calanus pacificus* as a function of diatom cell concentration. (After Frost, 1972)

■ CHAPTER SUMMARY

• Zooplankton can aggregate by swimming, but a local phytoplankton patch is a balance between population growth and dispersal by turbulence.

• For many decades the spring diatom increase has been explained

by the Sverdrup model: Because the development of water-column stability of shelf waters in spring keeps phytoplankton near the well-lit surface and increasing light allows greater primary production,

phytoplankton reach great abundance. This phytoplankton bloom, however, collapses as surface nutrients are depleted and some dead phytoplankton sink to the bottom. The following winter, nutrients do return to the surface, and phytoplankton growth begins anew. A newer model suggests that in winter, phytoplankton start to increase in abundance because grazing is at very low levels and phytoplankton cells start to accumulate at depth. The phytoplankton bloom continues into spring as light increases, the water column stabilizes, and phytoplankton continue to grow faster than they can be grazed by zooplankton.

- Photosynthetic pigments such as chlorophyll *a* capture visible light in the action spectrum of light that is used in photosynthesis. Light is attenuated in the water column through absorption and scattering. Photosynthesis increases with increasing light intensity, up to a plateau, and is then inhibited by high light intensity.

- Autotrophic forms use light and inorganic nutrients alone in photosynthesis, whereas auxotrophic forms require some biological product and heterotrophs absorb biological materials from other organisms. Much particulate organic matter can arise from the collision of polysaccharides in phytoplankton.

- Nitrogen—available as ammonium, nitrate, and nitrite—is essential for the production of amino acids. Microbes regulate this nutrient through nitrogen fixation, nitrification, and denitrification. Phosphorus is needed for ATP production in the cell and occurs in the water usually as phosphate.

- Much nitrogen and phosphate comes from deeper waters, stimulating

new production, but nutrients recirculate in the surface waters from zooplankton excretion, giving rise to the regeneration production component of productivity. The ratio of nitrogen to phosphorus is often constant, owing to the influence of phytoplankton uptake–decomposition processes. In the short term, nitrogen appears to be the limiting factor outside a few low-productivity regions.

- Iron is crucial in the synthesis of cytochromes, ferredoxin, and proteins. In areas remote from land, iron is the limiting element, as it is less available and primary production is therefore lower. Silicon, too, may arrive by wind, and so areas remote from land are depleted of diatoms. Vitamins and other trace substances are also important for growth.

- Nutrient uptake in phytoplankton varies with taxonomic group, cell size, and microturbulence. The uptake increases with increasing nutrient concentration before leveling off. Environments low in nutrients select for phytoplankton that are efficient in taking up nutrients at low concentration, whereas near-shore environments, which are higher in nutrients, select for phytoplankton with higher total nutrient uptake.

- A number of phytoplankton species have toxic substances and also can rapidly increase in numbers, causing harmful algal blooms. Species of dinoflagellates produce toxins that inhibit sodium channels and cause

harm to a variety of plankton, but also to humans through shellfish consumption and direct wafting of toxin-bearing aerosols to shore. Harmful algal blooms have increased in frequency throughout the world's oceans in recent decades.

- The succession of phytoplankton follows the seasonal availability of nutrients and the excretion of toxins. Even with succession and limited nutrients, though, many species coexist within the same water column, even on the microscale. The coexistence of photosynthetic and heterotrophic microbial groups presents an ecological paradox.

- Overall, nutrient concentration and turbulence interact to determine patterns of dominance by different phytoplankton groups. Dominance during phytoplankton succession tends to proceed from groups favored by high nutrients and high turbulence to those favored by low nutrients and low turbulence.

- Bacteria in the water column can rapidly take up organic substrates. In the microbial loop, these dissolved substances are efficiently converted to a particulate form available to small-bodied consumers, such as protistans and other smaller zooplankton.

- Zooplankton abundance usually follows phytoplankton abundance in coastal and shelf planktonic systems. However, zooplankton grazing does not always control phytoplankton abundance.

■ REVIEW QUESTIONS

1. In winter, which factors prevent a phytoplankton bloom from occurring? What factors are not limiting at that time?

2. Why does a spring phytoplankton bloom finally collapse?

3. Why does the peak of phytoplankton production not necessarily coincide with the maximum day length in the summer solstice, even if light is important for photosynthesis?

4. In very shallow estuaries, where do nutrients come from in summer to fuel phytoplankton growth?

5. What is the relationship of photosynthesis to light intensity?

6. Why do phytoplankton blooms occur occasionally in the fall?

7. Why is nitrogen an essential nutrient element for phytoplankton, and in what major forms does it occur chemically in the water column?

8. Why is the spring freshet important in nutrient supply to an estuary?

9. What is the difference between new production and regeneration production?

10. How may microbial organisms influence the availability and forms of nitrogen for phytoplankton?

11. Why is turbulence such an important driving force in determining dominance of certain phytoplankton groups? What disadvantage might arise as turbulence becomes strongly reduced?

12. What is the microbial loop, and what is its importance in nutrient cycling in the water column?

13. Why may different forms of phytoplankton dominate during different times of the year?

14. How may resting stages contribute to the eruption of nuisance blooms such as red tides?

15. Why do you think harmful algal blooms are becoming more and more common in coastal waters of the world?

16. What would be a good experimental test of Behrenfeld's argument that plankton blooms fail to begin in winter because zooplankton are so sparse? Explain your answer.

17. What would be good evidence, in your opinion, of the importance of the effect of marine grazers on phytoplankton abundance?

18. What is a mechanism by which plankton may be concentrated in the surface waters by wind action?

19. What factors may cause planktonic animals to migrate through the water column, and what are the possible adaptive advantages for such migrations?

20. In some marine habitats copepods undergo strong diurnal migrations, whereas in others migrations are very weak. Why may this be so?

21. What are two means by which plankton defend against predation?

22. Many tropical zooplankton have elaborate projections and spines, which have been interpreted in two distinct ways: as a defense against predators and as a means of increasing drag to slow sinking from surface waters. How would you test these two general hypotheses for a copepod that has elaborate spines on its first antennae? What is another hypothesis to explain this? (Hint: Think in terms of feeding.)

Visit the companion website for *Marine Biology* at www.oup.com/us/levinton where you can find Cited References (under Student Resources/Cited References), Key Concepts, Marine Biology Explorations, and the Marine Biology Web Page with many additional resources.

0.01 0.02 0.05 0.1 0.2 0.3 0.5 1 2 3

Productivity, Food Webs, and Global Climate Change

In this chapter, we shall discuss how productivity is measured, how it is controlled, and how consumption links the marine biota into a web of food interactions. We will concentrate on productivity and food webs in the water column. Production in the water column accounts for most of the biological production in the sea, even though certain more locally restricted benthic environments, such as sea grass beds and coral reefs, are far more productive per unit area. Water column primary production, moreover, is the basis of the food web of most of our important fisheries. Finally, we will relate all we have learned to the impact of global climate change on current and future oceanic productivity and elemental cycling.

Productivity and Biomass

■ **Productivity expresses the rate of production of biological materials, and biomass refers to the amount of biological material present at any one time. Both measures are expressed as a function of ocean area.**

Productivity is the amount of living tissue produced per unit time. It is often estimated in terms of carbon contained in living material and expressed as grams of carbon (g C) produced per day, in a column of water integrating production in 1 square meter of cross section (g C m^{-2} d^{-1}), from the surface to the seabed. An alternative often used is to express productivity per unit volume of seawater (g C m^{-3} d^{-1}). The amount of living material present in the water column at any one time is the **biomass** (expressed as g C m^{-2}). **Primary production** is that part of the production ascribed

to photosynthesis. **Secondary production** refers to production by the organisms that consume the phytoplankton; those that consume the organisms responsible for secondary production are said to be engaged in **tertiary production**. As we discussed in Chapter 11, **nutrients** are those constituents required by photosynthetic organisms. **Limiting nutrients** are those that are in potentially short supply and may limit phytoplankton growth. **Autotrophic** phytoplankton can photosynthesize and produce all necessary constituents of the cell by simply using light and inorganic nutrients. **Auxotrophic** phytoplankton require some sort of organic nutrient, such as a fatty acid or vitamin, synthesized by another organism. **Heterotrophic** organisms consume organic matter exclusively; they include animals and some algae that are sometimes heterotrophic. **Saprophytic organisms** decompose organic matter and include the bacteria and fungi.

Food Webs and Food Chains

■ **A food web is a complex diagram of feeding interactions. A food chain is a linear sequence, often a simplification of a food web that reveals which organisms consume which other organisms in an environment.**

A **food web** diagram shows the overall connection pattern of feeding among organisms, or who eats whom. **Figure 12.1** shows such a diagram for the North Sea, leading finally to feeding by the herring *Clupea harengus*. The feeding relationships are a bit complex (even this diagram is a very reduced set of the total interactions), but they can be simplified to the linear **food chain** on the left-hand side of the

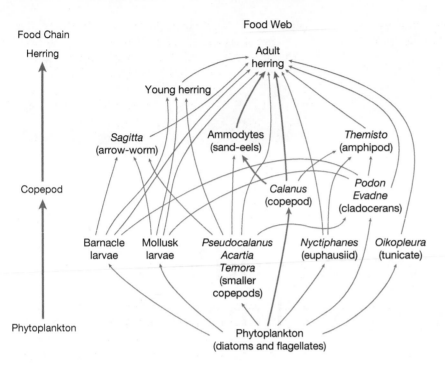

FIG. 12.1 Food web constructed from the feeding relationships of the North Sea herring *Clupea harengus* during different life history stages. A simplified food chain leading to the adult herring is diagrammed at left. (Modified after Russell-Hunter, 1970)

figure, which takes into account the main species through which organic matter cycles. Each main species represents a **trophic level**, which is defined as a species or group of species that all feed on one or more other species (which can be grouped into a lower trophic level). Species in one trophic level may be consumed by species in a higher trophic level. Nearly all oceanographic studies of productivity must focus on the abundant species and can produce only a simplified representation of a complex food web. However, it is important to recognize the actual complexity. For example, species may change levels in the food chain at different stages of the life cycle, as they get larger and seek larger prey. Other predators may feed in the water column and on the seabed. Adult codfish are predators on herring and large shrimp, but when they are small (<50 cm long), they feed on copepods and other planktonic crustaceans. In the Gulf of Maine, cod are now severely overfished, but they once fed and controlled many benthic consumers. Consumers may feed at several levels in the food chain. Copepods, for example, feed on phytoplankton, but they also feed on smaller zooplankton, fish eggs, and particulate organic matter.

Transfer Between Trophic Levels

■ **Transfer from one trophic level to the next is not complete.**

Not all the production from one trophic level is transferred entirely to the next. To estimate the potential production at the top of a food chain, such as fish production, we must tabulate the losses. They result from essentially two factors.

1. *Not eaten.* Some proportion of a given trophic level evades consumption through escape, unpalatability, or unavailability. Phytoplankton with large spines or toxins are avoided by zooplankton. Phytoplankton cell size may be too small, or too large, to permit ingestion.

2. *Inefficient conversion.* Some proportion of the food is not ingested and may sink from the water column and decompose; another fraction that is ingested is not converted into growth and egested.

We can construct a budget for ingested food as follows:

$$I = E + R + G$$

where I is the amount ingested, E is the amounted egested, R is the amount consumed in respiration, and G is the amount used in growth; G can be partitioned between somatic (body) growth and reproduction. Such a budget is usually constructed in terms of energy units, such as calories. Losses consist of $E + R$. Not all food can be digested and assimilated, and some, therefore, is egested (e.g., skeletal material) or not ingested at all. Some of the energy obtained in food is lost as respiration and hence is not available to the next trophic level.

The incompleteness of transfer up a food chain can be estimated in terms of **food chain efficiency**, *Eff*, defined as the amount of energy extracted from a trophic level divided by the amount of energy supplied to that trophic level. **Growth efficiency**, which is the proportion of assimilated food used in growth, can be in a range as high as 30–45 percent, but food chain efficiency is usually far lower, often in the range of 10 percent. Higher-latitude planktonic systems may have much higher food chain efficiency.

■ **Food chain efficiency can be used to calculate the potential fish production at the top of a food chain.**

Food chain efficiency can be used to estimate the potential fish production at the top of the food chain. If B is the biomass of the phytoplankton, and n is the number of links between trophic levels, then the production P of fish is

$$P = BEff^n$$

A change in Eff from 0.1 to 0.2 would magnify by a factor of 16 the estimate of fish production at the fifth trophic level!

Keeping the potential errors in mind, J. A. Gulland (1972) estimated the potential annual yield of common commercially exploited fishes at 100 million tons. This figure is disturbing because the 1970 yield was 55.7 million tons, and the world catch increased at the time about 7 percent per year. Gulland's estimate suggests that even in 1972, the world's fisheries would soon reach the limits of exploitation. In 2000, the United Nations Food and Agriculture Organization estimated the world catch as exceeding 100 million tons, with widespread declines noted in scores of fisheries. The recent worldwide decline or standstill in fisheries may be the realization of Gulland's prediction (see Chapter 21), but we can also take a food chain approach to classify marine systems and their potential for fish production.

Patterns of Food Chain Variation

■ **Food chains can be classified on the basis of oceanographic conditions, which helps determine the number of trophic levels.**

John Ryther (1969) classified marine planktonic food chains into three basic types (**Table 12.1**). The **oceanic system type** has five trophic levels, with a low annual primary production of about 50 g C m⁻² y⁻¹. The **shelf type** has three trophic levels, and primary production is about 100 g C m⁻² y⁻¹. The **upwelling type** occurs in areas such as the Peru Current and in the Antarctic and may have only two trophic levels. Upwelling provides higher and more continuous nutrient supply, leading to a primary productivity of about 300 g C m⁻² y⁻¹.

Table 12.1 shows the potential fish production of these three types. By far, the greatest potential for fish production lies in the upwelling system. The high potential of upwelling systems is enhanced by a greater food chain efficiency, which is related to the ease of ingestion and assimilation of large diatoms by planktivorous fishes. The low primary productivity and the large number of food chain transfers greatly reduce the fishery potential of the oceanic system.

Why do some food chains have many levels, whereas others are so short and simple? Temporal environmental stability and a stable water column may promote the survival of complex oceanic food webs (**Figure 12.2**). In nearshore and upwelling systems, on the other hand, strong temporal changes in temperature and mixing from upwelling would tend to collapse a complex multilevel food chain. In upwelling systems, the water column is typically unstable with strong vertical mixing. This instability may prevent the addition and sustainability of more trophic levels.

The number of levels may require some rethinking; the microbial loop discussed in Chapter 11 would suggest that more input to food webs occurs from the bottom up, which might affect overall estimates. On the other hand, the increase would involve at the most an increase of input of 20 percent—and likely less.

■ **Is there a limit to the number of links in a food chain? The total number may be limited by the structure of the food chain, the possible energy that can be transported through many links, or to a possible instability of large food chains.**

Two interacting processes can regulate food chains. **Bottom-up control** involves effects of fluctuations of species at a given trophic level on the levels above. If an influx of nutrients allows a large phytoplankton bloom, and if the bloom allows zooplankton herbivores to increase, then we are speaking of a bottom-up effect. Such effects have been seen in the northeastern Pacific in the form of strong correlations between chlorophyll concentrations estimated from satellite data and abundances of zooplankton and fish (Ware and Thomson, 2005). As an alternative, a population of consumers might strongly regulate species at a lower trophic level. For example, in many areas of the North Atlantic, cod feed on shrimp and strongly reduce their numbers. Cod, therefore, are exerting a **top-down control** in this food chain. The sum of effects caused by strong increases or declines in a species at the top of a food chain is known as a **trophic cascade** effect.

Interactions within food chains can be very strong. An interesting conclusion develops from considerations of fairly short food chains with strong top-down control (the interested student should consult Polis and Strong, 1996). With an even number of trophic levels and strong top-down controls, primary producers will be relatively rare. Think about it. Suppose we have phytoplankton consumed by copepods, which are consumed

TABLE 12.1 Some Characteristics of the Three Principal Types of Marine Planktonic Food Chain

FOOD CHAIN TYPE	PRIMARY PRODUCTIVITY (g C m⁻² y⁻¹)	NUMBER OF TROPHIC LEVELS	FOOD CHAIN EFFICIENCY (%)	POTENTIAL FISH PRODUCTION (mg C m⁻² y⁻¹)
Oceanic	50	5	10	0.5
Shelf	100	3	15	340
Upwelling	300	1.2	20	36,000

Source: Ryther, 1969.

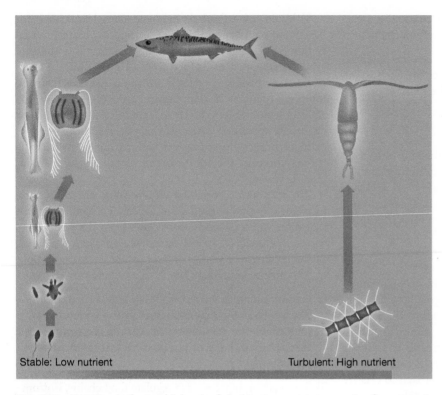

Stable: Low nutrient   Turbulent: High nutrient

FIG. 12.2 Variation in the trophic levels of planktonic ecosystems, ranging from oceanic environments to coastal environments to turbulent, upwelled high-nutrient environments. (After Landry, 1977)

by zooplanktivorous fish, which are consumed by larger piscivorous fish. When the piscivores start feeding, they reduce the numbers of zooplanktivores, which increases the copepods, which decreases the phytoplankton. With an odd number of trophic levels, say without the piscivorous fish, the phytoplankton will be abundant because their consumers, the copepods, will be reduced in numbers by the planktivorous fish. Clearly this is a dynamical system, as populations will fluctuate. But still we can conclude that there may be inherent differences that emerge just from the structure of food chains. These conclusions work only if we can prove that consumers have important effects in food chains. There is good evidence to support this. The other question, of course, is what determines the length of food chains?

Food chains are often an abstraction of reality, but it is not common to find more than four to five trophic levels in oceanic or lake food chains. Why should this be so? Two alternative hypotheses have been proposed, but we are not sure whether either or both are good explanations. The **trophic hypothesis** argues that there is a maximum number of trophic links through which energy can travel. After all, with an ecological efficiency of 10 percent, only 0.01 percent will reach a fifth trophic level. This may set a limit to the upper levels, and therefore this hypothesis argues that control is a **bottom-up process**. The **food chain stability hypothesis** argues that longer food chains are inherently unstable. If changes occur at one level, they will be propagated to the other levels. For example, if a population at one

trophic level becomes extinct, it will cause species at levels above to become extinct. This might prevent stable food chains from becoming too long. This idea is based on some theory and simulations, but many factors weaken the idea. Most important, the fact that species in upper levels of a food web may feed at many levels beneath could dampen fluctuations since the top predator has other options that will prevent extinction if its immediate food source on the trophic level below is depleted. Feeding on many food sources is known as **omnivory**, and this feature of many species will reduce the strong effects of fluctuations of a species in a given level of a food chain.

■ **Strong food web interactions in trophic cascades can result in strong regime shifts that are driven by human and climatic forcing.**

Strong interactions in the form of trophic cascades lead to two important predictions about marine food webs. First, **strong fishing of top predators may have strong effects on the structure of entire food webs.** As we mentioned earlier, fishing on cod has an immediate impact on shrimp in the North Atlantic. Strong overfishing of cod has resulted in major reorganizations of coastal systems in the northwest Atlantic (see Hot Topics Box 17.1). Second, **climate forcing may favor changes in the members of some food webs,** which can result in strong reorganizations by means of the trophic cascade. For many years, South San Francisco Bay was dominated by benthic suspension-feeding bivalves, which fed and greatly reduced the phytoplankton.

But after 1999, phytoplankton in San Francisco Bay rapidly increased and benthic suspension-feeding bivalves, which crop the phytoplankton, nearly disappeared (Cloern et al., 2007, 2010). The decline of benthic suspension feeders was caused by a sudden increase of demersal predators, including shrimp, crabs, and bottom-feeding flatfish, which increased at the same time and greatly reduced the bivalves, which in turn resulted in reduced grazing and increase of the phytoplankton. James Cloern and colleagues discovered that this **shift in ecosystem regime** was strongly correlated with the Pacific Decadal Oscillation, a major Pacific Ocean decadal swing in air pressure that influences climate throughout the northern Pacific, which may have brought normally cold-water predators from the north and then into San Francisco Bay. Increased coastal upwelling may have provided food to oceanic populations of predators such as bottom fish and crabs, which then entered San Francisco Bay and devastated the bottom populations of bivalves (see Figure 3.6). Such regime shifts have been noticed in many other water bodies and testify to the common strong links within food webs, causing trophic cascade effects.

Measuring Primary Productivity

Primary productivity must be known before we can estimate the potential production of fish at the top of a food chain. Marine ecologists have attempted to develop accurate and relatively simple field-oriented techniques. **Gross primary productivity** is the total carbon fixed during photosynthesis. We are more interested, however, in **net primary productivity**, which is the carbon that is fixed but not respired, and therefore available to higher trophic levels. Some of the primary production is respired away, leading to a somewhat lower net availability to higher trophic levels.

Oxygen Technique

■ **The oxygen technique relies on the fact that oxygen is released in proportion to the amount of photosynthesis.**

Because oxygen is released during photosynthesis, changes in oxygen concentration can be used to estimate primary productivity. A water sample is first collected and the zooplankton are strained from it, using a 150–300 μm plankton net. The remaining water is placed in an oxygen-tight biological oxygen demand (BOD) bottle and incubated in the light for several hours. Oxygen is measured at the start and finish of the experiment. Dissolved oxygen is measured either by the **Winkler method,** which is based upon chemical titration, or by a polarographic **oxygen electrode**, which measures an electric current that is proportional to the dissolved oxygen concentration.

During the course of the experiment, phytoplankton and bacteria are respiring in the bottle, reducing the oxygen concentration. The change in oxygen concentration is explained, therefore, in terms of **an addition from photosynthesis and a subtraction from respiration.** (For more, see **Going Deeper Box 12.1.**) We can estimate the loss from respiration by incubating a bottle that is completely covered in black tape so that no photosynthesis occurs. This **light–dark bottle method** can be applied by incubating in various levels of artificial sunlight or by actually dropping a string of bottles over the side and incubating them in natural light (**Figure 12.3**). With the artificial light technique, it is necessary to measure the light intensity as a function of depth so that the shipboard laboratory measurements can be used to calculate the amount of photosynthesis in the natural environment. Practically speaking, respiration is often not actually measured but is usually taken to be 10 percent of the total oxygen increase (which is based on measurements in rapidly growing laboratory cultures).

Radiocarbon Technique

■ **The radiocarbon technique uses the radioactive isotope ^{14}C as a tracer in uptake of bicarbonate during the process of photosynthesis.**

In this method, performed in the lab under optimal field light conditions, bicarbonate ion is labeled with the radioactive isotope of carbon, ^{14}C. (The common stable isotope is ^{12}C.) The uptake of carbon is measured directly, using ^{14}C-labeled bicarbonate as a tracer. First, a sample of seawater is inoculated with a small amount of radioactive bicarbonate solution, and the phytoplankton are allowed to photosynthesize for a specified number of hours. After this, the phytoplankton are trapped on a filter, and the radioactivity is then measured in an instrument known as a scintillation counter. Productivity is proportional to the percentage of the radioactive label taken up by the phytoplankton. (For more, see **Going Deeper Box 12.2.**)

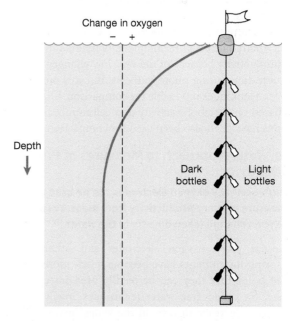

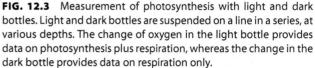

FIG. 12.3 Measurement of photosynthesis with light and dark bottles. Light and dark bottles are suspended on a line in a series, at various depths. The change of oxygen in the light bottle provides data on photosynthesis plus respiration, whereas the change in the dark bottle provides data on respiration only.

GOING DEEPER 12.1

How to Calculate Productivity Using the Oxygen Technique

If I is the initial amount of oxygen in both a light bottle and a dark bottle (see **Figure 12.3**), P is the oxygen produced in photosynthesis, and R is the amount consumed in respiration, we can show that, after a specified amount of time, the amount of oxygen in the light bottle, L, is $L = I + P - R$. The amount in the dark bottle is $D = I - R$ because there are no additions from photosynthesis. Therefore, the amount of oxygen in the light bottle, minus the amount in the dark bottle, is

$$L - D = (I + P - R) - (I - R) = P$$

It is desirable to convert photosynthesis from units of oxygen to units of carbon in the photosynthetic product (e.g., sugars). To do this, we must know the ratio of the molecules of oxygen liberated during photosynthesis to the molecules of carbon dioxide assimilated. This photosynthetic quotient depends on the type of photosynthetic product. The photosynthetic quotient is 1 if hexose sugars are produced, 1.4 for lipids, 1.05 for protein with ammonia as the nitrogen source, and 1.6 for protein with nitrate as the nitrogen source. Thus if the PQ is greater than 1, we are getting less photosynthetic output per unit of oxygen output. For example, the PQ increases when NO_3 is used as the nitrogen source instead of ammonium. In the case of one nitrate, eight atoms of oxygen have to be reduced to convert nitrate to ammonium for the photosynthetic process, which costs energy that is taken away from carbon dioxide uptake, and increases the PQ from 1.05 (ammonium use) to 1.6 (nitrate use).

The respiratory quotient is the ratio of molecules of carbon dioxide liberated during respiration to the molecules of oxygen used in respiration. The following relationships hold for gross primary production (GPP), respiration (R), and net primary production (NPP):

$$GPP = \frac{375\,(L-D)\,X}{PQ}$$

$$R = 375\,(I-D) \times RQ \times X$$

$$NPP = GPP - R$$

where I equals the initial oxygen content of the water added to the light and dark bottles, D equals oxygen in the dark bottle after a selected time period, L equals oxygen in the light bottle after the same time period, X equals the depth of a 1 m^2 cross-section water column, RQ is the respiratory quotient, PQ is the photosynthetic quotient, and 375 is a conversion factor (mg oxygen to mg carbon) when the photosynthetic quotient is 1.

The radiocarbon measurements of photosynthesis are usually lower than oxygen estimates (Peterson, 1980). Phytoplankton may excrete some of the photosynthate as it is produced, and the radiocarbon technique therefore provides a good estimate of the production of particulate matter from photosynthesis. It is also possible that respiration is high, which results in a large loss of ^{14}C during the experiment. The radiocarbon technique is preferable in waters of low productivity because of the very low changes in oxygen concentration during photosynthesis. But it is preferable to do both techniques if possible for comparison. Other methods based on metabolic activity and cell number change in cultures have been also been used for comparison.

Free-Water Approach to Measures of Primary Productivity

■ **High-accuracy oxygen electrodes can be used to measure primary productivity from measures of oxygen change taken directly in the water.**

In recent years, oxygen electrodes and other oxygen-measuring instruments have become much more accurate, to the point that they can be readily used instead of ^{14}C uptake. So-called **free-water techniques** involve placing an oxygen electrode directly in the water, instead of using sealed bottles, and following changes in oxygen in the water column by use of a recording or transmitting device (Cole et al., 2000). During the day, oxygen change is explained by primary production minus daytime respiration, but diffusion of oxygen out of the surface water into the air

must be accounted for. At night, no photosynthesis occurs, so a drop in oxygen can be explained by total water column respiration and diffusive loss of oxygen to the atmosphere. Gross primary production, net primary production, and total water column respiration can therefore be calculated by use of the recording oxygen electrode. This free-water method is preferable because it measures processes directly in the natural environment.

The Pump-and-Probe Fluorometer

■ **The amount of photosynthesis can be estimated by measuring the fluorescence obtained from phytoplankton active photosystem II centers that have been exposed to a sequence of light flashes.**

Most estimates of photosynthesis are tedious and often involve incubations on shipboard. A device known as the **pump-and-probe fluorometer** (Falkowski et al., 1992), or fast repetition rate fluorometer, allows in situ measures of photosynthesis. The technique depends on an important aspect of photosynthesis, which involves the transfer of electrons between different parts of the photosynthetic apparatus. It also depends upon the fact that photosynthetic pigments fluoresce when exposed to ultraviolet light. A total phytoplankton sample fluoresces when exposed to UV light, but the total amount of fluorescence corresponds to the total amount of photosynthetic pigment, not necessarily to the active photosynthesis. To actually probe the dynamic process of photosynthesis, the phytoplankton sample is exposed sequentially to UV flashes. The details are beyond

GOING DEEPER 12.2

Using the Radiocarbon Technique to Estimate Productivity

Gross productivity, GPP, can be estimated as follows:

$$GPP = \frac{(R_L - R_D) \times W \times 1.05}{R \times N}$$

where W is the weight of bicarbonate in water (mg C m^{-3}), N is the number of hours of the experiment, R is the counting rate expected from the entire amount of ^{14}C added to the sample, R_L is the counting rate of

a light-bottle sample, R_D is the counting rate of the dark-bottle sample, and 1.05 is a constant allowing for the different rate of uptake of ^{14}C relative to the more common isotope ^{12}C.

The radiocarbon technique estimates gross productivity, but only if measurements are made quickly. If measurements are made over several hours or more, some dissolved organic carbon, including the radio-labeled ^{14}C, leaks from the phytoplankton cells.

the scope of this text, but the response of the sample to the strength and the timing of the flashes allows calculation of the target area used by the phytoplankton cell for the absorption of light energy and the total photosynthetic rate. It is estimating the number of active photosystem II complexes (see Going Deeper Box 11.1).

This technique is complex in principle but simple in practice. Indeed, it allows us to deploy an instrument in the water to measure photosynthesis directly and has been used to study primary productivity in the water column, but also in the surf zone and in benthic photosynthesizers such as algae. The instrument can be gradually lowered through the water column to get an estimate of photosynthesis. In many cases, more has to be understood about the physiological and photosynthetic processes within algal cells to make a good quantitative estimate of primary production.

Optical Data Within the Water Column

- **Data collected by instruments on ocean moorings can be used to estimate continuous changes in primary production.**

One of the great disadvantages of the methods just discussed is the need for continuous and direct sampling of water from vessels if we wish to understand changes over time. This problem has led to the development of autonomous instrumentation that can be placed on open-ocean moorings. For example, one can sample various optical properties of seawater in combination with physical properties such as current speed, temperature, and salinity. The bio-optical measurements consist of two main types. First, **fluorometry** measures the fluorescence yielded in response to a pulse of ultraviolet light. This can be calibrated to calculate the concentration of chlorophyll *a*. Second, a high-accuracy spectral radiometer can be used to collect the intensity of light over the wavelength spectrum of light. Both instruments can collect data and transmit information to a satellite. The data can be used to connect light, chlorophyll abundance, and the physical data to an estimate of productivity, based upon previously established mathematical models. This type of approach has been tested by comparison of direct measures of primary productivity with the estimates derived from the bio-optical instruments on moorings at sea. The

results show good matches between direct measures and estimates from data collected by moored instruments (Honda et al., 2009).

Satellite Color Scanning and Productivity Models

- **Satellite color scanners can crudely estimate relative standing stocks of phytoplankton, which can in turn be used with ocean surface data and productivity models to estimate changes in primary production.**

Our data on primary production of the world's oceans come from literally tens of thousands of measurements made from ships, using either seawater incubated on ship or samples placed directly within the water column. Ship time is very expensive, however, and we need a means of monitoring primary production more rapidly over large geographic scales. The only hope of achieving this is through the use of **satellite data**. Satellites can be outfitted with color scanners, and these have already been used to survey the ocean (**Figure 12.4**). The Coastal Zone Color Scanner, operating from 1978 to 1986, measured the visible part of the spectrum of light that left the water (radiance). A Japanese sensor operated in 1996–1997, and the Sea-Viewing Wide Field-of-View Sensor (SeaWiFS) has been in operation since 1997 under the supervision of the U.S. National Aeronautics and Space Agency. It is a visible light scanner capable of covering 90 percent of the earth's surface every 2 days. The Moderate Resolution Imaging Spectroradiometer (MODIS) is a satellite scanner designed to measure fluorescence, which is a more direct means of assessing chlorophyll distributions in the ocean and is meant to be a successor to SeaWiFS. Both MODIS and SeaWiFS are crucial tools in tracking global temporal and geographic variation in primary production.

The water surface is bombarded by **solar irradiance**, which includes a spectrum of photons from ultraviolet to visible to infrared wavelengths. Any satellite measures the **radiance** that leaves the water, which is a function of reflected solar irradiance and light emanating from the ocean itself. With algorithms used to relate radiance to ocean features such as chlorophyll concentration, ground-truthing measurements of incident solar irradiance, the vertical

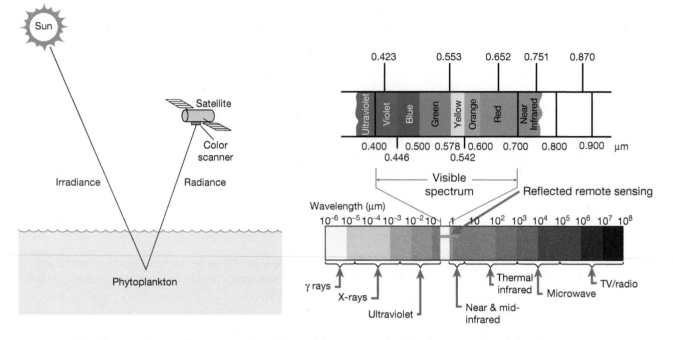

FIG. 12.4 Color detectors in a satellite measure the radiance of the ocean, or the light that is re-radiated after the sun's light encounters and partially penetrates the ocean (the light that reaches the ocean is called the irradiance).

structure of the water column, and sea temperature, it is beginning to be possible to estimate net primary production on a global scale.

There are several challenges with using satellite data. First, one is attempting to gather data on light received over a spectrum of wavelengths and to relate this radiance to the concentration of light-harvesting pigments in seawater. A set of equations that relate the concentration of photosynthetic pigments to the light received by the satellite is required. An index of chlorophyll density can, therefore, be developed from the satellite's reception of different wavelengths of light. However, there is a saturation effect, which prohibits the estimation of light-harvesting pigments above a few milligrams per cubic meter; thus, water in phytoplankton-rich coastal seas and bays cannot be measured accurately in this way. A satellite can "see" only partly into the water column; in some parts of the ocean, much of the chlorophyll is beyond the reach of detection. Given the overall errors, this problem is usually a secondary one.

With the aid of the satellite's color scanner, we obtain an estimate of photosynthetic pigment concentration. However, the relationship between pigment concentration and primary production varies considerably throughout the ocean. We therefore need to find a relationship that ties together pigment concentration and primary production and is specific to a geographic region. Usually in a given region, there is a relationship between (1) the ratio of primary production to phytoplankton biomass and (2) irradiance, or light that hits the ocean surface. It is the dream of proponents of the satellite approach that a measure, independent of local on-the-ground adjustments, will some day be developed.

There is an overall inverse relationship between water temperature and nitrate concentration in the open ocean. In warmer seas, vertical water exchange is more limited and nutrients are relatively depleted; primary production is, therefore, reduced. Because satellites can also estimate sea-surface temperature, this relationship can be used in a local calibration. Even though photosynthesis increases with increasing temperature, the relationship of sea-surface temperature to nutrient concentration is stronger, so overall, primary production decreases with increasing temperature.

A number of models have been developed that relate satellite-collected data to primary productivity. These models are not new but have acquired much more importance now that global-scale data can be collected so efficiently. **Depth-integrated models attempt to produce an estimate of net primary productivity (NPP) that accounts for all of the photosynthetically active depth of the water column.** The main objective is to relate NPP to a surface measure of phytoplankton biomass, such as can be estimated from satellite data. The parameters found to be most important are as follows (Behrenfeld and Falkowski, 1997):

- *Surface phytoplankton biomass.* This is estimated from radiance, which depends on a good model relating radiance to chlorophyll in the surface waters.
- *Photic zone depth.* The relationship of light penetration to depth varies with location, latitude, and other variables and is important because of its effect on photosynthesis.
- *Photosynthetic efficiency.* This variable gives the efficiency of photosynthesis, which is the carbon fixed per milligram of chlorophyll. This will vary with the dominant phytoplankton species, nutrient conditions, temperature, and many other variables. In some models, this is taken to be a constant number, but it is likely that imprecise estimates

of this variable are a major cause of inaccuracy in satellite-based estimates of NPP.

- *Day length.* This variable changes with season and latitude and provides crucial information on light availability.
- *Irradiance.* Light intercepted by the sea surface varies with season, latitude, and weather and should strongly influence the energy available for photosynthesis.

How well do satellite data correspond to measures from shipboard? Thus far, a number of studies show a statistically significant correlation between chlorophyll *a* estimates from direct measurements on shipboard and SeaWiFS satellite analyses (Moore et al., 1999). On the other hand, estimates from shipboard samples are still far more accurate. Satellite measurements show a great deal of variation, probably because of the difficulties mentioned earlier in this section. Satellite-based estimates of productivity are correct within a factor of 2 at best, when compared to shipboard measurements using the carbon-14 technique (Campbell et al., 2002). Much of the variation is related to different equations that have been developed to relate NPP to satellite measurements, based on the variables we have discussed. Errors were especially large in the Southern Ocean and in the Bering Sea. In any one place, the accuracy based on satellite measures is not encouraging and shipboard-based measurements are still far more accurate. Still, satellite-based data provide excellent maps of the globe and changes with season (see later).

More recently, a carbon-based model has been developed, based on the knowledge that chlorophyll/carbon ratios in marine phytoplankton are a reflection of physiological conditions and might provide additional information on the physiological state of the phytoplankton (Behrenfeld and others, 2005). Chlorophyll is measured by a satellite scanning photometer that is sensitive to wavelengths of 443 and 672 nm corresponding to chlorophyll. Cellular carbon is very strongly correlated with a backscattering signal from the sea surface that is detected at a different wavelength of 660 nm. As chlorophyll increases relative to cell carbon, one may conclude that phytoplankton cells are responding to conditions by increasing the capacity for photosynthesis. We will discuss in the section on world productivity how this might strongly affect world estimates of variation in primary production.

Geographic Distribution of Primary Productivity

Regional Variation in Productivity

■ **Continental shelf and open-ocean upwelling areas are among the most productive, owing to winds that move surface water offshore and bring nutrient-rich water from below.**

Continental-shelf upwelling regions have high productivity because of a consistent wind parallel or at a slight angle to the coast. Owing to the Coriolis effect, surface waters are deflected to the right in the Northern Hemisphere and

to the left in the Southern Hemisphere. As the spring and summer seasons progress, the axes of the trade winds shift, resulting in the movement of the principal location of upwelling toward the pole.

The movement of water toward offshore results in replacement by cooler nutrient-rich water from the bottom. The bottom nutrients originate from sedimentation of phytoplankton to the seafloor. The regeneration of nutrients to the surface fuels high primary production, and the great fisheries of the world are located in upwelling regions. The California, Peru, and Benguela (northward along the west coast of Africa) currents, for example, are eastern boundary oceanic currents.

Upwelling can occur from as deep as 200 m but usually occurs from 100 m depth or less (e.g., off the Peruvian coast). In this region, dense standing crops of phytoplankton and anchovy occur. The anchovy further serves as food for tuna, and these species form two of the world's great fisheries. Anchovy populations also feed large populations of birds, whose guano production covers the famous guano islands off the coast of Peru.

Upwelling in Antarctic seas and in the eastern equatorial Pacific also fuels high phytoplankton production. In some areas, upwelling is seasonal. In the monsoon regions of southern Asia, periods of high production alternate with low production seasons. In upwelling areas, production can exceed $1.5–2$ g C m^{-2} d^{-1} for protracted periods. Strong upwelling can also occur adjacent to submarine ridges as well as in areas of strong currents, such as the Faroe–Iceland Ridge, where production equals 2.5 g C m^{-2} d^{-1}.

■ Coastal areas are nutrient rich and productive.

Because of the shallowness of the water and the regeneration of nutrients from the bottom, waters close to shore are generally highly productive. At the shelf–slope break, intrusions of nutrient-rich slope water can fuel production in shelf waters. Much of the inner-shelf phytoplankton is not consumed by the zooplankton, and they sink to the bottom. On productive shelves, much of the food of the benthos comes from such sedimentation of uneaten phytoplankton. Some of the surface phytoplankton over the shelf–slope break may be exported to deep-sea bottoms, but probably far less material reaches the abyss.

Estuaries in Georgia are normally rich sources of nutrients, but the water is usually turbid and the depth of active photosynthesis is relatively shallow. Outer-shelf waters are clear, but nutrients are low and production is relatively low. Inner-shelf surface waters combine the best of both worlds: The water is relatively clear and nutrient rich. The combination causes higher primary productivity than is the case for either estuarine or outer-shelf surface waters (**Figure 12.5**). Outwelling of dissolved nutrients from human-derived sources has greatly increased primary production on the inner shelf in much of the world.

In shallow coastal waters, sea grasses and seaweeds may locally be the dominant forms of primary production. In the tropics, turtle grass beds (*Thalassia testudinum*) are very common in waters of a couple of meters depth or less;

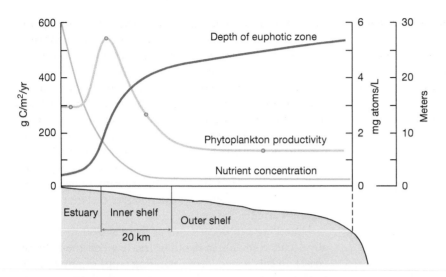

FIG. 12.5 Geographic variation in euphotic zone depth, nutrient concentration, and primary productivity on a transect from the coast of Georgia, United States, to the outer continental shelf. (After Haines, 1979, with permission of the Plenum Publishing Corporation)

eel-grass (*Zostera* spp.) dominates in higher latitudes. Such beds have primary production often greater than 1,000 g C m⁻² y⁻¹. Kelp forests dominate the shallow coastal waters of wave-swept, high-latitude hard bottoms and match sea grass beds in production (see Chapter 17).

■ At convergences and fronts, nutrients are concentrated and primary productivity is high.

A **front** is any rapid geographic change of seawater properties. Horizontal variation in hydrography has a significant effect on patterns of production. Surface convergences concentrate nutrients and plankton. Seabirds, fishes, and fishing boats congregate at convergences to take advantage of the food supply. Such fronts may be temporary (as are intrusions of slope water moving onto the shelf) or long-lived (as are some fronts off the shelf–slope breaks near Nova Scotia and near the southern Bering Sea). In general, fronts increase primary productivity when nutrient-rich water is upwelled from a deep source and is transported into a relatively stable shallow part of the water column, such as a warm surface layer. This is commonly the case in the surface waters over the shelf–slope regions, where deep nutrient-rich water seaward of the continental slope moves upward over the shelf–slope boundary region to fuel primary production in the surface waters.

In the southern Bering Sea, a series of fronts divides the ocean into distinctive food webs. A middle-shelf zone supports the production of large diatoms that are not grazed completely and settle to the bottom and provide food for benthic invertebrates. The outer-shelf phytoplankton are grazed in the water column efficiently, and the high fish production supports large populations of seabirds and mammals.

■ Central oceans and gyre centers are nutrient poor and relatively barren of primary productivity.

Productivity is extremely low between latitudes 10 and 40 within tropical gyres, such as the Sargasso Sea and the

North Pacific gyre. The nutrient regime is poor but is compensated by the year-round growing season and the great euphotic depth. Primary production is about 50 g C m⁻² y⁻¹. An important feature of such regions is the permanence of the thermocline. The density stratification of the water column reduces the potential for nutrients to be regenerated from deeper water. Near the Bermuda islands in the North Atlantic, the thermocline breaks down during the winter and early spring, resulting in upwelling and higher rates of primary productivity than in surrounding deep-water areas.

Productivity by Region

■ Different oceans have different overall levels of primary productivity, which are determined by latitude, ocean basin shape, wind-driven surface currents, and the influence of surrounding continents.

ATLANTIC OCEAN The North Atlantic and the South Atlantic are both characterized by large circulating gyre current systems. Eastern boundary currents, such as the Benguela Current off the coast of southwest Africa, produce a large upwelling system in the southern Atlantic. The Brazil Current moves southerly along the coast of South America. The center of the South Atlantic Ocean at midlatitude thus consists of a large counterclockwise eddy. A similar clockwise eddy is found in the North Atlantic, in the Sargasso Sea, where daily primary production is less than 0.1 g C m⁻² d⁻¹.

PACIFIC OCEAN The Pacific Ocean is divided into the North Pacific and the South Pacific, and its pattern of water circulation and production is similar to that of the Atlantic, with low production in the gyre centers and higher production on shelf waters. The northern part of the North Pacific differs from the Atlantic in that there is no noticeable burst of phytoplankton biomass in the spring, although primary productivity does increase. It is believed that zooplankton predation in this part of the Pacific prevents phytoplankton biomass from accumulating, but iron limitation also contributes. We have already discussed the great production in the Peru upwelling system. There is also an upwelling system off the west coast of North America. Overall, the Pacific is more productive than the Atlantic.

INDIAN OCEAN The rate of primary production in the open waters of the Indian Ocean is fairly low (about 0.2 g C m^{-2} d^{-1}), as in the South Atlantic. North of the equator, however, very high rates are found off the coasts of Somalia and India, and in the Arabian Sea. Winds called monsoons shift seasonally. Production rates vary with alternating periods of upwelling and influx of nutrient-poor surface water from offshore. During periods of upwelling, organic matter settles slowly in the thermocline region, and bacterial respiration causes the development of strong oxygen minimum layers.

ANTARCTIC OCEAN One of the more productive areas of the world, the Antarctic supports a dense planktonic diatom flora, and production is probably 1 g C m^{-2} d^{-1} during the production season of about 100 days. A slow upwelling of nutrient-rich water fuels the high productivity. Phytoplankton production can proceed under the ice, and ice-associated algae are a locally important source of food for krill (see Chapter 19). The Antarctic is a major location for satellite-based studies of production (Moore and Abbott, 2000).

ARCTIC OCEAN Production can be high in the Arctic Ocean in summer. Year-round figures are limited because of the short production season (remember that there is no light for 6 months of the year) and the ice cover, which intercepts much light. Nutrients are at high levels, and summer melting stabilizes the water column, especially

in the upper 10 m, resulting in considerable primary production, much of which is consumed by zooplankton in shallow layers (Olli et al., 2007). Primary production is of the order of 15 g C m^{-2} y^{-1} (Gosselin et al., 1997). This is much larger than had been believed to be the case, but the contribution of ice algae has proven to be important. As summer ice continues to decline, in future decades primary production from ice algae will decline, but light is stimulating more production in shallow open waters as summer sea ice disappears entirely and surface water temperatures increase with global warming. The Arctic Ocean is warming at twice the average global rate and the length of the plankton growing season is steadily increasing.

EQUATOR Upwelling a few degrees on either side of the equator fertilizes tropical ocean waters in both the Pacific and Atlantic. In the Pacific Ocean, for example, there are two westward-flowing currents, one on either side of the equator. The Coriolis effect causes a net southerly transport of surface waters south of the equator and a net northerly transport north of the equator. The movement of surface waters away from the equator in the eastern Pacific causes upwelling of nitrogen-rich water from greater depths. Productivity varies around 0.3–0.4 g C m^{-2} d^{-1}. In the western Pacific, nutrients are far lower in concentration.

WORLD PRODUCTIVITY Figure 12.6 shows the overall pattern of productivity in the ocean based on shipboard measurements. Net primary productivity in the ocean comprises about half of the earth's total. The overall world production is computed on the basis of oxygen evolution measurements (gross production) to be 100–200 g C m^{-2} y^{-1}; on the basis of radiocarbon estimates (net production estimate), the computed range is 25–35 g C m^{-2} y^{-1}. John Ryther (1969)

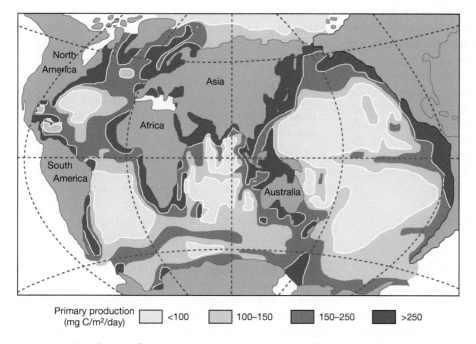

Primary production (mg C/m²/day): <100   100–150   150–250   >250

FIG. 12.6 Distribution of primary production in the oceans. (After Koblentz-Mishke et al., 1970)

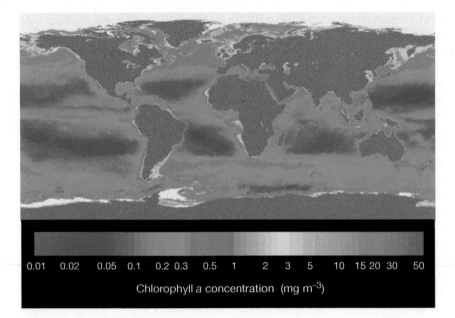

FIG. 12.7 Worldwide, year-round integrated estimate of chlorophyll concentration derived from data collected by the SeaWiFS satellite. False colors are used to represent the data, according to the accompanying scale.

has estimated that a realistic world average is about 50 g C m^{-2} y^{-1}. Overall, satellite-integrated scans of chlorophyll between 1978 and 1986 (**Figure 12.7**) conform well to this regional map, which is based on thousands of "ground-level" estimates. Satellite data suggest that world oceanic net primary productivity is in the area of 140 g C m^{-2} y^{-1}, which is a good deal larger than some previous estimates (Field et al., 1998). Any of these estimates can be used to show that all the nitrogen and phosphorus in the ocean must be used and recycled many times over. The supplies of nitrogen and phosphorus, therefore, limit total world productivity.

Recently, estimates of large-scale regional differences in primary productivity have been made using the chlorophyll/carbon approach, and some interesting differences have been found from the chlorophyll-only approach. As we mentioned earlier, chlorophyll-to-carbon ratios give us insight on the physiological state of the phytoplankton, since higher chlorophyll-to-carbon ratio indicates that the phytoplankton cells are increasing the capacity for photosynthesis. Behrenfeld and colleagues (2005) have used this approach and found that primary production in the equatorial regions may be much higher than indicated by depth-integrated models based upon chlorophyll alone.

Global Climate Change and the Global Carbon Pump
Cycling of Carbon and the Carbon Pumps

Carbon cycling is of great importance in the ocean because carbon dioxide interacts with seawater and dissolved substances. As a nutrient element, carbon is generally overlooked because it is not limiting to primary production relative to the obvious nutrients, nitrogen, iron, and phosphorus. But

because carbon reactions affect seawater chemistry, and because carbon dioxide influences global climate, we are very interested in its cycling through the ocean and the atmosphere. Also, there might be some surprises ahead with regard to the effects of increased dissolved carbon dioxide on primary productivity.

Inorganic carbon (e.g., dissolved CO_2, bicarbonate ion—HCO_3^-, carbonate ion—CO_3^{2-}) is the dominant form of carbon in the ocean; most of it occurs in deep water, beyond the photosynthetic zone. Only a small fraction of the carbon occurs as **organic carbon** whose chemical origin is biological, in forms such as proteins, carbohydrates, and degraded compounds of once-living organisms such as fragments of proteins, known as peptides. The shallow part of the ocean has both inorganic and organic carbon, whereas the deep ocean has much less organic carbon.

■ **Carbon dioxide increase in the past 200 years from human additions has led the earth to a carbon dioxide level in the atmosphere not seen in at least 400,000 years.**

We have records of atmospheric conditions over the past 400,000 years because of ice cores that preserved evidence of atmospheric conditions as the ice was being deposited. The ice core records show that atmospheric CO_2 was high during interglacial periods and low during glacial periods. This suggests that CO_2 changes drove the climate changes, but the question is how the ocean fits into the equation. Carbon dioxide changes have been documented in ice cores, and there is a strong correlation between carbon dioxide and climate. The crucial and intriguing result from ice cores is the variation of aluminum. When aluminum increases, carbon dioxide decreases in the cores. One might expect aluminum to be carried in dust, which would be high during glacial periods when the atmosphere is very dry and allows wind transport of dust to great distances from continents. Aluminum is known to be correlated with iron, and therefore iron increases may have increased the abundance of nitrogen-fixing phytoplankton and biological productivity in the surface waters, which resulted in sinking of carbon. This would have reduced atmospheric CO_2 and caused global cooling. As wind sources decreased, the reduction of windborne iron to the open ocean would have reduced ocean nitrogen fixation, and CO_2 would have built up in the atmosphere. This would suggest a cycle of change of carbon dioxide between glacial and interglacial cycles that would affect changes in atmospheric CO_2 and temperature.

It is of current interest to climate change expectations at present that the changes have been asymmetrical: The transition from glacial to interglacial times has been much faster than the reverse. We therefore have great concern about the results of industrial activities of the past century, since atmospheric carbon dioxide now exceeds all past glacial–interglacial records by more than 100 ppm.

It is very important to realize that the oceans contain more than 50 times the carbon than the atmosphere can contain. When each of the glacial periods ended, carbon dioxide was added to the atmosphere-ocean system, but most of it entered the ocean. In the past century, the additions of human-contributed carbon dioxide will therefore have likely entered the ocean, and it is of great interest to know the fate and effect of this carbon. Previously, carbon dioxide has reacted with carbonate in the ocean to produce a series of products, including bicarbonate ion, HCO_3^-, that buffered the ocean. This has not changed the ocean very much, but there is a danger that too much CO_2 will acidify the ocean, affecting fundamental biological processes. As we have discussed, phytoplankton productivity in the ocean is likely limited by nitrogen, phosphorus, and iron, so carbon dioxide additions may not stimulate oceanic productivity very much.

■ **Carbon is transferred downward into the ocean by means of a solubility pump and a biological pump. Fluctuations in these processes may be crucial in affecting global climate.**

Deeper in the ocean we find that the concentration of inorganic carbon increases. Why should this be so? Two processes, which we can liken to **pumps**, draw carbon to these depths. First, the **solubility pump** results from the greater solubility of carbon dioxide in seawater at colder temperatures, which is characteristic of deeper waters. Carbon dioxide dissolving into the ocean at the surface will tend to sink to deeper waters. In Chapter 2, we discussed thermohaline circulation, which brings cold and often saline high-latitude surface waters to great depth. While much of the deep water returns to the surface, the cold deep water retains much inorganic carbon. Models of global warming suggest that such sinking may be reduced, which will reduce the sinking and retention of carbon at depth. The increase of sea surface temperatures by global warming will reduce the solubility of CO_2 in surface waters.

Primary production withdraws CO_2 from the surface waters, which allows more CO_2 to be removed from the atmosphere. As we have discussed, a fraction of the production is never grazed and sinks to the depths, while some more material is consumed by zooplankton and then sinks to deeper waters in the form of fecal pellets. Many planktonic organisms secrete calcium carbonate skeletons, which sink and bring carbon to deep waters. Also, the collision of macromolecules forms marine snow, which also may sink. Respiratory biological processes would gradually oxidize some particulate material to inorganic carbon. It is believed that about one-quarter of the total surface primary production sinks toward the deep, which together is the product of the **biological pump.** Not all of this material reaches the deep, however, because some of it is broken down completely by bacteria in midwater and may be returned to shallow water by turbulence. At present, the tropics are a major location for formation of particles owing to the great abundance of nitrogen-fixing cyanobacteria. A good deal of this production sinks to the deep ocean. The tropical ocean is therefore a major carbon sink. Major rivers in the tropics, such as the Amazon, add nutrients to the ocean, which further drives production and potential storage of carbon in the deep ocean beneath tropical surface waters.

Global warming at the sea surface may reduce the rate of transport of carbon to the deep sea, which means that the deep ocean will not be as large a storehouse of carbon. Some circulation models suggest that if warming continues, there may be a strong reduction of the thermohaline convection system and downward transport of carbon will be reduced suddenly. This will result in accelerated warming or other surprising localized climate changes in different parts of the ocean. Unfortunately, current models do not provide enough precision to give us an exact roadmap for future change. Nutrient inputs to the world ocean are increasing, ocean circulation may change from warming, and shifts in wind patterns might even change, which would affect the input of iron from wind sources on ocean productivity.

Climate Change: Some Records and Effects on Oceanic Processes

■ **Ocean transparency measurements suggest that carbon dioxide increase has not had an apparent impact on primary production over the past 70 years, except in the Arctic Ocean. But there is evidence for a more recent decline of production that may have been caused by increased ocean stratification.**

Human activities add about 7 gigatons of carbon to the atmosphere each year. Phytoplankton fix 35–50 gigatons, so they could exert a considerable effect on the global carbon cycle, especially with regard to the human addition. If the phytoplankton increase in production and sink to the seabed, where the carbon becomes buried in the sediment or stored in deep water for 10^2–10^3 years, a considerable amount of anthropogenically derived carbon could be removed from the atmosphere and surface waters. So the question arises: Has phytoplankton productivity changed over the time during which industrial and other CO_2 release activity was on the rise?

There is good evidence for swings in production over decades. A study by E. L. Venrick and colleagues (1987) showed that chlorophyll *a* concentration increased over 2 decades in the open-ocean North Pacific gyre. During this period, winter winds increased, and this probably caused upwelling of nutrient-rich water from below the thermocline. Decade-level oscillations such as the North Atlantic Oscillation are known to impact regional temperature changes and can cause swings in ocean currents, plankton, and fish and whale populations.

Satellite data will eventually be crucial in estimating long-term effects of global warming. We have two sets of records from the Coastal Zone Color Scanner (CZCS, 1979–1986) and the Sea-viewing Wide Field-of-View Sensor (SeaWiFS). We have a number of years of data from the SeaWiFS (1997–2008), which can collect light data that are clearly related to other means of estimating productivity, such as ocean-surface data and indices of productivity developed from independent estimates of ocean changes. An analysis of the SeaWiFS data set of less than 10 years cannot show a long-term trend, but the data analysis demonstrates that sea temperature warming results in greater ocean stratification, which in turn results in reduced primary productivity. As stratification increases, we expect less nutrient replenishment from deep water, which explains the reduction of productivity (Behrenfeld et al., 2006).[1] Chlorophyll in the North Pacific between latitudes 36 and 46 has declined over 1997–2013, which can also be explained by increasing stratification. But chlorophyll in the northwest Pacific Ocean has increased, perhaps due to ocean warming in formerly cold and less productive waters (Siswanto et al., 2016).

Data collected by satellites on a global scale show the intimate relationship between large-scale fluctuations and climate, water stability, and primary productivity. Behrenfeld and colleagues (2006) examined global satellite data on ocean chlorophyll and primary productivity estimates over a period of 8 years and found an inverse relationship between estimates of water column stratification and productivity. World ocean productivity increased from 1998 to 2000 and then decreased from 2000 to 2006. The overall variation is explained mainly by changes in productivity of the broad expanse of tropical ocean production, which is strongly affected by climate change. A climate index that is a good estimator of the degree of oceanic stratification is directly related to productivity: The more stratification, the less productivity. Satellite-integrated data show an inverse relation between changes in sea-surface temperature (SST) and productivity. As mentioned earlier, stratification inhibits replenishment of nutrients from deeper water. In areas of the ocean (**Figure 12.8**) where SST has increased, productivity has decreased.

Longer-term data are difficult to obtain because widespread studies of primary production, even from shipboard, were done only from the 1960s onward. There is a simple substitute, however. In the nineteenth century, Secchi, an astronomer-monk, devised a white disk, which he lowered into the water until it could no longer be seen by the unaided human eye from shipboard (= the Secchi depth). Today, Secchi disks are about 20 cm wide and are painted in black and white quarters. The Secchi depth is a surprisingly good index of chlorophyll, which is corrected for such factors as latitude and season, which affect the local light intensity. Falkowski and Wilson (1992) compiled Secchi disk data from about 1920 to 1990 in the North Pacific Ocean and found no significant changes in Secchi depth, which suggests that there have not been long-term changes in primary production. The shortcoming of this conclusion is that it cannot be checked with other methods of estimating chlorophyll concentration or productivity. Another more recent study by Boyce and colleagues (2010) used a combination of transparency-based and in situ chlorophyll data and concluded that phytoplankton has declined in eight of 10 oceanic regions since 1899. This study has been strongly criticized because it used only one global calibration between transparency, chlorophyll, and primary productivity, so at present there is no general agreement about long-term trends. In more recent decades, the combined satellite data set shows a decline in the North Pacific of about 9 percent since the 1980s (Gregg et al., 2003), and as mentioned, there is strong satellite-based evidence for a recent decline (**Figure 12.8**).

Productivity has increased in one ocean, the Arctic. Annual primary production increased 30 percent for the period between 1998 and 2012 (Arigo and van Dijken, 2015). This increase appears to be linked to global warming

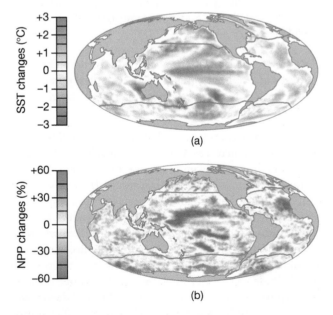

FIG. 12.8 Global changes in (a) sea-surface temperature and (b) primary productivity for the warming period 1999–2004. Temperature increases result in increased ocean stratification, which reduces nutrient replenishment from deeper water and therefore may have reduced primary productivity. (After Behrenfeld et al., 2006)

[1] An attempt has been made to unify CZCS and SeaWiFS data analyses, and the results show a 6 percent decline in ocean primary productivity since the 1980s (Gregg et al., 2003). This may be a response to global warming, but localized trends are complex.

and the reduction of sea ice. Ice extent has declined, resulting in warmer surface waters and a longer growing season.

Some Other Aspects of Greenhouse Gases

■ **Carbon dioxide increase may select for different dominance patterns in nutrient use among phytoplankton species.**

The apparent lack of carbon limitation on primary production would suggest that changes in primary production would stem only from indirect changes, such as increases of upwelling or changes of wind that might provide more windborne iron from the continents. What might be the effects of additions of carbon dioxide to the water column that are expected from industrial activities on the planet? While total carbon is not limiting, we must keep in mind that dissolved CO_2 is now much lower in concentration than bicarbonate ion. Some phytoplankton that take up CO_2 directly might simply be limited in the rate that they can get it from the water column by cellular processes. Danish scientists Hein and Sand-Jensen (1997) performed a direct test by altering natural seawater to increase dissolved CO_2 in waters from the low-nutrient part of the central North Atlantic. They found an increase of productivity of about 15–19 percent, which is modest. They also pointed out, however, that carbon dioxide increase might shift oceanic phytoplankton abundances toward species that are more limited by the rate of CO_2 uptake. Coccolithophores have been shown to respond positively to additions of carbon dioxide to the water and may benefit from the carbon addition as opposed to being negatively affected by low pH.

■ **Some relief from global climate change might be obtained by stimulating production by iron addition in high-nutrient, low-productivity areas of the ocean, resulting in sinking and carbon sequestration for hundreds or thousands of years in the seabed.**

We have mentioned the reduction of the effectiveness of the carbon pump, as industrial carbon dioxide additions to the atmosphere continue to occur and oceanic stratification increases. This will be compensated somewhat by possible increases in biological production and therefore the biological pump.

Can we take advantage of the biological pump by human intervention? In principle, it might be possible to add nutrients to some parts of the ocean where a nutrient is in short supply and net export to depth is likely. The likely target would be so-called high-nitrogen, low-productivity regions, such as parts of the Antarctic Ocean, where nitrogen and phosphorus appear to be abundant but primary production is limited because of low iron supply. As we discussed earlier in this chapter, a number of areas of the ocean are remote from the principal source of iron, which arrives by wind from the continents. All we need to do is add iron! Data from cores taken in Antarctic glacial ice show a relationship between aluminum and atmospheric carbon dioxide. When aluminum is up, carbon dioxide is down. Because aluminum is often correlated with iron, we may have found the smoking gun of global climate control. An iron increase

may stimulate primary production, which would increase the import of carbon dioxide from the atmosphere, in order to fuel photosynthesis. In this scenario, as organic carbon particles sank into the ocean, carbon dioxide would decrease and the greenhouse effect would decline as well.

An experiment performed in 1993 (Behrenfeld et al., 1996) in the equatorial Pacific showed that iron addition strongly stimulated photosynthesis, so at least the hypothetical effect of iron on a large scale has been confirmed. A similar experiment in the Antarctic also showed that iron additions stimulated particle sinking, but the size of the blooms and the degree of sinking would not likely have a major impact on sequestering carbon in the deep ocean (Buessler et al., 2004). The biological details may also be important, since some types of phytoplankton (e.g., nitrogen-fixing bacteria) are far more dependent on iron than other groups. Should such a program of iron addition be carried out on a massive scale? Is there any danger? Will stimulation of the phytoplankton cause nuisance algal blooms? Would the Antarctic phytoplankton be dominated by forms too small to be eaten by krill? These issues cannot be settled with our current poor understanding of the mechanisms that determine the relative abundance of the different major phytoplankton groups.

■ **Dimethyl sulfide is a natural by-product of phytoplankton metabolism and may strongly affect regional oceanic climate.**

Because of the strong evidence that phytoplankton productivity and downward transport of carbon may have an impact on global climate, we are beginning to look much more closely at the dynamics of phytoplankton and possible impacts on atmospheric processes. One process now believed to be important is the production of dimethyl sulfide (DMS) by ocean phytoplankton. A number of common phytoplankton species synthesize dimethylsulfoniopropionate (DMSP), which is implicated in osmotic balance, defense against grazing, and other processes. DMSP may be converted to DMS when phytoplankton cells are lysed by zooplankton grazing or by viruses. Virus attacks are important in geographically large-scale blooms of the coccolithophore *Emiliana huxleyi*, and the lysing of this species results in local large-scale production of DMS. Wind acting on the sea surface causes aerosols and DMS to be transferred into the atmosphere. Here, DMS is oxidized to sulfate, which is effective in cloud condensation. The increase of clouds may have an important effect on regional and even global climate. Although this process may be locally important, so far no general correlation has been found between DMS and cloud cover in specific regions of the ocean (Quinn and Bates, 2011).

■ **Ocean acidification may cause major reorganizations in the plankton by eliminating some calcifying plankton, especially those that precipitate aragonite.**

In Chapter 3 we discussed an important effect of additions of carbon dioxide to the atmosphere, caused by human activities of the past century or more. Carbon dioxide addition will cause a steady decrease of pH, which will inhibit those organisms that make skeletons from calcium carbonate. One form of calcium carbonate, **aragonite**, is more unstable than the other form, **calcite**. Thus, plankton that precipitate aragonite are more vulnerable.

A study by Orr and colleagues (2005) predicts that by 2050, the surface high-latitude ocean will become undersaturated with regard to aragonite, which will create major difficulties for pteropods, an important component of the zooplankton and a major source of food for larval and juvenile salmon in the northeast Pacific Ocean. A direct experiment using conditions predicted to be reached by the year 2100 resulted in incomplete pteropod skeletons under predicted decreases in pH conditions. Carbon dioxide conditions close to the present day appear to cause some incomplete calcification in some coccolithophore species, and experiments using conditions corresponding to projected increases by the end of the century cause poor calcification of individual plates and even missing plates in the spherical skeleton (Gattuso et al., 1998). Unfortunately, the relationship between degree of saturation and carbonate precipitation is nonlinear, so it is not possible to make generalizations about responses to lowered pH. Right now, the information we have for the pteropods is discouraging because lowered pH has an accelerating effect on reduction of calcification.

Given the enormous blooms of coccolithophores and other species seen in the ocean, a projected negative effect of ocean acidification of calcification is very worrying. But although there is some evidence that lowering of pH negatively affects calcification of some coccolithophorid species, an increasing-carbon-dioxide world might favor other species, particularly the widespread *Emiliana huxleyi*. Field studies of deposits of coccolith plates suggest that coccolithophorid plates have been increasing in size over the past 200 years or so, and a laboratory study demonstrates that increased carbon dioxide actually increases population growth of *E. huxleyi* (Iglesias-Rodriguez et al., 2008). On balance, the experimental addition of carbon dioxide, calibrated to expectations from global change models, actually enhances photosynthesis and shell deposition by coccolith plates of this dominant species. Results have been variable for other species. Still, these results show that effects of carbon dioxide increase can be quite variable and cannot be taken for granted.

■ CHAPTER SUMMARY

- Productivity is the rate of production of biological materials, whereas biomass is the amount of biological material present at any one time. Both are expressed as a function of ocean area.

- A food chain is a linear sequence that reveals which organisms consume which other organisms in an environment from top carnivore to primary producer. A food web is a fuller diagram of tropic interactions involving a network. A food chain's efficiency depends on the number of trophic levels and on the completeness of transfer from one level to the next. It can be used to calculate the potential fish production at the top of a food chain. Tropical food chains have more tropic levels and are less efficient than those in upwelling areas.

- Productivity can be measured by oxygen production during photosynthesis, using ^{14}C uptake and by remote sensing. In this satellite technique, a radiometer detects chlorophyll, which is converted to productivity by means of fitted equations.

- Primary production is high in upwelling regions, coastal and continental shelf regions, and in some oceanic boundary areas known as fronts, where nutrients are brought to the surface. Gyre centers have very low primary productivity.

- The shallow ocean has abundant organic and inorganic carbon, whereas the deep ocean has mostly inorganic carbon. Human activity in the past 200 years has led the earth to a CO_2 level not seen in at least 400,000 years.

- Carbon is transferred downward into the ocean by means of two kinds of carbon pump, and both critically affect global climate change. Colder, deep waters act as a *solubility pump* bringing carbon dioxide downward, while phytoplankton sinking from the surface to the deep-sea bed act as a *biological pump.*

- Increased sea-surface temperature might reduce the action of these pumps. Global warming may, therefore, have a negative feedback effect on planetary temperature, causing still more greenhouse gases. Productivity and the biological pump may be stimulated by artificial additions of iron.

- Satellite data suggests that global warming has caused increased warming of the tropical ocean surface, which is increasing water column stratification and reducing nutrient supply and primary production.

- Carbon dioxide addition to the ocean may have significant effects on the plankton, especially those that have a skeleton of aragonite. Carbon dioxide can also stimulate photosynthesis, so the overall effects of additions are not yet clear.

■ REVIEW QUESTIONS

1. Distinguish between autotrophic, auxotrophic, and heterotrophic.

2. What is the difference between a food chain and a food web?

3. What is the effect of increasing numbers of trophic levels on the productivity of the organisms in the top level?

4. Why is ecological efficiency in a food chain usually much less than 100 percent?

5. What gains and losses must be accounted for in the oxygen approach to measuring primary productivity?

6. Why are dark bottles used in the oxygen technique for measuring primary productivity?

7. In the radiocarbon technique of measuring primary production, why must a correction be made for the uptake of ^{14}C relative to ^{12}C?

8. Over the years, the oxygen production method has lost out to the radiocarbon method of measuring primary productivity. Why might this have happened?

9. What contributes mainly to the high primary productivity of coastal areas?

10. Current estimates of primary productivity using satellite imagery are very poor, yet satellite data may still be very useful relative to ground-based measurements. Why?

11. Why is it important to have a global estimate of primary productivity? What may be the advantage of having total estimates of productivity on scales more regional than that of the global ocean?

12. How does the vertical structure of the ocean change when SST increases? How does this change affect primary productivity?

13. Suppose you have a food chain with three levels: fish, zooplankton, and phytoplankton. What happens to phytoplankton if fish exert a strong top-down control and are suddenly fished out of the system?

14. Consider a food chain with three trophic levels and a food chain efficiency of 10 percent. What is the factor of increase at the top trophic level if the food chain efficiency is 20 percent?

15. Satellite sensors will continue to be crucial in global assessments of productivity. Does this mean that we can eliminate most of our sampling from ships and sensors attached to moorings? If so, why? If not, then why not?

16. Why might warming of the surface ocean have an effect on primary productivity. What will the direction of the effect be?

Visit the companion website for *Marine Biology* at www.oup.com/us/levinton where you can find Cited References (under Student Resources/Cited References), Key Concepts, Marine Biology Explorations, and the Marine Biology Web Page with many additional resources.

CHAPTER 13

Benthic Microorganisms, Seaweeds, and Sea Grasses

Most marine naturalists tend to focus nearly all their attention on animals; organisms such as whales and crabs seem to capture the imagination. Numerically, though, bacteria rule the ocean, both on the sea floor and in the water column. These organisms are crucial in the recycling of organic matter through decomposition and chemical conversions of a variety of substances, often in extreme environments. The intertidal and shallow subtidal seabed is often covered with seaweeds and sea grasses. These organisms are fascinating in their own right, but they also form the base of major coastal food webs. The purpose of this chapter is to introduce the diversity of microorganisms, algae, and plants that live on the seabed.

The Big Picture: Domains of Life

■ **All living organisms can be divided among three basic domains, on the basis of DNA sequence relationships.**

Although our notions of the overall organization of life first began with an analysis of overall morphological structure and an attempt to find structural similarities of major groups of organisms, current ideas about the major divisions of life are now based mainly on the use of DNA sequences and their change among species, features of cells, and arrangements and functions of genes.[1] In past classifications, a distinction was made between **prokaryotes** (often called bacteria), which include organisms usually of single cells that lack a nucleus, a nuclear membrane, and cellular organelles (e.g., mitochondrion), and **eukaryotes** (including

plants, fungi, animals, and protists), which have a nucleus, a nuclear membrane, and cellular organelles. It has been difficult to classify prokaryotic microorganisms, whose overall morphology was not sufficient to determine evolutionary relationships. Overall cell shape and morphology, such as rod-shaped or spherical, could evolve to converge to similar structure in many distantly related groups. Carl Woese and colleagues (1990) used data mainly from DNA sequencing of ribosomal DNA genes to demonstrate that there were three major divisions of life, which he termed **domains**: Archaea, Bacteria, and Eukarya.

Archaea have characteristics of prokaryotes but are united by similarities in sequences of 16S ribosomal DNA. Archaea have a number of distinctive features, such as a **distinctive lipid cell membrane** and a **unique ribosomal RNA**, even though Archaea in most other respects morphologically resemble members of the Bacteria in lacking a nucleus and occurring in similar cell shapes. Archaea in some ways resemble Eukarya in RNA polymerases and promoters, and in patterns of transcription and translation. They include many organisms that specialize in deriving energy by oxidizing or reducing substances, often in chemically extreme environments lacking oxygen (e.g., **methanogenic bacteria** that metabolize carbon dioxide and hydrogen into methane in anoxic waters or sediment pore waters), or having very high temperature (e.g., **thermacidophiles**, that live in high-temperature, acidic environments, as in hot springs and hot vents—see Chapter 18). While these so-called **extremophiles** are prominent Archaea have been found in more typical oceanic surface waters in recent years.

Bacteria also have characteristics of prokaryotes but form another evolutionarily distinct domain, based on data

---

[1] See bonus chapter, "Molecular Tools for Marine Biology," online.

from DNA sequences. They are single-celled and vary in shape from rods to spheres, and some, such as the familiar spirochetes, are spiral. Many bacteria species break down organic matter and are therefore vital in recycling nutrients in decomposition cycles. This domain also includes the photosynthetic cyanobacteria, which are widespread and important in oceanic primary production, and many species that are the source of disease in other organisms, such as syphilis and cholera.

Eukarya have a nucleus, a distinct nuclear membrane, and cell organelles, such as mitochondria and plastids (in plants and algae). Cell division involves separation of DNA in the form of chromosomes, aided by the micromechanical action of microtubules. While Archaea and Bacteria are far more abundant numerically, eukaryotic organisms reach far larger body sizes, from the paramecium to blue whales to trees. Included in this domain are the protists, fungi, algae, plants, and animals. In this chapter we will discuss bacteria, algae, and marine plants. We discuss invertebrate animal Eukarya in Chapter 14. We discuss some methanogens in Chapter 15 and extremophile Archaea in Chapter 18.

Bacteria

■ Bacteria occur as single cells and are the most numerous organisms in the sea. They are crucial in decomposition.

Bacteria are certainly the most numerous organisms in the ocean. They are prokaryotic and occur as single cells or as colonies. Most bacteria are tiny—usually only 0.2–2 µm in diameter. They are usually spherically shaped (coccus) or rod-shaped (bacillus), although this is also true for groups in the Archaea. In laboratory culture, the capacity for reproductive increase is fantastic, and cell doubling may take place every 20 minutes. In the ocean, bacteria rarely achieve such high rates, and doubling can be as slow as once a day. Bacteria occur free in the water column, attached to particles both in the water column and in sediments, and also in the pore waters of sediments. Large numbers of bacteria are also found in the guts of marine animals, and the green turtle, *Chelonia mydas*, has a special bacterial flora in the hindgut, which breaks down the cellulose in the turtle's sea grass food.

Most bacteria are important in the decomposition of organic matter in the sea. Such bacteria may absorb across the cell wall dissolved substances such as sugars. They may also produce enzymes externally. These break down organic material, which in turn is transported into the cell. Some bacteria can attach to particles and produce a network that allows enzymes to be trapped in an enclosed space so that they can work efficiently to break down organic matter.

■ Bacteria reproduce asexually, but DNA exchange as a form of sexuality is known.

Bacteria gain energy and grow and then reproduce by a form of fission. In the laboratory, population growth involves uptake of nutrients (lag phase), rapid growth (log phase), and nutrient depletion and low population growth (stationary phase). In some bacteria, reproduction involves budding. Sex as we see it in higher plants and animals is unknown in bacteria, although several mechanisms of genetic exchange, including transduction by means of bacteriophages and more direct exchanges by means of conjugation, allow transfer of DNA between cells. Bacteria may largely be clonal, but it is widely accepted that transfer of DNA between species is an important mechanism creating genetic variation, which increases opportunities for evolutionary change.

■ Bacteria are capable of gaining energy with many mechanisms of metabolism.

We can distinguish among bacteria by the way they gather or manufacture food. Recall that **autotrophic** organisms are those that can manufacture food if provided only with inorganic nutrients (e.g., phosphate, nitrate). A variety of bacteria and all higher plants are autotrophic and photosynthetic: They need light energy to aid in the process of manufacturing sugars. Many other bacteria are **chemoautotrophic**: They oxidize or reduce some substance (e.g., sulfate) to obtain energy, which is required to reduce carbon dioxide in order to manufacture sugars. Finally, many bacteria are **heterotrophic**: They obtain their food from the environment. In heterotrophs, the bacterium uses a substrate as an electron acceptor and generates energy in the process. In aerobic bacteria, oxygen is an electron acceptor, whereas in anaerobic environments, electron acceptors may be nitrate or sulfate. Aerobic bacteria use oxygen to break down organic compounds for energy. Other anaerobic forms use **fermentation**, the use of oxidation-reduction reactions with carbohydrate or similar substrates to generate energy for ATP production and produce products such as lactate and alcohol.

Chemophototrophic bacteria use light and nonorganic compounds to generate energy. A number of bacteria are **chemolithotrophic** and gain energy by oxidizing or reducing noncarbon substrates. For example, the photosynthetic **purple sulfur bacteria**, which occur as mats on the surface of the soft-sediment intertidal, oxidize hydrogen sulfide in sediment pore waters to produce elemental sulfur. The chemolithotrophic colorless sulfur bacteria (e.g., *Beggiatoa* sp.) also live as mats on sediment surfaces in contact with anoxic pore waters, gain energy by oxidizing hydrogen sulfide to elemental sulfur, and may oxidize the sulfur to sulfate. Sulfate-reducing bacteria are another common bacterial type in sediments, living at the interface between surface oxygen-bearing pore waters and deeper anoxic pore waters. Sulfide-oxidizing bacteria are common symbionts with many animals living either in anoxic environments (e.g., sediments, anoxic waters) where sulfide is common or in specialized deep-sea environments where sulfide comes from geothermal sources (see Chapter 18).

Of course, many species of bacteria are implicated in disease and are capable of penetrating living tissue of plants and animals. Unfortunately, our knowledge of marine bacterial disease is very poor, but extensive research is being done on coral disease (see Chapter 17).

■ **Cyanobacteria are photosynthetic cells or chains of cells, but the benthic forms usually live in association with anoxic sediments.**

Cyanobacteria (members of the group Cyanophyceae, also known as blue-green algae) are a widespread group of bacteria that occur as free-living forms but also in symbiotic associations with marine benthic plants and animals. Cyanobacteria are photosynthetic, but many also are capable of **nitrogen fixation**, the conversion of gaseous nitrogen into ammonium ion, and synthesis of amino acids (see discussion in Chapter 8). In benthic habitats, cyanobacteria may consist of spherical cells that are 1–25 mm in diameter. Commonly, they are multicellular (**Figure 13.1**) and are arranged in rows of cells known as **trichomes**. **Filaments** are trichomes or groups of trichomes enclosed in sheaths, and many cyanobacteria occur as long or branched filaments. Many cyanobacteria have **akinetes**, resting spores that allow overwintering or sometimes survival during several years of unfavorable conditions. Nitrogen fixation proceeds in enlarged cells known as **heterocysts**, and the enzyme nitrogenase enables the fixation of nitrogen. Cyanobacteria tend to favor environments where there is a local absence of oxygen because nitrogen fixation is favored under these conditions. Low oxygen levels can sometimes be achieved merely by growth in the form of a bacterial mat, whose interstitial water is anoxic in the interior of the mat. Cyanobacteria are often found in black crusts on the upper parts of rocky shores and also on mudflats, where grazing animals are uncommon. Cyanobacteria also live in association with the rhizomes of many sea grasses, and the latter benefit from the uptake of nitrogen into the rhizome system.

In areas with human-derived nutrient input—including iron, phosphorus, and nitrogen—very large blooms of benthic cyanobacteria, such as the filamentous *Lyngbya* (**Figure 13.1**), have developed, and these may smother soft bottoms and cover and kill sea grasses. In such blooms in Australia and other parts of the world, people have complained of itching, burns, and severe skin lesions after coming into contact with *Lyngbya*.

Eukarya

■ **The protists and allies include organisms that are at the cellular level of organization only, sometimes with tissues, but may also consist of multicellular forms.**

It is extremely difficult to organize for presentation the wide range of photosynthetic eukaryotic organisms that live on the seabed, but we will use a practical approach. We will first discuss a major group of the Eukarya, which include photosynthetic and heterotrophic single-celled organisms, fungi, and the algae, or seaweeds. The important distinction of these groups is the overall degree of biological organization. These groups have species that are often in the cellular form only, but some have discrete tissues. Eukarya encompasses an enormous range of morphologies and ecological types. They include single-celled forms such as diatoms and multicellular forms such as seaweeds. The taxonomy is very confused, and the most recent agreement does not even recognize a formal name to include all of these organisms, although groupings that have common ancestry have been defined. Further explanation is beyond the scope of this text, so we will avoid formal names altogether. The major groups we will discuss include photosynthetic single-celled forms, the largely saprophytic fungi, and multicellular algae.

Single-Celled Forms of Eukarya

DIATOMS

■ **Diatoms occur as single cells or chains of cells, are photosynthetic, and have a cell wall that is impregnated with silica.**

Diatoms are ubiquitous on intertidal rocks and soft bottoms and are common on the shallow-water seabed. One can often see a golden-brown mat of diatoms on intertidal sand flats. Planktonic forms have already been discussed in Chapter 8 as plankton, which are usually centric diatoms. On the seabed, **pennate diatoms** tend to dominate. This group includes relatively elongate bilaterally symmetrical cells (**Figure 13.2**), as opposed to the more disk-shaped cells (centric diatoms) in the plankton. Diatoms have a cell wall

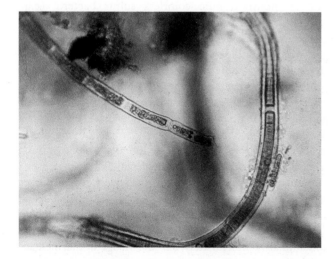

FIG. 13.1 The benthic cyanobacterium *Lyngbya*. (Photograph courtesy of Edward Carpenter)

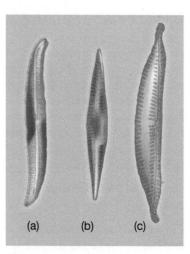

FIG. 13.2 Examples of benthic diatoms: (a) *Gyrosigma*, (b) *Navicula*, and (c) *Amphora*. (After Dawson, 1966)

impregnated with silica. Reproduction is mainly asexual and involves cell division. Some diatoms are motile and can move several millimeters. Commonly, intertidal species glide toward the surface at the time of low tide to maximize exposure to light. Some intertidal diatom species, however, glide below the sediment surface when low tide occurs in midday, to avoid desiccation. Although most occur singly, some exist as colonies, which are interconnected by a muci-laginous substance.

Fungi

■ **Ubiquitous and usually filamentous groupings of cells, fungi are very important in the decomposition of particulate organic matter.**

Fungi are either unicellular or multicellular, are eukaryotic, and occur ubiquitously in the ocean. Still, they are very intolerant of sodium, and most terrestrial fungi cannot live in seawater. Those living in seawater have mechanisms to transport sodium from within the cells. Fungi are relatively simple morphologically, and the cell walls are composed of chitin, as opposed to the cellulose found in plants. Fungi include marine yeasts and the filamentous-celled group known as **ascomycetes**, which include species that are important in decomposition processes. Cells of ascomycete forms usually grow into filamentous structures called hyphae and they appear under a microscope as a tangled mass of hairlike threads, known as the **mycelium**. They are heterotrophic and are important in the decomposition of plant matter in salt marshes and seaweed beds. Fungi can penetrate particulate organic matter, such as decaying sea grass, and commonly invade wood that has been transported to the seabed. They also can occur on inorganic surfaces, such as sand grains on a beach. Fungi that live on the surface of seaweeds and sea grasses probably survive by absorbing dissolved substances from the water. When grazing animals damage sea grasses or marsh grasses, fungi usually invade the wounds and can cause great damage to the plants.

■ **Fungi reproduce by means of fruiting bodies, which form and release spores.**

Fungi can reproduce asexually by forming and releasing spores, which settle on a substratum and form a new fungal mass. But filamentous fungi also may reproduce sexually by forming a **fruiting body** only a couple of millimeters long, within which sexual **spores** are produced and released. The spores are haploid, and two from different mycelia will fuse to form a founding cell of a new mycelium.

■ **Fungi are major sources of damage and occasionally disease to marsh grasses and sea grasses.**

Many species of fungi are decomposers, and they colonize and attack dying and dead marsh grasses (**Figure 13.3**) and sea grasses. The invasion of fungi into wounded grasses tends to accelerate decline of the grasses. This effect is likely to be widespread in both intertidal marsh grasses and subtidal sea grasses. In southeastern U.S. salt marshes (see Chapter 16, section on salt marshes), the snail *Littorina irrorata* scrapes on shoots of the common cordgrass *Spartina*

FIG. 13.3 A fungal invasion of a wound in the marsh grass *Spartina alterniflora*. Grazing snails scrape the grass blades, causing wounds (dark vertical lines) that are invaded by the fungi, which in this case are transmitted from the elongate snail fecal pellets on the grass blades. (Courtesy of Brian Silliman)

alterniflora and facilitates invasion by fungi, which in turn is used by the snail for food. This results in a mutually beneficial relationship between snail and fungi (Silliman and Newell, 2003). Some fungal species are pathogens and attack sea grasses. This is discussed further in Chapter 17 in the section on sea grass communities.

■ **Fungi live in a mutualistic association with algae to form lichens, which live in a band in the very high intertidal zone.**

It is common to see a brownish band in the upper intertidal of rocky shores, above the highest occurrence of barnacles. This band is a **lichen**, a mutualism between a fungus species and a photosynthetic alga, which may come from one of several algal groups (e.g., green algae, brown algae) or may be a cyanobacterium. For example, one intertidal lichen species consortium is a symbiosis between a brown alga and a fungal species. Cells of the two groups grow together in a complex weave that is believed to form a mat biomechanically evolved to resist wave damage (Sanders et al., 2004).

■ **Fungi also live in a mutualistic association with roots of some marsh grasses and sea grasses.**

Arbuscular mycorrhizal fungi live as mutualists in association with the roots of plants. Some live in the soil next to the roots, while others live within root cells. With the involvement of a third mutualistic partner, a bacterium, the consortium increases the rate of uptake of nutrients by the plant roots. These associations have been found in salt marsh grass species and in the common reed, *Phragmites australis*, usually in the mid to upper intertidal. No mycorrhizal fungi have been found so far among the roots of sea grasses such as eel grass. It may be that the occurrence of low oxygen and marine salinities act in combination to inhibit mycorrhizal growth (Nielsen et al., 1999).

Seaweeds

GENERAL MORPHOLOGY

■ **Seaweeds are eukaryotic, multicellular, photosynthetic, and usually attached to a substratum. They take up nutrients from the surrounding water, and do not have the extensive support structures or other adaptations needed for life in air.**

Seaweeds are considered by many to be allied with single-celled photosynthetic Eukarya but are multicellular, usually attach to a substratum, and are large in size, generally greater than 1 cm long. They vary greatly in form, from amorphous films to elaborate structures meters in length (**Figure 13.4**). Seaweeds must gather nutrients, and they also require light, but they lack most of the other familiar terrestrial constraints we associate with higher plants living in air. Water uptake and transpiration are not a problem except perhaps in higher intertidal species. Seaweeds gather nutrients in solution from the overlying water, and they therefore lack complex root systems. They also live in seawater, which is much denser than air. Hence seaweeds do not have nearly the degree of support tissue found in terrestrial plants. They must grow upward toward the light, of course, but they can do this to some extent by making structures within them

less dense than seawater. In many species, floats filled with gas suspend the plant in the water column.

■ **A seaweed usually occurs as a thallus, consisting of holdfast, stipe, and blade.**

An individual seaweed life-history stage, attached to the substratum, is the **thallus**, which can range in form from a tarlike crust to a thin green sheet to an erect simple filamentous branching structure to a more elaborate organism, complete with highly differentiated structures for light gathering, reproduction, support, flotation, and attachment to the substratum (**Figure 13.5**). Smaller seaweeds tend to be simple chains or sheets of cells, whereas larger forms are characterized by considerable morphological differentiation, serving the combined functions of support, protection against drag, and reproduction. The bulk of the thallus has no distinct anatomical subdivisions, such as vascular tissue, that might be found in a higher plant. Instead, the interior of the thallus usually consists of a complex structure resulting from the growth of cells into many different planes.

The thallus may be attached to the substratum by a **holdfast**. Seaweeds attach to a rocky substratum by means of secretion of a mucilaginous material but may penetrate the minerals of the rocky surface and even erode it by penetrating between mineral grains. Holdfasts may occur as a simple disclike structure, as in *Fucus* (**Figure 13.6**); a group of hairlike structures that are rooted in the sand, as is the case in *Penicillus*; or an elaborate and strong branching structure that attaches to a rocky surface, as in kelps or in the sea palm *Postelsia palmaeformis*.

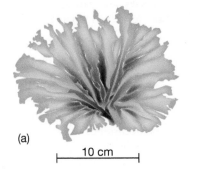

(a)

10 cm

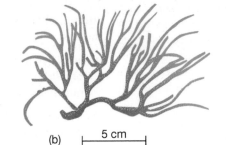

(b)

5 cm

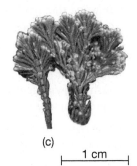

(c)

1 cm

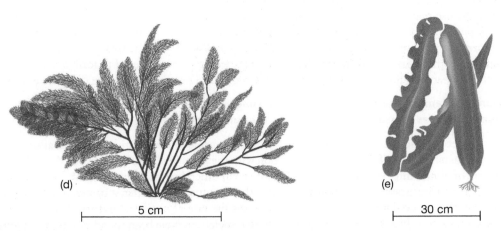

(d)

5 cm

(e)

30 cm

FIG. 13.4   Some of the diversity of seaweed form: (a) the green seaweed *Ulva*, (b) the green seaweed *Codium fragile*, (c) the red coralline alga *Corallina*, (d) the red seaweed *Polysiphonia*, and (e) the brown seaweed *Laminaria*.

FIG. 13.5 The brown seaweed *Durvillaea antarctica* on the coast of Chile, showing holdfast, stipe, and a frond. (Photograph by Consuelo Montero)

FIG. 13.6 The intertidal brown seaweed *Fucus gardneri* often dominates protected rocky shores of the Pacific Coast of the United States and Canada. The bulbous pneumatocysts provide buoyancy. (Photograph by Jeffrey Levinton)

The thallus may be further specialized into individual regions. The **stipe** (see **Figure 13.5**) is usually tubular and flexible; it connects to the holdfast and allows the thallus to bend over in a strong current. The stipe may be quite weak, or it may consist of a complex flexible network in species that live in strong wave surges. A flattened section connected to the stipe, known as the **blade** or **frond**, is specialized for light capture, whereas other parts are specialized for reproduction or flotation. The leafy blades or fronds differ considerably from terrestrial leaves: They are symmetrical and do not have a top side (specialized for light capture) that is differentiated from a bottom

FIG. 13.7 Buoyant pneumatocysts of a number of individuals of the kelp *Nereocystis leutkeana*, floating at the surface, near Bamfield, British Columbia. (Photo by Jeffrey Levinton)

side (specialized for gas uptake and release), as is the case for leaves of terrestrial plants.

In the smallest seaweeds, growth can occur by cell division nearly anywhere in the thallus. Larger seaweeds, however, have specialized zones of cell division, known as **meristems**. Photosynthesis occurs mostly on the blade, but other parts of the thallus also have chlorophyll and can photosynthesize. In some seaweeds, specialized floats, known as **pneumatocysts**, keep the blades of the attached thallus suspended in seawater (**Figure 13.7**). Pneumatocysts occur commonly in kelps, which keep their blades near the surface atop an extremely long stipe, but they are also present in sargassum weed and allow it to remain afloat in its pelagic habitat of the Sargasso Sea.

Combinations of the form elements just mentioned permit seaweeds to adopt a truly amazing range of morphology, ranging from calcareous crusts to filmy growths to massive kelps with stipes several tens of meters in length. Many types of seaweed are also capable of synthesizing a wide array of **secondary compounds** that function in deterring predators (Amsler, 2008; Hay and Fenical, 1988). Many are similar to those found in terrestrial plants and include polyphenolics and aromatic organic compounds. Animal grazers often avoid these species, which may have toxic effects. Some smaller benthic animals hide among chemically defended seaweeds because many grazers will avoid them because of their association with the toxic hosts. The seaweed *Desmarestia* makes sulfuric acid, up to 18 percent of the seaweed's dry weight, which deters urchin grazing.

The great diversity of form makes it difficult to classify seaweed success in particular habitats in simple terms of form variation. Fast-growing seaweeds with fewer structural and chemical defense adaptations tend to be good colonizers of disturbed habitats. Also, crustose seaweeds, such as coralline red algae, are often the survivors after grazers remove the softer and filmier forms. Erect species living in wave-swept or high-current environments tend to have robust and strong thalli.

VARIATION AMONG SEAWEEDS IN PHOTOSYNTHETIC PIGMENTS, LIFE CYCLES, AND STORAGE PRODUCTS

■ **Seaweeds are classified by the different pigments used in gathering light for photosynthesis, by their storage products, and by the types of flagellae in their spores.**

Like most algae, seaweeds have a range of photosynthetic pigments that are specialized to capture light energy over a broad range of the visible spectrum. Different groups of seaweed can be identified by their range of pigments and are commonly known as green, brown, or red algae. Also, carbohydrates are stored in different forms. Although all algae have cell walls composed primarily of cellulose, the main groups of seaweeds can be distinguished by other constituents, especially storage products, which often have commercial value. Finally, spores may be flagellated or not (e.g., red seaweed spores lack flagellae), and number and insertion also differ among the groups. The major differences are summarized in **Table 13.1**.

■ **Seaweeds depend on light for photosynthesis and can adjust to the decrease in light with increasing depth.**

Solar irradiance decreases exponentially with depth. Seaweeds usually cannot survive when light energy is less than 1 percent of the irradiance at the surface. In temperate shelf waters, this is about 15 m, but in clear tropical waters, seaweeds may be found at depths over 100 m. Large brown seaweeds such as kelps usually have far higher light needs to maintain their large thallus and blades.

TABLE 13.1 Some General Characteristics of the Major Seaweed Groups

SEAWEED GROUP	PHOTOSYNTHETIC PIGMENTS	STORAGE PRODUCTS	CELL WALL
Green algae	Chlorophylls *a* and *b*	Starch	Cellulose (not all)
Brown algae	Chlorophylls *a* and *c*, fucoxanthin	Laminarin, mannitol	Alginate
Red algae	Chlorophylls *a* and *d*, phycoerythrin, phycocyanin	Floridian starch	Agar, carrageenan

Seaweeds differ in how they use absorbed light of different wavelengths. The portion of the range of wavelengths of incident light used in photosynthesis is known as the **action spectrum**. Green algae use red and blue light the most and green light the least, whereas red algae use green wavelengths the most. Browns are similar to greens in this respect, with a little more use of the red end of the visible light spectrum.

Not only does the total light irradiance change with depth, but also blue-green light tends to penetrate to much greater depths than does light from the red part of the visible spectrum. It therefore stands to reason that algae might adjust to changing light in two ways. As total light irradiance decreases, algae increase their total pigment concentration. Deeper-dwelling seaweeds tend to have lower compensation light intensities (see discussion of compensation light intensity in Chapter 11) (Johansson and Snoeijs, 2002). The finger-shaped *Codium* is optically black, which means that it can absorb all incident light. It can therefore do rather well at great depth in low light (Ramus et al., 1976) even though its action spectrum makes it appear to be more efficient at higher light. In effect, the thallus has such high concentrations of pigment that it can capture all light.

A second possible means of adapting to changing light levels at different depths would be to alter the proportions of different light-harvesting pigments to adapt to the spectral composition of light as it changes with depth. Seaweeds at depth might be expected to have pigment proportions that would maximize the capture of light under the differing spectral conditions there. The evidence does not support this response by individuals of a given species as an adaptation to depth. However, there are some clear patterns *among* species. Rhodophyta absorb light more efficiently at the green and blue end of the spectrum and tend to live in deeper water than green algae, which tend to absorb more efficiently in the red end of the spectrum and live more frequently in shallower waters.

■ **Seaweeds have a complex life cycle, with differently shaped thallus stages alternating with dispersing stages.**

Like many marine organisms, seaweeds have complex life cycles, often with more than one macroscopic form. The gametes may be motile in seawater, and many species have mobile spore stages, which facilitate dispersal to new sites.

Sexuality and life cycles of seaweeds are very complex. Often, a haploid (*N* chromosomes) generation alternates with a diploid (2*N*) generation. Attached large stages may be haploid or diploid. The **gametophyte** stage produces gametes, which may be motile (as in brown algae) or nonmotile (as in red algae, whose gametes may still be planktonic and hence able to be carried by water currents). Gametes are usually formed and released from gametangia, which may be single cells or elaborate structures. The zygote, formed from the fusion of two gametes to make a diploid **sporophyte**, may be an attached plant or a motile flagellated form. Seaweeds may be monoecious (hermaphroditic) or dioecious (separate sexes).

Figure 13.8 is a diagram of some of the common life histories of benthic seaweeds. As can be seen, there are a great number of possibilities. Most notable is a lack of

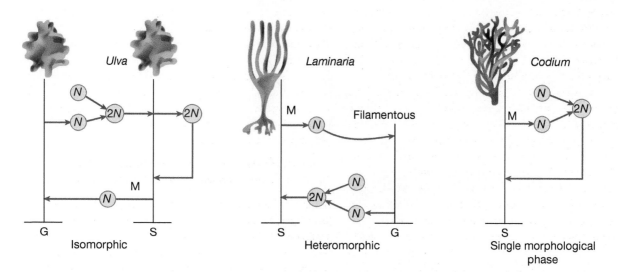

FIG. 13.8 Variation in the life histories of seaweeds. Diploid (2*N*) or haploid (*N*) status is indicated. G = gametophyte; S = sporophyte; M = meiosis.

consistent relationship between haploid or diploid condition and morphological form of the vegetative stages.

■ **Why do seaweeds have complex life cycles in the first place? It may have to do with alternative performance in differing microenvironments.**

There is a great deal of diversity in the relative size and form of the sporophyte and gametophyte stages. In some seaweeds, such as the sea lettuce *Ulva*, the stages are identical. In the Laminariales (kelps), the sporophyte is a large individual, whereas the gametophyte is a small filamentous form. The sporophyte and gametophytes are so different in some seaweeds that members of the same species were once erroneously assigned to different genera. It is not completely clear why such a difference has evolved, but many of the contrasts in forms between sporophyte and gametophyte allow different life stages to dominate in different seasons or to allow resistance against herbivores in one stage and an ability to grow rapidly in another stage. These differences may allow greater exploitation of different microhabitats and may therefore increase the ecological success of the given seaweeds. For example, the non-upright morphs, especially the crustose and boring forms, may be less likely to be grazed. By contrast, the upright morphs might be more vulnerable to grazing by herbivores, but also might be more productive and might produce more offspring when herbivores are less abundant. The combined life history might be a bet-hedging strategy to maximize survival and productivity (Lubchenco and Cubit, 1980).

TYPES OF SEAWEED

Green Seaweeds (Chlorophyta) Green seaweeds have photosynthetic pigments similar to those of higher plants, and they also have the same storage product, starch; indeed, a green algal group may have been the ancestors of higher plants. Most species live in fresh water. There is a fantastic array of morphologies, ranging from single-celled forms (e.g., the flagellated *Chlamydomonas*) to filmy seaweeds (e.g., *Ulva*, **Figure 13.4a**). Along the shore, the sea lettuce *Ulva* is a familiar sight, and it occurs as mats on many sand- and mudflats. The genus *Ulva* has many species, but all share a delicate leafy morphology, which is very susceptible to grazing by snails and other herbivores. *Ulva* has relatively few chemical defenses that deter feeding. Instead, it uses a rapid growth strategy to persist. Filmy green seaweeds are often associated with polluted environments and brackish water. They have a limited ability to store nutrients and so are associated with polluted environments where organic nutrients are in rich supply.

Codium fragile (**Figures 13.4b** and **13.9**) is a common green seaweed that has invaded coastal waters of the northeastern United States, destroying oyster beds by covering and sometimes smothering the oysters. On the large scale of oceans, *Codium fragile* was transported along with oysters transplanted for mariculture. Once arrived in a new area, the seaweed's dispersal is due mainly to the spread of flagellated spores that settle on rocks and oyster shells.

Brown Seaweeds (Phaeophyta) Brown seaweeds dominate low intertidal and shallow subtidal environments in all latitudes. The species with largest thalli seem to thrive in colder-water, nutrient-rich environments. They range from small filamentous forms to kelps, which include the largest seaweeds in the ocean (*Macrocystis*), often longer than 15 m. They include the large kelps (e.g., *Laminaria*, **Figure 13.4e**, and *Macrocystis*; see Chapter 18), sargassum weed, and the intertidal *Fucus* (**Figure 13.6**), which has

FIG. 13.9 The green seaweed *Codium fragile*, attached to a stack of the snail *Crepidula fornicata*, and stranded on a beach on Long Island, New York. (Photograph by Jeffrey Levinton)

FIG. 13.10 Red coralline algae on a rocky shore near Bamfield, British Columbia. (Photograph by Jeffrey Levinton)

bulb-shaped reproductive structures. Brown seaweeds get their color from xanthophylls and carotenes, which may mask the green of chlorophyll in them. Brown seaweeds can be far more morphologically differentiated than green seaweeds and usually have distinct holdfast, stipe, blades, and reproductive structures.

Brown seaweeds contain **phycocolloids**, which are colloidal agents used in many foods and other products for human consumption. The various phycocolloids serve to cushion the seaweed against wave shock and to help in water retention. Many species of brown seaweed are harvested for the extraction of **alginate**, used in toothpaste, pills, and salad dressing. Kelp harvesting of the dominant *Macrocystis pyrifera* is well developed on the west coast of the United States, and grazing sea urchins were once actively destroyed in California to maximize kelp growth. During World War I, kelp was harvested for fertilizer, but potash and acetone were extracted from brown seaweeds to manufacture a smokeless gunpowder known as cordite. Later, harvest was for alginate. Today, kelp harvest in California is greatly restricted and involves clipping of the upper fronds only, which allows a sustainable harvest by permitting regeneration of new fronds.

Red Seaweeds (Rhodophyta) The photosynthetic pigment phycoerythrin gives red seaweeds their color and nearly masks the green chlorophyll. Red seaweeds are found in shallow water and in the intertidal zone, and their phycoerythrin is often bleached by the sun, causing the seaweeds to be any of a variety of colors. They are by far the most diverse group of seaweeds in the ocean and are especially species-rich in temperate and tropical waters. They also are capable of surviving at far greater depths than members of the other seaweed groups.

The life cycle is complex and variable among species, but the common *Porphyra*, the seaweed used in sushi, has three distinct morphological stages. A haploid thallus stage (which is harvested) can produce haploid spores that grow into new thalli, or it can produce male or female gametes. The female gametes are retained on the thallus and are fertilized by motile male gametes. The fertilized carposporangia (diploid, since it is a product of fertilization) after mitosis produces spores that bore into shells and form the filamentous conchocelis stage. The conchocelis stage releases haploid spores that form the thallus stage, completing the cycle. At first, seaweed specialists thought that the very different conchocelis stage was a different species!

Red seaweeds (**Figure 13.4c, d**) have a variety of forms, and their holdfasts may consist of a single cell, spread out into a pad, or of many cells. Some groups (e.g., *Gracilaria* and the Irish moss *Chondrus crispus*) are stringy or filamentous in appearance. *Porphyra* is thin and filmy. The **coralline algae** (e.g., **Figures 13.4c** and **13.10**) secrete calcium carbonate in the cell walls and, depending on the species, grow as upright branching forms or as crusts. In the tropics, some coralline algal species build up wave-resistant structures, and a red algal ridge over a meter high is found on the windward side of many a Pacific coral reef.

Red algae often grow rapidly, and some species contain substances of great value for food production. Carrageenan is a series of gel-forming polysaccharides used as a structural element in many seaweeds. It has been obtained for centuries, mainly from Irish moss, and is used to thicken foodstuffs such as cream cheese, ice cream, and toothpaste. Agar is extracted from *Gracilaria* and other seaweeds and is used as a culture medium for the growth of bacterial strains. Agar is also employed in a variety of foods, including cake icing.

Sea Grasses

■ **Sea grasses are true flowering plants, but the flowers are simplified and the pollen is transported by water.**

Higher plants evolved on land, but descendants of some terrestrial flowering plants were able to reinvade the sea. Sea grasses occur in shallow temperate, subtropical, and tropical waters. Sea grasses, such as the Pacific *Phyllospadix* (**Figure 13.11**), may grow in wave-swept surf, but most others tend to grow in relatively quiet, shallow, subtidal soft bottoms. Their maximum depth is affected by light, so turbidity may greatly reduce their maximum depth. The eelgrass *Zostera marina* is widespread in shallow bottoms throughout the Atlantic, eastern Pacific, and Mediterranean. Other *Zostera* species can be found throughout coastal waters of the world. The turtle grass *Thalassia testudinum* (**Figure 13.12**) dominates shallow soft bottoms throughout the Caribbean. We discuss the ecology of sea grass beds in Chapter 17.

Sea grasses are flowering plants and probably evolved from terrestrial ancestors. The flowers are simple, and their pollen, which may be elongate and specialized to stick to the appropriate target, floats along until encountering a receptive stigma (**Figure 13.13**). Their elongate shape encourages them to rotate while moving in a current, which increases the probability of encounter. Eelgrass seeds move a short distance before setting and germinating, but turtle grass produces a fruit that can move over relatively long distances.

■ **Sea grasses usually extend populations by asexual growth, via a subsurface rhizome system.**

In many sea grasses, sexual reproduction and seed set represent a minor part of population increase. Sea grasses such as eelgrass and turtle grass have a complex **rhizome system** (a series of interconnected stems) that connects shoots beneath the sediment surface. Populations of sea grasses often extend mainly by rhizomal extension. Nutrients are obtained mainly from the pore waters of the substratum, although some sea grasses have symbiotic nitrogen-fixing bacteria in their rhizome systems, which aid in the uptake of nitrogen.

Sea grasses are usually tough, owing to the high cellulose content, and palatability varies greatly among species (See Chapter 17).

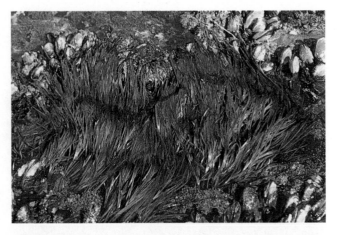

FIG. 13.11 The surf grass *Phyllospadix* sp. on the outer coast of British Columbia. (Photograph by Jeffrey Levinton)

FIG. 13.12 A turtle grass (*Thalassia testudinum*) bed, Florida. (Photo by Jeffrey Levinton)

FIG. 13.13 Pollen release by sea grass, *Zostera marina*. (Photo by Paul Cox)

■ CHAPTER SUMMARY

- All living organisms can be divided by evolutionary affinities based on DNA sequence relationships into three domains: Archaea (single-celled organisms, often living in extreme environments); Bacteria (single-celled organisms, including many forms that break down organic matter and derive energy by oxidative-reductive reactions, transforming nitrogenous compounds, or causing disease); and Eukarya (broad variety of organisms defined by having a cellular nucleus, including protists, fungi, plants, and animals).

- Bacteria are the most numerous organisms in the sea. They lack a true cell nucleus, cell organelles, and mitosis. Bacteria reproduce asexually but may exchange DNA. They are crucial in decomposition. Most are heterotrophic, but a number can photosynthesize or produce energy through oxidation-reduction reactions.

- Cyanobacteria are photosynthetic cells or chains of cells, but the benthic forms usually live in association with anoxic sediments. They are photosynthetic, but many are capable of nitrogen fixation.

- Diatoms are photosynthetic microorganisms whose elongate (pennate) forms dominate the benthos. Their cell wall is impregnated with silica. Reproduction is mainly asexual and involves cell division. Some diatoms are motile and can move several millimeters within an hour.

- Ubiquitous and usually filamentous chains of cells, fungi are very important in the decomposition of particulate organic matter. They are eukaryotic and may occur as single cells or in masses of filamentous cells called mycelia.

- Seaweeds are multicellular forms and usually connect to a substratum. Seaweeds take up nutrients from the surrounding water and do not have adaptations needed for life in air. An individual, or thallus, consists of a holdfast, stipe, and blade. Seaweeds are classified by the different pigments used in gathering light for photosynthesis, by their storage products, and by the types of flagellae in their spores.

- *Chlorophyta* (green seaweeds) have photosynthetic pigments similar to those of higher plants and also the same storage product, starch. They include single-celled and multicelled forms. *Phaeophyta* (brown seaweeds) dominate low intertidal and shallow subtidal environments in all latitudes and include the largest of all seaweeds, the kelps. *Rhodophyta* (red seaweeds, colored by phycoerythrin) comprise by far the most diverse group, including calcareous algal species and filamentous forms.

- Sea grasses are true flowering plants, but the flowers are simplified and the pollen is transported by water. They are in very shallow depths and are strongly light-limited. Sea grasses usually extend populations by asexual growth, via a subsurface rhizome system. Some but not all species are poorly digestible by grazers.

■ REVIEW QUESTIONS

1. What evidence has been used to classify the major divisions of living organisms?

2. Why does it matter ecologically that there are many different types of bacteria?

3. How do fungi participate in cooperative relationships with other organisms?

4. What types of marine habitats have few or no fungi?

5. What are the major body parts of seaweeds?

6. What is the function of pneumatocysts in seaweeds?

7. What traits are used to classify seaweeds?

8. What are some economic uses for seaweeds?

9. By what principal means do sea grasses reproduce?

10. What special interaction do cyanobacteria have with nitrogen in seawater?

11. What is the alternation of morphologies and chromosome number in the life cycles of seaweeds?

Visit the companion website for *Marine Biology* at www.oup.com/us/levinton where you can find Cited References (under Student Resources/Cited References), Key Concepts, Marine Biology Explorations, and the Marine Biology Web Page with many additional resources.

CHAPTER 14

The Diversity of Benthic Marine Invertebrates

There is a fantastic diversity of marine life, especially among the invertebrates, animals without backbones. If you go to a tide pool, without much effort you could find snails and sea slugs (Phylum Mollusca, Class Gastropoda), chitons (Phylum Mollusca, Class Polyplacophora), barnacles (Phylum Arthropoda, Subphylum Crustacea), bivalve mollusks (Phylum Mollusca, Class Bivalvia), worms of various sorts (Phylum Annelida, among others), sea stars (Phylum Echinodermata, Class Asteroidea), perhaps sea urchins (Phylum Echinodermata, Class Echinoidea), and more! Not to mention the seaweeds! This diversity is bewildering, and you may someday want to take a course in invertebrate zoology to study these animals in greater detail. For now, this chapter will serve as an introduction to many of the invertebrate groups you may encounter on the seabed and shores and which figure importantly in the benthic habitat (Chapters 15–19). In this chapter, we will cover invertebrates, down to the taxonomic level of class for the most part, which will give an adequate introduction to marine benthic animal diversity. In **Hot Topics Box 14.1**, we discuss the deep evolutionary structure of the animals and the Cambrian Explosion of animal life about 540 million years ago.

A number of evolutionary changes allowed the evolution of morphological and behavioral complexity. **Grade of construction** is especially important because it indicates the diversity of structure. Some animals merely consist of a **single cell** and, therefore, cannot be very complex or even very large. Others are just **groups of cells**. A few phyla possess a **tissue grade** of construction, wherein the body consists of layers of distinct cell types. This allows for some subdivision of labor among tissues. The most complex grade

of construction is the **organ grade**, where tissues combine to form organs such as a liver or large intestine, which can also be combined into systems of organs. Here there is a large degree of flexibility in using different cell types (skin, secretory, nerve, etc.) in combination to serve a complex function.

Complexity and different degrees of evolutionary transformation can also be found in terms of symmetry and organ types. Many invertebrates are **radially symmetrical** (**Figure 14.1**), which simply means that they possess no true front or rear. A sea anemone is an example of a radially symmetrical invertebrate, which can capture food from all directions. By contrast, most animals are **bilaterally symmetrical** (**Figure 14.1**): They have an anterior-posterior plane of symmetry and have anterior, posterior, dorsal, and ventral parts. Bilaterality allows directional movement and requires processing of information by anterior structures such as eyes. Bilateral forms, therefore, usually have a head with a brain and anterior sense organs.

Another indication of complexity in invertebrates is the presence of a **coelom**, or internal body cavity. A "true" coelom arises within the embryonic mesoderm, whereas a "false" coelom, or pseudocoel, has a different embryonic origin. The embryonic origin of the coelom is important as a tool in inferring evolutionary relationships among groups. Coeloms allow compartmentalization of the body and specialization of function. They are also sometimes used as fluid-filled chambers to form a fluid skeleton, which becomes stiffened by pressure and aids in functions such as swimming and burrowing.

Before proceeding with this chapter, I would like to apologize for omitting many phyla that would only increase

HOT TOPICS IN MARINE BIOLOGY

Where Did All This Invertebrate Diversity Come From?   14.1

A rocky tide pool usually contains a fantastic diversity of creatures. Flatworms, anemones, hydroid cnidarians, crustaceans, snails, bivalve mollusks, annelids, nematodes, sea stars, small fishes, and more can be found with just a bit of effort. Did all these diverse creatures even have a common ancestor, and how we would be able to tell? We also have a rich fossil record of most of these creatures. Did they arise all at once? Unfold slowly? What regulated the pace of their appearance?

Many clues come from traits shared by different groups. Morphological, genetic, and cellular information unites the sponges, Cnidaria, and all higher animals as the Metazoa. The bilaterally symmetrical animals, the Bilateria, share a common ancestor that must have all had a specific set of developmental genes, three distinct tissue layers, and a number of other shared characters. Embryology is a powerful tool for establishing relationships within the Bilateria. The mechanism of early cell cleavage allows most of the invertebrate phyla to be divided into two great groups, the protostomes (including flatworms, annelids, mollusks, and arthropods) and the deuterostomes (including echinoderms and chordates, which include human beings). Within each of these two groups, members may look very different as adults, but their early embryology unites them. The protostomes are nearly all characterized by a peculiar pattern of cell cleavage that makes the embryo asymmetric (**Box Figure 14.1**). The early cleavage in deuterostomes is, by contrast, symmetric about a polar axis. Also, the deuterostomes are distinguished by a pattern of invagination of the early ball of cells, known as the blastula. The blastula invaginates, leading to the gastrula stage, and the opening formed to the outside of the invagination eventually forms the anus. The mouth forms elsewhere on the embryo. In protostomes, the blastopore divides to form both mouth and anus. The sponges are the oddest and most remote of phyla; they are ancestral to all animals, including Cnidaria, Protostomes, and Deuterostomes. They seem to bear almost no relationships to the other phyla and probably arose from a colonial flagellate ancestor. Cnidarians (jellyfish, corals) also are difficult to place using embryological criteria, although they have various tissues (e.g., nerves) that can be related to those of most other invertebrate phyla.

Evolutionary trees of the animals are based mainly on DNA sequences, organized by sophisticated algorithms that group animal phyla into an evolutionary tree, which is still controversial but does retain a basic stability that conforms to the protostome-deuterostome split. The traits discussed in this chapter and many others lead to the evolutionary

tree shown in **Box Figure 14.2**. The ancestor must have been a eukaryote, meaning that it had a true nucleus, nuclear membrane, and organized chromosomes. The protostomes and deuterostomes share a fundamental bilateral body plan, which is only vaguely present in other phyla, such as Cnidaria. They also share a number of important genes that initiate crucial development events, including bilateral and anterior–posterior axes, vision, circulation systems, neurological structure, and the development of a number of other key structures. Within the protostomes, DNA evidence demonstrates that there is a major split between a group of phyla that share an exoskeleton that sometimes allows molting (Ecdysozoa: including Arthropoda, Onycophora, Nematoda), and a second group including a diversity of phyla (Lophotrochozoa: Mollusca, Annelida, Platyhelminthes, Nemertea). The Cnidaria and the other bilateral phyla are united by the presence of radial cleavage, which we conclude to have been transformed evolutionarily into spiral cleavage with the rise of the protostomes. Given that the Cnidaria and deuterostomes both have radial cleavage, we assume that the deuterostomes retained the ancestral condition of cleavage and the protostomes evolved asymmetric cleavage.

The Cnidaria can also be related to the protostomes and deuterostomes on the basis of molecular sequence similarities. The genome of the supposedly primitive sea anemone *Nematostella vectensis* (Putnam et al., 2007) is very complex, and the Cnidaria (grouped with the Ctenophora–comb jellies) are the evolutionary sisters of the lineage containing all the bilaterians. This anemone's genome contains vertebrate-like introns and a gene-linkage pattern also quite similar to vertebrates. Genes involved with cell adhesion, cell signaling, and synaptic transmission are already present in the anemone, suggesting that the genome was already complex and modified in various ways in descendants.

From what ancestral group did most of the invertebrates arise? Our current evidence suggests that sponges are a sister group to the Cnidaria and all bilaterians (e.g., annelids, mollusks, vertebrates). A group called the choanoflagellates are likely the closest relatives to animals and have many expected ancestral traits of the Metazoa, including the lines leading to sponges, cnidarians, and the bilaterians. Choanoflagellates are common in marine and fresh water and are cells that strongly resemble the collar cells of sponges. One would have to argue that at some time in the deep past a group of cells, perhaps all daughters of the same parent cell, kept some contact,

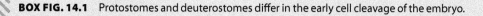

Radial cleavage
(Deuterostomes)

Spiral cleavage
(Protostomes)

BOX FIG. 14.1   Protostomes and deuterostomes differ in the early cell cleavage of the embryo.

continues

Where Did All This Invertebrate Diversity Come From? *continued* **14.1**

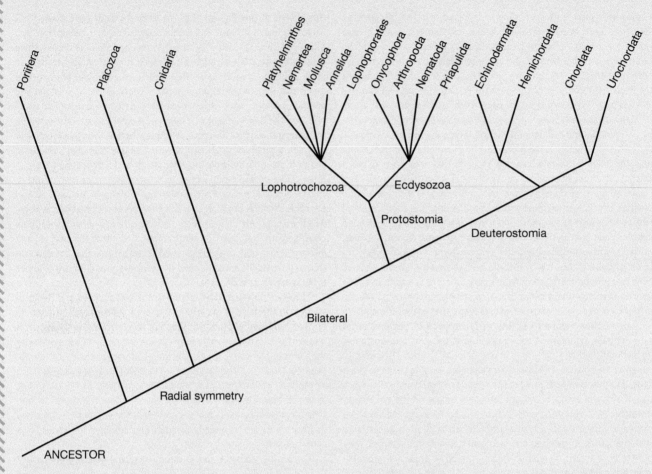

BOX FIG. 14.2 An evolutionary tree of the animal phyla. Labels on the tree indicate traits that unify groups. The lophophorate phyla (brachiopods, bryozoans) are allied through molecular evidence with the protostomes.

and this spurred the rise of a multicellular organism. Cell aggregation would be advantageous in increasing survival by diversifying function in specialized cells. Molecular biologist Nicole King and colleagues (2003) discovered a series of tyrosine kinase genes in some choanoflagellates, which are also found in bilaterian animals. These are important cell-signaling molecules previously thought to be present only in Metazoa. It is likely that this is evidence for an evolutionary and functional relationship between choanoflagellates and Metazoa. There is a viable alternative hypothesis: Choanoflagellates are derived from some group of sponges and are therefore related but not ancestral to sponges.

Others believe that a flat, wormlike organism named *Trichoplax* (Phylum Placozoa; **Box Figure 14.3**) may be a model for an early bilaterian ancestor. It is multicellular and reproduces by fission and by

budding. It can change shape much like an amoeba but is moved along on the substratum by an external layer of cilia.

In recent years, modern DNA sequencing techniques provide an independent perspective on the relationships among the phyla. Those with more sequences in common probably had a more recent common ancestor. By and large, some recent studies have confirmed the relationships established by more traditional methods of comparative morphology and comparative embryology, but some surprises may lie ahead. Many morphological traits could be similar simply because otherwise unrelated organisms evolved in similar environments. Thus far, some interesting conflicts have been generated and resolved by the analysis of DNA sequences. The velvet worm *Peripatus*, classified in the phylum Onycophora, possesses morphological traits linking the annelids and the arthropods. DNA sequencing, however, has allied

Where Did All This Invertebrate Diversity Come From? *continued*   **14.1**

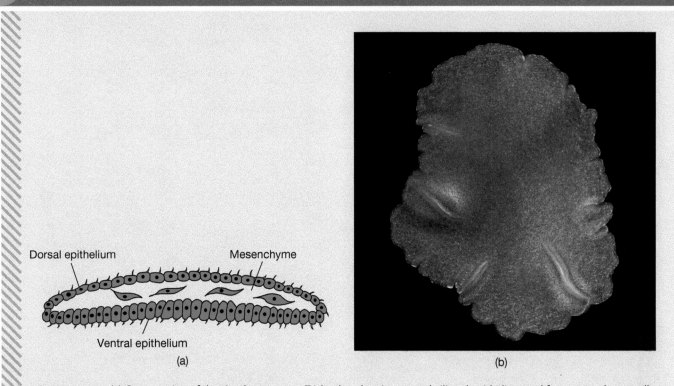

BOX FIG. 14.3   (a) Cross section of the simple metazoan *Trichoplax*, showing ventral ciliated epithelium and free mesenchyme cells. (b) Photo of *Trichoplax adhaerans*. (Photograph by Ana Signorovitch.)

the onycophorans far more closely with the arthropods, which have also been allied to the nematodes. DNA sequencing also placed into question the previous alliance of brachiopods and other lophophorates with deuterostome phyla, which has been the point of view supported by embryological evidence. Most biologists now accept that the lophophorates are protostomes.

DNA-based phylogenies have also given rise to a new perspective on the relationships of the phyla within the protostomes, which can be divided into two groups of phyla mentioned above, the Lophotrochozoa (includes Annelida, Mollusca, lophophorate phyla, Sipuncula) and the Ecdysozoa (includes the Nematoda, Arthropoda; **Box Figure 14.2**). A study by Dunn and colleagues (2008) shows that spiral cleavage arose only once. This type of cleavage in early embryogenesis unites the protostome phyla.

We still cannot be sure that the invertebrates all arose at just one time, but the fossil record tells a very exciting story indeed. Fossils of nearly all the living phyla first appear in the Cambrian period, starting about 542 million years ago. Current radioactive dating evidence suggests that the entire Cambrian explosion of bilateral animal life may have had its beginnings in a period of 12 million years or less in the Early Cambrian. At that time, there was a fantastic diversity of

creatures, perhaps much more so than any later time in the history of life. The great paleontologist C. D. Walcott discovered the Burgess Shale in British Columbia, Canada (**Box Figure 14.4**), in which were the remains of many fantastic and unknown creatures, beautifully preserved as organic coatings. Many other faunas have been discovered in southern China, Greenland, and elsewhere. The fossil evidence, taken at first hand, suggests a "big bang theory" of metazoan evolution. Molecular estimates of the time of divergence of the protostomes and deuterostomes precede the Cambrian by over 100 million years or so. Is there a missing period of Precambrian evolution? The fossil record strongly supports a much more sudden appearance in the Early Cambrian. The interested student should read Erwin and Valentine (2013).

If one goes back a bit further in geological time, one encounters the Ediacaran faunas (**Box Figure 14.5**), first discovered in Australian rocks over 650 million years in age and therefore somewhat older than those of the earliest Cambrian period, which began about 542 million years ago. These have impressions of soft-bodied creatures that have little or no resemblance to anything in the younger Cambrian rocks. Were they an early experiment in evolution, unrelated to the living invertebrates?

continues

HOT TOPICS IN MARINE BIOLOGY

Where Did All This Invertebrate Diversity Come From?   *continued*   14.1

BOX FIG. 14.4   The Cambrian world. The shallow marine environment represented by the Burgess Shale and other Cambrian sites harbored a wide range of the earliest metazoan creatures on earth. (From Levinton, 1992, courtesy of *Scientific American*)

Where Did All This Invertebrate Diversity Come From?   *continued*   **14.1**

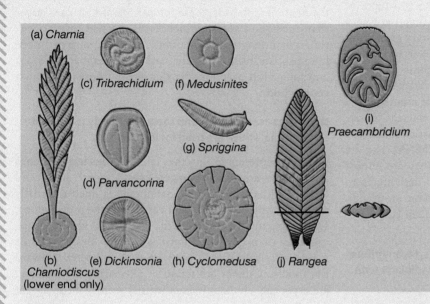

BOX FIG. 14.5   Some of the peculiar soft-bodied fossils of the Ediacaran period, before the Cambrian. Could these be an early "experiment" in evolution, independent of invertebrates living today? (After Glaessner and Wade, 1966)

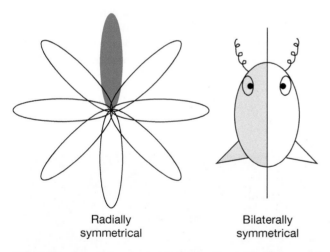

Radially symmetrical

Bilaterally symmetrical

FIG. 14.1   A radially symmetrical organism looks the same in many directions relative to a center (one radial element is filled in blue), whereas a bilaterally symmetrical animal has a front and a rear, an arrangement that promotes the evolution of sensory structures and a head.

the student's appreciation of the diversity of the invertebrates. The worm-shaped Priapulida are relatively minor today, comprising only about 15 species. Yet they have a venerable fossil record, stretching back to the Cambrian period, nearly 525 million years ago. DNA evidence places them within the Annelida. The Loricifera is a relatively recently described living phylum, and it has an eversible

spiny head. Its relationships are poorly understood, but it has some resemblance to nematodes, priapulids, and kinorhynchs (yet another group we will not cover). The most recent phylum to be described is the Cycliophora, tiny animals living among the mouthparts of lobsters. The many phyla provide much information on the diversity of morphology and resemblances among the phyla. The interested student should consult one or more of the invertebrate zoology texts cited in the references list on the book's website, especially Ruppert et al. (2003) and Brusca et al. (2016). The Tree of Life project (www.tolweb.org) is also a very useful resource for finding the latest information on evolutionary relationships.

Kingdom Protista: Single-Celled Organisms

BASIC FACTS

Taxonomic Level: kingdom Protista; grade of construction: in effect, single autonomous cells; symmetry: variable; type of gut: none; type of body cavity other than gut: none; segmentation: none; circulatory system: none; nervous system: none; excretion: diffusion from cell surface.

Other Features: autonomous single cells, reproducing by fission.

Number of Species: over 30,000.

■ **Protists are neither animals nor plants, but free-living, one-celled organisms.**

Protista include amoebas, ciliates, and a host of other organisms that consist of one cell living freely in the environment. All have a nucleus, cytoplasm, several types of cell organelle, and a cellular membrane. They are found abundantly in pore waters among sedimentary grains and are attached to the sediment and to hard surfaces. Most feed on very fine-grained particulate organic matter or bacteria, but some are carnivorous and feed on other Protista. Some ciliates (**Figure 14.2a**) have mouth openings lined with cilia that trap particles, whereas amoeboid forms surround and engulf food particles. Protists also include a number of macroalgae. The Protists are a difficult taxonomic grouping because they are almost certainly not a unique evolutionary group deriving from a single ancestor. Other competing taxonomic groupings, such as the Protozoa, have the same problem (Cavalier-Smith, 2003).

■ **Amoeboid forms include naked amoebas, Foraminifera with calcium carbonate skeletons, and Radiolaria with silica skeletons.**

Amoebas and their allies all share the ability to stream cellular protoplasm and to form extensions, or **pseudopodia**, which can surround and engulf food items. Forms lacking an external skeleton can adopt many shapes and can move in many directions. They usually live in pore waters or on the sediment surface. Foraminiferans (**Figure 14.2b**) construct beautiful chambers of calcium carbonate, which may be laid out in a row or in a spiral. The chambers may also be ornamented with spines, which deter predators from ingesting the organisms. Foraminiferans are usually less than 1 mm in length—but some such as the deep-sea xenophyophorids are colonial and quite large

(**Figure 14.3**). Protoplasm streams out through holes in the hard-walled chambers, and food particles that stick to the pseudopodia are captured and engulfed. Some foraminiferans have symbiotic algae, known as zooxanthellae, which transfer sugars to their protistan hosts. Like the xenophyophorids, these foraminiferans also are larger than most, and some extinct forms reached several centimeters in length. Radiolarians, another group of amoeboid protists, have elaborate hard skeletons made of silica and long, thin pseudopodia that extend outward from a central body.

■ **Ciliates are elongate, and their outer body is covered with cilia, which are used for locomotion.**

Ciliates, represented by the familiar freshwater genus *Paramecium*, are usually elongate, with a mouth at one end (**Figure 14.2a**). The cellular membrane is lined externally with cilia, whose coordinated beats propel the protist along. Depending upon the species, the mouth is of varying shape and may or may not be lined with cilia. Beating of the mouth cilia brings food particles such as bacteria toward the mouth. Ciliates are found in all marine habitats and are especially abundant as interstitial animals in sediment pore waters.

■ **Flagellates also are elongate, but they are propelled by one or a few flagellae instead of by many cilia.**

Flagellates are similar to ciliates, except long flagellae, usually very few in number, are embedded externally in the cellular membrane, and beat to move the organism. Like ciliates, flagellates usually have a mouth and consume bacteria or small organic particles. Some are carnivorous and feed on other protists. Many photosynthetic phytoplankton are flagellates, and these are discussed in Chapter 8.

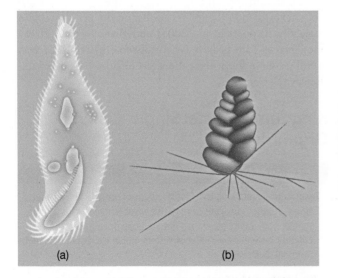
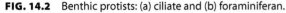

FIG. 14.2   Benthic protists: (a) ciliate and (b) foraminiferan.

FIG. 14.3   A xenophyophorid, a foraminiferan from the deep sea that agglutinates sedimentary particles. This one can be seen in the core (7 cm diameter) from which it was extruded. (Photograph courtesy of Lisa Levin)

Phylum Porifera: Sponges, Simplest of Animals

BASIC FACTS

Taxonomic Level: phylum Porifera; grade of construction: cellular, with no distinct tissues or organs; symmetry: variable but usually asymmetrical or radial; type of gut: none; type of body cavity other than gut: none; segmentation: none; circulatory system: none; nervous system: none; excretion: diffusion from cell surface.

Other Features: flagellated cells called choanocytes drive water through pores and cavities.

Number of Species: 8,000.

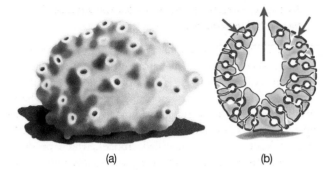

(a)   **(b)**

FIG. 14.4 Simplified diagrams of a sponge: (a) a whole animal and (b) a cross section of a colony, showing chambers lined with collar cells.

■ **Members of the phylum Porifera possess structures consisting of groups of flagellated cells, which move water and food particles into open chambers.**

Sponges, which comprise the phylum **Porifera**, are extremely simple animals that lack organs, do not have a gut, and have poorly developed tissue layers. They have several different specialized cells that are grouped to perform different functions. Sponges have an internal space, which is made up of a series of pores, canals, and chambers (**Figure 14.4**). Simple flagellated **collar cells**, or **choanocytes**, line the chambers and beat water through the entire system. The simplest of sponges is a two-layered animal built around one chamber, which is lined with collar cells. The outer part of the animal is a layer of epithelial cells interspersed with pore cells, through which water flows with food particles to the flagellae of the collar cells. Food particles, principally bacteria, are trapped on the flagellae and engulfed by the collar cells. Another cell type, the amoebocyte, is involved in transport of digested food to other cells in the sponge body. Amoebocytes also can transport waste products. While nearly all sponges consume bacteria and smaller phytoplankton, such suspended food is very scarce at deep-sea bottoms. A highly modified group, the Cladorhizidae, are adapted to consume much larger prey than other sponges, and trap small crustaceans by means of filaments bolstered by hook-shaped spicules. The evolutionary changes involved in adopting this new morphology included the loss of the typical sponge system of pores, canals, and chambers, including the loss of choanocytes (Vacelet and Boury-Esnault, 1995).

Although water enters through hundreds or thousands of pore cells, it leaves the central chamber through one exit, called the **osculum**, which may occur singly or as several openings. The action of thousands of collar cells moves water and concentrates it so that the wastewater leaves at a fairly high velocity through the osculum and is carried away by water currents (see Chapter 6 for a discussion of the water transport of sponges). All sponges (except one small carnivorous group) either have a simple one-chamber design or consist of hundreds to thousands of chambers in a complex array. Whereas some sponges consist of tiny filmy colonies, many are quite large and have distinct forms. Basket sponges, for example, which have a distinct vase shape, may be a meter high. For support of the colony, sponges secrete a fibrous protein known as **spongin** between the two main cell layers. Many sponges also secrete interlocking networks of needle-shaped **spicules**, which may be siliceous or calcareous. Different spicule types can be used to identify the sponge.

Sponges are capable of both asexual and sexual reproduction. Colonies can extend themselves over a surface by means of asexual reproduction. The relatively low level of organization allows extension of the colony in any direction. The eggs are usually retained within the body, to be fertilized by planktonic sperm. On coral reefs, whole sponge populations have been observed to produce a fog of sperm over the bottom, timed with the lunar cycle.

Sponges may seem to be easy targets for a predator, but many contain very toxic compounds that are distasteful and that can harm an attacker. Some of these compounds, especially those produced by some tropical sponges, are caustic to the touch. Nevertheless, sponges are often consumed by fishes.

Phylum Cnidaria: Hydrozoans, Jellyfish, Anemones, and Corals

BASIC FACTS

Taxonomic Level: phylum Cnidaria; grade of construction: two tissue layers; symmetry: radial; type of gut: blind gut; type of body cavity other than gut: none; segmentation: none; circulatory system: none; nervous system: network of nerve cells; excretion: diffusion from cell surface.

Other Features: two basic stages—sessile polyp and swimming medusa, both tentaculate; includes anemones, corals, sea whips.

Number of Species: 9,000.

- **The cnidarians are all built around a common cup-shaped polyp body plan, which has a ring of tentacles and a digestive tract with one opening.**

Although the cnidarians have a large range of morphologies, all are built around a basic individual, a **polyp**. Although some are highly specialized, most are cup shaped and have a single mouth–anus opening, surrounded by a ring of tentacles. The tentacles are usually lined with several different types of **nematoblast** (also called cnidae), which produce structures called **nematocysts**, which are coiled within the nematoblast cell. Upon contact with prey, a mechanoreceptor on the nematoblast is stimulated and the nematocyst everts explosively, thrusting either a sticky extension or a sharp barb at the prey. Species may have more than one type of nematocyst. The nematocysts may be: (a) mucus coated and entrap the prey; (b) long and thin and used to lasso prey; or (c) harpoon-shaped and used to stab prey and inject toxins, which can stun and even kill prey animals. After the nematocysts have everted and made contact with prey, the tentacles draw the food item through the mouth and into the digestive cavity, where digestive enzymes are secreted. After a period of time, digested material crosses the gut cavity wall, and undigested remains are extruded back through the mouth.

This basic body plan (**Figure 14.5**), a **polyp**, can be used by attached benthic creatures, such as sea anemones, whose tentacles extend upward into the water column. The plan also serves, however, in free-swimming jellyfish in a **medusa** form; the tentacles of the animals hang down, and the body is free living; muscular, rhythmic contractions allow a medusa to expel water downward and thus swim upward. In groups like corals and sea anemones, the polyp is the main form in the life cycle, whereas in the true jellyfish, the medusa form is the main adult stage. In the hydrozoans like *Tubularia* (**Figure 14.6**), the life cycle consists of an alternation between small colonies of polyps and small medusa jellyfish stages (**Figure 14.7**). The jellyfish stage produces gametes, which form zygotes in the water column and develop into larvae, which settle and produce the polyp stage.

Many cnidarians are colonial. In the case of corals, some anemones, and the polyp stages of hydrozoans, all the polyps

FIG. 14.6 The hydroid *Tubularia*.

are essentially alike and have feeding tentacles. Many cnidarians, however, are polymorphic, and different polyp types perform functions that contribute to the functioning of the colony. Hydrozoans often have separate feeding and reproductive polyps. In the feathery *Plumularia*, the reproductive polyps are found along a central axis, and the feeding polyps are found on the projections from the central axis. The Portuguese man-of-war *Physalia physalis* (see Figure 8.12h) belongs to a group known as the siphonophores and has several distinctive types of polyp. One highly specialized polyp forms a float, from which the rest of the colony dangles into the water beneath. A disk underneath contains a common digestive cavity used by the whole colony.

- **Cnidarians are divided into Hydrozoa, Scyphozoa (true jellyfish), and Anthozoa (corals, anemones).**

The class **Hydrozoa** are the simplest cnidarians, and most have a complex life cycle, with an alternation of a benthic colony and a planktonic small-jellyfish medusa stage (**Figure 14.7**). The benthic colonial stage may be polymorphic and may consist of both feeding and reproductive

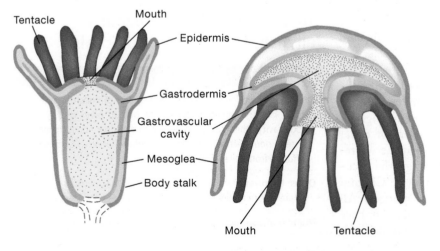

FIG. 14.5 Basic body plans of a cnidarian, including the polyp (left) and medusa (right) stages.

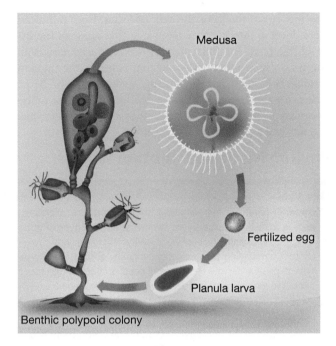

FIG. 14.7   Life cycle of a typical hydrozoan.

FIG. 14.8   A sea anemone, *Anthopleura xanthogrammica*, member of the Anthozoa, about 8 cm in diameter. (Photo by Jeffrey Levinton)

polyps, all connected by living tissue. The polymorphism is developed especially well in species of *Hydractinia*, which usually lives as a coat on the snail shells occupied by hermit crabs. Polyps are specialized into feeding polyps, spines, reproductive polyps, and protective dactylozooid polyps. Some hydrozoans are bush shaped, whereas others live as layers of polyps, attached to rocks or seaweeds. In a typical hydrozoan, reproductive polyps bud off small jellyfish-shaped individuals, which are known as medusae, and these medusae spawn gametes into the water column. The gametes form a larva called a **planula** (**Figure 14.7**), which swims for a period and settles to form the benthic colony stage.

The class **Scyphozoa** are true jellyfish, which were discussed more completely in Chapter 8. They look like a very enlarged version of the medusa stage of a hydrozoan, although their digestive cavity is far more complex: They are much larger (sometimes a meter in diameter), and the periphery of the body is often covered with sensory structures. The jelly layer is also far thicker than in hydrozoan medusae. Unlike the Hydrozoa, planktonic scyphozoans have a life cycle that usually has only a very small benthic polyp stage. Jellyfish shed gametes into the water and the embryo develops into a planula larva, which settles on the bottom to form the diminutive benthic stage. The benthic stage, in a process called strobilation, produces immature medusae, which swim off into the water column and mature into adult jellyfish.

Scyphozoans are all carnivores; they are propelled by rhythmic contractions of muscle rings that force water out of a structure known as a bell, named after its shape. After the contraction, the animal starts sinking, and tentacles draping from the bell can touch prey and fire off the nematocysts. In many species, nematocysts merely ensnare prey with mucus, but a number of species can inject virulent toxins. Another class, the Cubozoa, includes the sea wasps and box jellyfish, that are distinctly square in cross

section. Some species are among the most virulent species in the sea and can kill humans with their toxic nematocysts. *Chironex fleckeri* has a bell approximately 30 cm wide and occurs in the northern coastal waters of Australia; its venom is strongly neuro- and cardiotoxic to humans.

The **Anthozoa** include the anemones, corals, and sea fans (**Figure 14.8**). In these groups, the polyp stage dominates the life cycle, and there is no medusa stage. Gametes are shed by the animals, fertilization occurs, and the resulting embryos develop into swimming planula larvae. These larvae settle and metamorphose into the polyp stage, which feeds on zooplankton. Anthozoans have an element of bilateral symmetry because a slitlike mouth is elongate. There is still a ring of tentacles, however, that catch differently sized prey depending upon tentacle morphology and overall polyp size. The tentacles surround the polyp and have radial symmetry. Many anemones are solitary, though some consist of many hundreds of individuals that arise by fission from one original animal. Stony corals and soft corals usually consist of hundreds to thousands of polyps, interconnected by soft tissue. Stony corals secrete calcium carbonate and may have quite large and branching skeletons, or may be mound shaped. The great diversity of stony-coral forms gives coral reefs their spectacular appearance (see Chapter 17).

Phylum Platyhelminthes: Flatworms

BASIC FACTS

Taxonomic Level: phylum Platyhelminthes; grade of construction: organs derived from three tissue layers; symmetry: bilateral; type of gut: blind; type of body cavity other than gut: none; segmentation: absent; circulatory system: none; nervous system: small bundles of nerves (ganglia), two ventral nerve cords; excretion: excretory organs in many species.

Other Features: flattened free-living worms, often with tubular pharynx to gather food; also, many parasitic species; representatives include the common freshwater planaria and the tapeworms.

Number of Species: 12,000.

■ **Flatworms are truly bilaterally symmetrical, with anterior–posterior differentiation. They also have distinct organs.**

Flatworms represent a major transformation in evolution. They are bilaterally symmetrical, which means that their body exists in mirror images about a long anterior–posterior axis. They also have distinct anteriors and posterior, a morphology that gives the opportunity for consistently directed forward motion. Such a motion requires the development of a battery of sensory structures for smell, sight, and touch to allow for interpretation of the environment as the flatworm moves along. As the name indicates, the flatworms are wormlike, although the free-living forms are usually flattened. Movement is accomplished partially by means of ventral cilia in the small forms but also by rhythmic contractions of an outer layer of circular muscle and an inner layer of muscle that parallels the line of the body. Unlike the cnidarians, flatworms have distinct organs, or structures that consist of more than one kind of tissue. Examples are the excretory organs. Unlike the cnidarians, flatworms have tissues and organs that derive from three original cell layers (endoderm, mesoderm, and ectoderm) instead of from two. Flatworms have a central nervous system, with concentrations of nerve cells in the head that could be considered a brain.

Free-living flatworms (**Figure 14.9**) are flattened and usually have a muscular pharynx, which can be protruded to suck fluids from prey such as crustaceans and annelids. The animal has a blind digestive cavity, and undigested material must be egested from the same opening through which food enters. The larger forms have eyespots and sensory tentacles. Flatworms are found in crevices, under rocks, and sometimes on bare sediment surfaces. While many are cryptically colored, many tropical species are brightly colored to advertise danger and contain toxic defense compounds.

The marine flukes parasitize many vertebrates and invertebrates. Many have complex life cycles with several hosts (see Figure 4.11). Many flukes have one stage that infests a mollusk or annelid. This stage produces free-swimming individuals that enter the water column and may burrow into the legs of ducks or the skin of fishes. Eggs produced there develop into another free stage that reenters the invertebrate host. The fluke *Austrobilharzia variglandis* is an example of such a flatworm fluke that burrows into the skin of ducks or human bathers, causing "swimmer's itch." More serious are the flukes causing the debilitating disease schistosomiasis. The fluke *Schistosoma mansoni* moves between humans and a freshwater snail as hosts.

Phylum Nemertea: Ribbon Worms

BASIC FACTS

Taxonomic Level: phylum Nemertea; grade of construction: organs derived from three tissue layers; symmetry: bilateral; type of gut: complete, with anus; type of body cavity other than gut: rhynchocoel surrounding proboscis; segmentation: absent; circulatory system: present; nervous system: small bundles of nerves (ganglia), two nerve cords; excretion: excretory organs in many species.

Other Features: elongate free-living worms, with complete gut; carnivorous, using barbed proboscis to kill prey.

Number of Species: greater than 800.

■ **Ribbon worms have a proboscis and a complete gut and are mobile carnivores that burrow through the sand.**

Nemerteans are long, flat, carnivorous worms with remarkable powers of contraction. Some species can extend to over 10 m in length and can rapidly contract to just a few centimeters. They resemble flatworms to a degree and have an externally ciliated body; but they have a complete gut, with an anus, a circulatory system, and a pointed anterior with an elaborate proboscis for catching prey (**Figure 14.10**). Ribbon worms move with the aid of ciliary activity or by coordinated contractions of the body wall muscles.

When the animal is inactive, the proboscis is inverted and housed in a cavity separate from the mouth. The proboscis is surrounded by a fluid-filled cavity known as a rhynchocoel. When the ribbon worm detects a possible victim, the muscles contract the rhynchocoel and fluid pressure shoots out

FIG. 14.9 A free-living flatworm, *Prosterceraeus vittatus*, which lives in northern European waters. (Courtesy of Peter Glanvill)

FIG. 14.10 A large (>1 m long) nemertean worm *Parborlasia corrugatus* found on the Antarctic shelf. (Courtesy of Richard Aronson)

the proboscis, which is usually armed with a stylet that punctures the prey and injects a venom. The proboscis can also draw the prey to the worm's mouth.

Phylum Nematoda: Roundworms

BASIC FACTS

Taxonomic Level: phylum Nematoda; grade of construction: organs derived from three tissue layers; symmetry: bilateral; type of gut: complete; type of body cavity other than gut: pseudocoel; segmentation: absent; circulatory system: present; nervous system: small bundles of nerves (ganglia), two nerve cords; excretion: special excretory cells.

....

Other Features: small free-living and parasitic worms with only longitudinal muscles, circular in cross section; covered by cuticle; both free-living and parasitic.

....

Number of Species: 12,000.

....

■ **Many nematodes live free in all marine environments, using longitudinal muscles that work antagonistically against a fluid-filled body with a rigid wall. Free-living nematodes may be carnivorous or plant eating; some consume organic matter from sediment.**

Roundworms (**Figure 14.11**) are among the most widespread of all marine invertebrates and often have population densities of millions per square meter of soft bottom. The free-living forms are generally small, usually less than 3 mm in length. They are cylindrical and have a rigid outer organic case known as a cuticle. Growth cannot be continuous because of this structure, and the animal sheds, or molts, the cuticle four times in the adult life cycle. Unlike flatworms

and ribbon worms, roundworms have longitudinal muscles only. These muscles work antagonistically against the body, which is made rigid by the cuticle and a fluid-filled cavity. The fluid-filled cavity is known as a pseudocoel because it is not located within the mesoderm, like a true coelom. The working of the longitudinal muscles pushes against sand grains and moves the animal along. Out of the sediment, the worms appear to thrash about wildly.

Depending on the species, nematodes have a wide variety of feeding habits, reflecting the different teeth and rods that project from the cuticular hard surface of the mouth. Some species can pierce algal cells and suck out the juices, and others can consume tiny invertebrates. The latter have the best developed teeth. Other species ingest mud and absorb organic matter from the sediment. When considering molecular sequence data, Nemertea are closely allied with Annelida and Mollusca.

Phylum Annelida: Segmented Worms

BASIC FACTS

Taxonomic Level: phylum Annelida; grade of construction: organs derived from three tissue layers; symmetry: bilateral; type of gut: complete with anus; type of body cavity other than gut: coelom; segmentation: present; circulatory system: closed system; nervous system: brain, with nerve cords and bundles (ganglia); excretion: excretory organs in most segments.

....

Other Features: segmented worms, with great diversity of head and locomotory appendages; includes earthworm, sandworm, and lugworm.

....

Number of Species: 12,000.

....

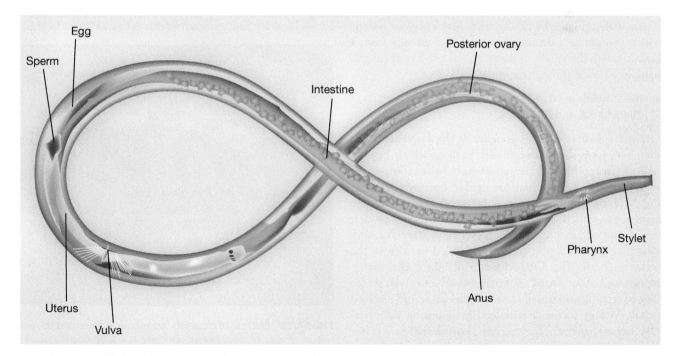

FIG. 14.11 Cutaway diagram of a typical free-living nematode; such animals are found in densities of millions per square meter of mud. (After Barnes, 1987)

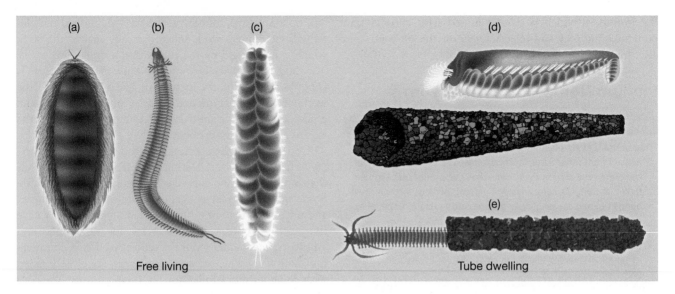

FIG. 14.12 Some of the diversity of polychaete forms: (a) *Aphrodite*, (b) herbivorous *Nereis vexillosa*, (c) *Harmothoe*, (d) deposit-feeding *Pectinaria* with tube of sand grains, and (e) *Onuphis*.

■ Annelids are worms divided into segments, with a tubular gut from mouth to anus and three distinct embryonic tissue layers.

Annelids include earthworms, marine polychaetes, and leeches, and all share a wormlike form, with a mouth and anus connected by a tubular gut. The body is divided into distinct **segments**, and digestive glands, reproductive organs, and locomotory appendages are usually repeated in each segment. The segmentation of musculature allows localized contraction, and the worm can crawl and even swim by sequentially moving appendages on individual segments. The body contains an interior fluid-filled space, or coelom, and the body wall muscles act against this fluid, which serves as a hydrostatic skeleton. In most of the annelids, only the head area and anal segments are differentiated (**Figure 14.12**). Annelids have a nervous system, which consists of a double nerve cord that extends from a brain through the length of the body. The nerve cord coordinates locomotion through the entire body.

■ The annelids are divided into the Polychaeta, Oligochaeta, and Hirudinea.

The most diverse class of annelids is the **Polychaeta**; its members have distinct segments, each of which bears a pair of **parapodia**. The parapodia have bristles known as setae (**Figure 14.13**), which are used in burrowing, crawling, and swimming. Some polychaetes have elaborate head areas, which may have a specialized proboscis that everts and seizes prey. Some polychaetes have feathery gills that have cilia and collect suspended food from the water column. The **Oligochaeta**, including the earthworms and other forms abundant in salt and fresh water, lack parapodia and usually have smaller setae and reduced heads. Marine forms usually feed by ingesting sedimentary organic matter. The leeches (**Hirudinea**) lack setae and use external suckers to attach so that the worm can move along by muscle contraction in the body wall. Most leeches are parasites that use their suckers to feed on a host's body fluids.

■ Annelid locomotion depends on layers of longitudinal and circular muscles working against a rigid fluid, compartmentalized among many segments.

In most annelids, the body wall consists of an outer cuticle and inner skin. Beneath is a circular muscle layer and a longitudinal muscle layer. By combining contractions, these two muscle layers work in opposition against the fluid-filled segment and can change its shape. In the polychaetes, this action moves the parapodia, whose setae press against the substrate and move the worm in a tube or burrow or along the surface. As the body wall muscles contract, the worm must also coordinate the protrusion and withdrawal of the parapodia and setae. Lacking parapodia, oligochaetes use setae and wriggling to burrow through the sediment. The segmentation allows localized movements, but this is absent in the leeches.

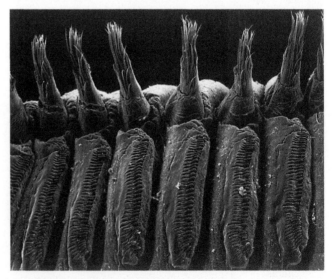

FIG. 14.13 Electron micrograph (ventral view: anterior is toward left of photo) of *Sabellastarte magnifica*, a tube-dwelling polychaete. The long capillary setae (protruding upward from the parapodia) aid the worm in moving along within its tube. The shorter hooked setae below are used to hook into the tube wall to prevent removal by a predator. (Courtesy of S. A. Woodin)

- **Annelids adopt a wide range of living positions and include free burrowers, infaunal tube dwellers, and epifaunal tube constructors. They may be carnivores, deposit feeders, and suspension feeders.**

The form and locomotory abilities of annelids have allowed them to evolve a wide range of living habits. Many annelids live freely in the sediment and may either be specialized carnivores or sediment-ingesting deposit feeders. Species of the sandworm genus *Nereis* have an eversible proboscis, which in some species is used for seizing prey and in others for tearing and ingesting fragments of seaweeds. A large number of species live in tubes or nearly permanent burrows. The western North Atlantic bamboo worm *Clymenella torquata* may be up to 25 cm in length and lives in groups, head down in a mud tube. The animal ingests sediment at depth and defecates on the sediment surface. Tube dwellers move rhythmically to irrigate the burrow and get rid of waste products and draw in fresh seawater.

A large number of species of annelid live in tubes but protrude a feeding organ into the water column to collect plankton. *Spirorbis borealis* secretes a small spiral tube that is cemented usually to the fronds of seaweeds. Ciliated tentacles protrude from the tube opening, and the cilia beat and drive water past the tentacles. Phytoplankton are collected and transported on a ciliated groove to the mouth. Species of the parchment worm *Chaetopterus* make a U-shaped, paper-like tube and create a current that is passed across a sheet of mucus stretched between a pair of specialized parapodia (see Figure 15.12a). This worm has extremely specialized body parts with very different appendages. It is so delicate that it cannot possibly live outside the tube.

Phylum Sipuncula: Peanut Worms

BASIC FACTS

Taxonomic Level: phylum Sipuncula; grade of construction: organs derived from three tissue layers; symmetry: bilateral; type of gut: complete with anus; type of body cavity other than gut: coelom; segmentation: absent; circulatory system: large coelomic cavity bathes most tissues; nervous system: brain with single ventral nerve cord, with branches; excretion: paired organs.

Other Features: worm shaped, living in sediment and among rocks, feeding on sediment with protrusible organ called an introvert.

Number of Species: about 300.

- **Peanut worms live in burrows in soft sediment and in rock crevices. They gather food by means of an introvert that has branched tentacles.**

Sipunculans are wormlike but are not segmented. They can be found in soft sediments and in crevices, but they are commonly seen living in the mud trapped in empty snail shells (**Figure 14.14**). They use fluid pressure in a large coelomic cavity to protrude the introvert, whose tip usually contains branched tentacles for feeding. Peanut worms feed on the organic matter in sediment. The introvert can be rapidly withdrawn by muscle contraction, giving the

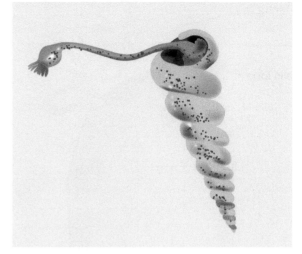

FIG. 14.14   A sipunculan, or peanut worm, protruding from the mud in a snail shell.

animal a bulging appearance. Molecular data suggest that Sipuncula are located within the group Annelida phylogenetically. This would suggest that ancestors of sipunculans were segmented but that segments were lost in the evolutionary lines leading to Sipuncula.

Phylum Pogonophora: Gutless Wonders

BASIC FACTS

Taxonomic Level: phylum (might properly belong within the phylum Annelida) Pogonophora; grade of construction: organs derived from three tissue layers; symmetry: bilateral; type of gut: none; type of body cavity other than gut: coelom in several sections; segmentation: present in one region; circulatory system: dorsal and ventral blood vessels, with part of the dorsal vessel muscularized into a heart; nervous system: brain with rudimentary nerve cords; excretion: paired organs.

Other Features: solitary worm-shaped individuals with no gut; rely on symbiotic bacteria for nutrition.

Number of Species: about 100.

- **The members of the phylum Pogonophora are generally deep-sea species that lack a gut and depend on symbiotic bacteria.**

If they did not exist, the Pogonophora would have to be invented to satisfy our imagination because they surely contradict our intuition of what an animal should be. DNA evidence suggests that they should not be an evolutionary separate clade, because reconstructed trees place them within the phylum Annelida. Generally, they are long and slender worms (**Figure 14.15**). An anterior head bears sometimes as many as 200 tentacles that protrude into the water. This is set aside distinctly from a trunk region, whose anterior is ringed with hard plates, somewhat similar to the setae of annelids. A third section is segmented, and each segment has several stiff rods similar to setae. This arrangement allows the animal to anchor itself in a tube.

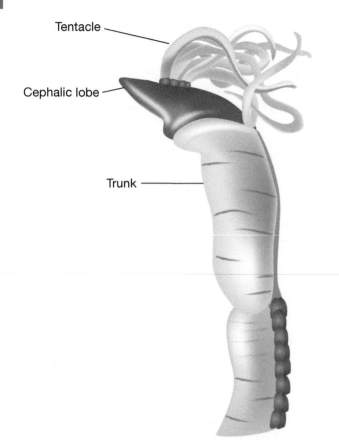

Tentacle

Cephalic lobe

Trunk

FIG. 14.15 A pogonophoran worm. (After Meglitsch and Schram, 1991)

The most conspicuous feature is what is missing: a gut. For years, zoologists thought that pogonophorans could survive only by absorbing dissolved organic matter from seawater, and DOM may well be part of the nutrition of this animal. However, it was discovered that a part of the trunk, the **trophosome**, contains large numbers of internal symbiotic bacteria. The animal may derive nutrition by digesting the bacteria or, alternatively, dissolved substances may leak from the bacteria.

Recently, gigantic relatives of pogonophorans were discovered near hot vents adjacent to volcanically active parts of the seafloor (see Chapter 18). These **Vestimentifera** can be as much as a meter in length and a centimeter in diameter. They secrete tubes, often 2 m long. They, too, have symbiotic bacteria living within the body. The bacteria are chemosynthetic and depend on sulfide delivered by the worm's circulatory system, which contains a special hemoglobin with a binding site for sulfide. The worms have a ciliated free-living trochophore larva, whose morphology suggests a possible evolutionary relationship with the annelids.

While Pogonophora and Vestimentifera are morphologically distinct, it is not so clear that they are separate evolutionary lineages. Molecular phylogeny analyses suggest that these groups are part of the evolutionary lineage defined by the annelids (Halanych et al., 2002), and all are grouped now in an annelid family, the Siboglinidae (Pleijel et al., 2009). This is a general theme for many groups formerly considered separate phyla such as Sipuncula.

Phylum Mollusca: Shelled Invertebrates (Mostly)

BASIC FACTS

Taxonomic Level: phylum Mollusca; grade of construction: organs derived from three tissue layers; symmetry: bilateral; type of gut: complete, with anus; type of body cavity other than gut: coelom; segmentation: absent; circulatory system: usually open to large coelomic cavity; nervous system: brain, with nerve cords and bundles (ganglia), brain very well developed in squids and relatives; excretion: excretory organs.

Other Features: typically externally shelled, mantle secretes shell; respires with ctenidium; includes clams, snails, squids, octopus.

Number of Species: more than 100,000.

■ **Mollusks have a head–foot complex, a mantle that usually secretes a calcium carbonate shell, and a gill, suspended in a mantle cavity, which is used for respiration and commonly for suspension feeding.**

The mollusks are one of the most successful phyla, and there are over 100,000 living species. All are believed to have evolved from a primitive form with a head–foot complex, a cap-shaped shell, and a posterior gill that could be used both for respiration and to collect phytoplankton on ciliated tracts. The mollusks have a coelom, or fluid-filled body cavity, used in clams as a hydraulic device to burrow into the sediment. Most have a shell, secreted by the mantle. From this basic form sprang the supposedly ancestral **Monoplacophora**; the **Gastropoda** (snails), with their usually coiled shell and varied feeding types; the **Bivalvia** (clams, oysters, and mussels), with usually two symmetrical shells and a variety of burrowing and epifaunal forms; the **Cephalopoda** (squids and octopus), with arms, and the ability to move rapidly; the **Polyplacophora** (chitons), which have a flattened foot; and the **Scaphopoda**, which have a tusk-shaped single external shell.

■ **Members of the class Bivalvia are distinguished by two symmetrical shells connected in a hinge region. The mantle secretes the shell, and a gill, or ctenidium, usually helps in respiration and collects phytoplankton on ciliated tracts.**

The Bivalvia (**Figure 14.16**) are distinguished by two shells, or valves, usually mirror images of each other, hinged together by calcified teeth and by a tough organic ligament, which tends to spring the shell open. Muscles counter this springing force and draw the valves closed. The mantle lines the valves and secretes new shell. Most bivalves have a powerful foot, which can be filled with fluid. The foot probes into the sediment and expands, whereupon the shell is drawn in behind. In some epifaunal forms such as mussels, the foot is strongly reduced and merely secretes byssal threads that glues the animal to the bottom.

Bivalves have a **mantle cavity**, and water is drawn through siphons across a **gill** or ctenidium (Figure 15.14). Phytoplankton are collected on the gill, and ciliated tracts pass the food to a **palp**, which also is ciliated and passes particles into the mouth and digestive system. The siphons

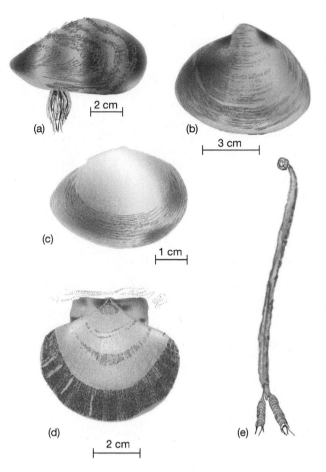

FIG. 14.16 Some of the diversity of bivalve mollusks: (a) mussel *Crenimytilus*, (b) suspension-feeding clam *Mactra*, (c) deposit-feeding clam *Macoma*, (d) scallop *Pecten*, and (e) boring bivalve *Bankia*.

directly to the mouth. Burrowing bivalves vary in shape, depending on their speed of burrowing. Rapid burrowers tend to be elongate and smooth, whereas poor burrowers are stubbier and may be ornamented. Although most bivalves are burrowers, some (e.g., mussels) live on the surface, and the shipworm *Teredo* bores into wood by means of mechanical abrasion and use of the wood-digesting enzyme cellulase (**Figure 14.16e**). These bivalves have symbiotic nitrogen-fixing bacteria that can convert nitrogen into useful forms for the bivalve. The borer *Lithophaga* can penetrate limestone and has special glands that secrete acid.

- **Members of the class Gastropoda have a flattened foot, usually a cap-shaped or coiled shell, and a mouth apparatus known as a radula. They are characterized by a twisting of the body, known as torsion.**

Gastropods (**Figure 14.17**) are usually distinguished by a flattened **foot** and a coiled shell, which can range from the high-spired (tall and pointy) type to low and flattened forms. The shell has one opening, and the animal often has an oval and stiff organic door, known as an **operculum**, which can seal the animal in the shell, protecting it from predators and from drying out. Within the shell, the soft body is twisted, so that both the anus and mouth can connect to the single shell opening. This condition is known as **torsion** and is unique to the gastropods.

Gastropods have a mouth apparatus that is distinguished by a **radula**, or tooth row, which moves back and forth over the food. The teeth differ according to the food source. In herbivorous snails such as periwinkles and limpets, the tooth row scrapes seaweeds and attached microalgae from the rock surface (**Figure 14.18**); in carnivores, the teeth are often fewer and stronger. Some species of the cone shell *Conus* have special modified radular teeth that inject a deadly poison, which is immediately fatal to the prey (and sometimes to humans!). Some snails (e.g., moon snails and dog whelks) have a mouth apparatus that can drill holes into calcium carbonate shells.

connect the infaunal bivalves to the sediment surface. Some bivalves feed directly on the organic matter in sediment. In clams of the superfamily Tellinacea, the siphons are separate, the ventral siphon vacuums the surface sediment, and particles are sorted on the gill. In nut clams (*Nucula*, *Acila*), palp tentacles collect sediment and the particles are passed

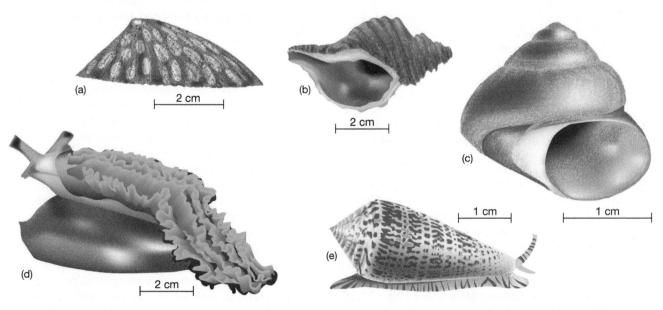

FIG. 14.17 Some of the diversity of gastropods: (a) limpet *Notoacmea*; (b) whelk *Neptunea*; (c) rocky-shore snail *Tegula*; (d) sea-slug *Elysia*, a shell-less gastropod; and (e) carnivorous *Conus*.

Sheltered   Exposed

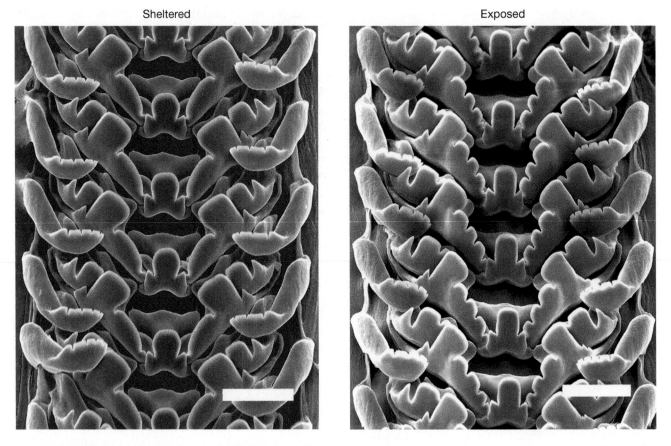

FIG. 14.18 Gastropods have a radula, which is a sliding belt of teeth. The intertidal periwinkle *Littorina obtusata* shows a plastic response in tooth development depending on whether snails are in sheltered areas where seaweeds are relatively soft (left), or exposed areas (right) where seaweeds are tougher. Central teeth are longest and lateral teeth the most numerous when snails are in exposed areas consuming tougher seaweed (see Molis et al. 2015). Scale bars are 50 μm. (Photographs by Ricardo Scrosati)

■ **Members of the class Polyplacophora (chitons) have a flattened foot, a radula, and eight dorsal articulated plates.**

Chitons (Class **Polyplacophora**) (Figure 14.19) are found commonly on hard substrates; they are oval shaped and flattened, and eight articulated dorsal plates cover the dorsal mantle. Chitons have a flattened foot that can adhere to the rock by suction, and a mouth with a radula that is generally similar to that of gastropods. Nearly all chitons are herbivores that feed on microalgae and seaweeds attached to rocks. A few species are grazers but also feed on small invertebrates.

■ **Members of the class Monoplacophora include simple animals with cap-shaped shell, posterior gill, flattened foot, and radula.**

The **Monoplacophora** are represented by just a few species, but they are believed to have traits very similar to those of the ancestor of the mollusks. The animals have a cap-shaped shell and a flattened foot, with a radula in the mouth apparatus. The taxonomic class was once thought to have only fossil representatives, but the group was "rediscovered" in a dredge sample taken in the Danish *Galathea* expedition of 1952. Many believe that the presence in the "living fossil"

Neopilina galathea of several examples of repeated structures (gills, muscles, auricles, nerves) links the Monoplacophora to other segmented groups, such as annelids.

■ **Members of the class Cephalopoda are distinguished by elaborate nervous and muscular coordination, the presence of grasping arms, and a carnivorous feeding mode.**

The **Cephalopoda**, which include squids, octopods, and the chambered nautilus, represent the peak of invertebrate evolution in terms of nervous organization and behavioral complexity. They have the basic molluscan body plan, including a mantle, external shell or internal shell remnant, and a foot that is modified into a water-squirting funnel. They differ considerably from the other mollusks, especially because the head–foot complex is usually in line with the rest of the body (dominated by the mantle) and because they have arms, often with suckers that can seize prey. All cephalopods have well-developed nervous systems and, with the exception of *Nautilus*, have an eye with cornea, iris diaphragm, lens, and retina. Octopods (**Figure 14.20**) are locally important predators and usually live in crevices. Individuals often specialize on specific prey, such as individual clam or crab species. Because cephalopods live

FIG. 14.19 The chiton *Catharina tunicata* at Point Lobos, California, about 10 cm long. (Photo by Jeffrey Levinton)

FIG. 14.20 The octopus is a major predator of benthic invertebrates, including crabs and mollusks. (Photograph by Paulette Brunner, with permission from Friday Harbor Laboratories)

mainly in the water column or are demersal, the subgroups are treated in more detail elsewhere (see Chapter 9).

■ **Members of the class Scaphopoda (tusk shells) have an elongate conical shell and live buried within the sediment, feeding on foraminiferans and other small animals.**

Scaphopods (**Figure 14.21**) number only a couple of hundred species and live in sand and mud. They secrete a tusk-shaped shell, usually less than 10 cm in length. They have a foot resembling that of bivalve mollusks, and they burrow into the substratum. Food is collected by means of tentacular structures, which are extended and probe into the sediment searching for small-animal prey. The animals have a radula, which pushes food into the mouth.

Phylum Arthropoda: Jointed Appendages

BASIC FACTS

Taxonomic Level: phylum Arthropoda; grade of construction: organs derived from three tissue layers; symmetry: bilateral; type of gut: complete with anus; type of body cavity other than gut: coelom; segmentation: present; circulatory system: usually open to large coelomic cavity; nervous system: brain, with nerve cords and bundles (ganglia), compound eyes and simple eyes; excretion: excretory organs.

Other Features: external cuticle, jointed appendages; includes horseshoe crabs, shrimp, crabs, sow bugs, insects.

Number of Species: over 1 million.

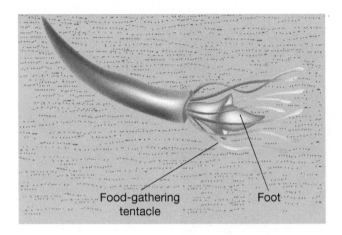

FIG. 14.21   A scaphopod in living position.

Food-gathering tentacle

Foot

■ **Arthropods are characterized by an external cuticle of chitin as well as by segmentation and jointed appendages.**

The phylum **Arthropoda** contains the largest number of living species and includes the insects, arachnids (spiders), millipedes, and centipedes. In the marine environment, arthropods are represented mainly by the subphylum **Crustacea**, which includes shrimps, crabs, and lobsters. Also present in marine environments is the class **Merostomata**, or horseshoe crabs, and the class **Pycnogonida**, or sea spiders. An extinct class, the **Trilobita**, dominated ancient shallow seas in the Paleozoic era.

All arthropods have an external flexible skeleton composed mainly of the polysaccharide chitin, although it is calcified in many species. The skeleton is jointed, and thinner spots allow flexing, especially at the jointlike areas in the limbs. The body is segmented but usually consists of a main cephalothorax (head–trunk) region and a multisegmented abdomen. The head contains elaborate sensory organs and mouth parts that process food. The thorax usually contains a series of paired walking limbs, which are very differently formed in different groups, depending on life habit.

Unlike some other phyla (e.g., annelids, mollusks), which use a hydrostatic coelom as a skeleton, the arthropods have a relatively rigid external skeleton. Muscles are inserted into the interior of the skeleton, and arthropod limbs move and operate as a series of lever systems. The skeleton allows for precise movement. But since the animal has a rigid outer case, it must periodically molt the external skeleton in order to grow. After molting, the animal takes on water and increases body volume. It then forms a new cuticle on the larger "frame." Different successive molts may vary in features such as swimming appendages and in terms of required foods. Most marine arthropods have a larval stage that is morphologically quite different from the adult stage, which in many species does not appear until after a major reorganization of the body, or metamorphosis.

Arthropods are sensory animals and usually have **antennae** that are covered with taste and smell receptors. A lobster has a set of antennae that is covered with tens of thousands of sensory nerves that can smell food at very low concentrations. Most have eyes of two types. **Compound eyes** consist of several units joined together and giving the appearance of a honeycomb. Each unit collects, focuses, and transmits light, and has separate nerve receptors. The animal pieces together the images, in mosaic fashion, to form a compound image. Some arthropods also have simple eyes, usually just a lens element and a single pit of sensory cells, that are usually capable of detecting only the presence and strength of light. These eyes are usually used for orientation. Arthropods have well-developed nervous systems, and many species are capable of exchanging salts with the water, through gills and excretory organs.

■ **The Trilobita (subphylum Trilobitomorpha) are an extinct class whose members had relatively unspecialized appendages and a compact body.**

Trilobites come from an oceanic world that no longer exists. They dominated the seas of the early Paleozoic era and survived until its end. *Trilobite* literally means "three lobes," signifying the three body sections (head, thorax, and telson) that characterize the group (**Figure 14.22**). Paleontologists have found a few specimens with the appendages intact, and they report that the trilobites were far simpler in structure and more unspecialized than their living arthropod relatives. Over time, arthropods have evolved more specialized and different types of limbs. In the Paleozoic era, trilobites dominated the marine arthropods, but became extinct at the era's end.

■ **The subphylum Chelicerata includes the horseshoe crabs, spiders, terrestrial scorpions, and pycnogonids (sea spiders). These animals are characterized by a first pair of movable claws and a division of the body into two general sections.**

FIG. 14.22   The trilobites are now extinct, but they were dominant invertebrates in early Paleozoic times. Pictured is *Cambropallas telesto* from the Cambrian of Morocco. (Photograph by Jeffrey Levinton)

Chelicerates are more familiarly represented by the terrestrial spiders and scorpions than by the marine horseshoe crabs and pycnogonids (sea spiders). Chelicerates are distinguished by a division of the body into two major sections. The first section has six pairs of appendages, but the first pair consists of chelicerae, or movable claws, quite different from the first appendages of the other major marine arthropod group, the crustaceans.

In the **Merostomata**, which includes the horseshoe crabs, the chelicerae and another pair of appendages handle and tear apart food before passing it to the mouth. The body is covered by a hard, calcified shield. There are two compound eyes on the dorsal surface. The second section of the body has several appendages, including highly modified, leafy book gills, which aid in swimming. The animal is often more than half a meter long and has a long tail spine, which can aid in righting the animal if it is flipped over (**Figure 14.23**). Horseshoe crabs prey on invertebrates living in sand- and mudflats. **Pycnogonids** (sea spiders) are far less conspicuous. They are spidery in appearance and have four pairs of walking legs. A sucking tube protrudes from the head. Sea spiders are often found perched upon colonial invertebrates and seaweeds, from which they suck fluids. Chelicerae are absent in many species of sea spider.

■ **The subphylum Crustacea is the largest marine arthropod group. Its members are characterized by a head with two pairs of antennae, three pairs of mouthpart appendages, and a trunk with several specialized appendages. The trunk is sometimes divided into a thorax and a posterior abdomen.**

The crustaceans number about 50,000 species and include the lobsters, crabs, copepods, and shrimp. All have a head with two pairs of antennae and specialized mouthpart appendages, which process food before it enters the mouth. The antennae are so densely covered with taste and smell receptors that they may be able to locate live food in the dark from several tens of meters downstream. All crustaceans have feeding, walking, or swimming appendages on the trunk. In the decapods (crabs and lobsters) there are

five pairs of appendages, but they are specialized differently in each group. In the crabs, the first pair consists of claws, and the remaining four pairs are walking legs. Lobsters (e.g., the New England lobster *Homarus americanus*) may have a large pair of claws and a couple of other pairs of legs armed with pincers that handle food, or they may have two or three anterior pairs of legs with small pincers (e.g., the spiny Pacific lobster *Panuliris interruptus*). In many species, the posterior part of the trunk contains a distinct area known as the abdomen. In the true crabs, this is bent over and more or less fused with the trunk, but in crayfish, the abdomen has appendages used for respiration and egg incubation and a couple of flattened appendages that can flex and move the animal rapidly backward. There is not enough space to describe crustaceans in detail, but **Figure 14.24** shows the range of diversity of body shapes and appendages.

Crustaceans have a wide variety of life habits (**Figure 14.24**). For example, species of the burrowing shrimp *Callianassa* live in a series of interconnected burrows, sometimes a meter below the sediment surface. The mole crab *Emerita* burrows rapidly on wave-swept beaches and migrates up and down with the tides. The dorsal-ventrally flattened isopods live on a wide variety of surfaces and sediments, as do the laterally compressed amphipods. Crabs and lobsters move along the surface, walking on their legs and leaping by rapid flaps of their posterior appendages. There are also planktonic crustacean groups, as we discussed in Chapter 9.

Benthic crustaceans, such as many of the crabs, are voracious carnivores, using their claws to seize, crush, and tear apart prey. The Atlantic and Gulf coast blue crab *Callinectes sapidus* (Figure 15.17c) is abundant in western Atlantic estuaries; during its annual migration up the estuary, the Chesapeake Bay population devastates bivalve mollusk populations. Other species are scavengers and even eat sediment, digesting the organic debris contained in it. Fiddler crabs (genus *Uca*; see Figure 7.1) feed on sediment by using the claw to scoop sediment into the mouth cavity, where the mouthparts separate microalgae and bacteria from the sand grains, which are then rolled into sand balls and deposited on the sediment surface. Males have a large claw that is used in waving displays to attract females and plays no role in feeding. Females are at a feeding advantage because they have two feeding claws. Other crustaceans, such as barnacles and copepods, filter phytoplankton and bacteria from the water.

Barnacles stand out as the most specialized of the crustaceans. Although the larvae are planktonic and typically crustacean, the adults are sessile and are enclosed by a series of calcium carbonate plates (**Figure 14.24a**). Midtrunk appendages, called cirri, collect food as the animal rapidly extrudes part of the body from the shell, and the feeding cirri comb phytoplankton from the water. Despite the overall sessile habit, they are very diverse. Acorn barnacles (**Figure 14.24a**) and their relatives the leaf barnacles (**Figure 14.24b**) live attached to rocks,

FIG. 14.23 A copulating pair of horseshoe crabs, *Limulus polyphemus*. (Photograph by Jeffrey Levinton)

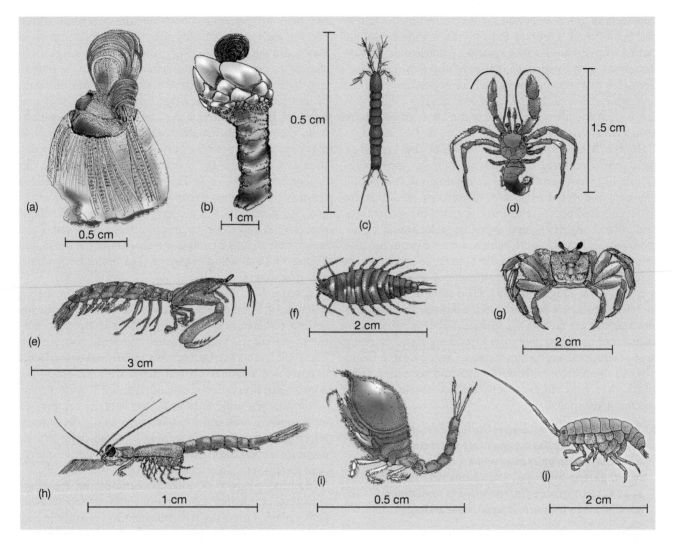

FIG. 14.24 Some of the diversity of benthic crustaceans: (a) acorn barnacle, (b) leaf barnacle, (c) harpacticoid copepod, (d) hermit crab (without shell), (e) mantis shrimp or stomatopod, (f) isopod, (g) brachyuran crab, (h) mysid shrimp, (i) cumacean, and (j) amphipod.

while goosenecked barnacles tend to attach to floating objects in the sea. Another group infests sea turtles and whales, and burrows into the surface with hooks or cements to the flesh. Boring barnacles bore holes into corals and mollusks. Finally, as we discussed in Chapter 4 (see Figure 4.10), rhizocephalans parasitize other invertebrates. Acorn barnacles are simultaneous hermaphrodites, but other groups vary in their sexuality.

The Lophophorate Phyla

- **The lophophorate phyla include the Bryozoa, Brachiopoda, and Phoronida. They are united by the presence of a looped feeding and respiring structure known as a lophophore.**

Three invertebrate phyla—Bryozoa, Brachiopoda, and Phoronida—are believed to be closely related because they all have similar looped feeding structures known as lophophores. A lophophore is ciliated and gathers suspended

food, mainly phytoplankton. The closeness of the phylogenetic relationship among the three phyla may be illusory, however, and the similarity may be more a result of convergent evolution of feeding structures.

Phylum Bryozoa: Moss Animals

BASIC FACTS

Taxonomic Level: phylum Bryozoa; grade of construction: organs derived from three tissue layers; symmetry: bilateral; type of gut: complete; type of body cavity other than gut: coelom, usually reduced; segmentation: absent; circulatory system: absent; nervous system: single ganglion with nerves branching throughout zooid; excretion: no special structures.

Other Features: colonial invertebrates, with small individuals (zooids) feeding with lophophore, growing in sheets or erect colonies; a colonial nervous system allows coordination of zooids.

Number of Species: about 4,000.

FIG. 14.25   (a) Closeup of zooids of the sheetlike bryozoan *Membranipora*. (b) The erect bryozoan *Tricellaria*.

■ **Bryozoans are abundant on hard surfaces and consist of colonies of small (<1 mm in width) individuals known as zooids.**

There are about 4,000 species of living **bryozoans**, and they occur throughout the world on hard surfaces, mainly in shallow waters. Bryozoans are abundant on the undersides of stony corals, on rocky surfaces, and even on the fronds of seaweeds. They are all colonial, and individual **zooids** (**Figure 14.25**) are much smaller than 1 mm in diameter. The soft parts of the animal are encased in an organic box, which may also be calcified. The animal feeds on suspended matter using a ciliated **lophophore**, which can be thrust into the overlying water by fluid pressure. Ciliary movement creates a feeding current that draws suspended particles to the lophophore. Ciliary currents then direct food to the mouth. Interspersed among the feeding zooids are other beak-like specialized zooids called avicularia, which help keep the colony unfouled by pinching at settling organisms.

■ **Bryozoan colonies can occur as sheets, erect colonies, or units connected by runners, known as stolons.**

Bryozoans have a wide variety of forms. *Electra*, *Membranipora*, and similar forms consist of a low-relief sheet (**Figures 14.25a, 14.26**). They usually live on flat surfaces. The peripheral zooids of some can produce spines to ward off predators. Many others, such as *Bugula*, grow as erect colonies (**Figure 14.25b**) and could be confused with cnidarians but for the characteristic lophophore. The colony often grows by continuous dichotomous splitting. A minority of bryozoans consist of colonies connected by modified zooids that form runners or stolons. This form may have the advantage of allowing rapid colonization of new microsites, but it is not nearly as common as the other two growth forms.

Phylum Brachiopoda: Lingulas and Lampshells

BASIC FACTS

Taxonomic Level: phylum Brachiopoda; grade of construction: organs derived from three tissue layers; symmetry: bilateral; type of gut: complete; type of body cavity other than gut: coelom, in two sections; segmentation: absent; circulatory system: heart with blood vessels; nervous system: rudimentary, but animals respond to light; excretion: no special structures.

Other Features: solitary individuals with two valves, attached to bottom with pedicle, feed with ciliated lophophore.

Number of Species: about 300.

■ **Brachiopods have two shells and use a pedicle to attach to the bottom; a lophophore allows them to feed on suspended matter.**

The living **brachiopods** represent a mere trace of their former abundance in the Paleozoic era, when they dominated shallow seas. Today, there are only about 300 species, which tend to live in cryptic coral reef habitats, some high-latitude shallow waters, and deeper waters. They are mainly sessile and live exposed on hard surfaces. They are quite distasteful to predators.

Brachiopods superficially resemble bivalve mollusks and have two calcareous valves. In brachiopods (**Figure 14.27c**), however, the valves are not the same: one usually has a perforation, through which a stalk, or pedicle, attaches to the bottom. Articulate brachiopods live on hard surfaces. The valves are articulated with a tooth-and-socket hinge, and the pedicle attaches to the rock surface. Inarticulates, such as *Lingula*, lack the articulate hinge system, however, and the valves can move more freely. The valves are connected by a complex

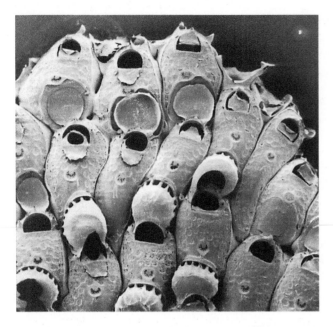

FIG. 14.26 Electron micrograph of the skeleton of the encrusting bryozoan *Fenestrulina*, showing pores through which feeding zooids emerge. Several egg-containing ovicells can be seen above the main zooid chambers. (Photograph by Sean Craig)

musculature, which in burrowing species allows them to move like two sliding sheets of cardboard and permits the animal to dig into the sand, valves first (see Figure 15.3). The animal assumes a living position with the valves up and pedicle down. When disturbed, the pedicle contracts and withdraws the valves below into the sand.

All brachiopods have a coiled lophophore, which projects outward as two symmetrical sections, usually supported by a calcareous loop. In some fossil brachiopods, this support was often in the form of a lovely spiral. The cilia on the

lophophore beat and create a current, which brings in particles through two lateral openings, across the two parts of the lophophore and toward a central area. Water and uningested particles exit through a central stream, between the valves.

Phylum Phoronida: Wormlike Animals with a Lophophore

BASIC FACTS

Taxonomic Level: phylum Phoronida; grade of construction: organs derived from three tissue layers; symmetry: bilateral; type of gut: complete; type of body cavity other than gut: coelom in several sections; segmentation: absent; circulatory system: blood vessels moving fluid with peristaltic action; nervous system: rudimentary, but giant neurons extend for the length of the animal to allow instant shortening; excretion: possible presence of organs of excretion.

- -

Other Features: solitary individuals, wormlike, living in vertical tubes with lophophore protruding into water above.

- -

Number of Species: approximately 10.

- -

■ **Phoronids are wormlike, with a lophophore that protrudes above the substratum.**

Phoronids (**Figure 14.27a**) comprise very few species and are occasionally abundant in sand flats of the Pacific coast of the United States and in some Atlantic coast sandy bottoms. The soft bodies of Pacific coast *Phoronopsis californica* and *P. viridis* are enclosed in a parchment-like tube that is buried in the mud. A feathery lophophore, which protrudes above the sediment surface, is a lovely orange in *P. californica* and green in *P. viridis*. As in the case of bryozoans and brachiopods, cilia circulate water across the

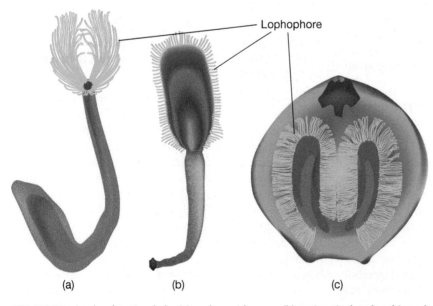

Lophophore

(a)   (b)   (c)

FIG. 14.27 Lophophorate phyla: (a) a phoronid worm, (b) an inarticulate brachiopod, and (c) opened articulated brachiopod; lophophore feeding organ shown in all three groups (light green).

lophophore, and mucus-laden cilia capture particles and transfer them in ciliated tracts, eventually to the mouth.

Phylum Echinodermata: Animals with Fivefold Symmetry

BASIC FACTS

Taxonomic Level: phylum Echinodermata; grade of construction: organs derived from three tissue layers; symmetry: radial, sometimes combined with bilateral, organized around an oral-aboral axis; type of gut: blind sac with very reduced anus, or complete with anus; type of body cavity other than gut: coelom; segmentation: none; circulatory system: usually open to large coelomic cavity; nervous system: major nerves extending from nerve ring, no brain; excretion: diffusion.

....

Other Features: A spiny skin enclosed an internal skeleton of interlocking calcium carbonate plates; feeding and locomotion on tube feet connected to water vascular system; common pentameral symmetry, includes starfish or sea stars, sea cucumbers, sea urchins.

....

Number of Species: approximately 6,000.

■ **Echinoderms are exclusively marine and have an outer skin that encloses a skeleton of interlocking ossicles.**

Members of the phylum **Echinodermata** (Figure 14.28) live only in the ocean; most are incapable of living even in estuaries. There are about 6,000 species, including the sea stars (Class **Asteroidea**), sea urchins (Class **Echinoidea**), brittle stars (Class **Ophiuroidea**), sea cucumbers (Class **Holothuroidea**), and crinoids (Class **Crinoidea**). The echinoderms have a unique adult radial symmetry, and structures are often repeated in multiples of five (e.g., the arms of sea stars). Echinoderms have an external leathery skin, which encloses an internal skeleton of interlocking calcium carbonate plates known as ossicles. In sea urchins, this system forms a calcareous ball, with openings for the mouth and other structures. The ball encloses the soft tissues, including digestive system and reproductive system. In sea stars, the ossicles are less well fused, and the animal is flexible. On the outer surface of sea stars and sea urchins, specialized groups of ossicles may form many pedicillariae (**Figure 14.29**), which can pinch and protect against predators or fouling organisms.

■ **Both locomotion and feeding in echinoderms are based on the water vascular system, which uses water pressure to operate many tube feet.**

Echinoderms all have a remarkable network of canals, pressure relief valves, and tubular suckers known as the **water vascular system** (Figure 14.30). Water is exchanged across a dorsal sieve plate, the **madreporite**. Combined with the action of relief valves, movement of water can create increases of pressure or suction in various parts of the canal system. The animal connects with the outside world through

FIG. 14.28 The diversity of echinoderms: (a) crinoid, (b) asteroid sea star, (c) ophiuroid brittle star, (d) holothuroid (sea cucumber), and (e) echinoid (sea urchin).

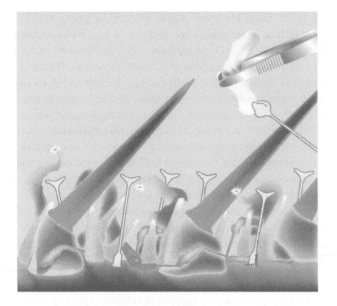

FIG. 14.29 A tube foot of a sea star is held near a sea urchin, whose spines bend away and whose pedicillariae pinch at the tube foot in defense. (After Feder, 1972)

thousands of **tube feet**, which can apply suction and also secrete mucus. On each tube foot is an **ampulla**, which looks like a rubber medicine-dropper bulb and functions by contracting and forcing fluid to expand the tube foot. When the tube foot contracts and withdraws, fluid is returned into the expanding ampulla. At the tip in some species is a sucker that sticks to the surface with the aid of mucus. Consider a sea star as an example (**Figures 14.28b, 14.30**). The madreporite is located in the dorsal surface in the central disk area. The tube feet number in the thousands and are located on the ventral surface. Tube feet are expanded and withdrawn

in coordination, and this allows the animal not only to pull itself along with thousands of tiny coordinated movements but also to feed. In some groups, such as sea urchins and sea lilies, tube feet are used only to collect food and are not used for locomotion.

All echinoderms are built around a general body plan of nearly radial symmetry, with five (or multiples of five) **ambulacral areas** radiating from the central mouth. The ambulacral areas are lined with tube feet, which are reduced to varying degrees. In the sea star, the ambulacral areas are the ventral parts of the arms, but in sea cucumbers, the areas are aligned parallel to the tubular body, with two on the dorsal surface and three on the ventral surface.

■ Sea stars (Asteroidea) are usually carnivorous, and many feed by extruding the stomach and digesting prey, mostly outside the body

Sea stars (Class **Asteroidea**) are ubiquitous throughout the ocean and nearly all are carnivorous. All have a central disk area, from which several arms radiate. Sea stars have remarkable regenerative capabilities and can produce a complete new set of arms and disk as long as some major portion of the central disk survives an attack. The ventral surfaces are covered with tube feet, which can attach to a prey item. Depending on the species, a sea star may attack anything it can handle or may be specialized to specific prey, some even to other sea stars. The asteroid sea star *Luidia sarsi* uses its arms to leap onto brittle star prey. Leaping can be induced artificially by soaking cotton with extracts of brittle stars and placing the bait near the *Luidia*. Some species simply swallow their prey whole. Clam and scallop eaters attach to the valves of bivalves by means of the tube feet and use force, perhaps over a period of hours, to pull

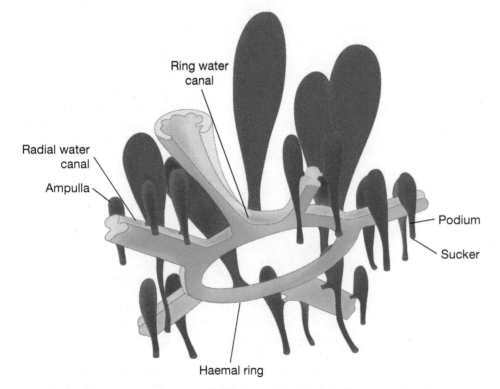

FIG. 14.30 The water vascular system and the operation of the tube feet in an asteroid.

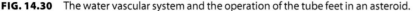

the shells apart a few millimeters. A sea star then pushes its stomach out through the mouth and into the bivalve and presses it against the soft tissues of the bivalve, secreting enzymes and digesting the prey. The animal then withdraws the stomach and sucks in the digested material.

- **Echinoids (Echinoidea) are usually covered with spines and may live on the surface, using the teeth of a structure called an Aristotle's lantern to scrape algae, or burrowing in the sand, feeding on sediment or suspended matter.**

The sea urchins, sand dollars, and heart urchins (Class **Echinoidea**) have an internal test of interlocking ossicles. The outer skin is covered with pinching **pedicillariae** (**Figure 14.29**) and spines, which also protect against predators. The spines of urchins are usually long and thin, but sand dollars and heart urchins are covered with a carpet of very short spines that are mobile and aid in burrowing. The Caribbean urchin *Diadema antillarum* has spines as long as 50 cm, which are very sharp, have thousands of tiny barbs, and produce great pain if touched.

Echinoids have a remarkable jawlike structure called an **Aristotle's lantern** (**Figure 14.31**), constructed from specialized ossicles connected by ligaments and moved by muscles. Five teeth come together at one point in the structure, allowing the animal to tear apart seaweeds or to scrape microscopic algae from a hard surface. Some urchins can actually scrape a depression in rock and nestle in it. The teeth are steadily worn down and replaced by secretion of calcium carbonate from dental sacs. Food is taken into the digestive system, and undigested remains are passed through the anus, on the dorsal surface. Nearly all sea urchins feed on attached algae, although some can catch drifting pieces of seaweed by means of long tube feet extending from the dorsal surface. Heart urchins and sand dollars are burrowers and feed on sedimentary organic matter, which they collect by means of modified tube feet on the ventral surface. Particles are passed from tube foot to tube foot and eventually to food grooves and to the mouth.

- **Sea lilies and feather stars (Class Crinoidea) are characterized by a cup-shaped body, with upward-reaching arms that catch zooplankton on specialized tube feet.**

Crinoids include the unstalked feather stars and stalked sea lilies (**Figure 14.28a**). All have a basic cup-shaped body.

FIG. 14.31   The Aristotle's lantern of a typical echinoid.

The arms extend upward into the water column and are covered with pinnules, which are in turn covered by sticky tube feet that catch zooplankton from moving currents. Feather stars attach temporarily to the bottom with arm-like cirri, but the animals have a stalk in early life. Feather stars can move into and out of crevices, and often are found at night feeding on the surface. They are sometimes common in coral reefs but also occur in higher latitudes. Sea lilies have a long stalk that is embedded into the bottom and often an umbrella-shaped array of arms. The stalk is composed of connected ossicles that resemble a stack of poker chips. Sea lilies tend to live in deep water today, but in the Paleozoic era they were a major dominant of shallow-water seas.

- **Sea cucumbers (Class Holothuroidea) are tubular and have a crown of tentacles, which either feed on sediment or are directed upward to feed on zooplankton.**

Sea cucumbers (Class **Holothuroidea**) are shaped as their name suggests (**Figure 14.28d**), and most have reduced ossicles, which makes them very soft and flexible. They move along by means of muscular contraction of the body wall and with the aid of the tube feet of three ambulacral areas that press against the bottom. The mouth has a crown of tentacles, consisting of highly modified tube feet. In cucumbers like the Pacific *Parastichopus californicus*, the tentacles press against the sediment, which is passed into the mouth and through the gut. Other species point the tentacles upward and collect suspended matter. *Leptosynapta* occurs on the Atlantic and Pacific coasts and lives burrowed in sand, feeding on the organic matter in it. The burrower *Molpadia oolitica* lives upside down in a vertical burrow, feeds on mud, and extrudes feces on the sediment surface.

Sea cucumbers react to predators (or to marine biologists collecting them) by evisceration: rupturing the mouth or anal opening and expelling nearly all internal organs. The eviscerated animal can then escape and regenerate. Tropical sea cucumbers can also deter predators by means of highly toxic substances located in the surface skin.

Sea cucumbers have a unique pair of **respiratory trees**, which are highly branched outgrowths of the hind gut. The cucumber draws in water through the anal opening and contracts the hind gut, driving water through the branches of the respiratory trees. Then, the water is expelled through the anal opening. Species of crab, protozoan, and fish live as commensals among the respiratory trees of some tropical sea cucumbers. The fishes leave through the anal opening to forage outside.

- **Brittle stars and basket stars have a central disk and distinct separate flexible arms that move the animal along the bottom without the use of tube feet.**

Brittle stars and basket stars (Class **Ophiuroidea**) are generally small, although some basket stars are as large as a meter across. The disk is quite distinct from the arms, and the tube feet are not involved in locomotion. The arms are very flexible and muscular and move the animal along the sediment surface. The arms are far more flexible than those of regular sea

FIG. 14.32 An oral view of the eastern Pacific basket star *Gorgonocephalus eucnemis*. (Photograph by Jeffrey Levinton)

Phylum Chordata: The Sea Squirts

BASIC FACTS

Taxonomic Level: subphylum Urochordata; grade of construction: organs derived from three tissue layers; symmetry: bilateral; type of gut: complete, with anus; type of body cavity other than gut: coelom; segmentation: none; circulatory system: heart with vessels; nervous system: brain with nerve cords; excretion: diffusion.

Other Features: benthic sea squirts have a barrel-shaped body, with incoming and outgoing siphons; tadpole larvae have features allying this group with vertebrates.

Number of Species: about 1,200.

stars, and the animals are much faster. Ophiuroids (**Figure 14.28c**) usually have only five arms. However, in basket stars the five arms are branched and subbranched, and the animals appear to have coiled tentacles (**Figure 14.32**). Basket stars often spend the day in crevices, with the arms wrapped in a compact ball. At night, they stretch out the arms and use their tube feet to trap zooplankton. Brittle stars may live within the sediment or on the surface, occasionally draped over erect colonial animals. They may be suspension feeders, sediment eaters, or carnivores. Some brittle stars live with the disk positioned below the sediment surface, the arms projecting into the water column. Particles are trapped on mucus suspended between modified spines, and tube feet remove the material and pass it to the mouth. Carnivores capture prey by looping the arms and bringing the captured victim to the mouth.

■ **Sea squirts are barrel-shaped animals that filter water through a mucus sheet. A tadpole-shaped larva relates them to the vertebrates.**

The subphylum **Urochordata** includes the benthic sea squirts (**Figure 14.33**) but also the planktonic salps, which were discussed in Chapter 8. The sea squirts are found on hard surfaces nearly anywhere in the ocean and often are the dominant form of benthos. They have a barrel-shaped body, with an incoming and an outgoing siphon on top. The outer part of the body is sometimes covered with a tough tunic made of cellulose. To avoid predation and fouling, some have evolved the ability to concentrate toxic heavy metals and even to secrete sulfuric acid. The inner part of the barrel is lined with a lattice-shaped pharynx, which is covered with a ciliated layer, coated with mucus. Water currents bring plankton to the mucus net, which can filter particles as small as 1 μm in diameter.

The larvae of sea squirts (**Figure 14.33b**) are especially exciting creatures because they demonstrate an evolutionary

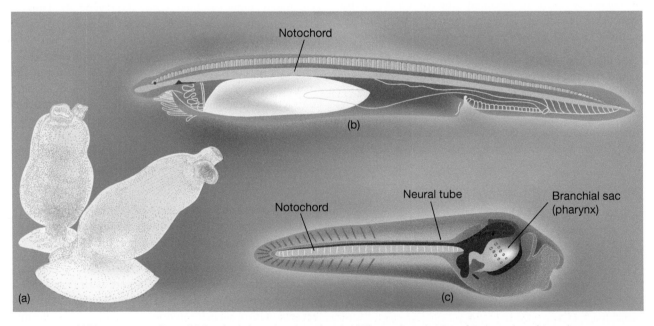

FIG. 14.33 (a) The sea squirt *Ciona*. (b) A tadpole larva (ca. 1 mm long). (c) The sea lancelet *Branchiostoma* (amphioxus).

relationship with our evolutionary branch of the phylum Chordata, which includes the vertebrates. The larvae resemble tadpoles and have a trunk and tail. The tail is stiffened with a rod, called a notochord, and has a dorsal tubular nerve. The animal also has gill slits. These traits are characteristic of vertebrate embryos. They also ally sea squirts with another common marine animal, the sea lancet (**Figure 14.33c**), which belongs to the subphylum Cephalochordata. The sea squirt larva does not feed and usually has a larval life of hours and cannot disperse very far.

■ CHAPTER SUMMARY

- Benthic invertebrates with *radial symmetry* have a center of symmetry. *Bilateral symmetry* (symmetry about a plane) is associated with directional movement and anterior sense organs. Some have a coelom, a body cavity that may allow compartmentalization.

- Protists are neither animals nor plants but usually free-living, mostly one-celled organisms. They include amoeboid forms but also colonial forms such as macroalgae.

- Porifera, or sponges, are cellular in organization, including flagellated collar cells, which move water and food particles into open chambers. Water enters through hundreds or thousands of pore cells but leaves central chambers through just one or several oscula. Sponges have an organic skeleton of spongin but also silica skeletal elements known as *spicules*.

- Cnidaria are built around a radial cup-shaped polyp and medusa body plan, with a ring of tentacles and a digestive tract with one opening. All have stinging nematoblasts that puncture, ensnare, or stick to prey. Benthic cnidarians include colonial hydrozoans, corals, and anemones.

- Platyhelminthes, or flatworms, have bilateral symmetry and distinct organs. They include free-living forms that feed with a muscular pharynx and glide along the bottom using ciliary motion as well as parasitic forms that infest other marine species.

- Nemertea, or ribbon worms, have a proboscis and a complete gut and are mobile carnivores that burrow through the sand.

- Many nematodes use longitudinal muscles that work against a fluid-filled body with a rigid wall. Free-living nematodes may be carnivorous or plant eating; some consume organic matter from sediment.

- Annelids are worms divided into segments, with a coelom, tubular gut from mouth to anus, and three distinct embryonic tissue layers. Annelids include free burrowers, infaunal tube dwellers, and epifaunal tube constructors. They may be carnivores, deposit feeders, and suspension feeders.

- Sipuncula, or peanut worms, live in burrows in soft sediment and in rock crevices. They gather food by means of an introvert with branched tentacles.

- Pogonophora are generally deep-sea species that lack a gut and depend on endosymbiotic bacteria. Molecular evidence suggests that they are part of the Annelida. Members or close relatives include the volcanic vent-associated Vestimentifera.

- Mollusks have a head–foot complex, a mantle that usually secretes a calcium carbonate shell, and a gill used for respiration and commonly for suspension feeding. They include the single cap-shelled Monoplacophora,

the Bivalvia, the Gastropoda, the Polyplacophora (chitons), the Cephalopoda, and the Scaphopoda.

- Arthropods are characterized by an external cuticle of chitin, usually compound eyes, segmentation, and jointed appendages. Marine forms include the Chelicerata (horseshoe crabs and pycnogonids) and the Crustacea (copepods, amphipods, isopods, crabs, barnacles, and shrimp).

- Lophophorates are united by a suspension-feeding structure, the ciliated lophophore, and include Bryozoa, Brachiopoda, and the Phoronida, which are wormlike, burrow in sediment, and protrude the lophophore above the sediment–water interface.

- Echinodermata have an outer skin enclosing a skeleton of interlocking ossicles. Both locomotion and feeding in echinoderms depend on water pressure to operate the tube feet. They include the Asteroidea (sea stars), Echinoidea (sea urchins and relatives), Crinoidea (sea lilies and feather stars), tubular Holothuroidea (sea cucumbers) with a crown of tentacles, and Ophiuroidea (brittle stars and basket stars).

- Urochordata, or sea squirts, are barrel-shaped animals with incurrent and excurrent siphons that filter water through a mucus sheet. A tadpole-shaped larva relates them to other chordates, including vertebrates.

■ REVIEW QUESTIONS

1. What is the advantage of a radially symmetrical body plan? What is a disadvantage of such a body plan?

2. What is the function of collar cells in sponges?

3. What function does polymorphism serve in some cnidarians?

4. What is the major difference in overall body symmetry between flatworms and cnidarians?

5. What is the main feeding mode of nemerteans?

6. What is the distinction between locomotory mechanisms in polychaetes and oligochaetes?

7. What is the distinguishing characteristic of nutrition in the Pogonophora?

8. What are three means by which different bivalve mollusks may feed?

9. What structure is used by gastropods to scrape, puncture, or tear apart prey?

10. What is the disadvantage of the stiff exoskeleton in arthropods?

11. What are some of the different life habits of crustaceans?

12. What trait do the brachiopods, phoronids, and bryozoans share?

13. What general trait unites the members of the phylum Echinodermata?

14. How do echinoderms generally move about?

15. What characteristics unite sea squirts with the vertebrates?

16. Why is the Cambrian period special in the history of animal life?

17. What are the differences between protostomes and deuterostomes?

18. Under what ecological-evolutionary conditions might you find an animal that is barely recognizable morphologically as a member of its taxonomic group?

Visit the companion website for *Marine Biology* at www.oup.com/us/levinton where you can find Cited References (under Student Resources/Cited References), Key Concepts, Marine Biology Explorations, and the Marine Biology Web Page with many additional resources.

CHAPTER 15

Benthic Life Habits

Introduction

Benthic Size Classification

Because benthic animals are often collected and separated on sieves, a classification based upon overall size is useful. **Macrobenthos** include organisms whose shortest dimension is greater than or equal to 0.5 mm. **Meiobenthos** are smaller than 0.5 mm but larger than the **microbenthos**, which are less than 0.1 mm in size. Meiobenthos and microbenthos are often interstitial, living among sedimentary grains.

Feeding Classification

Deposit feeders ingest sediment and use organic matter and microbial organisms in the sediment as food. **Suspension feeders** feed by capturing particles from the water, usually phytoplankton and smaller zooplankton but possibly bacteria. **Herbivores** eat nonmicroscopic photosynthetic organisms, such as seaweeds. **Carnivores** eat other animals, but there is no easy way to classify some organisms, such as suspension feeders that ingest zooplankton. Finally, **scavengers** feed on remains of other animals and plants. Many deposit feeders also scavenge. A good example of such species are the fiddler crabs, which are normally deposit feeders but can also tear apart dead fish.

Life in Mud and Sand

Important Features of Soft Sediments

■ **Sediment grain size is an important determinant of the distribution of benthos and increases with increasing current strength.**

Sediment grain size is generally classified as in **Table 15.1**.

The **silt–clay fraction** is the mass percentage of sediments finer than 62 μm in diameter. The **percent clay** (particles < 4 μm) may also be useful in describing sediment properties relevant to benthic organisms.

The **grain size of sediment** is largely related to the current strength of the overlying water column. Because stronger currents can transport larger particles, median grain diameter increases in areas of high current velocity (see **Going Deeper Box Figure 15.1** for a discussion of the measure of grain size). Areas with strong currents also experience extensive erosion; sediment and fine particles will be transported away. In a weak current regime, fine particles can settle out of the water column and will remain.

■ **Sediment sorting and grain size angularity also reflect the hydrodynamic regime.**

Sorting is an estimate of the spread of abundance of particles among the size classes. A sediment is poorly sorted when most of the sediment is spread over a large range of size classes, whereas in a well-sorted sediment, almost all the weight is confined to a few size classes, with a well-defined

TABLE 15.1 Size Classification of Sedimentary Particles

CLASS	SIZE RANGE (mm)
Clay	<0.04
Silt	0.04–0.0625
Sand	0.0625–2.0
Gravel	2–64
Cobble	64–256
Boulder	>256

Measuring Grain Size of Sediments

Median grain size is the simplest way to represent particle size characteristics of soft sediments. By washing the sediment through a series of graded sieves, one can get the size class data to construct a histogram of sizes (**Box Figure 15.1a**). To accommodate a range of particle sizes of many orders of magnitude within one graph, we plot grain diameter in logarithmic form (log to the base 2, which means that a value of 1 in **Box Figure 15.1** corresponds to 2 mm, a value of zero corresponds to 1 mm, and a value of −1 corresponds to a diameter of 0.5 mm). The diagram is used to construct a cumulative weight graph, where the percent weights of the successive size classes are accumulated and cumulative percent weight is plotted as a function of particle diameter (**Box Figure 15.1b**). The median diameter M, which corresponds to Q_{50}, is the

particle diameter corresponding to 50 cumulative percent. Calculation of the 25 and 75 percent classes is also shown.

Sorting is a measure of spread among the grain sizes. This can be quantified by

$$S = \frac{Q_{25}}{Q_{75}}$$

where Q_{25} is the grain size corresponding to the 25 percent cumulative weight (**Box Figure 15.1b**) and Q_{75} is the same value for 75 percent. As S approaches 1, the sediment is all the same size class and is perfectly sorted. **Box Figure 15.1c** shows examples of sediments that have been poorly sorted and well sorted.

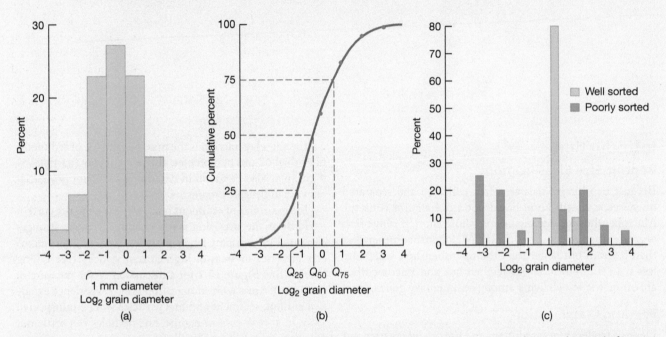

BOX FIG. 15.1 Graphical methods of presenting the particle size distribution of sediments. (a) Histogram, showing the weight frequencies of each particle size class as a function of the log of particle diameter (we use the log of particle diameter to be able to plot an enormous range of particle sizes on a manageable scale). (b) Cumulative frequency distribution curve, showing Q_{25}, Q_{50} (the median particle diameter), and Q_{75}. (c) Examples of a poorly sorted and a well-sorted sediment.

peak (see **Going Deeper Box Figure 15.1c**). A **well-sorted sediment** will be deposited in an environment with constant current strength. **Poorly sorted sediments** usually reflect a heterogeneity of sedimentary processes or origins of the sedimentary grains.

■ **In very shallow, sandy, wave-swept bottoms, currents generate ripples and bars, which create spatially varying microhabitat variation for benthic organisms.**

In areas of considerable current strength, surface sediment is eroded and transported continually, and a number of sedimentary structures may be established in equilibrium with

this transport. On a large spatial scale of tens of meters and kilometers, emergent and submerged **bars** may develop off-shore. On a smaller scale of meters and centimeters, sedimentary **ripple marks** commonly develop where sediment is in motion (**Figure 15.1**). In areas where currents are unidirectional, ripple marks are asymmetrical in cross section, with the steep slope facing downcurrent. Reversing tidal motion can reverse the form of ripples. By contrast, waves may produce sufficient oscillatory motion to generate symmetrical ripples.

Sedimentary ripples create a local microenvironment of their own, which strongly affects sediment stability and

FIG. 15.1 Geometry of a sand ripple in a unidirectional current. Note direction of sand ripple movement from upper right to lower left and the possibility of burial faced by invertebrates in its path. (Photograph by Jeffrey Levinton)

movement for organisms that are much smaller than the size scale of the ripples. For example, fine organic material tends to accumulate in the troughs, and deposit-feeding animals, therefore, are attracted to this microenvironment. By contrast, the crests of the ripples are relatively bare of this material and are also localized sites of erosion.

BURROWING

■ **Burrowers live in sediment ranging from packed sand to elastic mud to watery mud.**

Burrowers live in a wide range of sediments, whether they live in the intertidal or subtidal seabed. On one end of the spectrum, sediments are nearly pure sand particles in the range of 62.5–200 µm. Animals ranging from clams to larger polychaetes to crustaceans burrow into sand. The sand grains are piled onto each other and, because the sand grains are not compressible, the burrowers must exert forces to both compress water from the interstices of the grains and to displace sedimentary grains. Some clams can shoot a jet of water into the sediment just before burrowing, which liquefies the sand and makes it easier to displace.

Muds are a mixture of small mineral grains less than 62 µm, often of clays and organic particles and organic polymer coatings of the mineral grains. The organic polymers cause adhesion among the grains. This adhesion property makes the sediment behave like an elastic material, and a burrowing organ such as a bivalve foot or the proboscis of a worm can propagate cracks in the mud as the proboscis becomes engorged and thrusts into the sediment. As the cracks are generated, the sediment is locally weakened and the burrowing organ can continue to move through the sediment (Dorgan et al., 2005). As the water content of mud increases and the particle size becomes dominated by clay-sized particles, the sediment again changes in character. Sediment at this end of the spectrum is watery. Also, such sediments may consist of fecal pellets, which are aggregated groups of sedimentary grains, which increase the actual sedimentary grain size. As a burrowing

organ is pushed into the sediment, the sediment behaves like a viscous but watery material, simply pushing apart as the organism moves through the sediment. There is a reciprocal action since the burrowing act itself can increase the water content of muddy sediment by stirring in water from above the sediment surface as the animal burrows. As the burrowing organ pushes through, it may require less force to continue moving the burrowing organ. This property is known as **thixotropy**. In effect, the sediment becomes less resistant as you exert a concentrated shear force on it.

■ **Soft-sediment burrowers use hydromechanical and simple digging mechanisms to move through soft sediment.**

The initial displacement of sedimentary grains requires that a firm structure be pushed into the sediment with a sufficient force. To accomplish this, many burrowing organisms have a **hydrostatic skeleton,** which is a flexible tube that can be stiffened by the injection of fluid. In the case of bivalve mollusks, the foot is filled with fluid and becomes a digging device. The internal fluid is not compressible, and a set of longitudinal muscles usually operates in opposition to a set of circular muscles. The longitudinal muscles act to shorten the fluid-filled structure (e.g., a worm segment), and the circular muscles act to compress the body and lengthen the structure.

After the fluid-filled and rigid structure (i.e., bivalve foot) has pushed into the sediment, its distal end is engorged with fluid, creating an anchor. In order to propagate cracks into plastic muds, the tips of fluid-filled burrowing structures of many invertebrates are wedge-shaped. After penetration and formation of the anchor by widening of the fluid-filled structure, contraction of longitudinal muscles then pulls the rest of the body along (**Figure 15.2**). Within the sediment, a part of the body can be dilated to form an anchor, so that another forward part of the body can be extended forward by contraction of circular muscles. A series of dilations and extensions allows the animal to move within the sediment. This general principle applies to burrowing in mollusks, polychaetes, sipunculids, burrowing sea cucumbers, and other wormlike animals. There are some important detailed differences among these burrowing groups, however. Bivalves, for example, must rock the shell back and forth during burrowing, to allow the bivalve shell to work its way into the sediment.

The other major mode of burrowing involves the use of **mechanical displacement,** based on firm digging structures that act as spades and are moved by muscular action. A wide variety of crustaceans dig into the substrata by means of specialized digging limbs. For example, the mole crab, *Emerita talpoida*, has spadelike posterior appendages. Inarticulate brachiopods use a complex musculature to alternately push and rotate the two opposed valves through the sediment. The muscular rocking motion keeps the valves moving and constantly displacing sedimentary grains (**Figure 15.3**).

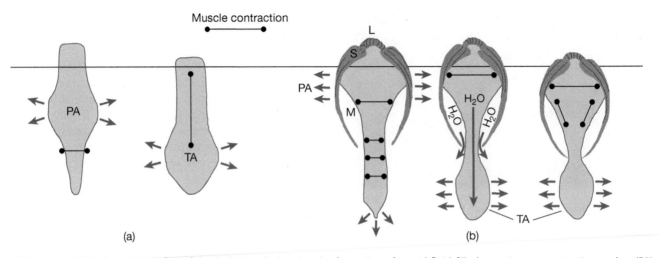

FIG. 15.2 (a) The burrowing of a soft-bodied animal, showing the formation of turgid fluid-filled mass into a penetration anchor (PA) and the dilation of a distal region, forming a terminal anchor (TA). Longitudinal muscles then drag the animal into the sediment. (b) How a clam uses its shell(s) and fluid-filled foot to burrow. Left: Clam is in sediment and presses shell outward, forming an anchor. At the same time, its fluid-filled foot thrusts into sediment. Middle: The foot fills with fluid at the tip, forming a new anchor. Right: Muscle contraction draws the shell together and drags it downward. (After Trueman, 1975)

FIG. 15.3 Burrowing inarticulate brachiopods have two symmetrical shells that are connected by a complex musculature. They burrow in the substratum by scissoring the shells back and forth, which shovels the sediment aside and pushes the animal downward. (Courtesy of Charles W. Thayer)

■ **Interstitial animals adapt to water flow and life in small spaces among particles by means of a simplified body plan, a wormlike shape, or by adhering to particles by means of mucus, suckers, and hooks.**

Interstitial animals are very small and move among sedimentary grains but do not displace them in bulk, as do burrowing animals.[1] Because they move through tight spaces, interstitial animals from many different phyla have evolved a wormlike shape and a simplified external body plan (**Figure 15.4**). Relative to their epibenthic relatives, for example, interstitial hydroids have reduced numbers of tentacles, which are important for capture of suspended prey from the water column. Smaller interstitial forms may be attached to sand grains by a variety of hooks and suckers.

The slender body form of some interstitial forms may be related to uptake of dissolved organic matter for food. Nematodes living in the low-oxygen parts of sediments (see the following) tend to be more slender than those living in the aerobic surface sediments.

Soft Sediments and the Role of Burrowers in Sedimentary Structure

THE SOFT-SEDIMENT MICROZONE

■ **Sediments consist of an oxygenated layer overlying an anoxic zone.**

If you dig into a protected sandy beach, you will first encounter light brown sediment but will soon reach a thin grayish zone and then a black layer with a rotten-egg-like odor. The changes in color and smell reflect a change of chemistry and microbiological processes. The light brown zone contains pore water with dissolved oxygen, whereas the black smelly zone is devoid of oxygen, and the gray layer is a transition zone between the two. The smell in the black zone derives from **hydrogen sulfide**, H_2S, which is generated by **sulfate-reducing bacteria**. Overall, the oxic–anoxic zonation results from a shifting balance between addition and consumption of dissolved oxygen in the pore waters. The boundary between the oxygenated zone and the anoxic zone is known as the **redox potential discontinuity (RPD)**.

---

[1] Interstitial animals are usually meiofauna, although benthos of meiofaunal size may also be epibenthos.

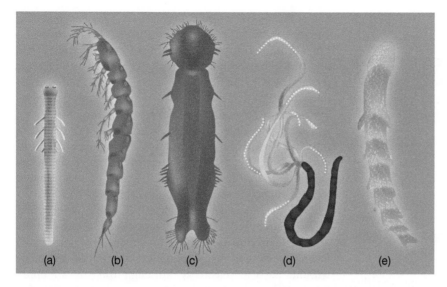

FIG. 15.4 The wormlike shape of interstitial meiobenthic animals of diverse phyla: (a) polychaete, (b) harpacticoid copepod, (c) gastrotrich, (d) hydroid, and (e) opisthobranch gastropod. (After Swedmark, 1964)

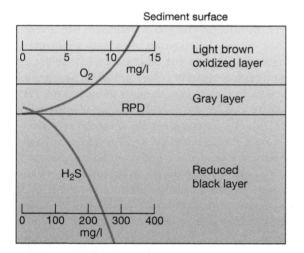

FIG. 15.5 Cross section of the sediment near the sediment–water interface, showing the redox potential discontinuity (RPD), which is a boundary between oxidative and reducing processes. The diagram shows the concentration of oxygen (above) and hydrogen sulfide (below).

It represents a sharp boundary between chemically oxidizing and reducing processes (**Figure 15.5**).

Near the sediment–water interface, oxygen diffuses or is stirred into the pore water from the overlying water column. In quiet areas, especially in organic-rich, fine-grained sediments, the transition to the anoxic zone can occur only a few millimeters below the sediment surface. Infaunal organisms may transport oxygen somewhat deeper by stirring the sediment or by irrigating their burrows. The combined actions of infaunal organisms may bring oxygen down to depths of several centimeters in muddy sediments that would otherwise have their oxygen content controlled by diffusion and never surpassing a few millimeters depth. Because of vertical burrowers, the RPD may not always be a horizontal surface but may be vertical in places, parallel with tubes and burrows (see later, **Figure 15.10**).

■ In sediments in quiet water, there is usually a vertical zonation of microorganisms.

Microorganisms are a crucial factor in determining the chemical conditions of the sediment, especially the pore water. Microorganisms help decompose particulate organic matter. The vertical gradient in oxygen, strongly affected by microorganisms, also affects the composition of the sediment microorganism community (see Fenchel et al., 2012). Aerobic bacteria and protists live near the sediment-water interface, but only anaerobic microorganisms can survive below the RPD. In order to obtain oxygen, nearly all animals living below the RPD must maintain contact with the sediment above the RPD by means of siphons, irrigated burrows, and tubes. It has been argued that a few metazoans, such as some nematodes, can survive without oxygen and that some macroinvertebrates can live for extended periods on the proceeds of anaerobic metabolism. T. Fenchel and R. Riedl (1970) first described the anoxic community, known as the thiobios. Some researchers, such as Riese and Ax (1979), have argued that this community does not really exist in truly anoxic sediments, but only in sediments of very low oxygen that are adjacent to anoxic microzones. Some protozoans, however, are clearly anaerobic and contain symbiotic anaerobic bacteria.

Microbial organisms may be autotrophic or heterotrophic. Recall that autotrophic organisms produce their own carbohydrates or sugars by means of either photosynthesis or chemoautotrophy. Photoautotrophy employs light as an energy source, whereas chemoautotrophy employs one of several chemical substrates (e.g., sulfate, hydrogen) to derive energy.

Figure 15.6 illustrates a generalized zonation of microbial communities in soft sediments. At the surface, aerobic photosynthetic microorganisms, such as diatoms and cyanobacteria, may predominate. These coexist with heterotrophic aerobic bacteria, which live in pore waters having dissolved oxygen and break down organic matter, and oxygen is the

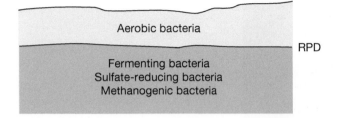

FIG. 15.6 Typical vertical zonation of bacterial components of quiet, muddy marine sediments.

terminal hydrogen acceptor in the decomposition process. In the deeper anoxic pore waters, however, heterotrophic microorganisms use a variety of other compounds as hydrogen acceptors. If the anoxic zone reaches the surface, then one often observes mats of photosynthetic bacteria such as purple sulfur bacteria, which use hydrogen sulfide as a reducing agent, producing elemental sulfur. But if the surface sediment is stirred by burrowers and pore waters contain oxygen, one then sees deeper into the sediment a series of anoxic bacteria beneath the surface aerobic sediment zone (**Figure 15.6**). Most notable of these are **fermenting bacteria**, which use organic compounds and produce end products such as fatty acids and alcohols, and the deeper **sulfate-reducing bacteria**, which reduce SO_4 to H_2S. The reduced compounds diffuse upward and are used by chemoautotrophic sulfur bacteria, which oxidize H_2S at the RPD region. Beneath the sulfate reducers are **methanogenic bacteria**, which grow successfully when sulfate is in short supply and break down organic substrates and produce methane as an end product.

Environmental constraints (e.g., the presence of oxygen) and energetic payoff combine to determine the successive dominance with depth of different heterotrophic bacteria groups (aerobic, fermentative, sulfate-reducing, and methanogenic). In the presence of oxygen, much more energy is obtained from the energy-efficient aerobic breakdown of organic matter by bacteria. Aerobic bacteria therefore are competitively superior in the microzone where pore waters have dissolved oxygen. Oxygen is lacking beneath this zone and different bacteria perform the energetically less efficient process of fermentation. The processes of sulfate reduction and methanogenesis are still lower in energy efficiency. Ultimately, heterotrophic bacterial activity is limited with depth by the lack of a food source. This can be shown by the steady decrease of substrate use by microbes with increasing depth into the sediment.

Deposit Feeding in Soft Sediments

■ **Deposit-feeding macrobenthic animals ingest sediment and derive their nutrition mainly from microalgae and particulate organic matter. Free-living sediment bacteria are digestible but not quantitatively important in the diet of larger macrobenthos.**

Deposit feeders are animals that ingest sedimentary material and derive their nutrition from some fraction of that material. Sediment is a complex mixture of inorganic

material, microorganisms, decomposing organic material, and pore water with dissolved constituents. Understanding the nutrition of these creatures, therefore, is a complex task, not at all like watching a caterpillar chew on a leaf! Deposit feeders tend to be more abundant in fine-grained sediments, because such sediments contain increased quantities of microorganisms, fine-grained particulate organic matter, and small ingestible inorganic particles. From the complex mixture of microorganisms, particulate organic matter, and dissolved organic matter, deposit feeders must obtain essential nutrients including amino acids, polyunsaturated long chain fatty acids, vitamins, and sterols.

Deposit feeders collect particles in a variety of ways that are associated with phylogenetic origins of the organisms and the environments within which they live (**Figure 15.7**). Representatives of many animal groups **swallow sediment** without particle selectivity, although there is an upper limit on the size of particle they can ingest. Many polychaetes have **tentacles**, which gather particles by means of a mucus-laden ciliated tract (**Figure 15.7a**). Sea cucumbers, such as the large northeast Pacific *Parastichopus californicus*, draw sediment into the mouth by means of a large crown of tentacles. Bivalves in the group Tellinacea use a separate inhalant vacuum hose **siphon** to suck up sedimentary grains (**Figure 15.7c**). In some other groups, the sediment is processed quite noticeably before a fraction is ingested by the deposit feeders. Many amphipods tear particulate material apart and ingest considerably smaller particles (**Figure 15.7e**). Fiddler crabs handle sediment extensively and ingest only the fine particulate organic matter; they reject the inorganic sand grains and drop them on the sand.

Deposit feeders differ in the depth of feeding below the sediment-water interface. **Surface browsers** use tentacles or siphons to collect surface sediment, which is rich in photosynthetic microbes such as diatoms. For example, spionid polychaetes have two tentacles that are pressed to the surface and thus collect both particles and benthic diatoms. At the other end of the spectrum, **head-down deposit feeders** (e.g., many vertical-tube-dwelling polychaete annelids) maintain their long axis vertical, consume particles at depth, and defecate at the surface.

A series of experiments gave us some important insights on how continental shelf deposit feeders deal nutritionally with the complex sediment to which they are exposed. In a classic series of experiments, B. T. Hargrave (1970) fed decaying leaves to a freshwater amphipod and found that its ability to digest and assimilate the material was very low, in contrast to its high efficiency at digesting bacteria. Similar results were found for several marine invertebrate deposit feeders of widely varying origin, including marine amphipods, gastropods, and sea cucumbers. The **microbial stripping hypothesis** states that particulate organic matter is relatively indigestible and that microbial organisms are, therefore, the main source of nutrition for deposit feeders. To be nutritionally useful for deposit feeders, therefore, **particulate organic matter (POM)** must be decomposed and converted by microbes into digestible microbial tissue. Particulate organic matter that derives from plants such as

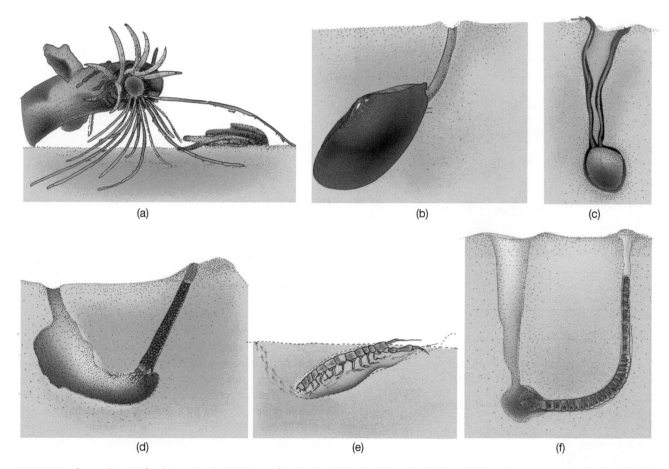

FIG. 15.7 Some deposit-feeding animals: (a) the surface tentacle feeder *Hobsonia*; (b) the within-sediment, tentacle-feeding bivalve *Yoldia limatula*; (c) the surface deposit-feeding siphonate bivalve *Macoma*; (d) the within-sediment-feeding Atlantic polychaete *Pectinaria gouldii*; (e) the surface-feeding *Corophium volutator*; and (f) the deep-feeding *Arenicola marina*. (Drawing of *Hobsonia* copied with permission from an original by P. A. Jumars)

sea grass is often rich in cellulose and indigestible, particularly because deposit feeders usually lack sufficient cellulase enzyme activity to digest the complex carbohydrates in the POM.

Particulate organic matter is decomposed by three processes, which often act simultaneously. **Fragmentation** involves the breakdown of large particles into smaller ones. This reduces the grain size and increases the surface area available for microbial attack. **Leaching**, the loss of dissolved materials from once-living organisms, is accelerated by mechanical fragmentation. Finally, **microbial decay** is the active use of POM nutrients by surface-bound microbes. As microbes—mainly bacteria—colonize, they enrich carbon-rich particulate organic matter with nitrogen (**Figure 15.8**). In intertidal environments, much decay can be attributed to marine fungi, especially in marsh grasses.

Grazing by deposit feeders on the benthic microbial community stimulates microbial productivity and, by extension, detrital decomposition (Figure 15.9). Oxygen consumption by microbes increases while consumers are grazing the organisms. The mechanism behind this stimulation is not well understood. Grazing may reduce the standing crop of bacteria and select for metabolically active cells with higher cell division rates. Grazing by some deposit feeders may break up particulate organic matter and make it more accessible to bacteria.

■ Particulate dead organic matter is also important in the nutrition of many deposit feeders.

Although microbes may be efficiently digested by deposit feeders, some low-level digestion and assimilation derives from the more refractory POM. Although the digestion and assimilation of particulate organic matter may be inefficient, POM is usually more abundant than microbes. Sediments in sea grass meadows contain large amounts of decaying sea grass, and deposit feeders cannot help but ingest much of this material. Thus, a poor rate of uptake may be balanced by the sheer abundance of the poor food source. Many other sources of POM exist in marine habitats, particularly the rain of recently dead phytoplankton, known as **phytodetritus**, in shallow embayments and on the continental shelf. Deposition of phytoplankton cells in the spring is a major source of sediment protein and probably sterols and vitamins, because the material is fresh and easy to digest. But as time goes by, the phytodetritus is degraded and becomes more refractory and probably is more difficult for deposit feeders to digest and assimilate (Mayer and Rice, 1992). Near shore, seaweeds may provide

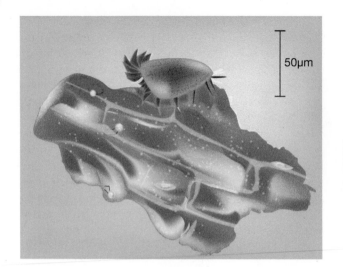

(a)

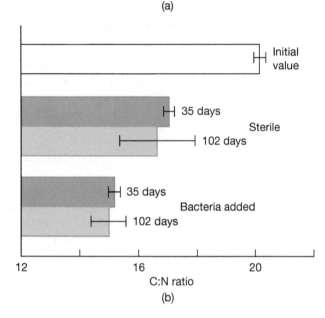

(b)

FIG. 15.8   (a) A piece of turtle grass (*Thalassia fesfudinum*) detritus and its microbiota. (Redrawn from Fenchel, 1972) (b) Change in the carbon-to-nitrogen ratio in particulate organic matter over time, with and without the presence of bacteria. (Modified from Harrison and Mann, 1975, © Blackwell Scientific Publications, Ltd.)

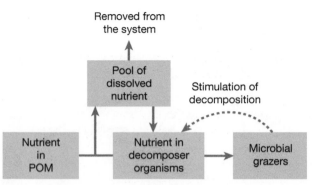

FIG. 15.9   The relationship between deposit-feeding microbial grazers and microbial decomposers, particulate organic matter, and the dissolved nutrients in the pore waters of sediments. Microbial grazers remove microbes, but they also stimulate decomposition by increasing microbial activity and by tearing apart particulate organic particles, which exposes more surface area to mechanical attack. (Modified after Barsdate et al., 1974)

a significant input of POM. As it turns out, seaweed detritus is far more digestible to deposit feeders than is sea grass detritus, and seaweed detritus can fuel deposit feeder population growth. In kelp forests, the rain of decomposing seaweeds supports large populations of benthic suspension feeders. Sea grass detritus may be rich in carbon, but other nutrients—including essential amino acids, vitamins, sterols, nitrogen, and phosphorus—may be in short supply.

Some quantitative estimates set some limits on the value of certain food sources. A number of studies suggest that the *typical abundance of bacteria in sediment can supply only a small fraction* of the energy requirements of a macrofaunal deposit feeder such as a polychaete or a bivalve mollusk. These estimates apply only to bulk calculations because specific fatty acids, amino acids, and vitamins may be necessary for specialized nutritional needs. Rich diatom mats

may be a more adequate food source, but only in intertidal and very shallow subtidal sediments, and these cannot be very important in even the relatively shallow waters of bays and estuaries, below the light compensation depth. In estuaries and on the continental shelf, the spring diatom increase is often followed by a sinking of phytodetritus, and this POM may be crucial in fueling the productivity of the deposit-feeding benthos. Microbial groups that are poorly known in sediments may produce essential fatty acids for deposit feeders. We are very ignorant of important microbial groups that may produce such substances.

In contrast to macrofauna, meiofauna probably depend mainly on a combination of bacteria and fine-grained particulate organic matter. Because of their small size, they cannot feed on particles much larger than 10–30 μm but can gain nutrition successfully from the bacterial fraction of the sediment.

■ Deposit feeders use a cocktail of enzymes and compounds with surfactant properties to digest organic matter from ingested particulate material.

A simple measure of carbon or nitrogen of sediments does not necessarily tell us how much food is available for deposit feeders. Much of the carbon may be bound up as indigestible complex carbohydrates such as cellulose, and nitrogen may also be bound up as indigestible nitrogenous compounds. Exposure of low-density organic particles from sediments to digestive enzymes shows that the particles are strongly enriched in protein and peptides, which confirms the idea that fine low-density particles are a possible food source for deposit feeders (Mayer et al., 1993).

Many larger deposit feeders employ **surfactants** (compounds with detergent properties that reduce the surface tension between water and hydrophobic organic compounds) in high concentration, which probably enables the stripping of organic material from particle surfaces (Mayer et al., 1997). This activity may allow some deposit feeders to digestively attack relatively refractory organic material in sediments. The surfactants include branched saturated or unsaturated fatty acids that are linked to amino acid residues (Smoot et al. 2003).

■ Microbes and particles comprise a complex renewable resource system for deposit feeders.

As we have discussed, many sediments are dominated by POM, but some consist mainly of microbes and particles. In intertidal soft-bottom flats, microbes such as benthic diatoms are the main food source for surface feeders. In these cases, microbes may be (1) free-living among the sedimentary grains, (2) attached to sedimentary grains, or (3) living as a mat on the sediment surface. Because the microbes themselves seem to be limited by some resource, the abundance of microbes at any one time is a balance between the microbial population growth rate and the grazing rate. At high grazing rates, the steady-state abundance of diatoms is kept at a low standing stock.

Some deposit-feeding invertebrates, such as polychaetes and gastropods, consume fine particles and bundle them into fecal pellets that are often not reingested. Deposit feeders may live in a mixture of fecal pellets and fine particles, ingesting only the latter. In some cases, the deposit feeder may try to get rid of the pellets. For example, the Pacific ampharetid polychaete *Amphicteis scaphobranchiata* has a specially modified branchium that flings fecal pellets out of the animal's feeding reach.

When such behavior is not possible, the deposit feeder must wait for the pellet to break down into its constituent particles before it will reingest the sediment. In crustaceans, pellets are often surrounded by a distinctive coating, and in mollusks, the sediment is bound together by mucus. As the pellets break down, they are probably colonized by microbes, so there is a value to having the particles sequestered as fecal pellets for a time. Presumably, the nutritive value of a new fecal pellet is far less than that of one that has had some time to simultaneously break down and be recolonized by microbes. In such a system, there will be an analogous equilibrium abundance of ingestible particles, which is determined by the competing rates of pelletization and pellet breakdown. Mud snails of the genus *Hydrobia* slow down feeding and may emigrate from microsites with fully pelletized sediments.

In some cases, feces may be enriched in organic matter, relative to the surrounding sediment. This occurs when the feeder is very selective and may even involve enrichment by microbial action in the hindgut of the deposit feeder. In such cases some deposit feeders ingest their own feces or the feces of other species and are **coprophagous**.

■ Many benthic animals do not feed directly on microorganisms but have symbiotic chemoautotrophic bacteria, which derive energy from dissolved ions in seawater.

Although many benthic animals feed actively on sediment, or on suspended organic matter (see section on suspension feeding), a large number of species depend on symbiotic bacteria, which may live intracellularly or in chambers in various organs, depending upon the group. Many bivalve mollusks, for example, have bacteria living intracellularly

in their gills. These bacteria oxidize reduced sulfur compounds. The oxidation processes provide energy, which is used by the bacteria to manufacture ATP, which in turn is used in bacterial cellular metabolism. Some species of the infaunal bivalve genus *Solemya* have a very small gut or lack one entirely. These forms rely exclusively on symbiotic sulfur-oxidizing bacteria. The animals are also tolerant to sulfide, which normally is quite toxic, especially to animals that use oxygen in metabolism. Mussels living near hydrocarbon seeps have intracellular bacteria in the gills. These bacteria rely on methane from the seeps for nutrition and energy. The bivalves rely exclusively on the bacteria for nutrition. This life habit is especially prominent in some deep-sea environments that are poor in organic matter but rich in sources of oxidizable sulfur compounds (see Chapter 18).

Burrowers and Sediment Structure

■ Deep feeders cause overturn of the sediment and strongly affect the soft-sediment microzone.

Sediments are strongly altered by burrowing animals. Donald C. Rhoads (1967) investigated the properties of burrowed sediments and found their mechanical properties to be quite different. The production of fecal pellets may increase the grain size of the sediment. A sediment with abundant deposit feeders that is wet-sieved tends to be dominated by fecal pellets, which are often about 50–150 μm in size. If the same sediment is placed in a blender and sieved, one finds that its constituent particles are closer to 50 μm or less. Burrowing, deposit feeding, and production of fecal pellets tends to make the sediment in the top few millimeters very watery, sometimes over 90 percent water. Recently, elegant sensors have been developed to measure changes in pressure within sediments. Using pressure sensors, the effect of daily and tide-related changes in burrowing intensity can be related to changes in pore water properties of sediments (Woodin et al. 2016).

■ Head-down deposit feeders create biogenically graded beds.

Many deposit feeders feed in a head-down position and defecate sediment at the surface (**Figure 15.10**). Head-down deposit feeders tend to ingest particles that are smaller than the average size for the sediment. Such animals may select fine particles and transport them to the surface, leaving a lag deposit of coarser material at depth. For example, the bamboo worm *Clymenella torquata* usually does not ingest particles greater than 1 mm. In poorly sorted muddy sediments, dense populations produce a **biogenically graded bed**, with small particles concentrated at the surface (**Figure 15.11**) (Rhoads, 1967). Such biogenically graded beds can be easily detected by walkers who suddenly encounter squishy sediment.

■ Hydrodynamic forces at the sediment–water interface cause sediment transport, which often induces switches from deposit feeding to suspension feeding.

Sediment-dwelling invertebrates often live on tidal flats where current energy causes extensive lateral particle transport with each tide and storm. If water turbulence

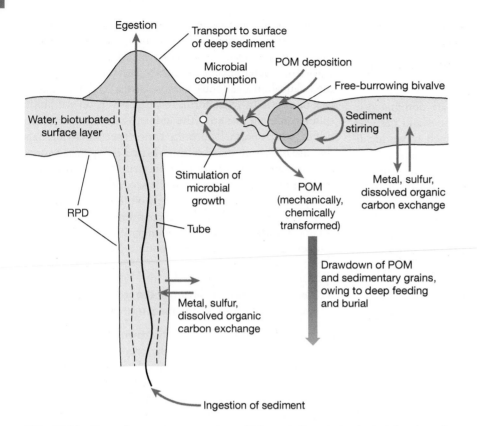

FIG. 15.10 General processes occurring within a sediment dominated by deposit feeders, including various transport processes. In deposit feeding, particles are taken up by a feeding organ, and some of them may be rejected before entering the gut. Particles may be packaged in fecal pellets, which are egested. As the pellets break down, the sedimentary grains are recolonized by microbes, which may be ingested and assimilated as the particles are ingested once again.

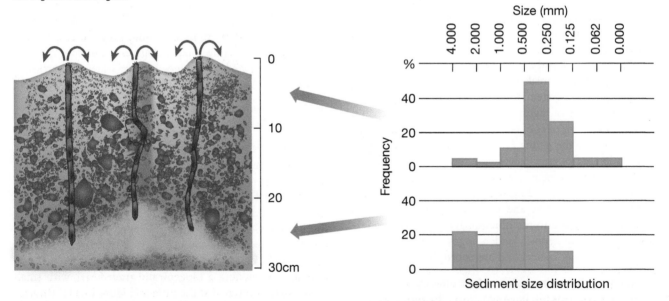

FIG. 15.11 Biogenically graded bed. Left: vertical reworking of intertidal sediments by the tube worm *Clymenella torquata*. Right: change in the vertical distribution of particle size as a result of vertical reworking of the sediment. (After Rhoads, 1967)

is sufficiently strong, deposit feeders may change their behavior significantly. In moving waters, some of the normally deposit-feeding tellinacean bivalves switch to suspension feeding by keeping the siphon within the burrow. This may be a reaction to particle saltation, which would bring the particles to the feeding organ passively.

Spionid polychaetes have cilia-covered tentacles, which in quiet water pick up particles from the sediment surface. If current speed is increased sufficiently to transport particles above the bottom, these worms deploy their tentacles in an erect spiral, which then serves as a suspension-feeding organ.

■ **Deposit feeders can optimize their intake by adjusting food particle size and gut passage time.**

Natural selection can be expected to optimize food choice and feeding rate to maximize fitness. Sediments with fine particles tend to be richer in microbial organisms, owing to the greater surface area per unit volume of the fine particles relative to coarse particles. If the expected surface-area relationship holds, then deposit feeders should select for fine particles.

Feeding rate and gut passage time may also be regulated according to food value. There may be an optimal feeding rate simply because feeding too quickly reduces the time available for digestion, whereas feeding too slowly may waste valuable time that could otherwise be applied to feeding on new material. This idea could be tested by consideration of foods of differing quality. If the cost-benefit approach is appropriate, deposit feeders should feed more rapidly on richer sediments. This has turned out to be true in several experiments on polychaetes, for which feeding and gut passage is steady.

Suspension Feeding

■ **Passive suspension feeders collect food by means of morphological structures that protrude into the flow and capture particles.**

Passive suspension feeders such as gorgonian corals and sponges commonly protrude a feeding organ into a mainstream current and collect particles as they encounter the feeding collection device. In a moderate unidirectional current, the best strategy for a colonial suspension feeder would be to deploy a network whose plane is perpendicular to the flow. But not all benthic passive suspension feeders adopt a vertical planar form, nor are their feeding structures always oriented upstream. Many suspension feeders are colonial and are bushlike (e.g., the hydroid coelenterates) or simply form a thin layer over the substratum (e.g., many sponges and corals). Although this may seem to be an inefficient way of feeding in a unidirectional current, a multidirectional orientation serves well when the current flow is complex. In many benthic habitats, water motion is oscillatory; the water just sloshes back and forth over the bottom. In other cases, tidal currents cause a complex reversal of flow. Under such circumstances, a bush-like shape will gather more food and oxygen than will a planar shape with individuals pointing upstream.

■ **Active suspension feeders are under some constraints similar to passive suspension feeders, but they also generate their own water currents to channel and ingest particles.**

Active suspension feeders create a feeding current to take in planktonic food. In many polychaete annelids and bivalve mollusks, ciliary currents draw particles toward the cilia, which capture particles and transport them down ciliated tracts (**Figure 15.12b, c**).

Processes near cilia are at low Reynolds number, and the cilia must directly reach out and capture particles. (See the discussion in Chapter 7 on planktonic larval ciliated feeding for more on this subject.) As transport occurs, the tracts reject unsuitable food particles. Many intertidal acorn barnacles use a different active strategy: The thoracic limbs move actively to capture particles that are drawn to and processed by appendages surrounding the mouth. Barnacles can adjust the orientation of the thoracic appendages at different flow velocities. At low velocity, the feeding cirri face into the current and capture particles. If the current passes a threshold velocity, the cirri are suddenly reversed and pointed downstream, to minimize drag and to capture particles passively. Barnacles living in wave-swept environments tend to have shorter and stouter cirri, which strengthen the feeding organs and reduce drag.

Many suspension feeders live infaunally and semiinfaunally in soft sediments. For example, the polychaete *Chaetopterus* lives in a U-shaped burrow (**Figure 15.12a**), and specialized parapodia drive an inhalant current into the tube. A sheet of mucus stretched between another pair of specialized parapodia captures particles, and this sheet is periodically rolled into a ball and passed to the mouth. The siphonate infaunal bivalve mollusk *Mercenaria mercenaria* creates a current by means of a ciliated gill. Water is drawn into an inhalant siphon, and the cilia strain and transport particles to a ciliated palp, which can reject poor food particles. Acceptable particles are then ingested and enter the gut (**Figure 15.12c**). Most soft-sediment suspension feeders rely on phytoplankton for food. In coastal waters, large numbers of detrital particles are in the water column, and these are digested poorly. Benthic algae, however, are often resuspended, and these may be an important food for benthic suspension feeders.

■ **Suspension feeders must be able to avoid clogging from heavy particle loads.**

When water moves above the surface of a soft sediment, the erosive power of fluids eventually saltates particles into suspension. For suspension feeders, this process dilutes their plankton food source with unwanted inert particles such as sand grains. Higher particle loads usually clog suspension-feeding organs, such as ciliary tracts and siphons. At very high water velocity, sediment moves laterally and ripples form. As crests and troughs alternately pass over a suspension-feeding animal, it becomes difficult for the animal to maintain a stable feeding position. Water eddies often form in the trough of a ripple, which creates a complex flow pattern.

Infaunal animals have a variety of means of dealing with increasing particle flux near the sediment–water interface. Some suspension-feeding siphonate bivalves have a ring of papillae at the siphon opening, which filters out sand grains. The inhalant siphon of some tellinacean bivalves is lined with papillae, which can help in rejecting unwanted sand. Most marine bivalves can "sneeze," or suddenly expel water and an overload of sand through the inhalant siphon. Even suspension feeders on hard surfaces may suffer clogging when sedimentary loads become high. When sediment is deposited on their colonies, some corals can produce mucus, which transports the clogging material off of the colony.

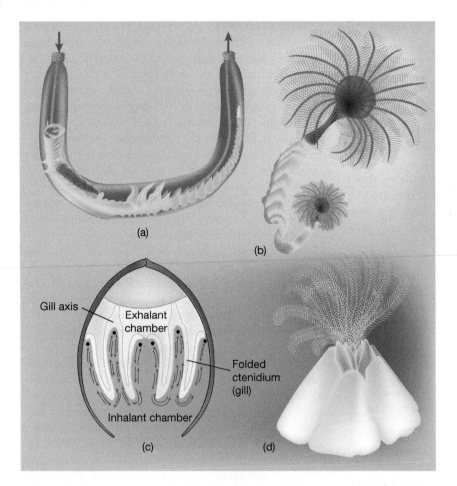

FIG. 15.12 Some suspension-feeding invertebrates. (a) The active suspension-feeding parchment worm *Chaetopterus*. (b) The suspension-feeding polychaete *Serpula*, which uses ciliary currents to draw particles to tentacles. (c) Cross section of a bivalve mollusk, an active suspension feeder (arrows denote ciliated tracts transporting particles). (d) The acorn barnacle *Semibalanus balanoides* with cirri protruded like a basket, the concave side pointing into the flow, and particles trapped on feeding appendages, which are then withdrawn. If the flow increases beyond a point, the basket is reversed, to maintain stability in the flow.

PARTICLE SELECTIVITY BY SUSPENSION FEEDERS

■ **Many suspension feeders can select for nutritionally valuable particles and reject poor particles before they enter the gut.**

Particle selectivity would be strongly adaptive to allow suspension feeders to reject nonnutritive particles such as resuspended sand grains, even before they enter the gut. Otherwise, the bulk of the gut's volume will be filled with nonfood material.

Ciliary suspension feeders, such as polychaetes with tentacle crowns and bivalves with large gills, collect particles on the ciliated tracts. Can such organs reject particles of low nutritive quality? We know most about suspension-feeding bivalves such as mussels and oysters. These bivalves bring material into the mantle cavity and capture particles, probably mostly by direct interception on cilia in the gills. Then these particles are transferred, via a ciliated groove, to another highly ciliated organ, known as the palp. The palps are very strongly folded and ciliated

and often can reject particles that are poor in nutritive quality or even toxic.

■ **Selectivity after particle collection can be studied with a surgical endoscope.**

Where are the sites of selectivity? Bivalves have long been known to select for nutritionally valuable particles, by simply comparing the organic content of particles that are rejected and released back on the sea bed with the usually higher organic particles that are passed to the mouth and ingested. But the mechanisms of particle selection are particularly difficult to study because it is hard to observe within the mantle cavity of a bivalve. The use of a surgical endoscope (**Figure 15.13**) helped solve this problem. This instrument is a glass lens, no more than 2 mm wide, that can be inserted within the bivalve's mantle cavity to observe movement of particles with video and even to direct collection of particles using a micropipettor. In the mussel *Mytilus edulus,* particles were captured on the gills and transported

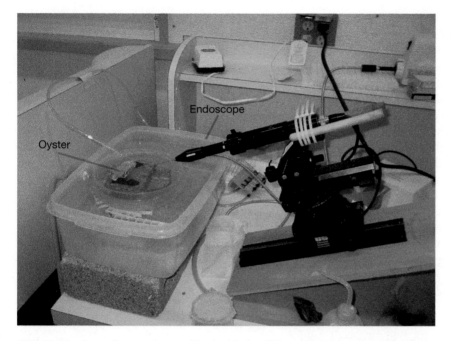

FIG. 15.13 An endoscope inserted into a bivalve. This surgical telescope allows us to observe how internal mantle cavity organs of a bivalve function in processing particles without surgically altering the mantle organs, such as gills and palps. (Photograph by Jeffrey Levinton)

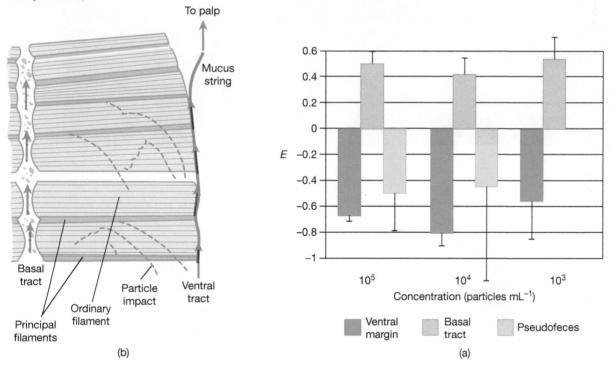

FIG. 15.14 (a) The gill surface of an oyster. Note the different directions in which particles are transported by ciliary tracts. Particles transported to the basal tract are transported to the palp and then to the mouth, where they are ingested. Particles transferred to the ventral tract are eventually rejected by the palps as pseudofeces. (b) Selectivity for algae relative to non-nutritive particles at total particle concentrations of 10^3–10^5 particles per milliliter. Positive values indicate selectivity for algae. Note positive values for basal tract, which delivers particles to the mouth. Algal particles are depleted in ventral margin, where particles are rejected as pseudofeces, which also have less algae than was proportionally fed to the bivalve.

to a ciliated tract on the ventral part of the gill. There the particles were enmeshed in mucus and transported to the palps, which were capable of sorting and rejecting poor particles and transporting more nutritive particles to the mouth (Ward et al., 1998a, 1998b).

The structure of oyster gills (**Figure 15.14a**) is rather different from that of many other bivalves. The gill is folded, and the "valleys" of the folds are lined with cilia that beat dorsally (toward the bivalve hinge). The "hills" also are covered with cilia and can either transfer particles dorsally (refer to the figure or you will become confused!) or ventrally. Some particles were captured by cilia on the hills and were transported to a ventral tract, enmeshed in mucus, and rejected. But others were trapped by other cilia in the lower folds, transported to a dorsal tract, and eventually transported to the mouth (**Figure 15.14a**). Thus oyster gills could sort particles on the gill itself, which was not true of mussels.

A creative experiment demonstrated the amazing extent to which selectivity could occur. Sandra Shumway and colleagues (1985) pioneered the use of the **flow cytometer**—a laser-based device that analyzes fluorescence and other characteristics to differentiate particle types. They fed mixtures of algal cultures to see whether species of algae were preferentially ingested, which could be told by comparing available food particles in the water with the material rejected in the pseudofeces. On comparing samples of algal mixtures, it was clear that a variety of bivalves were selectively rejecting certain phytoplankton species and preferentially retaining others for ingestion.

This result was applied to the oysters. Ward et al. (1998a) sampled the particles that had been sorted to the dorsal and ventral tracts. One could sample these two tracts and compare them with the particles in the water, representing the available food. Ward and colleagues fed the oysters *Crassostrea virginica* and *C. gigas* with aged cordgrass (*Spartina alterniflora*) detritus, which they had ground to match an alga (*Rhodomonas*) in particle size. The detritus was very poor in nutrients and essentially indigestible. The results were immediately apparent, especially because *Rhodomonas* is vivid red. The particles in the dorsal tract were clearly red and the ventral tract was tan, the color of the cordgrass detritus. This demonstrated that it was the gill doing the sorting, not the palp, as in most bivalve species. The flow cytometer was able to count particles (**Figure 15.15**) that were selected for comparison with those that were rejected. In other bivalve species, the gill just captures particles, which are transported in ciliary tracts to the palp, where selection occurs. In all cases, bivalves can distinguish

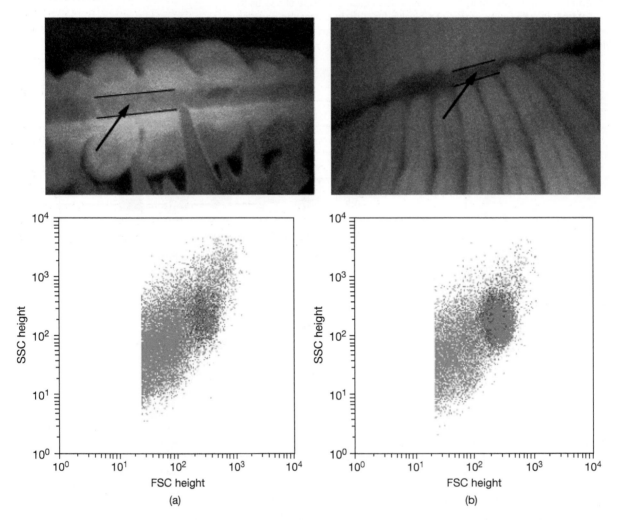

FIG. 15.15 Plot showing distinction between two particle types, the red alga *Rhodomonas lens* (red areas of graphs) and cellulose-rich particles derived from the decomposition of the cordgrass *Spartina alterniflora* (blue areas of graphs): FSC, forward light scatter (estimate of particle size); SSC, side scatter (estimate of index of refraction). (a) Particles sampled from a ciliated tract (arrow in photo) that collects rejected particles; note the enrichment of cordgrass detritus. (b) Particles sampled from another ciliated tract (arrow in photo) that collects particles that will be passed to the palp and then ingested. Note the increased abundance (big red dot) of *Rhodomonas lens* and relatively smaller amount of cordgrass detritus.

nutritive particles and eventually reject poor particles *before* they enter the mouth. This is a tremendous advantage to the bivalve, since the gut does not have to waste time and energy on poor food particles, which are abundant in the environment.

- **Suspension feeders may live in current regimes that deliver particles in uniform currents, but sometimes flow and particle supply direction may be very complex.**

Suspension feeders such as barnacles and sea squirts may live in widely different current regimes depending on their living position relative to the sediment–water or rock–water interface. A directional current may become far more irregular near the seabed. This has effects on species of different sizes and especially on species that may experience complex flow as small juveniles but strong directional flow after the juveniles grow to be larger adults. Small colonies of the tropical sea fan *Gorgonia* have an irregular shape and no preferred orientation, in contrast to larger colonies, whose fan shape is oriented approximately perpendicular to unidirectional currents. Near-bottom currents are erratic in direction, owing to irregularities of bottom topography and surrounding erect organisms. As the colony grows, it changes its orientation and grows into a fan shape that faces perpendicular to the mainstream current. This will maximize food particle interception.

This difference of adaptation to unidirectional and variable current direction can be seen among species of feather star crinoids. Feather star species found in crevices generally experience multidirectional currents and have their pinnules arranged in four rows at approximate right angles, which maximizes food capture from several possible directions. By contrast, the erect Caribbean feather star *Nemaster grandis* protrudes strongly into unidirectional currents and has its pinnules arranged in a plane, which maximizes capture under these circumstances. In the brittle star *Ophiothrix fragilis*, tube feet arise from either side of the tentacle and are also arranged in a plane (**Figure 15.16**). Food particles are captured by the tube feet and are compacted into a mucus-clad bolus that is passed down the arm.

Benthic Carnivores

- **Carnivory relies on mechanisms of prey search, location, seizure, and ingestion.**

Carnivorous animals hunt and eat other animals (**Figure 15.17**). Defining benthic carnivores is not entirely straightforward because those that eat zooplankton are as much suspension feeders as they are carnivores. Of necessity, most carnivores are mobile and have a variety of means of prey detection. Many species are capable of detecting soluble substances emanating from the prey. The European sea star *Astropecten irregularis* moves along the sediment surface but can detect its prey within the sediment. Many carnivores orient to upstream prey. Specialized bivalve mollusks known as septibranchs detect prey by chemical means. A specialized pumping septum moves suddenly, expels water through the exhalant siphon, and draws water plus prey into the inhalant siphon (**Figure 15.17b**).

Vision is a common means of prey detection. Bottom-feeding birds, crabs, fishes, and cephalopods such as cuttlefish all detect prey visually. Visual detection is usually accompanied by sophisticated and rapid eye–motor coordination. The oystercatcher *Haematopus ostralegus* can dash onto an open mussel as a wave recedes and plunge its beak into the mussel, severing the prey's adductor muscles, thus making it helpless. More rarely, the oystercatcher hammers with its bill and crushes the shell. In either case, the bird assesses the size of the mussel and tends to take prey that are larger than average. Lobsters and crabs use both chemical detection and vision in predation, and can rapidly attack and immobilize prey.

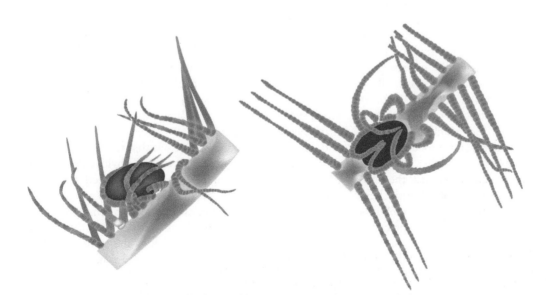

FIG. 15.16 The food-collecting wave of a suspension-feeding brittle star, showing the planar arrangement of the pinnules. (From Warner and Woodley, 1975)

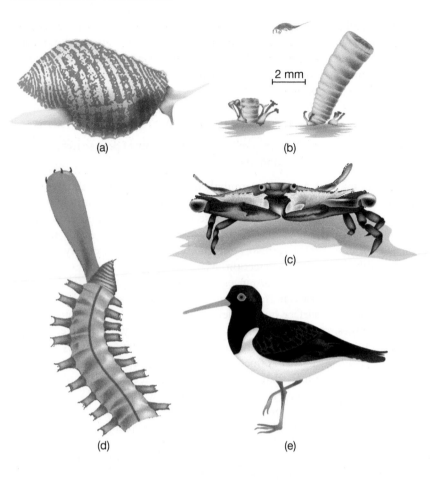

FIG. 15.17 Some marine benthic carnivores. (a) Gastropod *Nucella*, which uses a specialized radula and buccal mass to drill holes in barnacles and bivalve mollusks. (b) Bivalve mollusk *Cuspidaria*, which uses a pumping septum to suck up small prey. (c) Crab *Callinectes sapidus*, whose strong claw can crush mollusks. (d) Polychaete *Glycera*, which has a proboscis armed with hooks, used in seizing and tearing prey. (e) The oystercatcher *Haematopus ostralegus*, a predator on intertidal bivalve mollusks.

Odor detection is crucial in the behavior of many predators. Many prey organisms release signals such as excretory products into the water column, and predators use stereotyped movements to seek increased concentrations of the prey odors. In Chapter 6, we discussed the odor-seeking behavior of the blue crab *Callinectes sapidus*, which allows it to locate odors emanating from the exhalant siphon of buried clams. A **threshold concentration** of a signal molecule is required for detection. The relationship of this threshold to actual behavior is poorly understood in marine systems (Weissburg et al., 2004). On the other side of the relationship, prey also use odor signals to detect the presence of predators. Crabs are major predators of sessile benthos, and many are capable of cracking shells of mollusks. In the intertidal zone, snails and mussels can detect the odor of conspecifics being dismembered by crabs or even the odor of predatory crabs. As a quick response, they may move away. The mud snail *Tritia obsoleta* rapidly burrows into the sediment when it detects the odor of conspecifics that have been crushed by crabs. If predation persists in the area, some snails may devote resources to thickening their shells. This can be demonstrated by placing crabs in cages near to molluscan prey (Trussell and Smith, 2000).

The several strategies for attacking and seizing prey are obviously related to the mode of prey detection. Many predators are essentially sessile and must wait for prey to arrive. Anemones usually remain fixed to a hard surface and have access only to prey that swims or falls in contact with the tentacles. One large eastern Pacific anemone, the intertidal *Anthopleura xanthogrammica*, lives in low intertidal pools and depends for food upon mussels that fall from above.

The handling of prey varies with phylogenetic background because morphologies and neurodetection mechanisms are so disparate. Seizing prey involves some sort of appendage, such as a crab claw or a starfish arm. Many crab species have large crushing claws with denticles that enable handling of prey. Some crabs, such as the stone crab *Menippe mercenaria*, have robust claws and musculature and can crush thick-shelled mollusk prey. Others, such as the shore crab *Carcinus maenas*, are not terribly strong and have trouble crushing mussels unless they discover a weak spot in the shell. Some crabs repeatedly apply a crushing load to bivalves. Eventually, after several applications of pressure, the shell fatigues and can be crushed. Some tropical crabs can easily peel the shell of a snail to expose the soft parts. Many crabs and lobsters have a crushing claw, but the other claw is specialized for cutting tissue. Crushing claws have a high mechanical advantage and greater muscle mass (**Figure 15.18**). Spider crabs use pincers to rip apart seaweeds, sea stars, and other macroinvertebrate prey. Polychaetes, such as some species of *Glycera* (**Figure 15.17d**), have a protrusible proboscis with hook-shaped teeth; other polychaetes have large chitinous jaws that can tear prey apart.

In the gastropods, drilling is a specialized way of penetrating prey that have exoskeletons. This occurs in the prosobranch families Muricidae (*Urosalpinx, Murex*), Naticidae (*Polinices*), and Thaiidae (*Nucella*) and involves alternations of mechanical rasping and chemical secretions from an accessory boring organ. Some octopods drill holes in their molluscan prey and may inject a paralyzing venom through the hole.

Cone snails of the genus *Conus* consist of several hundred species living mostly in the tropics and subtropics. Venom, consisting of a cocktail of toxins, is produced by epithelial cells in a venom duct and injected into the proboscis by a squeeze of a specialized bulb. A highly modified radular tooth punctures the prey and delivers the venom (**Figure 15.19**). The shape of the proboscis varies with the method of capture and the type of prey. In some fish-hunting species, the snail has a part of the proboscis modified into a

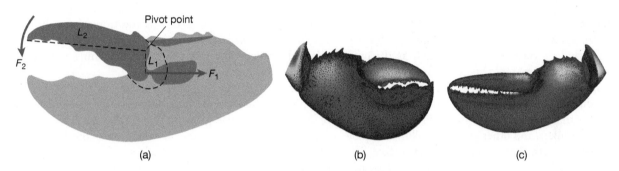

FIG. 15.18 Claws of the lobster *Homarus americanus*. (a) Features of the claw: forces and pivot of the claw apparatus. (b) The crusher claw. (c) The cutter claw. (After Elner and Campbell, 1981)

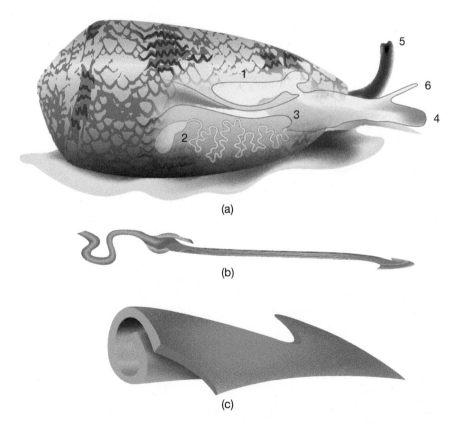

FIG. 15.19 (a) The stinging apparatus of a cone snail: (1) harpoon sac; (2) venom gland; (3) pharynx; (4) proboscis; (5) siphon; (6) eyestalk. (b) Harpoon-shaped specialized radular tooth. (c) Close-up of tip. (Courtesy of Dietrich Mebs)

lure, which attracts fish. The attraction is fatal because the disposable tooth harpoons the fish, paralyzes it, and reels in the prey, which is swallowed whole.

The toxin is usually a mixture of as many as 50 highly specific peptides, which are short chains of amino acids that usually attack a group of cellular ion channels, such as sodium channels. The peptides appear in the venom duct as precursors that are cleaved by enzymes in order to obtain final toxicity as they are injected into the prey. The combination of different venoms overwhelms the prey and immobilizes it almost immediately after the harpoon-like tooth is injected.

Benthic Herbivores

■ **Benthic herbivores are divided between microphages and macrophages.**

The food of benthic herbivores (**Figure 15.20**) can be divided by size class into two major categories. Benthic microalgae include a variety of groups, such as diatoms, cyanobacteria, and microscopic stages of seaweeds. These organisms may form a thin layer on a rock surface or on the surface of sediment. **Microphages** have a range of morphological features that allow them to graze efficiently on this layer. Chitons, limpets, and other grazing mollusks employ a radula, a belt of teeth that scrape along the surface. The movement of the subradular membrane over a cartilaginous portion of the buccal mass erects the teeth and scrapes them

over the surface. The radula and buccal mass are retracted, and food trapped on the teeth is delivered to the buccal cavity. This feature can be used on rocks, and limpet grazing scars are common on rocky shores. Radular scraping is also employed by gastropods feeding on soft-sediment surfaces. Some polychaetes can graze on sediment microalgae by pressing their tentacles onto the surface and collecting particles and microalgae, which are transported to the mouth by means of a ciliated tract.

A wide variety of herbivorous **macrophages** can tear apart and consume macroalgae and marine higher plants. Periwinkles, for example, use the radula to rasp and tear apart delicate seaweeds such as species of the sea lettuces *Ulva* and *Enteromorpha*. Sea urchins possess an Aristotle's lantern, which is a complex of calcareous teeth, ligaments, and muscles. This device can tear apart a variety of seaweeds, and some urchins are even capable of devouring relatively less digestible sea grasses, such as the tropical Caribbean *Thalassia testudinum*. Sea grasses are usually thought to be relatively indigestible, but a number of urchins and fishes regularly consume them.

Many crustaceans are also herbivorous. Many smaller amphipods and isopods feed on relatively soft seaweeds or on the microalgae growing on seaweed surfaces. A variety of fishes are also efficient herbivores. The jaw teeth of parrot fishes (Scaridae), which are fused into plates, are capable of cutting material from the surface of coral skeletons. Surgeon fishes (Acanthuridae) also can scrape

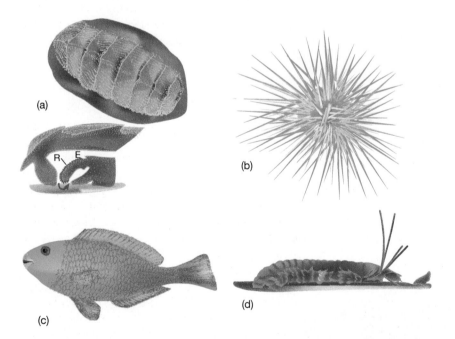

(a)

(b)

(c)

(d)

FIG. 15.20 Some benthic herbivores. (a) The chiton *Tonicella*, a scraper of microalgae; inset shows anterior sagittal cross section, indicating the action of the radular tooth belt in scraping algae from the substratum (R, radula; E, esophagus). (b) The sea urchin *Arbacia*, which uses a toothed Aristotle's lantern to scrape microalgae or to tear apart seaweeds. (c) A parrot fish, which uses specialized fused teeth to scrape algae from coral surfaces. (d) The nereid polychaete *Nereis vexillosa*, which tears apart sea lettuce with buccal hooks. (Copied from an original by K. Fauchald)

algae from corals, and the two groups are major causes of erosion on coral reefs. Even smaller invertebrates, such as isopods and polychaetes, have sufficiently strong mouth parts to tear apart algae. The buccal hooks of some species of the sandworm *Nereis* are employed in tearing apart soft green algae.

Although herbivores are usually mobile, many rock scrapers are capable of homing. A home base may provide a reference location, allowing efficient exploitation of the renewable resource of microalgae living on hard surfaces. The eastern Pacific owl limpet *Lottia gigantea* and the limpet *Patella longicosta* both defend a home spot, which often appears as a scar on the rock.

Although most benthic herbivores search for suitable food, some wait for the food to come to them. Many urchins capture drifting seaweed fragments on their dorsal spines, and dorsal tube feet transfer them toward the mouth. Sand-flat polychaetes such as species of *Nereis* and *Lumbrinereis* can drag seaweed fragments down into their burrows. In some cases, the downward dragging is incomplete, and some polychaetes practice farming by attaching fragments of *Ulva* to their tubes and letting them grow.

■ Some benthic herbivores can feed on highly indigestible plant material.

Most marine herbivores are restricted to relatively soft seaweeds and microalgae, with a minimum of relatively indigestible complex carbohydrates, such as cellulose. A small number of species, however, have adapted to such difficult food sources. Some invertebrates can bore into wood and digest it or may depend on the marine microbiota living in the wood. The wood-boring bivalves *Teredo* (shipworm) and *Bankia* scrape the wood particles, and symbiotic bacteria synthesize the digestive enzyme cellulase to attack the cellulose (**Figure 15.21**). The wood-boring isopod *Limnoria* can also digest cellulose, but it requires wood-boring fungi as a source of nitrogen. Wood-boring bivalves

derive their nitrogen from symbiotic nitrogen-fixing bacteria, since nitrogen is not present in sufficient quantities in the wood.

Sea grasses, such as eelgrass (*Zostera*), salt marsh cordgrass (*Spartina*), and Caribbean turtle grass (*Thallassia testudinum*), are relatively indigestible to most marine consumers because of the abundance of cellulose the grasses contain. Some small grazers consume the microalgal surface layer, but relatively few species can consume, digest, and assimilate material from the grass itself. As mentioned earlier, a few species of urchins can deal with turtle grass. Eelgrass and cordgrass are remarkable for the minuscule amount of grazing they experience from marine herbivores. An interesting exceptional species is the green turtle, *Chelonia mydas*, which can digest cellulose derived from turtle grass (Fenchel et al., 1979). It has a hindgut that bears a functional resemblance to the stomachs of ruminant mammals, such as cows and horses. The postgastric gut region is greatly elongated, and postgastric fermentation is facilitated by the presence of digestion-aiding symbiotic bacteria and protozoa. In the Caribbean, green turtles often feed within a restricted area, and cropping of leaves encourages new growth with lower concentrations of complex structural carbohydrates and higher nutritional content (Moran and Bjorndal, 2007). In effect, the turtles are prudently altering the sea grasses. Green turtles have also been found to be feeding either on sea grasses or seaweeds, and this may relate to the difficulty of switching microfloras in the hindgut to specialize on efficiently digesting both food types in rapid succession. It may be that the low grazing pressure on eelgrass and cordgrass is misleading. Valentine and Heck (1999) have argued that many tropical sea grasses are consumed by urchins and fishes to a far greater degree than has been appreciated.

■ Benthic plants have evolved both mechanical and chemical defenses to deter herbivory.

The sessile habit of benthic plants makes them very susceptible to herbivores, but many groups have evolved **mechanical and chemical defenses** to deter feeding. Many marsh grasses are difficult to tear apart when alive. The common cordgrass *Spartina* secretes silica particles, which can exceed 1 percent by mass and make grass blades difficult to chew. Many seaweeds, such as red calcareous algae, are calcified. It is common for these seaweeds to survive in areas of intense herbivory and to dominate hard substrata in both the tropics and high latitudes. In addition, many marine plants have evolved chemical defenses. The cordgrass *Spartina alterniflora* synthesizes cinammic acid esters of glucose, which are stored in vacuoles and deter herbivores. Lignins also deter predation by inhibiting digestion of herbivores. In the tropics, many seaweed species on coral reefs combine calcification and toxic compounds to deter herbivory. Calcified seaweeds always have such chemical defenses, suggesting that calcification is also an adaptation to deter herbivores, such as fishes and urchins (Paul and Hay, 1986).

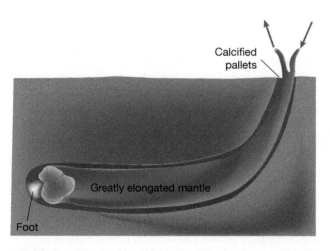

FIG. 15.21 Cross section showing the living position of the wood-boring bivalve mollusk *Teredo*. (After Trueman, 1975, *The Locomotion of Soft-Bodied Animals*, with permission of Edward Arnold [Publisher] Ltd.)

■ CHAPTER SUMMARY

- Epibenthic organisms live attached to the surface and usually protrude into the flow. Infaunal benthos live below the sediment–water interface. Demersal animals are mobile but associated with the seabed. Interstitial animals are elongate and live among sedimentary grains.

- Benthic feeders include deposit feeders, suspension feeders, herbivores, carnivores, and scavengers.

- Soft sediments are a mixture of inorganic particles, organic particles, and pore water. Grain size determines the distribution of benthos and increases with increasing current strength. In very shallow, sandy, wave-swept bottoms, currents generate ripples, bars, and other sedimentary structures.

- Sediments range from packed sand to elastic mud to more watery mud. Organic molecules may make sediment sticky and prone to cracking as burrowers move through. Because they move through tight spaces, interstitial animals from many different phyla have evolved a wormlike shape and a simplified external body plan.

- Sediments consist of an oxygenated layer overlying an anoxic zone. In quiet water there is usually a vertical zonation of microorganisms. In the oxygenated surface layer, aerobic microorganisms

most efficiently break down organic matter and coexist with aerobic photosynthetic microorganisms. In the anaerobic zone beneath, microorganisms may use substrates as hydrogen acceptors. Sulfate-reducing bacteria produce H_2S.

- Deposit feeders ingest sediment and derive their nutrition from microalgae, particulate organic matter, and to a smaller extent from bacteria. Particulate matter decomposes by fragmentation, leaching, and microbial decay. Burrowing and feeding by benthic animals stimulates microbial activity. Particulate dead organic matter is also important in the nutrition of many deposit feeders.

- Deposit feeders use a cocktail of enzymes and compounds with surfactant properties to digest organic matter. Many benthic animals do not feed directly on microorganisms but rather harbor symbiotic chemoautotrophic bacteria, which derive energy from dissolved ions in seawater.

- Deep feeders cause overturn of the sediment and strongly affect the soft-sediment microzone. Head-down deposit feeders create biogenically graded beds, with finer particles often transported toward the surface and deposited as feces. Deposit feeders can optimize their intake of food by

adjusting the particle size to be ingested and its gut passage time.

- Active suspension feeders create a current, usually with cilia, to draw particles toward a feeding structure. Passive suspension feeders protrude the body into the current and capture particles directly by impact without the aid of an active current.

- Suspension feeders may be capable of extensive particle selectivity, which allows the feeder to avoid ingesting too many particles of low nutritive content, which may be abundant in the water column.

- Organisms must adapt to strong laminar flow above the bottom, but near the surface they experience more complex flows. Epibenthos must have traits to minimize pressure drag by changing orientation.

- Benthic carnivores rely on prey search, location, seizure, and ingestion. Odor detection allows predators to cover large ranges, even if their traveling velocity is relatively low.

- Benthic herbivores are divided between microphages and macrophages. Some benthic herbivores can feed on highly indigestible plant material. Benthic microalgae, seaweeds, and sea grasses have evolved both mechanical and chemical defenses to deter herbivory.

■ REVIEW QUESTIONS

1. How do active and passive suspension feeders differ?

2. What type of hydrodynamic condition do well-sorted sediments reflect?

3. What is a burrowing anchor, and why is it required in a burrowing organism?

4. What do most interstitial marine animals have in common, in spite of being from quite different taxonomic groups?

5. What factors help to determine the depth of the redox potential discontinuity?

6. Why do bacteria of different types tend to dominate in muddy sediment at different depths below the sediment–water interface?

7. What is the microbial stripping hypothesis?

8. What are the components of decay of particulate organic matter in sediments?

9. How does a bivalve such as *Solemya*, which lacks a gut, manage to derive its nutrition?

10. Why and under what conditions do some benthic infaunal species switch between suspension feeding and deposit feeding?

11. How can sessile epibenthos reduce pressure drag?

12. Why are many suspension-feeding structures not simple sieves, whose interfiber distance can be used to predict the diameter of particles that can be captured?

13. What is the advantage to carnivorous crustaceans of having differentiated crusher and tearing claws?

14. Why is it possible for some marine animals to digest cellulose, which is nearly indigestible for most organisms?

Visit the companion website for *Marine Biology* at www.oup.com/us/levinton where you can find Cited References (under Student Resources/Cited References), Key Concepts, Marine Biology Explorations, and the Marine Biology Web Page with many additional resources.

CHAPTER 16

The Tidelands
Rocky Shores, Soft-Substratum Shores, Marshes, Mangroves, Estuaries, and Oyster Reefs

Coastal benthic habitats are among the most productive marine environments. They receive a high nutrient supply, which is influenced both by terrestrial nutrient sources and by strong coastal phytoplankton production. The richness of these habitats also makes them feeding and nursery grounds for migratory species, particularly fishes, crustaceans, and birds. The elevated nutrient supply and the visitations by predators often lead to cycles of population increase of prey species, followed by population crashes because of predation. The great abundance of organisms also leads to rapid depletion of resources and strong interspecific competition. Because tidelands are so accessible to ecologists, we know most about marine ecological processes from these habitats.

Rocky Shores and Exposed Beaches
Vertical Zonation and the Protected, Wave-Swept Coast Gradient

■ **Vertical zonation, the occurrence of dominant species in distinct horizontal bands, is a nearly universal feature of the intertidal zone, but many localities do not "obey" the rules.**

The **intertidal zone** is the shoreward fringe of the seabed between the highest and lowest extent of the tides. An important feature of the rocky shore intertidal zone and, to a lesser degree, the soft-sediment shore, is **vertical zonation**, the occurrence of dominant species in distinct horizontal bands (**Figure 16.1**). For example, a general pattern

of zones is found throughout temperate and boreal rocky shores (**Figure 16.2**). From highest to lowest, the zones are (a) a black lichen zone, (b) a periwinkle (littorine gastropod) zone with sparse barnacles, (c) a barnacle-dominated zone either overlapping with a mussel-dominated zone or with mussels below, and (d) a zone dominated variously, but usually by seaweeds. On North American rocky shores, mussels of the genus *Mytilus* dominate below the barnacle zone. Vertical zonation occurs on sand- and mudflats, but it is rarely as distinct as on rocky surfaces.

While overall patterns of zonation exist, on a gross scale, it is equally noticeable how often there is *not* a predictable pattern of dominance at different levels of the tide zone. Most commonly, one sees **patches** dominated by different species at the same tidal levels or patches unoccupied by organisms.

■ **Different assemblages occur in protected and wave-swept waters in the same area.**

Figure 16.2 shows an expansion of the vertical extent of organisms in wave-swept areas and differences in the types of organisms that occur in wave-exposed sites relative to protected waters. In some cases, wave-exposed sites have species that do not occur in protected waters. On the Pacific coast of North America, the large California sea mussel *Mytilus californianus* is found in exposed wave-swept coasts, but protected bays have a smaller species of mussel. In other cases, it is more a matter of relative dominance. In protected temperate waters of northern Europe and New England, seaweeds dominate rock surfaces, whereas barnacles and mussels dominate more wave-exposed open coasts (**Figure 16.2**).

FIG. 16.1 Zonation on a rocky shore in British Columbia. Seaweed zone below and mussel zone (dark blue color) above. (Photograph by Jeffrey Levinton)

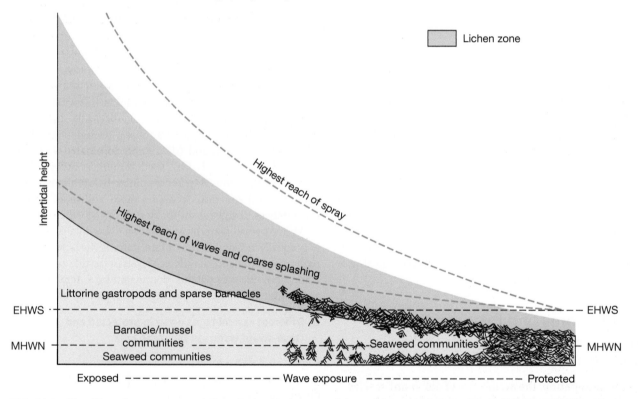

FIG. 16.2 The effect of wave exposure in broadening biotic zones of the rocky intertidal zone of the British Isles: EHWS, extreme high water spring tide; MHWN, mean high water neap tide. In quiet-water habitats, represented on the right of the diagram, the mid-intertidal zone is dominated by seaweeds, but with more wave exposure, barnacles and mussels come to dominate. The upper intertidal is dominated by herbivorous snails in all cases with a band of lichens toward the top. In strongly wave-swept habitats (left), the vertical zones are expanded. (After Lewis, 1964)

Factors Affecting Rocky Intertidal Organisms

■ **The intertidal zone is alternately a marine and a terrestrial habitat. At the time of low tide, heat stress, desiccation, and shortage of oxygen increase, and opportunities for feeding and respiring decrease.**

With increasing tidal height, more of the shore is exposed to air for a greater proportion of the day. The highest part of the intertidal zone is essentially a terrestrial environment, but organisms living at the low-tide level are less affected by aerial exposure. At low tide, intertidal sandy sediments may be quite dry, but finer-grained sediment retains water and many organisms can maintain activity within burrows and even at the sediment surface.

At the time of low tide, marine organisms face both heat and desiccation stress. On a hot day, invertebrates rapidly heat up, although they possess several adaptations to counteract thermal stress. If body fluids become warm past a critical temperature, physiological function and even the stability of proteins may diminish. In the worst case, intertidal organisms can literally dry up. In summer, this happens commonly to fragile species, such as thin green seaweeds. Organisms such as barnacles and mussels survive drier and more exposed areas because their shells can enclose water at low tide and seal the soft parts from desiccation. Organisms lacking external skeletons do not usually occur on open rocks exposed to the sun but are more common in moist cracks or in tidal pools (**Figure 16.3**).

Heat and desiccation stress vary on quite small spatial scales. As might be expected, sessile animals on sun-exposed and flat rocks will gain far more heat than those in a moist crack or in the shade. The timing of low tide during the day also can have profound effects on the heat inputs experienced by sessile intertidal organisms (Helmuth, 1999). In the outer, exposed coast of Washington, for example, spring low tides in summer come in the morning, when air temperatures are often only 15°C. In contrast, summer spring low tide in Puget Sound comes at midday, when air temperatures can surpass 25°C. Small-scale spatial differences in microhabitat and timing of low tides may, therefore, have far greater effects than overall annual variation and even broad-scale latitudinal variation, at least on the west coast of North America, where climatic latitudinal gradients are slight. On the east coast of North America, the latitudinal gradient is probably the main factor in heat stress differences between localities over long latitudinal-geographic distances.

Body size and **body shape** both influence the degree of heat and water loss. As body size increases, the surface area, relative to body volume, decreases, and this aids in reducing water loss because proportionally more of the body is not exposed to surface evaporation. However, the decrease

in surface area relative to body volume that often comes with increasing size is a *disadvantage* with regard to heat loss. Small animals, with their higher relative surface area, tend to lose heat faster. The combination of these two factors must strike a balance. An intertidal animal cannot be too small or else it will dry up in the sun. If it grows too large, however, it may not be able to dissipate heat fast enough through its body surface. Shape has an important effect. Long and thin organisms will dry up much more easily than spherical ones. This is one reason that a sea anemone contracts into a small equidimensional cylinder during low tide. The change in shape reduces surface area and water loss.

Intertidal invertebrates can avoid overheating by **evaporative cooling**, combined with **circulation of body fluids**. As a result of such processes, intertidal snails are usually cooler than inanimate objects of the same shape, size, and color. Higher-intertidal animals are better adapted to desiccation than lower-intertidal species. Movement of the cirri (feeding appendages) in intertidal acorn barnacles increases with increasing temperature but declines near an upper thermal limit (**Figure 16.4**). Upper-intertidal barnacles tolerate high temperatures better than do barnacle species found in the lower intertidal. Species living in the high-intertidal zone in the tropics tend to maintain coordinated ciliary motion at higher temperatures than do species living in the low-intertidal zone or subtidal zone.

Genetically based variation in shell color is related to reflection of sunlight. Along the east coast of the United States, for example, the mussel *Mytilus edulis* has light and dark shell color forms (see Figure 4.23), which are genetically determined. On the east coast of North America, the light-colored form is found more frequently toward the south (**Figure 16.5**). Dark mussels gain heat more rapidly and are superior in environments where cold air temperatures are common, whereas light mussels reflect heat in the sun and are superior in the south, where high-temperature stress is a problem.

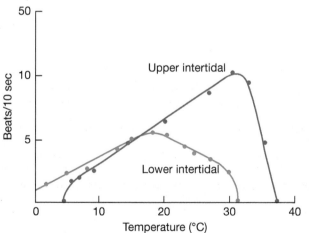

FIG. 16.4 Beating frequency of cirri (feeding appendages) in relation to increasing temperature in an upper-intertidal (red curve) and lower-intertidal (blue curve) barnacle species. (After Southward, 1964)

FIG. 16.3 A tidal pool on an exposed shore in Washington State. Note the presence outside the pool of desiccation-resistant barnacles and mussels, and more sensitive anemones, hydrocorals, and sea stars within. (Photograph by Jeffrey Levinton)

Species living in the high part of the shore retain cellular function at higher temperatures than those living in lower shore levels (**Figure 16.6**). In Chapter 5, we discussed the use of **heat-shock proteins** to maintain protein function under high temperature stress. Lower-intertidal animals and very shallow subtidal animals acclimate to reduced conditions of heat stress and tend to produce more heat shock proteins when exposed to temperature stress, as compared to higher-intertidal animals (Tomanek and Sanford, 2003). Species of snails living higher in the intertidal are superior in producing heat-shock proteins at higher temperatures than species living in the lower intertidal (Tomanek, 2002). We would expect that summer-acclimated individuals would be more tolerant to temperature stress, which can be seen in induction of heat-shock proteins. Buckley and colleagues (2001) studied the heat-shock protein Hsp70, which is induced under high temperature stress. For the west coast mussel *Mytilus trossulus*, mussels collected in February had a threshold activation of Hsp70 at 23°C, whereas mussels collected in August had a threshold temperature of 28°C.

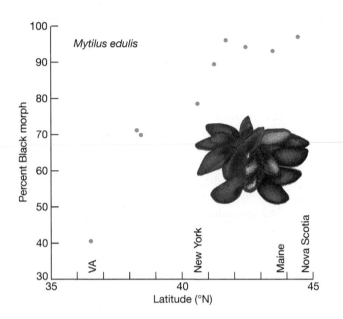

FIG. 16.5 Latitudinal variation in black and brown shell color forms of the blue mussel *Mytilus edulis*. The brown morph reflects solar radiation more efficiently and is favored in lower latitudes. (Data from Innes and Haley, 1977, and Mitton, 1977)

■ Climate change is altering the geography of temperature stress of intertidal communities.

Global climate change patterns include increases of sea-surface temperature, which is having strong effects on rocky-shore communities. A few decades ago, planktonic larvae of the blue mussel *Mytilus edulis* settled in the region and even south of Cape Hatteras, North Carolina. But the southern range end of this species is moving steadily northward as air and water temperatures have increased (Jones et al., 2009). Mussels transplanted to areas in North Carolina where mussels once thrived now experience catastrophic mortality. Community species membership is also changing as climate warms. In rocky shores of Monterey Bay, California,

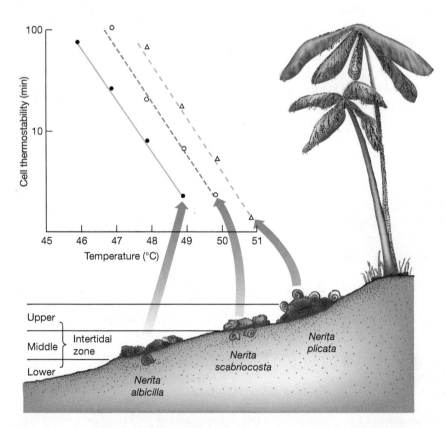

FIG. 16.6 Survival of ciliated epithelial cells in three species of the intertidal gastropod genus *Nerita* and the relation of temperature tolerance to position on shore. (After Ushakov, 1968)

five of nine species with a more high-latitude distribution decreased in abundance from the 1930s to the 1990s and eight of nine more southerly species are steadily increasing in abundance. Mean summer maximum temperatures increased 2.2°C over the period (Barry et al., 1995).

■ Living in the upper intertidal reduces the time of access to food and usually to oxygen in nearly all species, although the specific mechanisms differ.

Mobile carnivores, such as drilling gastropods and sea stars, must seize and consume prey, which at low tide exposes their soft body parts (e.g., foot of gastropods, tube feet of sea stars) to thermal and desiccation stress. Grazers such as periwinkles and limpets cease feeding at low tide on hot, dry days, although they may feed when the air is cool and moist. Suspension feeders obviously have no access to phytoplankton when exposed to air. Therefore, **higher-intertidal animals have reduced feeding periods and grow more slowly than do low-intertidal forms.** Upper-intertidal herbivorous gastropods may be able to compensate for the reduced feeding time by feeding more rapidly. Nevertheless, reproductive output is usually far less for high-intertidal animals than for low-intertidal members of the same species.

At low tide, marine organisms are exposed directly to air, or they are confined in burrows or with small volumes of trapped water (as in barnacles or mussels). As a result, most intertidal organisms face shortages of oxygen and buildup of metabolic wastes. **Reduction of metabolic rate** is a common means of reducing the need for oxygen during low tide. Some species have developed the capacity to breathe air. Mussels may gape at low tide and consume oxygen from the air, as long as their gills are moist. The eastern North American salt marsh mussel *Geukensia demissa* lives in the upper intertidal zone and probably obtains more oxygen from air than from water. Oxygen diffuses from the air across the moist surface of the bivalve's gills. Many high-intertidal crabs, such as many grapsid crabs, fiddler crabs and ghost crabs spend much of their active time in air, and such crabs have greater posterior gill surface area devoted to gas exchange with air rather than water (Takeda et al. 1996). Some high-intertidal crabs have membranous disks on their legs that are thin and very permeable to gas uptake when the crab is in air (**Figure 16.7**). Beneath the membrane is a complex of blood spaces, which takes up oxygen. Other air-breathing crabs (e.g., *Ocypode* and *Geograpsus*) have vascularized tissue—an analog of a lung—adjacent to the gill that can remove oxygen from air. Certain burrowing invertebrates, such as the lugworm *Arenicola marina*, have blood pigments, such as hemoglobin, which bind tightly to oxygen and release it for metabolic needs during low tide.

■ Mobile intertidal animals may remain fixed at one tidal level, becoming active at the time of low tide, while others migrate up and down the shore with each tidal cycle in order to maintain normal activity and to remain moist.

Mobile intertidal animals may maintain a fixed territory at low tide. The owl limpet *Lottia gigantea* maintains a territory

FIG. 16.7 The Australian high-intertidal sand-bubbler crab *Scopimera inflata* has a membrane on each leg (shaded blue) designed to exchange gas from the air for uptake into arterial blood. (After Maitland, 1986)

and remains in a dug crevice until high tide when it moves outward to graze within a territory, which it defends against other limpets. Other grazing limpets and snails spend low tide in moist cracks in the rock or among the byssal threads of mussels, departing for open areas during high tide. Mobile carnivores such as sea stars and gastropods actively move downward at the time of high tide to avoid drying up.

Many species have a series of responses designed to keep a position relative to the water level. In the case of rocky-shore mobile forms, this usually involves responses to light, gravity, and water that combine to keep the animal at the correct level. This can be illustrated by G. S. Fraenkel's work on the high-intertidal periwinkle *Littorina neritoides*. When submerged in seawater, *L. neritoides* is negatively geotactic. It is negatively phototactic when immersed and right side up, but positively phototactic when immersed and upside down. When the animal is moist but not submerged, it has no light response. **Figure 16.8** illustrates how these responses combine to lead *L. neritoides* to its high-intertidal splash zone habitat after it has been dislodged by waves or after a feeding excursion in the lower-intertidal zone.

As in the case of rocky-shore habitats, some ocean sandy-beach inhabitants continually live on the edge of death because they may find themselves in sediments too high and dry or too low and wave swept. Open beaches experience enormous erosive force. The eastern surf clam *Spisula solidissima* lives in such exposed areas and, when dislodged, is capable of rapid reburrowing. In the winter, when much of a beach is eroded, intertidal open-beach invertebrates often migrate to subtidal parts of the beach because the erosional forces of waves are too great to allow them to maintain position or to reburrow if dislodged by erosion.

As the tide falls, an infaunal sandy-beach species may find itself in the equivalent of desert sand. As the tide rises, waves may suddenly erode the sediment and make it impossible to maintain a burrow. This constant change is why very few macrobenthic animals live successfully in wave-swept beaches. A few species have evolved novel ways of migrating up and down the shore by periodically leaving the sediment, riding the swash, and reburrowing in an optimal moist but stable sediment. Species of the mole

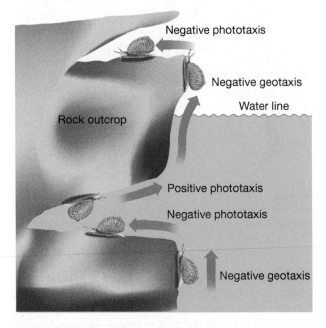

FIG. 16.8   The contribution of behavioral responses of the high-intertidal gastropod *Littorina neritoides* to the regulation of its vertical position on rocky shores. (After Newell, 1979)

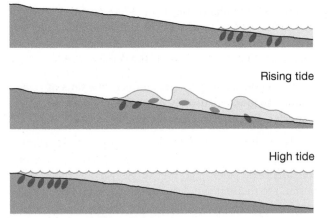

FIG. 16.9   Vertical beach migration of swash riders, such as the mole crab *Emerita* and the beach clam *Donax variabilis*. Note that animals seek sediments that are neither too dry nor too wave washed.

crab *Emerita* (*E. analoga* on the west coast of the United States and *E. talpoida* on the east coast) and several species of the wedge clam *Donax* have adopted the strategy of behaving as **swash riders**, leaving the sediment as they feel the pressure of an approaching wave (**Figure 16.9**). They then ride the wave to a higher position in the intertidal zone and rapidly burrow. When the tide falls, they reverse the process, which guarantees that they will always be located in a moist but not excessively wave-swept level of the beach. As the waves wash back down the beach, the mole crab extends its feathery second antennae, which trap phytoplankton. The mole crab has a streamlined shape and special digging appendages (**Figure 16.10**).

Wave Shock

- **Wave shock is a major factor determining the distribution and morphology of intertidal organisms.**

The impact of waves and the material that they carry (sand, pebbles, logs) are important in selecting morphological adaptations to intertidal life. We need only visit a rocky shore on a stormy day to witness the tremendous energy that is focused on the shoreline. Waves may rip organisms from the rocks, erode large volumes of sand, or propel a variety of projectiles to the shore.

Breaking waves can damage rocky-shore organisms in the following ways:

1. *Abrasion.* Particles in suspension or floating debris scrape delicate structures. Water turbulence may whip seaweeds and other erect organisms against rocks.

2. *Pressure drag.* The hydrostatic pressure exerted by breaking waves can crush or damage delicate and

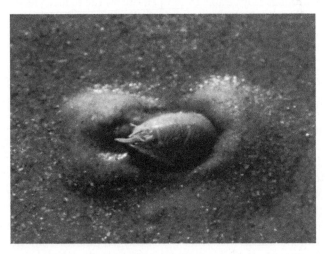

FIG. 16.10   The streamlined shape of the mole crab *Emerita talpoida* allows the animal to burrow into the sand rapidly. (Photograph by Jeffrey Levinton)

compressible structures, such as gas-filled bladders of seaweeds. Most intertidal organisms are liquid filled and, therefore, are relatively incompressible. Larger epifauna will experience greater pressure drag (see discussion in Chapter 6 on limpet size and wave pressure drag).

3. *Acceleration.* Water exerts a directional force against intertidal organisms. Larger organisms may also experience acceleration as their large mass is carried along by waves. Acceleration might also cause ripping of holdfasts or snapping of byssal threads of mussels.

- **Force transducers can be used to measure the force of wave shock.**

Organisms living in wave-swept coasts experience dynamically changing forces as waves collapse upon them. Organisms that project into the water will likely experience strong pressure drag, which is the difference of force upstream and downstream of the organism (see Chapter 6). To measure these forces, one needs to implant **force transducers**, which can measure force along the plane of the rocky surface and perpendicular to the rock. A force transducer is a beam that bends when water impinges on it. As the beam bends, the electrical resistance of a metal band on the beam changes, which causes a change in the voltage, which is proportional to force.

On a wave-swept rocky shore in Washington State, Mark Denny (1985) estimated pressures on a model of a limpet averaging about 0.6 N (N is a newton, a unit of force), but often reaching 3 N. Equally important was the unpredictability of the time of maximum force or the direction. Wave forces can dislodge a number of intertidal organisms and also cause cessations of movement by limpets and snails. Single organisms that project from the rock surface are more at risk than those living among other organisms. For example, when mussels are packed together into dense beds, the entire mussel bed can dissipate the wave along its surface, reducing the force on each individual. Mussels forming such beds are **foundation species**, and strongly affect the structure and physical conditions of associated species.

■ **Phenotypically plastic changes in form can reduce the risk of wave shock, but there are often trade-offs with other biological functions.**

While intertidal animals may be vulnerable to being torn from a surface or smashed by wave forces, many species have plastic responses that may reduce mortality. As mentioned, species living on wave-exposed coasts may have thickened skeletons and reduced delicate projecting spines. Barnacles experience increased drag even on the microscale and have feeding appendages and even penises that are shorter or thickened at the base. Emily Carrington (2002) found that mussels in the intertidal of Rhode Island were especially stressed in October, when food was low and water temperatures were still high. Because of this, fall storms were capable of rapidly dislodging large numbers of mussels from intertidal rocks (**Figure 16.11**). The mussels increase byssal thread production in winter, which increases the total attachment strength of the mussel to rocks. Seaweeds may be smaller in stature in open exposed coasts, which reduces drag, and more delicate and longer in protected waters. Larger seaweeds may bend over in a wave-swept surf zone, which reduces drag. But the very size of larger seaweeds causes the mass to accelerate, which can break off the thallus at the stipe.

These plastic responses, however, may come at a cost. The increase of production of byssal threads is physiologically costly to mussels and influences reserves available for somatic growth and gamete production. Seaweeds that grow to smaller sizes in open coasts may survive better, but they also produce fewer reproductive structures and there is, therefore, a trade-off of response to wave action and reproductive output.

CAUSES OF ZONATION

■ **Zonation results from preferential larval settlement and adult movement, differential physiological tolerance, and biological interactions such as competition and predation.**

(a)

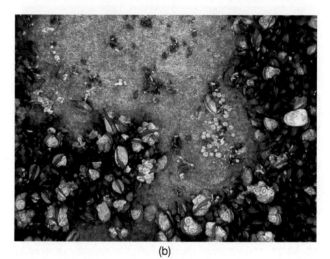

(b)

FIG. 16.11 Mussel dislodgement. The mussel *Mytilus edulis* changes its attachment strength to deal with seasonal changes in water currents and secretes more threads so that it can resist winter storms. But in early fall, it does not have sufficient reserves to attach very well. (a) Mussels in October 3, 2001. (b) Bare space on October 21, 2001, after mussels were stripped from a rock because of fall wave action. (Photographs by Emily Carrington)

Having examined the important physical factors in the intertidal gradient, we can ask the question: What causes zonation? The remarkably similar patterns of species vertical ranges, combined with the sharp boundaries between zones, seem to suggest that zonation can be explained with few factors.

Several major factors combine to form zones.

1. *Physiological tolerance.* Species found higher on the shoreline are generally more tolerant of desiccation, reduced feeding time, reduced access to oxygen, and extreme temperature.

2. *Larval and adult preference.* To some degree, larvae of sessile animals are able to locate the tidal height suitable for adults of their own species. Larvae of the common European Atlantic higher-intertidal barnacle *Chthamalus stellatus* settle in the high shore, although they also settle somewhat lower. Gregariousness (settling in groups, on adults of the same species) also causes

FIG. 16.12 (a) Intertidal rocks are often completely covered by organisms. This rock, on Tatoosh Island, Washington, is covered by barnacles (light) and mussels (dark). (b) Close-up of a mussel-dominated rock. (Photographs by Jeffrey Levinton)

FIG. 16.13 Predation line on rocky shores of England. Below this line (adjacent to the crack in the rock), the gastropod *Nucella lapillus* (white spots) can clear the rocks of its prey, the mussel *Mytilus edulis* (dark-colored areas). (Photograph courtesy of Raymond Seed)

preferential settlement on the "correct" level of the shore. In eastern Australia, larvae of the barnacle *Hexaminius popeiana* settle preferentially on adults of their own species (Coates and McKillup, 1995). If adults are transplanted above the typical shore level at which the adults usually live, the larvae settle there preferentially as well. Survival of recruits, however, is much lower than within the normal tidal level where the adults usually live. Mobile animals can adjust their tidal height by a combination of responses to light, gravity, and moisture, as discussed earlier.

3. *Competition.* Intertidal habitats, particularly rocky shores, may be severely space limited (**Figure 16.12**). Species capable of overgrowing or undercutting others may come to dominate that level of the shore in which they can do well physiologically. Mussels, for example, can usually move by forming and detaching byssal threads. They can climb on top and smother competitors.

4. *Predation.* Predators are often strongly limited by the time of immersion because carnivores such as sea stars and snails must be moist as they move to locate, seize, and ingest prey. This requirement usually limits

predation to the lower part of the shore and creates a refuge from predation above a certain shore height, where predators do not have enough time to capture prey. **Figure 16.13** shows a rocky-shore mussel bed on the coast of England. Dog whelks (*Nucella lapillus*) come out of moist lower-intertidal cracks during the rising tide and prey upon mussels. They must return to the cracks when the tide withdraws. Mussel beds on the outer Pacific coast, dominated by *Mytilus californianus*, often have a similar sharp lower limit, which is controlled by the carnivorous sea star *Pisaster ochraceus*. Robert Paine removed sea stars for over a decade and found that the mussel bed gradually extended lower in the tide zone.

Interspecific Competition

■ **Field experiments show the importance of interspecific competition for space on rocky shores.**

Field experiments were crucial in understanding the cause of zonation. On rocky shores of both the American and northern European sides of the North Atlantic, one often sees the zonation pictured in **Figure 16.14**. In the highest

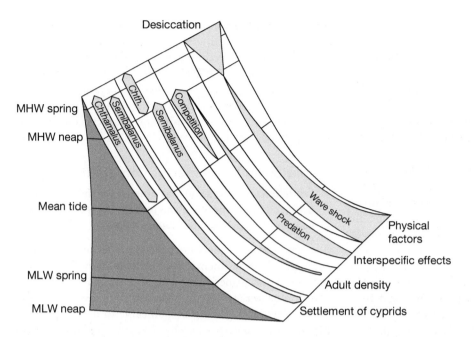

FIG. 16.14 Distribution on rocky shores of Scotland of adult and newly settled larvae of the barnacles *Semibalanus balanoides* and *Chthamalus stellatus*. Width of the bars indicates the relative effects of principal limiting factors. MHW, mean high water; MLW, mean low water. (Modified after Connell, 1961)

part of the shore, adults of the barnacle *Chthamalus stellatus* (*C. fragilis* in New England) dominate available space down to the approximate level of mean high tide. Below, adults of the white barnacle *Semibalanus balanoides* usually dominate. At the level of mean low water, however, barnacles are usually rare. What explains the distribution?

Joseph Connell (1961) studied this problem on a protected rocky shore in Scotland by selective removals of hypothesized competitors and by caging selected areas to exclude the common predatory gastropod, the dog whelk *Nucella lapillus*. First he considered larval recruitment. Cyprids of *Chthamalus* tended to settle high on the shore, but they also settled well within the range in which adults of the same species were rare. *Semibalanus* cyprids recruited in large numbers throughout the intertidal but failed to do very well above mean high water. Hence, there was a substantial area of overlap of larval recruitment, although differential larval recruitment explains some of the difference in adult vertical distribution.

Connell transplanted rocks downward with newly settled *Chthamalus* and found that these barnacles were rapidly overgrown and undercut by recruited *Semibalanus*, which are much faster growing, even causing intraspecific mortality. Above high water at neap tide, *Chthamalus* survival was greater, but the *Semibalanus* there died, owing to their poor survival in the dry reaches of the upper intertidal (**Figure 16.14**).

These experimental results show that the lower limit of *Chthamalus* on the shore is regulated by competition with *Semibalanus*, whereas the upper limit of *Semibalanus* is probably controlled by desiccation. The upper limit of the *Chthamalus* zone is also controlled by desiccation. Predation by the drilling gastropod *Nucella lapillus* controls the lower limit of *Semibalanus*. These results led to the generalization that **the upper limit of an intertidal species is regulated by physical factors, whereas the lower limit is regulated by biological factors** (e.g., competition).

Seaweed zonation may also be strongly controlled by interspecific interactions. On New England's protected rocky shores, the mid-intertidal is dominated by species of the brown seaweed *Fucus*, whereas the lower shore is usually covered by the Irish moss *Chondrus crispus*. Jane Lubchenco (1980) used experimental manipulations to show that the removal of the common herbivore periwinkle *Littorina littorea* and Irish moss resulted in complete coverage of the low shore by the normally mid-intertidal *Fucus* sp.[1] If the herbivore snails were left in place, *Fucus* sp. did not colonize the lower shore so successfully. In this case, the herbivore was functioning as a predator, much as the carnivorous gastropod *Nucella lapillus* functioned in the barnacle example mentioned earlier.

■ Climate change may tip the balance of biological interactions toward changes in intertidal community structure.

Competitive success on rocky shores involves the ability to overgrow a competitor combined with physiological tolerance to desiccation. The superiority of *Chthamalus* on the upper shore depends on its tolerance of stress relative to other species. Changes in climate would therefore interact with preexisting differences in adaptations to temperature and desiccation stress. A particularly severely cold winter in 1962–1963 in the United Kingdom caused extensive mortality in generally warm-adapted marine species. One surprise was a very high intertidal barnacle, noted for its resistance to heat. It showed no mortality, perhaps because of its general physiological resistance to conditions on the

---

[1] When a genus, for example, *Fucus*, is named as *Fucus* sp., we are not sure of the exact species identity.

high shore, which frequently involve cold as much as heat. In the past few decades, coastal waters have been increasing in temperature in this part of Europe and a number of lower-latitude species have been expanding north, including two species of warm-water barnacles (Hiscock et al., 2004). As we have discussed, competition between barnacle species can be intense, so what role will global warming play in reorganizing competitive outcomes in rocky intertidal communities?

Cape Cod is a latitudinal thermal threshold in eastern North America. Nearest the mainland, a canal connects the cold waters of northern Cape Cod Bay with the relatively warmer waters south of Cape Cod. South of the Cape, the barnacles *Semibalanus balanoides* and *Chthamalus fragilis* interact, much as Connell described interactions in Scotland between *S. balanoides* and *C. stellatus*. But in the north side of the Cape, waters are simply too cold for *Chthamalus*. David Wethey (2002) studied interactions between these two species along this special northern frontier. On the south side of Cape Cod, *Semibalanus balanoides* could not outcompete *Chthamalus fragilis* in the highest intertidal, which was too hot for survival. But on the northern end of the canal, the water and air temperatures were sufficiently colder to allow *S. balanoides* to outcompete *C. fragilis*, right to the top of the intertidal zone. Wethey was able to prove that the northern limit of *C. fragilis* was determined by competition as much as climate. He transplanted *Chthamalus* on plates to a locale 80 km to the north of the northern part of the Cape Cod Canal. Adults survived quite well in the absence of *Semibalanus*, which proves that the limit is not just controlled by temperature but by an interaction effect with interspecific competition.

It is interesting to speculate how regional warming, which is well documented in marine waters in this region, will affect range extensions. *Chthamalus fragilis* will probably extend northward, as will many other species by a variety of mechanisms. Some evidence shows that the northern limit of the fiddler crab *Uca pugilator* at Cape Cod is determined by larval survival. Increase of temperature may allow a range extension of this and all other similarly limited species. This may cause a reorganization of species interactions in Cape Cod Bay, just north of Cape Cod.

Predation and Interspecific Competition

■ **Predation (or herbivory) may ameliorate the dominance achieved by competition and may strongly affect species composition and species richness.**

The importance of predation has also been demonstrated through field experiments. Robert Paine studied this effect on west coast American rocky shores by removing the carnivorous sea star *Pisaster ochraceus* (**Figure 16.15**). The mussel *Mytilus californianus* soon came to dominate. Such experiments suggest a common effect: **Predation delays the competitive displacement of competitively inferior species by the competitive dominant.** On North Atlantic rocky shores, the dog whelk *Nucella lapillus* preys in the lower intertidal upon the competitive dominant barnacle *Semibalanus balanoides*. In waters of Great Britain, this predation opens up space for colonization by the competitively inferior *Chthamalus stellatus*. This general effect has also been observed among seaweed competitors: That is, competition is reduced by the introduction of grazers such as urchins and snails. On New England shores, snail and sea star predation on mussels permits the Irish moss *Chondrus crispus* to dominate lower-intertidal shores. Mussels tend to dominate available space on more exposed headlands where predators are eliminated owing to wave shock.

FIG. 16.15 The sea star *Pisaster ochraceus* is a top predator on eastern Pacific rocky shores. This sea star was turned over while attacking a mussel, *Mytilus californianus*. (Photograph by Jeffrey Levinton)

Species such as sea stars that prevent the monopolization of a habitat by preying on most potential competitors are known as **keystone species**. These species exert **top-down control** on ecosystems.

Like dog whelks, the sea star *Pisaster ochraceus* is limited by moisture and ascends with the rising tide to attack the lower edge of the mussel bed. At the time of low tide, sea stars can be seen on rocks below the mussel bed (**Figure 16.16**). Paine removed sea stars for many years, and the mussel bed gradually moved downward, as sea stars failed to remove the lowest level of mussels.

Although the sea star *Pisaster ochraceus* is a keystone species with respect to species attached directly to rocks, one must remember that the interstices of the mussel bed may contain large numbers of smaller animal species, such as polychaetes, barnacles, and smaller gastropods. Sea star predation may remove the competitive dominant and increase the number of coexisting large-bodied species attached to rocks; but by removing the mussel bed, it also acts in the same manner as clear-cutting a terrestrial forest—namely, causing the disappearance of many smaller dependent animal species.

Strong interactions come not only from animal predators but also from herbivores. On rocky shores, limpets are especially important in the mid-intertidal zone because they are very effective grazers of diatoms attached to rocks, coralline algae, and seaweeds. Simple removal experiments are very informative. Under normal circumstances, limpet-covered rocks are quite barren. However, Betty Nicotri (1977) removed limpets from the intertidal zone of San Juan Islands, Washington, and soon observed the development of a lush cover of diatoms. Limpet grazing has also been found to inhibit the growth of seaweeds on rocky shores in many regions of the world. Finally, limpets affect strongly the distribution of barnacles because the limpets bulldoze aside and ingest newly settled cyprid larvae. On rocks dense with limpets, it is common to observe (**Figure 16.17**) only a few very large barnacles that managed to survive the limpets (and other sources of mortality).

Predation may be so intense that the competitive dominant is rather rare. For example, several species of turf-forming coralline algae compete for space in the lower-intertidal zone of the outer coast of Washington. Experimental combinations of species produced a surprise: The dominant competitor was rather rare in the natural community. In a sense, the basis for its competitive superiority was the key to its population downfall. It grew rapidly enough to displace other competitively inferior species, but this also made it the favorite food of some common grazing chitons and limpets. As a result, the relatively slower growing and less preferred algal species dominated and occupied most of the space, which was monopolized by coralline algae (Paine, 1990). This result leads to some interesting conclusions. First, the complete monopolization of a resource does not necessarily indicate that predation is unimportant. Second, competition cannot be divorced from predation.

■ Spatial heterogeneity may strongly affect the pattern and intensity of predation on rocky shores.

We have discussed the decreasing effectiveness of mobile carnivores such as sea stars in the higher part of the shore. But even at the same level in the intertidal zone, spatial heterogeneity can strongly influence local patterns of predation. Cracks retain moisture, and many mobile carnivores spend the time of low tide there, with no need to retreat to the lowest part of the tide zone. It stands to reason, therefore, that prey items within easy reach of the cracks and pools will be taken with great frequency. In the intertidal zone of the west coast of the United States, species of the genus *Nucella* drill a variety of barnacle and mussel species.

FIG. 16.16 Mussel bed near Bamfield, British Columbia, showing abundant purple and orange sea stars *Pisaster ochraceus* below a *Mytilus californianus* mussel bed. (Photograph by Jeffrey Levinton)

FIG. 16.17 Limpets on a mid-intertidal rocky shore, San Juan Islands, Washington. Note the unoccupied space with sparse large adult barnacles, caused by intense limpet grazing, which consumes the diatoms but also bulldozes aside most newly settled barnacle larvae. (Photograph by Jeffrey Levinton)

They are often found abundantly at the edges of mussel beds and can stay in these relatively moist areas during low tide. When the tide rises, they move from these protected edge habitats out on the rocks to prey on barnacles and smaller mussels. If a large patch is opened by a storm,

however, the snails tend not to move across a large area of open rock. Thus, barnacles and mussels settling in the middle of the patch may occupy a refuge from predation.

While rock cracks and pools are relatively permanent, biological refuges such as mussel beds and seaweeds provide shelter and protection from desiccation and need not be permanent. Anthony J. Underwood (1999) was surveying a rocky shore in New South Wales, Australia, when a severe storm hit in 1974. Before the storm, large patches of the seaweed *Hormosira banksii* covered the rocks, and the carnivorous drilling gastropod *Morula marginalba* survived desiccation at low tide by retreating to the wet shelter of the seaweed canopy (**Figure 16.18**). Barnacle mortality was very high near and within the seaweeds, which was explained mostly by snail predation. In some cases seaweeds whip back and forth, scraping barnacle-settling larvae from the rocks, but that was not the case in this study. The seaweeds indirectly affected the barnacles by positively enhancing the local abundance of the snail. The storm, therefore, set a new pattern for predation interactions by removing the shelter for a major predator. It took about 5 years for the strongly disturbed patches of *Hormosira* to recover fully. We explore this further later when we discuss the issue of the scale of disturbance.

■ **Mobile predators can often detect newly recruited sessile benthos and may form strong aggregations and feed on new recruits.**

As we have stressed, larval recruitment is not homogeneous, either between sites or from one year to the next. Predators, such as sea stars, drilling snails, and crabs, would alternately be overwhelmed by food and starved if they did not attempt to find spatially concentrated new sources of food. Such concentrations are liable to develop in sites where larval recruitment was very high; perhaps the water flow was high and brought settling larvae to a given location. It is quite common to see abundant predators where recruitment is high. On rocky shores of Maine, where springtime is a wet, bone-chilling affair, large herds of the drilling snail *Nucella lapillus* move along on rock surfaces

(a)   (b)

FIG. 16.18   (a) The seaweed *Hormosira banksii* forms a canopy under which snails hide and keep moist at low tide. (b) The gastropod *Morula marginalba* is a major predator of invertebrates on rocky shores of southeastern Australia. This snail, in the process of consuming a barnacle, is about 1.5 cm long. (Photographs by Jeffrey Levinton)

FIG. 16.19 Aggregation of sea stars, *Pisaster ochraceus*, at the periphery of a mussel bed dominated by *Mytilus californianus*. (Courtesy of Thomas H. Suchanek)

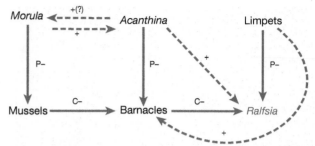

FIG. 16.20 Direct (solid arrows) and indirect (dashed arrows) interactions in an intertidal rocky-shore community in the northern Gulf of California. For example, limpets graze the seaweed *Ralfsia*, whose removal allows greater settlement by barnacles. Key: P−, a negative effect by a predator; C−, a negative effect by a competitor; one symbol alone specifies a positive indirect effect. (Drawn from descriptions in Lively et al., 1993)

at low tide, mowing down newly settled barnacle recruits. On the west coast of North America, large aggregations of the sea star *Pisaster ochraceus* can often be seen on the lower shore periphery of dense mussel beds (**Figure 16.19**). Working on the outer coast of British Columbia, Canada, Carlos Robles and colleagues (1995) found that sea stars aggregated around local areas of high mussel recruitment, consumed the juvenile mussels, and then dispersed.

Many intertidal organisms have behavioral and structural defenses against predators.

Predation can be very intense, but many potential prey species have evolved effective defenses against predators. Hiding in cracks, living high in the intertidal zone, and simply growing too large all provide effective refuges against predation. Many species, however, cannot escape predator contact. Sessile species and slow-moving mobile species cannot escape predators unless they are cryptic or have some sort of defense. Mussels and barnacles can deter predators with their shell, and large mollusks often have shells too thick to be penetrated by boring snails. The mussel *Mytilus edulis* is especially vulnerable to drilling gastropods, such as the eastern Atlantic dog whelk *Nucella lapillus*, but it has an effective defense: When a whelk wanders onto a group of mussels, the prey animals react by secreting byssal threads, which can entwine and trap the snail (Petraitis, 1987).

Strongly interacting species often cause indirect effects in rocky-shore food webs.

Rocky-shore food webs often consist of several strongly interacting predator and prey species; the latter also compete intensely for space. If these interactions are diagrammed by means of arrows, one can sometimes see **indirect effects**, which are effects that a species experiences because of its interactions with another species that also interacts with yet another species. Consider **Figure 16.20**, which partially depicts the interactions among sessile species and mobile carnivores in the northern Gulf of California (Lively et al., 1993). As can be seen, mussels compete with barnacles and the seaweed *Ralfsia*. The snail *Acanthina* consumes

barnacles, which in turn allows *Ralfsia* to expand, thus providing more food for grazing limpets. So, indirectly, *Acanthina* benefits *Ralfsia* and limpets. By grazing on *Ralfsia*, limpets allow barnacles to settle, which indirectly benefits *Acanthina*. Because the interactions among rocky-shore species are so strong, such indirect effects are common. It is even possible to imagine mutualisms. For example, the gastropod *Morula* consumes mussels, which otherwise overgrow barnacles. Consumption of mussels therefore benefits the barnacle-consuming *Acanthina*, but consumption of barnacles by *Acanthina* may accelerate the competitive dominance by mussels, a development that benefits *Morula*. Therefore, *Morula* and *Acanthina* are mutualists, via these indirect effects.

Disturbance and Interspecific Competition

Physical and biological disturbance often determines the species composition of intertidal communities.

The intertidal zone stands in the way of some of the most intense storms in the ocean. This is especially true of rocky headlands exposed to the open sea, where waves, boulders, and logs all crash on rocky shores. On the outer Pacific coast of North America, such storms commonly strip off the dominant mussels and seaweeds and thereby open space for colonization. The palm seaweed *Postelsia palmaeformis* (**Figure 16.21**) colonizes such newly opened spaces, mainly by dispersal of adult plants. Once it has colonized a newly opened area, the seaweed produces spores that move down the fronds and locally increase the patch size of the plant population.

Disturbance by storms on rocky shores has much the same effect as predation. It liberates space and allows inferior competitors to persist. Moderate disturbance should allow more species to coexist than strong storm disturbance, which would strip off most species. Storms may indirectly benefit anemones in tidal pools, which can eat mussels stripped from the rocks.

Soft-bottom areas, especially exposed beaches, are also disturbed continually. Waves and currents continuously erode open-ocean beaches, and very few animal species

FIG. 16.21 The brown palm seaweed *Postelsia palmaeformis* on an exposed rock on Tatoosh Island, Washington State. It colonizes exposed eastern Pacific rocky shores that have been recently disturbed. (Photograph by Jeffrey Levinton)

can survive the instability. In some locations, **biological disturbance** is an important factor. In eastern American protected beaches, the horseshoe crab *Limulus polyphemus* comes into shore to breed, and its burrowing activities can destroy the burrows of many species of invertebrates.

Disturbance and Succession in the Intertidal Zone

- **Succession of intertidal seaweeds may be irregular, but an overall spectrum of life histories begins with early successional good colonizers, which are prone to grazing, yielding in many cases in late succession to good space holders that are resistant to grazing and to competitors.**

Seaweed succession on rocky shores cannot be divorced entirely from animals because several animal grazers can affect succession, and some rocky-shore animals compete with seaweeds for space. Still, many sites are dominated by seaweeds. Consider the colonization of a recently constructed rock jetty on a North American coast. Succession is initiated by a disturbance, and continual disturbance maintains the earliest stage of succession. At first, nothing colonizes the boulders, but later a surface slime develops, consisting of bacteria and other microorganisms. After this, species of a green seaweed genus, *Ulva* or *Enteromorpha*, will most likely colonize. Limpets or periwinkles may graze the seaweed down to bare rock.

After several cycles of colonization and grazing, the brown seaweed *Fucus* (*Gigartina* or others on the west coast) may dominate the lower shore. Its later appearance does not depend on the previous appearance of any seaweed species. It recruits more slowly than green seaweeds, and it is not likely to dominate unless they have been grazed away. Once *Fucus* has appeared, however, greens will not likely dominate again unless and until *Fucus* is removed by a storm disturbance or by grazers. The periwinkles *Littorina* spp. are not very good at eating *Fucus*, owing to its relatively tough stipe and holdfast and, possibly, its poisonous compounds.

Despite a considerable degree of irregularity and the absence of a guarantee that macroalgal succession will move toward completion, certain properties of seaweeds reflect their position in the successional sequence. In early succession, seaweeds place a premium on rapid growth. Later in succession, however, resistance to grazing and the ability to combat competitors are more important, but the investment of seaweeds in tough tissues and various poisonous compounds imposes a cost that reduces growth rate. In California, the sea lettuce *Ulva californica* dominates early colonization, whereas the fucoid brown seaweed *Pelvetia fastigiata* dominates the late successional stages. Algal productivity and nutritive value (in kilocalories per ash-free unit dry weight) decrease from early to late successional species. Late successional seaweeds allocate more energy to the synthesis of support structures, attachment structures, and structures that fend off urchin grazers. Later successional forms are tougher and more resistant to wave shear. These features all contribute to persistence, the hallmark of late successional species.

- **The spatial scale of disturbance often results in different outcomes. Large-scale disturbances may result in unpredictable recolonization, leading to alternative stable states in large patches on rocky shores.**

As we have mentioned, the interactions on the individual level often lead to predictable outcomes when competitors and predators meet on a rocky shore. But disturbances over larger areas can produce very different outcomes of community structure.

Robert Paine and Simon Levin (1981) found that if patches opened by disturbance were relatively small, on the order of a square meter or less, the mussels would gradually move into the open space. But what if disturbances opened very large patches? Thomas Suchanek (1981) found that if storms opened large patches, then the immigration by adult mussels was too slow to seal up the patch (**Figure 16.22**). In these large patches, seaweeds and another mussel species, *Mytilus trossulus*, colonized the patch centers. Predatory snails in the surrounding California mussel bed would not move across open bare rock space in the large open patches. This situation allowed rapid growth of *M. trossulus*, and it was able to reach sexual maturity in a few months. Suchanek accumulated data on recruitment in California and discovered that *M. trossulus* recruited more in the winter, which might have been an adaptation to the opportunity to settle in open spaces in the mussel bed created by winter storms. To summarize, above a certain size threshold, disturbance patches were ecologically distinct from their smaller cousins.

We now have a fascinating possibility of a landscape that is not only patterned by disturbance but also *maintained* as a series of **alternative stable states**, which are differing ecological assemblages of interacting species that can occur under the same environmental conditions. Once the ecological assemblage develops, a stabilizing mechanism keeps the assemblage intact. This may make sense of one of our

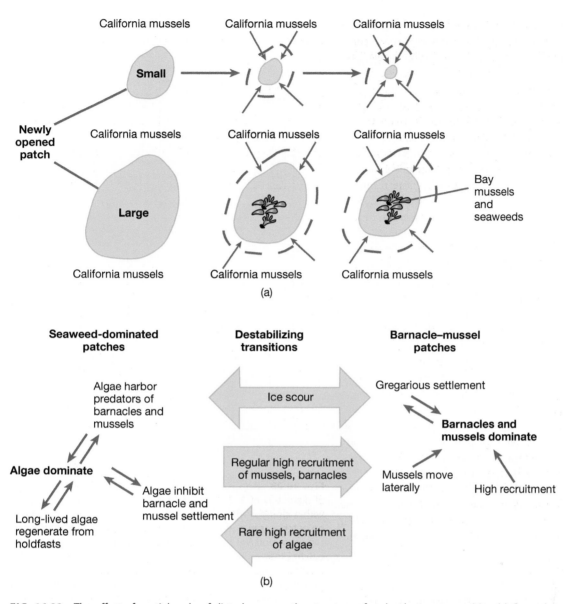

FIG. 16.22 The effect of spatial scale of disturbance on the structure of rocky-shore communities. (a) On rocky shores of the western United States, the size of a patch opened by disturbance strongly affects the future of the patch. If small patches are opened, mussels move in toward the interior by means of crawling, and the patch is closed. If the patch is large, however, the ratio of perimeter to interior area is smaller, which reduces access by the surrounding mussels and inhibits incursions of predatory snails. This gives enough time for colonization of the patch center by seaweeds and another species of mussel, which may persist. (b) Alternative stable states on the rocky shores of Maine, and the factors that perpetuate the states and the strong events that effect a transition from one state (seaweed dominated) to the other (barnacle and mussel dominated). (part b after Petraitis and Latham, 1999)

commonest observations about the marine environment: the coexistence of large patches with completely different dominating species, even when organisms occupy all of the space. How can this be, when intertidal experimentation tells us that space shortage should be a surefire indication of a predictable process of competition?

On protected rocky shores of Maine, one is struck by how different species may dominate different sites that appear to be similar in physical features. Overall, protected sites tend to be dominated by seaweeds and exposed sites tend to be dominated by barnacles and mussels. Protected areas tend to have more carnivores, which eliminates the barnacles and mussels, which in turn facilitates the dominance of

seaweeds. In exposed sites, wave action inhibits predation, which allows mussels and barnacles to dominate. That is all well and good, but in protected coasts, some areas are covered with barnacles and mussels, while others are covered exclusively by brown seaweeds, most commonly stands of *Ascophyllum* or *Fucus*. Peter Petraitis and students examined these assemblages and could not see any reason for competition or predation alone to lead to this sort of alternative dominance (Petraitis and Latham, 1999). The key to these alternatives was thought to be in rare events that could flip a site from one stable state to another.

How can these alternative states on Maine rocky shores be generated? **Figure 16.22b** summarizes the processes that initiate the formation and tend to maintain alternative stable states on rocky shores of Maine. Once seaweeds have colonized a patch, they provide moist shelter for predatory gastropods, which rapidly consume recruiting barnacles and mussels. To some degree, the complete rock cover by seaweeds prevents successful settlement of barnacles and mussels in the first place, and whiplashing by seaweed fronds also tends to damage newly settled barnacle larvae. On the other hand, barnacle–mussel patches are maintained by high recruitment and to a degree by lateral movement of mussels, which can fill in empty spaces (the eastern mussel *Mytilus edulis* is far more mobile than the western mussel *Mytilus californianus*). Most important, barnacles and mussels settle gregariously, which tends to perpetuate patches. Mussels also can resist predation when in a dense bed. Predatory *Nucella lapillus* snails can attack only from the periphery of the mussel bed because, with the aid of byssal threads, mussels can enmesh snails that venture closer.

It is essential for a major perturbation to occur to cause a shift from one patch type to the other. That is the important lesson of this research. Small-scale processes may be inferred by use of manipulative field experiments, but it is the longer-term and sometimes unpredictable disturbance events at larger spatial scales that have such an impact on community structure. Petraitis and Dudgeon speculate that **ice scour**, which occurs only every 10–20 years or so in Maine, might be the agent of patch opening. If so, one cannot predict the outcome of species interactions without an understanding of the history of a particular location.

Petraitis and Steven R. Dudgeon (1999) experimentally demonstrated the effect of large-scale disturbances by clearing patches at several spatial scales (1, 2, 4, and 8 m in diameter). They planted a small patch of mussels in the center of each cleared patch. A discontinuity was discovered between patch sizes of 2 and 4 m diameter: Mussel mortality dropped in a distinct step in the 4 and 8 m patches, relative to the 0, 1, and 2 m patches (**Figure 16.23**). It is possible that at this threshold of disturbance, mussels and barnacles can colonize and resist predation by the drilling snail *Nucella lapillus*. Indeed, mussel mortality from *Nucella* predation declined in a steplike fashion in experimental patches that were greater than 4 m in diameter. This explanation is similar to Suchanek's explanation of the role of large open patches in discouraging the movement of carnivorous *Nucella* snails and, therefore, reducing the mortality of colonists at the patch centers. A more recent study by Petraitis and Dudgeon (2005) showed that opening such large patches 8 m wide and greater resulted in the unpredictable colonization by barnacles or the seaweed *Fucus*. At a given site seaweed unpredictability in recruitment is much greater than for mussels or barnacles. Therefore, seaweed recruitment unpredictability combined with priority effects (first in holds the space) are the forces that matter the most in creating alternative dominance patterns in different areas of the Maine study sites (Petraitis and Dudgeon, 2015).

The idea of alternative stable states on rocky shores has been a source of controversy. Bertness et al. (2004) found far more determinism in the outcome of species colonization.

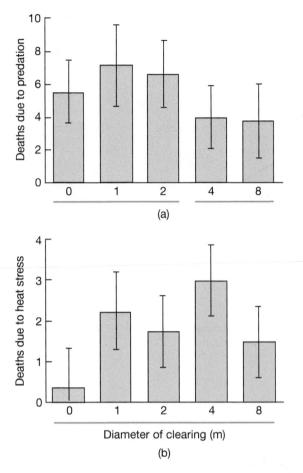

FIG. 16.23 An experiment on rocky shores of Maine. Patches of diameters of 1, 2, 4, and 8 m were scraped clear and planted with a clump of the mussel *Mytilus edulis*, including areas where no patch was opened up (0 m diameter). (a) Predation success by the gastropod *Nucella lapillus* is distinctly higher on 0, 1, and 2 m patches than on 4 and 8 m patches. The red bars unite patch size treatments that are statistically indistinguishable. (b) Mortality of mussels owing to physiological stress. Note that mortality in all cleared areas is indistinguishable statistically, but all cleared areas have greater stress-induced mortality than the uncleared areas. (After Petraitis and Dudgeon, 1999)

Unfortunately, this study cleared off much smaller patches, within the range that Petraitis and colleagues also found no effect of patch clearance. Still, this controversy is well worth discussing in class, using the recent literature.

Two Outside Forces: Phytoplankton Productivity and Flow

■ **Water flow and primary productivity by phytoplankton interact strongly to determine intertidal community dynamics.**

For many years, marine investigators have known that larger regional forces influence community interactions. We shall discuss estuaries later, but an important example emerges from studies of estuaries and bays, which have spatially and temporally varying degrees of water exchange with open coastal areas. J. D. Andrews (1984) found that a range of estuarine tributaries feeds into Chesapeake Bay.

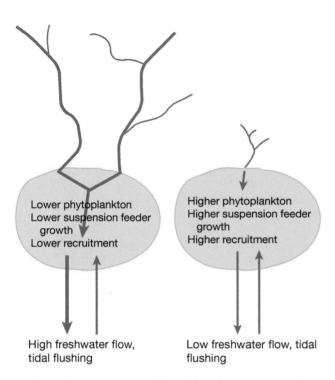

FIG. 16.24 The expected relationship between retention time of water in an estuary, water entering from the watershed, and recruitment rate of an estuarine invertebrate.

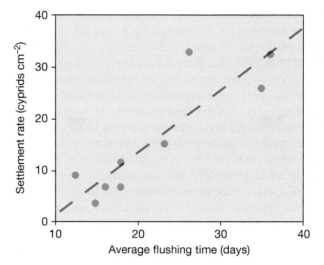

FIG. 16.25 Settlement rate of the barnacle *Semibalanus balanoides* in different years (dots) as a function of flushing time of Narragansett Bay, Rhode Island. When rainfall is high, fresh water enters the bay more frequently and flushing time decreases. (Modified from Gaines and Bertness, 1992)

The James River has relatively high flow and tends to have very poor larval sets of oysters, which increase a bit downstream as the river widens and current velocity declines. On the other hand, a number of smaller rivers have very poor exchange with the open parts of the bay, and in those tributaries there are commonly very dense sets of oyster larvae. These observations can be embraced under the concept of **flushing time**, which is the mean time that water and entrained larvae and phytoplankton are maintained in a bay (**Figure 16.24**). Apparently the flow regime, especially the flushing time of water in an area, is a good predictor of whether larvae will be kept in a bay system until settlement or will be flushed. If flushing time is short, then larvae and phytoplankton will be mainly flushed out and larval settlement will be poor. But if flushing time is long, then larvae will build up in the bay and settle in great densities.

In Narragansett Bay, Rhode Island, some years are characterized by a great deal of rainfall and river flow into the bay, which results in stronger flushing and mixing with coastal waters. In such years, barnacle sets of *Semibalanus balanoides* are quite low in Narragansett Bay. In years with low flow, the retention time of water and larvae is much greater and barnacle recruitment is increased (Gaines and Bertness, 1992) (**Figure 16.25**).

On more open rocky shores, there is the question of larval supply. In such an open system, water flowing from elsewhere is the only delivery mechanism for planktonic larvae and phytoplankton that could be consumed by sessile suspension feeders. As flow increases, one might expect higher larval supplies and more suspension-feeder food. In low-flow situations the larval supply would be low, and the

suspension feeders might be able to filter all the phytoplankton, thereby putting a cap on their food supply.

Regional differences in primary productivity should influence the pattern of rocky-shore species abundances. For example, striking differences were found in rocky-shore animal abundances on different sides of a headland along the coast of Oregon. The side with abundant rocky-shore benthos had strong upwelling and primary productivity. Subsequent investigations using satellite photos demonstrated that shores with abundant benthos are also characterized by abundant near-shore phytoplankton (Menge et al., 1997). Such effects have been demonstrated in widely dispersed regions, including New Zealand exposed rocky shores (Menge et al., 2003). This result shows that the **bottom-up control** of phytoplankton productivity can strongly influence community structure.

Combined with flow, larval supply often is a major determinant of community interactions.

Ecologists have thought that in most environments, the interactions of competition, predation, and occasional disturbance were the main determinants of species composition. But many of the dominant players have planktotrophic larvae that remain in the plankton for weeks. Is there always an adequate supply of larvae to permit the interactions to be predictable everywhere?

The roles of larval supply and transport can be strong controllers of regional differences in community composition (Underwood and Fairweather, 1989). This is likely to be a crucial factor on coasts where upwelling along the shore is prominent and carries planktotrophic larvae out to sea and far from parental sites. On the west coast of the United States, kelp forests separate the rocky coasts from the open sea. Where kelps are scarce, barnacle larvae appear to recruit more heavily. Kelp forests seem to act as larval filters, partially because they slow down the advance of shoreward water movement, which would otherwise carry larvae to the coast. However, kelp forests are also dense with planktivorous fishes, and these pick off much of the larval population that enter the relatively quiet waters.

Larvae will be lost to sea in upwelling situations, in estuaries where larvae stay in the buoyant seaward estuarine flow, and also on some open coasts where longshore currents are strong. Larvae released in a deep embayment may be trapped inside for weeks, and adjacent coastal embayments may never profit from the large flush of larvae in any one locale. In southeast Australia, a persistent current tends to transport surface water offshore; thus larval settlement success depends on local larval production and retention, not on regional supply of larvae from far down the coast. Such local retention has been found to be very common, and larval adaptations of coral reef fishes often take advantage of shifting currents to return to the site of larval origin.

It is easier to examine **recruitment**, which is the appearance of newly settled larvae as metamorphosed juveniles, as opposed to settlement of the larvae themselves, which are small and hard to observe directly. The classic experiment done by R. T. Paine demonstrated that removal of the predatory sea star *Pisaster ochraceus* resulted in dominance by the mussel *Mytilus californianus*, but this effect depended on a good recruitment season. As it turned out, many subsequent seasons were quite poor, and the experiment probably would not have "worked" in quite the same way had it been repeated in another year. For example, *M. californianus* can outcompete the brown seaweed *Hedophyllum sessile*, but an outer coast Washington State rocky shore was dominated by the seaweed for at least 20 years before a strong larval set of *M. californianus* occurred, resulting in mussel dominance (Paine and Trimble, 2004). This long-term lack of mussel recruitment was probably reversed by a warming event associated with the 1997–1998 El Niño. Realistically, most communities consist of species whose respective temporal variations may depend on very different factors, operating on many temporal and spatial scales. Many dominant species on the west coast, such as the California mussel, have long life spans, and this also may contribute to the relative dominance of

importance of interactions within rocky-shore communities, relative to fluctuations in larval supply.

Soft-Sediment Interactions in Protected Intertidal Areas
Zonation and Interspecies Competition for Space

- **Soft sediments are also physiologically stressful at low tide, but water retention reduces desiccation and temperature stress.**

Animals living in intertidal sand- and mudflats face temperature and desiccation stress as in rocky shores. Fine-grained sediment, however, retains water at the time of low tide, so burrowing is a strategy that successfully reduces desiccation and temperature stress. During low tide, animals can remain within cool and wet sediment by living in relatively deep burrows, where water is retained. As a result, many intertidal sand-flat invertebrates maintain burrows as much as 25–50 cm below the sediment surface (**Figure 16.26**). At the time of high tide, the animals either come to the surface or extend siphons, tentacles, or other feeding and respiratory organs to the surface. Shallow-burrowing eastern U.S. coast bivalves, such as the northern quahog *Mercenaria mercenaria*, deal with desiccation by tightly sealing the shell. Deep-burrowing clams, such as the razor clam, *Ensis directus*, and the soft-shell clam, *Mya arenaria*, are bathed in sediment pore water at greater burial depths and have longer siphons, with shells containing gapes that preclude the shell from being tightly sealed. In muddy sediments, water does not drain quite as easily, and small soft-bodied organisms may live closer to the surface.

- **Soft-sediment intertidal species compete for space and food by direct displacement and chemical secretions.**

Interspecific competition is also important in soft sediments because food and space for burrow occupation are often limited. The common mud snail *Tritia* (formerly *Ilyanassa*) *obsoleta* dominates eastern muddy shores of temperate

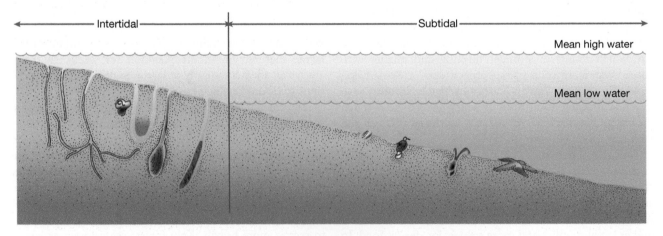

FIG. 16.26 Cross section of the bottom of intertidal and subtidal marine soft sediments. The depth of burrowing is deeper in the intertidal, where exposure to desiccation and temperature variation at the sediment surface is greatest. (After Rhoads, 1966)

North America. Its burrowing, surface deposit feeding, and probably ingestion of smaller benthic invertebrates usually results in the reduction of numbers of competing deposit-feeding polychaetes (Kelaher et al., 2003). In the last few decades, *I. obsoleta* was introduced and spread rapidly into San Francisco Bay. It has displaced the local mud snail *Cerithidea californica*, which now occupies a refuge in the upper part of the intertidal zone. The upper intertidal refuge for *C. californica* seems to be due only to the poor desiccation tolerance of *I. obsoleta*. Margaret Race (1982) built an enclosure spanning the intertidal zone, and the *I. obsoleta* was removed from the lower part of the shore. *Cerithidea* soon expanded its vertical range downward. Predation of *Cerithidea* eggs also contributed to the reduction of its ecological range.

In muddy subtidal bottoms, burrowing invertebrates compete for space within the sediment, and one often sees **vertical stratification of dominant species** within the sediment. Some species, for example, are free burrowers with no permanent stationary burrow, and their movement through the sediment disrupts the permanent burrows of other species (**Figure 16.27**). The limitation of space can be shown by a field experiment. If deep-dwelling, suspension-feeding clams are reduced in density, the remaining individuals grow faster and have greater survival. In Mugu Lagoon, California, the deep-dwelling clam *Sanguinolaria nuttali* is affected by the presence of other deep-dwelling species. Variation in density of a shallow-burrowing species, however, causes no effect (Peterson and Andre, 1980). Because all the species feed by means of a siphon connected with the surface, this experiment suggests that space

within the sediment, and not food, is probably the major limiting factor.

A limited number of benthic species, including some polychaetes, hemichordates, and a species of phoronid, produce high concentrations of **bromine-containing aromatic compounds**, which are toxic and are avoided by other infaunal organisms. These compounds are released into the sediment, a halo of brominated aromatics surrounds the worms. The sediment is smelly to humans. If these aromatics are experimentally injected into the sediment at the same concentrations found naturally next to chemical-producing animals, juvenile bivalves and nonbrominated polychaetes burrow much less than in aromatic-free sediment (Woodin et al., 1997). Presumably the bromine production defends against incursions by competitors, but it may also deter predators.

Food Supply in the Soft-Sediment Intertidal

■ **Inputs of resources may be spatially variable, which may affect interspecific interactions even at the same tide level and cause significant spatial differences in local dominance by species.**

On soft-bottom shores, the spatial distribution of food input may be quite uneven. The sea lettuce species of the genus *Ulva* are important sources of particulate organic matter, but *Ulva* usually occurs in discontinuous patches on a sand flat or mudflat surface (**Figure 16.28a, b**). Using a tracer of seaweeds in sediments can show this. Levinton and McCartney (1991) analyzed sediments in a sand flat in Washington State for the photosynthetic pigment lutein, which is a good tracer of the seaweed *Ulva*. Lutein was analyzed with a technique called high-performance liquid chromatography. The distribution of lutein in sediment was spatially very discontinuous and mostly corresponded to patches of *Ulva*, and was found only sparsely in sediments distant more than 1 meter from *Ulva* patches.

In mudflats near salt marshes in Long Island, New York, these *Ulva* patches go through the following stages:

1. Sediment is covered by live *Ulva*.
2. *Ulva* dies in late spring and decomposes.
3. Sediment gains particulate organic matter and becomes anoxic in patches.
4. Soon, the sediment is more oxygenated and colonized by populations of smaller polychaetes, which creates strong spatial variation in smaller worm densities (**Figure 16.28**).
5. In some cases, larger mud snails of *Tritia obsoleta* move toward the patches and negatively affect the smaller worm populations. This is a force in opposition to stage 4, and tends to reduce spatial heterogeneity. Field experiments show that the mud snails can detect the presence of patches of *Ulva* decomposing.

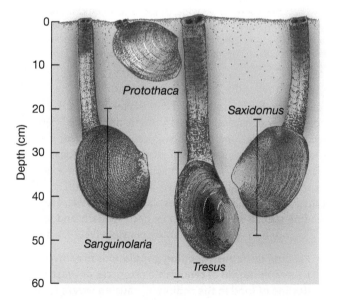

FIG. 16.27 Stratification of bivalve mollusks in intertidal sediments of Mugu Lagoon, southern California. *Protothaca staminea*, a shallow-burrowing clam (at top), is unaffected by the presence of deep-burrowing clams. Experiments demonstrate that *Sanguinolaria nuttalli* (lower left) is strongly depressed by other deep-dwelling clams. Vertical bars show vertical burrowing position. (Drawn following descriptions in Peterson and Andre, 1980)

■ **In soft sediments, deposit feeders appear to be food limited, whereas suspension-feeding populations are more variable and are not affected as much by population density.**

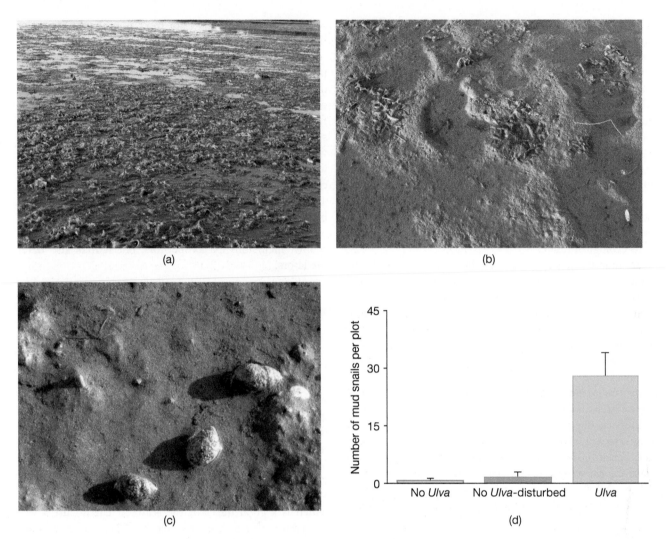

FIG. 16.28 The soft-bottom intertidal landscape often has spatially patchy inputs of food. (a) Patchy occurrence of the green seaweed *Ulva rotundata* on a mudflat in Long Island, New York. (b) Close-up of *Ulva* patches, about 15 cm wide. (c) Close-up of mud snails *Tritia obsoleta*. (d) Results of an experiment in which bare surfaces were compared with sediment into which *Ulva* pieces were buried; the sediment was also disturbed without inserting *Ulva*, as an experimental control. SE = 95 percent confidence limit. Results demonstrate that mud snails are attracted to plots with planted *Ulva* fragments. (From Kelaher and Levinton, 2003) (Photographs by Jeffrey Levinton)

Deposit feeders depend on microbial organisms and particulate organic matter for food. By contrast, suspension feeders feed on phytoplankton or resuspended particles. In reality, the difference is not so great because particulate organic matter may be deposited on the bottom from the water column, which is an especially important process in subtidal habitats. In the intertidal, however, the food source of deposit feeders may be dominated by microalgae such as benthic diatoms. Deposit-feeding populations may be living with a more stable food source, whereas suspension feeders may be exposed to much more variable phytoplankton and may have a food supply that may be quite variable, owing to the large spatial and temporal changes that occur in the water column (Levinton, 1972).

The hypothesis of deposit feeder stability versus suspension feeder variability has now been tested several times and has been confirmed for soft sediments in intertidal and very shallow subtidal areas. Einar Olafsson (1986) planted varying densities of the bivalve *Macoma balthica*, which suspension feeds in sand and deposit feeds in mud. Higher population densities reduced growth only in the deposit-feeding populations, which suggests that deposit-feeding food is limiting as a resource but the phytoplankton source was not limiting. This approach has also been extended to the community level in a study of tidal mudflats of the Dutch Wadden Sea (Kammermans et al., 1994). Deposit-feeding species were found to be both spatially and temporally more homogeneous than suspension-feeding species.

■ **Feeding by deposit-feeding populations may involve overexploitation of renewable resources or seasonal decline of food in the sediment, causing severe food limitation.**

Many intertidal deposit-feeding species graze diatoms from the surface of the sediment. If you add several surface/deposit-feeding clams to a tray of sand, you will find that the surface remains clean. Remove the clams and, as long as there is light, diatoms will grow profusely as a brown surface layer. Deposit feeders, therefore, are exploiting a renewable resource. A number of studies show that grazing by bivalves and gastropods reduces sediment microalgae to low levels and that animal growth is then inhibited. This is a mechanism for reduction of somatic growth, reproduction, and, therefore, population regulation of surface-grazing deposit feeders.

Many deposit feeders, such as small polychaetes and oligochaetes, appear in great numbers in the spring in temperate, protected tidal soft sediments of southern New England. Their generation times are short and populations usually build up, eventually declining by end of summer. Early spring sediments probably have a considerable amount of nutritive particulate organic matter deposited on the sediment surface. As the spring progresses, this material is eaten by deposit feeders and fuels population growth (Cheng et al., 1993). By the middle or end of summer, however, this particulate material either is all eaten or is degraded microbially into material that is relatively indigestible to deposit feeders. In response to this cycle, the common oligochaete *Paranais litoralis* reproduces by budding, and the population increases, reaching a peak in early summer (**Figure 16.29**). Then food quality declines and the population crashes. If the organic matter collected over the season is labeled with ^{14}C, one can observe a steady decline in absorption of the material when fed to the worms. This suggests that the material is becoming steadily less digestible.

■ Seasonal influxes of predators in the intertidal can devastate local soft-sediment communities.

Although many habitats have their indigenous permanent population of predators, some habitats experience seasonal invasions of predators that cause major reductions of prey populations. Shorebirds (**Figure 16.30**) often have favored feeding grounds, which they visit successively during their seasonal migration, that may cover thousands of kilometers. Along the coast of eastern North America, sites such as Plymouth Bay (Massachusetts), Cape May (New Jersey), and Jamaica Bay (New York) are famous for their periodically dense populations of shorebirds. Sanderlings (*Calidris alba*), semipalmated sandpipers (*Calidris pusilla*),

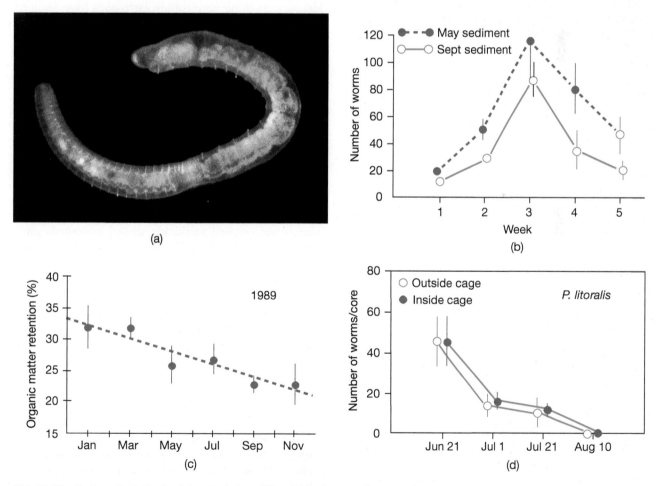

FIG. 16.29 Testing a hypothesis of the regulation of deposit-feeder populations by food limitation in Flax Pond, New York. (a) *Paranais litoralis*, which reproduces mainly by budding asexually. (Courtesy of Alexa Bely) (b) If food has entered the mudflat in a pulse, populations should increase and then crash, and sediment from the beginning of the cycle (May) should be capable of growing more worms than sediment taken from late in the population cycle (September). (c) If food is becoming less abundant, carbon in the sediment should become more and more difficult to assimilate during the spring–summer season. (d) If predation is not significant in causing population decline, then *Paranais litoralis* populations should decline at the same rate on the open mudflat and within predator-exclusion cages. (After Cheng et al., 1993)

FIG. 16.30 Laughing gulls, *Laros atricilla*, feeding on horseshoe crab eggs on a sand flat at Cape May, New Jersey. (Courtesy of Joanna Burger)

TABLE 16.1 Percent Mortality of Intertidal Mudflat Invertebrates Fed on by Migratory Shorebirds in Plymouth Harbor and Kingston Harbor, Massachusetts

SPECIES RANK IN JULY	PERCENT MORTALITY*
1	84
2	78
3–5	67
5	36

* Mortality is listed as a function of the prey species' rank at the start of the study.
Source: Data from Schneider, 1978.

short-billed dowitchers (*Limnodromus griseus*), and black-bellied plovers (*Pluvialis squatarola*) all prey on mud-flat invertebrates. David Schneider (1978) caged areas of a Massachusetts mudflat with dense migrating shore-birds, simply by placing a rope suspended from four short sticks. Within the enclosed area, the benthos survived for 2 months at the same densities. Outside the enclosure, the soft-bodied benthos (e.g., polychaetes) were totally consumed. The birds concentrated their predatory efforts on the most abundant benthic species (**Table 16.1**).

Similarly, migrations up Chesapeake Bay of the blue crab *Callinectes sapidus* obliterate the shallow-burrowing benthos (although in recent years this species has declined in abundance). Spring migrations of fish and crustaceans into shallow-water embayments and mudflats cause great decreases in benthic population size. In many cases, however, smaller predators—including gastropods, nemertean worms, and polychaete annelids are major predators.

Invasions and the Reorganization of Intertidal Communities

■ **Invasions of predators have had major effects on intertidal communities.**

Invasions are becoming a common story in marine communities, and they have been noticed especially in the intertidal zone. Two important predators have invaded both coasts of North America and have had major impacts on shore communities in New England, including strong impacts on the diversity of native species. The shore (also called the green) crab, *Carcinus maenas*, arrived in New England probably in the nineteenth century and became a dominant crab in intertidal and shallow subtidal sandy and rocky bottoms. It is not very abundant south of New Jersey, perhaps because of predation by the blue crab *Callinectes sapidus*. The Japanese shore crab, *Hemigrapsus sanguineus*, was first noticed in New Jersey in 1988 and by 2008 had spread from North Carolina to Maine. It is now the most common intertidal crab on intertidal cobble beaches of the middle Atlantic United States and has displaced the shore crab, perhaps by preying on juveniles (Lohrer and Whitlatch, 2002). Both crab species as adults are omnivores, feeding on macroalgae but also on mollusks such as snails and juvenile bivalves. The arrival of these species, particularly the shore crab *H. sanguineus*, seems to have had dramatic effects on the New England shore fauna. Intertidal rock-dwelling herbivorous gastropods have declined greatly in abundance, which has allowed greater opportunities for seaweeds to expand.

While direct predation has directly removed many juvenile snails, crabs also may exert a trait-mediated indirect effect by causing snails to feed less on seaweeds (Trussell et al., 2002). Unfortunately, the experiments that have been done involve placing crabs in containers around a plot with snails. It is by no means clear that the signal received by the snails (an odor of some sort) is scaled properly to real-world conditions. Also, when invasive crabs appear, prey snails and mussels respond by increasing the thickness of their own shells, and snails may even reduce the size of the aperture. Equally interesting is the response of the predator. Green crabs exposed to thicker-shelled gastropods grow more robust crusher claws.[2] We are thus witnessing an arms race between invasive predators and native prey. Comparisons between new and older invasions by alien crab species in northeastern North America suggest that prey snail species have quite variable responses to odors of recently invaded Japanese shore crabs, suggesting that evolutionary responses by snails have not yet stabilized into stereotyped responses. But older invasions, greater than 110 years by the European invasive crab *C. maenas*, show more stabilized responses of prey, suggesting that evolution has moved to the point that prey responses are highly evolved and have less flexible antipredator behavior (Edgell et al. 2009).

Carnivores are not the only invasive species that are potentially important in altering intertidal community structure in New England. For example, the herbivorous gastropod *Littorina littorea* greatly expanded its abundance in New England in the nineteenth century, spreading from Nova Scotia to southern New England in a matter of a few

---

[2] Green crabs have a thick claw, which is used for crushing prey, and a thinner claw used for slicing tissue and handling prey.

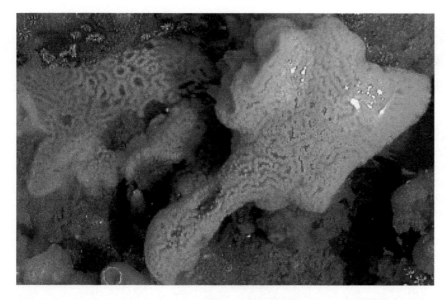

FIG. 16.31   The colonial sea squirt *Botrylloides diegensis* has invaded rocks of southern New England and displaced colder-water-adapted native species. Here, colonies of *B. diegensis* are overgrowing native *Ciona* sea squirts, colored white. (Courtesy of Robert Whitlatch)

decades by means of a planktotrophic dispersal stage. This species is a very effective grazer and has had great impacts on seaweed abundance. Ironically, as mentioned earlier, it now is being strongly affected by the invasive Japanese shore crab. The intertidal zone of New England is therefore impacted to a major degree by invasive species.

- ■ **Invasions of a number of sessile suspension-feeding species, combined with sea-surface temperature warming, have caused major reorganizations of intertidal communities.**

While invasive predators are likely to have cascading effects on target communities, sessile suspension feeders have also been spread throughout the world, probably both as hitchhiking adults on boats and other floating materials and as larvae brought in ballast water of ships. Such invasions appear to have accelerated in the past few decades,

but are certainly not new. In the 1940s the barnacle *Elminius modestus* invaded waters of Great Britain, all the way from Australia. Clearly, shipping was involved in the transport. The invasive barnacle tolerates a broad range of salinities and has become increasingly a dominant part of the protected rocky-shore biota of the United Kingdom (Allen et al., 2006).

Long Island Sound, New York, has been especially well studied, and a number of colonial ascidian species have invaded and risen to dominance in recent years. The invasive orange sea squirt *Botrylloides diegensis* (**Figure 16.31**) has become especially abundant, but at least three other invasive species dominate quiet-water areas. These invasives are more abundant in years of warmer temperature, whereas native sea squirts do better in colder years (Stachowicz et al., 2002). As climate gets warmer, this trend will only increase.

HOT TOPICS IN MARINE BIOLOGY

Sea Star Catastrophe: Disease, Its Spread, and Its Diagnosis   16.1

Marine biologists have gradually become aware of the tremendous ecological consequences of diseases in the marine environment. Consider the rate of spread and intensity of effects of past outbreaks. In the late 1930s, a catastrophic decline of sheepswool sponges was noticed in the Bahama island chain in the Caribbean. Especially striking was the speed of the decline in a year or so and the spread apparently by ocean currents from one island to the next (Smith, 1941). Overall, 70–95 percent of sponges disappeared from the Caribbean in 1938 (Galstoff, 1942). A fungal infection was believed to be the disease agent, but bacteria have been found that cause sponge declines (Webster, 2007). The widespread and abundant sea urchin *Diadema antillarum* largely disappeared from the Caribbean

basin within one year in the 1980s. Disappearances also occurred sequentially from island to island, following expected current patterns (Lessios et al., 1984). In the ocean, such rapid waterborne transmission is apparently common and spreads disease over thousands of kilometers within months. It is not clear whether the transmission occurs by disease organisms free in the water or perhaps by a vector such as planktonic larvae.

While water currents can rapidly spread disease, we cannot overlook the possibility that similar environmental changes in widespread and often scattered locations are facilitating disease outbreaks of formerly rare microbial disease agents. The enormous El Niños of 1997–1998 and

continues

HOT TOPICS IN MARINE BIOLOGY

Sea Star Catastrophe: Disease, Its Spread, and Its Diagnosis   *continued*   16.1

2015–2016 have brought extreme warm waters to coral reefs throughout the Pacific, which may facilitate the rise of a number of microbial pathogens. Not all disease microorganisms benefit from ocean warming, but widespread phenomena like ocean warming and El Niños likely will facilitate the growth and possibly spread of disease. On the U.S. east coast, warming has facilitated the northward expansion of protozoan parasites of bivalve shellfish species, making them major agents of mortality and physiological weakening.

Disease spread is especially effective ecologically when the victims are major players in ecological interactions. We therefore must focus on species that provide structure in benthic environments, such as sponges, corals, and seagrasses, and species that are strongly interacting predators in marine food webs. Many marine communities have their ecological dynamics dominated by trophic cascades, in which a top keystone predator's removal results in dramatic reorganizations of the food web

Sea star wasting disease (SSWD) usually makes its appearance as a series of whitish lesions, but eventually the disease causes arms to drop off and reduces the sea stars to a pile of slime and skeletal elements (**Box Figure 16.1**). In July 2013, a major mortality event was recorded in Howe Sound, British Columbia, but by 2014, mortality events were recorded as far south as Mexico off Baja California. Appearance and spread of the epizootic has been associated with increased sea-surface temperatures, and the recent major outbreak occurred during an especially warm winter of 2013–2014 on the west coast of the United States. An outbreak in Washington State coincided with positive summer temperature anomalies of 2–3°C (Eisenlord et al., 2016).

BOX FIG. 16.1 Progression of sea star wasting disease on the ochre star *Pisaster ochraceus* in the San Juan Islands, Washington. Sea stars were abundant at start of disease progression in May 2014. (a) Healthy sea stars (b) develop lesions (c) that can lead to arms detaching from central disk (d) prior to extensive tissue necrosis and (e) death. (From Eisenlord et al., 2016)

What causes SSWD? This was initially a mystery, because no great abundance of bacteria or other organisms was visible using standard microscopy. But exposure of tissue homogenates of symptomatic sea stars homogenized and then filtered through a 0.22 μm filter resulted in the SSWD syndrome after about 10–17 days, whereas heat-treated controls resulted in no infection. These results demonstrated experimentally that nondiseased sea stars could be infected with virus-sized material. It is challenging to go to the next step. Next-generation sequencing of viral extracts of infected sea stars must be employed, along with complex metagenomic statistical techniques to assemble a viral genome. Assembly of metagenomes from the sequencing process resulted in the deduction of the sequence of a near-complete virus of the type densovirus, which was termed sea star-associated densovirus (SSaDV) (Hewson et al., 2015). This virus is in far greater concentrations in sea stars with SSWD than in normal healthy sea stars. The diagnosis is clear, but the dynamics of spread along the coast and the role of ocean warming still need to be determined. We can expect more such outbreaks in the future in many marine species and communities.

Spartina Salt Marshes

Ecosystem Engineers, Geographic Extent, and Setting

■ **Spartina salt marshes are dominated by cordgrasses, which function as ecosystem engineers by binding fine sediment and causing the buildup of meadows above low water.**

Spartina salt marshes (**Figure 16.32**) develop in tidal areas of quiet water, where a variety of salt-tolerant grasses colonize the sediment and then trap fine sediment. A study of *Spartina* salt marshes on the Atlantic and Gulf coast of North America found a range of vertical sediment accretion rate of 0.09–1.78 cm per year (Turner et al., 2002). Characteristic of these grasses is an extensive **rhizome system** beneath the sediment surface, which takes up nutrients and is crucial in maintaining the structure of the marsh sediment and the entire salt marsh.

Spartina alterniflora, the dominant eastern and Gulf coast American species found lowest in the intertidal zone, must put up with long periods of immersion in saltwater, to which it is more tolerant than other grass species. The sediment pore water has little or no oxygen unless burrowing organisms aerate the sediment, and the root system of *S. alterniflora* connects to air by means of air pockets in the midcortex. Much of the leaf and stem section of the plant is highly vascularized. A cross section of the plant shows the large amount of open space near the surface, which is devoted to air and oxygen transport (**Figure 16.33**). Plants cannot use nutrients efficiently without oxygen, so this tissue allows a connection between the aerobic leaves and stems, which are surrounded usually by anoxic water.

Rhizomes extend laterally, and shoots grow above the sediment-water interface. Although these grasses develop flowers and set seed, the asexual spread of the rhizome system is usually the major form of local spread. Genetic analysis using enzymes demonstrates that marshes often

FIG. 16.32   The salt marsh at Herring Creek, near Harwich, on Cape Cod, Massachusetts. Tall form of *Spartina alterniflora* in foreground; short form behind. In the distance is the common reed *Phragmites australis* fringing the marsh. (Photograph by Jeffrey Levinton)

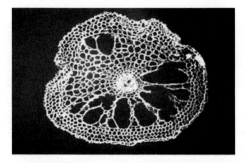

FIG. 16.33 The aerenchymal tissue allows *Spartina* to exchange gases, even when surrounded by an anoxic soil. The tissue in this photograph is visible as a series of circular passageways around the periphery. (Photograph courtesy of Mark Bertness)

consist of a few clones, or groups of genetically identical individuals, each of which has spread by asexual growth. The overall morphology of rhizome system and shoots creates a series of projections that form a baffle against water movement, which encourages sedimentation. The accumulation of sediments results in the formation of a meadow with a nearly horizontal surface. *Spartina* salt marshes are, therefore, an excellent example of the **ecosystem engineer** concept, in which one or a few species create a structural habitat on which many other species depend.

A single plant may colonize an open area of sediment (**Figure 16.34**), either by rafting or by setting of seed. After the clone has developed, the grass blades will develop a density sufficient to slow current speeds and accelerate the deposition of fine-grained sediment. This will gradually cause the development of a rising sediment surface. Thus, a salt marsh begins to spread and evolve into a meadow of sediment held together by dense grass stands. Eventually the sediment will be high in organic matter known as **peat**. Organic content varies a great deal and can range from

2 to 40 percent. Older marshes have thicker layers of peat. The meadow usually encloses a system of salt marsh **creeks**, which are often nursery grounds for juvenile crustaceans and fishes (**Figure 16.35**).

If there is no major change of local sea level, then the level of the sediment surface will rise, and the dominant plants will change gradually from low-intertidal grasses to terrestrial plants as more sediment accumulates and the surface of the sediment rises. Thus, a "mature" marsh has passed through the stages of (1) bare intertidal sediment, (2) early colonization of patches of grass, (3) extension of the grass patches and trapping of sediment, (4) gradual rise of the sediment surface and transformation into organic-rich peat, and (5) development of a higher marsh.

In North America, *Spartina* salt marshes are best developed on the east and Gulf coasts, and marshes of hundreds to thousands of acres can be found. *Spartina* marshes are also common in California, dominated by *S. foliosa*, and an American Atlantic species is an invasive element of the Pacific Northwest. Southern California salt marshes are spatially quite variable, perhaps because of the dry Mediterranean-type climate. *Spartina* marshes are also common in northern Europe. The most spectacular American marshes are found in the southeastern United States, especially in South Carolina, Georgia, and Gulf coast states. Salt marsh plants are salt tolerant, and *Spartina alterniflora* contains siliceous deposits, presumably to deter grazing by birds and mammals. The cellulose composition and the mechanically tough leaves seems to prevent much successful direct grazing. Usually much less than 10 percent of the leaf production is consumed by herbivores in the northeastern United States, but damage is greater farther south. The great majority decomposes and may support large populations of decomposing bacteria and fungi. Predation on flowers is often intense, and *Spartina* seed production is, therefore, often very limited.

FIG. 16.34   How it begins. Newly established seedlings of *Spartina alterniflora* on an open sand flat. (Photograph by Jeffrey Levinton)

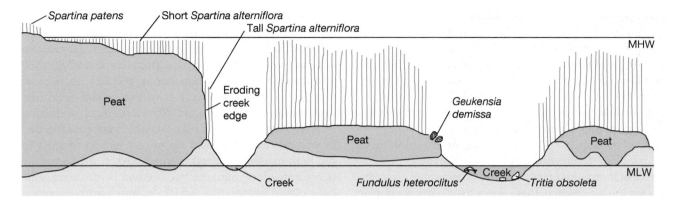

FIG. 16.35 Subhabitats of the *Spartina* salt marsh environment. A tall form of *Spartina alterniflora* is associated with the high nutrient supply of flowing creeks. (After Redfield, 1972)

Vegetational Zonation and Plant Interactions in Salt Marshes

■ **Vegetational zones in salt marshes develop from the interaction of competition and physiological ability to survive salt and drowning.**

Moving from the low-water mark in a tidal creek to the terrestrial environment, one encounters a vertical zonation of vegetation, each zone dominated by a different grass species (**Figure 16.36**). In most east and Gulf coast North American marshes, zones occur in the following order, from low to high intertidal: tall form (1–1.5 m tall) of *Spartina alterniflora*, short form (< 0.5 m tall) of *S. alterniflora*, *Spartina patens*, *Juncus gerardi*, and terrestrial shrubs. The border between zones is often quite sharp and at a predictable tidal height. For example, the zone boundary between *S. alterniflora* and the higher-intertidal

S. patens, approximately at spring high tide, is often used in legal disputes to define the marine–terrestrial border. Why are such zones present? Research by Mark Bertness (1991a) and colleagues showed that plants of the low marsh *Spartina alterniflora* grow well into the high marsh *S. patens* zone, if the latter species is absent. In other words, no physiological factor prevents *S. alterniflora* from invading upper levels. In its own zone, *Spartina patens* outcompetes *S. alterniflora*. If one transplants *S. patens* to within the typical *S. alterniflora* zone, however, the former does badly physiologically, owing to a relative intolerance to immersion in salty water and to drowning for more of the day at the lower-tide level. *Spartina patens* also lacks an efficient mechanism to survive in lower-intertidal anoxic sediments and, therefore, does badly within the lower-level *Spartina alterniflora* zone, where oxygen supply to roots is a major limiting factor. Successful dominance in a zone is

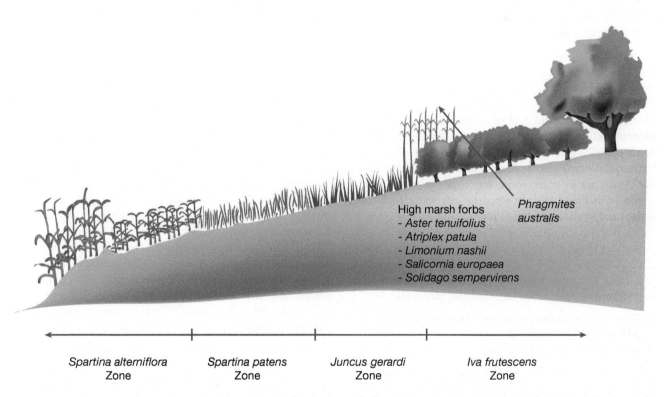

FIG. 16.36 Vegetational zonation in a southern New England salt marsh. (From Bertness et al., 2002, Copyright National Academy of Sciences, U.S.A.)

thus a combination of physiological tolerance and relative competitive ability. In the high marsh, *Juncus gerardi* outcompetes other grasses, although the grass *Distichlis spicata* is more capable of invading newly opened bare sediment patches that are often salty from evaporation. Ironically, *Distichlis* then shades the sediment, allows it to become moist and less saline, and thus facilitates the eventual invasion and predominance of *Juncus gerardi*.

Competition is therefore a major determinant of salt marsh plant dominance, but it works in reverse of dominance of invertebrate competition on rocky shores described earlier in this chapter. These are, after all, terrestrial plant species that are invading seawater. These plants tolerate salt, but the environment is more physiologically stressful as you go lower in the intertidal, because of lowered oxygen in the sediment, and high salt content of water and pore water. *Spartina alterniflora* predominates in the low zone partially because of its superior ability to survive these stresses. In higher zones, plants are good competitors because of the ability to produce a dense rhizome system, which outcompetes other species for space and nutrients.

In Florida salt marshes, *S. alterniflora* dominates the low marsh, and *S. patens* is found at higher levels that are inundated by seawater. However, large areas in this region at higher levels are often in lower-salinity water, and these areas are dominated by the black needle rush *Juncus romerianus*. Just a few centimeters of higher elevation allows the rhizome of this species to avoid exposure to salt, and this species dominates large patches. Still fresher-water expanses are dominated by the saw grass *Cladium jamaicense*, the dominant of expanses of Everglades marshes. Mangroves replace salt marsh as the dominant quiet saltwater shore habitat of southern Florida.

Spartina-dominated zones exert larger-spatial-scale impacts on adjacent ecosystems.

The presence of dense marsh growth exerts longer-distance influences on more distant ecosystems. We discuss below the potential for salt marsh systems to export nutrients, both dissolved and particulate, to nearby coastal systems, subsidizing shallow subtidal benthic systems and fueling coastal phytoplankton blooms. But the wave-buffering effect of dense *Spartina* grass growth also can influence inland, more terrestrial systems (van de Koppel and others, 2006). Without dense marsh grass, more terrestrial forbs and shrubs would be battered and drowned by salt water from wave action, and would succumb to physiological stress from salt water. But the absorption of wave shock by dense *Spartina* growth and peat development on cobble beaches of southern New England usually allows the landward development of a dense and more terrestrial plant forb community, consisting, for example, of sea lavender and glasswort.

Floating wrack often smothers plants and creates patches of bare sediment, which become salty and inhibit colonization for a time.

High-marsh zones often consist of acres of continuous tracts of one species of grass. Bare patches are common, however,

and sometimes span several meters across. Considering the lushness of a salt marsh, the patches are surprising. Mats of cyanobacteria cover some, and others are nearly abiotic, with layers of salt on the surface. But piles of dead *Spartina* leaves, or wrack, may accumulate in marshes where there is a limited water exchange with the coastal zone. The high-marsh bare patches are often a remnant of floating rafts of decaying *Spartina* shoots, which are concentrated by currents and then float up to rest on top of grass in the high marsh. The grass is smothered, and a bare zone is created. Once the area has become bare, strong sunlight causes evaporation, which in turn greatly increases the salt content of the sediment pore waters, and sometimes a layer of salt develops on the surface. Mark Bertness (1991a, 1991b) demonstrated that the salty water prevents seed germination and the bare patch is self-sustaining. If a plant can colonize, the shading reduces evaporation immediately, and the saltiness of the water decreases. Many of the grasses can extend from the edge of the patch through vegetative growth, but a strong rain may reduce the salt content, and seeds may then be able to germinate. The patch will be eventually covered by vegetation. The rapidly colonizing grass *Distichlis spicata* is best able to colonize these patches.

Salt marsh assemblages may exert positive and negative effects as ecosystem engineers.

In the past few years, ecologists have come to think of **ecosystem engineers** (this term is often freely exchanged with foundation species, which is often associated with important primary producers in a community) as exerting a positive effect on the presence of a number of codependent species. *Spartina* species create a relatively quiet sedimentary environment, which may enhance the presence of species dependent on soft sediment. The tight association of grass shoots also is known to protect predators such as crabs from moving high into the marsh, which results in a refuge for a large number of salt marsh invertebrates.

Although an ecosystem engineer often creates an environment that is strongly altered, thereby favoring growth of the engineering species itself and many other associates, the local microhabitat might be no longer suitable for other groups of species. An excellent example of this is the occurrence of *Spartina* patches and patches of the lugworm *Arenicola marina* in Netherlands salt marsh habitats. The work of van Wesenbeeck and colleagues (2007) demonstrated that patches of the cordgrass *Spartina anglica* had sharp borders with patches of bare sediment, dominated by the lugworm *Arenicola marina*, an active deposit-feeding polychaete, which lives in U-shaped burrows. Modifications of the sediment by these two different dominants causes mutually negative effects, resulting in occurrence of one or the other "engineer." The lugworms burrow in the sediment, which was shown experimentally to destabilize the sediment and increase mortality of planted *Spartina* seedlings. *Arenicola* worms, on the other hand, could not burrow effectively in the sediment dominated by *Spartina*, which was dense with rhizome material.

Salt Marsh Creeks and Mudflats

■ **Salt marsh systems include creeks and mudflats, which are often biologically diverse and abundant and are corridors between salt marshes and other habitats for many marine fish species.**

As mentioned earlier, salt marsh habitats are usually a series of broad *Spartina* meadows, alternating with salt marsh creeks (**Figure 16.35**). At a creek edge, a marsh may be at a standstill, eroding, or accreting in size. As the grass stands trap sediment, the marsh can grow over bare sediment and into a creek. Several species are often found in the high-intertidal zone, often at the creek edge. The marsh mussel *Geukensia demissa* lives semi-infaunally in the sediment and apparently aids marsh accretion by trapping sediment (**Figure 16.37**). It also enhances grass growth through the addition of organic-rich fecal material on the sediment surface. Fiddler crabs (genus *Uca*) burrow in the upper-intertidal zone and also are found on creek edges in marshes. The burrowing apparently enhances the growth of *Spartina*, perhaps by aerating the sediment. Normally, marsh sediment is anoxic, which inhibits the growth of fungi associated with plant roots, known as **mycorhizae**, which greatly increase the gathering of nutrients from the sediment. Crab burrowing helps aerate the sediment, which in turn increases mycorhizal growth.

The creeks themselves often have strong tidal flow, with bottoms of well-sorted sand or fine mud, depending on the degree of current strength. Marsh creeks often have dense soft-sediment faunas, including polychaete annelids and mollusks. In tidal creeks of northeastern North America, the mud snail *Tritia obsoleta* occurs in densities of hundreds per square meter. In the southeastern United States marshes, the marsh periwinkle *Littorina irrorata* feeds on the muddy surface, but climbs grass blades at high tide to avoid incoming predators. Many species of smaller crabs and shrimp are also abundant, especially the mud crab *Rhithropanopeus harrisii*. Smaller fish, such as killifish (species of *Fundulus*) and silversides (*Menidia menidia*), are also common and may attract predatory bluefish into the creeks. A variety of wading birds such as black-crowned night herons (**Figure 16.38**) and snowy egrets are also common. These birds stalk mobile forms such as crabs and shrimp. Diving birds such as terns and kingfishers also frequent the creeks, where they dive for smaller fish. Predators—including killifish, blue crabs, and other fish—are abundant in the creeks. Blue crabs move into the marsh grass to feed but cannot usually penetrate any farther than the tall form of *Spartina* adjacent to the creek.

The high density of prey invertebrates and protected waters makes tidal creeks ideal habitats for many species of fishes. Salt marsh creeks are usually regarded as nursery areas for juveniles of many fish species, but fish using marsh creeks are usually also found in associated coastal water habitats such as larger estuarine basins. The abundant predators on soft-sediment infauna create the potential for a trophic cascade. In the previous section on the soft-sediment intertidal, we discussed some of the dynamics of benthic invertebrate populations on mudflats, which are often associated with salt marshes. These include polychaetes, oligochaetes, deposit-feeding snails, and meiofauna, which are abundant in salt marsh flats of the east coast, Gulf coast, and in marsh flats in California. Many of these deposit feeders feed on microalgae, which creates a common food chain of microalgae → mudflat deposit feeders → fish–crab carnivores. Experiments on nutrient enrichment of salt marshes in New England suggest that this chain may not be tightly linked (Deegan et al., 2007). On the other hand, we mentioned earlier a very tight linkage between green seaweed input and invertebrate abundance.

Eastern American mudflats are dominated by polychaetes, oligochaetes, and bivalves and are strongly affected by inputs of particulate organic matter, especially from seaweeds. California salt marsh soft-sediment environments are less biologically diverse and dominated by oligochaetes (Levin et al., 1998; Levin and Talley, 2002). Spatial

FIG. 16.37 Population of the semi-infaunal marsh mussel *Geukensia demissa,* among *Spartina alterniflora.* (Photograph by Jeffrey Levinton)

FIG. 16.38 The black-crowned night heron *Nycticorax nycticorax,* a major predator of fish in salt marsh creeks. (Photo by Jeffrey Levinton)

variation of density and species occurrence is quite high. This difference in diversity from New England might be related to the very dry Mediterranean climate of the region. California salt marshes are highly impacted by human dredging and filling, and some active restoration projects are in process.

Spartina Marshes as Sources of Organic Matter

■ ***Spartina* salt marshes produce large amounts of particulate and dissolved organic matter, which may influence the food webs of salt marsh benthos and perhaps the food webs of coastal marine systems.**

In the late fall, leaves of the dominant lower-intertidal *Spartina alterniflora* senesce, turn a lovely yellow brown, and eventually sever from the main plant. Large amounts of floating material enter the marsh system, although the material is relatively concentrated in indigestible cellulose and takes some time to be decomposed by physical fragmentation, tearing apart and ingestion by detritivores, and bacterial decomposition. This material is probably a source of nutrition for a large fraction of the deposit feeders in marsh soft sediments. Minimally, the *Spartina* fragments are substrates for bacteria and fungi, which are consumed by mussels, deposit-feeding polychaetes, and gastropods. However, these animals can also inefficiently digest particulate organic matter and probably derive some of their nutrition from the *Spartina* detritus itself. Experiments show that additions of such particulate matter can stimulate somatic and population growth of salt marsh oligochaetes.

A minority of the *Spartina* production is consumed by herbivores in New England salt marshes, and little of the high-marsh and not all of the mid-marsh grass is consumed in the southeastern United States (but see later discussion for effects of a common snail). The salt marsh ecologist John Teal showed that in Georgia salt marshes, herbivores appeared to consume a minority of the plant production, the majority of which entered into the detritus food chain (**Figure 16.39**). Many believed that the large amount of

detrital material floated from salt marsh creeks into coastal continental shelf waters and was a major source of nutrition both for zooplankton and benthos of the continental shelf. This was known as the **outwelling hypothesis**. Further studies, however, showed that salt marsh plant detritus does not significantly contribute to coastal shelf secondary production but is mostly retained within marsh creeks or right near shore (see Chapter 12).

Although we have few studies to bolster this idea, it appears that export of particulate organic matter may be slight but the export of dissolved nitrogen sources may be considerable in some cases. Dissolved nitrogen may be produced during decomposition, diffused from sediment pore water into the water column, and then transported away by moving waters. A study of the enclosed marsh–creek system in Sippewissett marsh on Cape Cod, Massachusetts, shows that considerable amounts of dissolved nitrogen are exported to the adjacent Buzzards Bay, which has the potential to fuel a considerable amount of the primary productivity by phytoplankton (Valiela and Teal, 1979). But large net export was not found on the north shore of Long Island, New York (Woodwell et al., 1979).

Spartina Marshes as Trophic Cascades

■ ***Spartina* marshes in the southeastern United States appear to be controlled by top-down effects in a trophic cascade.**

The previous section portrays salt marshes as a source of food, which drives local and regional coastal ecosystems. But salt marshes have generally been envisioned as food-rich environments where direct grazing on *Spartina* was unimportant.

This concept of salt marsh structure has been stood on its head mainly by the creative work of Brian Silliman and colleagues. It was long known that carnivores were very important in southeastern salt marshes. Blue crabs attacked and sometimes consumed the marsh periwinkle *Littoraria irrorata*. But what ecological effect did the periwinkle have? Silliman noticed that snails appeared to be associated with wounds in *Spartina* plants, and these wounds were filled with fungi (see Figure 13.3). The scraping of the snail's radular teeth damaged the *Spartina* plants, which were attacked by fungi. Manipulative experiments (Silliman and Zieman, 2001) demonstrated that *Spartina* growth was strongly inhibited by the presence of the snails. The snails may not have been feeding directly on the plants but they were causing major damage by causing fungi to invade *Spartina* leaves, which damaged them extensively. Silliman, therefore, identified a major trophic cascade in low marshes of the Southeast: *Spartina* → snails → blue crabs (**Figure 16.40**).

The identification of this cascade has produced two very interesting interactions. First, the snails appear to have a facultative mutualism with the fungi. By damaging blades of *Spartina* grass, they facilitate fungal colonization, and the fungi is food for the snails. This is not as stable a relationship as other animal–fungi relationships, such as in leaf cutter ants, because the *Spartina* grass leaves soon die and are inhospitable for the snails. By contrast, leaf cutter ants

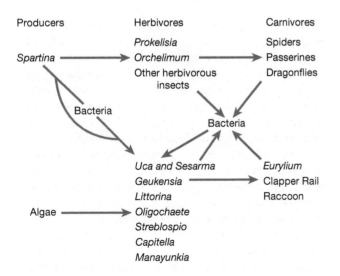

FIG. 16.39 The food web of a Georgia salt marsh. (After Teal, 1962)

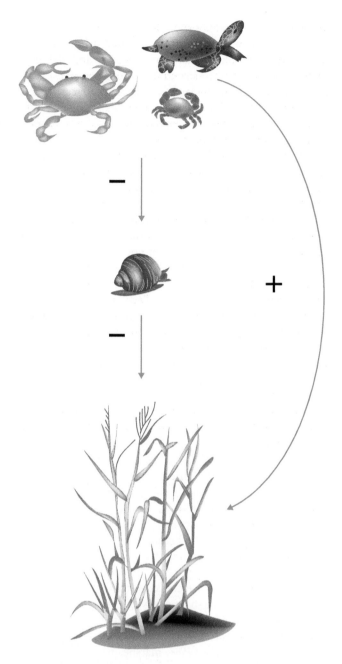

FIG. 16.40 A trophic cascade in salt marshes. Predators of the fungivorous snail *Littoraria irrorata* facilitate growth of *Spartina* because the snail strongly damages marsh plants. (From Silliman and Bertness, 2002, Copyright National Academy of Sciences, U.S.A.)

bring food (in the form of leaves) to a subterranean home for fungi.

Second, the trophic cascade has an important implication for fisheries management and conservation. Blue crabs have been exploited by commercial fishers down to very low numbers, and in the years since 2000, disease has strongly affected blue crabs in the southeastern United States. Predation on *L. irrorata* has, therefore, been depressed and may be contributing to increased mortality of low-marsh *Spartina alterniflora* populations in the Southeast in the past few years. Droughts have been very severe in the southeastern United States since about 2000 (e.g., 2006–2007, 2012), and a lack of moisture and heat stress have likely

been the major contributors to this die-off. Negative effects by *L. irrorata*, however, are adding to the dieback of low-marsh plants (Silliman et al., 2005).

■ **Grazing on salt marshes of the eastern United States probably increases in intensity toward the south, as evidenced by presence of grazing species.**

Grazing by herbivores on *Spartina* plants in northern salt marshes of the northeastern United States appears to be at a low level. In contrast, grazing in the southeast is more frequent, for example, from grasshoppers. Damage from snails is absent in the Northeast but severe in the Southeast. Like many grasses, *Spartina* and other salt marsh grasses have a number of defenses against predation. *Spartina alterniflora* has phenolic acids, mainly cinammic acid esters of glucose, which are stored in cell vacuoles. Other phenolic compounds are bound to the cell walls. Lignin is also associated with the cell walls and helps to deter digestion by grazers. Silica is also used to deter grazing. Steven Pennings and Brian Silliman (2005) found that lower-latitude *Spartina* leaves are less palatable than higher-latitude leaves. This may be an evolutionary response to the increased grazing pressure in more southerly areas.

Invasion of Salt Marsh Species

■ ***Spartina* species have been introduced, accidentally and purposefully, and have greatly modified shoreline environments throughout the world.**

Spartina species have been introduced in many parts of the world and, like the brooms in the story of the sorcerer's apprentice, have spread rapidly by vegetative growth while altering greatly the nature of shoreline habitats.

The English introduction is particularly fascinating (Thompson, 1991). *Spartina alterniflora* was introduced accidentally and hybridized with the English native *S. maritima*, which produced a form known as *Spartina townsendii*, which was believed to be sterile. Somehow, a chromosome doubling in this form produced the perfectly fertile *Spartina anglica*, which spread rapidly throughout the protected coastal regions of Great Britain. This new form displaced the native marsh species in many locations. It is exceptionally efficient at spreading by rhizomes. The spread of marshes resulted in the reduction of usable habitats for some birds and spawning fishes.

The supposedly sterile *Spartina townsendii* was imported to many localities around the world for shoreline stabilization, creation of duck habitat, and so on, and somehow the reproducing *Spartina anglica* was brought along, or arose independently in many localities. Now, *Spartina anglica* and other *Spartina* species are being regarded as a scourge in the Pacific from New Zealand to Washington State, where they are spreading rapidly to protected shores along the west coast of the United States. Its major means of spread appears to be by floating seeds.

The Atlantic *Spartina alterniflora* was introduced into the state of Washington probably as packing material for oysters in the 1890s, but was not noticed growing locally until the 1940s. In Willapa Bay, Washington, *Spartina alterniflora* is now displacing all viable oyster grounds and intertidal mudflat spawning sites for fishes, and may be creating habitat unsuitable for the local migrating birds. The marshes, spread mainly by rhizome extension, are also displacing eelgrass beds, which are believed to be important feeding and nursery grounds for commercially important fishes. Invasive *Spartina* marshes are being controlled by herbicides.

Filling, Erosion, Nutrient Addition, and Sea-Level Rise

■ Dredging and filling have destroyed many salt marshes.

Salt marshes have been vulnerable to shoreline development and in many areas of the United States have been mostly filled in or dredged away. In Palm Beach County, Florida, over half of the salt marshes were removed by development. Manhattan island's coastline was largely salt marsh before 1800, and today's shoreline is less than 5 percent salt marsh. Over 75 percent of southern California marshes have been removed, and many areas have no salt marsh left. Southeastern and Gulf coast salt marshes are quite large and remain the largest areas of salt meadow coverage in North America. Salt marshes have been very important in the development of the concept of **ecosystem services**: the monetary and sociological valuing of ecosystems for human needs, both economic and recreational.

Eugene and Howard T. Odum first pointed out the possible value of salt marshes as a food source for coastal fisheries, but this has been strongly questioned. Still, the approach of placing a value on the service of salt marsh ecosystems was extremely important. Salt marshes are known to be key nursery grounds for fish and shellfish, protectors against erosion, and also of obvious aesthetic importance. Salt marshes also can absorb nutrients to a degree, and especially can process much nitrogen by denitrification. Owing to the efforts of the Odum brothers, legislation protecting salt marshes was initiated in the 1950s, and many states now have rigorous protection of salt marshes from dredging and filling. Numerous attempts have been made at restoring salt marshes, with varying success.

■ Nutrient addition and near-shore development has caused measurable effects on salt marshes and associated fauna.

Salt marshes are commonly in areas occupied by human populations and therefore are vulnerable to nutrient additions from sewage pipes and groundwater. To some degree, salt marshes can absorb nutrients, especially nitrogen. **Denitrifying bacteria**, common in anoxic marsh sediments, can convert a great deal of human-derived nutrients into nitrogen gas. Because of this, salt marshes have been used to some extent as sinks for sewage, as opposed to dumping sewage directly into open bays and coastal waters.

Although marshes do have the capacity to absorb nutrients, some effects have been discovered that suggest direct impacts on salt marsh communities. Bertness and colleagues (2002) found that Rhode Island salt marshes show an increase of the extent of the low-marsh species *Spartina*

alterniflora in areas where shoreline development is greater. The upper limit of *S. alterniflora* increased, which may effectively squeeze out higher-marsh species and therefore decrease biodiversity. The spread of the common reed *Phragmites australis* in the high marsh is displacing other salt marsh plant species, which further threatens marsh plant biodiversity. *Phragmites* tends to occur in lower-salinity areas in New England, but in recent decades it has displaced stands of both freshwater and saltwater marsh dominant plants. This dominance is apparently the result of an invasion of a surprisingly competitive genotype (Saltonstall, 2002) that appeared in the United States in the early twentieth century. It spread from Europe or Asia and could be identified by a distinct molecular marker in herbarium specimens in U.S. museums, collected after 1910, but not before.

Coastal eutrophication may be a major cause of reduction of salt marshes throughout eastern North America. Retreat of the lower extent of the *Spartina alterniflora* zone is a major feature of salt marshes, leaving bare sediment or bare peat. Nutrient addition tends to cause *S. alterniflora* to devote more resources to shoots rather than the sediment-stabilizing rhizome system. As a result, moving water tends to rip out shoots and poorly developed root systems, resulting in the loss of *S. alterniflora* beds. The effectiveness of this hypothesized mechanism was supported by an experiment done on a Massachusetts salt marsh, involving additions of nitrogen to marshes along tidal creeks, done in comparison with control creeks that were not subjected to nutrient addition (Deegan et al., 2012). Creeks with nutrient addition showed strong retreat of *S. alterniflora* stands from the creek edges, whereas no effect was observed in the control areas. Nutrient-enriched creeks had sediment with higher-porewater content and increased slumping of sediment into the channels. After several years, more energy was devoted to shoots relative to rhizome, and more cracks in the sediment surface accumulated in the marsh creeks with experimental nutrient addition. Human nutrient additions therefore help to undermine the physical structure that stabilizes salt marsh sediments. Some *S. alterniflora* marshes have retreated because of grazing by herbivorous crabs (Schultz et al., 2016).

■ Salt marshes are very sensitive to sea-level fluctuations and may be affected by sea-level rise derived from anthropogenic global climate change.

Spartina salt marsh zonation is very closely tied to sea-level conditions. On the scale of thousands of years, salt marshes have had to keep pace with a rise in sea level for about 11,000 years since the end of worldwide continental glaciation. Fossilized accumulations of salt marsh peat have been used in combination with ^{14}C dating to show that the coastline of Connecticut submerged about 10 m in the past 7,000 years. Salt marshes have probably changed their areal extent substantially in response to fluctuations of sea-level rise and local basin shape. There is some concern that global warming of the past century will accelerate sea-level rise and perhaps cause damage to salt marshes by accelerating erosion or by drowning of salt-sensitive, high-marsh species. Erosion might increase because raised sea level would expose marshes to more frequent waves and current energy. Sea-level rise might also interact with the tolerance of different salt marsh species to submergence and high salinity. Initially, accelerated sea-level rise might increase the relative abundance of the salt-tolerant, low-marsh *Spartina alterniflora* and reduce biodiversity in marshes. This would combine with the effect of increased nutrient supply, which also tends to increase the dominance of *Spartina alterniflora*.

Hartig and colleagues (2002) studied the Jamaica Bay, New York, salt marshes and found a loss of about 12 percent in the period 1959–2000. Current marsh plant growth was typical of healthy marshes through the region. Nevertheless, small marsh islands were disappearing, and the fringes of marshes showed evidence of sediment slumping. Since the 1930s, sea-level rise has averaged about 0.2 cm y^{-1} in the New York region (this reflects a global rise), and projections of sea-level rise with global warming models suggest that marshes may not be able to keep pace with sea-level rise. Damage already done has included erosion, slumping, and dissection and fragmentation of marsh islands.

The growth and retreat of salt marshes may also be strongly affected by human land use patterns. After Europeans colonized the northeastern United States, much of southern New England's forests were cleared for agriculture. This resulted in erosion of agricultural lands providing sediment that could be bound up in salt marsh sediments. Forest clearing likely also increased the supply of dissolved nutrients to salt marshes from cleared fields and agricultural lands. Thus, salt marsh growth may have been stimulated by land clearing. In recent decades, however, agriculture in New England has declined and forests have returned, which may have choked off sediment supply. This may be a major factor in a widespread retreat of salt marshes that has been noticed in the Middle Atlantic states and New England (Kirwan et al., 2011). An overabundance of nutrient supply from human sources may have caused destabilization of marshes discussed above.

Mangrove Forests

■ **Mangrove forests are intertidal and emergent plant communities dominated by trees, which are rooted in marine soft sediment.**

Mangrove forests (also called **mangels**) are tropical and subtropical in distribution, and over 80 mangrove species can be found in Australia, the Americas, Asia, and Africa. Dominance, however, is usually confined to fewer than five species that occur in zones. Mangrove growth in lower latitudes is continuous, and mangrove trees tend to grow taller there than mangrove trees in higher-latitude mangrove forests. Mangroves have abundant and diverse marine and terrestrial animal life. Large numbers of falling leaves provide a continual localized source of detrital material.

Mangrove forests are found along quiet-water tropical and subtropical marine coasts in water temperatures greater than 20°C, but no less than 16°C in the coldest month (**Figure 16.41**). They range in size from enormous tracts of forest, mudflats, and creeks covering an order of 10^2–10^3 km^2 to tiny cays in shallow seas such as the Caribbean. They are dominated by shrub- or treelike mangroves, which are rooted in anoxic muddy sediment that is waterlogged with seawater. Once established, they greatly decrease wave energy of the shorelines on which they live. Waterlogging is a major physiological problem for mangroves, especially because the sediment pore water is often anoxic. Mangrove belowground tissue is, therefore, subjected to long periods of exposure to anaerobic conditions, which slows nutrient uptake and allows the accumulation of toxins such as hydrogen sulfide, methane, carbon dioxide, and reducing metals.

FIG. 16.41 Mangroves along a tidal creek in Palmar, Ecuador. (Photograph by Windsor Aguirre)

Exposure to decomposing bacteria is also a problem, which may explain the high tannin concentrations in mangrove tissues that function to protect against bacterial invasion. Mangrove species have evolved independently from ancestors in a number of plant evolutionary groups, but are united in their tolerance of waterlogging and salinity stress.

Mangroves are adapted to the anoxic sediments by air-projecting and shallow roots.

Mangroves are usually broadly rooted but only to a shallow depth. This may be a response aimed at avoiding exposure to deeper-lying anoxic sediments. Above the water level, mangroves are in many ways typical terrestrial shrubs, with trunks, stems, leaves, and flowers. Their root system, however, is adapted to the anoxic sediment, and all mangrove species have root extensions that project into the air so that the underground parts of the plant root system can obtain oxygen.

The variety of root morphologies maintained by a single tree allows differentiation of function. Mangroves can have prop roots, structures that extend midway from the trunk and arch downward for support; roots that direct upward into the air (knee roots or larger pneumatophores, depending on the species); and finer roots for gathering nutrients (**Figure 16.42**). Oxygen is gathered and directed into the highly chambered upward-directing roots, which transport oxygen to belowground tissues. This assures aerobic metabolism of the plants within the anaerobic environment of the sediment. The formation of pneumatophores in the genus *Avicennia* is induced by anoxia in the sediment.

Mangroves are salt tolerant.

Mangroves live at the edge of the marine environment, and the roots penetrate into a surface sediment layer that is of high salt content, known as the **vadose layer**. The salinity of the vadose pore waters is usually less than full-strength seawater because of rainfalls. Mangroves are quite salt tolerant. They have a variety of mechanisms for excluding salt. One group of species has **salt glands** that secrete salt from the leaves (**Figure 16.43**). In the morning it is common to see small dots of moisture where the salty drops have been excreted. One often sees leaves covered with salt crystals after a few hours in the sun. In another group of mangrove species, roots are capable of reducing salt uptake to a degree by an ultrafiltration system. A membrane-bound ion channel that can exchange H^+ for Na^+ accomplishes this. Species using ultrafiltration also store Na^+ in vacuoles. Because the mangrove circulatory system fluids contain less salt than in the sediment pore waters, there is always an osmotic gradient to maintain. This is quite costly, energetically speaking.

Mangrove species show vertical zonation, which is strongly affected by seedling dispersal and invertebrate predation on seedlings.

Mangrove forests can be divided into a series of zones, with different tree species dominating with increasing distance from the shoreline. Landward of the mangrove forest one tends to find typical terrestrial trees, which differ depending on location. Although subtropical forests may have only one mangrove tree species, many tropical forests have several dominants, which are usually zoned in abundance with increasing distance from the shoreline. In southern Florida and in the Caribbean, the red mangrove *Rhizophora mangle* dominates the seaward part of mangrove forests and is the first species to colonize an unvegetated shoreline. It has prop roots, which extend into the water, and tolerates full-strength seawater and tidal inundation. The black mangrove *Avicennia mangle* lives shoreward of red mangroves and tolerates only occasional seawater inundation, usually at highest high tides. Landward may be

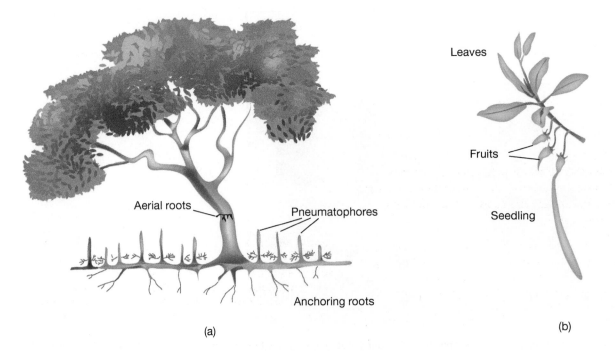

Leaves

Fruits

Seedling

(a)

(b)

FIG. 16.42 (a) A typical mangrove tree, showing the root systems. (b) Fruits and seedling. (b after Tomlinson, 1986)

Aerial roots

Pneumatophores

Anchoring roots

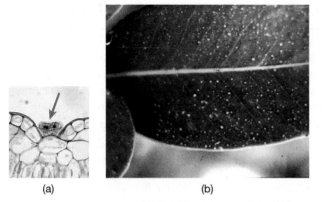

FIG. 16.43 (a) A vertical section through a mangrove leaf, showing the salt gland (arrow). (Photograph by Peter Saenger) (b) A mangrove leaf with numerous excreted salt crystals. (Photograph by Robert Twilley)

found the white mangrove, *Laguncularia racemosa*, which is rarely inundated by seawater.

Red mangrove seeds germinate while still attached to the parent plant. Seedlings develop and dangle from the parent until they either drop into the mud, like darts, or float away in the water (**Figure 16.42b**). Those that float away are finally carried by winds to another muddy shore, where they root in the sediment. A seedling coat is shed, giving the seedling negative buoyancy, which causes it to drop to the bottom. Survival of seedlings is strongly affected by predation due to grapsid crabs, which may cause total mortality of newly recruited mangrove propagules. The extent of predation may determine the species composition of individual mangrove forests. Ellison and Farnsworth (1993) found that crab predation of seedlings was especially high under existing mangrove forest canopies. Seedlings in areas where canopy was removed experimentally had faster growth rates and leaf production rates and apparently passed a threshold where crab predation did not occur. Seedlings of the red mangrove *Rhizophora mangle* tend to invest more heavily in roots and stems; this may explain its ability to rapidly colonize open, well-lit environments at the shoreline. This strategy tends to minimize crab predation and suggests that **top-down control** is important in mangrove communities in explaining species distributions. The black mangrove is far less successful in resisting predation.

■ Mangroves and upland vegetation compete for space, which is probably determined from differences in salt tolerance and local precipitation and evaporation in the vadose layer.

Mangroves are in competition with more terrestrial vegetation. The spatial transition between mangroves and upland coastal hardwood hammocks is often abrupt and complex. Hammocks may generally be landward of mangroves, but hardwood hammocks may also consist of a series of islands, slightly topographically higher than surrounding vegetation. Hammock species in Florida consist of species intolerant of salinity. While this would exclude them

physiologically from saline soils, mangrove trees grow quite well in upland habitats with fresh water in soils. Therefore, mangroves must be excluded from these habitats by competition with upland hammock species. Sternberg and colleagues (2007) have argued that the border resulting from competition between mangroves and hammocks is unstable and based on transpiration, which is the removal of water from the soil, transport through the tree's circulatory system, and transport to the atmosphere through leaves. They argue that saline water inhibits transpiration of hardwood hammock species, which live better in fresh water. A drought or more tidal inundation tips the advantage toward mangroves, which transpire well and thus bring very saline water from deep in the soil into the vadose layer. But a modest increase in salinity within the upland hammock species tends to reduce transpiration in hardwood hammock species, which also reduces movement of saline water into the vadose layer and allows the hammock species to persist. This process maintains the two species types as separately occurring groups with a sharp boundary between them (**Figure 16.44**).

■ Mangrove sediments have abundant and diverse invertebrate populations, and particulate organic matter is important in the economy of mangrove communities.

The prop roots of mangroves that extend into open sea water usually support a diverse invertebrate and seaweed community. The flat tree oyster *Isognomon alatus* attaches to the roots (**Figure 16.45**), as do a variety of crabs, shrimp, and barnacles. A number of species also live on the trunks

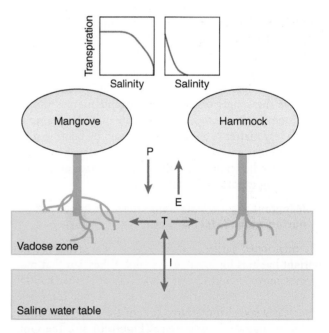

FIG. 16.44 Water uptake and transpiration in mangroves and upland hardwood hammock trees. The vadose soil layer usually has lower-salinity water and overlies a body of saline ocean water. Transpiration of mangroves is less sensitive to salinity. P = precipitation; E = evaporation; T = water uptake and transpiration; I = infiltration between the vadose layer and the layer below. (From Sternberg et al., 2007)

FIG. 16.45 Wood tree oysters attached to Caribbean mangrove roots. (Photograph by Robert Twilley)

and leaves, including barnacles and snails. A number of crab species move cyclically between the trees and mudflats, where they feed at low tide when predators are absent. The Caribbean tree crab *Aratus pisonii* is often an abundant mangrove leaf herbivore, but is only a sporadic defoliator. Unlike other marine habitats, mangrove forests are also terrestrial habitats, and southwestern Pacific mangrove forests may have large populations of herbivorous insects, monkeys, and bats.

Detritus feeders dominate the roots and sediments in the seaward part of the mangrove forest. In the inner part of the forest, mudflats are rarely inundated, and many of the dominant species live in air for extended periods of time. The mudskipper *Periophthalmus* is a small Indo-Pacific fish with excellent eyesight, capable of slithering on the surface of the intertidal mudflat and even climbing low branches. Fiddler crabs in these habitats release larvae during the rare times of tidal inundation. The soft-sediment fauna often contains crustacean species capable of deposit feeding and scavenging. This includes a number of large crabs that are exploited commercially throughout the tropics. This abundance is probably related to frequent leaf falls, which enhance the supply of particulate organic matter into the ecosystem. Stable isotopes demonstrate that a large amount of this leaf matter is consumed and used by benthic mangrove mudflat animals. Dissolved nutrients, such as N and P, appear to be retained within mangrove systems, so there is no strong export to coastal systems.

Mangrove shallow waters and creeks are important nursery grounds for fisheries.

We mentioned earlier that salt marsh tidal creeks are important habitats for juvenile and adult fishes, but that most species move in and out of creeks and use other habitats as well. This is likely true for mangroves, but some species clearly spend important parts of their lifetime wholly within mangrove creek systems (**Figure 16.46**). The Goliath grouper, *Epinephelus itajara*, spends as much of its first 5 years within Florida mangrove creeks and emigrates to more open water at about 1 m body length (Koenig et al., 2007). Food of common mangrove species can be related to mangrove fishes through a tracer such as ^{13}C, a stable isotope of carbon. Abundances of this isotope can be related

to a standard, and deviations from the standard can be studied in fishes and compared with potential food items. For example, a study in Tanzania by Lugendo and colleagues (2007) demonstrated that the isotopic signature of mangrove fish generally resembled that of invertebrates living in mangrove creeks. Mangroves are believed to be extremely important in sustaining coastal fisheries in the tropics.

Mangrove forests are very endangered throughout the tropics because of shoreline development and the dredging of mangroves for the use of shrimp farms.

Like most coastal vegetation, mangroves have been removed to make way for human habitation. Over half of the world's mangrove habitat has been eliminated. The largest loss of mangroves, however, has been for the establishment of shrimp farms. Mangrove trees have been removed on a large scale, and have been dug out and converted into basins for shrimp farms (**Figure 16.47**). Besides the loss of mangrove habitat and loss of feeding areas for migratory birds, this activity also causes the loss of invertebrate food for fishes, subsurface salinization of otherwise fresh waters, and loading of organic matter causing anoxia. Mangrove deforestation is contributing strongly to a decline in coastal fisheries, which has been documented in Thailand, a major location of shrimp farms (Barbier and Strand, 1998). It is also a major problem in the Americas. Owing to crowded conditions, disease is a major source of shrimp mortality, and farmers respond by using high concentrations of antibiotics. These, too, are being released to freshwater canals and coastal waters.

Sea-level rise from global warming is a great threat to mangroves but particularly upland hammocks, which are salt intolerant.

Global warming is a major potential source of sea-level rise, owing to thermal expansion of seawater and melting of glacial ice. As sea level rises, the frequency of incursions of seawater in coastal storms increases. This is especially a problem in the Gulf of Mexico, where hurricanes are frequent and the tidal range is very small. On the west coast of Florida, negative effects of saline water are greatest in areas with more frequent seawater flooding events. A combination of flooding, causing hypoxia, salt stress, and incursions of salt-tolerant plants has strong negative effects on salt-intolerant species such as *Sabal palmetto*, which is common throughout the Gulf of Mexico. Such plants lose their ability to regenerate and then may die several years later. This is especially worrisome because such upland hammock species are important in protecting coastlines from flood damage from hurricanes. A drought in 2000–2005 has caused major declines in cabbage palm and southern red cedar, and it is likely that sea-level rise is exacerbating the losses (Desantis et al., 2007).

Estuaries

Estuaries are environments where oceanic seawater mixes with freshwater input from a discrete source, such as a river. Because of this, estuarine biology is intimately connected

FIG. 16.46 Mangroves line the intertidal zone of tropical tidal creeks, as shown here along Estero Pargo in Terminos Lagoon, Mexico. (Photograph by Robert Twilley)

FIG. 16.47 Aerial photo showing patches of mangrove forest and excavated shrimp farms in Palmar, Ecuador. (Photograph by Windsor Aguirre)

with the **watershed**, or the surrounding land that provides water input to the estuary. Estuaries vary greatly in size, from small tidal creeks fed by rivers with watersheds that are only a few square kilometers to the relatively enormous Chesapeake Bay estuary, whose watershed is about 180,000 km². The interface of the watershed with the main course of the estuary provides an opportunity for the entry of nutrients, pollutants, and other substances through tributaries and general flow down slopes into the estuary.

Much of our habitat coverage above applies as coverage of subhabitats of estuaries, depending on their location. Therefore, much of our coverage here is an overview of overall estuary properties.

■ **Review: Estuarine structure is controlled by seaward flow of fresh water combined with tidal mixing.**

The input of fresh water through rivers causes an overall **estuarine flow**, sending lower-density water at the surface toward the ocean. There is a compensating deeper upriver flow of higher-salinity oceanic water beneath the buoyant surface flow. When the flow is strong and the tidal mixing is small (see Figure 2.27), estuaries are **highly stratified**, with a distinct low-salinity layer flowing toward the sea, and a compensating movement of higher salinity below, upriver from the ocean. From the air, one can readily see the movement of low-salinity surface water from the Mississippi

River into the Gulf of Mexico. Most estuaries are **partially mixed** from tidal action and wind. There is a stretch near the opening of the estuary where mixing occurs, but still the salinity in this zone is always lower at the surface than at depth. In very small and shallow estuaries, wind tends to fully mix the water to the bottom, creating a **vertically homogeneous** estuary.

■ **Estuaries range from open marine to a range of successively decreasing salinity zones, to tidal fresh water and associated creeks and marshes, to fresh water.**

Figure 16.48 shows a series of salinity ranges and associated salinity-related habitats. The topography of the watershed, the slope and size of the river(s) feeding into the main part of the estuary, the size of the main estuary channel, and tidal flow all contribute to determining the salinity structure. The Hudson River estuary has a discrete main stem, with relatively small tributaries entering throughout its length. The slope of the main part of the Hudson River is low, and a large part of its upriver length is tidal but fresh water. By contrast, the Chesapeake Bay estuary consists of five major river systems that feed into a central estuary (see Figure 2.26). Overall, **river discharge** is also a major influence on salinity transitions, and spring increases in river flow move fresh water farther down the estuary. The main part of the estuary is strongly affected by wind and tidal action, but the rivers entering the main part of the estuary are quite variable and may not exchange as strongly with the main estuary in summer because of low river discharge, low tidal flow, and narrow openings into the main part of the estuary. In the main part of the estuary, low flow in summer combined with nutrient addition may result in sluggish circulation and low oxygen concentrations. Strong rains have very strong effects on salinity structure, and hurricanes often dump large amounts of fresh water on an estuary, causing a general decrease in salinity throughout the estuary. Storms also may cause powerful episodes of erosion and seaward sediment transport.

■ **Estuaries are geologically ephemeral but abundant in nutrient supply and biological production.**

Because rather small changes of sea level can completely fill or empty out an estuarine basin, estuaries are geologically impermanent features of the coastline. As sea level rises, the sea may flood river valleys; but such basins will also empty as sea level drops, such as during a worldwide increase in glaciers, which lock up the water of the ocean as ice. It is believed that the specific form of an estuary will rarely exist for more than 10,000 years or so. On the other hand, rivers are surprisingly ancient, and therefore some type of estuary may exist for far longer periods.

Despite the ephemeral nature of estuaries, they are among the biologically richest habitats in the world. Supplies of nutrients from freshwater sources and recycling of nutrients from the seabed as in denitrification combine to

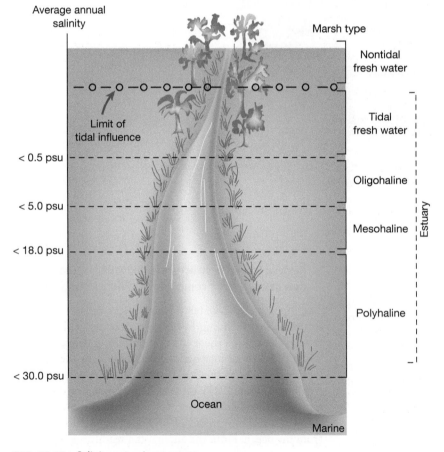

FIG. 16.48 Salinity zones in an estuary.

support high levels of primary production, which in turn supports large numbers of estuarine benthic invertebrates, fishes, plants, and birds. In eastern North America, estuaries such as the Hudson River and Chesapeake Bay support major fisheries, which have been strongly impacted by habitat destruction, overfishing, and high nutrient inputs. The Chesapeake Bay system had enormous oyster beds (they are now depleted owing to overfishing and disease) and populations of blue crabs (also overfished). The Hudson River has seasonally large populations of fishes such as shad and striped bass. The bottom is alive with annelids, mollusks, crustaceans, and insect larvae (in the freshwater parts of the estuary).

The decreased salinity at the headwaters of estuaries can reduce the number of marine species.

The most noticeable gradient in an estuary is that of decreasing salinity as one goes upstream. Marine species, as discussed in Chapter 5, generally can tolerate fluctuations in salinity, but their tolerance is often exceeded when salinity falls below 10–15 psu. Some major invertebrate groups, such as echinoderms (sea urchins, sea stars), tend to drop off in estuaries, due to a general incapacity to evolve resistance to lowered salinity. Others, such as crustaceans, are capable of good regulation in the face of osmotic stress and are often quite abundant in estuaries. In estuaries with some degree of vertical density stratification, the salinity is greater on the bottom than at the surface, where freshwater flow moves lower-salinity water downstream (see Chapter 2). As a result, marine bottom species can often penetrate an estuary farther upstream than surface planktonic organisms can. Infaunal species experience less salinity variation than do epifaunal species over a tidal cycle in a very-well-mixed estuary because of the buffering effect caused by sediment pore waters that exchange water slowly with the overlying water column (**Figure 16.49**).

The second major estuarine salinity transition is the critical salinity range of 3–8 psu. Many marine groups apparently find it hard to survive in this salinity range, even though many more species are capable of living either in fully fresh water or in waters of higher salinity. Along the estuarine gradient, species numbers are at a pronounced minimum in the critical salinity range (**Figure 16.50**). Mollusks may be incapable of cell volume regulation at salinities this low. Freshwater species, however, can regulate ionic concentrations and maintain a hyperosmotic state. They have lost the ability to regulate cell volume, however, and therefore cannot penetrate even the low salinities of the critical salinity range. The critical salinity is thus a no-man's land, hospitable to neither marine nor freshwater species.

Although salinity change is often a critical factor in limiting the range of marine species, many are capable of rapid regulation and adjustment to the changing osmotic stress of varying salinity. Many fish species are capable of extensive regulation of tissue fluids (see Chapter 5) and can swim across strong salinity gradients. Striped bass, salmon, and killifish are just a few examples of fishes that migrate from completely saline water to fresh water in a few weeks

or even days. Some crab species are also quite adaptable to changing salinity and perform migrations over nearly the same range of salinity traveled by fishes. In the Hudson River, males of the blue crab *Callinectes sapidus* can migrate from full-strength seawater in the lower New York Bay to fresh water in the upper reaches of the estuary.

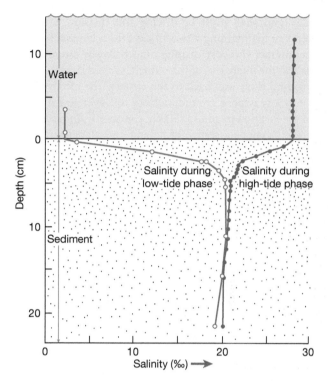

FIG. 16.49 Salinity variation in the water column and within the sediment in the small tidal Pocasset River estuary, Massachusetts. Note that the salinity varies a great deal in the water column, owing to tidal motion and freshwater flow. Within the sediment, however, the salinity is relatively constant and intermediate in value. (After Sanders et al., 1965)

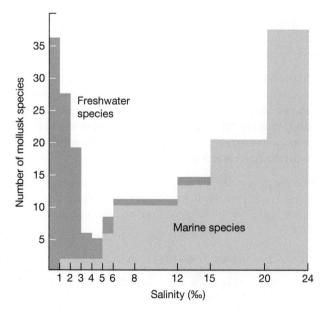

FIG. 16.50 Species richness along the estuarine gradient of the Randersfjord, Denmark. (After Remane and Schlieper, 1971)

■ **Overall, estuaries and shelf environments comprise a two-phase system that corresponds to life-history stages of many fish and invertebrate species.**

While many species, especially tidal freshwater groups, complete their entire life cycle within the estuary, a large number of species live a two-phase existence in the estuary and on the continental shelf. Clearly, differences between estuary and shelf must be driving this division, and the benefits of maintaining a two-phase life cycle must be able to counteract the cost of migrating between estuary and shelf and the likely loss of individuals, such as larvae, which fail to find their way back to the estuary. The estuary may serve as a spawning ground, nursery for juvenile fishes, or even the location of adult life, as in the case of *Anguilla* eels. The shelf may be a source of habitat and planktonic food for larvae of estuarine adults (as in blue crabs and fiddler crabs), or the location of adult feeding grounds (as in many anadromous fish species).

Changing ecological conditions within an estuary might cause strong alterations in behavior. These are most easily accomplished by nekton, which could relocate to areas with better conditions. For example, estuarine fish populations might fluctuate strongly; and there is good evidence for density-dependent effects, in which fishes grow and reproduce more poorly at high local densities (Craig et al., 2007). We would expect such species to move to more favorable habitats, reducing local densities. Changes in the structure of food webs might also stimulate such relocations. For example, invasive bivalves have reached high densities in some estuaries and have greatly reduced phytoplankton abundance (see later discussion on invasive species in estuaries). In San Francisco Bay, the invasion of the Asiatic clam *Corbula* (*Potamocorbula*) *amurensis* had such an effect, but there was a surprisingly small effect on planktivorous fishes, as might be expected from a competition-based expectation for limited food. Wim Kimmerer (2006) found that the northern anchovy *Engraulis mordax* simply moved to the higher-salinity parts of the estuary, where the Asiatic clam did not invade and had little effect on phytoplankton abundance.

■ **Some estuarine species are adapted to counteract the estuarine flow to the sea, in order to be retained within the estuary; others are broadcasted onto the shelf and return to estuaries at the time of metamorphosis, making the estuary a two-phase system in coordination with the continental shelf.**

Estuarine flow is usually seaward in the low-salinity waters at the surface. This net flow is known as **buoyant flow** and is especially strong in spring, following rains and snow melt

in watersheds in higher latitudes, known as the **freshet**. Small estuaries with extensive tidal flushing cannot support large nurseries of juvenile fishes, and retention adaptations probably would be insufficient to counteract the tidal exchange with the open sea. Larger estuaries with longer flushing times and vertical density stratification can support fisheries because of the reduced loss of larvae to sea. Within large estuarine systems, such as Chesapeake Bay, tributaries like the St. Mary's River have relatively low tidal exchange rates with the rest of the estuary. The reduced exchange tends to trap larvae and may be the reason for heavy larval sets of the oyster *Crassostrea virginica*. To counteract the estuarine flow that does exist, estuarine fish and invertebrate larvae have been observed to keep to the bottom during the ebbing tide, swimming actively at the surface with the incoming tide (see Figure 7.27). Menhaden are more easily netted during flood tides, indicating their adaptations for retention within the estuarine system. The mud crab *Rhithropanopeus harrisii* is found in greater abundance upstream as the larval life period progresses. As mentioned in Chapter 7, some species export larvae to adjacent coastal waters and depend on a variety of tidal and wind sources to enable them to reinvade the estuary. Estuaries are, therefore, for many species part of an interactive two-phase system, in coordination with the continental shelf (**Table 16.2**).

A number of other species live and mate within the estuary, but larvae move in the buoyant flow toward the shelf. Good examples are the blue crab *Callinectes sapidus* (discussed in Chapter 7) and species of fiddler crabs. As mentioned, the problem for these species is to return to the estuaries, which is aided in the case of blue crabs near Delaware Bay by seasonal changes in regional wind systems. When blue crab megalopae (last larval stage, competent for settlement and metamorphosis) enter the estuary, they are adapted to swim upward on the flood and go toward the bottom on the ebb tide, in order to assure entry into the estuary. Glass eel larvae of the American eel also take advantage of flood tidal streams to enter the estuary and move upstream after they move toward the inner shelf waters.

■ **Estuarine suspension feeders may control phytoplankton of shallow parts of large estuaries or in entire well-mixed estuaries; tidal exchange may be a driving force for suspension feeder food in smaller estuaries.**

Estuaries in temperate latitudes commonly have large populations of suspension feeders, particularly bivalve mollusks such as oysters, mussels, and burrowing clams. In small estuaries with strong tidal exchange, most phytoplankton supplied to these bivalve populations come

TABLE 16.2 Examples of How Life Cycles Fit Spatially into the Two-Phase Estuary-Shelf System

BIOLOGICAL SYSTEM	ESTUARY	SHELF
Crabs type 1 (fiddler crab, blue crab)	Adults: feeding, spawning	Larvae: feeding
Invertebrates type 2 (mud crab)	Adults and larvae	None on shelf
Anadromous fish (striped bass, shad)	Larvae and early juveniles: spawning	Adults: feeding

from the adjacent ocean. The **retention time** is the average number of days that a phytoplankton cell stays in such a small estuary before it is washed out to the adjacent shelf. This value can be compared with the turnover time, which is the number of days it takes for a population of bivalves to completely filter a water column. If the turnover time is less than the retention time, then it is possible for the bivalves to completely filter the water column. For example, if the turnover time is 10 days and the retention time is 60 days, the water column will have been filtered 6 times before an average phytoplankton cell has the chance to be mixed back with the adjacent ocean.

When the estuaries are relatively well mixed by wind and tide, bivalves may have access to the entire water column, develop large populations, and greatly depress phytoplankton densities. This has been observed in two cases where invasive suspension bivalve species have appeared. The Asiatic clam *Corbula amurensis* has had a strong effect on the low-salinity phytoplankton in the very shallow San Francisco Bay. The zebra mussel *Dreissena polymorpha* invaded the tidal freshwater part of the Hudson River estuary in the late 1980s and has had a major impact on phytoplankton populations. In years of high mussel population density, well over 90 percent of the potential phytoplankton biomass is removed. If the estuary is very stratified, bivalves on the bottom may not have access to all of the phytoplankton and may not completely deplete the phytoplankton from the estuary. Also, bivalves may not be active during earlier and cooler parts of the spring when phytoplankton blooms may occur.

■ **A combination of historical high nutrient inputs and removal of predators has created great ecological instability from a combination of bottom-up and top-down processes.**

Estuaries are ecologically and economically valuable and have been densely settled by human populations. Owing to human influence (see Chapter 22), estuaries have had major additions of nitrogen and phosphorus. This has greatly stimulated primary production of phytoplankton, which may greatly decrease water clarity. Much of the phytoplankton is not consumed by zooplankton or benthic suspension feeders and is broken down by bacteria, which reduces oxygen concentrations. This has had two main effects on estuarine environments. First, submerged aquatic vegetation (often called SAV), an important structural habitat in estuarine habitats, is strongly affected by low light conditions caused by the shading of concentrated phytoplankton. While eutrophication increases nutrients, which might stimulate eelgrass plant growth, the effect is more than compensated by the negative effect of strong growth of competing seaweeds, stimulation of epiphytes that grow on and intercept light from eelgrass, and the increase of phytoplankton in the water, which intercept light (Short et al., 1995). We have discussed the capacity of salt marshes to absorb nutrients and to recycle N to the atmosphere by means of denitrification. Estuaries exposed to nutrient enrichment today have a reduced means of nitrogen absorption, and **the growth of phytoplankton and epiphytes impedes eelgrass growth**. In the western part of Long Island Sound estuary, there is virtually no eelgrass left, which correlates strongly with high nutrient inputs. Eelgrass populations are still relatively abundant in the eastern third of Long Island Sound. Nutrient loading into Waquoit Bay on Cape Cod has been studied extensively, and declines of eelgrass can be related directly to increases of housing and nutrient loads (Short and Burdick, 1996). Eelgrass declines in Chesapeake Bay may be greater than the losses seen during the major eelgrass epidemic in the 1930s (Orth and Moore, 1983).

Phytoplankton in the water column that is not grazed can die and sink as **phytodetritus** to the bottom. The nature of phytodetritus can be tracked by analysis of photosynthetic pigments, which help to identify the phytoplankton types that are sinking. In the Baltic Sea, the bulk of the surface production sinks, with a successive dominance of diatoms, dinoflagellates, and cyanobacteria, from spring to late summer. Large deposits of phytodetritus results in colonization of benthic surface-feeding deposit feeders, but past a threshold, phytodetritus will overwhelm the bottom, resulting in bacterial decomposition and anoxia at the sediment surface. Such conditions can be found in late spring and summer in Chesapeake Bay, in western Long Island Sound, and in the Baltic Sea.

Along with bottom-up effects, overfishing has caused the loss of many top predators, causing trophic cascade effects from the other end of the food chain. Many top carnivores have been exterminated due to overfishing, in the same way that any possibility of top-down control by oysters and other suspension feeders has been eliminated. The oligohaline Lake Pontchartrain has lost over 96 percent of its apex predators, alligator gar and bull shark, since the early 1950s. We might not be surprised that this has initiated a trophic cascade effect through the estuarine food chain. The decline of apex predatory sharks in southeastern U.S. waters has been implicated in the increase of their prey, the rays. Cow-nosed rays have increased greatly in North Carolina estuarine waters, and these have caused large-scale mortality of their molluscan prey, especially the bay scallop *Argopecten irradians* (Myers et al., 2007). With the disappearance of scallops, the rays are digging out infaunal clams, thereby destroying more scallop sea grass habitat. In the southeastern United States, blue crabs have been severely overfished and have been recently decimated by a combination of drought and disease caused by the parasitic dinoflagellate *Hermatodinium perezi*, which proliferates in the crab's hemolymph. Drought in the Southeast since about 2000 has increased salinities in estuaries, which has in turn increased the exposure of estuarine blue crabs to the parasite. Food web effects of this decline are not well understood.

■ **Large estuaries are targets for biological invasions and occasional strong ecological alteration, although the rate of invasion is quite variable.**

Large estuaries experience large environmental changes. Interannual changes in the entire watershed are likely to result in strong changes in rainfall or snowfall, and this will influence the salinity structure, temperature structure, and nutrient input of estuaries. As a result, organisms in estuaries are the targets of enormous physiological changes. A major storm may change the salinity from mesohaline to freshwater for days. Large estuaries are commonly major locations of human habitation, and therefore are strongly disturbed by pollution and shoreline habitat alteration. Of course, large estuaries are also major ports and hence targets for the introduction of alien marine species.

The combination of habitat instability, biological change, and continuous introductions of alien species makes estuaries likely hotspots for biological invasions of aquatic organisms. Human activities continuously sever the natural barriers to dispersal across large oceans. In the nineteenth century, most ballast in ships was solid material, which facilitated invasions of terrestrial plants. In the latter half of the twentieth century and beyond, the increase of ship speed combined with the use of water for ballast has exacerbated the rate of introduction of species from afar, especially because larvae and microorganisms can survive in the ballast water. Many sessile organisms arrive attached to ships' hulls. But the increased rate of introduction does not guarantee success. Introductions may fail for several reasons:

1. *No suitable habitat.* Many introduced species cannot survive the target environment because of inappropriate temperature, salinity, or substrate needs.

2. *Shipping practices.* Ships that bring species that can survive in the target region may dock in the wrong place. In Chesapeake Bay, many ships dock in salinities far too low for survival of the marine species found in ballast tanks or attached to hulls.

3. *Too small an invasive population.* The invasive population may be too small to survive random changes encountered and various challenges, such as finding appropriate larval settling sites and reproducing successfully into the next generation.

4. *Inappropriate dispersal strategy.* Invasives that survive a first generation may have a dispersal stage that precludes survival into successive generations. Planktonic larvae increase dispersal distance, but the lack of appropriate adaptations to the local hydrographic conditions might result in washout to sea or to inappropriate sites. Lack of dispersal might result in local extinction.

Despite the impediments to successful establishment, estuaries have been invaded extensively. Over 200 species have invaded San Francisco Bay, which is the most affected estuarine system in the United States (Cohen and Carlton, 1998). Its biota is completely dominated by invasive species, ranging from salt marsh domination by east coast *Spartina* cordgrass to benthic domination by the Asiatic clam *Corbula amurensis.* A 200-year record shows that San Francisco Bay invasions began to accelerate at the beginning of the twentieth century and far exceed that of other water body systems that have been studied (**Figure 16.51**).

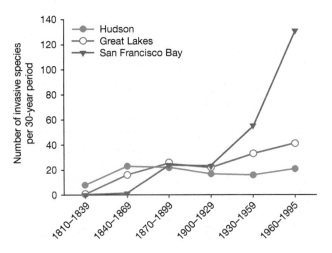

FIG. 16.51 Pattern of invasion of the Great Lakes, freshwater tidal Hudson River, and San Francisco Bay since the early nineteenth century. (From Strayer, 2006)

The ecological consequences of successful invasions are quite variable. The Hudson River tidal freshwater estuary has been invaded by over 100 species, many of which have had no strong ecological effect. The vulnerability of estuaries to ecologically significant invasions might result from three important factors:

1. *Frequent environmental overturn.* Environmental disturbances in estuaries commonly result in large-scale overturns. The timing of such overturns might increase the vulnerability of estuaries to invasion by ecologically important aliens.

2. *Ecological vacancies.* Estuaries have relatively low diversity, and some habitats may be open, remaining to be occupied. The successful invasion of the zebra mussel *Dreissena polymorpha* in the tidal freshwater Hudson might have been encouraged by the lack of common sessile suspension feeders that could attach to shallow subtidal hard bottoms.

3. *Competitive superiors.* Some invasives may simply be competitively superior to natives. The mud snail *Tritia obsoleta* successfully invaded San Francisco Bay and forced the native mud snail species into small high-intertidal refuges where the invader could not occupy.

It is likely that a number of newly appeared pathogens and toxic algae in estuaries are alien species, but we often cannot exclude the possibility that they were simply rare formerly and increased because of a recent environmental change.

Oyster Reefs

■ **Oyster reefs occur in estuarine environments throughout the world.**

Oyster reefs, especially on the east and Gulf coasts of the United States, are intimately connected with estuaries, but once were found abundantly in estuaries and along open coasts and estuaries on both coasts of North America and

FIG. 16.52 An intertidal oyster mound of the eastern oyster *Crassostrea virginica* in South Carolina (left) with a closer look at a cluster of oysters (right). (Photos by Loren Coen)

South America, northern and southern Europe, Australia, and east Asia. In the tropics, oysters are associated with mangrove lagoons, but in the temperate zone, estuaries and protected coasts are the most important habitats. Because oysters are a highly prized food, oyster reefs were heavily exploited in the eastern United States as soon as the first European settlers arrived. Subtidal oyster reefs and intertidal populations were especially prolific on the Gulf and eastern coasts of North America, occupied by the eastern oyster *Crassostrea virginica* (**Figure 16.52**).

Because oysters produce planktonic larvae, which settle, metamorphose, and cement themselves on hard surfaces and especially other oysters, they accumulate to form intertidal bars and shallow subtidal mounds or reefs. Their accumulations comprise a matrix strongly dominated by a single species, but otherwise resemble coral reefs in their ability to produce a mound of material dominated by calcium carbonate (shell material).

Because oysters comprise a group of valuable resource species, they have been exploited, farmed, and moved among 73 countries for hundreds of years. This makes oyster reefs a target for overexploitation and transport of alien species, including disease organisms, around the globe. Oysters have also been a major food source for many nations. It is hard to believe that oysters over 30 cm (12 inches) in shell height were once common in enormous stretches of reef in the waters of New York Harbor and that New Yorkers—rich and poor alike—ate oysters on average once a week. Overexploitation was already apparent by the Revolutionary War, and the reefs were replenished by seeding with newly settled oysters until they also were overexploited by the early twentieth century, when pollution eliminated the rest.

■ Oysters are ecosystem engineers that construct reefs by means of larval settlement, attachment, and accumulation of oyster shell.

The accumulation of oyster shell into oyster reefs fits our definition of oysters as ecosystem engineers. Oyster reefs may protrude 1–2 m or more above the soft estuarine bottom and extend laterally for hundreds of meters. Such structures alter local currents, and the oysters themselves alter the regional ecosystem by feeding on phytoplankton, depositing organic material on the bottom, and creating a three-dimensional structure that attracts many other species.

■ Oyster reefs attract many sessile, mobile benthic, and demersal species.

Oyster reefs are complex three-dimensional structures, with an irregular upper surface created by the living oysters and an intricate set of interstitial spaces that may cause deposition of sediment, but also allows the colonization of a large number of species, thus increasing local biodiversity. Of course, if an oyster reef is absent, another set of species occupying bare sediment bottoms would be present, but the oyster reefs present hard surfaces, which allow a wide variety of sessile and hard-bottom-associated species. Most important, oysters attract predators. For example, the eastern oyster *Crassostrea virginica* in New York waters will attract predatory sea stars *Asterias forbesi*, the oyster drill *Urosalpinx cinerea*, various species of crabs, and occasional bird predators. But a wide variety of sessile and mobile benthic invertebrate species also are attracted to living oysters in reefs. Some fish species, such as the oyster toadfish *Obsanus tau*, lay eggs on oyster shells, and hatched juvenile toadfish feed on crabs. Many wider-ranging fish swim among oyster reefs, searching for smaller fish prey species. Oyster reefs are therefore foci of biodiversity.

■ Oyster reefs thrive in relatively shallow water and when the reef shell is in high relief on the seabed.

Oysters are suspension feeders and feed on phytoplankton, usually with cell size above 3–5 μm in diameter. They can live in waters of medium oxygen content but do poorly in waters of very low oxygen, like most other marine invertebrates. As we discussed earlier, estuaries tend to be vertically stratified, because of flow of low density-low salinity water and solar warming in late spring and summer. The development of stratification makes water circulation often sluggish at greater depths, which tends to result in lowered oxygen concentrations. Phytoplankton also tend to be more concentrated near the surface in a highly stratified estuary in summer. Thus, oyster reefs do more poorly at greater depth.

Oyster reefs also grow and survive better when the oyster shell accumulation puts the living population of oysters in more shallow water and away from the seabed, where highly turbid water can be resuspended and negatively affect the suspension feeding oysters by clogging the suspension feeding gills and depositing mud in the interstices of the reef. High-relief oyster reefs 1–2 m above the bottom therefore grow more vigorously than oyster reefs where shells accumulate in a pile just a few inches above the bottom (Lenihan and Peterson, 1998).

■ **Oyster reefs may greatly influence estuarine ecosystem processes and perform important ecosystem services such as reducing phytoplankton density and enhancing nitrogen processing.**

The potential reduction of phytoplankton by bivalves has been of great interest, since estuarine bivalves such as oysters have declined greatly in estuaries and coastal waters due to habitat destruction, overfishing, and disease. In Chesapeake Bay, the eastern oyster *Crassostrea virginica* was once abundant and a major fishery, but the oyster populations have shrunk to only a few percent of their densities in the nineteenth century (Rothschild et al., 1994). Roger Newell (1988) linked the power of oysters to filter the water column with declining water quality and ecosystem changes in the Chesapeake in the past few decades. Phytoplankton densities have increased and oyster densities have declined, which suggests a linkage. Newell used data on filtration rates and estimates of historical oyster densities to conclude that oysters filtered the entire Chesapeake estuary in 2–4 days. Oysters would not only have cropped down the phytoplankton, they may have made feces that were buried in the sediment and, therefore, removed nutrients from the estuarine system (Newell, 2004). This conclusion about the beneficial effects of oysters has been used to spur oyster restoration in the Chesapeake and many other water bodies.

Newell's top-down hypothesis both has exciting connections to the idea of trophic cascades and leads to a conservation management plan. Because bivalves in other

estuaries, such as San Francisco Bay and the Hudson River, have had strong effects on the phytoplankton when they invaded, we have at least seen a possible model of the effects of bivalve restoration in other estuarine regions. Oysters therefore perform an important ecosystem service, since reduction of phytoplankton increases water clarity, which enhances bottom growth of sea grasses. Reduction of phytoplankton also reduces the probability that they will be broken down by bacteria, causing reduced oxygen in the water column (see Chapter 22).

It may be that the estimate in the Chesapeake gives too much credit to the oysters' ability to filter water columns in many estuaries, including the main stem of the Chesapeake estuary. Models of estuarine circulation show that oysters in Chesapeake Bay may have easy access to phytoplankton only in shallow estuarine tributaries (**Figure 16.53**). In the main part of the bay, spring phytoplankton blooms tend to develop but are not mixed very well to where oysters are located. Gerritsen and colleagues (1994) concluded that oysters and other suspension feeders could not effectively filter water in the main part of the bay. The relatively small vertical mixing reduces such access of phytoplankton in surface layers to suspension feeders on the deeper bottom of the main part of the bay. Also, oysters are not very abundant in these deeper waters. Still, oysters in tributaries and shallows would be able to filter the water, which would have a positive impact on water clarity and would help sea grasses to grow more efficiently in clearer water. If suspension-feeding mollusks such as oysters are placed in experimental containers with the eelgrass *Zostera marina*, phytoplankton densities decline and eelgrass production doubles (Wall et al., 2008).

Pomeroy and others (2006) argued that oyster feeding rates were not as high as Newell asserted, because Newell's model included only very high summer feeding rates, while ignoring lower rates, expected in late winter and spring when the main spring phytoplankton bloom occurred. This controversy is still quite active, and the interested student should read further about this hypothesized connection between grazing and ecosystem health.

Oysters may also strongly influence nitrogen cycling in estuaries by means of their feeding and removal of phytoplankton from the water. Especially in areas of high

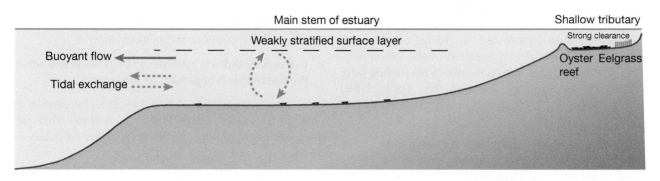

FIG. 16.53 Arrangement of estuarine habitats and diagram of water exchange. Abundance of oysters in shallow water tributaries results in strong water clearance of particles, which benefits eelgrass population growth. This effect may not be very strong in the main stems of estuaries, where the water is weakly to strongly stratified and oysters in deeper bottoms have less access by suspension feeding to clearance of phytoplankton in surface waters.

phytoplankton concentration, oysters regulate the amount of material ingested by rejection of particles by the gills, producing pseudofeces (see section "Particle Selectivity by Suspension Feeders" in Chapter 15). Ingested food passes through the gut and emerges as feces. The sum of production and deposition of feces and pseudofeces is known as **biodeposition**. This material may accumulate on the bottom and therefore remove nitrogen temporarily from the water column, thus reducing the nutrient supply that might otherwise subsidize plentiful phytoplankton, which decays and reduces dissolved oxygen. A microbial process known as denitrification would act in anoxic sediments to convert the particulate nitrogen eventually to nitrogen gas. Thus, by enhancing biodeposition, oyster reefs can strongly affect nitrogen removal and improve water quality (Newell et al., 2005).

■ **Oyster reefs in eastern North America and the Gulf coast succeed in relatively low salinities, where disease and predators are excluded but salinity is still high enough to allow physiological functioning, feeding, and growth.**

The eastern oyster *Crassostrea virginica* forms reefs with the most vigorous growth, survival, and recruitment in the approximate salinity range of 10–20. This optimum is an excellent illustration of the interaction of opposing forces to determine the distribution of abundance of an important ecosystem engineer (**Figure 16.54**). In waters with higher salinities, a number of important marine predators are abundant, especially the oyster drill *Urosalpinx cinerea* and the sea star *Asterias forbesi*. Both of these species do poorly in low salinity but can be devastating to oyster populations in the open marine salinities near the mouth of the estuary. Oyster diseases (discussed shortly) also thrive in higher salinities and do not infect oysters very much in salinities less than 10. At the other end of the salinity spectrum, oysters grow, survive, and reproduce increasingly poorly as salinities in spring and summer dip below 10. At these low salinities, oysters expend considerable energy to regulate osmotic balance within cells (see Chapter 5), and may even die when salinities are below 2–3 in the warmer months (oysters can often overwinter in fresh water for months). Thus, the most vigorous oyster growth is at the intermediate salinities of 10–20.

■ **Strong trophic cascades may influence species abundances on oyster reefs.**

Do oysters and oyster reefs have strong interspecies interactions, and do these interactions affect the estuarine ecosystem? As mentioned earlier, oyster reefs are mainly a matrix of oyster shell and live oysters with a large number of sessile and mobile benthic species that depend on the oyster matrix for their survival. The interstices and sides of the reef are relatively quiet-water sites where small nekton can live and reproduce. In these ways, oyster reefs are comparable to coral reefs, discussed in Chapter 17.

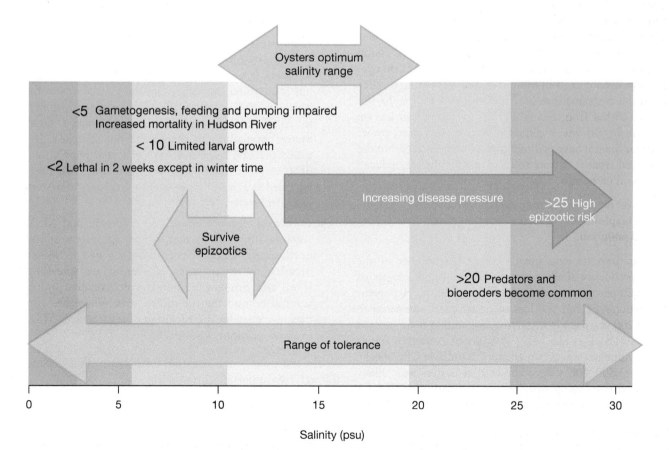

FIG. 16.54 Effects of various factors affecting survival and growth of the eastern oyster *Crassostrea virginica* in different salinity ranges. (Courtesy of Adam Starke)

There are two main interaction loops to examine. First, oysters themselves may filter out phytoplankton and therefore affect the general estuarine ecosystem. Filtration of phytoplankton would cause competition with other suspension feeders, such as other estuarine bivalves, and also with planktonic phytoplankton consumers such as copepods.

Second, oyster reefs also consist of a complex association of species that might comprise a strongly linked food web of predators and prey. On the small scale, oysters themselves are habitats for large numbers of smaller sessile species. A number of tube-building polychaetes encrust the outer shell of the eastern oyster, and the polychaete *Polydora websteri* builds tubes on the inner part of the shell and causes damage to oysters, which usually deposit shell material that encases the tube, forming a blister. Oysters are also substratum for barnacles, sponges, and other species. In most of these cases, the sessile species are commensals, and therefore the oyster reef is a focus for biodiversity.

A simple food chain can be described linking oysters to fish carnivores. Oysters are consumed often in great numbers by the oyster drill *Urosalpinx cinerea*, which can drill a hole in the shell. Juvenile oysters are often consumed by the mud crab *Panopeus herbstii*, which are in turn consumed by the oyster toadfish *Opsanus tau*. The oyster toadfish, often called the "oyster cracker," therefore might benefit oysters by preying on mud crabs. This is of interest because the oyster cracker depends on the oyster: It lays eggs on the undersides of live oyster shells. But Jonathan Grabowski (2004) found that nearly all of this predation effect was a **trait-mediated indirect interaction**, involving predator avoidance behavior of the mud crabs instead of direct predation. Grabowski constructed artificial "reefs" and found that the indirect beneficial effect of oyster crackers on oysters was even greater in more complex reefs of high relief. In simple reefs, toadfish preyed directly on mud crabs, but crabs also hid from the fish and therefore preyed less on oysters. In the more complex experimental reef, oyster crackers did not have as direct access to attack crabs, but the crabs hid from predators still and this resulted in reduced predation on oysters.

■ **Oyster reefs are in strong decline because of overexploitation, habitat disruption, disease, and pollution.**

Human impacts have greatly reduced a number of the most productive shallow-water ecosystems, with losses of about 20 percent for coral reefs and as much as 50 percent for mangroves. But the loss of oyster reefs compared to about 200 years ago is staggering, amounting to 85 percent or more (Beck et al., 2009). A study of American native exploitation of oysters in prehistorical times in Chesapeake Bay (~3,500–400 y ago) shows that oysters were likely exploited sustainably with no strong negative effects, for example, on oyster size (Rick et al., 2016).

Oyster reefs are essentially gone in locations such as Australia, much of Europe, and the northeastern Pacific coasts of the United States and Canada. In terms of their ecological engineering effects, oyster reefs are often *functionally extinct*. South American shellfish reefs are in much better condition, and a number of species of the oyster family Ostreidae occur there. The Olympic oyster *Ostrea lurida*, which once dominated beaches and estuaries from Vancouver Island to northern California, are now largely gone. Some populations have been found in isolated locations such as southern Vancouver Island and Oregon.

In the case of the eastern oyster *Crassostrea virginica*, three major impacts have caused declines and threatened recovery:

1. *Overexploitation.* Oysters have been exploited by dredging and the insertion of tongs into oyster reefs since the time of European arrival to the eastern and Gulf coasts of the United States. As a result, oyster reefs were overexploited, even by the end of the eighteenth century in New York waters. In Chesapeake Bay, commercial catches are less than 1 percent of the levels in the late nineteenth century.

2. *Habitat destruction by shoreline alteration and pollution.* Oyster habitat has been removed, especially along shorelines. In New York Harbor, oyster beds existed in a stretch that was essentially continuous in harbor coastlines, extending up the lower Hudson River. But alteration of the shoreline by straightening and adding rock borders has eliminated much of this habitat. Pollution from sewage has also caused degradation of local waters by reducing oxygen and adding toxic substances.

3. *Disease.* The eastern oyster is infected by two protozoan parasites. One, known as Dermo, *Perkinsus marinus*, invades oyster tissues, especially in higher salinities. It was first identified in the Gulf of Mexico but is common throughout oyster populations as far north as Massachusetts. The other protozoan, known as MSX, *Haplosporidium nelsoni*, originated from oyster populations in the Pacific but somehow invaded the east coast and has devastated populations of *C. virginica*. Both disease organisms have been regarded as a major impediment to recovery of oysters in coastal waters.

4. *Climate change and disease.* Both Dermo and MSX seem to thrive in warmer waters. As waters of the northeastern United States have warmed in recent decades, the occurrences of Dermo are spreading (Ford and Smolowitz, 2007), which reduces oyster survival beyond the age of 2–3 years. There has also been evidence for the evolution of resistance to MSX in some locations.

The most dangerous situation for oyster reefs is decline through two important thresholds:

1. *Sustainability by larval recruitment.* Populations of oysters are so rare in some areas, such as New York Harbor, that oyster recruitment is nearly absent. Adults are so rare that not enough larvae are being produced to sustain a population. Although a few oysters

have been found in parts of New York Harbor, large areas such as Jamaica Bay have virtually none, indicating that there is not a large enough source population to produce the larvae needed to replenish losses through mortality. This is clearly the case, since juveniles and adults that are transplanted into Jamaica Bay, New York, grow vigorously and spawn, so local conditions would support oysters if larvae were available.

2. *Balance of shell production and degradation.* The success of oyster reefs on submerged mud bottoms depends on the accumulation of shell, on which oyster larvae and many other dependent species can settle and attach. But if recruitment rates are low, then oyster shell will decompose faster than shell growth can replace it; the oyster mound will degrade and disappear into the sediment (Mann and Powell, 2007). As we discussed, oyster reefs thrive when a large shell pile exists that raises oysters above the turbid near-bottom waters. The question is whether oyster reefs can be restored after this threshold is breached. In some cases, however, shell piles survive for many decades in some localities.

Attempts at oyster restoration are being made throughout the world, but so far success has been limited. A notable success was achieved in the Great Wicomico River in Chesapeake Bay (Schulte et al., 2009). The success was greatest when shell was piled to great heights. Oyster recruitment of course depended on a source of larvae, which was still available in this estuary. Over several years, shell accumulated, multiple-year classes of oysters developed, and the oysters were not wiped out by disease. This study holds great hope for other attempts at restoration of oyster reefs, which are numerous and widespread. A large number of restoration projects are now actively being pursued, most notably an approximately 200 hectare restoration in Harris Creek in the Choptank River of Maryland.

■ CHAPTER SUMMARY

- Vertical zonation is a nearly universal feature of the intertidal zone. Different assemblages occur in protected and wave-swept waters in the same area, and wave shock is an important force in this environment.

- The intertidal zone is alternately a marine and a terrestrial habitat. With increasing height above the low-water level, animals are exposed to air for a greater proportion of the day. The changes cause heat stress, desiccation, shortage of oxygen, and reduced opportunities for feeding.

- Mobile intertidal animals may remain fixed at one tidal level, becoming active at the time of low tide. Others migrate up and down the shore with each tidal cycle in order to remain moist. Changes in form can reduce the risk of wave shock, but there are often trade-offs.

- Zonation results from preferential larval settlement and adult movement, variation in physiological tolerance, and biological interactions, such as competition and predation. Keystone species and wave shock on rocky shores exert strong effects.

- Water flow delivers phytoplankton and oxygen, while primary productivity strongly regulates growth rates and abundances of benthos. Larval supply is often a major determinant of community interactions.

- Zonation on soft shores is not as strong. Soft-sediment intertidal species compete for space and food by displacement and chemical secretions.

- Resources may vary even at the same tide level. Deposit-feeding populations may cause overexploitation of renewable resources or seasonal decline of food in the sediment.

- Seasonal influxes of predators, such as birds and crabs, can devastate local soft-sediment communities. Invasions of alien species into the intertidal have become frequent in recent years.

- *Spartina* salt marshes are dominated by cordgrasses, which function as ecosystem engineers by binding fine sediment and causing the buildup of meadows above low water. Shoots above the surface are connected below by a rhizome system. Sediments are usually anoxic, thereby limiting oxygen and nutrient uptake. Some co-occurring animal species facilitate marsh growth by oxygenating the sediment and adding nutrients.

- Vertical zones in salt marshes develop from competition and the ability to survive salt and drowning. Floating wrack often smothers plants and creates patches of bare sediment. Biologically rich creeks and mudflats are corridors between salt marshes and other habitats.

- *Spartina* salt marshes produce large amounts of particulate and dissolved organic matter, which may influence the food webs of salt marsh benthos and perhaps coastal marine systems. *Spartina* species have been introduced throughout the world, with major invasive effects. Marshes have often been destroyed by human activity and sea-level change.

- Mangrove forests are biologically diverse intertidal and emergent woody plant communities. Mangroves are adapted to the anoxic sediments by air-projecting and shallow roots. They are salt tolerant, and one group has salt-excreting glands in leaves. Mangroves show zonation with increasing distance to freshwater and terrestrial environments. Colonization may be strongly affected by crab predation on seedlings.

- Mangroves and upland vegetation compete for space. Mangrove sediments have diverse and abundant invertebrate populations, and creek systems are nurseries for fishes. Mangrove forests are endangered in the tropics by shoreline development, dredging for shrimp farms, and sea-level rise.

- Estuaries are intimately connected to their watersheds, which supply both water and nutrients. Estuaries range from open marine to successively decreasing salinity zones to tidal fresh water and associated creeks and marshes to fresh water. Decreased salinity at the headwaters can reduce the number of marine species.

- Estuaries and shelf environments comprise a two-phase system. Some species can counteract the estuarine flow to the sea; others return from the sea at metamorphosis.

- High nutrient inputs and the removal of predators have created ecological instability. Tidal exchange may be a driving force for suspension feeding in smaller estuaries. Large estuaries are targets for biological invasions.

• Oysters occur in reefs, which are structures that increase local biodiversity. Oyster feeding and biodeposition may strongly affect the overlying water column and nitrogen cycling in some estuaries. Pollution, disease, and overexploitation of oysters has resulted in widespread degradation of oyster reefs throughout the world.

■ REVIEW QUESTIONS

1. Why are vertical zones expanded on wave-swept coasts relative to protected shores?

2. How may an intertidal animal prevent overheating upon exposure to air and sunlight?

3. Do swash riders manage to survive the tide and waves of an exposed beach?

4. What factors contribute to determining zonation on a rocky shore?

5. Why is zonation often not as clear-cut on a soft-sediment intertidal environment as on rocky shores?

6. How do vertical burrowing differences affect coexistence of potentially competing burrowing species?

7. What is character displacement, and what evolutionary process may it reflect?

8. How does predation alter the course of interactions among major space keepers on a rocky shore?

9. How does the size of a disturbed patch influence the subsequent course of colonization and dominance by sessile marine invertebrates?

10. Why are so few species able to live in the transition between estuarine and freshwater habitats?

11. How does invasion by *Spartina* lead to a rise of the sediment level in a protected coastal flat?

12. Why do *Spartina* marsh plants violate the general intertidal rule that an upper limit to a species distribution is determined by physiological problems?

13. Why has *Spartina* had such an important impact on shorelines to which it has been introduced?

14. What is the main means by which *Spartina* extends its population on a tidal flat?

15. Why are mangrove roots important to the invertebrates of mangrove forests?

16. Why do you think many residents of coastal areas greatly regret that mangroves have been removed from many tropical coasts, beyond just the loss of interesting species?

17. The marsh snail *Melampus bidentatus* lives abundantly only in the upper part of salt marshes. What are two hypotheses for why this might be so? How would you test these hypotheses?

18. Rocky shores are often dense with living sessile animals and plants. Large stone jetties are built along the coast, especially along sandy beaches, yet large populations of barnacles and mussels do not necessarily rapidly colonize the boulders of these jetties. Why do you think this is so?

19. Open, exposed, rocky coasts often have mollusks with thick shells. What are two different reasons why this might be so? How would you test these two explanations?

20. The southeastern marsh snail *Littoraria irrorata* has been described as a fungal eater. How would you test this claim directly?

21. Many conservationists claim that oysters, once restored, will protect shorelines from storm surges. How would you test this idea?

Visit the companion website for *Marine Biology* at www.oup.com/us/levinton where you can find Cited References (under Student Resources/Cited References), Key Concepts, Marine Biology Explorations, and the Marine Biology Web Page with many additional resources.

CHAPTER 17

The Shallow Coastal Subtidal
Sea Grass Beds, Rocky Reefs, Kelp Forests, and Coral Reefs

In this chapter we discuss a series of shallow, coastal subtidal environments that have several features in common. Grass beds, rocky reef–kelp forest systems, and coral reefs are the most productive and diverse subtidal benthic environments. They are dominated by highly productive benthic primary producers, which also contribute to the structural habitat and qualify as ecosystem engineers: Sea grasses, kelps, and corals all strongly alter water flow and provide a highly altered habitat that allows the presence of many dependent species.

Sea Grass Beds

- **Sea grasses are marine angiosperms, or flowering plants, that are confined to very shallow water and extend mainly by subsurface rhizome systems within the sediment.**

In soft sediments, from the low-water mark to about 3–5 m, broad flats are often inhabited by sea grasses and a rich associated biota, from the tropics to as far north as Alaska and northern Norway and as far south as Chile, southern Africa, and Australia. There are only 50–100 species of sea grasses, as opposed to the far more diverse submerged flowering-plant flora in fresh water. **Sea grasses dominate communities and are foundation species,** whose subsurface structure, upright posture above the sediment surface, and current baffling combine to influence hydrodynamic conditions, microhabitat structure, and food supply within the sea grass bed.

In the temperate zone, the eelgrass *Zostera marina* forms thick beds in sediments ranging from sand to mud

(**Figure 17.1**). In the southeast American tropics, the turtle grass *Thalassia testudinum* dominates and covers very shallow flats of carbonate sediment. Like *Spartina* marshes, sea grasses have a large **rhizome system,** which is a root network within the sediment, capable of taking up nutrients. As the rhizome system extends laterally, **shoots** may be sent up. The shoots lack the stomata of terrestrial plants and have a thin cuticle, which allows gases and nutrients to diffuse into the leaves from the water. The rhizomes also take up dissolved nutrients, mainly from pore waters in sediments. While the rhizomes are important in nutrient uptake and reproduction, they also require resources to maintain, which may be a problem in low light conditions.

FIG. 17.1 A bed of *Zostera marina* in Padilla Bay, Washington. Blades of this sea grass are 50–100 cm high. (Photograph courtesy of Charles Simenstad)

Sea grasses can extend by means of rhizomes, but they are flowering plants, and pollen moves between plants in the water currents (see Figure 13.13). Fancy flowers in the terrestrial environment are associated with attraction of pollinators such as insects, so sea grass flowers are rather simple and quite inconspicuous. Sea grasses set seed, which are borne by water currents. Local population spread by flowering is less important in sea grasses, which appear to reproduce more by asexual spread through the rhizome system.

Sea Grass Growth Conditions, Succession, and Production

■ Sea grasses grow best in conditions of relatively high light and modest current flow.

Sea grasses grow densely in shallow water only—rarely found in depths greater than 3 m. Light strongly limits photosynthesis, and the lower depth limit is usually strongly related to light irradiance. Sea grasses generally form linear leaves, with the growing meristem region at the base. Leaves extend from the rhizome system toward the surface. **Epiphytes**, such as bryozoa and microalgae, grow on the leaves and may strongly reduce light capture and photosynthesis. Sea grass meadows appear to develop more extensively under modest current conditions, which may reflect the delivery of dissolved nutrients. Sea grasses obtain a portion of their nitrogen by means of nitrogen-fixing bacteria on the rhizome, but most nitrogen is probably taken up from the water. Sea grasses are affected adversely under high flows of 0.5 m s^{-1}, owing to sediment erosion and shear stress on the leaves.

■ Sea grass beds most easily colonize sediment after a successional sequence featuring a previous colonization by seaweeds.

Sea grass may extend laterally into bare sediment by means of the rhizome system. Bare sediment pore water, however, is generally low in nutrients. The rhizome system aids in transport of nutrients from the already established bed. Colonization of new areas by seedlings is difficult unless the sediment is already physically stable and the pore water rich in dissolved nutrients. This can be accomplished by the presence of seaweeds, which stabilize the sediments and add nutrients in forms such as ammonium. Thus, bare sand may change, by succession, to a sea grass bed (**Figure 17.2**). Mature sea grass beds usually consist of coexisting sea grasses and seaweeds. Caribbean turtle grass (*Thalassia testudinum*) beds often have dense coexisting growths of the **calcareous** alga *Halimeda*, whose mounds stimulate turtle grass growth, perhaps by release of carbon dioxide (Kenworthy and Reid, 2003). *Halimeda* also provides microhabitat for numerous small grazing invertebrates. This suggests the potential role of a diverse system of species in total primary and secondary productivity; more species may be able to occupy more types of microhabitats and, therefore, enhance total productivity.

Eelgrass, *Zostera marina*, grows more rapidly in muddy relative to sandy sediments. Leaves are wider and add more leaf area over time in mud than in sand. This is likely related to nitrogen supply from sediment pore waters. Ammonium concentrations are higher in muds, and higher ammonium is associated with more rapid shoot growth.

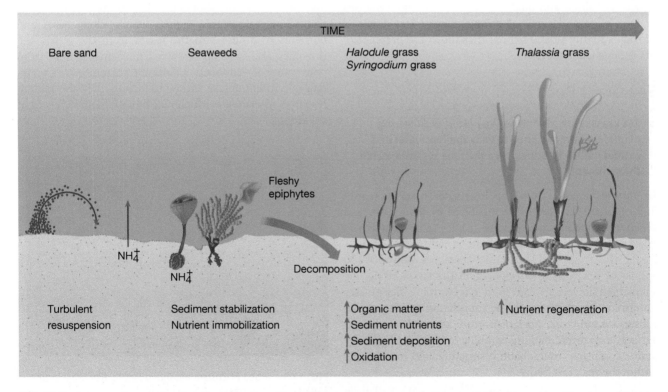

FIG. 17.2 Scheme of succession from bare sand to a *Thalassia testudinum* sea grass bed in coastal Florida waters. (Courtesy of Susan Williams)

■ Sea grass beds have high primary production and support a diverse group of animal species.

Shallow subtidal sea grass beds are among the most productive of all marine communities. The grasses themselves grow rapidly, but there is a large variety of associated animal and algal species. Sea grasses form dense meadows and are often colonized by a wide variety of fouling organisms, such as hydroids, sponges, bryozoans, and seaweeds. Many invertebrate species depend upon sea grasses. The common Atlantic bay scallop *Argopecten irradians*, for example, has a planktotrophic larva that recruits to sea grasses (**Figure 17.3**). The early juvenile stage attaches by byssal threads to the grass blades of *Zostera marina* and remains attached for several months before living free on the bottom. The attachment places the juveniles out of the reach of predatory crabs.

■ Sea grass beds reduce current flow, deter the entry of some predators, and may enhance the growth and abundance of infaunal suspension feeders.

Eelgrass beds are often quite dense, with thousands of shoots per square meter. Water currents are greatly reduced within sea grass meadows, and a number of species of burrowing invertebrates increase greatly in abundance, relative to nearby bare bottoms. The reduced flow may encourage settlement of swimming larvae. Some experiments show that larval settlement is greater when simulated sea grass bottoms of soda straws have intermediate densities (**Figure 17.4**). The density of blades, combined with the density of the root system, tends to deter some carnivores from entering the beds. Large carnivorous crabs, snails, and flatfish have some trouble entering eelgrass, which protects invertebrates, such as bivalves, within the bed.

The rhizome system is especially important in stabilizing the sediment within the sea grass bed. This stabilization tends to perpetuate the survival of the entire sea grass system. Because such stable sediment beds are in very shallow water adjacent to the shoreline, sea grasses provide important ecosystem services by protecting human shoreline development against erosion and shoreline wave damage.

Larger suspension-feeding invertebrates, such as the clam *Mercenaria mercenaria*, tend to be larger and grow faster just within the eelgrass than outside the bed in bare sand or muddy sand. There are two likely explanations. First, suspended food may be more abundant within the eelgrass bed. Because of the reduced current speed, resuspended material may be retained and ingested by the suspension feeders. Much of the suspended food of clams on open sand flats comes from resuspension of benthic diatoms. Second, predation is reduced within the eelgrass beds, which allows bivalves to grow older and larger.

If suspended phytoplankton comes laterally from outside the bed, the reduction of current flow into a sea grass bed might reduce the supply of phytoplankton food for suspension feeders deep into the bed. Allen and Williams (2003) tested this hypothesis in San Diego Bay by adding

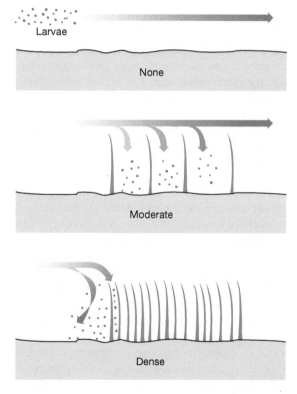

FIG. 17.4 An experiment showing the possible effect of sea grass beds on larval settlement. Larvae are carried over a bottom with no impediments to flow (top). A low density of soda straws reduces flow and causes dense larval settlement throughout the imitation grass bed. Finally, a dense array of straws cuts flow so much that only a little settlement at the edge of the "bed" is permitted. (Figure after work described by Eckman, 1983)

FIG. 17.3 The scallop *Argopecten irradians* lives as a juvenile attached to eelgrass blades, which reduces predation by crabs.

phytoplankton above the interior of a *Zostera marina* sea grass bed, to see if it would enhance the growth of an invasive mussel, *Musculista senhousia*. Mussels grew 50 percent faster when phytoplankton food availability was locally increased.

■ **Benthic suspension feeders may enhance sea grass growth through feeding and biodeposition, but burrowing shrimp may exert negative effects on sea grasses.**

Suspension-feeding bivalves deposit feces and rejected material, known as pseudofeces, to the bottom. Such **biodeposition** provides a benefit to sea grasses in the form of nutrient enhancement. Peterson and Heck (2001) experimentally added the mussel *Modiolus americanus* and demonstrated that mussel biodeposition enhanced the productivity of the sea grass *Thalassia testudinum* in a Florida bay. Because sea grasses are so limited by light, bivalves may also enhance the growth of sea grasses by filtering out locally dense phytoplankton and reducing turbidity within the sea grass bed (Wall et al., 2008). This effect is important, especially where nutrient addition in polluted estuaries and coastal lagoons enhances local phytoplankton abundance.

Turtle grass meadows in the American tropics and subtropics are often interspersed by broad open sand and mudflats dominated by calcareous sediment. Several species of the burrowing shrimp genus *Callianassa* are especially abundant, and two species form large, often conical mounds of sediment following extensive bioturbation of up to about 2.5 kg per day. Experimental transplants of these shrimp demonstrate that this activity has strong negative effects on growth and survival of turtle grass, which may result from the turbidity generated by bioturbation, which reduces light input to the grasses and may even smother them (Suchanek, 1983).

■ **Sea grasses are grazed to variable degrees.**

Some sea grasses are apparently unpalatable. The eelgrass *Zostera marina* in northerly latitudes is not consumed by any major herbivores. But this is not the case for sea grasses in subtropical and tropical habitats. A few sea urchins and the green turtle *Chelonia mydas* consume the Caribbean turtle grass *Thalassia testudinum*, which is rich in cellulose. Both green turtles and some urchin species have symbiotic gut bacteria that digest cellulose. Green turtles seem to behave like prudent farmers. They do not overgraze the turtle grass, but instead nip the leaf tips, which encourages regrowth of newly grown grass, which is easier to digest and has a lower content of complex carbohydrates and tannins. In Australia, dugongs, an important marine mammal grazer, graze eelgrass, resulting in the preferential growth of a more palatable grass, *Halophila ovalis*, which grows rapidly and has less fiber and higher nitrogen (Preen, 1995). In northwestern Australia, dugongs avoid predators like tiger sharks and often are forced to graze in deeper, less productive grass habitats.

In the tropics, sea grasses are often associated with coral reefs, and smaller, so-called patch reefs usually stand as islands in the middle of sea grass meadows. **Halos** bare of turtle grass often surround patch reefs (**Figure 17.5**). The

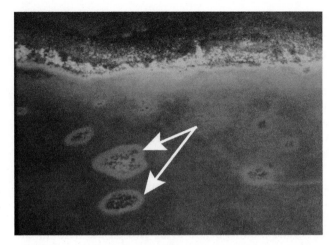

FIG. 17.5 Aerial photo of an area of patch reefs in Tague Bay, St. Croix, U.S. Virgin Islands, in the 1970s. Halos were created by grazing urchin *Diadema antillarum* that hide in the patch reef during the day, but depart and graze on the surrounding sea grasses at night, when visual predators cannot spot them. A disease in the 1980s reduced populations of these urchins throughout the Caribbean. (Photograph by Thomas H. Suchanek)

urchin *Diadema antillarum* was responsible for the halos. By day, the urchins hide in crevices on the patch reef, but at night they move out onto the surrounding sand flats and consume turtle grass. To avoid predators, they must return to the patch by morning, so usually only a limited area around the reef is denuded. These urchins nearly disappeared from the Caribbean following a devastating disease in the 1980s (see section on coral reefs) and have recovered in scattered locations. Still, other urchins remain important grazers on Caribbean sea grasses.

Today, grazing is reduced in many tropical sea grass beds because major grazers such as green turtles are absent or greatly reduced in abundance. Green turtles have extended hindguts with a microbiota that is adapted to digesting plant material. The turtles specialize on either high-cellulose grass or low-cellulose seaweeds. Such alternations of behavior may cause local strong differences in grass and seaweed abundance, so we now have only a limited idea as to how they affected Caribbean sea grass meadows where they were once very abundant. Other missing elements of the grazing biota include many grazing waterfowl and now-rare marine mammals, such as manatees in the tropical Atlantic and dugongs in the southwest Pacific. All of these were probably much more abundant and, therefore, were more influential consumers of sea grasses. Still, it is not clear that *Zostera marina* in temperate latitudes was ever grazed to any degree, and this may reflect differences in nutritional value or perhaps the absence of suitable grazers. The lack of grazing in higher latitudes has a parallel in fish: Herbivorous fish are far less frequent in higher latitudes than in the tropics. In warmer waters such as the Caribbean and the Mediterranean, fish grazing has a major impact on sea grass abundance. It may be that nitrogen content determines the degree of herbivory (the interested student should read Heck and Valentine, 2006).

Seaweed Interactions and Trophic Cascades in Sea Grass Beds

■ **Sea grass meadows are diverse in lower latitudes and interact strongly with coexisting seaweeds, which are grazed by many species.**

In higher latitudes in the Northern Hemisphere, sea grass meadows contain just one grass species, but at lower latitudes in the Caribbean basin, sea grass meadows typically have over 10 species. Some species tend to dominate when disturbance and grazing pressure are low and have lower growth rates and high biomass, including a well-developed rhizome system. Other species grow more rapidly and tend to be easier to graze by urchins and vertebrates.

Seaweeds grow among the grass blades and are an extremely important component of the biota. This is true of lower-latitude diverse sea grass beds and higher-latitude grass beds of lower diversity. Eelgrass (*Zostera marina*) in northern latitudes is not consumed to any degree, but seaweeds among the grass blades are consumed by a wide variety of herbivores, including crabs, snails, and urchins. These herbivores are, in turn, consumed by a number of benthic predators, such as blue crabs, *Callinectes sapidus*, and pipefish and seahorses. In turn, fishes consume the benthic predators. That is the important potential trophic cascade component of the system, which we will discuss shortly. But there also is a crucial horizontal interaction between eelgrass and seaweeds.

■ **Nutrient addition and loss of suspension feeders both have a negative impact on sea grass.**

At least two dynamic processes can produce dominance by seaweeds over sea grasses: nutrient addition and reduction of suspension feeders. As we mentioned, seaweed decay and even biodeposition by suspension feeders could provide nutrients that encourage the colonization and growth of sea grasses. But this does *not* happen when plant nutrients are in high concentrations. Seaweeds tend to respond much more than eelgrass to increased nutrients. Therefore, in areas adjacent to human habitation, sea grass tends to contain dense growths of rapidly growing seaweeds. Also, nutrients tend to increase the growth of epiphytes directly on the sea grass blades, which negatively affects sea grass growth. Second, processes that reduce suspension feeders might cause local increases in phytoplankton in sea grass beds, which might shade them out and favor seaweeds. In recent years, overharvesting of suspension feeding oysters and other bivalves and the increasing occurrence of nuisance phytoplankton blooms have caused drastic reductions of suspension-feeding bivalves such as scallops, clams, and mussels. Such phytoplankton bursts can shade seagrasses and suppress their growth. Many smaller grazers of seaweeds live among, and, therefore, depend on, bivalve beds, and are no longer available to keep seaweeds and epiphytes in check. Seagrasses can thus be smothered by seaweed and epiphyte growth.

■ **Predation is important in sea grass beds, but it is not clear whether there is a strong top-down trophic cascade.**

As we have discussed, grazing is variably important in sea grass beds, and may often make a major difference in the occurrence of various plant species. It is not so clear whether there are strong grazing interactions that result in different types of plant and animal assemblages, however. The same appears to be the case for carnivores. Predation on animals is clearly a major factor in the relative abundance of species in sea grass beds. For example, in the southeastern United States, the pinfish *Lagodon rhomboides* is a major predator on small grazers such as amphipods within sea grass beds. A visual predator, it prefers species that are more mobile and aggregated. On the other hand, many amphipods and other mobile small crustaceans can hide among sea grass blades. Wide-bladed sea grasses such as the turtle grass *Thalassia testudinum* comprise better hiding places than narrow-bladed sea grasses such as *Halodule wrightii* and *Syringodium filiforme* (Stoner, 1980). While pinfish are carnivorous, they also feed on epiphytes and therefore do not necessarily produce only negative effects on sea grass by consuming epiphyte grazers like amphipods (Heck et al., 2000). The feeding of pinfish on more than one trophic level blurs the simple idea of a trophic cascade with strong top-down effects.

Decline of Sea Grasses

■ **Disease was a major cause of eelgrass decline. Now many sea grasses are declining because of pollution and water turbidity.**

Sea grasses have declined greatly in abundance, and a disease nearly collapsed one species in the Atlantic. In the early 1930s, eelgrass meadows declined rapidly in the bays of Europe and the eastern United States, though for some reason the Mediterranean was spared. During this period, many species dependent on eelgrass declined precipitously. For example, settling larvae of the bay scallop *Argopecten irradians* must attach their byssal threads to grass blades, and this important commercial shellfish population went into a tailspin. Many other benthic species declined in abundance, and certain soft-bottom habitats that were protected by eelgrass beds, now exposed to the brunt of the sea, were eroded away. It was only in the late 1940s and early 1950s that extensive recovery occurred. Because the pathogen does poorly in low salinity, eelgrass survived in refuges in the low-salinity parts of estuaries, which allowed repopulation of broader areas.

Unfortunately, the cause of this disaster will never be known with certainty, but recent eelgrass epidemics have been studied properly and the cause attributed to a pathogenic strain of the heterokont *Labyrinthula zosterae*. Such species are active sources of potential mortality. Jensen and others (1998) found that the turtle grass *Thalassia testudinum* had resistance to *Labyrinthula* infection by means of a flavone glycoside that reduced attachment to the leaves. A more recent study shows that the eelgrass *Zostera marina* is still infected by the *Labyinthula zosterae* but with no apparent ill effects. It may be that eelgrass evolved resistance to a pathogen that was once a regional killer (Brakel et al., 2014).

Sea grass is otherwise declining at this time, especially in the Atlantic and Caribbean. Nutrients from sewage tend to increase phytoplankton density, and this decreases the amount of light that can reach the bottom. Once abundant, the eelgrass *Zostera marina* is nearly gone from Chesapeake Bay and other areas in coastal eastern U.S. waters. Nutrient input into shallow bays also encourages the colonization and spread of rapidly growing seaweeds, such as the sea lettuce *Ulva* spp., which choke out sea grass beds.

However, there is an alternative top-down hypothesis for the decline in Chesapeake Bay. Increased fishing may have reduced grazing on the epiphytes that live on the sea grass blades. This may have resulted in an explosion of epiphytes and a choking off of light to the sea grass. In basins where grasses have been shaded to extinction, scallops have also decreased greatly. Disturbance of the shoreline has increased sedimentation, which also has caused sea grass decline. Finally, overfishing of top carnivores has likely had major effects on sea grass communities, but they have

not been measured completely. Human disturbance from dredging and boat propellers probably also exerts strong negative forces on sea grasses. In the past few years, remediation of nutrient inputs and especially seed additions (see Orth et al., 2012) into areas such has Chesapeake Bay has resulted in local recovery of some eelgrass beds. Restoration of suspension-feeding bivalves in shallow areas should greatly reduce phytoplankton loads and benefit sea grasses further.

The Rocky Reef–Kelp Forest System

In the following two sections we will cover subtidal rocky reefs and kelp forests. In many instances, these two habitats are part of one system, strongly interlinked by common top predators, grazers, transport of plant detritus, and even species occurrences. In **Hot Topics Box 17.1**, we discuss the complete reorganization of a rocky reef–kelp forest

HOT TOPICS IN MARINE BIOLOGY

Reorganization of a Rocky Subtidal Ecosystem: A Cod and Lobster Tale
17.1

Kelp forests abound in the northeastern Atlantic coast of the United States. What we see today is an erratic abundance of one common species of urchin and a number of seaweed species, often dominated by kelps of the genus *Laminaria*. These kelps come to dominance when grazing is low.

When you read historical descriptions of fishing in New England and New York waters in the seventeenth and eighteenth centuries, a feeling of disbelief sets in. From Long Island, whalers rowed boats just a mile or so offshore to hunt their prey, which included sperm whales. Many remarked that you could literally walk on the cod in Boston Harbor; they were that dense. After all, Bartholomew Gosnold named Cape Cod in 1602 for the "vexing" abundance of cod he saw from his ship. Of course this is no longer the case, and a series of remarkable studies in the Gulf of Maine have charted the ecological effects of human fishing over the past centuries.

Ted Ames, a fisherman and trained biologist, gave shape to our knowledge by interviewing other fishers and learning about historical spawning and fishing sites in Maine. With this information he was able to reconstruct a remarkable map of about 260 cod inshore fishing and 91 inshore spawning sites in the 1920s and earlier (**Box Figure 17.1**). Offshore sites also harbored feeding cod. The abundance of cod could also be documented in Native American kitchen middens from 200 to 4,000 years ago. The subtidal rocky reef and kelp communities are undergoing an ecological revolution in the Gulf of Maine because of the loss of the cod.

The Gulf of Maine was described as a case of trophic-level dysfunction (Steneck et al., 2004), which sounds a bit like a psychological situation, but refers instead to the human removal of crucial parts of the trophic cascade. For thousands of years, abundant cod, haddock, and wolf fish in the Gulf of Maine functioned as top carnivores, keeping crabs and lobsters in check. Of course, no observations or studies of

ecological interactions were made, so we can infer only a long-term stability, somewhat bolstered by evidence we see in the absence in coastal Native American middens of abundant fish prey: crabs and lobsters. Urchins were probably not very common. Offshore, where spotty concentrations of cod and wolf fish can be found, urchins are quite rare and kelps are common, which makes sense from the trophic cascade perspective. At least through the 1930s, cod were abundant inshore and kelps dominated the bottoms. Predatory finfish were abundant through the 1960s before fishing finally overwhelmed the Gulf of Maine stocks. Starting in the 1940s, lobster populations and landings increased dramatically, which probably was a response to the relaxation of fish predation. Once this happened, a second phase of reorganization commenced.

In the 1930s, fishing in the Gulf of Maine became more efficient: Fish could be located more rapidly, and onboard refrigeration allowed rapid processing of the catch. From the 1960s to the 1970s, stocks were exploited partially by foreign fleets and the fish declined precipitously. Following the extension of the U.S. fisheries zone to offshore (see Chapter 21), overfishing by American fleets continued and the cod—the Gulf of Maine's top predator—declined to almost nil (**Box Figure 17.2**). The second phase began: Urchins increased in numbers, denuded nearly all kelp beds, and left a barrens zone, dominated by calcareous algae. The kelps were gone, and associated species of fish and invertebrates were missing. Cod had been replaced by some smaller species of fish.

Phase three, removal of the urchins, followed. After 2 decades of urchin superabundance, a fishery developed in the late 1980s. The green sea urchin *Strongylocentrotus drobachiensis* was exploited by divers and mainly exported overseas. Urchins were extirpated in just a few years, and by the 1990s seaweeds had recruited back to shallow hard bottoms, but finfish remained uncommon. A number of areas

HOT TOPICS IN MARINE BIOLOGY

Reorganization of a Rocky Subtidal Ecosystem:
A Cod and Lobster Tale   *continued*   **17.1**

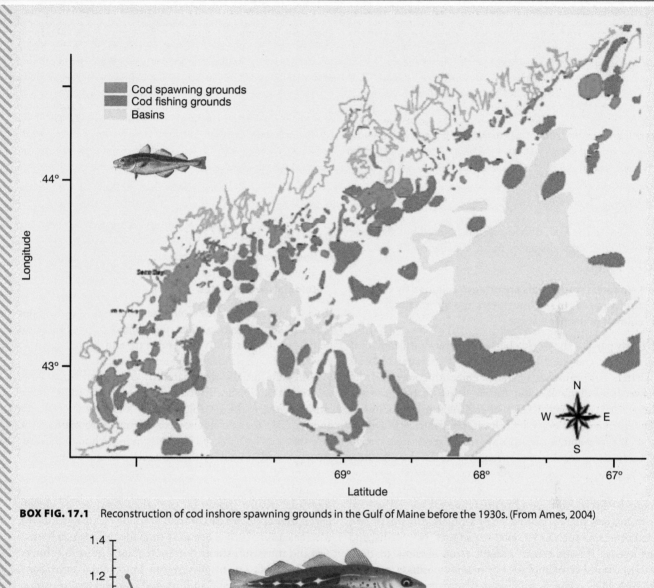

BOX FIG. 17.1 Reconstruction of cod inshore spawning grounds in the Gulf of Maine before the 1930s. (From Ames, 2004)

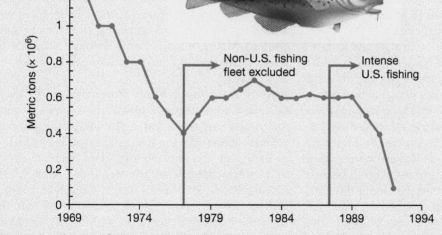

BOX FIG. 17.2 Change in fisheries take of cod from the western Atlantic. (From Steneck and Carlton, 2001)

continues

were closed to urchin fishing, but abundant crabs and even amphipods preyed on newly recruited urchin juveniles, so the luxuriant growth of macroalgae continued. The macroalgae whip around and may create an unstable surface for moving sea urchins when they are tiny juveniles.

A major lesson has been learned from this story, pieced together by an inquisitive fisherman and also by a scientific team of investigators from institutions all over the Gulf of Maine and beyond: Overfishing not only results in depletion of the fishery stock, but it may have drastic effects on the ecosystem because of the strong trophic links. These changes snowball and have many other impacts. Can public policy and private fishing practice allow the return of the cod, and can this ecosystem be restored? Warming of the Gulf of Maine may preclude recovery of the cod. But only time will tell. At least we now know the dimensions of a problem that is far more complex than anyone ever might have predicted. Restoration must involve management of the entire ecosystem.

ecosystem in the Gulf of Maine, all facilitated by successive harvesting of crucial species in an overall trophic cascade in this system.

■ **Kelp forest–rocky reefs are often dominated by kelps and seaweeds in shallow waters and by epifaunal animals in deeper waters.**

In many parts of the world on hard substrata with cool temperatures, a traverse from shallow to deep water involves a switch of dominance of major epibenthic species from seaweeds in shallow water (often kelps, to be discussed shortly) to coverage by colonial epifaunal invertebrates in deeper water. Obviously, light limitation restricts seaweeds to shallower water. In these locations seaweeds are a system of foundation species, that strongly alter local water flow and also provide a continuing source of detrital particles, used by many invertebrates. As the water deepens to below about 10–20 m, dominance of attachment sites to the hard bottom gives way to dominance by animals, principally sessile organisms such as colonial sea squirts, bryozoans, and anemones. Throughout a transect from shallow to deep water, a number of mobile organisms dominate, ranging from mobile predatory fish and invertebrate predators such as sea stars, snails, and nudibranchs to grazers such as urchins and abalones on the U.S. Pacific coast.

Figure 17.6 shows variation in macroalgal and invertebrate cover in a depth transect on hard substrata in Friday Harbor, Washington State, USA. Seaweeds dominate the surface in the shallows, but there is a switch to dominance by sessile invertebrates at a depth of 15–20 m, and invertebrates dominate strongly by 20–25 m (Britton-Simmons et al., 2009). Calcareous algae constitute an important component of the sessile biota in shallow waters. They often dominate when colonial invertebrates die off because urchins move in (predatory sea otters are rare in this area) and graze the surface for colonizing seaweeds, making it difficult for colonial invertebrates to recruit again. This shifts the advantage to calcareous algae.

Subtidal Rocky Reefs

■ **Rocky subtidal reefs harbor abundant communities of algae and invertebrates and are often dominated by colonial invertebrates.**

Rocky subtidal reefs are common off rocky coasts, often beneath kelp forests as just mentioned. Our discussion here is confined mainly to temperate and higher-latitude environments. Subtidal rocky outcrops occur in the tropics, but they are usually dominated by corals and other large calcareous organisms. In the temperate zone and in higher latitudes, the rocks are colonized by a variety of seaweeds and animals that do not secrete calcium carbonate at the rate that can be achieved in the tropics, although calcareous algae and animals with calcareous skeletons are usually very abundant.

Subtidal rocky reefs differ in an interesting way from intertidal rocky shores in that they tend to be **dominated in many sites by colonial animals** (see **Figure 17.6**), whereas animals that occur as individuals, such as barnacles and mussels, usually dominate rocky intertidal shores. Subtidal reefs are often covered by colonial bryozoans, hydroids, sponges, and solitary and colonial sea squirts, as well as by a variety of forms that are individuals, such as anemones, bivalves, and barnacles. In the rocky subtidal coast of Maine, mussels and anemones dominate many areas, so the subtidal colonial dominance pattern is not always found.

Why are colonial forms more common in the subtidal? The preponderance of colonial animals on subtidal rocky reefs may relate to the physiological limitations of colonial animals in the intertidal zone. Individuals such as barnacles and mussels can close themselves at the time of low tide and survive desiccation well. Most colonial groups—such as ascidians, sponges, bryozoa, and hydrocorals—are not capable of closing off soft tissues from the extreme temperature and salinity changes and desiccation of the intertidal. On exposed rocky shores of

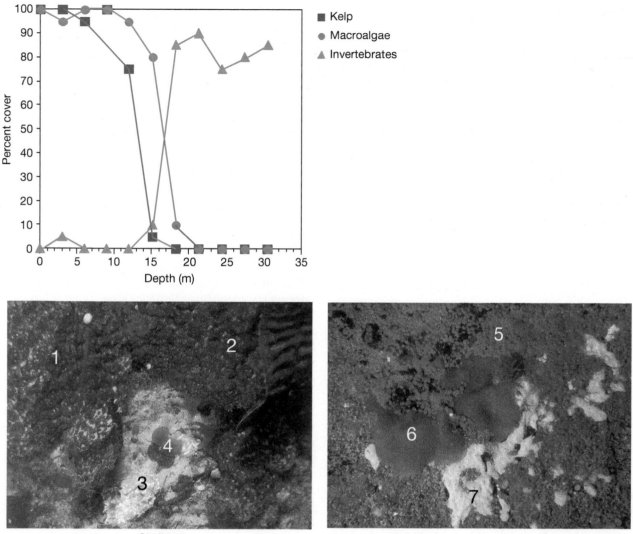

FIG. 17.6 Abundance of kelps, macroalgae (kelps plus other seaweeds), and sessile invertebrates on a transect with increasing depth in Lopez Island of Washington State. Photo on lower left shows dominance of macroalgae in shallow water, and photo on lower right shows domination by colonial sea squirts. (1) Kelp *Agarum fimbriatum*; (2) kelp *Saccharina latissima*; (3) crustose coralline alga; (4) fleshy red seaweed; (5) sea squirt *Aplidium* sp.; (6) sea squirt *Didemnum* sp.; (7) sea squirt *Metandrocarpa taylori*. (Data courtesy of Kevin Britton-Simmons, and photographs by Kenneth Sebens)

the Pacific northwestern United States, the calcareous hydrocoral *Allopora* can be found abundantly, but only on the lowest level of the shore. It grows well in subtidal rocky areas. Salinity variation may also be an important factor, as many colonial invertebrates are not well adapted to strong salinity variation. On rocky surfaces near Vladivostok on the Russian Pacific coast, only mussels are found abundantly in the upper meter or two. Below this depth one finds exceptionally rich epibenthic invertebrate assemblages of species that lack external and sealable shells, including sea stars, bryozoans, and ascidians. In the upper meter or so, the water temperature and salinity vary a great deal seasonally. Salinity can be below 20 owing to freshwater sources, and the surface-water temperature can surpass 15–20°C. In contrast, water below this surface layer is predictably cool and of open marine salinity all

year round. Mussels in New Zealand fjords (**Figure 17.7**) also dominate in a shallow low-salinity layer (Smith and Witman, 1999).

Colonial organisms are modular and consist of interconnected and usually identical units. The ability of a modular organism to spread along a surface makes it more resistant to total removal by predators and perhaps to competition as well. A predator, such as a grazing snail, may take only a portion of a colony, but the penetration of a mussel shell means death to the individual. One can imagine a group of predators consuming portions of prey colonies, the remainder of which grow back to cover the space. Such a growth advantage over individuals might make a modular organism an eventual winner in competition for space. That may explain the strong subtidal rocky reef dominance of colonial ascidians seen in **Figure 17.6**.

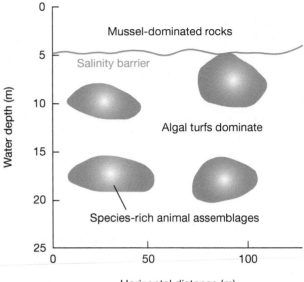

FIG. 17.7 Scheme of biotic patches on subtidal rock walls of New Zealand. The upper water layer is dominated by mussels, which are physiologically more tolerant. The algal turfs are sites of disturbance by landslides. (Drawn from descriptions in Smith and Witman, 1999)

■ **Rocky reefs often are very patchy, with alternations of rocks dominated by rich invertebrate assemblages and turf-forming calcareous red algae.**

Epibenthic organisms of rocky subtidal reefs are very patchy, at scales from centimeters to hundreds of meters. This seems to characterize such reefs, whether from New Zealand or New England or the Pacific Northwest in the United States. It is common to see large patches with rich assemblages of sessile marine invertebrates, especially on vertical or near-vertical rock walls. Within these patches space is clearly very limited, suggesting the possible importance of interspecific and intraspecific competition. This is especially interesting because species with extremely different morphologies and life histories are competing for the same space (**Figure 17.6**). The other common patch type is a turf that is dominated by red calcareous algae. This turf is quite distinctive because of the reddish tint to the crust on the bottom. While sessile invertebrates are found in such patches, they are decidedly rarer than in the other patch type. On more horizontal surfaces, seaweeds predominate in shallow waters, except when they are kept in check by large populations of urchins.[1]

The **alternation of invertebrate and calcareous algal patches** appears to be maintained by a number of processes. Disturbance, both biological and physical, appears to select for the calcareous algal patches. In rocky reefs of New England, urchins are associated with patches of calcareous algal turfs, which are more resistant to the urchin herds and their

scraping of the bottom than the more delicate hydroids, ascidians, and other colonial invertebrates.

The distribution of subtidal rock wall faunas in New Zealand integrates nicely a set of physiological effects, disturbances, and the dispersal biology of the dominants of the animal-rich patches (**Figure 17.7**). Mussels dominate a surface low-salinity layer of 0–5 m depth, possibly because they are physiologically more tolerant than colonial invertebrate species. It is also possible that the strong pycnocline presents a barrier for dispersal of larvae from deeper water to the surface layer. In deeper waters, one finds high-diversity (28–32 species) patches, dominated by bryozoans, sponges, and ascidians, which are interspersed with low-diversity patches dominated by encrusting calcareous algae.

The deeper low-diversity crustose algal patches are maintained by disturbance. Landslides are quite localized, but when they occur, they destroy delicate invertebrates, favoring the encrusting calcareous algae. What is less clear is whether the colonial animal–dominated patches are on a path headed for extinction. What maintains them? Franz Smith and Jon Witman (1999) placed settling plates within and at various distances from the diverse patches and found that settlement was much higher within the patches and trailed off with distance from the patch. Therefore, larval recruitment was sufficient to perpetuate the local patches. On Massachusetts subtidal rock surfaces, many of the animal species with short larval lives were found most abundantly within a few centimeters of the rock walls (Graham and Sebens 1996). This provides a continuous supply of short-lived swimming larvae. It may well be that gregarious settlement also contributes to localized high recruitment. The patches are likely self-seeding, which would prolong the survival of stable animal-rich patches, as long as there is no major disturbance. In effect, the rock wall large patches are relatively closed systems.

■ **Subtidal rock wall patches of animals often are short on space, suggesting the importance of competition.**

Competitive interactions on subtidal rocky reefs are poorly understood, and investigations are hampered by the need to understand the competitive abilities of a wide variety of species with different ecologies. While animal patch edges allow growth into empty space (or at least space dominated by turf algae), the middle of patches is characterized by crowding. All of the space on pilings in South Australia was occupied and overall the dominant species did not change over a 2-year survey (Kay and Butler, 1983). But if you looked at individual quadrats, a considerable proportion of the total space was lost by some species, to be taken over by others. It was as if the community was following the Red Queen's advice to Alice in Wonderland: Move as fast as you can, just to stay in the same place.

Competitive success must be related to the ability to overgrow neighbors and to resist overgrowth by other species. As mentioned earlier, many if not most species in such communities spread along the rock surfaces asexually, which makes spread at the expense of other species crucial

---

[1] Harvesting of urchins in recent years by people has led to enormous increases of seaweed abundance and major reorganizations of subtidal hard-surface communities throughout the world.

in competitive success. Kenneth Sebens (1986) followed rock walls in northern Massachusetts by photographic plots for 2 years and was able to determine competitive interactions in a complex fauna. A species of ascidian, a sponge, an octocoral, a red crustose alga, and a mat of amphipod tubes dominated the rock wall. Space occupation was generally positively correlated with competitive ability, either in displacing neighbors or by holding out against incursions by neighbors. Rapidly growing species were good at overgrowing but were not good at resisting overgrowth themselves. Larger and thicker colonies, somewhat slower growing, were usually at the top of the competitive hierarchy. The variety of competitive mechanisms may therefore render the outcome of competition unpredictable.

In Chapter 4, we discussed alternative stable states, which John Sutherland used to explain changes of dominance in subtidal hard surfaces in coastal North Carolina. Often certain species would colonize a bare surface that no other species could then invade. Such **priority effects** led to strikingly different patterns of dominance in different experimental settling tiles, depending on the time of year the tile was placed. Colonization of bare space, combined with the ability to stand off many newly entering species, may be the key to much of the coexistence or persistent dominance of some patches by one species.

Many of these colonial species are passive suspension feeders, and the flux of particles is therefore crucial for feeding. The question of whether food limitation influences colony growth rate arises. Beth Okamura (1992) demonstrated that lateral movement of water across a bryozoan colony resulted in depletion of food. Downstream zooids of a bryozoan received less food than upstream zooids.

■ Rocky reefs are grazed more intensely, mainly by sea urchins, on horizontal benches.

Competitive interactions on rocky reefs are mediated by predation and grazing. Because space is the principal limiting factor, all processes that liberate space are opportunities for strong shifts in dominance. A good example is grazing by sea urchins, which are often abundant on rocky reefs. However, they often are much more abundant on horizontal surfaces because currents tend to dislodge them when they move on open exposed vertical surfaces. If colonial invertebrates arrive first, then urchins might not have any effect, although urchins are known to be omnivorous and may consume some invertebrates and even smaller mussels. But if colonial invertebrates die back, urchins will move in, exploiting newly colonized seaweeds. In doing this, they will most probably favor a shift toward dominance by turfs of calcareous algae on the horizontal benches.

Kelp Forests

■ Kelp forests are dominated by a few species of brown seaweeds of the group Laminariales with fantastic growth rates.

Kelp forests are dominated by brown seaweeds of the group Laminariales and are among the most beautiful marine habitats (**Figure 17.8**). They are found throughout the world in shallow open coastal waters, and the larger forests are restricted to temperatures below 20°C, extending to both the Arctic and Antarctic Circles. A dependence on light for photosynthesis restricts them to clear shallow water, and they are rarely found much deeper than 5–15 m. The kelps have in common a capacity for some of the most remarkable growth rates in the ocean. In southern California, the giant bladder kelp *Macrocystis pyrifera* can grow 30 cm per day, and may grow 25 m, from the bottom to the surface, in 120 days. Species of *Macrocystis* are widespread and occur on the west coasts of North America, Australia, New Zealand, on both coasts of South America, and in South Africa. Somewhat smaller laminarians, such as species of *Laminaria* and other large kelps, extend the range of kelp forests to Alaska and the northwestern Atlantic. Kelp forests are, therefore, among the most widespread and productive of coastal marine

FIG. 17.8 A kelp forest in the Aleutian Islands: *Cymathere triplicata* (foreground); *Alaria fistulosa* (rear). (Courtesy of David Duggins)

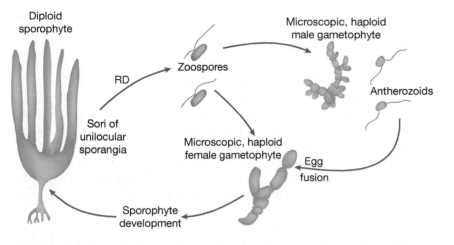

FIG. 17.9 The life cycle of a typical kelp of the genus *Laminaria*. RD = reduction division. (After Dawson, 1966)

FIG. 17.10 The Atlantic kelp *Laminaria longicruris* showing its stipe and broad frond. (Photograph by Jon Witman)

habitats. There is far more diversity of kelps in the Pacific basin than in the Atlantic.

Kelps differ widely in size and growth rate, so the "forests" actually range from beds of relatively simple single-bladed seaweeds of a few meters depth, as in the Atlantic *Laminaria* beds of Nova Scotia, to vast stands of seaweeds that extend from holdfasts on the deep (25 m and more) bottoms to stipes and blades floating at the surface (as in the *Macrocystis* and *Nereocystis* beds of the Pacific coast from California to Alaska). Kelps usually have a complex life cycle (**Figure 17.9**), which alternates between a large asexual sporophyte and a tiny filamentous gametophyte. The sporophyte is the "kelp," which consists of a **holdfast** for attachment to the bottom; a **stipe**, which looks much like a stem and is strong and flexible; and a leaflike **blade**, the main site of photosynthesis and growth. The blade tips are often eroded rapidly, but growth at the base replaces the tissue. In the smaller Atlantic *Laminaria*, the stipe is the only support for the plant, aside from water currents (**Figure 17.10**). In larger Pacific, South Atlantic, and

Indian Ocean kelps (e.g., *Nereocystis*, *Macrocystis*), the stipe is quite long, and large floating air bulbs, or pneumatocysts, support long blades, forming a **canopy** above the seabed.

■ Kelp forests are biologically diverse and support many seaweed and animal species.

The seabed beneath kelp forest canopies often supports rich communities of seaweeds and animals. To use the forest analogy, the kelp forest floor often has an understory consisting of a wide variety of seaweed species. To some degree, certain kelp forests have several distinct canopy layers, each dominated by a different kelp species. In rock crevices and in open areas, a number of grazing species are very common. In the *Laminaria* kelp beds of Nova Scotia, urchins are quite abundant, as are suspension feeders such as the blue mussel *Mytilus trossulus*. The lobster *Homarus americanus* is abundant and preys mainly on mussels, but it may also consume other mollusks and urchins. In eastern North Pacific kelp forests, grazers include a number of species of abalones, limpets, and sea urchins (**Figure 17.11**). In dense kelp forests, kelp erodes and fragments into minute particles, which support dense populations of suspension feeders such as mussels. When kelp is dense, the urchins have a particularly interesting behavior of trapping drifting fragments of kelp and other seaweed fragments on the dorsal spines and tube feet. This material is then transported to the oral surface and ingested. Urchins feeding in this way are usually rather sedentary and wait for the material to come to them. In areas with steep slopes, much of the drifting seaweed detritus may be transported to greater depths and provides food for urchins there as well.

Although hunting drove it to extinction in the eighteenth century, Steller's sea cow (*Hydrodamalis gigas*) was once a major browser on the upper kelp canopy throughout the north Pacific Rim. Another animal associated with kelp, the sea otter *Enhydra lutris*, is a major predator on sea urchins, mollusks, and fish (**Figure 17.12**). It dives to the bottom, removes a prey animal, and brings it to the surface. It often smashes the prey open with a rock and then floats on its back at the surface while eating. The otter was hunted

(a)

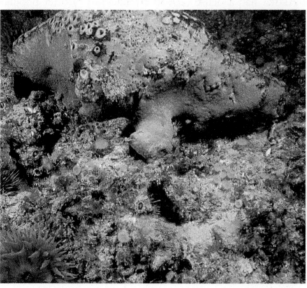

(b)

FIG. 17.11 Invertebrates in Alaskan kelp forests. (a) The sea urchins *Strongylocentrotus purpuratus* (purple), *S. franciscanus* (red), and *S. droebachiensis* (light-colored); (b) a hard bottom, showing some of the diversity of benthic invertebrates. (Photographs by David Duggins)

for fur and was reduced to just a few populations, mainly in the northern Gulf of Alaska and a remote area in central California. It was nearly driven to extinction, but was protected after 1911 and is now increasing rapidly through natural population increase and reintroductions. As we shall see, sea otters are a major top-down driving force of trophic cascades in eastern Pacific kelp communities.

Factors Affecting Kelp Communities

■ **Kelp communities are often strongly affected by a combination of storms, presence or absence of sea otters, and behavioral changes in herbivores.**

Kelps are attached to the bottom by means of sturdy holdfasts, but they project upward into very rough surface seas.

FIG. 17.12 The sea otter *Enhydra lutris* bringing a sea urchin to the surface to bash and eat. (Photograph by James A. Mattison III)

In the winter on the Pacific coast of North America, storms coming from the open sea can rip kelps from their holdfasts, and those with floats and blades at the surface are the most vulnerable. In quiet waters, the upper-canopy kelps are competitively superior because they have first access to the sunlight. However, lower-canopy kelps survive better after a storm, and it may be difficult for the larger kelps to colonize and grow from beneath the dense canopy of smaller kelps. The periodic El Niño events (see Chapter 3) that affect the eastern Pacific may thus cause a kelp forest to shift toward domination by lower-canopy species. In the extreme case, all kelps may disappear from a region, especially because El Niño also brings water that is too warm for the kelps to survive very well.

Storms can initiate changes in urchin behavior that can strongly affect the later history of a kelp forest. When the bottom is stripped of kelps, large aggregations of urchins move along the bottom, and newly recruiting seaweeds are the preferred food, because drift algae are no longer available. As a result, urchins scrape the bottom, **coralline algal turfs** take over, and **barrens**, or bare bottoms, develop (**Figure 17.13**). Harrold and Reed (1985) found that the barrens remain devoid of dense kelp until there is a good year for macroalgal recruitment. Then, some seaweeds grow sufficiently to escape urchin devastation. Once this threshold has been breached, dense kelp can develop, and drift algae become abundant once more. As a result, the urchins adopt a more sedentary behavior and cease to devastate new seaweed recruits by marauding over the rock pavement (**Figure 17.14**). This only reinforces the redevelopment of the kelp community. Trophic structure, therefore, is strongly affected by extrinsic factors, such as storms, and intrinsic factors, such as changes in sea urchin behavior. Sometimes storms can strip the bottom of the roving urchins themselves and allow the colonization of kelp. But it seems clear that there is a sudden transition in regime between kelp forests and barrens area (Filbee-Dexter and Scheibling 2014). **Figure 17.15** summarizes these interactions. A worldwide analysis of kelp forests (Filbee-Dexter and Scheibling, 2014) suggests that the urchin barrens–kelp dominance states are always discontinuous despite major differences

FIG. 17.13   An urchin barrens. (Photograph by David Duggins)

in dominant elements, such as top predators (e.g., cod and other fish in the northwest Atlantic, sea otters in the northeastern Pacific). Evidence in many regions supports the idea that each state is locally stable and requires a major change (e.g., sudden increase of urchin density or sudden loss of urchins owing to top predators or urchin overfishing—see **Hot Topics Box 17.1**) to shift from one regime (e.g., urchin barrens) to another (e.g. kelp domination).

North Pacific kelp forests can be simplified to a few main trophic levels. Although sea otters (*Enhydra lutris*) were once actively hunted to extinction in many East Pacific kelp beds, they still have large populations in some parts of Alaska, and they have recolonized some California beds. **Sea otters sit atop a strongly linked trophic cascade.** They prey on urchins and abalones, the principal grazers in the system. In turn, these species, as well as others, graze on attached seaweeds and crustose coralline algae.

Sea otters can completely change the structure of a kelp forest. In the 1970s, Amchitka Island in the Aleutian Islands of Alaska had a dense (20–30 animals km^{-2}) otter population, low urchin density, and a lush cover of kelp (Estes and

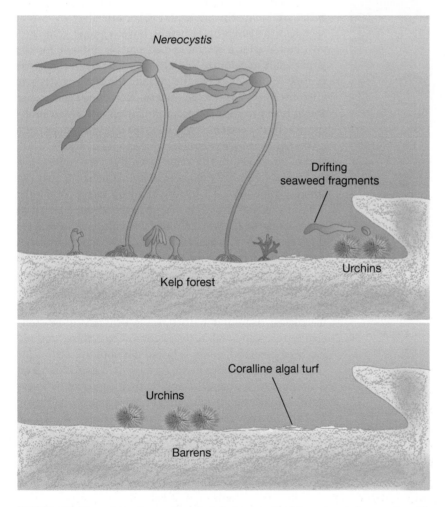

FIG. 17.14   Two alternative regimes in kelp forests. (Top) Luxuriant kelp growth, with urchins that are sedentary and trap drift algae on the dorsal surface. (Bottom) A storm has stripped the kelps, and urchins rove about the bed, denuding the bottom of all potential new recruiting kelps.

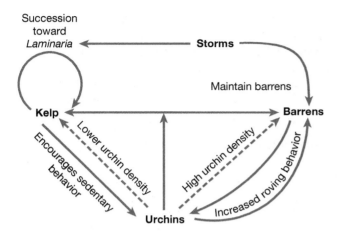

FIG. 17.15 The interactions of urchin population density, urchin behavior, and storms as they affect the character of a kelp forest.

Palmisano, 1974). By contrast, Shemya Island lacked otters, had a high density of sea urchins, and lacked macroalgae except in the lower-intertidal zone. Otters, therefore, controlled the pattern of macroalgal productivity by mediating the herbivore trophic level, which initiated a trophic cascade to the seaweeds. On a smaller spatial scale, the carnivorous sea star *Pycnopodia helianthoides* preys on patches of urchins, a practice that may allow locally dense patches of kelp to develop.

As sea otters in many sites in Alaska recovered in the 1970s and 1980s from their former low abundance, the resulting effects on kelp communities were striking. Sea urchin population densities declined by 50 percent in the Aleutian Islands. In southeast Alaska, kelp growth and increase in numbers were very rapid, but the removal of only about half of the urchins in the Aleutians resulted in a smaller increase of kelps. Why the smaller response in the Aleutians? Estes and Duggins (1995) collected urchin fragments in sea otter feces and found that the otters rarely took urchins as small as 15–20 mm in test diameter. High recruitment at the Aleutian sites almost guaranteed the presence of large numbers of such small urchins, however, which kept up the grazing rate on kelps despite abundant otters.

Sea otters in high densities promote increasing abundance of kelp. This effect is so strong that kelp productivity in the presence of otters rises from a range of 25–70 g C m^{-2} yr^{-1} to 313–900 g C m^{-2} yr^{-1}, which represents a 4.4–8.7 teragram increase in carbon storage in the shallow ocean. Such a large carbon reservoir would help ameliorate carbon release to the atmosphere and has been estimated to be valued at U.S. $205–408 million on a European Carbon exchange (Wilmers et al., 2012). Such carbon storage might also be enhanced in other trophic cascade systems that affect primary producer biomass at the bottom of the food chain.

■ Predation from an offshore source has introduced a new trophic level to some eastern Pacific kelp forests.

James Estes and his colleagues (1998) followed in the 1990s a later precipitous decline in sea otters in Alaskan coastal waters (**Figure 17.16**). Attacks by killer whales on sea otters were observed regularly. In Clam Lagoon, a bay in the Aleutian Islands, sea otters were inaccessible to cruising killer whales and their numbers were unchanged between 1993 and 1997. In nearby Kuluk Bay, otters were exposed to open-coastal killer whales, and the otters declined 76 percent over the same period. In most cases, killer whales work together in groups to hunt salmon schools. However, there have been many observations of individual killer whales that hunt marine mammals. Energetic calculations suggest that a single killer whale would have to kill 1,825 sea otters each year to maintain its metabolic requirement. Estes and colleagues estimated that 6,788 otters a year were dying in Kuluk Bay. The mortality could be explained, therefore, by 6,788/1,825, or 3.7 killer whales, maybe even fewer when energetic requirements for reproduction are factored in.

Killer whales shifted their feeding preferences to sea otters, probably because of a precipitous decline in alternative prey, pinnipeds, including harbor seals and Steller's sea lions, which began in the 1970s. Pinniped declines can be traced to declines in North Pacific fish stocks, which have also caused a precipitous decline in fish-eating seabirds. The apparent reorganization of the North Pacific ecosystem is still more difficult to explain, but an increase in sea-surface temperature over the past few decades or perhaps strong fishery pressure in the North Pacific may be at least partly responsible.

The otter decline has had predictable effects on Alaskan kelp forests (**Figure 17.16**): Sea urchin density and grazing of kelp increased over 90 percent, and kelp biomass correspondingly decreased. In this one area, the kelp forest had four trophic levels. With three levels, a top carnivore like the sea otter will cause urchins to decline and kelp to be abundant. But the addition of a predator to the top of this food chain of a fourth level should cause indirect negative effects on the otters, since a decline of kelp should cause a decline of urchins and, therefore, a further decline of sea otters, unless they can feed and thrive on an alternative food, such as fish (**Figure 17.17**).

While conservation efforts initially resulted in the increase of both sea otter population size and geographic extent in California, a decline has been noticed in the past couple of decades. Central California populations are especially in decline but the reasons are not clear. Diseases that afflict small mammals such as cats may be spreading to sea otters. Otters in California appear smaller and less healthy than before. Conservation efforts on the sea otter must therefore be greatly increased, despite initial promising changes for the better.

■ In Alaskan kelp forests, succession depends on the interplay of grazing pressure, disturbance, and competition for light.

The strong interaction between carnivorous sea otters, grazing sea urchins, and kelps is now well known. Sea otters are effective predators and reduce urchins rapidly, allowing kelps to grow and become abundant. Although

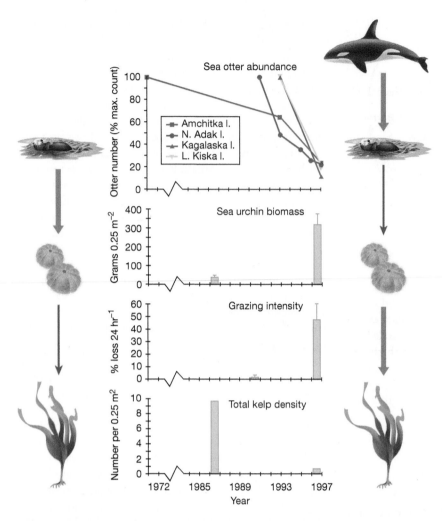

FIG. 17.16 Changes over time in abundance of sea otters, urchins, and kelp. (After Estes et al., 1998)

FIG. 17.17 A sea otter *Enhydra lutris* with greenling fish prey. (Photograph by Si Simonstad)

there are many alternative paths, consider the case in which an urchin population collapses because of otter predation or a disease affecting the urchins. Such a collapse can be simulated by an experimental removal of urchins, as performed in a southern Alaskan area rich in urchins and barren (at first) of kelps (**Figure 17.18**). After only 1 year, the experimental plots were dominated by several species of

kelps living at several levels above the bottom. An upper-canopy annual species, the bull kelp *Nereocystis luetkeana*, was abundant at first. Eventually, the lower-story perennial *Laminaria groenlandica* dominated the forest. By covering the bottom, it shaded out juvenile seaweeds of other species and prevented annual species from reinvading and growing above the lower story. In the initial stages of succession,

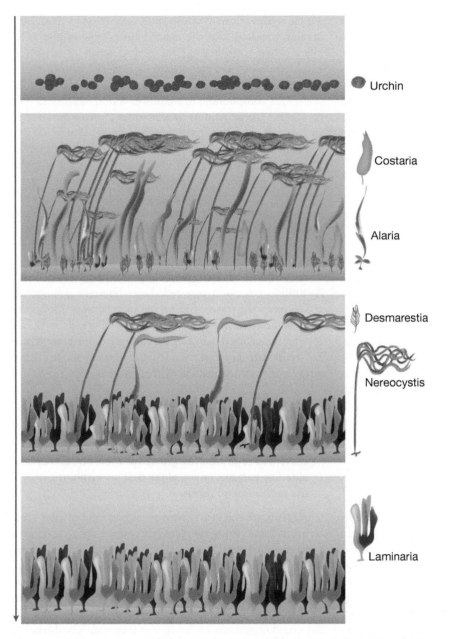

Urchin

Costaria

Alaria

Desmarestia

Nereocystis

Laminaria

FIG. 17.18 Succession in an Alaskan kelp forest. Eventually (bottom), the kelp *Laminaria groenlandica* dominated the forest and prevented taller species from reinvading by shading out juveniles. (Courtesy of David Duggins)

occupation of many layers above the bottom and rapid colonization tended to increase greatly the diversity of algal species. However, in the end *Laminaria* came to dominate and reduce kelp species diversity (Duggins, 1980).

In lower-latitude California kelp forests, a larger diversity of predators beyond sea otters exerts top-down effects.

Most of our evidence for strong top-down control of kelp forests comes from studies in Alaskan systems. In California, however, there is evidence that a number of other species are influential predators on sea urchin and snail species, which, in turn, affect kelp abundance. Predators include starfish, such as the voracious large sea star *Pycnopodia helianthoides*, and several species of crabs. While

P. helianthoides can kill a wide variety of prey, it prefers the purple sea urchin *Strongylocentrotus purpuratus* because its defense, using pedicellaria, is less effective than the movable spines of the red sea urchin *S. franciscanus* (Moitoza and Phillips, 1979). This very diversity of predators may influence grazing by providing multiple signals that tend to suppress urchin grazing behavior. Multiple predators may also increase the spectrum of effectiveness on the spectrum of grazers of kelps. Byrnes and colleagues (2006) found that carnivore diversity was negatively correlated with herbivore abundance and positively correlated with kelp biomass. They performed experiments using multiple carnivores and found that the degree of grazing was not clearly correlated with grazer abundance but that experimentally increased diversity of carnivores on herbivores tended to result in increased kelp growth. The abundance of kelp in field sites in central California increased with increasing carnivore diversity. This suggests that carnivore species diversity itself may provide a deterrent to herbivory, which constitutes a **trait-mediated indirect effect**. No single carnivore species affected all prey species, so overall carnivore diversity, using several hunting styles, was of instrumental importance in causing the top-down indirect effect on kelp growth.

■ Kelps vary widely in their susceptibility to grazing.

Because of the high density and effectiveness of grazing by sea urchins, kelps would be expected to have mechanisms of resistance to herbivory. This is not realized entirely, as some kelps, such as species of *Laminaria*, are not mechanically resistant to grazing, nor do they harbor high concentrations of compounds that would deter grazing. It may be that their extremely rapid growth reflects devotion of resources to escaping predation by a strategy of rapid colonization, growth, and reproduction. Kelps such as the enormous *Macrocystis pyrifera*, off California, and *Nereocystis luetkeana*, found from Alaska to the Pacific Northwest, are also palatable. On the other hand, once one of these kelps grows beyond a threshold size, the large stipe and float lifts the more vulnerable fronds from the seafloor, making them nearly inaccessible from urchin grazing.

Chemical deterrents to grazing are employed by some kelps. We expect kelps and other seaweeds in kelp habitats that are less erect to be better defended. Species of the seaweed genus, *Desmarestia*, found in Pacific kelp forests, synthesize sulfuric acid, which can erode the Aristotle's lantern of sea urchins and discourage grazing. In Chilean kelp forests, *Macrocystis* plants cannot recruit and grow unless they settle within an area enclosed by *Desmarestia* plants.

Other kelps are capable of producing high concentrations of compounds that discourage grazing. The Pacific kelp *Agarum fimbriatum* has high concentrations of polyphenolics (believed to be the marine equivalent of terrestrial plant tannins), which discourage urchins from chewing on the fronds. Bryozoans nevertheless seem to be able to settle and even prefer to live on this seaweed, and the coverage further defends the kelp against grazing. Stands of *Agarum* can be found in the Pacific Northwest, usually in deeper waters than other kelps, and probably are the result of their chemical defense, leaving *Agarum* to dominate locally. Roving bands of urchins tend to avoid stands of *Agarum*.

Some interesting depth and geographic patterns of chemical defense by kelps may be explained by the presence or absence of important elements of the sea otter–urchin–kelp trophic cascade we discussed earlier. Estes and Steinberg (1988) found that deeper-occurring kelp species in the northeastern Pacific tended to be better chemically defended than shallower species. This might be related to the historically great abundances of sea otters, which tends to keep urchin populations at low levels in shallow water,

which in turn allows shallow-water kelps to grow with little grazing pressure. On the other hand, deeper-water kelps like *Agarum* were beyond the diving access of sea otters and, therefore, there was strong natural selection to evolve antigrazing mechanisms such as high concentrations of poisonous compounds. In shallower water, *Laminaria* outcompetes *Agarum*, but in deeper water, *Agarum* survives by virtue of its resistance to urchin grazing.

A similar argument can be made on a larger geographic scale. Australasian kelp communities simply lack a major top carnivore like sea otters. As a consequence, kelps are under much more grazing pressure than in the shallow kelp forests of the northeastern Pacific. Northeastern Pacific kelp forests without otters have very high rates of herbivory because the dominant shallow-water kelps have not evolved strong chemical defenses. But Australian kelps do not suffer such strong herbivory (Steinberg et al., 1995). Australian kelps and seaweeds have deterrent compounds at 5–6 times the concentration found in comparable northeast Pacific species. Herbivores in Australia, moreover, had apparently evolved some resistance to these compounds and tended to graze seaweeds with them at much higher rates than northeastern Pacific herbivores.

Coral Reefs

■ Coral reefs are constructional wave-resistant features that are built by a variety of species and are often cemented together. The growth of these structures is aided by zooxanthellae, algae that are symbiotic with the reef-building corals.

Coral reefs are the most diverse and beautiful of all marine habitats. Large wave-resistant structures have accumulated from the slow growth of corals. The development of these structures is aided by algae, known as **zooxanthellae**, that are symbiotic with reef-building corals. Coralline algae, sponges, and other organisms, combined with a number of **cementation** processes, also contribute to reef growth. The dominant organisms are known as **framework builders** because they provide the matrix for the growing reef. Corals and coralline algae precipitate calcium carbonate, whereas the framework-building sponges may also precipitate silica. Most of these organisms are colonial, and the slow process of precipitation moves the living surface layer of the reef upward and seaward.

Coral reefs are usually topographically complex (**Figure 17.19**). Much like a rain forest, a reef has many strata, as well as areas of strong shade due to the overtowering coral colonies. Because of the topographic complexity, thousands of species of fish and invertebrates live in association with reefs, which are by far our richest marine habitats. In Caribbean reefs, for example, several hundred species of colonial invertebrates can be found living on the undersides of platy corals. It is not unusual for a Pacific reef to have several hundred species of snails, 60 species of corals, and several hundred species of fish. Of all ocean habitats, reefs seem to have the greatest development of complex biological associations.

FIG. 17.19   To the left is the branching elkhorn coral *Acropora palmata*; in the right foreground is the massive coral *Montastrea annularis*. The photograph was taken on the north coast of Jamaica before a combination of hurricanes and loss of a major grazer in the early 1980s resulted in a phase shift toward a bottom dominated by coral rubble and macroalgae. (Photograph by Philip Dustan)

■ **Reef-building corals belong to the calcium carbonate secreting Scleractinia. Hermatypic corals contribute most to reef growth and have abundant endosymbiotic zooxanthellae.**

While many members of the phylum Cnidaria (see Chapter 14) occur on coral reefs, reef building is due mainly to some members of the calcium carbonate secreting cnidarian order **Scleractinia**. While some species are solitary and consist of

FIG. 17.20   Close-up of polyps of a scleractinian coral. (Photograph by Robert Richmond)

a single polyp and skeleton, most are colonial and consist of hundreds to thousands of polyps (**Figure 17.20**) that are interconnected by living tissue. In all cases, the polyps collect zooplankton by means of **nematocysts** on their tentacles and digest their small prey in a blind gut (**Figure 17.21**). As we will discuss shortly, the symbiotic zooxanthellae contribute to nutrition. **Hermatypic corals** have large numbers of zooxanthellae and are the important contributors to reef growth.

Reef-building corals can be divided into **massive** and **branching forms**, although there are intermediates between these two types. Massive corals are mound shaped and often irregular. They tend to grow slowly, not usually increasing in any linear dimension much more than 1 cm per year. Branching coral colonies usually (**Figure 17.19**) tend to grow rapidly, on the order of 10 cm y^{-1}. The more rapid linear growth sometimes allows branching corals to spread rapidly on the reef. Storms sometimes break up branching corals, but the fragments may be able to start new colonies.

The Crucial Role of the Coral-Zooxanthellae Symbiosis

■ **Zooxanthellae are symbiotic with many invertebrates, and they are crucial as a source of nutrition for reef-building corals.**

The remarkable mutualism between hermatypic corals and the photosynthetic **zooxanthellae** is the driving force behind the growth of coral reefs. Because of the symbiosis with zooxanthellae, hermatypic corals respond to the environment

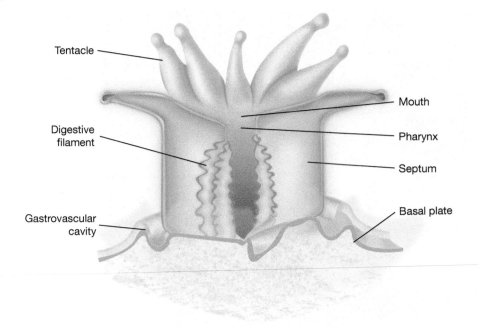

Tentacle

Mouth

Digestive filament

Pharynx

Septum

Gastrovascular cavity

Basal plate

FIG. 17.21   Diagram of a polyp of a scleractinian coral.

in many ways that are reminiscent of plants. As a result, the growth of corals associated with zooxanthellae is strongly light dependent, as is the overall growth of reefs. Coral reefs are among the most productive marine benthic habitats, mainly owing to the photosynthetic activities of zooxanthellae.

Zooxanthellae are specialized single-celled algae, and are dinoflagellates that live intracellularly within membrane-lined vacuoles in the endodermal tissues of scleractinian corals, concentrated in the tentacles. They are coccoid (oval shaped) and lack the typical biflagellated form of free-living dinoflagellates. The zooxanthellae-coral interaction is so interactive that the whole system has traits that are subject to evolution and probably strong disruption under stress. Such an interactive unit between a microorganism and a host is known as a **holobiont**.

Zooxanthellae are also found in a wide variety of other marine species, including foraminifera, sponges, sea anemones, jellyfish, octocorals, flatworms, and the so-called killer clams of the genus *Tridacna*. They are classi-fied within the genus *Symbiodinium*, but species concepts and identifications are very unclear. Molecular sequences have been used to identify at least seven separate evolu-tionary lines, or **clades**, of zooxanthellae. Corals typically contain only four of these clades, and a given coral may have members of more than one clade in its tissues. All four clades are found in the Caribbean, but only two are found in the Indo-Pacific. There is some evidence that in-dividual clades (this is a very loose term when applied to what we know about zooxanthellae) have different light and thermal adaptations, and this may explain the differ-ential occurrence of the four clades in deep- and shallow-water Caribbean coral species or in populations of the same coral species at different depths. There does not seem to be any evidence for correlated speciation in coral hosts and zooxanthellae.

Zooxanthellae also live free outside of corals, but their outside distribution is not well understood at all. But stud-ies of free-living forms show that clades found within corals are also found free living in the substratum surrounding corals (Granados-Cifuentes et al., 2015). When corals reproduce asexually by fragmentation, the polyps inherit zooxanthellae. But eggs produced in sexual reproduction often do not have zooxanthellae, so presumably they must be picked up from the environment. There is some evidence for attractions of zooxanthellae in the water to corals, and it is also possible that corals take up zooxanthellae by feeding. Zooxanthellae have been found in ciliates associated with diseased coral. In the case of **coral bleaching**, when zoo-xanthellae are lost from corals, the recolonization process must be influenced by the pool of zooxanthellae that are on the sea bed.

When stressed, corals appear to expel their zooxanthel-lae, a phenomenon known as **bleaching (Figure 17.22)**, which has been observed when corals are stressed by high temperatures and stricken by disease. Under normal cir-cumstances, corals appear to expel zooxanthellae that are in poor condition. But when corals are subjected to stress, zooxanthellae in good condition also apparently depart or are expelled by the coral. Little is known about the exact condition of zooxanthellae during expulsion, but a study by Strychar and colleagues (2004) showed that the cells were either dead or programmed to die as they were expelled. It is possible that a bleaching coral might be able to reab-sorb other zooxanthellae, but the most probable explana-tion of bleaching is a stress reaction when zooxanthellae begin to die. Temperature stress apparently causes increases in oxidizing molecules, which causes a breakdown of cellu-lar communication between host and endosymbiont, lead-ing to expulsion of the zooxanthellae (Vidal-Dupiol et al., 2009). After expulsion there is evidence that the immune

FIG. 17.22 (a) A healthy colony of the Caribbean elkhorn coral *Acropora palmata*. (b) A bleached colony of the same species. (Photographs courtesy of James W. Porter)

system of the coral is suppressed for some time (Pinzon et al., 2015), which might make it prone to disease. This area is still wide open for further research.

■ **Zooxanthellae provide nutrition and aid in calcification of their coral hosts. The benefit obtained by zooxanthellae may relate to protection from grazing and access to nutrients from coral excretion.**

Zooxanthellae probably benefit from the coral association by obtaining protection from grazing and by access to nutrients derived from coral excretion. The benefits of zooxanthellae to corals have been the subject of great controversy over the years. The possible benefits of zooxanthellae to hermatypic corals include the following: (1) Zooxanthellae aid in the removal of dissolved excretory products, (2) they provide oxygen via photosynthesis, (3) they manufacture carbohydrates and essential amino acids that can be used for coral nutrition, (4) they enhance coral calcification, and (5) they aid in the synthesis of lipids. The first two hypotheses relating to waste removal and oxygen supply are unlikely to sustain the mutualism, since hermatypic corals typically live in wave-washed environments, where moving waters carry away excretory products and bring ample dissolved oxygen. We therefore focus on the remaining factors.

Corals commonly live in tropical oceanic water of low nutrient concentrations, so ultimately the advantage of zooxanthellae could be the gathering of nutrients into a localized space and provision of various products to the coral. Zooxanthellae take up nitrogen and reduce the overall concentration in coral tissues, creating a concentration gradient that stimulates the further uptake of nitrogen, either actively or passively by diffusion. When enriched with dissolved nitrogen, zooxanthellae respond by increasing in abundance and nitrogen content, and the use of radiotracers demonstrates that dissolved ammonium appears to find its way directly to zooxanthellae (Swanson and Hoegh-Guldberg, 1998).

Zooxanthellae are photosynthetic, and the carbohydrates they manufacture are transferred to the corals. This can be shown by labeling bicarbonate with radioactive ^{14}C, which is taken up by the zooxanthellae during photosynthesis. Eventually, a considerable amount of this carbon can be detected in the animal tissues by a technique known as autoradiography. Slices of corals are placed on radiation-sensitive film, and coral tissues that take up radiolabeled carbon from the zooxanthellae release radiation, which reacts with the film. Studies on zooxanthellae productivity show an excess of primary production over zooxanthellae respiratory needs in shallow-water corals. Corals, therefore, benefit from the association by getting carbohydrate. Zooxanthellae also synthesize amino acids that are used by the coral.

All hermatypic coral species depend on the zooxanthellae for food, but they also capture small animals by use of their tentacles and nematocysts (**Figure 17.20**). In the Pacific coral *Stylophora pistillata*, the carbon received from the symbiotic zooxanthellae must be supplemented by nitrogen, which probably comes from feeding on zooplankton. The nitrogen is essential for amino acids, the building blocks of proteins. Corals capture microzooplankton from moving water, but their efficiency of capture declines with high current speeds. The partial dependence on zooxanthellae has also been demonstrated for species of the bivalve *Tridacna*. The dependence of these clams on the symbiotic zooxanthellae is supplemented by suspension feeding upon phytoplankton. The contribution of suspension feeding to overall nutrition, however, appears to decline in larger-sized species of *Tridacna*. There are groups of soft corals, such as members of the family Xeniidae, that have lost the ability to feed and appear to depend entirely on zooxanthellae.

The presence of zooxanthellae enhances calcification in hermatypic corals. By using a radioactive isotope of calcium, ^{45}Ca, Thomas Goreau and Nora Goreau (1959) demonstrated that calcium uptake is greater when hermatypic corals are exposed to light. Shade or photosynthetic inhibitors reduce the rate of skeletal growth. By removal of carbon dioxide, zooxanthellae may shift the carbonate–bicarbonate–carbon dioxide interactions toward conditions favorable for calcium carbonate secretion. Zooxanthellae may also enhance calcification by removing phosphate, which inhibits calcification, or by aiding in the secretion of the organic matrix upon which the calcium carbonate is deposited.

It is also believed that zooxanthellae influence lipid production in corals. Lipids constitute about a third of the dry weight of corals and serve both as structural support and perhaps as a source of energy. Lipogenesis is strongly accelerated in light, which is likely due to the uptake of acetate by zooxanthellae, followed by the synthesis of lipids.

Factors Limiting Reef Growth

■ Reef development is limited by the presence of relatively high temperature, open marine salinity, available light, and low turbidity.

Coral reefs are confined to tropical and subtropical waters, generally between 25° N and 25° S. The high calcification rates required for vigorous coral growth are limited to warm waters. In general, coral reefs are found in open marine salinities. Reefs are rarely found in tropical estuaries.

Next to temperature, light is probably the most important limiting factor to well-developed coral reefs because of the symbiosis between reef-building corals and zooxanthellae. Because light intensity decreases exponentially with increasing depth, active reef building ceases below depths of 25–50 m, at least as far as coral growth is concerned. Deeper than this, cementation and growth of sponges permit some reef accretion. Basically, however, coral reef growth is a light-limited process.

Although light is essential for the growth of hermatypic corals, the ultraviolet part of the spectrum is potentially dangerous, especially in clear tropical ocean water, where light penetrates to great depths. Because of the dependence of the zooxanthellae symbiosis on light, reef corals thrive in shallow well-lit waters where ultraviolet radiation is intense. A variety of UV-absorbing materials have been discovered in corals. When corals are exposed to photosynthetically active radiation combined with UV, corals produce more of the UV-absorbing materials. Some corals have conspicuously dark pigments that also are believed to absorb ultraviolet radiation. Especially effective compounds are **mycosporine-like amino acids**, found in several species, which appear to protect the zooxanthellae from ultraviolet light damage. These compounds have been synthesized in the laboratory and have been patented for use as a human sunscreen.

Turbidity (number of particles per unit water volume) and **sedimentation** both have adverse effects on reef-building corals. Turbid waters intercept light and reduce photosynthesis. Sedimentation tends to smother coral colonies and inhibit feeding and the extension of the polyps' crowns of tentacles. Blankets of sediment also may encourage the growth of disease-causing bacteria. Corals with relatively flat surfaces (e.g., brain corals) tend to produce large amounts of mucus, which traps the sedimented particles and transports the material off the colony. Erect corals intercept less material and usually have a lower capacity to produce mucus. The overall effect of turbidity and sedimentation reduces the development of corals. For example, the coast of eastern Venezuela is very turbid owing to river input and coastal sediment, and coral development there is quite poor. Coral growth is better developed in the clearer coastal waters of western Venezuela.

■ Coral reefs live exposed to high wave energy, but strong waves can break coral colonies and limit reef growth.

Because coral reefs require clear water and are constructional topographic features, they tend to be located in areas of high wave energy. Moving water brings nutrients and zooplankton to the corals, and is therefore beneficial. The exposure to waves also has its disadvantages. Erect branching forms, such as the Caribbean elkhorn coral *Acropora palmata* (**Figure 17.19**), are often greater than 2 m across and live in reef crest zones that must withstand severe wave shock. Storms, such as cyclones in the Pacific and hurricanes in the Atlantic, often topple coral colonies and may exert massive destruction of coral reefs.

■ Reef development is a balance between growth and bioerosion.

As was just mentioned, reef organisms precipitate calcium carbonate, which enables reef growth. In Jamaica, the growth rate was estimated to be about 1 cm per thousand years, both upward and seaward. Cementation is the chemical process that solidifies the reef by precipitation of calcium carbonate in crevices and cracks.

Coral reefs are continually under attack by **bioerosion**. Many species of invertebrates bore into coral skeletons. Parrotfish and surgeonfish rasp away at coral surfaces, and urchins enlarge crevices and bore into colony bases. A group of sponges (Family Clionidae) penetrate coral rock by chemical attack. Although destructive in the long run, the erosion provides living and hiding crevices for many reef species. The collective effect of bioerosion of reefs produces tons of fine-grained sediments, which may be transported from the reef to deeper water, sometimes to depths of 1,000 m or more.

■ Coral reefs can be divided into atolls and coastal reefs.

Coral reefs often have complex histories, but they can be divided into **atolls** and **coastal reefs**. Atolls are horseshoe- or ring-shaped island chains that cap an oceanic island of volcanic origin. You can think of an atoll as resembling a truncated cone capped by a necklace of islands. They are open-ocean structures, not usually found near a continental coast. Although most are in the Pacific, a few are in the Caribbean. Coastal reefs are elongate structures that border a continental coast. They may be enormous structures, like the 2,000-km-long Great Barrier Reef off the east coast of Australia. These reefs have an overall architecture in the form of a **platform reef** with zonation that is determined by thousands of years of reef growth, producing a distinct

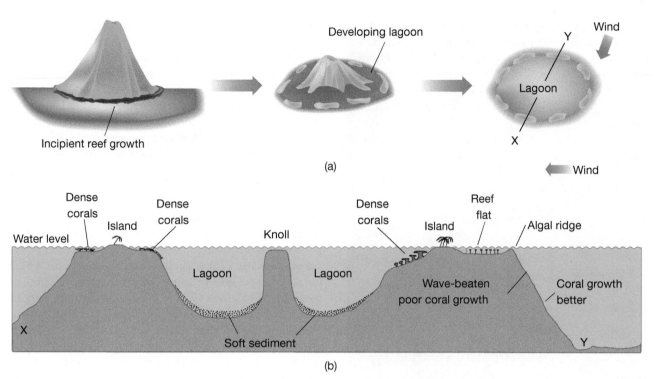

FIG. 17.23   (a) Darwin's theory of the origin of atolls. (b) Cross section showing major atoll subhabitats.

topography, as is found in many sites in the Caribbean (see upcoming description for Jamaica) and the Indo-Pacific. But corals may also just comprise a **fringing reef**, which is a relatively thin layer of corals on a subtidal coast such as an oceanic volcanic island like Bonaire in the southern Caribbean basin or on the coast of Israel and Egypt, in the Red Sea. Small growths of corals and associated calcifying organisms often growing on sand with the area of approximately 10–100 square meters are known as **patch reefs.**

Atolls have a unique origin, which we discussed briefly in Chapter 1. Charles Darwin reasoned that atolls developed when coral grew upward from the top of a sinking volcano on the seafloor. At first, a volcano grows and reaches the surface. Coral then grows around the fringe of the volcanic island. As the volcanic island submerges, coral reefs continue to develop and grow upward. This forms the island ring, leaving a lagoon in the center (Figure 17.23a). After many millions of years of upward growth, there should be a great thickness of coral rock capping the sunken volcano. In the 1950s, Harry Ladd and colleagues drilled a borehole in Enewetak Atoll and had to penetrate 1,400 m of reef rock before they struck volcanic rock. The reef dates back to Eocene times, or 40–60 million years ago. Thus, Darwin was correct about the origin of these reefs. He erred, however, in believing that the crust beneath all coral reefs is subsiding; this is not true for many coastal reefs.

Atolls usually reside in a stable wind system, and **Figure 17.23b** shows a typical cross section. The side facing the wind, the **windward side**, is usually strongly affected by wave action. Corals facing the sea do not grow very well, and a red algal ridge usually accumulates on the seaward side of a broad reef flat. The ridge is caused by algal precipitation of calcium carbonate and trapping of sediment

by coralline red algae. The ridge protects a broad reef flat, which has abundant small colonies of corals. Large coral colonies grow somewhat better at depth, but coral growth is much lusher on the **leeward side** of the open-ocean part of the reef and within the lagoons.

Coastal reefs parallel shorelines and have diverse origins. Part of the Great Barrier Reef develops by means of the growth of corals on subsiding rock, but this is not universal. Because of the combination of sea-level rise and fall correlated with Pleistocene glacial retreats and advances, the topography of coastal reefs is often the net result of the interaction upward and seaward growth of coral as sea level rises, and erosion, as sea level falls. The erosion often creates hollows and caverns, through which seawater may gush as the waves beat on shore.

■ **Both atolls and coastal reefs have prominent depth zones, each of which has a different set of dominant framework-building coral species.**

Depth zonation is a major feature of coral reefs, whether they are atolls or coastal reefs. If you swim from shallow to deep reefs, you will cross a succession of distinct bottoms, or zones, each dominated by a different species or group of coral species. Although the exact explanation for dominance by certain species is often unknown, the following factors must be important: (1) wave and current strength, (2) light, and (3) suspended sediment. Competition among species often leads to dominance by single competitively superior species. Because of the strong dependence of reef-building corals on light and space for attachment, one species may displace another by overgrowth. **Figure 17.24** shows a mound-building coral being overgrown by a branching coral. On Pacific reefs, species of the

FIG. 17.24 The shallow-water mound-building coral *Montastrea annularis* being overtopped by the elkhorn coral *Acropora palmata*. (Photograph courtesy of James W. Porter)

genus *Pocillopora* are superior at overgrowing other species and come to dominate shallow reef zones (**Figure 17.25**).

Coral reefs usually protect an inner **lagoon** (**Figure 17.26**). The lagoon has a soft-sediment bottom and is dominated by sea grasses, urchins, sea cucumbers, and sparse corals and sea whips. As one moves seaward in Caribbean reefs, different corals dominate, resulting in a series of zones. The lagoon is protected by an intertidal reef flat, which is dominated by small corals and colonial anemones. Seaward is the reef crest, which is dominated by large colonies of the elkhorn coral, *Acropora palmata*. In strong unidirectional currents, the branches grow to point into the current, to reduce tensile stress.

At slightly greater depths of Caribbean coastal reefs is a **buttress zone**, dominated by the massive coral *Montastrea annularis*. These corals form large mounds, which form **spur and groove topography**, or an alternation of mounds and channels (**Figure 17.26**). The channels may be grown over by coral, forming long caves. Below the buttress zone, a broad low-relief bottom is covered with thickets of the staghorn coral, *Acropora cervicornis*. Because of recent outbreaks of disease, much of the shallow reef zonation, especially of staghorn and elkhorn corals, has been strongly disrupted in the Caribbean.

Deeper than this zone, corals are much more sparse, although there is a zone of platy corals of a species that is closely related to the shallow-water *Montastrea annularis*. This coral looks like a tree fungus, and its flat aspect may serve to capture whatever little light exists at these depths. Seaward of this zone in Jamaica, the reef drops

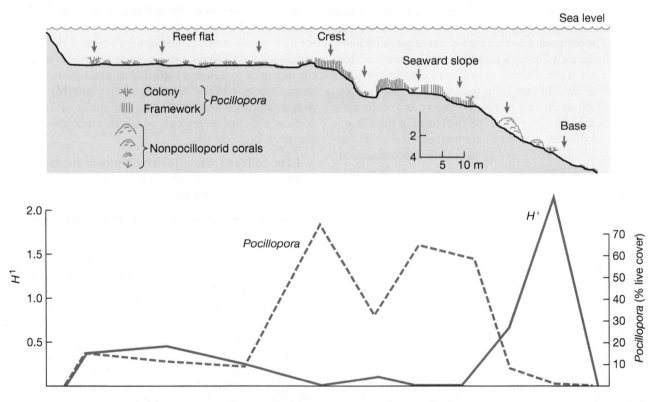

FIG. 17.25 Percentage cover of the coral *Pocillopora damicornis* on a coral reef on the Pacific coast of Panama versus diversity of all corals. Diversity, *H*, increases with increasing number of species and with increasing evenness of distribution of numbers among the species. Hurricanes have strongly altered these reefs in recent years. (After Glynn, 1976)

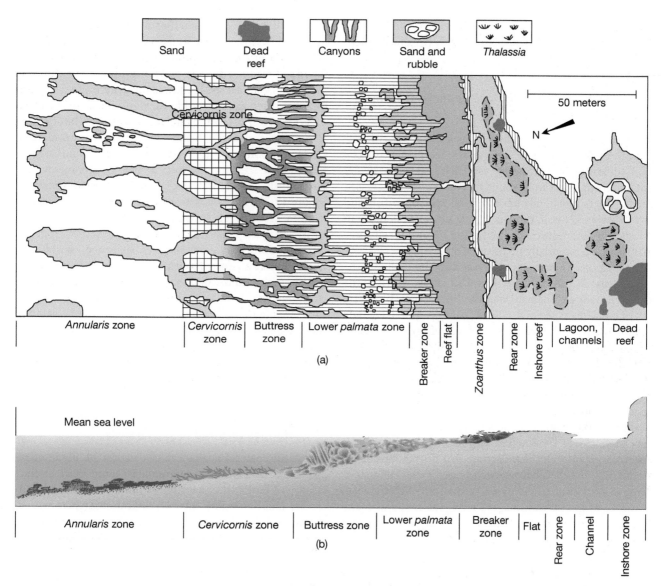

FIG. 17.26 (a) Aerial map view of coral reef environments of the north coast of Jamaica before an ecological overturn in the late 1980s. (After Goreau and Land, 1974) (b) Cross-sectional view of depth zonation of the coral reef at Discovery Bay, Jamaica. (After Goreau, 1959)

nearly vertically, and only sponges and the cementation of downward-sliding sediment contribute to reef growth. In other areas of the Caribbean, corals grow in great density on near-vertical deeper walls, usually down to depths of ca. 30 m.

Deep reefs in both the Atlantic and Pacific are generally poorly known, because access is limited to submersibles and highly specialized diving techniques. It has been suggested that deeper reefs, while living in very low light, might be a source of larvae and population replenishment for shallow reefs, many of which are endangered. This idea is supported by finding shallow-water coral species at surprisingly great depths, surpassing 40 m. Recent genetic evidence suggests that depending on the species, deep populations may or may not be highly genetically divergent from shallow-water conspecifics, which suggests that deeper-reef populations have species-specific potential to replenish shallow reef populations (Bongaerts et al. 2017).

Coral Reef Distribution and Biodiversity

GEOGRAPHIC DISTRIBUTION OF REEFS AND TYPES OF REEF

■ **There are two distinct biogeographic realms, Pacific and Atlantic.**

As reef corals and their associated faunas evolved, most of the world's tropical oceans were interconnected. The Atlantic and Pacific were connected by a broad seaway known as Tethys, or the Tethyan Sea, which dried up toward the end of the Miocene epoch, roughly 10 million years ago. Even at the beginning of the Miocene, divergence began between the Pacific and Atlantic, but by the end of the Miocene, faunas in the Atlantic had diverged from those of the Pacific, although there still was a connection through what is now Panama, because the Isthmus of Panama did not arise until 3 million years ago. Before the Isthmus of Panama was uplifted, many groups with Pacific affinities lived in the Caribbean Sea. After the rise of the isthmus, the Pacific forms in the Caribbean largely became extinct.

CORAL DIVERSITY

■ Coral reefs are diverse, but Indo-Pacific reefs are more so than in the Caribbean.

In general, the number of living species is about twice as great in Pacific reefs as in the Atlantic reefs, with a maximum diversity in the southwest Pacific (**Figure 17.27**) (more information can be found in Veron, 1995). Increased Pacific diversity extends beyond coral species to fish and invertebrates. Overall, there is a great deal of difference in the species composition of Pacific and Atlantic reefs, now that there are essentially no connections.

The region of maximum coral reef diversity in the southwest Pacific is known as the **Coral Triangle**, which is the area of Indonesia, Thailand, and the Philippines. Biodiversity of coral-reef species drops off from the Coral Triangle with increasing latitude north and south of the equator, and diversity also drops off east and west of the Coral Triangle, such that Hawaii, Panama, and the Red Sea are outposts of relatively low diversity. The southwest Pacific has the least environmental variation and has had the most continuous occupation of coral reefs over geological history. Regional climatic disturbance was also relatively slight, which may have reduced extinction within the Coral Triangle. Paleogeographic maps demonstrate that fluctuations of sea level during the Pleistocene epoch caused major changes in the distribution of land and sea in the Indo–West Pacific. At times of glacial maxima, sea level was low, and many areas between Australia and mainland Southeast Asia were mostly dry land. At times of high sea level, such as the present, most of this region is under water. Times of low sea level might have isolated coral populations and contributed to speciation (the interested student should consult Potts, 1983, and Veron, 1995).

Fossil evidence suggests that a series of major environmental shifts over the past few million years has caused extinctions in the Caribbean. There is no strong evidence for strong regional diversity gradients of species richness in the Caribbean, except for lower diversity in colder fringe locations, such as the Florida Keys and Bermuda.

We have the most quantitative diversity information about corals and the fishes associated with coral reefs. Coral reef fishes are quite diverse but are dominated by species that are strongly tied to reef environments. The surgeonfishes (Acanthuridae) and parrotfishes (Scaridae, **Figure 17.28**) feed on seaweed material on the hard substratum of the reef. Butterfly fish mostly feed on coral polyps (**Figure 17.29**). Many others seek shelter in the crevices afforded by coral colonies. Bellwood and Hughes (2001) found a strong correspondence between the diversity of corals and of coral reef fishes. They found the same latitudinal and longitudinal trends between fishes and the corals we mentioned earlier, but also found that habitat area was the major explanatory factor of fish diversity.

CORAL SPECIES: DISTINGUISHING GENETIC DIFFERENCES FROM PLASTICITY OF FORM

■ Coral reefs contain a large number of closely related species with strong phenotypic plasticity and require careful molecular and ecological study.

Before any careful ecological studies can be made of responses of corals to the environment, competitive interactions, and biological diversity, we must be able to identify species and determine their ecological differences. Corals present a difficult problem. We know there are many species, but growth forms of a single coral species vary in response to water energy, light, and competitors, usually on the same reef. In the Caribbean, the elkhorn coral *Acropora palmata* may occur in extremely different forms even over a short distance (**Figure 17.30**). In the past few years, a

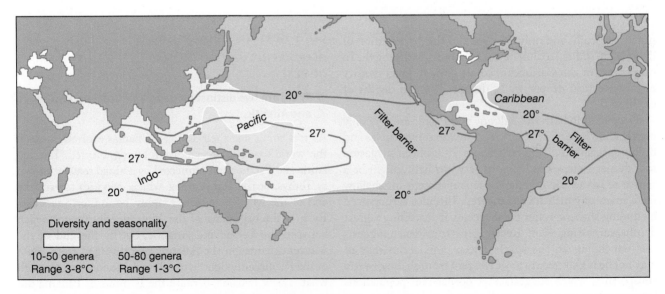

FIG. 17.27 Approximate limits of the tropical Indo-Pacific and Atlantic coral reef provinces compared with diversity at the generic level and minimum average sea temperatures. (After Newell, 1971)

FIG. 17.28   A queen parrotfish *Scarus vetula* in Bonaire, Netherlands Antilles (Caribbean), which feeds on algae attached to hard calcareous surfaces, showing fused teeth plates. (Photograph by Jeffrey Levinton)

FIG. 17.29   The foureye butterfly fish *Chaetodon capistratus* on the reef at Bonaire. It preys on coral polyps. (Photograph by Jeffrey Levinton)

toolkit of molecular and ecological components has been developed to distinguish between species differences and ecological plasticity, but we have not completed the task.

Marine species often have broad geographic ranges, perhaps because they experience stronger gene flow between populations than one sees in terrestrial environments. We have learned that planktonic larvae don't necessarily all move as far as their larval lives might predict in a linear current, but it does not take many exchanges of genes to prevent species from diverging over broad geographic expanses.

This does not mean, however, that the ocean has few species that are easy to identify. **Sibling species**, closely related species that are very similar in appearance, have been discovered in groups that are entirely conspicuous and with considerable dispersal potential. It is crucial to be able to distinguish among such species, since they may have different fixed specializations.

The problem of **distinguishing genetic differences from ecological plasticity** has been studied in recent years with a new set of tools. DNA sequences are being used

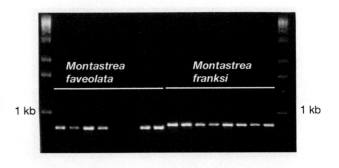

FIG. 17.31 A gel of nuclear DNA fragments that shows a distinct difference between two species of the Caribbean coral genus *Montastrea*. (Photograph provided by J. V. Lopez)

(a)

(b)

(c)

FIG. 17.30 Phenotypic plasticity in the Caribbean elkhorn coral *Acropora palmata*, ranging from (a) bidirectional flow environments to (c) more multidirectional flow. (Photographs by Phil Dustan [a] and Jeffrey Levinton [b and c])

to distinguish among species with extensive morphological overlap and plasticity. Careful morphological analysis and DNA research shows that there are three distinct species of the mound-building coral genus *Montastrea*: *Montastrea annularis*, *M. franksi*, and *M. faveolata* (Knowlton et al., 1997; Levitan et al., 2004). The three species can be distinguished by a combination of DNA sequencing (**Figure 17.31**) and by examining cross sections of the polyp skeletons (**Figure 17.32**).

It was particularly interesting that these species were distinct because they all participated in a mass spawning, Caribbean style. The team spent several nights collecting gametes at frequent intervals, studying fertilization,

gamete abundance, and gamete viability. The reproductive separation between the three species turned out to be complex. *M. faveolata*, the most molecularly and morphologically distinct species, hybridizes poorly with the other two, *M. annularis* and *M. franksi*, which hybridize rather easily with each other. *Montastrea annularis* and *M. faveolata* spawn over approximately the same hours in the late evening but have incompatible gametes. *M. franksi* spawns about 2 hours ahead. If you perform a cross in the lab, it is very compatible with *M. annularis*, but in the field, it has evolved a nonoverlapping spawning time. By the time *M. annularis* spawns, the *M. franksi* sperm are nearly all inviable. Overall differences in spawning time, sperm aging, dilution of gametes, and gametic incompatibility act in various combinations among the three species, making it unlikely that hybrid fertilization would occur. This story has even more complications, and the student is encouraged to read Fukami and others (2004) to see the role of geographic variation.

A fascinating area of research is how the species arose in the first place. One possibility is that a given species arose by being isolated for a time in a newly established reef system, but later the new species dispersed and commingled with the others. It may also be possible that a new species may be able to arise within the range of its closest sibling species. This would be extremely difficult because of the mass-spawning phenomenon. Perhaps a shift in timing of spawning in the evening occurs, as we have seen above. But what would drive this shift? For the foregoing example, it makes more sense that the shift was driven after the species formed, to avoid hybridization and production of hybrids with low fitness.

To make things even more complicated, hybridization *is* known to occur between closely related coral species in the wild. For example, Caribbean divers quickly learn how to distinguish the elkhorn coral, *Acropora palmata*, from the staghorn, *A. cervicornis*. The former can make large antler-shaped colonies, whereas the latter is often seen in the form of thicket-like growths. Vollmer and Palumbi (2002) began to notice a distinct intermediate morph between staghorn and elkhorn. Using DNA sequencing, they showed that the distinct morphs were F1 (first generation) hybrids, and the morphology was different depending on which species was

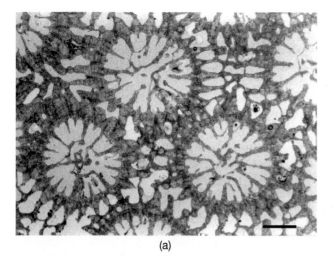

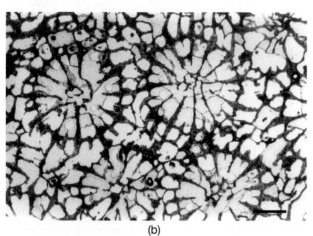

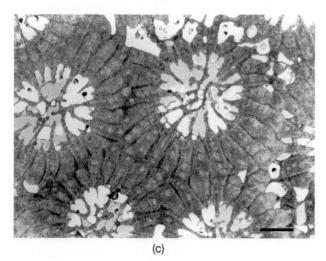

FIG. 17.32 Cross sections of the polyp skeletons of three sibling species of the *Montastrea annularis* complex on the Caribbean. (a) *Montastrea annularis*. (b) *Montastrea faveolata*. (c) *Montastrea franksi*. (Photographs by Nancy Budd)

the source of sperm (**Figure 17.33**). Without the genetic information, the morphs could have been mistaken for distinct species. In acroporid coral species, hybrids are often found in new and marginal habitats that are not suitable for either parent species, which suggests that new species of acroporid corals can evolve by hybridization and settlement in novel marginal habitats (Willis et al., 2006).

FIG. 17.33 Morphological variation and hybrids. Here we see the elkhorn coral *Acropora palmata* (right side), the staghorn coral *Acropora cervicornis* (center), and a hybrid morphology on the left. (Photograph by Hector Ruiz)

Mutualisms in Coral Reefs

■ **Coral reefs and nearby environments harbor some of the most remarkable mutualisms in the sea.**

Coral reefs are fantastically diverse, and many interdependencies have evolved between species. Many such dependencies begin with the protection that the crevices of a large sessile species can give to smaller mobile animals. In many sponges, the crevices are inhabited by a large number of species. In some cases, such commensal relationships have apparently evolved into mutualisms. For example, the Pacific branching coral *Pocillopora damicornis* harbors a group of species of crabs, shrimp, and fishes. One of the coral's principal predators is the crown-of-thorns sea star, *Acanthaster planci*. When the sea star mounts the coral, shrimps and crabs emerge from crevices within the coral colony and actively attack the sea star, usually causing a retreat.

On coral reefs, anemone fish (also known as clown fish, Family Pomacentridae) often live among the tentacles of large tropical anemones (**Figure 17.34**). This relationship is more clearly mutualistic. The fish do not induce the firing of the anemone's nematocysts. To avoid attack, a clown fish slowly rubs the anemone's mucus onto its body wall, so that the anemone no longer perceives the fish to be a prey item. The fish then lies among the tentacles, which would otherwise be fatal. The fish obtains protection from predators and also gets food items such as anemone tissue, wastes, and some prey collected by the anemone. Anemone fish also breed in microsites among the anemone tentacles. In the western Pacific anemone fish *Amphiprion percula*, a breeding pair is often accompanied by other nonbreeding individuals that are distinctly smaller and appear to not affect the breeding pair's activities (Buston and Cant, 2006). The breeding pair may be associated with each other for years. The anemone benefits by the fish's removal of

FIG. 17.34 A mated pair of the clown anemone fish *Amphiprion percula*, nestled among the tentacles of the anemone *Heteractis magnifica* in Madang Lagoon, Papua, New Guinea. (Photograph by Peter Buston)

necrotic tissue, and by the strong territorial behavior of the clown fish, which tends to scare off some potential anemone predators. Recent evidence even shows that anemone fish excretion transfers nitrogen to the anemone's zooxanthellae (Cleveland et al., 2011).

One of the most remarkable mutualisms in coral reefs is the cleaning mutualism between cleaner shrimp or cleaner fishes and a large number of fish species (see Figure 4.9). Cleaning organisms are strongly aggregated on reefs in "cleaning stations." Cleaner shrimp and fishes feed by picking ectoparasites off the fishes, which visit regularly. There is limited evidence that the absence of such cleaning fish increases the parasite load of fishes. The cleaning fish *Labroides dimidiatus* maintains cleaning stations that are visited by about 50 species of fishes each day. Clients recognize as a signal the undulating movements of the cleaning fish, whereas clients may also signal by opening the mouth. During cleaning, the black-colored Pacific surgeonfish *Acanthurus achilles* turns a bright blue. Such movements and color change suggest a complex set of recognition signals that have evolved to reinforce the interspecific relationship between cleaners and clients. Some of the interaction, however, is very fluid. The movements of cleaner fish are so stereotyped that it probably is not surprising that there might be deceptive interlopers. The fish *Aspidonotus taeniatus* mimics the undulation of *L. dimidiatus*, but instead of picking parasites, it attacks an approaching fish and bites its fins.

It is extremely important to understand that mutualisms are dynamic partnerships and that there are differences in the tightness of mutualistic relationships between species. In many cases, cleaner wrasses and gobies may bite living tissue off of their clients, instead of just ectoparasites. The reverse is also true. Clients may eat their cleaner fish in the

vicinity of cleaning stations. In the Caribbean, the sharknose goby, *Elacatinus evelynae*, may be found on sponges where they are not cleaner fish and feed on crustacea, but also on coral heads, where they mostly pick parasites and necrotic tissue from visiting fish. White and colleagues (2007) found that the gobies on coral heads grew more slowly and had a higher mortality rate than when on sponges. In this case, being a cleaner fish was not such a great deal, possibly because "clients" were eating the cleaners. The potential damage of cleaners to their clients also suggests that cleaner fish–client mutualisms are dynamic and have shifting costs and benefits. Presumably, the return of clients to cleaning stations is reinforced by the honesty of the cleaning species, which remove ectoparasites and refrain from biting the client. This dynamism is probably much more widespread than many appreciate and indicates that mutualisms can be successfully studied to see when benefits exceed costs, which would lead to more interdependency in a mutualism.

In many marine habitats, several animal species may share the same burrow, constructed by one of the species. On coral reefs, one of these relationships has apparently evolved into a mutualism. While diving on reefs, one can often see the head of a goby poking upward from a burrow in the sand. Beneath, a shrimp of the genus *Alpheus* can be found burrowing. In the reefs of Eilat, Israel, each of four shrimp species has its own companion goby that lives in the burrow. The shrimp has poor vision and relies on the goby to warn against predators. In three of the four cases, the shrimp alone excavates the burrow; but in one case the digging is shared by the two species. When digging sand, the shrimp exits from the burrow and communicates by pressing its antennae to the goby's tail. When an intruder approaches, the fish flicks its tail and the shrimp become motionless or retreats into the burrow. The fish then follows if the danger is severe.

Defenses Against Predation

■ Predation is intense on coral reefs, and many species have evolved strong defenses.

Coral reefs are topographically complex with many crevices. Mobile prey must eventually expose themselves to predators, if only at night. Sedentary species such as corals also must eventually become exposed to predators, such as butterflyfish, as they protrude feeding structures into the water. We have discussed in Chapter 15 the highly specialized lure and piercing radular tooth of cone snails, which allows attraction of prey and rapid attack of invertebrates and even fishes. Frogfish (family Antennariidae) have evolved to look like the irregular topography and color of the reef but can stalk prey (crustacea and fish) and rapidly engulf prey in just a few milliseconds by sucking them in as they enlarge the mouth by expelling water through the gills. At night, moray eels emerge from their crevices and actively hunt fish prey, so not all predators just sit and wait for prey to come to them.

Such interactions have led to extensive antipredator adaptations. Many reef species have acquired strong chemical defenses against predators and fouling organisms. The tropical Caribbean tunicate *Phallusia nigra* has sulfuric acid-filled vacuoles. Fishes usually refuse to eat these sea squirts, which are also loaded with other unpalatable substances, including the toxic metal vanadium. Saponins (triterpene glycosides) are produced by many sea cucumber species, and fish usually die if they feed upon them. The sea cucumbers are also unpalatable. The Caribbean gorgonian *Plexaura homomalla* is comprised of about 5 percent by wet weight of the hormone prostaglandin. This hormone in *P. homomalla* is about five orders of magnitude higher in concentration relative to other gorgonian species. Fishes that try to eat extracts of the sea whip soon vomit their meal. Prostaglandin is not distasteful, but the fish learn to avoid ingesting pellets that contain the hormone because they associate vomiting with the pellets. (This is analogous to getting sick after eating at a bad restaurant and deciding not to return.) Surprisingly, the flamingo tongue *Cyphoma gibbosum* (gastropod) and the fire worm *Hermodice carunculata* (polychaete) have somehow managed to circumvent the toxic effects of prostaglandin to eat the gorgonian. The fire worm is very exposed to predators while feeding on gorgonians, and it has very painful setae that project from its parapodia (**Figure 17.35**). *Cyphoma* species are commonly found on gorgonians that are well defended with chemicals (**Figure 17.36**). How the predators detoxify their prey is often a fascinating and unresolved problem. Butterfly fish employ the CYP 450 oxygenase system for detoxification when feeding on gorgonians and hard corals (see Maldanado et al., 2016).

Many species with chemical deterrents smell bad, and predators can probably detect them at a distance. The most poisonous marine organisms are often conspicuous, as opposed to being cryptic. The poisonous tunicate *Phallusia nigra* is black and conspicuous against a usually white coral reef or sand background. Many toxic tropical species are bright red or yellow. The Panamanian tunicate *Rhopalea birklandii* is acidic and a bright electric blue.

The extent to which chemical defense influences herbivory in coral reefs is still not very clear. Many herbivores, such as parrotfish and sea urchins in coral reefs, are generalists and can consume most or all of the plants despite production of deterrent compounds. It is hard to separate the effects of mechanical from chemical defense without experimentation, but current work supports the conclusion that grazing fish can detect the presence of seaweed and sponge deterrent compounds.

Crypsis and nocturnal activity are also major adaptations to avoid predators. The Caribbean peacock flounder *Bothus lunatus* is active during the day but can rapidly change its dorsal color pattern to resemble clean white bottoms or dead coral covered with dark spots, making it inconspicuous to predators (see Figure 4.4). Many other species, including corals, are active and feed at night, which allows them to avoid visual predators.

Reproduction and Mass Spawning in Coral Reefs

■ Corals reproduce by a wide range of mechanisms.

Corals in reefs are typically simultaneously hermaphroditic or occur as separate sexes. Gamete production, however,

FIG. 17.35 The fire worm *Hermodice carunculata*, feeding on a gorgonian on a coral reef in Bonaire, Netherlands Antilles. It defends against predators with sharp setae that project from the parapodia and are quite painful, even to human touch. (Photograph by Jeffrey Levinton)

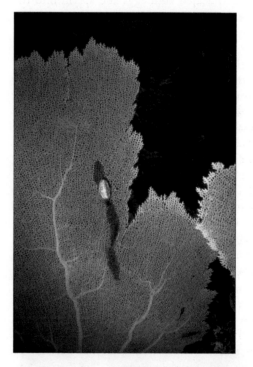

FIG. 17.36 The reef-dwelling gastropod *Cyphoma* sp. grazing on a gorgonian. Cyphoma species can survive toxic compounds produced by many gorgonians and can deploy the toxics in their colorful mantle, which is wrapped around the shell. (Photograph by Robert Richmond)

also has two possible modes: **brooding**—fertilization by planktonic sperm of sessile eggs, followed by release of planktonic planula larvae; or **broadcast-spawning** of gametes from both sexes, followed by formation of planula larvae. Planktonic planulae spread in currents, which allows individuals to colonize new environments. There is no strong association of these modes of sex type or dispersal type with phylogeny or obvious aspects of morphology.

■ **Most broadcast-spawning corals spawn gametes in the water column, often in synchronized multispecies mass spawning at night. This is also true of species of other taxonomic groups.**

In Chapter 7, we discussed larval dispersal and mentioned that some Australian reefs were found to have large populations of planktonic larvae, which tended to settle on the patch reef from which they were derived. Self-seeding of the reef was promoted by currents, which tended to create eddies, which in turn brought larvae back to the reef. There has been very little information until recently concerning the important question of whether corals spawn gametes into the water column or whether gametes are brooded. It was thought that gamete brooding was the rule until both Australian and American divers made some remarkable observations. In Australia, divers observed massive spawning by over 30 species a few nights after full moons in spring. (Keep in mind that this discovery could be made only by doing large numbers of scuba dives at night!) Nearly all species were simultaneous hermaphrodites, and most coral species released both eggs and sperm into the water column, where fertilization occurred (**Figure 17.37**). Since then it has been realized that well over 100 species participate in the spawning event. Caribbean corals with zooxanthellae seem to respond to blue light at 480 nm, which is the wavelength of maximum transparency of tropical ocean waters. This sensitivity may allow them to detect variations in lunar-reflected light, and perhaps stimulate spawning a few days after full moon (Gorbunov and Falkowski, 2002). The response is apparently due to the presence of cryptochromes, blue-light photoreceptors found in corals and many other organisms. Gene expression of coral cryptochromes increases when stimulated by blue light (Levy et al., 2007).

Despite the tropical habitat, coral mass spawning is highly seasonal, although the time of spawning seems to differ by locale (e.g., spawning on the Great Barrier Reef of Australia occurs for a few nights in the Southern Hemisphere spring in November between last quarter and full moon, and mass spawning occurs 7–8 days after the August

full moon on the Flower Garden Banks of the northern Gulf of Mexico). Mass spawning may not be universal, but there are not yet enough data to describe regional variations in this phenomenon.

There are two questions raised by mass spawning. First, why would all individuals of any one species spawn simultaneously? Second, given the multispecies phenomenon, why should many species spawn in the same evenings? In Chapter 7, we discussed shedding of gametes and simultaneous spawning. It makes sense that population-level simultaneous mass spawning within a species would increase the probability of fertilization. Fertilization success increased dramatically (Babcock and Mundy, 1992; Oliver and Babcock, 1992) on major spawning nights, which suggests that timing of sperm concentration in the water column regulates fertilization success to a strong degree. Corals spawn in evenings during slack neap tides at times of low wave action. This timing might minimize the spread of gametes by turbulence, increasing fertilization rates. It also might reduce damage to early coral embryos, which are very delicate and could be damaged by turbulence. But what about multiple-species spawning? This evolved behavior might overwhelm potential predators with too many gametes to possibly eat. The multispecies mass spawning is coincidental because there is only disadvantage in crossbreeding with other species. Most likely, all species independently evolved a spawning time to coincide with some predictable time marker, such as the full moon. Such timing may also relate to favorable current conditions, which maximize the encounter of gametes.

■ **Known strong dispersal potential of coral larvae presents a conundrum. Is the great potential for long-distance dispersal realized in widespread mixed populations?**

We do not know enough about the implications of mass spawning and larval dispersal in corals. Winds can blow most of the larvae produced in a mass spawning out to sea. Larval life span of planula larvae can be longer than 100 days but is usually only a few days before larvae are competent to settle and metamorphose. Many coral species have planktonic larvae that become negatively buoyant after a few days and appear to prefer crustose coralline algae upon which to settle (Heyward and Negri, 1999).

The fecundity of a local population may also be important in explaining local larval recruitment. Terry Hughes and colleagues (2000) found that the number of settling recruits in a given sector of the Great Barrier Reef was weakly correlated with the number of larvae, so the source population does matter in localized recruitment.

Some coral larvae have the potential for long-distance dispersal. The larval stage of the common Pacific coral *Pocillopora damicornis* can last 105 days or more before settling (Richmond, 1987). *P. damicornis* releases larvae that are endowed with more energy reserves, and therefore a longer period of larval settling competency and larval zooxanthellae also contribute to larval nutrition and longevity. Larvae derived from broadcasted planktonic gametes have been studied less, but they have considerable larval

FIG. 17.37 A Pacific *Acropora* sp. beginning to spawn. (Photograph by Robert Richmond)

competency periods ranging from 26 to 56 days. This would still allow considerable dispersal. Such long-lived larvae suggest that extensive dispersal may connect widely separated island populations across the Pacific (Wilson and Harrison, 1998). There is evidence for regional genetic differentiation of animal populations in coral reefs over hundreds to thousands of kilometers, but it is not always clear whether this is due to isolation or local selection for different genetic types. Molecular evidence suggests three distinct clades of *Pocillopora damicornis* in the tropical eastern Pacific that are likely isolated from the rest of the tropical Pacific Ocean (Pinzon and LaJeunesse, 2010). This suggests that, while occasional transport of larvae connects the rest of the Pacific with the tropical eastern Pacific, there is isolation between *P. damicornis* and other coral species from conspecific populations in the rest of the Pacific, but the ranges of genetically homogeneous populations are very large.

Biological Interactions and Community Structure of Coral Reefs

■ Space is limiting on coral reefs, and coral species compete for space by overgrowth, shading, and aggressive interactions.

A reef often gives the impression of Darwin's famous "tangled bank," where interactions for space among competing species must be strong. Large numbers of species coexist on coral reefs, and space is often fully occupied. Most of the dominant invertebrates on reefs are colonial invertebrates. Many species spread by overgrowing their neighbors. Because of the symbiosis with zooxanthellae, reef corals are strongly light limited, and erect corals can reduce the growth rate of neighbors by shading.

Coral reef ecologists Thomas and Nora Goreau discovered a great paradox on the coral reefs of Jamaica. After measuring the growth rates of many species of corals, they found that the most rapidly growing species did not necessarily dominate the reef. The relatively slow-growing, mound-building corals were often dominant. Why?

Judith Lang (1973) made a pioneering discovery that helped account for this phenomenon. Having observed that some reef coral colonies often have bare zones between them, she placed two species of the solitary and slow-growing coral genus *Scolymia* side by side. After several hours, one of the two forms extruded mesenterial (digestive) filaments through openings in the polyp wall and extended them toward the other coral. Within 12 hours, the mesenterial filaments had completely digested part of the second individual, leaving the underlying skeleton exposed. Areas of the victimized coral that were beyond reach seemed to be quite healthy. **Interspecific digestion** was found to be an effective means of competition (**Figure 17.38**).

Further research demonstrated that there is an aggression hierarchy that is related to growth rate. The most aggressive species capable of digesting their neighbors are in general the slowest-growing forms. Some slow-growing species also have long sweeper tentacles that damage neighbors within reach of the stinging cells. Rapid growth can result in competitive success but it is not necessarily a

FIG. 17.38 Interspecific digestion of the coral *Scolymia cubensis* (left) by the coral *Scolymia lacera*. Note bare skeleton to the right of *S. cubensis*, whose soft tissues have been digested away. (Photograph courtesy of Judith Lang)

guarantee of competitive success among corals. The race is not always to the swift.

On coming into contact, members of the *same* species of coral usually fuse. The interaction of competition with fusion between neighbors is an important aspect of studying the success of colonial organisms. Many colonies of scleractinian corals are probably chimeras of genotypes that have met and fused. Fusion may enhance survival by reducing the perimeter open to attack by competitors and predators.

In Caribbean coral reefs, hundreds of species of invertebrates coexist on the undersides of the skeleton of the platy coral genus *Agaricia* and in coral reef caves, even though nearly all the space is occupied. By searching for instances of one species overgrowing another, it is possible to develop a table showing which species routinely outcompete others for space. The competitively successful species consist mainly of colonial ectoprocts and sponges. Why did one or a few species not win? Predation is not as important in this environment. It is possible that there is no clearly competitively dominant species in the first place. The species with the lowest growth rate, for example, may have another means to allow competitive success, such as production of a poison. In Guam, a species of sponge in the genus *Dysidea* produces a sesquiterpene that appears to cause tissue death in competing sponges and is likely a mechanism of competition for space known as **allelopathy**. Thacker and colleagues (1998) demonstrated this by placing either crude sponge extracts or purified sesquiterpene in an agar strip next to a likely competing sponge species. As we discussed earlier, slower-growing corals can outcompete faster ones by interspecific digestion or by deploying specialized sweeper tentacles. This leads to standoffs, and to coexistence of competitors. In some cases, even different conditions may reverse the outcome of competition. In the cryptic invertebrate communities under platy corals, cheilostome bryozoans may or may not be able to overgrow other species, depending on the orientation or the degree to which a competing colony is fouled by other organisms.

Larval recruitment in this habitat is minimal, making unlikely the rapid colonization and rise to dominance by single species.

In summary, complexity and diversity of competitive interactions tends to slow down the rate of competitive dominance by any one species, either on well-lit surfaces or in more shaded undersides of platy corals. Among corals, cases of dominance by single species may occur in the absence of disturbance and predation. Some zones, as we discuss below, do show dominance by one or a few species of corals.

Herbivores exert strong effects on the species composition of reefs.

Hard surfaces free of coral are potential sites for seaweed growth. Large numbers of herbivores are common on reefs and can exert strong controls on the abundance of both plant and animal species. Many species of fish are herbivorous, including the parrotfishes (Scaridae; **Figure 17.28**) and surgeonfishes (Acanthuridae). Parrotfishes have highly fused beaklike mouths, whereas surgeonfishes have sharp teeth that scrape algae. Urchins are especially efficient and can denude a reef of its sea grasses and seaweeds. Experimental removals of the common Caribbean urchin *Diadema antillarum* are followed by colonization and heavy growth of algae. In the early 1980s, a disease caused mass mortality throughout the Caribbean basin. This die-off was followed by extensive seaweed growth in shallow parts of the reef, where the urchins were abundant. Deeper-living herbivorous fishes moved into shallower water to take the place of the urchins. In islands where herbivorous finfish were severely overexploited by fishing, the growth of algae was especially severe.

Coral reefs probably have strong top-down effects exerted by resident and mobile fish predators.

Coral reefs have both resident and mobile predatory fish that probably exert strong effects on coral fish populations. Resident predators such as small groupers hide among crevices and attack smaller fish such as butterflyfish and damselfish. But larger predators such as jacks and sharks move among reefs and may exert strong effects. It is difficult to estimate the effects of such predators, but Hixon and Carr (1997) performed a field experiment that manipulated the local abundance of the blue chromis, *Chromis cyanea*, a damselfish that tends to stay put in a very small area of a reef. Mobile predators such as the bar jack *Caranx rubber* aggregated at experimental patch reefs with higher densities of blue chromis, but predation was highest at a patch reef when both mobile jacks and resident small predatory groupers were present. Coral reef fish recruitment may be strongly affected by transient aggregations of mobile predators, since the effect of predation by visiting mobile predators increases when local prey density increases.

Large and mobile predatory fish probably exert strong top-down effects on community structure of coral reefs. At present, many coral reefs are severely overfished, so our understanding of these effects is very limited. In the Caribbean, the Nassau grouper, *Epinephelus striatus*, can successfully attack and consume a wide variety of fish, most importantly parrotfishes, which are important grazers on seaweeds. Grazers reduce macroalgal cover and may facilitate coral recruitment. Relationships, however, can be complex. In a marine preserve, Nassau groupers increased in abundance, and one might expect that parrotfishes would decline and macroalgae would increase. However, the part of the parrotfish population that caused most of the grazing consisted of individuals that were too large to be killed by the groupers. As a result, adding a top predator in the system caused only a small change in grazing, less than 10 percent (Mumby et al., 2006).

A study of reefs south of the Hawaiian Islands confirms the likely importance of top-down effects in coral reefs. Sandin and colleagues (2008) studied two pairs of reef tracts. One set was adjacent to islands populated by humans, and fishing resulted in the relative rarity of larger top predators such as sharks and an abundance of smaller planktivorous and seaweed-grazing fishes. More remote reefs with no human fishing had an abundance of sharks and other top predators. Top-down effects in these reefs resulted in greater coral recruitment and lower incidences of coral disease, although the specific mechanisms are not well understood.

A biological invasion by the lionfish, *Pterois volitans* (**Figure 17.39**), has provided us with an experimental opportunity to see the effects of sudden increased predation that may disrupt trophic structure in Caribbean coral reefs. Lionfish are native to the Pacific and are strikingly beautiful scorpionfishes with distinctive red stripes and sharp venomous spines. They are kept by aquarists throughout the world. Two closely related species of lionfish were released probably from aquaria during the 1970s through 1990s, but definitely at the time Hurricane Andrew hit southern Florida in 1992. They spread southward into the Caribbean and are now abundant in the Bahamas, Hispaniola, and Jamaica, but have also been found in the farthest southern limits of the Caribbean basin, such as Bonaire just off the coast of Venezuela. They have planktonic eggs and larvae, and juveniles have also dispersed northward in the Gulf Stream as far as Nantucket near Cape Cod, but they cannot survive the winters. They are effective predators on fish, and a field study using patch and artificial reefs with and without experimental additions of *P. volitans* demonstrated that fish recruitment was strongly depressed by lionfish (Albins and Hixon, 2008). Prey such as damselfish, which do not seem to recognize the lionfish as predators, swim right up to them and are rapidly attacked and consumed. Lionfish have venomous fin spines and cryptic red striped coloration that protect against their own predators, at least in their native range. They also blow directed jets of water at prey fish, which likely confuse naïve Caribbean potential prey (Albins and Lyons, 2012). It seems likely that it is too late to control this species in the Caribbean, although divers find them relatively easy to catch and isolated islands such as Bonaire have active eradication programs in progress.

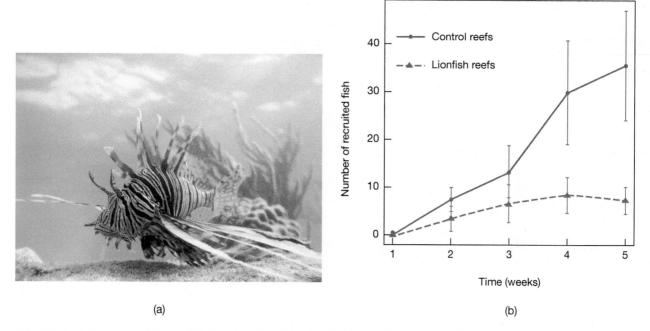

(a)

(b)

FIG. 17.39 (a) Invasion of the lionfish *Pterois volitans* into the Caribbean. (Photograph by Tim Pusack) (b) Effect of presence of lionfish on recruitment of coral-associated fish in an experiment in which lionfish were present on some patch reefs and absent in others.

Disturbances: How Storms, Ocean Warming, and Disease May Lead to the Death of Coral Reefs

■ **Coral reefs are often subject to extensive disturbance from tropical storms. In the eastern Pacific, El Niño events are also major sources of large-scale disturbance.**

In the popular literature, reefs are often pictured as completely benign environments, where exquisite biological interactions have evolved over eons of constancy. Although some of the most remarkable symbiotic relationships are to be found in coral reefs, the environment is far from benign and has experienced many catastrophic changes. First, rises and falls of sea level due to glacial advances and retreats over the last several hundred thousand years have been a source of profound disturbance; many reefs have failed to keep up with rising sea level and have been drowned. At the other extreme, falling sea level has caused major changes in sedimentary regimes and extensive erosion. On smaller time scales, reefs are often located within very strong tropical storm belts. In the Caribbean, many reefs have been pounded by hurricanes, and cyclones do similar damage in the Indo–West Pacific region.

The north coast of Jamaica was hit by Hurricane Allen in 1980 and by Hurricane Gilbert in 1988. These two hurricanes caused massive damage to erect coral colonies and broke them up in pieces to the point that they were vulnerable to overgrowth by algae and smothering by sediment. Following its first appearance on the Caribbean coast of Panama in 1983, a disease nearly eliminated throughout the Caribbean the grazing urchin *Diadema antillarum*, resulting in algal overgrowths and inhibition of coral

settlement. Before these major events, overfishing had already greatly reduced grazing fish, so *D. antillarum* was effectively the only major grazer before the disease struck. So it was the combination of hurricane damage and removal of the only abundant grazer that caused a flip to a new regime that was very unfavorable to coral colony colonization and growth. A formerly exquisite coral reef on the north coast of Jamaica was devastated and thus far has been recolonized by very few large colonies; recovery will take many years. Staghorn corals are beginning to grow, but the coral-eating snail *Coralliophilia* may prevent the small colonies from enlarging. Localized growth of corals has occurred, especially in sites where urchins have come back.

In the Pacific, cyclones are often devastating, and enormous brain corals are often toppled easily. Outbursts of predators can also be a major source of disturbance. In the 1960s, the crown-of-thorns sea star *Acanthaster planci* (**Figure 17.40**) increased in numbers in many Pacific reefs, and corals were devastated. Before this time, the sea stars were relatively low in abundance and largely nocturnal. As the outbreaks occurred, the sea stars fed on corals during the day and were abundant enough to eliminate large areas of coral. Strong outbreaks occurred in the 1980s. Many areas of the Great Barrier Reef still show the effect and outbreaks are still on the rise.

It has never been clear why this species increased so rapidly. Endean (1977) noted the removal by shell collectors of its major predator, the giant triton *Charonia tritonus*. However, *C. tritonus* does not seem to be a major source of sea star predation. Charles Birkeland (1982) argued that storms have caused washouts of nutrients, which in turn have increased larval recruitment by the sea stars, owing to an increased food supply. Starfish outbreaks in the Great

FIG. 17.40 The crown-of-thorns sea star *Acanthaster planci* on a reef of the west coast of Hawaii. (Photograph by Jeffrey Levinton)

Barrier Reef, which have been numerous, also followed large increases of nutrient discharges from rivers in eastern Australia in recent decades (Brodie et al., 2005). Phytoplankton has increased in the season when the starfish larvae disperse and in areas where outbreaks occur. This would suggest that larvae are typically limited by starvation, but this appears not to be the case (Olson, 1987). Still, nutrient supply might enhance larval growth and metamorphosis rates. In any case, the devastation was real, and it will take many years for corals to recover completely from the algal-dominated bottoms that the sea star outbursts left behind. Correlations between human population density and starfish abundance suggest that overfishing by humans has removed fish predators of young starfish, allowing local population explosions (Dulvy et al., 2004).

In the eastern Pacific, El Niño events can have drastic effects. As we discussed in Chapter 3, these periodic episodes bring warm surface water poor in nutrients to the shallow coastal zones of the eastern equatorial Pacific. In 1982–1983, the El Niño was particularly severe, and the increase in water temperature caused massive mortality among the dominant corals of Costa Rica, Panama, and especially the Galápagos Islands. Many dependent species also declined, and the lovely reefs were replaced by turfs dominated by filamentous algae and eroded dead coral rock. High-temperature excursions are now becoming commonplace and worldwide warming is interacting with periodic peaks of temperature associated with El Niño. See discussion of global climate change at the end of this chapter.

■ Disease can devastate the dominant species of a coral reef, which can result in major changes in reef community structure.

To further weaken the image of stable coral reefs, we now know that disease can rapidly destroy important species in coral reef systems. We mentioned earlier the catastrophic

decline throughout the Caribbean in the 1980s of the sea urchin *Diadema antillarum*, which is a major grazer on Caribbean coral reefs. This species is normally very abundant and conspicuous, so its rapid disappearance over the span of 1 year was noticed and studied intensively (Lessios, 1988). Although the pathogen was not conclusively identified, it was clear that the pattern of initial spread followed the path of surface-water currents, so the presumed pathogen was waterborne. Since other urchins did not suffer, the pathogen is likely to have been host specific. Mortality was nearly 100 percent in many locales, starting with coastal Panama. A catastrophic decline of several species of sponges occurred in the 1930s in the Caribbean (Smith, 1941). A fungal parasite probably caused the disease, though increased salinity may have exacerbated mortality. Here, too, the sequence of local mortality could be predicted by water current patterns. In the 1990s, another fungus, *Aspergillus sydowii*, spread rapidly in the Caribbean and caused tissue lesions and mass mortalities of sea fans (Nagelkerken et al., 1997).

The geographically large scale of these declines suggests that near-to-complete extinction of a widespread species can occur, simply by waterborne transmission. Some of these species, like the Caribbean sea urchin *Diadema antillarum*, exert strong effects on communities, and their disappearance may change completely the relative abundance of most other species. The characteristically tremendous surge in urchin population growth is usually followed by a decline of seaweeds. Similarly, sudden mortality is usually followed by lush seaweed growth. Considering the structural complexity that a seaweed population can bring to an otherwise featureless seabed, one can well imagine the importance of disease in the economy of marine communities.

Ecosystems that are strongly disturbed might be more vulnerable to disease than those that are intact. For example, in the terrestrial environment, vegetated areas are known to reduce the spread of pathogenic bacteria, and are often planted near sewage treatment areas. Vegetation may filter toxins and pathogens, and natural substances produced by plants might detoxify or kill microorganisms. Sea grass beds have also been discovered to reduce pathogenic bacteria within the beds and in adjacent coral reefs. This was supported by sampling *Enterococcus* bacteria near Indonesian islands with poor sewage control. Areas without seagrass in adjacent shallow waters had three times as much *Enterococcus* and those with sea grass, and nearby coral reefs adjacent to sea grass beds had half the *Enterococcus* of coral reefs near seabeds without sea grass beds (Lamb et al. 2017). These results suggest that sea grass beds are absorber-filters of pathogens and benefit the adjacent coral ecosystem.

In recent years, a number of diseases have been recognized in corals. In the Caribbean, the very shallow reef, dominated by elkhorn (*Acropora palmata*) and staghorn (*Acropora cervicornis*), has been devastated by disease. These two dominant coral reef framework building species have nearly disappeared as shallow-water dominants from much of the Caribbean basin.

Unfortunately, we are not in a position to understand the diseases completely. In most cases we cannot isolate and culture the disease organisms, nor can we typically use a lab culture to reinfect a coral and obtain the same symptoms. It is possible that some of these syndromes are opportunistic colonizations of corals that are already dying: Disease is often correlated with other signs of weakening such as bleaching. Still, the following syndromes have been noticed in a number of coral species, especially in the Caribbean, where the geographic extent and frequency of disease is on the increase in the past decades. Given the long-term observations by divers, it is unlikely that these diseases were common before a few decades ago.

White band disease. This is probably caused by a bacterium, whose invasion results in a white band adjacent to dying coral tissue. It has been associated with acroporid corals in the Caribbean (especially the elkhorn coral *Acropora palmata* and the staghorn *A. cervicornis*), which have been disappearing from shallow reef areas throughout the Caribbean, changing the entire nature of coral reefs there (Aronson and Precht, 2001). The disease appears to invade from base to tip and kills entire colonies. The application of antibiotics arrests the progression of WBD, which suggests a bacterial cause to the disease (Sweet et al., 2015).

White plague. This is a complex of diseases. White plague I is associated with the appearance of dense occurrences of an unidentified white coccoid bacterium. Degradation occurs on the scale of mm d^{-1}. White plague II is caused by the bacterium *Aurantimonas coralicida*, is infectious, and spreads much more rapidly. In this case, there is much more certainty about the microbial cause of the disease. An outbreak in the 1990s caused extensive mortality of many species of corals in the Florida Keys. A third variant, white plague III, advances very rapidly

and has killed colonies of the dominant coral *Montastrea annularis* in the Caribbean. There are other variants known as well, and it is likely that the symptoms unite the disease more than the bacterial causes, which are probably quite different.

White pox disease. This disease causes large white lesions and has caused high mortality in the elkhorn *Acropora palmata*, especially in the Florida Keys. Sutherland and colleagues (2011) demonstrated that a human pathogen, probably released in sewage in the Florida Keys, is the likely cause of this disease.

Yellow band disease. A consortium of species of *Vibrio* attacks zooxanthellae within gastrodermal tissues of coral, leaving yellow blotches on the coral colonies. Genetically similar consortia of bacteria have been found in both Atlantic and Pacific species of corals (Cervino et al., 2008).

Black band disease. This is a consortium of microorganisms, dominated by cyanobacteria, that form a mat; it tends to affect nonacroporid corals and leads to sulfide accumulation and toxicity in corals. A densely interwoven black mat of filaments develops, which eventually separates the coral tissue from its underlying calcium carbonate skeleton. Other bacteria then invade, and the coral tissue dies, leaving a bare skeleton.

Coral species differ strongly in susceptibility to black band disease, but the condition is widespread and can kill corals rapidly. In Florida Keys waters, the disease has steadily become more widespread, perhaps as corals become physiologically weaker and more susceptible to disease. Bleaching and black band disease have both been noticed more frequently in Atlantic corals, and many believe that this reflects greater stress. Other diseases have also been identified, sometimes in areas remote from obvious localized sources of human impacts. The interested student should consult the summary by Sutherland and others (2004) and Sato et al. (2016).

Like many other organisms, corals have known defenses against disease, including production of mucus to slough off invading microbes, wandering phagocyte cells that engulf and encapsulate microbes, and antimicrobial substances. Unlike vertebrates, corals lack acquired immunity.

■ Because corals are foundation species, death from disease and coral bleaching should have strong negative effects on dependent species.

Like salt marsh plants and sea grasses, corals are foundation species or ecosystem engineers, structuring a habitat within which thousands of other species reside. Corals are food for a number of predators and interact with mutualists, but also provide a structural habitat for many species. In recent decades, warming, disease, and major losses of keystone species raise the question of whether one or more combined disturbances will propel coral reefs into a new state or regime that is hard to reverse. Feather stars, feather duster worms, and anemones with their clown fish mutualists are examples of dependent species that would disappear if corals die. Fish cleaning stations are usually located in crevices of coral colonies, and other crevices provide shelter for hunting moray eels, and for a large number of fish that seek hiding places from predators.

Experiments and observations of the effect of loss of essential coral habitat are interesting. Booth and Beretta (2002) found little effect of coral bleaching on the loss of coral reef fish. Their results, however, may apply only to the short term, as they surveyed larger long-lived fishes, which might not be immediately affected. On the other hand, fish larval recruitment to recently dead coral skeletons was not terribly different from live corals, but fish larvae recruited poorly to long-dead coral skeletons that were colonized by algae (Feary et al., 2007). Bellwood and colleagues (2006) found major overturn in the composition of smaller fish species following bleaching events on the Great Barrier Reef. In the western Pacific, a major bleaching event was not followed by major declines in fisheries, although juveniles became far less common (Graham et al., 2007). Effects of coral bleaching and death are therefore probably severe, but might not be observed for several years following the event. This has not been true in the extreme bleaching event on the northern Great Barrier Reef in 2016. Bleaching has occurred in nearly all reef localities, and many reef-associated fish species are missing.

■ The decline of corals in the Caribbean from disease and pollution has resulted in a general increase of abundance of sponges.

Increasing seawater temperature, disease, bleaching, and pollution have all caused major declines of coral cover, especially in the Caribbean. Vulnerable acroporid elkhorn and staghorn corals have disappeared at many sites, and coral cover has declined in polluted areas such as the Florida Keys and parts of the Bahamas, but the decline really has also occurred in unpolluted areas of the northern Caribbean region that have experienced high sea temperature and influxes of coral disease. Joseph Pawlik (2011) and colleagues discovered that coral decline has been accompanied by increases of sponges. Rapidly growing sponges often come to dominate, but these species usually synthesize low concentrations of chemical defense compounds and are vulnerable to sponge predators such as some angelfishes and filefishes. Some sponge species have high concentrations of chemical defense compounds and come to dominate when such predators are common. Either way, many Caribbean reefs are now becoming dominated by sponges instead of corals.

■ Sponge growth and expansion on coral reefs may be affected by food supply.

The factors facilitating sponge expansion in recent decades are not well understood, but we need to understand them since a major ecological transition to sponge increase is occurring. Sponge abundance and diversity in the Caribbean tends to increase with depth, as does the abundance of picoplankton (one micron or less in size), the presumed size of plankton eaten by sponges. Does this mean that food is limiting to sponge growth and diversity? We have discussed the role of predation as a top-down factor. We already know that sponges are consumed by a wide variety of fish and turtles. Bite marks are common on sponges. The issue of food limitation is not so clear. But is there a bottom-up factor as well?

A study by Lesser and Slattery (2013) strongly suggests that food is limiting to sponge growth. Food consumption and sponge growth were measured in shallow and deep sponges by comparing the available food taken up by the sponge and the water released through the sponge osculum, after the water had been filtered (see discussion of sponge feeding in Chapter 14). Growth of sponge colonies was also measured. Deeper sponges grew faster. Sponge abundance and size increased with depth, which fit with the presence of more picoplankton food in deeper water. Shallow-water sponges cleared less picoplankton than at deeper sites. Predation at this site in Belize had no trend with depth. The data therefore supported the presence of bottom-up effects on sponge growth.

These results are convincing but still controversial, since alternative explanations have been suggested that, overall in the Caribbean, food supply is not related to sponge abundance (Pawlik et al., 2015). This controversy requires more work, done consistently in many localities, to determine whether food supply is universally limiting to sponge growth in coral reef habitats.

■ Owing to strong regional disturbances and other factors, coral reefs may exist in distinctly different community regimes.

If the water is oxygen rich, moderately warm, and free of very strong storms, coral reefs will grow vigorously in waters of moderate to strong wave action. The diversity of corals creates a landscape within which hundreds of species of fish and invertebrates can reside. Such a reef is, in the large sense, a stable state, supporting further growth of corals and allowing continued residence of hundreds to thousands of dependent species. But what if there are significant disturbances? The case of Jamaica, mentioned earlier, suggests that it was propelled into a new regime. Hurricanes, the disease-driven collapse of the herbivorous urchin *Diadema antillarum*, and previous overfishing of algal-grazing fish by humans have combined to convert a coral world to a seaweed world. Shading by seaweeds inhibits coral settlement, and a lack of grazing encourages algal growth. If we examine coral cover in the 1970s before the storms, it was usually in the range of 20–80 percent, and algal cover was usually a few percent. But after the hurricanes and *Diadema* die-off, coral cover was just a few percent and algal cover was always greater than 75 percent (**Figure 17.41**). Hughes (1994) concluded that coral reefs of the north coast of Jamaica have been forced into a completely different regime, as disturbance combined with the removal of grazers. The difficulty in recovery is to establish conditions in which coral recruitment is successful. Edmunds and Carpenter (2001) noted that recovery of *Diadema* in some areas of Jamaica has resulted in increased algal grazing that favors coral recruitment. In the past 5 years, students from my home university, Stony Brook, have noticed some recovery while diving on Jamaican reefs at Discovery Bay. Coral recruitment has

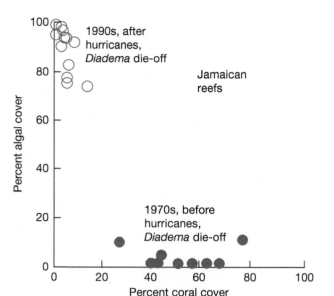

FIG. 17.41 A phase shift in coral reef community structure. Reef areas have either high coral cover combined with low algal cover or high algal cover combined with low coral cover. Solid circles correspond to sites on the north coast of Jamaica in the 1970s (before hurricanes and *Diadema* disappearance), whereas open circles refer to sites in the 1990s (following the changes). (After Hughes, 1994)

been increasing (Knowlton, 2001), which provides hope for the reestablishment of vigorously growing reefs.

The phase shift in Jamaica is not an isolated occurrence. Graham and colleagues (2015) examined the strong effects of the 1998 El Niño–related strong bleaching event in coral reefs of the Indian Ocean. In many reefs, warming events were followed by widespread bleaching, often lasting a year or more with over 90 percent loss of coral cover. Of 21 reefs examined, 12 recovered to previous high levels of coral cover. But 9 failed to do so, and formerly coral-covered bottoms were now covered by seaweeds. Recovery was favored in deeper-water reefs where fish abundance was high and nutrient concentrations fairly low. Thus, regime shifts to seaweed-covered bottoms were favored where warming was more intense in shallow water and reduced grazing and high nutrient input with high light intensity favored algal growth on coral rubble (see Ledlie et al., 2007).

Large-scale analyses of coral and algal cover around the world shows that most coral reef regions do not show the strong bimodality of either complete algal cover or coral cover. Instead we mostly see a continuum (Bruno et al., 2009). Within the Caribbean, sites other than Jamaica have declined less in coral cover relative to algal cover (Côté et al., 2016). But as future warming events are more intense, such regime shifts seem more and more likely to occur.

It has been argued that the Caribbean might be more vulnerable to such a phase change than the Pacific province. The Caribbean has fewer species of coral, and the loss of a single dominant species might result in the loss of an entire structural element from the reef. For example, the elkhorn coral *Acropora palmata* has steadily disappeared from many Caribbean coral reefs, and there simply is no other species with similar morphology to take its place.

In the southwestern Pacific, however, there are many species of acroporid corals, so there is some redundancy to the system.

Synthesis: How the Major Factors Interact to Produce Changes in Coral Reef Community Structure

■ **Different scenarios illustrate how coral reef dominance might shift in response to different conditions of storms, disease, warming, and overfishing.**

Our discussion of coral reefs produces a rather complex set of interactions between the major components, but we still can see strong patterns emerge when parts of the system are perturbed. So let's examine some scenarios of change relative to an "undisturbed" condition (**Figure 17.42**). We will consider what we have learned from Caribbean systems, where corals, sponges, and seaweeds change dominance depending on environmental changes.

Scenario 1: Undisturbed. In this case, corals are competitively dominant, and predators on sponges (spongivores) and corals (corallivores) are abundant. Herbivores such as fishes and urchins are also abundant and keep seaweeds at low abundance. A top predator, such as large Nassau grouper, is abundant and preys successfully on herbivorous fish and fishes that feed on corals and sponges. Corals under these "normal" circumstances tend to dominate the benthic layer. This was largely the case in the 1970s and earlier in Caribbean systems.

Scenario 2: Heavy storms and loss of herbivores. This scenario corresponds to the condition in Jamaica (**Figure 17.42**) in which herbivory was reduced by feeding on fishes and storms caused major mortality on corals. Here, seaweeds come to dominate, and corals and sponges are relatively sparse owing to breakup by storms and overgrowth by abundant seaweeds.

Scenario 3: Strong stress from coral disease and regional warming. In this case, corals are reduced in abundance and are outcompeted by sponges. Herbivorous fishes and urchins are still present, so seaweeds are kept in check.

Scenario 4: Strong overfishing. Here, top predators and fish that prey on benthos are strongly overfished. Seaweeds increase in response to reduced herbivory and overgrow corals. Sponges also benefit and increase in abundance.

■ **Coral reef areas might deteriorate to a tipping point, where colonization by coral larvae and recruitment of coral reef fish might decline precipitously.**

An array of processes—thermal stress, disease, overfishing—can nearly eliminate the coral cover from an area of reef. Is recovery possible? For a devastated reef to recover, coral larvae must resettle in the area and produce viable colonies. Formerly common fish must return to areas where their coral habitat has largely disappeared. Dixson and colleagues (2014) examined colonization of coral larvae and fish in a reef area along the coast of Viti Levu, Fiji, in the Pacific. Some areas were strongly disturbed by local

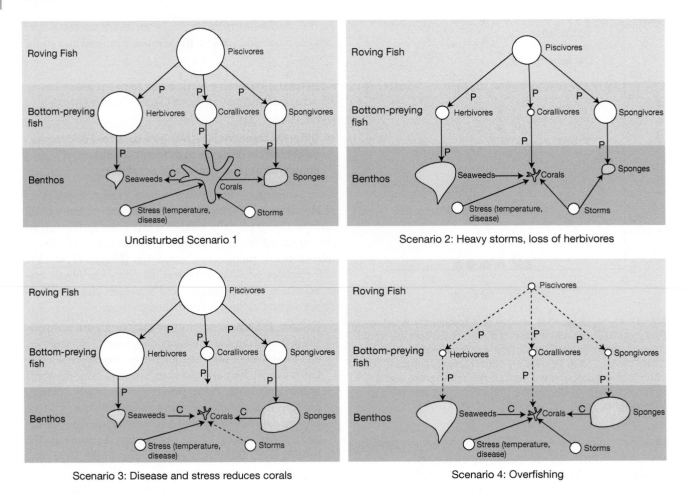

FIG. 17.42   Scenarios showing response of benthic dominants on coral reefs to various environmental changes. P → corresponds to predation; C → corresponds to competition.

fishing, and corals and coral reef fish were very depleted. Field experiments demonstrated that coral larvae failed to recruit in such sites and preferred water taken from where corals were abundant. Juvenile coral reef fish also avoided depleted sites but actively swam toward odors that emanated from local rich coral areas. These experiments suggested that coral destruction could pass a tipping point where an area would no longer be colonized by new coral recruits or juvenile coral reef fish. In some cases, this problem can be rectified by transplanting corals into barren areas. In the Caribbean, disease has caused nearly the loss entirely of acroporid corals, which formerly dominated the shallowest zones of the reef. Corals are now being grown on floats and planted into shallow areas in the hopes that they will spread and reproduce (**Figure 17.43**).

Global Warming and Ocean Acidification

■ **The global warming trend in recent decades is strongly associated with stress on corals, especially widespread bleaching events.**

The role of increased temperature on coral condition has been appreciated for many years, following Peter Glynn's

important work concerning the effects of the 1982–1983 ENSO event on coral reefs of Panama. High temperatures surpassed the physiological limits of many of the local coral species, causing extensive coral bleaching. Bleaching events have increased in frequency all around the tropical world since the 1970s. While coral bleaching (**Figure 17.22**) occurs in response to a number of physiological stresses, so-called mass bleaching, in which zooxanthellae are expelled but coral tissue remains alive (at least for a while), usually seems to be a response to temperature stress. At least 70 percent of mass-bleaching events in the 1990s were related to higher-than-average temperatures (Glynn, 1993). Mass-bleaching events, which have been more frequent in recent decades, were rare to absent on the Great Barrier Reef before 1979. In the western Pacific, a major bleaching event occurred in 1998, with mass bleaching and mortality in a number of islands. In 2005, an unprecedented increase of sea temperature occurred throughout the Caribbean (Eakin et al., 2010), and bleaching was higher in areas where the temperature increased the most (**Figure 17.44a**). Temperatures in the Caribbean have been increasing since 1985 (**Figure 17.44b**). Donner and colleagues (2005) used models of climate change predicting sea-surface temperature and concluded that dangerous conditions for mass-bleaching events would be annual or biannual events in the next 30–50 years. Reefs in the western

FIG. 17.43 (a) Caribbean fragments of the coral *Acropora cervicornis* grown on floating plastic frames off the bottom on the island of Bonaire. (b) Grown fragments have been connected and planted on the seafloor at 2 m depth. (Photographs by Jeffrey Levinton)

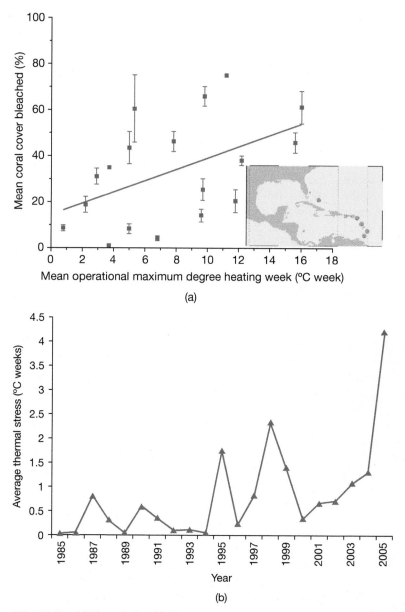

FIG. 17.44 (a) The relationship between mean percent coral cover that was bleached at a number of Caribbean coral reef sites and the mean maximum temperature for the hottest week at a given locality in 2005. (b) Thermal stress from 1985–2005 averaged over a group of stations in the Caribbean. (Courtesy of Mark Eakin and NOAA)

Pacific are particularly vulnerable. Hoegh-Guldberg undertook a similar exercise, using another index of bleaching and also data on various bleaching events. The models, combined with temperature thresholds, predicted widespread bleaching in 30–50 years (Hoegh-Guldberg, 1999). These general climate change effects are likely to be more important in reducing coral cover than localized human disturbance (Bruno and Valdivia 2016).

Coral species do not all respond to thermal stress in the same way, nor do reefs recover at the same rate. It is therefore difficult at this time to make specific predictions about the rate of coral reef loss from temperature stress. For example, coral species living at lower latitude have larvae that are more tolerant of high temperature than larvae of those species from higher latitude. This suggests that evolution to optimize performance at different temperatures has occurred in corals and there might be some room for evolutionary response to global warming (Dixon et al., 2015).

Without any check on carbon dioxide emissions into the atmosphere, the future of coral reefs will involve more long-term anthropogenic ocean warming, combined with the periodic stress of ENSO events. The interaction of warming and ENSO may be a factor in the extensive occurrence of coral bleaching across the Pacific that has arisen with the 2015–2016 ENSO event. A recent air survey of the Great Barrier Reef in the winter of 2016 showed that 81 percent of over 500 surveyed reefs in the northern sector were severely bleached. Bleaching was less strong in the central (33 percent severely bleached) and southern parts (1 percent severely bleached) because of persistent cloud cover. But the central part of the Great Barrier Reef started experiencing severe bleaching in 2017. Bleaching was often uneven, and some coral species have suffered more than others (**Figure 17.45**). This major event builds on a report

FIG. 17.45 A diverse assemblage of corals on the northern Great Barrier Reef, Australia, was heavily bleached in 2016 due to thermal stress. The most susceptible species are flat tabular *Acropora hyacinthis* and *A. cytherea*, and branched bushy corals (multiple species of *Acropora* and *Pocillopora*). (Courtesy of Terry Hughes)

of strongly reduced coral calcification rates in the Great Barrier Reef in recent years (De'ath and others, 2009) of 14.2 percent since 1990. The cause of the decline is unknown but may be related to increasingly frequent high ocean temperature stress.

The only escape from the effects of global warming would be for coral populations or their zooxanthellae to evolve increased thermal tolerance. As we mentioned in the section on coral reefs, some corals living in thermally extreme environments are resistant to higher temperatures, so there is evidence for evolutionary response to higher temperature. When the common Pacific coral *Acropora millepora* is exposed to high temperatures, there is a shift in dominance from one strain of zooxanthellae to another of higher thermal tolerance (Berkelmans and van Oppen, 2006). A reshuffling of strain dominance might help in forestalling bleaching and coral death from warming as a result of climate change. But coral reefs that have experienced repeated bleaching appear to have corals with the same strain of zooxanthellae.

■ **Acidification presents another threat to corals as ocean pH declines in the coming decades, but preliminary evidence suggests a homeostatic mechanism to buffer pH.**

We discussed ocean acidification in Chapter 3, and described how ocean pH is being driven down by a steady increase of human-derived carbon dioxide into the atmosphere. This increase is calculated to have already reduced the world ocean's pH by about 0.1 unit. Increases of atmospheric carbon dioxide will reduce pH and steadily reduce dissolved carbonate ions available for calcification. Calcium carbonate occurs in two forms: in a more stable calcite and in the less stable form of aragonite. Coral skeletons are composed of aragonite and are therefore among the most vulnerable of marine organisms to increases in acidity.

Even one of the more modest scenarios for carbon dioxide increase would make the Southern Ocean undersaturated with respect to calcium carbonate in the year 2100. Going Deeper Box 3.2 shows how to calculate a parameter Ω (omega), which equals 1 when the water is saturated with respect to a given mineral. But it turns out that Ω must be 3.3 or more for aragonite to be able to be produced by corals, which is close to the present levels. The current carbon dioxide concentration in the atmosphere is 380 ppm, and the threshold value of $\Omega = 3$ for corals that will limit calcification will be reached when that concentration reaches 480 ppm. Even the most conservative carbon dioxide model with no corrective action will bring the value below 3 by the year 2100.

Gattuso and colleagues (1998) found in the Pacific coral *Stylophora pistillata* that calcification increases nearly threefold in the laboratory when Ω increases from 0.98 to 3.90, the latter value of which applies to the present day in the tropical ocean. This means that further increases of carbon dioxide in the atmosphere are liable to have drastic effects on corals. But the negative effects may already have begun. A study of corals of the genus *Porites* on the Great Barrier Reef demonstrated a reduction of calcification over the last 16 years, resulting in a reduction of skeletal density of 0.36 percent per year and a reduction of linear growth rate of 1.06 percent per year. Together, these numbers suggest a reduction of calcification of about 1.3 percent per year (Hoegh-Guldberg, 1999). As acidification increases, experiments demonstrate that some coral species may be more tolerant than others, so ocean acidification may also cause reorganizations of abundances of different coral species before pH levels become intolerable for all corals.

A cleverly designed field experiment, however, suggests that corals may be resilient to some degree of acidification. Georgiou and colleagues (2015) were able to manipulate pH in the field 0.05–0.25 units below ambient near corals on a Great Barrier Reef flat and found no ill effects on the coral species *Porites cylindrica*. The calcifying fluid within the coral remained unaffected, perhaps regulated, at pH 8.4–8.6, suggesting that the coral was resilient to external acidification.

■ CHAPTER SUMMARY

- Sea grasses are marine *angiosperms*, or flowering plants, confined to shallow water and extending mainly by rhizome systems within the sediment. They grow best in high light and modest current flow. Sea grass beds most easily colonize sediment after previous colonization by seaweeds. They are diverse in lower latitudes and interact strongly with macroalgae, but are in widespread decline because of eutrophication.

- Sea grasses are grazed by sea turtles, urchins, and other species, and are a foundation species for a diverse group of animal species. They deter the entry of some predators and may enhance the abundance of infaunal suspension feeders within the bed.

- Subtidal rocky reefs and kelp forests often comprise a related pair of local systems. They are often dominated in shallow waters by kelps and seaweeds and by epifaunal animals in deeper waters.

- Subtidal rocky reefs harbor abundant communities of algae and are often dominated by colonial invertebrates. They are frequently very patchy, with alternations of rocks dominated by rich invertebrate assemblages and turf-forming calcareous red algae. Animals are often short on space, suggesting the importance of competition. Grazing, mainly by sea urchins, is more intense on horizontal benches.

- Kelp forests are dominated by a few brown seaweeds (Laminariales) with large thallus and high growth rates. A typical life cycle alternates between a large asexual sporophyte and a tiny filamentous gametophyte. Kelp forests support many plant and animal species.

- Upper-canopy kelps have first access to sunlight, but lower-canopy kelps survive better after a storm. In the northeastern Pacific, sea otters sit atop a strong trophic cascade, but in southern California other predators are also important. Succession depends on the interplay of grazing pressures, disturbance, competition for light, and colonization of urchins. Some kelps have developed chemical defenses to grazing.

- Coral reefs are built by a variety of species cemented together. Development is limited by warm temperature, open marine salinity, available light, and low turbidity. Reefs are composed of corals, algae, sponges and other calcareous organisms, and calcium carbonate. Symbiotic algae (*zooxanthellae*) contribute to coral nutrition and are essential for high calcification. Hermatypic corals contribute most to reef growth.

- Reefs are wave resistant and topographically complex, with many microhabitats. They are more diverse in the Indo-Pacific than the Caribbean. Coral reefs contain many closely related species and require careful molecular and ecological study.

- Coral atolls, more common in the Pacific, are ring-shaped island groups arising from oceanic volcanoes that reach the surface but slowly subside. Both atolls and coastal reefs have prominent depth zones, each with a different set of dominant species.

- Corals in reefs are hermaphroditic or occur as separate sexes. Gametes may be produced by (1) internal fertilization, followed by the release of planula larvae, or (2) broadcast spawning, followed by the formation of planula larvae. Most corals spawn gametes in the water column, often in synchronized multispecies mass spawning.

- Herbivores exert strong effects on the composition of reefs mediating interactions between corals and macroalgae, but top carnivores such as roving fishes appear to exert strong top-down effects. Colonial species compete for limited space by overgrowth, shading, and aggressive interactions. Coral reefs are often subject to disturbance from disease, tropical storms, and El Niño events in the eastern Pacific. Diseases are becoming more prevalent.

- Coral reefs and nearby environments harbor some of the most remarkable mutualisms in the sea. Predation is intense, but many species have evolved strong defenses. These include retreating to protective crevices, nocturnal activity, and poisonous defense compounds.

- Coral reefs have faced increasing stress from rising sea temperatures, and bleaching events have become more frequent since the 1970s. Ocean acidification is another outcome of climate change that may severely impact coral reefs in coming decades. Aragonite-secreting corals are much more vulnerable to acidification than many other calcifying marine organisms.

- Coral reef restoration may involve three cornerstone activities: establishment of marine protected areas; restoration of predators and grazers; direct transplant of coral species decimated by disturbance and disease.

■ REVIEW QUESTIONS

1. How do sea grasses spread to occupy more space?

2. Which would be better for a sea grass and under what circumstances: spreading seed or growing vegetatively?

3. Why are sea grasses important for the existence of other marine species?

4. Why did eelgrass decline so rapidly in the 1930s? What was the consequence for near-shore marine animal populations?

5. Why is sea grass grazed relatively little? What characterizes sea grass species that are grazed?

6. Why do you think urchins are less effective in regulating species composition on vertical surfaces of rocky reefs, relative to horizontal surfaces?

7. Why are kelps important in the survival of many benthic invertebrate species?

8. What would happen if a sea otter population were removed from a kelp forest rich in otters, kelps, urchins, and abalones?

9. How do kelp forests affect the local hydrodynamic regime, and how may this affect benthic invertebrate larval recruitment?

10. What is a barrens, and why might it be a stable state in a kelp habitat?

11. What characteristics determine whether a given coral species is likely to be a hermatypic coral?

12. What processes determine whether and at what rate an entire coral reef will grow upward and seaward?

13. What evidence is there that zooxanthellae provide nutrition to hermatypic corals?

14. What evidence is there that space is a limiting factor in interactions among sessile species in coral reefs?

15. How may corals protect against incursions by other corals into their space?

16. Why may it be adaptive for many species of corals to have synchronized mass spawnings? What problems are created by multispecies mass spawning?

17. Many people once thought that coral reefs were rather stable environments and used this stability to explain the high species diversity found in reefs. Is this explanation valid? Why or why not?

18. What factors may have resulted in such high coral diversity in the area of the Coral Triangle?

19. Why do bleached corals look so white in color?

20. What wide-scale phenomenon in the Pacific Ocean can teach us a great deal about the effects of future warming trends on coral reefs?

21. What can cause coral bleaching to be a geographically widespread process?

22. How might ENSOs and global climate change interact to greatly increase problems for coral reef survival?

23. Are there keystone species in coral reefs?

24. Discuss: When an El Niño causes decline of local coral populations, replenishment of populations in the short run must come from local populations.

24. Why might human fishing have more dramatic effects on communities with tighter trophic linkages?

25. Why do the interactions of major factors cause more profound changes to ecosystems than just the action of one factor alone?

26. If ocean acidification continues, what major group of animals may come to dominate coral reefs?

Visit the companion website for *Marine Biology* at www.oup.com/us/levinton where you can find Cited References (under Student Resources/Cited References), Key Concepts, Marine Biology Explorations, and the Marine Biology Web Page with many additional resources.

CHAPTER 18

Benthos from the Continental Shelf to the Deep Sea

THE SUBTIDAL LANDSCAPE

Sampling the Subtidal Soft-Bottom Benthos

■ **Past sampling of continental shelf and deep-sea bottoms involved bottom dredges and grabs, which often failed to give a complete picture of sea-bottom diversity.**

Understanding of the subtidal seabed is hampered by how remote the seabed is. In the 1870s, the sailing ship H.M.S. *Challenger* circumnavigated the globe and provided the first worldwide description of bottom organisms. To do this, a dredge was cast over the side, attached to thousands of meters of piano wire, and dragged along the bottom. It was then winched to the surface, which after hours brought a bag of sediment plus animals to be sifted and picked. The sample came up completely scrambled, with no idea of the micro-environment from which it came.

Our knowledge of the shelf and deeper seabed was advanced with the development of better bottom samplers. An ideal sampler should (1) sample a large area of bottom; (2) sample a defined area and uniform depth below the sediment–water interface; (3) sample uniformly in differing bottom substrata; (4) have a closing device, to avoid washout of specimens as the sampler is brought to the surface, sometimes for a distance of thousands of meters; and (5) bring up a sediment sample intact, so that the living positions of the animals may be examined directly. It is also important for bottom samplers to hit the bottom gently. Most devices push water ahead as they hit the bottom, creating a **bow wave** that erodes the sediment and scares the mobile organisms.

Dredges are heavy metal frames with cutting edges designed to dig into the sediment. An attached burlap or chain bag is attached at the rear and collects sediment as the device is dragged along the bottom. The **anchor dredge** has a control plane that constrains the dredge to bite to a defined depth (**Figure 18.1**). The area sampled can be calculated from the volume of sediment collected, divided by the biting depth. **Sleds** are dredges with ski-like runners that permit the device to dig in only a few centimeters. They are usually not quantitative, but they can collect large amounts of material. This is a virtue in the deep sea, where animals are often sparse.

Gravity and spring-loaded **grabs** sample a precise area of bottom with two or more sharp digging sections. As the **Petersen grab** hits the bottom and the supporting wire has some slack, the hook, whose support depends on the wire's tension, releases and allows a chain to pull the two sections closed (**Figure 18.2a**). This device is efficient in shallow (25–50 m) mud bottoms. The **Smith–McIntyre grab** is a heavy, spring-loaded device that digs efficiently in both muds and sands.

The **box corer** is a useful deep-sea gravity corer that is guided into the bottom by a movable plunger mounted on a frame. When the frame hits the bottom, a spade is released that digs into the sediment and closes the bottom of the corer as the frame is lifted by a wire (**Figure 18.2b**). It takes a sample of defined area and depth.

■ **Submersibles, remote underwater vehicles, underwater video cameras, and other remote devices are crucial tools in investigating the seafloor.**

These samplers have a number of disadvantages. Samples cannot be located exactly as the sampler hits the seabed.

FIG. 18.1 A deep-sea anchor dredge used to sample soft bottoms. The metal plane forces the dredge to bite at a defined depth. The sediment is collected in the bag to the right, brought to the surface, and sieved for organisms. (From Sanders et al., 1965)

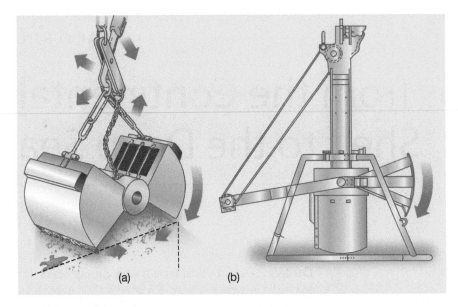

FIG. 18.2 Some benthic sampling devices. (a) The Petersen grab taking a sample from the seabed. (b) A box corer developed to take samples from the deep-sea bottom. (After Hessler and Jumars, 1974)

Also, it is difficult to control the spatial distribution of a series of repeated samples at the same general location. Drift of the vessel and currents make the exact sampling locations very indefinite. One literally cannot see what one is getting until the sampler is brought onboard ship. Finally, most samplers can take only a small amount of bottom.

A number of **remotely operated vehicles (ROVs)** have been developed that can navigate with great precision (**Figure 18.3**; see also Figure 1.7). They have manipulator arms with samplers attached and visual data can be collected with high-definition video cameras. Vehicles such as these were used to seek out and explore the sunken ocean liner *Titanic*. The *Ventana*, an ROV operated by the Monterey Bay Aquarium Research Institute, is equipped completely with sampling gear, video, and other instrumentation and has been completely renovated and remodeled several times (see Figure 1.7).

Human-occupied submersibles are among the most valuable vehicles for underwater exploration. The *Alvin* (see Figure 1.6), operated by the Woods Hole Oceanographic Institution, is capable of diving to depths of 4,500 m. The submarine usually has a pilot and a scientist onboard and is equipped with cameras, manipulator arms, and observation lights. The manipulator arms have permitted the establishment of many experiments at great depths. It is possible, using accurate satellite-based navigation on

shipboard in conjunction with submerged acoustic transponders, for the *Alvin* to establish an experiment, and to return later to the same site.

Ship-deployed devices have been used in recent years to rapidly construct a map of the seabed, and changes in surface properties of the bottom can be related to benthic communities that differentially affect sediment structure. **Multibeam sonar** is especially useful. A sonic signal scans the bottom as the ship is moving, allowing a signal to be gathered from a spreading beam 10–20 degrees from the ship's keel. A map of the seabed can be constructed and benthic samples can be collected to understand the changes in seabed surface features.

Sediment Type and Benthic Distribution

Suspension Feeders Versus Deposit Feeders

■ **Suspension-feeding benthic animals dominate sandy sediments, whereas deposit feeders dominate muds.**

Most of the subtidal seabed consists of soft sediments. **Deposit feeders generally dominate muds**, whereas **suspension feeders dominate sands**. Muds contain small sedimentary particles, which reflect a quiet water environment, and it is here that fine-grained organic matter tends to

Suspension feeders appear to do best in well-sorted sandy substrata. Suspension feeders depend on currents to deliver planktonic food (mainly phytoplankton). Sandy bottoms have faster currents and therefore probably have greater access to phytoplankton. Benthic suspension feeders may also feed on dead phytoplankton sinking from the overlying water. Collisions of large macromolecules in the surface water may also produce particles (marine snow) that sink through the water column and are consumed both by microorganisms and animal consumers.

■ Suspension feeders function poorly in muds, owing to the clogging effect of resuspended particles and to the destabilizing effect of deposit feeders on the sediment.

Why are suspension feeders relatively rare in muddy substrata? The answer has to do with particles suspended near the sediment–water interface. Water is often turbid just above muddy bottoms. Bottom currents over the muddy seafloor erode and suspend fine particles especially easily. The feeding and burrowing activities of deposit feeders further increase the water content of the sediment and convert the sediment to mainly fecal pellets, which can be eroded relatively easily by bottom currents. If a vane is moved over the bottom, it causes more resuspension of sedimentary particles than is observed when deposit feeders are absent. This effect is due mainly to the higher water content and relatively low stability of the deposit feeder–dominated sediment (**Figure 18.4**).

Both the instability of the muddy sediment and the near-bottom turbidity generated by the action of bottom currents over soupy muds have a strongly negative effect on the feeding efficiency, growth, and survival of benthic suspension feeders. Because deposit feeders make the sediment watery and unstable, they only further deteriorate the environment of suspension feeders. The negative effect exerted by deposit feeders on suspension feeders is an example of **trophic group amensalism** (amensalism is a negative effect

FIG. 18.3 Launch of a small remotely operated vehicle, operated by Friday Harbor Laboratories, equipped with real-time video. This vehicle is very useful for transects in shallow water. (Photograph by Jeffrey Levinton)

settle from the water column. After the time of the spring diatom increase (see Chapter 11), many diatoms often settle to quiet bottoms in estuaries and inner continental shelves. The abundance of deposit feeders correlates best with the percent clay size (particles < 4 μm in diameter) fraction of the sediment. This fraction is a good indicator of the settling of fine-grained organic particles, which are a good food source for deposit feeders.

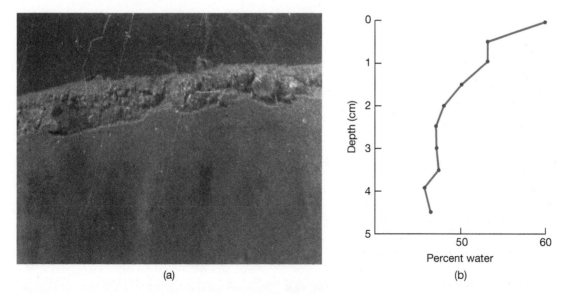

(a)

(b)

FIG. 18.4 The surface sediments of muddy bottoms have a high water content, owing to the burrowing and feeding activities of invertebrates. (a) A photo cross section of the bottom, showing a water surface layer. (b) Diagram showing the high water content of the surface sediments. (Photograph by Jeffrey Levinton)

of one organism on another; a trophic group is a group of organisms that feed in the same way).

Suspension-feeding organs tend to be clogged at high particle concentrations in the water. **Figure 18.5** illustrates this effect on the suspension-feeding estuarine clam *Rangia cuneata*. The **condition index** (the ratio of dry tissue weight in grams to shell cavity volume in cubic centimeters) is a general measure of nutritional state. Over a 2-year period, the condition index was greater for clams living in sands than for those living in muds. Suspended solids were greater in bottom waters over muds, relative to sands.

Over muddy bottoms, the weak bottom currents resuspend particles continually, and well over 90 percent of the fine particles spend some time resuspended in the water column. This probably enhances bacterial decomposition of the organic material associated with the resuspended matter (**Figure 18.6**). In late spring and summer, the shallow shelf water column becomes stable, owing to the presence of low-density warm water above and cooler water below. During that time, suspended fine particles are often trapped in a layer over muddy bottoms. Divers

often call such layer a "false bottom" because it appears to be the seabed as the diver descends. Such near-bottom, high-turbidity layers can often be traced over broad areas of estuarine and shallow shelf bottoms. In these areas, the near-bottom turbidity is too great for suspension feeders to grow well or even survive.

The Seabed Landscape

■ Modification of the sediment by burrowers and digging predators causes local patches.

A traverse along the coastal rim of the "oceanic bathtub" leads from coral reefs to mangroves to mudflats to rocky shores. In the deeper parts of the ocean (the continental shelf, slope, and the deep sea), the soft-sediment seabed appears at first glance to be far more homogeneous, even dull, relative to the shallow parts. Most of the seafloor is covered with soft sediments, ranging from clean sand to fine muds. The impression of homogeneity is reinforced by superficial visual inspections of the seabed. When the first photographs (**Figure 18.7**) and, later, videos, were taken of the bottom, one could see traverses over endless expanses of muds and sands, sometimes covered by mobile organisms such as brittle stars, bottom-living shrimps, and benthic fishes. The homogeneity is a bit of an illusion because much activity may occur beneath the sediment surface. Many soft-sediment animals actively burrow and strongly modify the sediment and can convert an otherwise spatially homogeneous sedimentary environment into one that is discontinuous.

Figure 18.8 shows a cross-sectional view of a mound made by the burrowing sea cucumber, *Molpadia oolitica*, which is common on the subtidal muddy seafloor of New England, north of Cape Cod. The animal creates fecal mounds protruding a few centimeters above the seabed. These attract smaller suspension-feeding bivalves, amphipods, and polychaetes, which live on the flanks of the

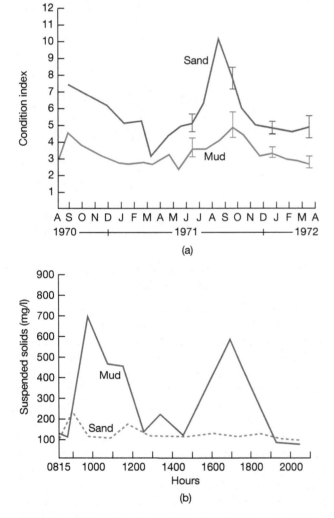

(a)

(b)

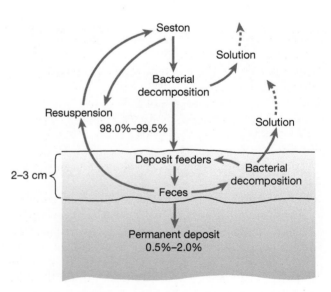

FIG. 18.5 (a) Condition index of the clam *Rangia cuneata* in sand and mud (vertical bars indicate 95 percent statistical confidence limits). (b) Suspended solids in the water column during a day over sandy and muddy bottoms. (After Peddicord, 1977)

FIG. 18.6 Resuspension and deposition of fine particles and relation to decomposition of particulate organic matter over muddy bottoms in shallow-water marine environments. (After Young, 1971)

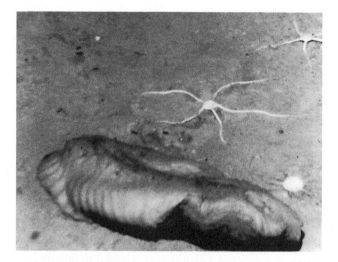

FIG. 18.7 Photograph of the soft-bottom seabed at 1,800 m depth. A sea cucumber is in the foreground, and brittle stars are to the right rear. (Picture taken by J. F. Grassle, courtesy of Woods Hole Oceanographic Institution)

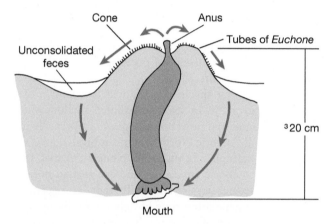

FIG. 18.8 Cross section of the sediment showing the feeding position, surface cone, and overall microtopography generated by the burrowing sea cucumber *Molpadia oolitica* in Cape Cod Bay, Massachusetts. (From Rhoads and Young, 1971)

mound. Between the mounds, deposit-feeding animals decrease the sediment stability and make the bottom inhospitable to suspension feeders. In effect, a population of *Molpadia oolitica* creates a small mountain range, with distinct "mountain villages," as opposed to the "valley communities." In the deep sea, similar mounds are created by echiurid worms and several species of polychaetes, which generate a landscape of hills and valleys in the seabed.

Larger mobile animals feed on benthic organisms and thereby increase the patchiness of the seafloor. Skates and rays, for example, leave large depressions in the sediment as they feed on benthic fish and burrowing invertebrates, and create a series of hills and valleys. The mysticete gray whale *Eschrichtius robustus* is a major predator of benthic amphipods in the Bering Sea, the west coast of North America, and other geographically scattered soft bottoms, and it turns over the sediment extensively as it feeds on the bottom. Such mobile predators not only kill some of the benthos, but also completely disrupt the local sedimentary regime and the living positions of the survivors.

■ Bottom currents disturb the sediment and generate a series of microenvironments.

Bottom currents create a second source of spatial patterning. Currents exert a shear stress on the seabed, and sedimentary particles are bounced and dragged along the bottom or propelled into the mainstream. **The movement of sedimentary particles often results in a variety of sedimentary structures,** including **ripple marks** if the current moves in a single direction for any length of time. Observations of a deep-sea site on the continental rise of the western North Atlantic Ocean show that erosion and deposition are both very active processes, with large-scale and sudden changes of benthic populations (Aller, 1998). The local environment on the crests and the troughs of the ripples can be quite different when compared with the needs of differing benthic creatures. Fine-grained organic detritus often accumulates in the ripple troughs and is of use to detritus feeders. The crests are a bit more exposed to the currents, which is of advantage to suspension feeders. If the current is particularly strong, then the crests migrate rapidly, sometimes on the scale of centimeters per hour. Under these circumstances, animals requiring a stable seabed to maintain a permanent burrow do not survive very well.

Certain benthic groups, such as clams and polychaete annelids, tend to be abundant even in deep-sea high-energy bottoms. Following storm events, however, members of meiofaunal groups, such as harpacticoid copepods, are often eroded away. To experimentally test the hypothesis that bottom currents altered meiofaunal and macrofaunal communities, David Thistle and Lisa Levin (1998) placed a flume on the seabed at about 583 m depth in a Pacific locality. The flume altered local bottom currents by guiding water through a narrow passage, which increased the velocity at the seabed surface (remember the principle of **continuity,** discussed in Chapter 6). Following storms, meiofauna were depleted relative to control sites, where currents were not altered experimentally.

■ Organic material is distributed discontinuously, generating local sites of high food value.

Certain parts of the seafloor comprise a sort of organic trash heap. When a large fish dies, it often falls to the seabed, where it commences to decompose. Soon thereafter, the decomposition consumes local oxygen, and the surrounding sediment becomes anoxic. Near the coastal zone, currents transport a surprising amount of plant matter to deep-sea bottoms. For example, large fragments of kelp have been collected or photographed at depths of 2,000–3,000 m on the seafloor adjacent to the southern California coastline. Here, the continental shelf is only a few kilometers wide so it is quite easy for such material to be swept out to the deep sea. North of St. Croix, in the U.S. Virgin Islands, large fragments of the turtle grass *Thalassia testudinum* can be found in deep-sea sediments. Again, the seaward extent of the shallow continental shelf is quite small, and only a few kilometers from the shoreline there is a rapid plunge to the deep-sea bottom.

Such carcass and detrital falls may have a positive effect on deep-sea benthos. As will be discussed further, the deep sea is a world of very low food supply. These falls, therefore, may be the source of quite patchy organic matter inputs, which can be eaten by scavengers or may gradually be broken down into particles of a size suitable for deposit feeders. Alternatively, bacteria may decompose the material and deposit feeders may derive nutrition by eating the bacteria in the sediment.

The impact of food falls was first discovered by the use by Scripps oceanographer John Isaacs of a "monster camera" that was placed over a bucket of fish scraps placed on the deep-sea bed. Very rapidly, large populations of fishes and a very large amphipod were attracted to the scraps and consumed them (**Figure 18.9**). This experiment demonstrated that there are deep-sea organisms adapted to moving toward such fish falls, which are rare riches in an otherwise poor environment.

■ Patchy dispersal and recruitment also create spatial change in benthos on the seabed.

Planktonic larval dispersal is another source of patchiness. Owing to currents, predation, and other factors such as bottom substratum, temperature, and salinity, larval settlement is very patchy. As a result of patchy settlement, a series of samples taken along a transect of seabed may have quite different assemblages of animals.

There are very good and very bad years for larval survival. Thus, every few years, an extremely large and localized larval recruitment may colonize a small area. This colonization may survive for many years, leaving a strong impact on the local animal community. This is probably a widespread phenomenon, but it is most noticeable in the echinoderms, which are relatively large and, therefore, conspicuous organisms. Brittle stars and sea urchins often arrive suddenly and in great numbers, lasting for many years. Through

FIG. 18.9 A shark attracted to a bait bucket placed on the deep-sea floor. (Photograph by John Isaacs, courtesy of Scripps Institution of Oceanography)

bottom sampling, they have been identified often as a series of patches of differing age, depending upon the time at which a larval swarm settled successfully in a given spot. Therefore, where planktonic larval dispersal dominates benthic recruitment, the seabed is a mosaic of settlement of dominant year classes of larval settlement.

■ Succession of subtidal soft sediments involves colonization of a disturbed and abiotic substrate by rapidly colonizing surface dwellers, followed by colonization and dominance of deeper-burrowing species.

Marine soft sediments are well known to have cycles of disturbance and succession. Lack of dissolved oxygen in sediment pore water and continual erosion are two major sources of stress, and a disturbance required to initiate succession usually involves these two factors. In Long Island Sound, New York, winter storms commonly erode the bottom down to water depths of approximately 10 m. The top few millimeters or centimeters of surface sediments are removed, exposing a sulfide-rich anoxic layer. The removal of burrowing organisms and the increase of sulfide make the sediment inhospitable for a wide variety of benthic infauna, and larval recruitment of many species is inhibited (Marinelli and Woodin, 2002).

Some pioneering species are adapted to colonize disturbed subtidal muds. In strongly polluted shallow-water habitats, the small red polychaete *Capitella* can rapidly invade. The small Atlantic North American bivalve *Mulinia lateralis* is an example of a successful pioneer colonist. It is well adapted to anoxic conditions and can sustain normal activity after several days of anoxia. It feeds on suspended matter. It has a generation time of only 2 months and can rapidly produce a set of offspring. Species adapted to colonizing disturbed bottoms feed either on suspended matter or on surface detritus. Deeper feeding is precluded, owing to the anoxic hydrogen sulfide-rich pore waters.

Later-colonizing, deeper-burrowing colonists (**Figure 18.10**) continue the process of sediment detoxification by burrowing and by circulating water through tubes and siphons that extend to the sediment–water interface. These species, which include gastropods, bivalves, and polychaetes, tend to be characterized by lower colonization potential, smaller investment in reproduction, and lowered mortality rates, probably stemming from their burrowing below the sediment–water interface.

THE DEEP-SEA GRADIENT
The Shelf–Deep-Sea Gradient

■ A transect from the shelf to the deep sea defines a shift from shallow-water bottoms with high biomass and productivity to remote deep-sea bottoms with low biomass and productivity.

Once the problem of quantitative sampling was solved, it was possible to understand the change of animal population density from the shelf to the deep-sea abyssal plain. The

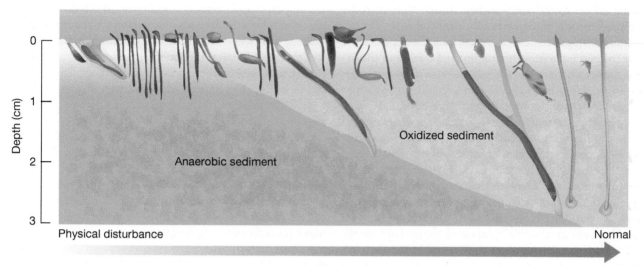

FIG. 18.10 Dominants of early (left) and late succession (toward right) in soft sediments of southern New England. Note the transition from dominance by surface forms to deeper-burrowing species. (After Rhoads et al., 1978)

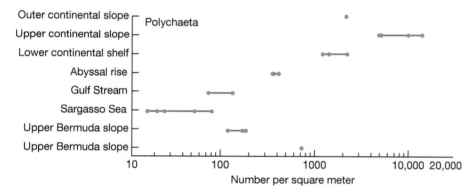

FIG. 18.11 The number of polychaetes found in different depth zones, along the Gay Head–Bermuda transect. (From Sanders et al., 1965)

standing stock of animals, measured in terms of numbers or biomass, declines extensively with increasing depth. Howard Sanders and colleagues (1965) established a transect from Cape Cod to Bermuda and showed that numbers declined substantially as one moved seaward of the continental slope. Abyssal densities of polychaetes are 100-fold less than those on the upper continental slope (**Figure 18.11**). The same pattern applies for meiobenthos, whose biomass decreases toward the deep sea. The question raised is: Why is benthic biomass so reduced in the deep sea?

Input of Organic Matter

■ **The input of organic matter to the deep-sea floor is low because of the distance from the coast and the great depth through which organic particles must travel, but localized areas of the seabed receive pulses of organic matter.**

The supply of food to the seabed depends greatly on production in the water column above. Primary production is great in shelf waters and in upwelling systems, and some of

the production inevitably leaves the surface and reaches the sea bottom. In the deeper-water parts of the seabed, animals depend on this deposition because light at such depths is too low for indigenous photosynthetic production. Near shore on the inner continental shelf, primary production is high, and bottom deposition is also great. There is also a peak of production over the continental shelf–slope break, where nutrient-rich waters intrude onto the shallow waters of the shelf. In the open ocean, seaward of the shelf–slope break, primary production is much reduced, which in turn reduces the supply of organic matter to the bottom. Importantly, a very small fraction of the primary production on the continental shelf is likely exported across the shelf-slope break to be deposited on the deep-sea floor (Falkowski et al., 1988). This greatly reduces the organic matter supply from the surface waters of the very-low-productivity gyre centers. In the inner continental shelf, for example, 30–50 percent of the phytoplankton may reach the bottom. However only about 2–7 percent of the surface production reaches the bottom beneath the gyre center of the North Atlantic. Because the primary production in the open-ocean surface waters is less than that of the inner shelf, the supply of organic matter to the deep-sea bed is much less than the supply to bottoms of the inner continental shelf.

Along with production at the surface, the depth of the water column influences the amount and quality of the organic matter that reaches the bottom. The deep sea averages about 4,000 m in depth, and organic particles may take weeks or even months to reach the bottom. During this period of descent, bacteria attack the particles, and only very refractory particulate organic matter eventually reaches the bottom. While localized pulses occur, overall there is a strong relationship between deposition of organic matter to the deep-sea floor and benthic biomass. In the western North Atlantic, surface primary production is strongly positively correlated with deposition on the sea floor (Johnson et al., 2007). Input of particulates to the sea floor was the driving force of biomass of the deep-sea benthos. Deep-sea bottoms that are deep, beneath surface waters of low primary productivity, and very remote from the continental margin (e.g., deep-sea bottoms below subtropical oceanic gyre centers) are generally depauperate in supplies of rich POM. In the Antarctic, meiobenthic biomass declines tenfold from deep-sea bottoms with relatively high food input relative to bottoms with low food supply.

In some regions, organic material falls to the bottom rather rapidly, especially in oceanic surface areas where primary production is high and there is, therefore, a large source of sinking particulate organic material from the surface. This rapid sinking is well known in the northeast Atlantic, for example, west of Ireland, where oceanic primary production is relatively high. Such rapid transfer occurs over bottoms that lie beneath surface waters with strong seasonal pulses of primary production (Billet et al., 1983; Graf, 1989). There is evidence that benthic animals and microorganisms in the deep-sea sites beneath these rich surface waters respond to seasonal deposition of organic matter by consuming sediment immediately after the organic matter reaches the bottom. Smaller benthic creatures such as foraminifera process the material as it reaches the seabed. Sediment bacteria are probably equally responsible for processing of sedimentary particulate organic material by decomposition. Mobile benthic organisms, such as holothurians and ophiuroids, move along the bottom and locate patches of phytodetritus that has reached the seabed. Seasonal studies of these animals in the northeast Atlantic using time-lapse benthic cameras show movements of populations of these mobile creatures and grazing of sedimented phytodetritus, which reaches the bottom following the spring phytoplankton bloom (Bett et al., 2001). Some material, moreover, is consumed by midwater creatures, which in turn transfer organic matter deeper by defecating relatively nutritious material.

■ The decline of input of organic matter to the deep-sea bottom is accompanied by strong changes in feeding types of benthic animals there.

The reduced organic input to the deep sea also affects the relative abundance of the major benthic feeding types. With increasing depth, suspension feeders become relatively less abundant, whereas deposit feeders become the dominant feeding type in fine-grained sediment. The lack of phytoplankton makes the deep sea inhospitable for phytoplankton-feeding suspension feeders typical of the shelf, but deposit feeders can still ingest the sediment and extract the organic matter that settled from above to the seafloor. One bivalve has been found to have an extraordinarily long gut relative to its close relatives of the same size in shallow water. This may be a reflection of the very low food quality, as well as the need to digest as much as possible from the food as it passes through the gut. Benthic carnivores become less abundant as one samples from the shelf to the mid-ocean deep-sea abyssal plain. This is probably due to the general reduction of animal biomass in the remote abyss, which makes it difficult for the carnivores to find prey. The very rare carnivores may also have difficulty in finding mates.

■ The deep-sea bed consists of sediment with little organic matter, and microbial activity is very low relative to upper-slope and shelf bottoms.

The input of organic matter from the overlying water column to the seabed declines steadily as we go seaward of the continental shelf–slope break. In inner continental shelf bottoms, the organic content of the sediment is in the range of 2–5 percent. In the open ocean, the seabed is very low in organic content, between 0.5 and 1.5 percent. The situation reaches an extreme in open-ocean abyssal bottoms beneath water columns of very low primary production, such as beneath gyre centers. Here, the sediment is so low in organic matter (<0.25 percent) that it is oxygenated and reddish, called red clays. One may expect that such bottoms would be very low in microbial and animal biomass and activity.

The morphology and activity patterns of some deep-sea animals reflect the low food supply. Many deep-sea bottom fish are sluggish and have low muscle mass relative to animals of similar size in shallow water. Metabolic rates of deep-water fish are very low. Many abyssal fishes and fishes living in the deep-water column seem adapted for the consumption of the rare meal that comes along. Unfortunately, we know far too little about the actual activity patterns of deep-sea fishes. I. G. Priede and colleagues (1991) put acoustic transmitters in baits placed on the seabed of the North Pacific and North Atlantic oceans, at depths of 4,000–6,000 m. Grenadier fish consumed the baits and moved at speeds of 0.11 m s^{-1}. The fish seemed to be quite active and probably moved rapidly toward the bait, rather than sluggishly waiting until they smelled the rotting fish fall. Even though meals may be rare in the deep sea, animals may of necessity move briskly to take advantage of the events.

The impoverished nature of the deep-sea bed was highlighted no better than by a shipside accident in 1968, which caused the sinking of the deep-sea submarine *Alvin* to the seafloor at a depth of 1,540 m. Luckily, no one was hurt seriously, but some Woods Hole Oceanographic Institution scientists were unlucky enough to have left their lunches in the submarine that went over the side. In a remarkable salvage operation, the *Alvin* was recovered about a year later, and the food (thermos with bouillon, apple, and sandwiches) showed little decomposition. Dr. Holger Jannasch tasted the bread, and it was edible. The soup, initially prepared from canned meat extract, was not decomposed very much after 1 year. When such food was kept in a refrigerator at 3°C, bacterial attack was immediate and starch and protein fractions spoiled in just a few weeks. Therefore, the low temperature of the deep sea (2–4°C) was not the main factor that retarded decomposition. This accident stimulated serious scientific studies of the rate of decomposition at deep-sea bottoms.

The impression of a low rate of decomposition in the deep-sea floor was only strengthened by other investigations. Oxygen consumption of deep-sea sediments was measured by placing a bell jar on the seabed and inserting a polarographic oxygen electrode in the water enclosed by the jar. Oxygen consumption of the abyssal seabed was 100-fold less than on continental shelf bottoms. The lowered oxygen consumption reflected the severely reduced animal and microbial activity in deep-sea sediments. Bacterial substrates labeled with radioactive carbon are taken up by bacteria at a rate usually less than 2 percent of the rate measured in shallow waters.

The mechanisms accounting for the lowered rate of decomposition are not understood completely. The low temperature of the deep sea cannot explain the difference because, at the same low temperature, the shallow-water measurements of decomposition rate are much higher. There may also be a direct effect of the great pressures on the deep-sea bottom (the pressure at 4,000 m, e.g., is about 400 surface atmospheres). Deep-sea bacteria are known to be barophilic—that is, they function best under high pressure—but still, their maximum rates of decomposition may be less than those of shallow-water bacteria. As we shall see, however, the pressures in the deep sea do not necessarily retard metabolic activity by a direct physical limitation such as pressure.

Environmental Stability in the Deep Sea

■ **As we go from the shelf to the deep sea, we pass from physically variable environments to those that are stable.**

The continental shelf is a physically unstable environment. In the midlatitudes, especially, there are strong seasonal changes in temperature, salinity, dissolved oxygen, and light. In North Carolina, for example, coastal water temperatures vary from winter minima near 3°C to 30°C in summer. At these latitudes, light also varies extensively, owing to changes in day length and the angle of the entry of sunlight into the sea surface. At higher latitudes, temperature is a bit more constant through the year, though low. Day length, however, is even more variable; consider the midnight sun of the Arctic versus the sunless days of the Arctic winter. The only real exception to the general rule of shallow-water variation is in the tropics, where the water is rather warm and stable all year round. The same holds for day length.

Decadal and even sub-decadal scales impose strong changes in conditions on continental shelves. The Pacific Decadal Oscillation and the North Atlantic Oscillation are examples of decadal-scale changes in temperature and wind patterns that bring strong changes to the shelf (also to the open ocean). The **Pacific Decadal Oscillation**, with a 20- to 30-year cooling-warming cycle, has strong impact on west coast U.S. fisheries such as salmon and, when combined with another cycle, the **North Pacific Gyre Oscillation**, can even be related directly to shifts of marine bottom communities within coastal areas such as San Francisco Bay (Cloern et al., 2010; see Chapter 3). On longer time scales, the shelves and estuaries of the world are less stable than deeper water habitats. Because of the advance and retreat of the glaciers over the last 2 million years or so, large amounts of seawater alternately have been locked up in glacial ice or liquid water at higher sea-level stands. This difference is sufficient to have caused sea-level fluctuations of 100–200 m. During the last glacial maximum, about 11,000 years ago, global sea level was more than 100 m lower than it is today, making the continental shelves far narrower. Some of the major estuarine systems, such as Chesapeake Bay, San Francisco Bay, and the Baltic Sea, did not even exist. At present, estuarine environments are far more environmentally variable than open-coast systems.

The variability of climate on the inner continental shelves and estuaries is often increased by the dominating effect of continental processes. In North America, weather systems move from west to east, so continental effects are stronger on the east than the west coast. Thus, seasonal

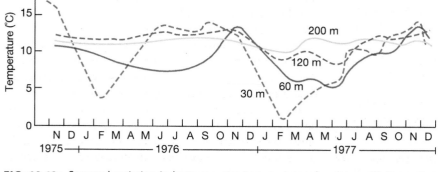

FIG. 18.12 Seasonal variation in bottom-water temperatures decreases with increasing depth. (Courtesy of D. F. Boesch)

temperature change in the shallow ocean is more extreme on the east coast of the United States than on the west coast.

As we move to greater depths, the variability in the physical environment is dampened considerably. **Figure 18.12** shows seasonal variation in bottom temperatures at varying depths on the continental shelf and upper continental slope off the coast of New Jersey. Note how much more constant the 200-m depth is, relative to the 30-m depth. This trend continues as we move to deep-sea bottoms. At the base of the continental slope, toward the abyss of 4,000 m, the temperature and salinity variations are minuscule. Over the year, temperature in the abyss varies less than 1°C. The deep sea is thus a physically constant environment on the scale of years and even centuries.

Deep-Sea Biodiversity

■ **Despite the poor food conditions, the deep sea in the vicinity of the continental slope and continental rise actually has more soft-bottom species than do corresponding environments on the inner continental shelf.**

The remoteness from inner-shelf sources of organic matter, combined with the great depth of water through which organic matter passes and is decomposed, makes the deep sea an impoverished environment. Animal density is very low. Until the late 1950s and early 1960s it was generally believed that the deep sea also had very few species. This belief was based mainly on deep-sea dredge samples, which pulled up very few individuals and very few species. The impression was an illusion. Howard Sanders, Robert Hessler, and their colleagues began a sampling program in the deep North Atlantic, but they used much higher-quality samplers than had been used previously (Sanders et al., 1965; Sanders and Hessler, 1969). The large deep-sea sled they used could collect a great deal of sediment and was equipped with a closing device, which tripped just as the sampler was raised from the seafloor, to prevent washout. As a result, they collected far more small animals than had ever been collected before, with the exception of an earlier Soviet expedition. Previous samplers had no closing device, and most small animals were washed out of the sampler during its ascent through several thousand meters of water column.

The sampling showed that the deep sea at depths of 1,500–2,000 m, despite its low animal density, has large numbers of species relative to similar muddy-bottom habitats on the continental shelf (Sanders and Hessler, 1969). Grassle and Maciolek (1992) reported that North Atlantic muddy bottoms at depths of 1,500–2,500 m had at least 1,500 species!

It is important to realize that estimates of deep-sea diversity are not straightforward. First of all, samples may have large differences in numbers of specimens. This makes it difficult to know the number of species that come from the locality from which the sample is extracted. We might expect that the more specimens we have, the more species we collect. This positive trend between specimen number and number of collected species would continue with increased sample size up to a plateau; eventually, more

individuals would reveal no new species. But such a plateau has yet to be reached in benthic samples from the deep sea. Even with hundreds of samples, one still finds many species with only one individual, suggesting that there are many rarer species that have not yet been detected. Is there something unusual about genetic mechanisms that generate new species in the deep sea? Ron Etter and colleagues (1999) sequenced DNA of deep-sea mollusk species and found that DNA differences between species were similar to that of shallow-water species.

Before estimating species diversity from a site, deep-sea biologists use a large number of samples to construct a **rarefaction curve**, which relates the number of species collected to the number of specimens. Then, a sample can be standardized to a standard number of specimens collected. This allows one to compare samples with thousands of specimens to others with just a few and get an estimate of species richness. Most studies now use the so-called **expected number of species**, which is the expected number of species to be found in a sample of n individuals that are selected at random from a larger collection containing N individuals, S total species, and a defined distribution of abundance among the species (the interested student should consult Hurlbert, 1971).

■ **Biodiversity of soft bottoms increases with increasing depth, but then declines as depths surpass about 2,000 m.**

Figure 18.13 shows trends in numbers of species as a function of water depth, with samples standardized to a sample size of 50 specimens. For a wide variety of animal groups, diversity first increases to a maximum at a depth of about 2,000 m; at greater depths the diversity decreases (see Rex, 1981, and Rex and Etter, 2010). The increased biodiversity at intermediate depths has been found in temperate and tropical latitudes, although areas with strong upwelling can depart from this pattern.

It is possible that in depths much greater than 2,000 m, the scarcity of food in such remote abyssal bottoms outweighs the effects of increased environmental stability. Population sizes may be so low that extinction is more likely, thus lowering diversity. This hypothesis is consistent with the especially strong decline in carnivores found in remote abyssal bottoms. Carnivorous gastropods decline more rapidly than deposit-feeding snails, as one samples with increasing seaward distance from the shelf–slope break. By contrast, the more abundant microscopic Foraminifera increase in diversity over the same gradient. This may be explained by their greater population sizes, which lower the probability of extinction.

Ron Etter and Fred Grassle (1992) discovered that high-diversity sites at intermediate depths of about 1500 m also have greater sedimentary particle diversity. The particle diversity may reflect a diversity of ingestible particles or even a diversity of hydrodynamic regimes, both of which may support a greater diversity of ecologically different species. The only problem with this hypothesis is that the same diversity gradient occurs in many groups that are not deposit feeders and are not particularly dependent on sediment grain size. It is also possible, as will be described in the next section,

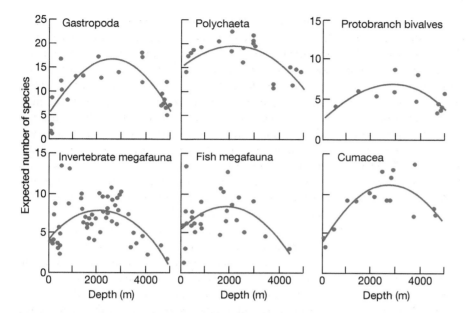

FIG. 18.13 Variation in species diversity of various benthic animal groups as a function of increasing water depth. Species number is an estimate for a sample of 50 individuals. (Data compiled by Rex, 1981)

that other types of environmental variation are important in this depth zone that affect groups beyond deposit feeders.

- **Because of strong regional and depth differences in environment in the slope and bathyal zones of 200–4,000 m, there is much more regional difference and depth difference in species occurrence and genetic differences within species than in the deeper abyssal zones of greater than 4,000 m.**

You should not visualize the deep sea as one vast and monotonous realm. This is especially true of the continental slope to bathyal depth range of 200–1,000 m. A number of environmental variables show very strong geographic and temporal changes. For example, the continental slope within this range is dissected by enormous canyons, where very large amounts of sediment move suddenly into deep water from slope depths. The canyons also can isolate different parts of the slope from each other, which might promote the origin of species and geographic genetic differentiation between different locations on the slope and even the continental rise. The sediment that is eroded and deposited within the bathyal realm also creates strong differences of sediment type over relatively short distances. Finally, the history over the past 2 million years of glacial retreats and advances resulted in strong fluctuations of sea level of the order of a fall and rise of 120 m, respectively. During low stands of sea level during glacial maxima, erosion on the continental slope was heightened. Ocean thermohaline current structure was greatly altered relative to high stands, which also affected temperature change in the slope-bathyal zone. The bathyal zone has therefore fluctuated a great deal and is far more regionally variable than the abyssal depths deeper than 4,000 m. In a few words, the bathyal zone is a world of spatial and temporal change, relative to the more homogeneous deeper abyssal zone.

The depth difference in variability has had a major impact on the benthic invertebrates. Slope-bathyal communities have greater numbers of species than shallower shelf and deeper abyssal depths (**Figure 18.13**), and differences between areas in species occurrence is also greater within this zone than in the deeper abyssal depths (but not necessarily on the shelf). We have only limited information on genetic differences within species, but there appears to be greater genetic variation over the slope-bathyal depth and continental rise zones than over much larger spatial areas in abyssal bottoms (Etter et al., 2005; also see Jennings et al., 2013). In these zones, depth-related genetic differences may be caused by ecological changes with depth that enhance isolation and geographic speciation. The impact of environmental change in the slope and bathyal zones is therefore reflected in natural selection that results in genetic differences over these depths.

- **Despite the apparent similarities in sediment properties, there is a prominent latitudinal diversity gradient in deep-sea benthos.**

There is no obvious difference in sediment properties among latitudes. Indeed, the differences are much more pronounced with changes of depth, since both coarse and fine sediments and mixtures are found on the continental shelf. We might therefore expect no latitudinal gradient in species diversity, but this is not so. Michael Rex and colleagues (1993, 2000) compiled data for deep-sea bivalves, gastropods, and isopods and found clear latitudinal diversity gradients, especially in the North Atlantic (**Figure 18.14**). South Atlantic species diversity declined with increasing latitude, but the pattern was much less clear.

These results suggest that there must be something more global driving deep-sea diversity because the latitudinal pattern parallels a shallow-water latitudinal gradient.

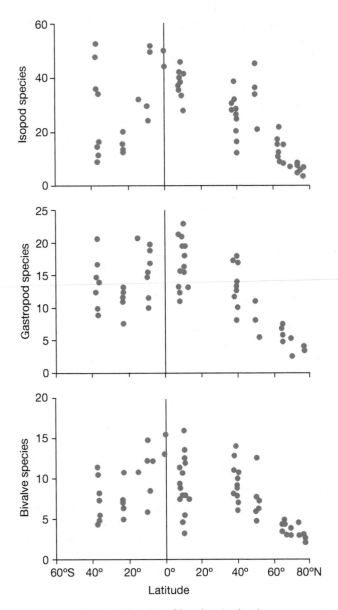

FIG. 18.14 Species diversity of benthos in the deep sea versus latitude for Isopoda, Gastropoda, and Bivalvia in the Atlantic. (Data from Rex et al., 1997)

As we discussed earlier, the deep sea is not disconnected from the surface, as we once thought. Seasonal primary production at the surface in some areas is often exported to the deep-sea bed in rapid pulses. It may therefore be that patterns of longer-term change at the surface affect the deep sea. This does not answer the important questions of how and why, however. We discuss the general problem of regional variation of diversity in Chapter 20.

■ **Deep-sea biodiversity may have been promoted by a stable environment, although disturbance and microhabitat heterogeneity also may have promoted diversity.**

The high biodiversity of the deep sea, seaward of the shelf-slope break, was a great surprise to ecologists. Despite the absence of light, very low food supplies, and very low microbial activity, fantastic numbers of species have been recovered. At first, one tends to conceive of the deep sea as a monotonous muddy bottom, which causes us to ask: How can so many species coexist in what appears to be a continuous environment with no variation? One might think that the reduced food would cause strong interspecific competition, resulting in competitive extinctions and low diversity. The deep sea is not homogeneous, however. We have discussed the strong effects of currents, which disturb the sediment and reduce populations, much in the same way that disturbance affects rocky shores. This might promote the coexistence of species to some degree. But there is no reason to believe that such disturbance is any more prevalent than on the continental shelves, where biodiversity is less. Muddy-seabed animals also produce tubes, burrows, and mounds. This would surely increase local environmental heterogeneity and promote an increase of microhabitats that would allow more species to coexist. But again, one cannot see why bioturbation, and bioturbation-promoted sediment structures, would be less on the shelf.

To understand the high biodiversity of the deep sea, we have to think about the process of speciation. Deep-sea species mostly have reduced dispersal mechanisms relative to their closest shelf relatives, which tend to have planktonic larval dispersal. The deep-sea dominance of reduced dispersal might help in promoting isolation, which is necessary for speciation.

Sanders and colleagues explained the high diversity in terms of the constancy of the deep sea. Because the deep sea was environmentally stable and perhaps an older habitat, species may have lower extinction rates and may accumulate to higher diversities than on shelf habitats. On the shelf, the habitat is less stable, and large-scale fluctuations in sea level may have contributed to increased extinction rates and lower speciation rates. If we go back just 8,000–10,000 years, the world was much more covered by continental glaciers, sea level was lower by about 150 m, and most of the continental shelves of the world were exposed to air. Such disturbance over millions of years may have caused extinctions on the shelves, but the deep sea of course has never been subjected to such massive disturbances and, as mentioned, speciation may be enhanced in the slope-continental rise region by depth-related ecological changes. We know that in the last few million years there have been significant extinctions of some of the shelf macrofauna, but we do not have a good record of deep-sea macrofauna on the same time scales. The solution to the high diversity of the deep sea still eludes us.

Deep-Sea Island Hotspots of High Diversity

COLD-WATER DEEP CORAL MOUNDS

■ **Deep- and cold-water hard surfaces often contain habitats dominated by stony and soft corals, ranging from a few colonies to enormous mounds of calcium carbonate sediment.**

In Chapter 17, we discussed the ecology of coral reefs, which are confined to open marine warm and shallow-water marine habitats. Communities with corals as a structuring element are also found in deep cold waters throughout the world over a depth range of 30 to over 1,000 m. They have been found in environments as varied as canyons in the continental slope, open-ocean seamounts, and in a large number of deep and hard substratum environments. Much of what we know comes from observations and sampling made by deep-sea submersibles. At first, many were impressed with the large colonies of gorgonians, which were meters tall; but later submarine expeditions revealed a more diverse fauna in many locations on enormous mounds of calcareous sediment. They are especially abundant in the northeast Atlantic but are also widely distributed in the Gulf of Mexico, the northern Pacific, the northernmost coast of South America, and off northwest Africa. They are extraordinary for their diversity and apparent support of some rich demersal fisheries. The latter may be their undoing.

The occurrence of these **deep-water coral mounds** makes them devoid of plants and algae but dominated by animals, particularly scleractinian calcium carbonate–secreting corals (without zooxanthellae), sea fans, and sea whips. The colonies provide a complex three-dimensional habitat that provides shelter and habitat for many other invertebrates and fishes. The coral species contribute calcium carbonate to the seabed, with the interaction of boring organisms such as boring sponges. The result ranges from small aggregations of corals, to mounds over 300 m in height and a few kilometers across, consisting of the accumulation of coral-derived sediment. For example, in the Porcupine Trough, a deep-sea area off the southwest of Ireland, deep-water coral mounds occur at depths of 500–1,200 m, may be as much as 350 m above the surrounding seabed, and have complex topography. These mounds may interact with surface currents, bringing organic matter downward from surface waters, providing a source of food to the deep coral mounds

(Soetaert et al., 2016). Thickets of corals are found on the top of the mounds, and they appear to be as complex as shallow-water coral reefs, though not as diverse. Coral mounds dominated by *Lophelia pertusa* are found throughout the world at depths from 200 to over 1,000 m (**Figure 18.15**).

Deep-water coral mounds are probably initiated by special sedimentary conditions that allow corals to colonize bottoms with hard substratum. Corals might have colonized submarine deposits of rocks that were delivered by sea ice during glacial maxima during the Pleistocene. As the glaciers retreated, currents eroded the fine particles, leaving a lag deposit of cobbles and boulders. Conditions that promote sedimentation during glacial advances in the Pleistocene would have been detrimental to coral mounds, which would be eventually buried in soft sediment. **We therefore have a cycle of mound development and burial that is related to glacial cycles.**

Mounds are more common in parts of the ocean with higher surface primary productivity. It is likely that sinking particulate matter is the main food of corals and other species, and this community, therefore, is similar to deep-sea soft-sediment animal communities that depend on the drizzle of organic matter from the surface.

■ Deep-water coral mounds are extremely diverse.

Despite the lack of contribution from photosynthetic organisms at these depths, coral mounds can be very diverse. Suspension feeders dominate the tops of mounds in the northeastern Atlantic, but the flanks of the mounds have soft sediment and are dominated by deposit feeders. A study of northeastern Atlantic deep coral mound tops found over 1,300 species, including corals and other cnidarians and a wide array of associated invertebrates and fishes. J. Murray Roberts, a major investigator in this environment, informs me that this number should be revised to over 2,000 species.

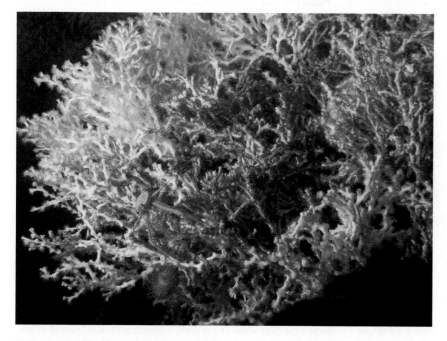

FIG. 18.15 The deep-water coral *Lophelia pertusa* with squat lobster and sea urchin. (Photograph by Steve W. Ross and others)

Corals and other cnidarians are the foundation species for this community. They grow slowly and live a long life. Colonies of one living black coral (*Antipatharia*) species were estimated by the ²¹⁰Pb method to be 200–500 years old (Williams et al., 2007). The foundation species of this community, therefore, are comparable to ancient terrestrial trees as structural elements in these communities. Black corals of the genus *Leiopathes* are especially fascinating, and colony ages have been estimated by means of radiocarbon dating to over 2,000 years (Roark et al., 2006) (**Figure 18.16**). They may well be the oldest living organisms in the ocean. Sea fans and scleractinian corals have a large number of associated and probably commensal brittle stars, crabs, and shrimp. It is not clear what interdependencies among species may exist, beyond the provision of protective and supportive habitat by the corals. Mobile species probably use this habitat for shelter, support, and food. A number of larger mobile fish feed in this habitat, especially species of the rockfish genus *Sebastes*, which feed on smaller invertebrates.

What is the biogeographic structure? Are all mounds alike? It is too early to say, but many of these species other than the main framework dominants are localized to a region. DNA markers demonstrate that *Lophelia* coral populations show strong differences between open-ocean and fjord populations in Norway, and there is evidence for population differences along the coast (Le Goff-Vitry et al., 2004).

Dominant species differ from region to region, but one or a very few species are the foundation for much of the community and only six coral species comprise the foundation globally for these coral mounds (Henry and Roberts, 2016). In the northeastern Atlantic, the Gulf of Mexico, and the Mediterranean, species of the coral *Lophelia pertusa* form colonies over a meter wide, which break down to coral rubble, and large numbers of species are associated with the rubble and the living coral colonies. Somewhat shallower reefs (ca. 70–100 m deep) dominated by species of the coral *Oculina* occur off the Florida continental shelf; they grow as large as 2–3 m wide.

The diversity of this community is striking. John Reed (2002) reports this level of diversity from about 42 *Oculina* colonies living in the Florida reefs: 230 mollusk species; 50 species of decapod crustaceans; 47 species of amphipods; 21 echinoderm species; 15 species of pygnogonids; 23 families of polychaetes; and numerous sipunculids, nemerteans, isopods, tanaids, ostracods, and copepods. Shallow-water tropical reefs are probably orders of magnitude more diverse.

■ Deep-water coral mounds support important fisheries, which attract damaging trawlers.

By now, coral mounds are being discovered throughout the world ocean, but their large fish populations are also being targeted by deep-sea bottom trawlers, whose gear drag along the bottom and destroy large parts of the coral-dominated seabed. This destruction was discovered partially by Norwegian fishers, who noticed that fishing declined as the corals were being destroyed by fishing gear. Sophisticated sonar-based means of locating the mounds makes it easy for trawlers to locate coral-dominated mounds. The slow growth of many of the corals makes recovery time far too long to allow a policy of destruction and protection, which would encourage renewal of the reefs. Fishing has damaged as much as 30–50 percent of Norwegian coral mounds. The Norwegian government issued protections in 1999 for about 1,000 km² of coral mounds, which was followed by protection of deep-water coral mounds in other countries, including the United States. Coral mounds on the high seas are not protected because they are in international waters and such protection treaties are difficult to establish (see Chapter 21 for some recent progress).

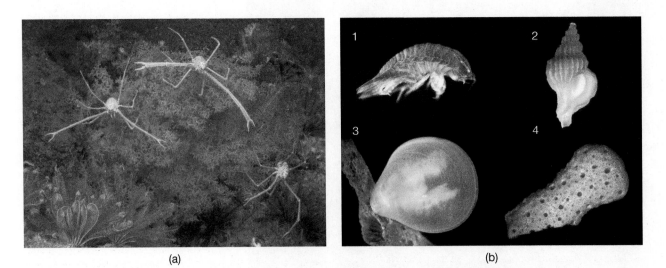

(a)   (b)

FIG. 18.16 Some organisms found on deep-sea coral mounds. (a) Large antipatharian coral (probably *Leiopathes*) on a northeast Atlantic carbonate mound. (Image courtesy of AWI & I. Fremer) (b) These examples show fauna from a giant carbonate mound in the northeast Atlantic: (1) isopod *Natatolana borealis*, (2) gastropod *Boreotrophon clavatus*, (3) brachiopod *Macandrevia cranium*, (4) hydrocoral *Pliobothrus symmeticus*. (Images courtesy of L. A. Henry, Scottish Association for Marine Science)

■ **Deep-water coral mounds are affected by the ability to secrete aragonite and may be in a danger zone because of increasing carbon dioxide in the atmosphere.**

Like shallow-water scleractinians (see Chapter 17), deep-water corals secrete aragonite skeletons. This can be a strongly limiting factor because increasing depth and pressure make it more difficult to secrete calcium carbonate in the form of aragonite. There is a maximum depth, the **aragonite saturation horizon (ASH)**, beneath which aragonite-based organisms cannot live. In the North Atlantic, the ASH is greater than 2,000 m, but in the Pacific, it is far shallower, sometimes near the surface (50–600 m). This fits with the abundance of deep-sea coral mounds in the North Atlantic but the near absence of such deep communities in the Pacific. When deep-water coral mounds occur in the Pacific, the species secrete calcite, which is the more stable form of calcium carbonate (Guinotte et al., 2006). For example, in the Aleutian Islands of Alaska, the ASH is less than 150 m, and deeper coral communities are dominated by octocorals, stylasterid corals, and sponges, all of which secrete calcite and not aragonite. The pattern is not perfect. The ASH in the Antarctic is rather deep, but deep-water coral mounds are not found there.

The question arises as to whether increased carbon dioxide in the earth's atmosphere will reduce the ASH in the Atlantic and cause the extinction of some coral mounds dominated by scleractinian aragonitic corals such as *Lophelia*. The shallower ASH in the Pacific suggests that this might be a problem, but we do not have enough information yet to be sure of the likely effects.

Seamounts

■ **Seamounts are relatively isolated elevated areas, usually of volcanic origin, rising 1 km or more above the seafloor and whose top does not reach the sea surface.**

Thousands of submarine seamounts don't reach the surface;[1] they harbor a diverse biota on their summits (Clark et al., 2010). Seamounts are widespread through all oceans (**Figure 18.17**) and may rise from the deep-sea abyssal plain, although these are less common than those that are located near more typical sources of volcanism, such as oceanic island arcs or midoceanic ridges. About half of all seamounts are in the Pacific Ocean. Many rise with their summits within the photic zone, but the majority are in deeper waters and therefore do not have access to photosynthesis as a source of production.

■ **Seamounts are mapped and observed with multibeam acoustic methods, video taken by ROVs, and direct sampling.**

During World War II, a major oceanic bathymetric exploration in the Pacific resulted in the discovery of hundreds of flat-topped seamounts that were named **guyots**. Since that time, satellite-based methods and the use of multibeam sonar has located thousands more. Researchers use

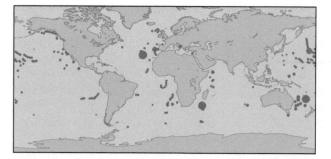

FIG. 18.17   Global distribution of seamounts. (From Clark et al., 2010)

FIG. 18.18   Color visualization of a multibeam sonar survey of the Brothers Seamount complex near New Zealand. (Courtesy of CenSeam, New Zealand)

different types of vehicles, including manned submersibles, ROVs, and autonomous underwater vehicles, to create visuals of the topographical surface (**Figure 18.18**) and details of the tops and sides. All have recovered spectacular photographs and videos of a rich biota, including corals and other invertebrates and large schools of fishes.

■ **Seamounts are dominated by a large number of sessile invertebrates, but also by a diverse array of fishes and mobile bottom invertebrate species.**

Seamounts have a diverse biota of invertebrates and fishes (**Figure 18.19**). Many seamounts have the coral faunas we described earlier as cold-water deep coral mounds, with some coral species that are nearly global in distribution. Coral species are especially abundant on the generally hard bottoms, but invertebrates from many phyla are found as well, along with a diverse group of fishes, both associated with the bottom and mobile species with broad geographic distributions. On the Norfolk Rise, east of Australia, over 500 species have been found on a series of seamounts, about 100 from nearby Lord Howe Island and neighboring seamounts, and nearly 300 species from a series of seamounts south of Tasmania, southeast Australia (de Forges et al., 2000). So far, 168 larger invertebrate species have been found at the Davidson Seamount in the northeastern Pacific, southeast of Monterey Bay (McClain et al., 2009). Therefore, many seamounts are diversity hotspots, with much

---

[1] Sea mounts rise at least 1 km from the sea floor.

(a)   (b)

FIG. 18.19   Photographs of the Davidson Seamount, located in the northeastern Pacific off the California coast near Monterey Bay. (a) The sea fan *Paragorgia* sp. (b) Feather star *Florometra* sp. with several species of soft corals. (Courtesy of Monterey Bay Aquarium Research Institute)

higher species diversity than habitats at the same depths, for example, on the continental slope.

Seamounts are not only diverse, but they can have very large populations of fishes. Fish may be attracted to the complex surface of seamounts, with corals and other epifauna. Locally, crustacea may move down during the dawn in a diurnal vertical migration and can be trapped near the surface of a seamount, which in turn attracts large numbers of fish.

- ■ **Seamounts often have hard substrata on top because deep-sea currents are focused on their summits, preventing accumulation of soft sediments.**

Because seamounts rise relatively steeply from the seafloor, they often project into complex current systems, which are often of high velocity and sweep along the summits of the seamounts. The strong bottom currents leave either bare hard bottoms or beds of soft sediment that usually have a rippled surface. On Cross Seamount, a location in the central North Pacific Ocean, the strong currents are tidal in origin (Noble and Mullineaux, 1989). These currents strongly influence the settlement of planktonic larvae, which are a very common ecological factor in these habitats. While larvae may preferentially select some hard substrata, such as iron-manganese nodules on the bottom, they also have behavior that allows them to drop into the boundary layer to select sites for settlement and metamorphosis in an active current environment (Mullineaux and Butman, 1990). While many seamounts have hard bottoms, some, particularly flat-topped guyots, are in relatively quiet-water conditions and therefore have soft sediments with a burrowing fauna not very different from that found at the same depths on the continental slope.

- ■ **Seamounts have been claimed to have highly endemic faunas, but this is not clear.**

One might expect seamounts, or at least groups of seamounts, to be very isolated and therefore contain species that are endemic to one or a group of neighboring seamounts. Because they usually rise directly from the deep-sea floor, one might imagine that larval transport might be insufficient to connect seamounts. The opposite impression, however, comes from reports of small parts of seamount faunas, which tend to show widespread occurrence of seamount species, and a minority of species endemic to one seamount system (Rogers, 1994). This generalization is controversial. On the one hand, de Forges and colleagues argued that seamounts near southeastern Australia had large numbers of species not found on other seamounts. But McClain and fellow researchers (2009) found that about 70 percent of the species found on Davidson Seamount were also found on other seamounts and in other nonseamount habitats nearby in California, so there might be a regional biota, but not one specialized to seamounts themselves. This interesting subject is far from settled, but Clark and colleagues (2010) argue that widespread species are the rule. There are some very isolated seamounts in the Pacific and even in the Caribbean basin that must be examined to see if isolation is as rare as McClain and his coworkers assert.

Many benthic species on seamounts have planktonic larvae. Many of these larvae may be lost to sea, or they might colonize other seamounts. However, the interaction of regional currents and tidal currents with seamounts also may create eddies, even circular eddies that would prevent larval spread away from the seamount and might create a mechanism of retention of larvae at the seamount (Mullineaux and Mills, 1997).

- ■ **Seamount biotas are endangered from overfishing and undersea mining.**

Seamounts have dense populations of sessile and mobile invertebrates and fishes, and therefore they attract large numbers of large mobile species of fishes and crabs. Large schools of fish near seamounts can be fished by trawling and can easily be located by sophisticated sonar systems available to oceangoing fishing fleets. The effect of fishing is devastating because many fishes associated with deeper seamount summits live in a relatively constant and relatively food-poor environment, where natural selection causes evolution of life histories with slow body growth and low reproductive rates. In New Zealand, the orange roughy *Hoplostethus atlanticus* (**Figure 18.20**) has been severely overfished from seamounts over 600 m deep (Clark, 2001). Its value as a fishery was virtually unknown before the 1970s, and then fishing caused a collapse of known populations on seamounts near Australia and New Zealand. The characteristics of this and other seamount species suggest that such concentrated fisheries are unsustainable without very strong limits on fishing.

Because seamounts are usually of volcanic origin, the bedrock may be a source of metal deposits that can be exploited by undersea mining and recovery technology that was not available even a couple of decades ago. Iron-manganese crusts and nodules are common on seamounts with no strong volcanic activity, but a wide variety of hydrous metal deposits with manganese, iron, and other metal sulfides are found in seamounts in volcanically active regions such as island arcs of the southwest Pacific and in volcanically active midoceanic ridges. A number of these areas have been licensed for mining in the New Zealand region, and many open-ocean seamounts are not under any national jurisdiction. Active research is underway to investigate the biological richness and vulnerability of seamounts to mining.

Hot-Vent Environments, Cold Seeps, and Whale-Fish Falls

■ **Hot vents, located near areas of submarine volcanic activity, spew out hot, sulfide-rich water and support unique biotic assemblages.**

The deep-sea bed gives the impression of monotony and poor productivity. Most deep-sea bottoms are low in production when contrasted to the upper continental slope and the continental shelf. In the 1970s, the submarine *Alvin* descended to a midoceanic ridge to examine the generally steep and rocky bottom environments. The expedition came upon a series of fissures from which spewed hot water of partial volcanic origin.

The midoceanic ridges are volcanically active, and the molten rock heats seawater in crack systems, often to a superheated state of several hundred degrees. The underwater landscape often consists of a series of "smoking chimneys" emanating hot water that is loaded with high concentrations of various metals and dissolved sulfide (**Figure 18.21**). The discoveries of hot vents were followed by the discovery of a fantastic assemblage of animals associated with the hot vents, although in not-so-hot waters (the animals live in temperatures ranging from cold ambient to over 16°C). In contrast to adjacent cold-water deep bottoms, the rocky surfaces surrounding the hot vents are covered with animals, including large limpets, clams, and mussels. Zoarcid fish and octopods move along the bottom searching for prey. The most curious and spectacular organism is a large tubeworm, *Riftia pachyptila* (**Figure 18.22**). The worm is often 1 m in length and secretes tubes up to 3 m long, which are rich in chitin and protein. Several species of tubeworms have now been discovered and classified in the group Vestimentifera, which has affinities to both the Pogonophora and the Annelida, and DNA evidence shows that they arose within the Annelida as a very specialized group. The growth rate of these worms is very rapid. Sequential dives on a newly established volcanic terrain demonstrated (Lutz et al., 1994) that the worms could grow tubes over 1.5 m long in just 18 months! The tubeworms are probably a foundation species and allow other species to occur and may even facilitate the colonization of other dominants, such as bivalves.

The community presents a paradox. How can such a productive biota occur in the midst of general trophic poverty? The food supply from surface waters is clearly insufficient to maintain these rich animal communities. The **dissolved sulfide** emanating from the vents seems to be the answer.

FIG. 18.21 Volcanic activity results in superheated water to be expelled from fissures. Here the water is 350°C! Locale is 21° N on the East Pacific Rise. In 1993–1994, scientists were lucky enough to catch an eruption in action and were able to follow the development of volcanic deposits and biological colonization. (Courtesy of the Woods Hole Oceanographic Institution)

FIG. 18.20 An orange roughy at a seamount near New Zealand. (Courtesy of Malcolm Clark)

(a)

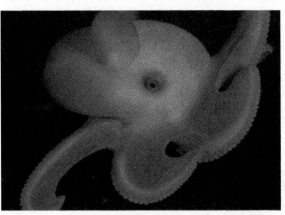

(b)

(c)

(d)

FIG. 18.22   A hot-vent fauna, in the tropical eastern Pacific: (a) the vestimentiferan tubeworm (red animals) *Riftia pachyptila*, with zoarcid fish; (b) undescribed "Dumbo" octopus; (c) galatheid crab; and (d) seascape showing vestimentiferans with predatory crabs. (Photographs courtesy of Richard Lutz)

Certain bacteria derive energy from the oxidation of sulfide. In the vicinity of the vents, where sulfide-rich water is spewed out, bacteria that derive energy from the oxidation of sulfide live as mats on rocky surfaces and on the surface of soft sediments. These bacteria are grazed by a variety of consumers, including crabs, shrimp, limpets, and bivalves. Galatheid crabs are especially abundant at vent peripheries and consume bacterial mats (**Figure 18.23**). Sulfide-dependent bacteria also are found in "bacterial snow" that has been observed in the water column coming from vent waters. This material is also a source of nutrition for consumers.

FIG. 18.23 Galatheid crabs are often abundant near hot vents, and here we see rows of them lining a fissure at a hot vent on the East Pacific Rise. (Courtesy of the Woods Hole Oceanographic Institution)

However, many (and perhaps most) vent animals do not seem to feed directly from the external medium. The Vestimentifera, for example, have no obvious gut. Instead, they have a specialized organ, the trophosome, which has dense concentrations of chambers that contain sulfide-oxidizing bacteria. The animal, in turn, can digest the bacteria and derive nutrition. *Riftia* has a specialized hemoglobin, which binds to both sulfide and oxygen (Zal et al., 1998). Two amino acid sites on the hemoglobin code for the amino acid cysteine, which binds to sulfide. The bound sulfide is transported to the trophosome, where it

is used by the symbiotic bacteria. There are several species of *Riftia* and also a number of species of symbiotic bacteria. But when DNA sequences are used to construct evolutionary trees for both groups, there is no correspondence, which suggests that speciation of *Riftia* or bacteria has no influence on each other (Nelson and Fisher, 2000). Bacterial species may be transferred from the environment to their hosts, but little is known about this subject. The lack of evidence for cospeciation of bacteria and animal host is similar to what we have seen in other systems, such as corals and their zooxanthellae symbionts.

The bivalve mollusks *Bathymodiolus thermophilus* (**Figure 18.24**) and *Calyptogena magnifica* have symbiotic sulfide-oxidizing bacteria, which detoxify sulfide and fix carbon, living in specialized cells of the bivalve's gills. The bacteria use sulfide as an electron donor and obtain carbon from CO_2. Evolutionary analysis of DNA sequences demonstrates that bacterial strains from all bivalves investigated with sulfide-oxidizing bacteria have the same evolutionary origin and are related to a free-living chemoautotrophic bacterium with the same metabolic functioning (Distel et al., 1994). Some species also have methanotrophic bacteria, which convert methane to energy-bearing carbon compounds useful to the bivalves. The bacteria can be identified by unique DNA sequences and also by specific polar lipid fatty acid profiles that differ between bacterial metabolic types (Colaío et al., 2007). Fatty acids in the bacteria can be traced to bivalve tissues (Saito and Osako, 2007).

This discovery of symbiotic bacteria was made in one of the most inaccessible environments on Earth, and certainly

FIG. 18.24 A bed of the deep-sea mussel *Bathymodiolus thermophilus*, which has symbiotic bacteria. The chamber in the background was deployed by the deep-sea submersible *Alvin*. (Photograph by Richard Lutz)

one of the most newly discovered. However, many other benthic animals seem to have symbioses with bacteria. In shallow subtidal and intertidal muddy sediments, bivalves of the genus *Solemya* live in Y-shaped burrows. Species of this genus usually have no gut or a reduced gut and have symbiotic bacteria on the gills. The bacteria function similarly to those on bivalves living near hot vents: They detoxify sulfide and fix carbon (Scott and Cavanaugh, 2007). Individuals gain oxygen by ventilating the upper arms of the "Y," but they obtain sulfide from the lowest arm of the "Y" in the anoxic sediments at the bottom of the burrow. In tropical seagrass beds, anoxic sediments contain burrowing lucinid bivalves, which also have symbioses with sulfide-oxidizing bacteria.

Many of the most exciting discoveries made at the vents have been microbiological. Many species, as we have mentioned, live in symbiosis with animal hosts, but others are free living and are adapted in very surprising ways. One of the most exciting is the discovery of a photosynthetic anaerobic bacterium. Yes, photosynthetic in an oceanic realm with no visible light! This is a green sulfur bacterium that oxidizes sulfur and obtains carbon from carbon dioxide in the process. But it has a photosynthetic pigment that absorbs the radiative energy from geothermal radiation, coming from the vents (Beatty et al., 2005). The radiation is most intense at wavelengths of about 750 nm, which corresponds to thermal radiation. The energy source is likely an auxiliary source to energy derived from oxidation of sulfide. On the other hand, such a life-form greatly expands the possibilities of life on earth and in extraterrestrial environments where typical solar energy is not available. Other bacteria are also capable of living within cracks in the rock at very higher water temperatures. These so-called **extremophiles** belong to the microbial groups Archaea and Bacteria and are adapted to living in extremely high temperatures combined with high metal concentrations within the vents.

Hot vents were discovered over 30 years ago, but new and bizarre organisms continue to be discovered. In 2005, at a vent near Easter Island in the tropical Pacific, the French scientist Michel Segonzac descended to the bottom in the *Alvin* and found a bizarre crab, distantly related to the galatheids mentioned earlier, at a depth of about 2,200 m. Molecular sequence studies show that it is a member of a new taxonomic family, but most interesting are the strange claws that are covered with hairs, which are in turn covered completely by filamentous bacteria (**Figure 18.25**). The hairy look inspired the Latin name *Kiwa hirsuta* and the common name "Yeti crab" after the Abominable Snowman. But what function do the hairs serve? Some have speculated that the hairs are grown to farm the bacteria, which are eaten periodically by the crab. Maybe so, but the crab was discovered while it was eating a broken mollusk. Some recent isotopic evidence shows that the body of another species of *Kiwa* living near a methane seep resembles the bacteria on its hairs. The crab appears to be farming bacteria on the hairy appendages, and the bacteria are periodically combed off of the appendages by a specialized mouth appendage. The crabs swing their claws rhythmically, perhaps to increase exposure of the bacteria on the hairs to more methane (Thurber et al.,

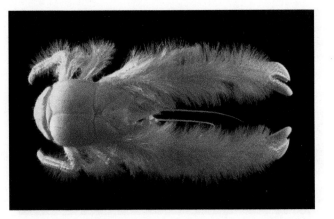

FIG. 18.25 *Kiwa hirsuta*, a bizarre crab about 15 cm long with hairs festooned with filamentous bacteria. This crab was found near a hot vent, in the vicinity of Easter Island in the eastern Pacific. (Courtesy of Michel Segonzac)

2011). Recently, a newly discovered hot-vent community in the Antarctic was found to have dense populations of yet another species of *Kiwa* (Rogers et al., 2012).

■ Hot-vent faunal habitats are ephemeral, and species disperse along ridge systems often by means of planktonic larvae.

The vent faunas are also fascinating because they are dependent on hot water sources that are temporary. Hot vents may last no more than hundreds of years or a few thousand years at most. Early in a vent's history, vestimentiferans dominate, but as the vents begin to cool down, bivalves become more common. Some "dead" vents have been encountered with piles of shells nearby.

The question then arises of how vent animals disperse from one vent system to the next. Although there is not yet an accurate map of the distribution of the vents, they are probably often hundreds and even thousands of kilometers apart. Despite the distance, genetic differentiation among marine invertebrates is very slight along thousands of kilometers of the same ridge system. There must be extensive dispersal to homogenize the distant populations. The degree of genetic differentiation of populations of the same species between separated ridge systems is quite high, however, indicating that connections between populations occur only along the axes of ridges.

Most of the vent species probably have lecithotrophic larvae, which in shallow waters would not be expected to disperse very far. Deep-sea biologist Richard Lutz has suggested that the coldness of the deep sea may greatly slow down larval development rate, increase larval life, and, therefore, allow more time for dispersal. Larvae may be entrained in the plumes of water that rise from hot vents, and these plumes may disperse laterally for distances of kilometers. Direct pump sampling by Lauren Mullineaux and colleagues (2005) found high concentrations of larvae in midoceanic ridge valleys that might direct them to new vent habitats. As was discussed in Chapter 7, some of the vent animals probably have planktotrophic larvae that swim to surface waters, a circumstance that raises the enigma of how a larva can find the surface and then find a highly localized vent biota of only a few hundred square meters at most. Some have suggested that certain vent organisms may be able to disperse to nonvent habitats, such as fish falls or whale carcasses. After settlement, mobile species can move as adults to adjust to local thermal conditions and even to reduce competition with other species (Bates et al., 2006).

■ **Some cold-seep areas in the Gulf of Mexico and the Florida escarpment are rich in hydrocarbons and also have rich benthic communities based on local food sources.**

Cold seeps are areas of seabed where hydrocarbons such as methane or sulfides occur at temperatures comparable to that of seawater, are abundant in rock fissures, and rise to the seafloor. They are associated with releases of methane from submarine rocks and sediments. Along the continental slope of the Gulf of Mexico, they arise from deposits of petroleum and other hydrocarbons. But on the Florida escarpment, they derive from sulfide-rich brines. They are found worldwide in all types of continental margin, in depths of 400–6,000 m. They provide another hotspot of nutrition in the deep sea for bottom organisms, and many organisms related to those found near hot vents are found adjacent to the seeps (Sibuet and Olu, 1998). Organisms similar to those found in hot vents dominate the seep areas, including vestimentiferan worms and bivalves, all with symbioses with bacteria. In the Gulf of Mexico, the common bivalve *Bathymodiolus childressi* maintains a symbiosis with methanotrophic bacteria, which react oxygen and methane to form formaldehyde, which is then converted into organic compounds. Bacteria are located in specialized bivalve cells on the gills known as bacteriocytes. A few bivalve species have both sulfur bacteria and methanotrophic bacteria. In the Gulf of Mexico, methane is abundant and bivalves with methanotrophs dominate the bottom. Methane is correlated with low oxygen, so these environments are often stressful for the invertebrates that live there.

The food supply from cold seeps is usually far less than around hot vents. Although vestimentiferan worms are found there, individual growth rates are far slower than near hot vents, and worm size is much smaller, usually growing less than 1 cm in length per year. Cold seeps often have very low oxygen concentrations, and many invertebrates have hemoglobin, which is adapted to bind to oxygen at very low oxygen concentrations (Hourdez and Weber, 2005).

Cold seeps are usually distinct environments from hot vents, but there are locations such as a deep-sea site off the Pacific coast of Costa Rica where methane seeps are mixed with hot vents.

■ **Fish and whale carcass falls also are a localized deep-sea environment where species successively colonize and dominate as the carcasses decompose.**

Large fish and whales die and sink to the bottom, creating very patchy and localized sources of decaying organic matter. Marine mammal carcasses appear to be a series of ephemeral islands throughout the deep sea, supporting a correspondingly ephemeral group of colonizing species populations. There are four stages in the decomposition and colonization of the carcasses (Smith and Baco, 2003):

1. *Scavenger stage.* Dead whales are visited by sharks, bony fish, and crabs and torn apart (a year or so).
2. *Enriched sediment stage.* Invertebrates living on enriched sediment from organic material derived from decomposing flesh (up to 2 years).
3. *Sulfide stage.* Appearance of organisms that rely on hydrogen sulfide emitted from sediment as organic matter decomposes and sediment continues to be anoxic. Includes chemoautotrophic bacteria and animals such as vesicomyid clams with symbiotic sulfide-oxidizing bacteria.
4. *"Reef" stage.* Remaining bones provide supporting substratum for epibenthic suspension, feeding animals.

Greg W. Rouse and colleagues (2004) made a spectacular discovery on a whale found on the bottom of Monterey submarine canyon in California. Species of *Osedax*, a worm related to Vestimentifera, has a large posterior ovisac (sac with eggs) connected to a vascularized system that roots within bone marrow of the fallen whale's skeleton (**Figure 18.26**). The rooting system has cells within which are located bacteria known to be involved in the degradation of organic matter. This arrangement of rooting and symbiotic bacteria is unique in the animal kingdom. The original described species consisted of these larger individuals, which were females, with numerous tiny males found within the female's tube. Other species of *Osedax* have since been discovered in waters deeper than 300 m, and some do not have dwarf males. Six species have been described so far in Monterey Canyon alone and are found in all stages of whale carcass degradation just described. In the past few years a number of stranded whales have been moved to experimental sites at various depths, and the stages of degradation and colonization are being studied.

■ **Hot vents, cold seeps, and fallen carcasses demonstrate a surprising diversity on highly localized and often ephemeral habitats.**

Hot vents, cold seeps, and fallen carcasses all have in common a trophic source that is apparently conducive to selecting for macrofaunal-bacterial symbioses. The details differ with regard to food source and chemistry, but the animal hosts have specialized enclosing cells for the bacteria and rely on the bacteria's respective abilities to either degrade organic matter or fix carbon by means of chemoautotrophy. The habitats also appear to be impermanent. Hot vents tend to last no more than a few hundred years and probably cease to exist fairly suddenly. Carcasses undergo a natural degradation cycle and eventually cannot support species with the specialized symbioses we have discussed. Hydrocarbon seeps must last decades to hundreds of years in some cases, given the slow growth and the size of vestimentiferan

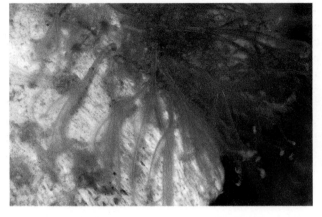

(a)

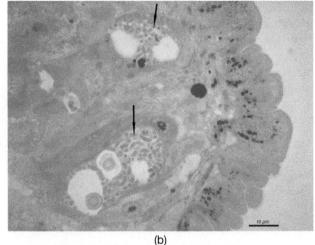

(b)

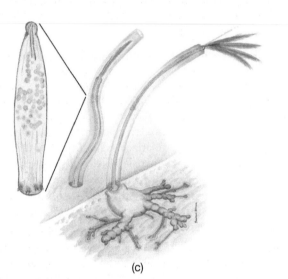

(c)

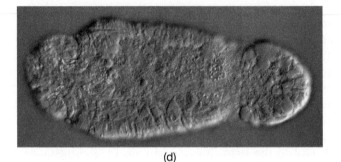

(d)

FIG. 18.26 The boneworm *Osedax*. (a) Live specimens of an undescribed species emerging from the bone of a gray whale from 600 m depth; (b) section through a "root" of a female *Osedax roseus* with arrows pointing to two bacteriocytes filled with bacteria; (c) diagram of *O. rubiplumus*. On the left is a blowup of a dwarf male (actual size ≤ 1 mm long), and the lines converge where males tend to be found in the female tube. The female "trunk" is shown in contracted state, about 5 cm long. Plumes in red. Bones are in cross section to show ovisac and roots filled with bacteria; (d) scanning electron micrograph of dwarf male. (Photographs by Greg Rouse; diagram painting by Howard Hamon)

worms that have been found there. In all cases, the habitats blink on and off in a spatially complex manner, and most resident invertebrate species have planktonic larvae that colonize new patches.

Despite the patchy and ephemeral nature of these habitats, they support a surprisingly large number of species restricted to a given habitat type. For example, Tunnicliffe and colleagues (1998) report that of the 433 species found in hot vents, 366, or 84 percent, are restricted to hot vents. At cold seeps, 47 of 136 species are restricted. Whale falls have the least restricted fauna. Only 21 of 407 species have been found to be specific to this habitat (Smith and Baco, 2003), which would fit with its ephemeral nature.

■ **The vents and other habitats demonstrate that the deep sea is not constrained to be in slow motion.**

The evidence from colonization and growth of organisms on hot vents and other hotspots of diversity demonstrate

that animal growth rates and microbial activity are not inevitably slow in the deep sea. Rather, when growth is slow, the rate of food or nutrient supply is low. Thus, deep-water coral mounds show very slow coral growth rates, which are likely related to the low rate of supply of particulate organic matter from the surface. On the other hand, other species growth rates on coral mounds may be rapid. There are other habitats that violate the general picture of a deep sea in slow motion. As we discussed earlier, experimental introductions of barrels of rotting fish are usually immediately followed by recruitment of various scavengers, such as amphipods and fishes, to the rotting fish. Ruth Turner has found rapid colonization and rapid growth by boring bivalves on pieces of wood she placed in deep-sea habitats (Turner and Lutz, 1984). The bivalves derived their carbon from the wood and obtained nitrogen from symbiotic nitrogen-fixing bacteria. Cold-seep bottoms are rich in hydrocarbons, and these are dominated by rich faunas of invertebrates. Thus, depth or pressure per se cannot explain slow rates of bacterial and animal activity in more typical deep-sea bottoms.

■ Microbial activity and benthic creatures in trenches and other areas with high organic matter input confirm that organic input is a major influence on deep-sea activity.

We discussed earlier the strikingly low rate of microbial activity on remote deep-sea abyssal bottoms, to the point that human foodstuffs can last a year with relatively little decomposition. The soft sediment habitats in the remote deep sea suggest that this environment is in slow motion. But there is little evidence that depth per se is the cause of such slow activity. Earlier, we mentioned deep-sea habitats in the North Atlantic in which deep-sea animals show activity just days after phytoplankton blooms occur at the surface (e.g., a few hundred kilometers west of Ireland). Apparently organic matter input is the driver of the degree of microbial and animal activity on the deep-sea floor. This theme extends to trenches, which can reach depths of 11,000 m. Many trenches, and some other deep-sea bottoms, are sites of high deposition rates of organic matter. For example, bait placed at a depth of about 5,000 m in the northeast Pacific Ocean attracted the largest amphipods ever seen at the time, of approximately 9 cm in length (Hessler et al., 1972). Large amphipods in great densities were later found commonly in deep sea trenches, at depths of 9,000–11,000 m, reaching lengths of 29 cm, where a typical amphipod in shallow depths of the United States would be 1–2 cm at most. Microbial activity in the bottom of the Challenger Trench (near Mariana Islands, southwestern Pacific) was twice that of nearby abyssal bottoms (Glud et al., 2013). The high microbial activity and high benthic densities of large detritus-eating invertebrates is a reflection of organic matter input, which fits with the other islands of high biological activity we have described, including hot vents, cold seeps, and whale carcass falls. The slow motion of so much of the deep-sea bottom is a reflection of low organic matter input, arising from remoteness of organic matter and nutrient sources.

■ A bacterial realm has been found deep below the deep-sea floor.

Investigations at the deep-sea floor have greatly expanded our knowledge of the diverse ecological adaptations of bacteria, ranging from intimate mutualisms with many animal species at hot vents to species that are surprisingly tolerant of extreme environments of very high temperature and metal content. But more startling surprises have recently emerged from investigations of sediment cores taken beneath the deep-sea floor. Viable bacteria have been found through the cores, down to depths greater than 500 m below the abyssal floor surface (Parkes et al., 2004). Cells occur in pore waters that have been estimated to be 60–100°C, and the DNA sequences of some match those of other known microorganisms in the Archaea that live in extreme environments such as hot vents. Bacterial activity has also been discovered within deep-sea sediments that contain ancient

brines and even at the interface of the soft sediment on the deep-sea bottom and rocks below. Bacteria in this thick layer of sediment beneath the seafloor may constitute as much as half of the living biomass on the planet! Similar evidence of widespread microbial activity has been found in samples from oceanic crust basalts, which lie beneath sediments on the flanks of midoceanic ridges. Sequencing of microorganisms in basalt cracks show widespread chemoautotrophy involving methane cycling and sulfate reduction in these anoxic microenvironments (Lever et al., 2013). Like the deep soft sediment environments below the sea floor, these microenvironments are global in extent.

Pressure Change

■ Pressure increases of approximately 1 atmosphere in every 10 m depth strongly affect the biochemistry of deep-sea organisms.

The great pressure of the water column also exerts strong effects on the functioning of deep-sea organisms. The average depth of the ocean is about 4,000 m, and the pressure is 400 atmospheres. Some species live over depth ranges of greater than 1,000 m, and, therefore, members of the same species can experience pressure differences of as much as 100 atmospheres. Increased pressure causes a decrease in concentration of dissolved oxygen. What are the effects of such differences?

Aside from living at high pressure, most deep-sea organisms live at low temperature, ranging from 1°C to 4°C. Temperature has a strong effect on enzymes, which catalyze important reactions in cells. An enzyme binds to a substrate and other molecules and catalyzes a reaction in which the substrate is converted to a product. The efficiency of the enzyme depends on its ability to bind and then detach rapidly. At low temperature, this process slows greatly, and the time of binding to substrate and other molecules increases. This reduces enzyme-binding efficiency.

The effects of pressure are not well known and have been investigated mainly on the functioning of fishes and on gene expression in bacteria. The simplest experiment that can be done is to subject a shallow-water organism to increased pressure in the laboratory. When this is done to fish, their behavior is changed markedly. The reaction is similar to that of the strong effects of low oxygen. Fish lose balance and quickly become immobile. But deep-sea fish appear to swim normally and show no such effects, which suggests that they have evolved mechanisms to counteract the effects of pressure. Pressure should affect the functioning of enzymes and the structure of membranes.

Pressure affects the volume of biologically active molecules such as enzymes. The simplest effect is in changing the volume of the voids in enzymes, in which substrates and enzyme cofactors attach briefly during the reaction process. This effect tends to decrease the efficiency of enzyme catalysis. In deep-water fishes, enzymes that are less sensitive to pressure change have evolved. Increased pressure tends to harm the functioning of enzymes derived from shallow-water fish, whereas enzymes taken from deep-water fishes manage to function at much higher pressures (Siebenellar, 1991). Enzymes of deep-sea fishes are structurally far more stable than those of shallow-water species. This may be an adaptation to prevent high rates of protein turnover, which would be maladaptive in the food-poor deep-sea environment. These properties may also apply to proteins of deep-sea bacteria, which, as discussed earlier, seem to metabolize very slowly. Pressure tends also to compress membranes, and cell membranes taken from deep-water fish species are designed to remain fluid at pressures much higher than membranes from shallow-water related species. Deep-water fishes also appear to have cell-signaling proteins adapted to high pressure. Joseph Siebenaller (2003) found that G proteins, which couple cell-surface membrane receptors to other proteins, function well under high pressure in deep-sea fishes but respond more poorly to pressure in a shallower-water fish species. Adaptations to high pressure are, therefore, present both at the level of enzyme and cell membrane structure and in retaining the function of membrane signal transduction at high pressure.

Pressure does not seem to cause the low metabolic rates observed in deep-sea species. This seems to be more closely related to a set of adaptations to reduce activity, since food is so rare and often of such poor quality in the deep sea. Muscle-related enzymes, such as lactate dehydrogenase, are much lower in activity in deep-sea fish species relative to shallow-water relatives. Muscle-related metabolic power is very low, much lower than one would predict from the low temperature of the deep sea. The lowered metabolism is therefore an adaptation to low-food conditions, not a response to pressure, temperature, or other physical conditions.

Pressure adaptations have also been investigated in bacteria. Cell membrane function appears to evolve to keep its relatively liquid functional state in deep-water bacteria (DeLong and Yayanos, 1985; Yano et al., 1998). Enzymes in deep-sea bacteria are functionally different from shallower-water forms and may be activated by increasing pressure, but our understanding is still very incomplete. Recent research has focused on molecular approaches to study gene expression (see bonus chapter, "Molecular Tools for Marine Biology," online). In bacteria, functional approaches have been taken at the level of an operon, which is a genetic unit including genes for protein synthesis, combined with a promoter that initiates gene expression. Some operons have been shown to vary in expression with pressure.

■ CHAPTER SUMMARY

- Dredges of the deep sea by remote sampling cannot give a complete picture of seafloor biology. Submersibles, remote underwater vehicles, underwater video cameras, and other devices now allow direct visual observation, precise location of samples, and experimental manipulation. Cabled observatories will soon allow repeated data gathering from specified sites.

- Suspension-feeding benthic animals dominate sandy sediments, whereas deposit feeders dominate muds. Shelf and deep-sea bottoms are mainly sand and mud, but population densities are patchy. Organic material is distributed discontinuously, generating sites of high food value, including seaweed and animal carcasses.

- Patchy dispersal and recruitment also cause spatial uncertainty in subtidal soft sediments. Rapid colonization by surface dwellers is followed by the dominance of deeper-burrowing species.

- In contrast to shallow-water bottoms with high biomass and productivity, deep-sea bottoms are characterized by low biomass and productivity. That includes a reduction of benthic invertebrate carnivores, but microbial activity is also very low. The input of organic matter to the deep-sea floor,

especially beneath gyre centers, is low because of the distance from the coast and the great depth through which organic particles must travel. However, localized areas of the seabed receive pulses of organic matter.

- Thanks to a stable environment, the biodiversity of soft bottoms increases with increasing depth to the continental rise, despite reduced food conditions. It declines as depths surpass about 2,000 m seaward toward abyssal bottoms, where especially poor food conditions probably cause local extinctions.

- Deep- and cold-water hard surfaces contain habitats dominated by stony and soft corals, ranging from a few colonies to enormous mounds of calcium carbonate sediment, at depths ranging from 30 to over 1,000 m. These deep-sea coral mounds are diverse, and colonies live for several hundred years. They also support fisheries and attract damaging trawlers.

- Seamounts are found worldwide and often have a rich epibenthic invertebrate community on the seamount summits, often associated with rich fisheries.

- Hot vents, located near volcanically active oceanic ridges, spew out sulfide-rich water. The sulfide is used by free-living bacteria, which are consumed by

abundant vent animals. Other bacteria, also using sulfide, live in symbiotic association with vestimentiferan worms, bivalves, and other groups. This habitat is very productive and some organisms grow very rapidly. Species disperse along ridge systems, often by means of planktonic larvae.

- Cold seeps are areas of the seabed where abundant hydrocarbons or sulfides rise from rock fissures to the seafloor. The fauna also includes invertebrates with bacteria mutualisms with many similarities to the fauna of hot vents.

- Species successively colonize and dominate fish and whale falls as the carcasses decompose. Hot vents, cold seeps, and fallen carcasses all demonstrate a surprising diversity on highly localized and often ephemeral habitats.

- Pressure in the deep sea increases by approximately 1 atmosphere for every 10 m depth, strongly affecting the biochemistry of organisms. Enzymes of deep-sea fishes are structurally far more stable than those of shallow-water species. Cell membranes taken from deep-water fish species are designed to remain fluid at pressures much higher than membranes from shallow-water species.

■ REVIEW QUESTIONS

1. What general sediment type is dominated by deposit feeders? By suspension feeders?

2. How do deposit feeders contribute to making their benthic environment relatively inhospitable to suspension feeders?

3. What are three mechanisms that might create patchiness of benthic organisms on the seafloor?

4. What is an advantage of using remotely operated submersibles, relative to more standard bottom samplers such as dredges?

5. What feeding type of benthic organism dominates the deep sea? Why?

6. How does benthic biomass change as one goes from the continental shelf to the deep-sea floor of the open sea?

7. What evidence is there for a low rate of organic matter decomposition in the deep sea?

8. What are the changes in variability of the physical environment as one passes from shallow shelf habitats to those of the abyssal deep-sea floor?

9. Why are hot vents places of rather high benthic biomass and secondary production, despite their location in the deep sea?

10. How does benthic diversity change over a gradient from the inner continental shelf to the abyssal deep-sea floor?

11. Why do hot vents present a paradox in terms of some species with planktotrophic larval dispersal?

12. Why is the activity of deep-sea fishes so low?

13. Deep-water coral mounds have been surveyed to have over 1,300 species.

What does this make you think about the role of zooxanthellae in promoting shallow tropical coral reef diversity?

14. What general pattern leads to the expectation that animals in the deep-sea soft bottom have a poor food supply?

15. Why might you expect that species on a seamount might be found on that seamount alone, and not elsewhere in the sea?

16. It has been suggested that low rates of microbial activity are caused by inherent limitations of some aspect of depth, such as pressure. What evidence contradicts such basic physiological limitations on rates of microbial activity?

17. What morphological adaptations are known in deep-sea animals that allow them to be hosts of symbiotic bacteria?

Visit the companion website for *Marine Biology* at www.oup.com/us/levinton where you can find Cited References (under Student Resources/Cited References), Key Concepts, Marine Biology Explorations, and the Marine Biology Web Page with many additional resources.

CHAPTER 19

Polar Marine Biology

Introduction

■ **Polar ecosystems are influenced by strong seasonality, especially of ice cover, and a diverse phytoplankton assemblage fuels the polar food webs. Carnivores exert top-down effects, and overfishing of carnivores must have had strong effects on polar food webs.**

The hallmarks of polar ecosystems are cold temperatures; strong seasonal change of light, temperature, and especially ice cover; and a generally rich photosynthetic base of the food web with a surprisingly abundant and diverse phytoplankton community. Seasonal ice cover expansions in winter and contractions in summer characterize both Arctic and Antarctic communities, and microalgae are known to reside in fissures and cracks in the ice. As the polar spring commences, the ice melts and retreats, light increases and enters the newly ice-free water column, and nutrients are released. The combination of these events sets off the spring phytoplankton bloom (Smetacek and Nicol, 2005).

Polar systems are also dominated by a large array of carnivores, especially mammals and birds that feed heavily on fish, squid, krill, and crustaceans, depending on the locality. For example, the Antarctic is famous for its dense populations of penguins, which in turn are preyed on by carnivores such as leopard seals and killer whales in water and by giant petrels and skua on land. The famous polar bears, and also killer whales, prey on Arctic sea birds and seals. Closer to the base of the food web, Arctic carnivores tend to feed on fish and benthos, whereas Antarctic carnivores tend to feed on krill and squat lobsters. While phytoplankton are clearly the base of food webs and exert bottom-up effects on grazers such as krill,

top-down predators such as baleen whales must depress krill populations, which would result in a positive effect on phytoplankton growth. Thus, polar food webs are very dynamic.

Historically, we can guess that polar ecosystems were complex and likely different from today. There is no simple answer to the question of whether food webs were dominated by bottom-up processes (nutrient supply, phytoplankton fluctuations) or top-down processes (fluctuations in top carnivores). In the ocean, large mammals tend to be carnivores, either at the top level (e.g., killer whales hunting seals) or intermediate levels (e.g., baleen whales hunting krill and small fish). But historically, human exploitation greatly reduced the abundance of such marine mammals. In the twentieth century, the discovery of large Antarctic whale stocks (see Chapter 21) resulted in catastrophic overfishing declines. In the Northern Hemisphere, the Steller's sea cow *Hydrodamalis gigas* and the European populations of the gray whale *Eschrichtius robustus* were both driven to extinction by humans. These declines must have strongly influenced food-web structure.

■ **Climate change is causing strong reorganizations of polar ecosystems, especially where warming waters are causing increased seasonal retreat of sea ice.**

Ocean warming is having clear effects on the organization of polar ecosystems, which we will discuss below. In the Arctic Ocean, sea ice has steadily declined, occurring at a loss rate of 9 percent per decade (Comiso, 2002). Melting has also increased from the Greenland ice sheet. This causes increase of light penetration in summer and great influxes of fresh water at the surface, which tends to increase water stratification. Similar ice retreats in the Antarctic, especially along the Antarctic Peninsula, have caused changes in food webs and especially in populations of some species of penguins (see discussion below). Warming of the circumpolar current around Antarctica is causing reductions in sea ice where the current impinges on the limits of sea ice, such as at the West Antarctic Peninsula where basal melting of floating ice has greatly increased in recent decades (Pritchard et al., 2012). Such effects have not occurred in other parts of Antarctica, although there is worry about the near future collapse and breakup of ice sheets.

Arctic Marine Systems

Arctic Marine Ecosystem Structure

- **Arctic environments are characterized by cold seawater temperatures, seasonal changes in sea ice, and high productivity, which influences the subtidal benthos by deposition of organic matter.**

The Arctic region traditionally has had a terrestrial definition: the northern limit of the tree line, where tundra environments begin. The southern limit can also be placed at the Arctic Circle, at 66° N latitude. Marine environments include the large Arctic Ocean and adjacent lower-latitude water bodies such as the Norwegian Sea and Barents Sea in the far North Atlantic, and the Beaufort Sea, north of the Bering Sea in the Pacific.

These environments share three important traits that strongly influence both surface waters and the benthos below. First, of course, is the large expanse of surface oceanic ice, which covers most of the Arctic seas in winter, but melts back to varying degrees during the Arctic summer. The ice layer is quite thick and often strongly erodes the seabed of the shallow continental shelf. As the winter ice is formed, resuspended sediment is incorporated, only to be released again upon melting to the water column during the polar spring and summer. This process results in the **release of nutrients** that help to fuel phytoplankton blooms. These blooms are the base of a highly connected food web. Phytoplankton are consumed by copepods and other zooplankton, which are consumed by baleen whales and a variety of fish species including juvenile cod, arctic char, and capelin. These fish are consumed by a number of species of seal, which are consumed by killer whales and polar bears (**Figure 19.1**). Ice algae and phytodetritus fuel the great abundance of Arctic benthos, which are consumed by several species of benthic carnivorous fishes. The large open expanse of ice reflects sunlight, and this of course greatly cools the surface waters beneath. As we will discuss shortly, the summer cover of ice has steadily diminished as global climate change has strongly affected this region.

Second are the steady low sea temperatures of this very high-latitude region. Temperatures in the Arctic range from −2°C (the freezing point of seawater) to usually not much more than 6°C in the Arctic summer.

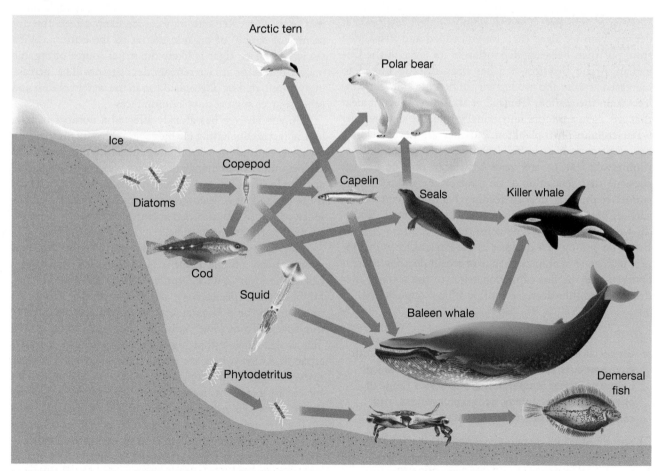

FIG. 19.1 Some of the connections in a typical Arctic food web.

Third, high productivity of Arctic waters and inputs of algae bound in ice results in high inputs of organic matter to the benthos, resulting in apparent high benthic productivity and strong benthic-pelagic coupling.

- **High primary productivity of open-water phytoplankton and ice algae provide an abundant source of organic matter that is deposited on the seabed and subsidizes benthic communities.**

Arctic shelf environments have very high productivity, although over a relatively short season. Much of this water column productivity, mainly of diatoms, provides food for copepods, which are extraordinarily abundant. Most of the biomass of the widespread copepod genus *Calanus* is in Arctic waters. The copepods, in turn, supply abundant food that supports very large populations of zooplanktivorous fish such as herring, but also are important food for bowhead whales and ctenophores. Herring can be consumed by larger carnivorous fish species, but the decline of sea ice (see following) may reduce their availability to ice-edge predators such as seals, polar bears, and gulls. The diatoms that are not consumed by zooplankton become phytodetritus, which is deposited on the seabed, guided by current systems, which might direct phytodetritus laterally sometimes for hundreds of kilometers. Deposition rates are high in Arctic waters. This regime is quite different from the Antarctic, where primary productivity is generally lower and benthic invertebrates are of much larger body size and probably of greater individual longevity than in the Arctic benthos.

Polar regions have a special supply of algal detritus that derives from **ice algae** that live embedded and suspended from polar ice. Seasonal deposition of ice algae in the Canadian Arctic Beaufort Sea has been directly linked to increased activity and productivity of benthic animal populations on the seafloor (Renaud et al., 2007). It is not clear that ice algae are more nutritionally valuable than regular water-column phytoplankton, but they are nevertheless a source of deposit of useful organic matter to the bottom (Sun et al., 2007). Fatty acids have been used as a tracer showing that deposition of ice algae on the seabed is consumed mainly by deposit-feeding benthic invertebrates, such as bivalves and polychaetes (McMahon et al., 2006). This rich benthos may be consumed by bottom-feeding fishes. As summer ice declines in the Arctic Ocean in concert with general climate change of recent decades, there is concern that the flux of phytodetritus to the seafloor may decrease and thus decrease benthic animal populations.

Climate Change and Arctic Systems

- **Global climate change has had strong effects on Arctic environments, resulting in reduced summer sea ice, higher seawater temperatures, and a wide range of effects on the physiology, geographic ranges, and community structure of Arctic biotas.**

The extent of sea ice cover in the Arctic has declined steadily (**Figure 19.2**), from about 9×10^6 km² in September 1953 to less than 4×10^6 km² in September 2007 (Stroeve and

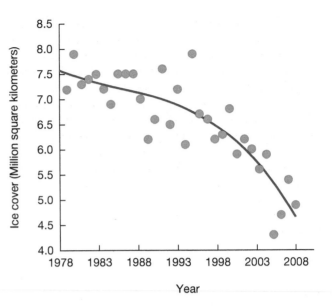

FIG. 19.2 Changes in extent of Arctic sea ice in September from 1978 to 2010, with curvilinear fit of line to data. (After Stroeve et al., 2012)

Serreze, 2008). The decline of sea ice affords greater sunlight penetration into the water column, which has caused surface sea temperature to rise in winter and summer. Warming has also caused thermal stratification of the Arctic Ocean water column. Primary production in the water column has increased significantly, partially from the retraction of ice and mainly from the longer growing season for phytoplankton that now exists. Retraction of sea ice likely affects the dynamics of supply of organic matter to the bottom. With sea ice cover, ice algae is likely the major source of organic matter, but as the sea ice retreats, deposition will be increasingly dominated by phytoplankton in the water column and other sources such as zooplankton feces.

Climate change has already affected a number of biological factors, including (1) biogeographic range, especially for lower-latitude species that are invading the Arctic and the beginning of exchange of species between the Atlantic and the Pacific Oceans; (2) seasonal changes, especially for those dependent on sea temperatures and presence or absence of ice; (3) species dependent on ice; and (4) species adapted to cold temperatures.

The warming of Arctic and subarctic waters has had strong effects on the benthic and planktonic communities of the high-latitude North Atlantic and Barents Sea. High-latitude species have been disappearing from the waters of northern Norway and have been replaced by benthic species from lower latitudes (Edwards et al., 2009). This would suggest that high-latitude species in the subarctic to Arctic waters are under thermal stress. A similar story emerges for phytoplankton and zooplankton: Arctic species are in decline in the Barents Sea and southerly temperate species are expanding into the subarctic realm.

There is only scattered information on the possible effects of high temperature stress as Arctic Ocean temperature continues to warm. As we discussed in Chapter 5, species living at high latitudes have evolved to function at low ocean temperatures and may not do well as temperature increases. We discussed in Chapter 5 the strong effect of increased temperature on the circulatory system of fishes and the increased synthesis of molecules such as heat-shock proteins. These responses may strongly affect species as polar sea temperatures increase, but we as yet know few specifics. The Antarctic ice fish *Trematomus* has been discovered to show more gene expression of heat-shock proteins when exposed to high temperatures (Buckley and Somero, 2009), but the relation of this response to population decline or mortality is not clear. A recent study of four invertebrate species (Richard et al., 2012) suggests that there is some leeway beyond the usual seasonal highest temperatures of about 7°C in the Arctic regions north of the Atlantic. The urchin *Strongylocentrotus droebachiensis* and the gastropod *Margarites helicinus* were able to acclimate to 10.3°C, and some acclimation to higher temperature was found by the bivalve *Serripes groenlandicus* and the amphipod *Onisimus* sp. These results suggest that there is still some scope for thermal increase. It may be more likely that responses to warming will not directly involve upper thermal limits but will instead involve reorganizations of communities in response to changing sea ice, sea temperature, light, and introduction of exotic species that have important effects on marine food webs. Such reorganizations are well underway.

The loss of sea ice in the Arctic Ocean in summer will also have some indirect effects owing to opportunities for human exploitation. With the reduced sea ice, the Arctic will be much easier to exploit for oil drilling, increasing the chances of drilling accidents and transportation spills. Trans-Arctic commerce will also greatly increase in the future, which will expose the Arctic to more pollutants and even exhaust from oil-fired engines. Trans-Arctic commerce will also inevitably enable the invasion of species between Atlantic and Pacific oceans. This will not be the first time that such an interchange occurs. About 3 million years ago, an ice-free period resulted in widespread invasion of hard-substratum marine benthic invertebrates from the North Pacific into the North Atlantic Ocean. Already, the Pacific diatom *Neodenticula seminae* has been observed in the Atlantic Labrador Sea and has spread through the North Atlantic since the 1990s. Core records show that this is the first occurrence in the Atlantic in over 800,000 years (Reid et al., 2007).

There has been a great deal of focus on ice-dependent species. Polar bears, *Ursus maritimus*, move on large patches of ice, and their numbers have been declining in Arctic regions. Between 1982 and 2006 in the Beaufort Sea, polar bear body size has decreased. Number and survival of pups have decreased during years of lower ice extent (Rode et al., 2010), which suggests that food availability has decreased or energetic cost of obtaining food has increased as ice declines. The same holds for the Ivory gull *Pagophila eburnea*, an ice specialist whose colonies have been reduced in numbers in Canada and Greenland and perhaps in other parts of the Arctic. The decline of sea ice must also be affecting Arctic seals, *Phoca groenlandica*, which are losing suitable habitats for breeding and resting. After a major collapse

of sea ice near Svalbard, Norway, foraging effort increased in the ringed seal, *Pusa hispida*. As ice shifted northward, subadult seals had to greatly increase their swimming distances and diving depths to hunt prey, which suggests great increases in the energetic cost of foraging (Hamilton et al., 2015).

Our information is very limited, but the reduction of sea ice must be affecting the balance of organic matter supply from ice-associated algae and algae living freely in the water column, which must be affecting water column food webs and deposition of organic matter on the seabed.

A major effect on community organization has been documented in fjords near Svalbard, Norway, where seasonal sea ice has been reduced over the past few decades and sea-surface temperature has increased approximately 0.5°C per year (Kortsch et al., 2012). Observations for over 10 years after the 1980s showed no change on rocky reef bottoms in two fjords with dominance by calcareous algae, but soon macroalgal cover increased greatly and reached 80 percent in one of the fjords. The increase of macroalgal cover was followed by a large increase of benthic invertebrate abundance. This large-scale regime shift was attributed to the reduced sea-ice cover, which allowed light to penetrate, which allowed the expansion of light-sensitive seaweeds that also required higher sea temperatures to increase in abundance.

■ Incursion of warmer waters into the Barents Sea and reduction of sea ice has led to major expansions of krill, capelin, and cod.

Climate change, ocean warming, and sea ice retreat in the Barents Sea in the subarctic North Atlantic region has caused major reorganizations of seabed ecology as mentioned in the previous section. But it also has resulted in major expansions of lipid-rich krill, which are fed on by capelin, an abundant cool-water fish that is often a major food source for cod and many marine mammal species (Dalpadado et al., 2012). Cod populations in the Barents Sea have exploded in abundance, which has attracted destructive fishing effort (see Chapter 21). It may be that the explosion of cod has resulted in competitive reduction of access to krill and capelin by marine mammals such as minke whales and harp seals, as judged by their failure to increase in numbers like the cod (Bogsted et al., 2015).

■ On a larger geographic scale, disappearance of Arctic ice from climate change interactions with typical ocean climate fluctuations may be causing much broader ecological effects on Atlantic food webs distant from the Arctic Ocean.

In both the Pacific and Atlantic Oceans, there appear to be important interactions between long-term trends in anthropogenic climate change and natural fluctuations in annual- to decadal-scale climate oscillations. In Chapter 3, we discussed the potential interaction that can occur between a long-term trend and a shorter-time-scale oscillation. The oscillation can build on the longer-term trend and cause major shifts in ocean ecology.

The apparent enhancement by longer-term anthropogenic climate change on the Arctic Ocean, interacting with patterns reflected in a natural climate oscillation known as the **Arctic Oscillation,** is causing increased ice retreat during the Arctic Ocean spring and summer and also a reduced latitudinal temperature gradient. The ice retreat has, in turn, caused the export of ice melt–derived, lower-salinity water to the North Atlantic—for example, to the shelf waters of Nova Scotia and the Gulf of Maine, which have been termed Great Salinity Anomalies (Greene et al., 2013). The loss of sea ice also makes the surface waters of the Arctic Ocean more responsive to atmospheric wind patterns. Since these wind patterns regulate the export of fresh water out of the Arctic into the North Atlantic, the downstream propagation of great salinity anomalies (GSAs) during export-favorable, positive-phase AO conditions can impact shelf ecosystems thousands of kilometers away through changing water-column stratification processes (Greene et al., 2013). Stratification in these conditions is increased, which traps nutrient-rich and warmer water at the surface. In the Gulf of Maine and continental shelf off Nova Scotia, major ecological shifts in response to freshening shelf waters have been observed, with significant increases in the production of fall and winter phytoplankton and small copepod species; a rapid increase in the abundances of various forage fish species, including herring, sand lance, and capelin; and a dramatic burst in northern right whale reproduction (Meyer-Gutbrod et al., 2015).

This climate-driven, bottom-up forcing of shelf ecosystems reverses itself when wind patterns less favorable to freshwater export return to the Arctic. At present, it is unclear whether anthropogenic climate change will alter the timing or frequency of GSAs being released from the Arctic. However, it is expected that more fresh water will be exported out of the Arctic during such events in the future, especially with the accelerated melting of Greenland's ice sheet. Freshwater export on this scale has the capacity to completely change the food-web dynamics of the North Atlantic pelagic ecosystems.[1]

Antarctic Marine Systems

Antarctic Food Webs: The Central Role of Krill

■ **Krill occupy a central role in driving coastal Antarctic food webs: They interact with shifting sea ice and feed on algae, and they are consumed by a wide variety of predators.**

Like the Arctic, sea ice figures importantly in the dynamics of coastal Antarctic ecosystems. In most of Antarctica, glacial ice extends from land to floating ice that extends to sea, often for several kilometers and several meters thick. During the Antarctic summer, the ice retreats; it then extends seaward again during the cold and dark winter. Some ice forms as free-floating fragments of pack ice, whereas other ice forms as attachments to land-connected ice.

As the ice melts in Antarctic summer, large amounts of material is released into the water, as occurs in the Arctic. This includes vital nutrients for phytoplankton, such as iron. Also, as in the Arctic, large amounts of algae grow in cracks within the ice. This ice algae has high productivity, as is true in the Arctic. It is a small fraction of open Southern Ocean primary production on an annual basis but is very concentrated spatially and is often abundant in seasons when open-water phytoplankton are less so. Fissures in sea ice are often filled with seawater and high densities of diatoms. So, ice algae may have a crucial impact on grazers that might seek the food source when food is not abundant elsewhere. Even abundant guano from seabirds can be a major source of nutrients, especially phosphorus and nitrogen, which can fuel phytoplankton blooms. A similarly large nutrient impact can be caused by feces deposited by marine mammals in waters near the edge of the sea ice.

There also is evidence that phytoplankton that get trapped in sea ice can provide seed populations for more open-water phytoplankton blooms when the spring–summer phytoplankton bloom occurs. Ice-melt water produced in spring and summer tends to stabilize the water column seaward of the ice, and primary production is elevated there. Whalers hunted near the ice edge for krill feeders such as blue whales because they were known to congregate there, presumably for the krill. So why are the krill so tightly associated with the ice edge?

Krill, *Euphausia superba*, (see Chapter 8) are shrimplike herbivorous animals a few centimeters in length that often dominate the zooplankton of surface waters of high productivity such as upwelling centers of the coastal and open oceans. Krill is the Norwegian word for "whale food," and this tells us much about their distribution and importance in oceanic systems. They are especially abundant in the Antarctic and are crucial food for a wide variety of important populations of baleen whales, other marine mammals, and, crucially, penguins. Krill are in many ways the center of a strongly integrated food web ranging from phytoplankton (mainly diatoms) to krill to a large number of predators (**Figure 19.3**). As might be expected from the positive influences of sea ice and algae, **krill abundance is positively correlated with sea ice extent,** although the relationship is rather complex (Siegel and Loeb, 1995).

Recent warming trends on the Antarctic Peninsula would be expected to be detrimental to krill population size. If krill decline, we should expect a bottom-up effect on predators higher in the food chain. Krill are also a major fishery because of their high concentrations of omega-3 oils, and overfishing has been a major worry because of the potential indirect impacts on the species that depend on krill. Krill are still very abundant, however; but commercial fishing and seafood production have been hampered because krill decompose rapidly and the shells are difficult to remove. Still, the economics of declining world fisheries have resulted in increased fishing impacts. Concern for overfishing in the Antarctic began when krill fishing greatly increased

---

[1] The interested student should see Charles Greene's 2012 article in *Scientific American*, "The Winters of Our Discontent."

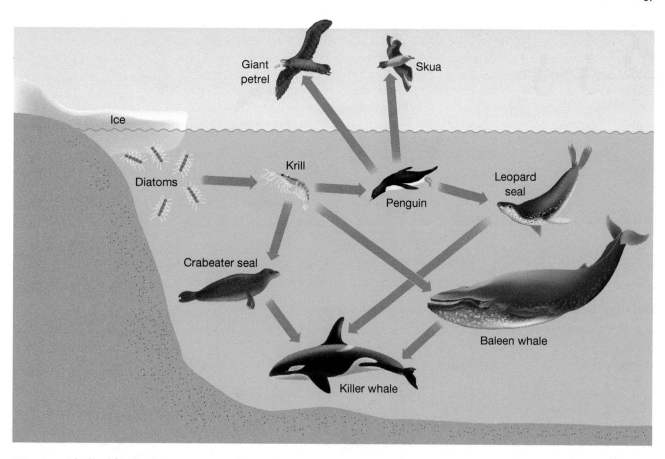

FIG. 19.3   Idealized food web in Antarctic surface waters.

in the 1970s, and a 1980 international treaty (Conservation of Antarctic Marine Living Resources) allowed the imposition of fishing limits. The fishing take in recent years has been under the limit imposed by a scientific committee. On the other hand, krill fishing has been on the rise since 2000 owing to uses for krill in the aquaculture industry, because extracts are useful in fish food and potentially in medical applications, including dietary supplements. A major conservation advance was the 2016 international treaty establishing an Antarctic **Ross Sea marine sanctuary**, which prohibits commercial fishing for 35 years except in a small adjacent area. This protected area, measuring approximately 1.5 million square km, is the world's largest ecological reserve and a major part of the Antarctic coastal ecosystem.

A key connection in this trophic cascade is the discovery of a strong association between krill and sea ice, creating a food chain involving ice algae, phytoplankton that grow beneath the ice and offshore of the ice edge, krill that consume algae, and penguins that live near the sea-ice boundary and hunt krill, often under the ice (**Figure 19.4**). Especially in the winter, the evidence suggests that larval krill depend on sea-ice algae, although adults are more abundant beyond the ice border (see Quetin et al., 1996). Under the sea ice may also be a refuge for krill from predators. Whatever the specific mechanism, one tends to find more krill where sea ice is more extensive in winter. A striking association between krill and sea ice was discovered using an autonomous underwater vehicle; echolocation surveys consistently showed strong

peaks of krill abundance under the ice, up to a distance of 11 km from the edge. Krill were far less abundant offshore (Brierley et al., 2002). Krill are strongly associated with sea ice and ice algae, and the dependence is strongest for the larval stages—and is especially important in parts of the year (e.g., November) when alternative foods are not common. The larvae appear to starve more readily than adults and, therefore, are very tied to the food available from the ice.

Krill recruitment success has fluctuated greatly in recent decades. Krill in the Southern Ocean declined dramatically in the period 1970–2003 (Atkinson and others, 2004), and it is likely that this is largely due to the decline in sea ice in this area. Krill abundance shows a strong positive correlation with the duration of winter ice, and krill also increase when ice reaches farther from the Antarctic continent (**Figure 19.5**). Ice algae have declined with the decline of sea ice, which is one important source of food for krill. As waters have warmed near the West Antarctic Peninsula, growth of smaller microflagellates has been favored, as has salps, which efficiently consume the microflagellates. A connection has also been discovered between whale excretion and nutrients supplied to phytoplankton. Whales may feed in deeper waters on krill and rise toward the surface and defecate, which supplies significant nutrients to phytoplankton.

Penguins are abundant in the Antarctic Ocean (**Figure 19.6**) and nearby latitudes, and are very dependent on both high-quality breeding areas on land/ice and oceanic feeding areas. Much of what we know about penguin

FIG. 19.4. Links among sea ice, ice algae, phytoplankton, krill, and penguins.

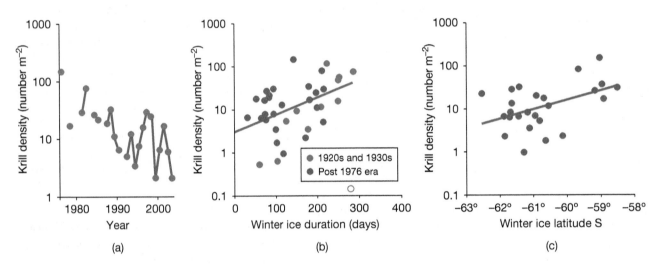

| (a) | (b) | (c) |

FIG. 19.5 (a) Change of krill abundance in a sector of the Antarctic Ocean near the Atlantic. Relationship between krill abundance and (b) duration of winter ice cover and (c) latitude to which ice extends. (After Atkinson et al., 2004)

| (a) | (b) |

FIG. 19.6 (a) The Adelie penguin *Pygoscelis adeliae*. (b) Adelie penguins swimming. (Photographs by Jerónimo Pan)

population change comes from studies of the Antarctic Peninsula (the peninsula on a map of Antarctica that seems to point to the tip of South America). In this region, sea ice has fluctuated considerably. Overall, true polar penguins are on the decline, although the details and dependencies are complex. Abundant species such as Adelie (*Pygoscelis adeliae*) and chinstrap penguins (*Pygoscelis antarcticus*) depend on krill for food, and their fate is tied directly to krill abundance and proximity. They have declined about 50 percent since the 1970s, and recruitment of Adelie juveniles to the parent population has declined by 80 percent (Lynch et al., 2012). Analysis of satellite images can now be done at the 0.5 m scale, and such analyses show that breeding areas for chinstraps have declined by at least half on Deception Island from 2003 to 2010 (Naveen et al., 2012). Fledging has not changed very much and tends to increase and decrease to the same extent for all species. In the summer breeding colonies, conditions do not seem to differ much for the species as well. This suggests that the problem is at sea and falls mainly on juvenile Adelies and chinstraps foraging in the winter. Adelies and chinstraps depend more on the ice and march more closely to the beat of the krill, which is becoming fainter. Chinstraps are less affected than Adelies by ice decline; they do better in open water and have had short-term intradecadal increases in past decades when sea ice retreated, relative to Adelie penguins. Adelie penguins are losers to climate change, since they are dependent on ground ice for nesting and have more trouble moving over larger expanses of open water to get to patches of krill offshore. In contrast, the Gentoo penguin, *Pygoscelis papua*, is a climate change winner. Its diet is more flexible than the krill specialist Adelie penguin, taking krill, fish, squat lobsters, and squid, and Gentoos can nest on ice-free cover, which increases their opportunities in a declining-ice world. The Gentoo penguin's larger body size allows deeper dives than those of Adelie penguins. Gentoo penguins have *not* declined in the past decades (Hinke et al., 2007).

Several seal species are important residents and ecosystem interactors in the Antarctic. Most notable are the crabeater seals (*Lobodon carcinophaga*), a true seal that lives along the entire Antarctic ice perimeter. They have specialized lobe-bearing teeth that are clearly adapted to filter large zooplankton (**Figure 19.7**), and they are major consumers of krill. Their abundance is not well known, but there are at least 7 million of them and likely many more. The leopard seal is a slender voracious predator that feeds on penguins and juvenile seals of other species. Several other species are found breeding on Antarctic shores.

The Antarctic Shelf Benthos: A Special Case of Isolation

■ **The Antarctic shelf benthos are isolated from the rest of the ocean's benthic fauna and have adapted to permanently cold waters.**

The shallow waters of the Antarctic Ocean are physically isolated from the rest of the world ocean, principally by the eastward-flowing circum-Antarctic current, which marks a large feature known as the Antarctic Convergence, or Polar Front. This feature suppresses exchange between the waters surrounding the Antarctic continent and the rest of the global ocean. Isotopic evidence suggests that the Antarctic became very cold during the Miocene epoch, more than 10 million years ago. At present, the temperatures in surface waters near the Antarctic Peninsula vary between –1°C in winter and +2°C in the Antarctic summer. These physical features already suggest that the Antarctic realm may have been long isolated both by geography and the extreme low temperatures (Clarke et al., 2004).

Resident organisms are distinctly adapted to the very low Antarctic temperatures, and the upper thermal limits of a number of species have been shown to be less than 10°C. We discussed in Chapter 5 the special adaptations of some Antarctic icefishes that use organic molecules as antifreeze to prevent cellular fluids from freezing.

■ **The Antarctic shelf benthos is dominated by a diverse fauna with unusual elements and lacks shell-crushing predators, such as crabs and lobsters.**

The continental shelf of Antarctica can be quite deep, extending to about 500–1,000 m. The benthic fauna of the Antarctic shelf has no major animal groups that are not found worldwide, but still the proportions are very different from most other oceanic shelves. Epifaunal soft-bodied suspension feeders such as sponges, bryozoans, hydrozoans, and ascidians are dominant in shallower waters of the shelf, less than 100 m deep. They create a biologically controlled landscape of mounds created by organisms on both soft and hard substrata (**Figure 19.8**). The dominance of passive suspension feeders may be related to a very low sedimentation rate, characteristic of the Antarctic shelf. This ecological aspect of the fauna seems very archaic and functionally resembles benthic faunas of the Paleozoic Era.

Often on soft bottoms of deeper shelf waters there is an abundant mobile fauna. Brittle stars, asteroid sea stars, isopods, gastropods, and polychaetes are also very common in the Antarctic benthos, even if the number of species of these groups is relatively low compared to other parts of the world's continental shelves. The abundance of echinoderms and isopods has reminded many researchers of a resemblance to benthic

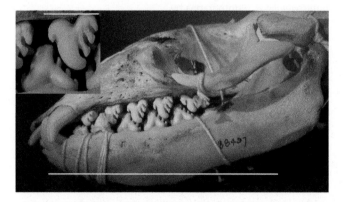

FIG. 19.7 Specialized teeth of the crabeater seal, adapted to strain krill from the water through the presence of krill-trapping spaces, while the seal keeps the teeth tightly occluded. Left view of skull (scale = 15 cm) and detail of teeth (inset, scale = 15 mm). (Thanks to American Museum of Natural History; photograph by Jeffrey Levinton)

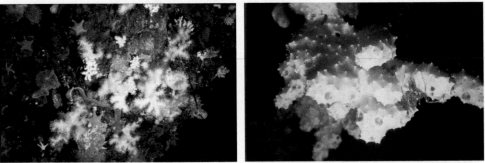

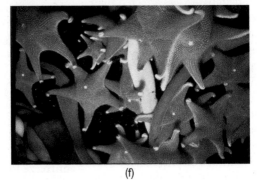

FIG. 19.8   Some of the Antarctic shallow shelf benthos: (a) isopod *Glyptonotus antarcticus*; (b) giant pycnogonid *Thavmastopygnon* sp.; (c) crinoid *Promachocrinus kerguelensis*; (d) cactus sponge *Dendrilla membranosa*; (e) vase sponge *Scolymastris joubini*; and (f) sea star *Odontaster validus*. (Photographs by Richard Aronson)

faunas of modern deep-sea soft bottoms. Some groups such as the crustacean group Pycnogonidae (**Figure 19.8b**) are much more common than in comparable biotas elsewhere.

But most fascinating is a general **lack of shell-crushing predators** such as crabs, lobsters, sharks, and fish. Predators instead tend to be slow-moving invertebrates such as asteroid sea stars. The Antarctic fauna has existed for millions of years without these predators. Considering this and the dominance by passive suspension feeders, it is as if the Antarctic is a refuge that ecologically resembles benthic communities of the Paleozoic Era, rather than more modern ecological structures found in the rest of the world ocean, where actively burrowing species dominate more in sediments and shell-crushing predators are present.

Have these differences arisen because of the isolation? The evidence is somewhat limited but fascinating. Occurrences of fossil-rich, shallow-water fossil deposits in the Eocene (ca. 56–34 million years ago) of the Antarctic show abundances of the crab and lobster predators that are now absent. Perhaps when cooling occurred in the Eocene to the Miocene, the Antarctic fauna acquired its unique and present characteristics. The Polar Front is believed to be about 25 million years old. There is a small amount of evidence showing some interchange of organisms coming from the north (Clarke et al., 2005), but the evidence for long-term isolation is compelling.

Another interesting feature of the Antarctic benthos is the much higher biodiversity found in the Antarctic shelf benthos relative to that of the Arctic. Apparently, the long isolation of the Antarctic has allowed more speciation and diversification. It may be that the Antarctic has existed for millions of years with less major disturbance than the Arctic. Also, the Arctic contains abundant shell-crushing predators of the kind (fish, crustacea) that are lacking in the Antarctic.

■ **Global climate change is rapidly increasing sea-surface temperatures around the Antarctic Peninsula, perhaps making the fauna vulnerable to invasion by formerly absent predators.**

As in much of the rest of the world's shallow ocean, seawater temperatures are increasing in Antarctica, especially in the area of the Antarctic Peninsula. Since 1945 the waters around the Antarctic Peninsula have increased about 2.5°C, which is several times the world average.

What might be the impact of temperature increases? First, consider the increase of boat traffic, especially of relatively fast ships. Fishing ships move continuously from north of the Polar Front into Antarctic waters, and these ships inadvertently carry large numbers of marine organisms, including planktonic larvae in ballast waters. Also, organisms are bound to float across this barrier occasionally on wood and other objects. In the past, it is likely that many species adapted to the warmer waters of the north arrived and died of the cold. Antarctic biologist Richard Aronson has speculated that many of the most mobile predators simply cannot function well in the frigid Antarctic waters (Aronson et al., 2007). But now, as the waters warm, predators such as crabs and lobsters may arrive and survive.

As discussed earlier, the shallower shelf biota of the Antarctic, dominated by epifaunal suspension feeders, resembles the ecological types that dominated the shallow seafloor during the Paleozoic Era. In the Mesozoic Era, these Paleozoic-type bottom communities were replaced by more mobile bottom invertebrates that defended themselves by burrowing into the sediment and producing skeletons that protected them against the skeleton-crushing sharks, rays, bony fish, crabs, and lobsters that became so abundant. As an example, surface-dwelling ophiuroids, or brittle stars, became rare when such predators rose to prominence in the Mesozoic. Paleontologist Geerat Vermeij (1977) termed this increase of predation the "Mesozoic marine revolution." The Antarctic benthos, although containing specific invertebrate

groups, many of which are quite different from those of the Paleozoic, still harbor in splendid isolation a vulnerable fauna. Right now, the dominant fishes are the notothenioids, which have small conical teeth and cannot crush skeletons, nor do they present a danger to the abundant surface epifaunal suspensions feeders, such as ascidians or brittle stars.

Low temperature may limit the ability of vigorous shell-crushing fishes to attack benthos. Crabs may now be limited from the frigid Antarctic waters by an inability to regulate magnesium ions at very low temperatures. But in recent years a number of direct observations of crabs have been made in Antarctic waters. A spider crab species from the Northern Hemisphere has been found in Antarctic waters, as has a king crab (family Lithodidae), native to the Southern Hemisphere. These crabs may have arrived as larvae in ballast water or as adults attached to ship surfaces. On the other hand, Griffiths et al. (2013) point out that there are some lithodid species that are endemic to Antarctica, suggesting that lithodids have been around for a long time and are not recent invaders. They also argue that the fossil record of lithodid crabs is too poor to make strong conclusions as to whether the crabs have been absent from Antarctic for millions of years. This interesting controversy needs to be resolved, and Griffiths et al. (2013) discuss crab diversity in the Southern Ocean at length.

Aronson's fascinating hypothesis argues that a local revolution in predation may begin soon on Antarctic shelves. The extent of Antarctic isolation and its possible vulnerability to invasion raises an interesting question. How connected are the oceans? Other sections of this text suggest that strong currents, like the western boundary currents (e.g., the Gulf Stream), can carry planktonic larvae a great distance. A study of deep-sea octopuses suggests that even the global ocean current conveyor belt system (see Chapter 2) can be a transport route for species dispersal. Strugnell et al. (2008) examined octopuses in the deep sea and studied their molecular evolutionary relationships. Strong evidence points to their origin in the Southern Ocean, and they may have spread to other parts of the ocean by the thermohaline circulation system, where water moves as water masses according to density. The timing of the origin (33 million years ago) and radiation (15 million years ago) of the deep-sea octopuses coincides with major changes in the global thermohaline system, with dispersal and species origins in the Atlantic, Pacific, and Indian Oceans.

■ **CHAPTER SUMMARY**

- Arctic benthic communities are strongly affected by continuous cold temperature, seasonal change in sea ice cover, and deposition of organic matter from the plankton and from ice algae embedded in sea ice.

- Fissures and cracks in polar sea ice harbor large amounts of ice algae, which are principally diatoms.

- During the polar spring and summer, melting ice releases both ice algae and

nutrients to coastal waters, which helps initiate strong coastal phytoplankton blooms.

- In the Arctic, high primary productivity of open-water phytoplankton and ice algae provide an abundant source of organic matter that is deposited on the seabed and subsidizes benthic communities.

- Global climate change is decreasing summer ice cover in the Arctic, and

major changes are occurring in sea temperature, input of organic matter to the bottom, and the organization of benthic communities.

- Decline of Arctic sea ice is having strong negative effects on a number of species of large predators that depend on ice as a place to roost and stage hunting trips, and for increased proximity to prey.

- On a larger geographic scale, disappearance of Arctic ice due to climate change may be causing much broader ecological effects on Atlantic food webs distant from the Arctic Ocean, because of melt-water spreading southward and increasing water-column stratification.

- Krill occupy a central role in driving coastal Antarctic food webs: They interact with shifting sea ice and feed on algae, and they are fed upon by a wide variety of predators. Decline in sea ice due to ocean warming is causing strong reductions of krill populations.

- Antarctic food webs depend on ice algae and phytoplankton blooms, which are in turn dependent on sea ice, which is declining. Phytoplankton is consumed by krill, which support large populations of penguins, baleen whales, and crabeater seals. These phytoplankton consumers support a rich array of carnivores, ranging from marine mammals to large predatory sea birds. Declines of krill and loss of ice cover has caused major declines of dependent Adelie and chinstrap penguins.

- The Antarctic shelf benthos is dominated by a fauna with unusual elements and lacks shell-crushing predators, such as crabs and lobsters. Climate change may allow the invasion by these predators, causing devastating effects on the local benthos.

■ REVIEW QUESTIONS

1. What general features of polar habitats suggest that they would be ecologically distinct from those at lower latitudes?

2. How does the seasonality of polar habitats affect polar food webs?

3. What is a major difference between Arctic and Antarctic regions that might strongly influence differences in oceanic circulation at the two poles?

3. What are the unique aspects of nutrient supply found in polar habitats?

4. How has climate change affected polar regions? What distinct effects have developed in the Arctic in comparison to the Antarctic?

5. How is krill abundance related to the extent of sea ice?

6. As ice recedes in the Arctic Ocean, what specific effects might occur that will affect Arctic food chains, in terms of top-down effects?

7. How might climate change affect the supply of phytodetritus to Arctic benthos?

8. Suggest a test that will estimate the degree of response of polar organisms to ocean warming.

9. What environmental changes in Antarctica are liable to have strong differential effects on different species of penguins?

10. How has fishing affected the structure of polar food webs?

11. As Antarctic waters warm, what major danger will be faced by subtidal epibenthic invertebrates in the future?

12. How has ice coverage changed in the Arctic Ocean in past decades?

13. How would you use an integrated approach studying molecular trees, biogeography, and the fossil record to understand whether Antarctica is an origin for global marine faunas or, instead, just a receptacle where species wind up, originating in other parts of the world's oceans?

Visit the companion website for *Marine Biology* at www.oup.com/us/levinton where you can find Cited References (under Student Resources/Cited References), Key Concepts, Marine Biology Explorations, and the Marine Biology Web Page with many additional resources.

CHAPTER 20

Biodiversity and Conservation of the Ocean

In this chapter, we shall consider patterns of biological diversity from the broader perspectives of evolution, extinction, and biogeography. Short-term ecological processes control the number of species in a region locally, but in the long term the relative rates of speciation and extinction explain the number of species. Regional variation in geography and climate influences the origin of species, and also the degree of geographic structuring of the world into assemblages of species. Finally, there has been an ever-increasing recognition of the importance of conserving the earth's biodiversity, and we will discuss in this chapter some of the important factors involved.

Diversity, Speciation, Extinction, and Biogeographic Factors

The Speciation–Extinction Balance

- **Although local patterns of species diversity are often explained in terms of short-term dynamic interactions, regional patterns are probably as much explained by the balance of speciation and extinction.**

In Chapter 4, we saw how a series of factors interacted to control the number of species in a community. We must take into account the origins and extinctions of species to explain variations in diversity on larger geographic and temporal scales. A predator may depress interspecific competitive exclusion and increase local diversity, but this process does not create species. Similarly, intense predation rarely drives a species to extinction over its entire geographic range. Rather, extinction usually follows larger-scale regional changes, such as a deterioration of climate.

Thus, an understanding of the factors that regulate the number of species over the long term leads to the need for an understanding of processes controlling speciation and the regional changes that cause extinction. We briefly discuss speciation in Chapter 4.

The sum of speciation tends to increase the total number of species in the world, unless there is extinction. **In the end, species diversity is a function of speciation minus extinction.**

- **Habitat change or destruction, widespread diseases, biological interactions, or random fluctuations of population size may cause extinction.**

Extinction, or the loss of species, may be caused by shifts in the environment, such as sudden temperature changes. Destruction of major habitats may also cause extinction. For example, most coastal lagoons are geologically unstable habitats and are destroyed after the passage of several thousand years. Any species restricted to a lagoon would be in similar danger of extinction unless they had sufficient geographic range over a large number of lagoons, leading to a high probability that some would survive in the long run. Another such example of habitat destruction would be marginal seas, such as the Mediterranean, that have experienced massive anoxia, nearly complete evaporation, and other effects so major as occasionally to make them nonmarine habitats.

Biological interactions may also be the cause of sudden extinctions. In Chapter 4, we discussed the rapid spread of disease in the ocean and the potential of disease to destroy even a widespread dominant species. The great *Zostera marina* eelgrass epidemic of the 1930s (see Chapter 17) caused the loss of an important foundation species upon which many marine invertebrates depended. For example, the eastern

American limpet *Lottia alveus* lived and fed on eelgrass, and its size fit closely to eelgrass leaves. The wasting disease in the 1930s resulted in the extinction of this utterly dependent commensal species, which has not been observed since the 1930s (Carlton et al., 1991). The invasion of competitors and predators to new regions may have a similar effect, although loss of habitat or regional destruction of water bodies is more likely the cause of most extinction.

The total number of species in a region, therefore, is the result of a net balance between the rate of species production and the rate of extinction. In unstable areas, where fluctuation between environmental extremes is the norm, newly isolated species are liable to become extinct, or are never established at all, owing to rapid shifts of climate and habitats. In such areas, the rate of extinction is likely to be high as well. Thus, the speciation–extinction balance tips toward low standing diversity. In more stable habitats, species might have a higher probability of survival when newly formed, thus increasing diversity. In some cases, immigration from another biogeographic region might add new species.

Biogeographic Factors

■ **Geographic isolation and major geographic gradients in temperature and salinity combine to determine provinces of statistically distinct groupings of species.**

The spatial arrangement of the continents and oceans, combined with the influence of the latitudinal gradient of temperature, organizes the world oceans into a series of distinct regions, characterized by geography, local circulation patterns, and water properties. Owing to seafloor spreading and continental drift over the past tens of millions of years, most coastlines are oriented approximately north–south. Thus, the geographic ranges of many shallow-water species are limited by temperature, which varies greatly along a coast with changing latitude. Many coastal biogeographic ranges are also influenced by geographic barriers, such as capes, which often coincide with temperature changes and isolation caused by current systems. For example, Cape Cod projects into the ocean and is a common termination of the coastal range of many species. It appears to be a barrier to dispersal, but sea-surface temperature also changes considerably as one moves from south to north of Cape Cod. The northern limit of the fiddler crab *Uca pugilator*, for example, may be limited by larval survival and perhaps an inability of adult individuals to molt at low temperatures north of the Cape (Miller and Vernberg, 1968; Sanford et al., 2006). Global warming, however, is allowing other species to spread north of Cape Cod, such as the mud fiddler crab *Uca pugnax*.

Because of the set of geographic and environmental barriers, the ocean can be divided into a series of **provinces**, or biogeographic regions with characteristic assemblages of species. Boundaries between provinces can be water-mass borders, major thermal discontinuities, points of land coinciding with thermal discontinuities, or boundaries between water bodies of differing salinity. Along the west coast of North America, for example, there is a series of provinces (**Figure 20.1**) whose boundaries usually coincide with

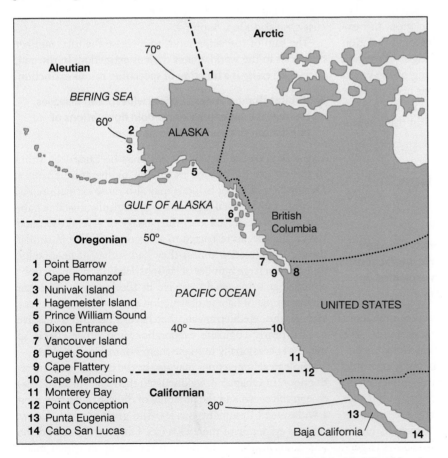

FIG. 20.1 Biogeographic provinces (borders shown as dashed lines) defined by the continental shelf mollusks of the northeastern Pacific. (After Valentine, 1966)

thermal breaks. For example, Point Conception in southern California marks a major shift from southern warm water to northern cold water. Although most boundaries coincide with major environmental shifts, currents may isolate one region from another, which is true through the summer at Point Conception. Multidecadal climatic oscillations such as the NAO and the PDO also periodically shift climatic zones north and south (see Chapter 3). Provinces are usually recognized statistically, not by unique assemblages of species. Many species have ranges that transcend provincial boundaries. Presumably, these are usually species with strong dispersal stages and broad thermal tolerances.

■ The relation of geography to speciation can be accomplished by relating evolutionary trees to patterns of geographic occurrence.

In Chapter 4, we discussed the construction of evolutionary trees. Such trees establish relationships of groups of species as being **monophyletic**, or arising originally from a single ancestral species. Many evolutionary lineages may occur throughout the world, but it is unusual for species to appear and randomly disperse, resulting in little or no relationship with geography. Usually, evolutionary trees can be related to the geographic barriers that would have influenced isolation of lineages. The study of relating evolutionary lineages to geographic relationships *within* a species is known as **phylogeography**. These relationships can be extended to the species level. **Figure 20.2** shows a hypothetical example of the origin of a set of species on an archipelago, as influenced by geography, colonization, and local evolution. At first an ancestral species is present and colonizes two nearby islands; then subsequent colonization occurs. With each colonization step, DNA sequence differences accumulate, which allows the construction of an evolutionary tree of the final result. It is important to realize that barriers may appear and later disappear, which allows species to disperse and intermix after their initial formation by allopatric speciation.

An important connection must be made between evolutionary trees constructed from DNA evidence and the absolute times of divergence. This is difficult without a timetable. The establishment of **molecular clocks** can make such connections. One way that such a clock can be established is by connecting speciation events to geological events whose age has been established by independent means. Such examples are unusual, and the appearance of the Isthmus of Panama has been used in many studies. For example, a few pairs of sister species (species most closely related and having the same ancestor) have been found on either side of the isthmus. Geological evidence shows that the isthmus arose and separated the Caribbean from the Pacific about 3 million years ago. We can compare the degree of DNA sequence difference to this time of separation to estimate the amount of sequence difference that arose per million years. Assuming that a given amount of DNA divergence occurs for a given gene type, this clock calibration has been used to estimate broader divergences. This is based on the idea that increased DNA difference occurred at proportionally older time periods, as calibrated by the

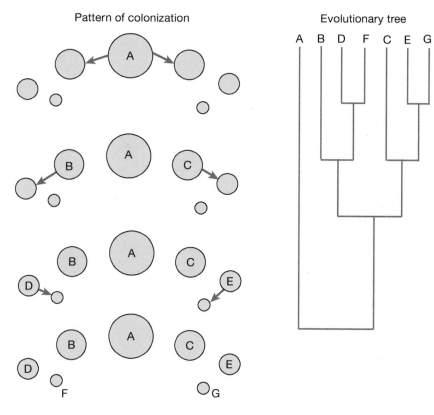

FIG. 20.2 Left: Species *A* occupies a central island, and successive colonizations of nearby islands and local divergence results eventually in the evolution of seven species in the time sequence from top to bottom. Right: Tree shows relationships between species, which could be constructed from similarities in DNA sequences.

Panamanian isthmus divergence. In the fiddler crab genus *Uca*, for example, a difference in sequence of sister species on either side of the isthmus has been used to calculate that the entire genus *Uca* diverged from an ancestor approximately 22 million years ago (Sturmbauer et al., 1996). This type of clocklike extrapolation can only be approximate, since it is unlikely that evolution is so clocklike, and many studies show strong divergences from a simple molecular clock. Still, this is a powerful tool to get general estimates of ages of evolutionary divergence.

■ **The geographic history of a region and major ocean basins is a crucial background for the evolution of marine species and the origin of marine biodiversity patterns.**

The ocean and its arrangement of coastlines have been dynamic throughout geological history. Continental masses have moved thousands of kilometers because of continental drift. Areas of the ocean have been connected and separated after global sea level has risen or fallen, respectively. Finally, local tectonic motions have caused rises of landmasses such as the Isthmus of Panama, resulting in separation of populations and rearrangement of current patterns.

These rearrangements are often an important backdrop for an understanding of the present distributions of marine biotas. In Chapter 2, we discussed the presence of marginal seas and the fact that many have very different recent geological histories, resulting in differences of environmental conditions and even existence in just the past few million years. The Mediterranean was completely dry just a few million years ago and was, therefore, recolonized in the more recent past as sea level rose. On a much more local scale, Long Island Sound, New York, was in the past 11,000 years completely covered by glacial ice, a freshwater lake, and (now) an estuarine water body. Its current biota must have derived by invasion from nearby marine coastal areas.

The shore marine biota of the North Atlantic is an excellent example of the impact of geological history on North Atlantic shore diversity and evolutionary history. First, consider an important event that occurred about 3.5 million years ago. At this time, the glacial era had not yet begun, so the Arctic Ocean was hospitable to a wide range of species. At this time, the Bering Strait, a present-day opening between Alaska and Asia, first opened up and allowed species from the North Pacific to enter toward the north. Something quite surprising occurred. A large number of rocky-shore species moved northward from the Pacific, across the Arctic Ocean, and successfully invaded the North Atlantic. At present, the European and American North Atlantic rocky-shore fauna, including mollusks, sea stars, and many other species, are very much dominated by this invasion through a **corridor** from the Pacific that existed a few million years ago but is now closed off. Nearly 300 molluscan species invaded across at that time, or descended from previous invaders. Geerat Vermeij (1991) reached this conclusion after an exhaustive study of the fossil and current distributions of mollusks in both ocean basins. With the increased melting of the Arctic Ocean sea ice in recent years owing to anthropogenic climate change, such a corridor is now being reestablished.

The story continues with the effects of the Pleistocene glaciation on the North Atlantic. For the past 2 million years, continental glaciers have advanced and retreated on the eastern and western sides of the North Atlantic. Evidence from continental glacial deposits tells a very different story on the American and European sides of the ocean. The American side was completely glaciated several times, southward to the current southern New England and New York coastlines. The rocky-shore fauna was probably nearly obliterated; but this was not true on the European side, where glacial deposits do not demonstrate such a major advance of the glaciers so far south (**Figure 20.3**).

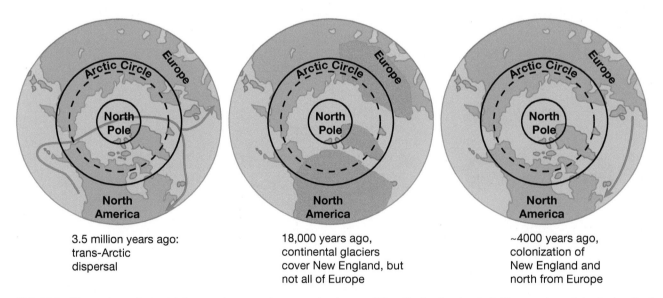

3.5 million years ago: trans-Arctic dispersal	18,000 years ago, continental glaciers cover New England, but not all of Europe	~4000 years ago, colonization of New England and north from Europe

FIG. 20.3 Three phases in explaining species occurring on rocky shores of New England today. Left: The opening of the Bering Strait 3.5 million years ago allows Pacific species to invade into the European and American sides of the North Atlantic through the Arctic Ocean, displacing local species. Center: Continental glaciation in the Pleistocene wipes out most species on New England rocky shores but has much smaller effects on European shores. Right: Species invade New England shores in the last few thousand years from the European side of the North Atlantic.

The difference on both sides of the Atlantic raises an interesting question. If most of the Atlantic New England coast was glacially covered and scoured, where did the current living fauna originate? The only logical conclusion is that somehow it came from European waters across the Atlantic Ocean in the past few thousand years since the glaciers retreated about 8,000–10,000 years ago. The Arctic was closed off, so our fauna could not have come again from the Pacific, as it did a few million years ago. How could we test the likelihood of such a process?

The record in the genes of marine invertebrate populations allows us a surprisingly good understanding of past population size and stability. If we sequence a gene in a population of a marine invertebrate species, we are likely to find genetic variability. The longer a population has been in any one place in great numbers, the more mutants will appear and be retained in that population. If a population is very small, a process called genetic drift will result in the random loss of many variants and genetic diversity will be small. Newly invading populations are often a small subsample of their progenitors, so the invaders will probably also have low variability. This leads to two predictions we can apply to the situation in the Atlantic:

1. If European populations have not been devastated by glaciations, they should have accumulated higher variability from mutation than North Atlantic American populations, which were more recently glaciated.

2. If American populations were devastated and recolonized from Europe, then they should have low variability and the dominant genetic variant should be traceable to Europe.

These hypotheses were tested on a number of species of invertebrates that live now on both sides of the Atlantic, and mostly the results fit the hypotheses. Wares and Cunningham (2001) sequenced the *Carbonic anhydrase 1* gene and looked for variants and generally found that populations of species on the European side had many variants coexisting within populations. Populations of the same species on the American Atlantic side had few variants whose DNA sequences could be linked to hypothetical European colonists. In an exceptional case of a barnacle species, there was evidence from unique American genetic variants that the barnacle species had survived the glaciation; we could see the trace in genetic variants that could not be traced to European origins.

■ **A geographic barrier may isolate populations of just one species; or it may isolate recently evolved closely related species or groups of species of a variety of distantly related groups.**

Although provincial boundaries can be recognized by the ends of the geographic ranges of marine species, one would expect that broader-ranging species might still be isolated and therefore have limited dispersal across the same boundaries. One might then expect some degree of genetic difference between populations of the same broadly ranging species on either side of a boundary. **Figure 20.4** shows differences in frequencies of major variants in mitochondrial DNA length polymorphisms along the east and Gulf coasts of North America. The Atlantic coast of southern Florida is known to be an important provincial boundary. Species with ranges that transcend

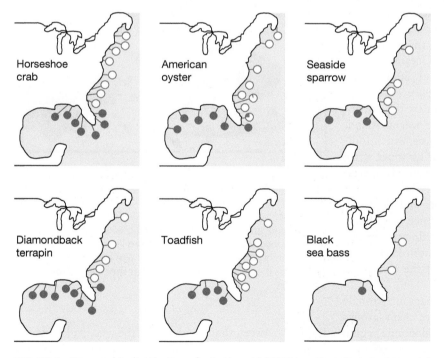

FIG. 20.4 Geographic distribution of mitochondrial DNA length variants in six coastal species. The pie diagrams for a given species at each locality show the relative frequencies of the two most common monophyletic groups of length variants (or groups of length variants with the same ancestor) for the given species. As can be seen, the South Florida region marks a major biogeographic discontinuity. (After Avise, 1992)

this boundary nevertheless are strongly differentiated genetically on either side of it.

Many other boundaries have existed long enough to cause the production of newly formed species on either side of the boundary. For example, the Isthmus of Panama arose by tectonic activity about 3 million years ago. This resulted in a major extinction in the Caribbean; but also, a number of species were each isolated into separated populations on either side of the isthmus, following a period of genetic differentiation and probably different selective forces on either side, owing to differences in sea temperature and other variables.

In the southwest Pacific, an archipelago comprised of islands of Indonesia, New Guinea, and Australia makes up a barrier that separates the waters of the western Pacific from those of the easternmost Indian Ocean in Southeast Asia. This is a complex region, and during times of glaciation, there were many land barriers when sea level was much lower. This area is known as the Coral Triangle. Paul Barber and colleagues (2006) investigated DNA sequences for the *cytochrome c oxidase* gene of three species of stomatopods (mantis shrimp) and found that they each shared a strong degree of genetic difference in the form of distinct evolutionary intraspecific gene lineages that reflected isolation across the Coral Triangle barrier during the Pleistocene low-sea-level stand (**Figure 20.5**). One finds the same pattern in each species with respect to the Indonesian archipelago: a distinct southern population in the Indian Ocean, a distinct population on the Pacific side, and a distinct central population in between.

A somewhat different pattern also reflecting isolation in this region was found for species of the tropical snail genus *Echinolittorina*. Some species of this genus tend to be found adjacent to continents, and patterns of speciation in the past few million years fit well with the presence of a barrier between the Indian and Pacific Oceans during a Pleistocene low level of sea stand (Reid et al., 2006). There is consistent evidence from DNA sequence differences that species arose as a result of isolation by the barrier. But other species that tend to live on more oceanic islands are not strongly different in DNA sequences, which indicates that there was interchange even during times when sea level was lower, when glaciers had advanced. Speciation did not occur in these latter groups. This pattern is not universal; however, a study by Barber and Bellwood (2005) used the rate of molecular divergence to estimate times of divergence of fish faunas. The divergence between a group in the Indian Ocean and the western Pacific was dated to be older than the appearance of the barrier we have discussed.

On a larger scale, broad expanses of oceans are strongly isolating because even planktonic larvae of many groups cannot traverse such a wide area. The midoceanic region of the Pacific Ocean is a major barrier. For example, Sturmbauer and others (1996) investigated the fiddler crab genus *Uca*, which is spread throughout the world tropics. Using DNA sequence data they constructed an evolutionary tree, which showed a strong break in evolutionary

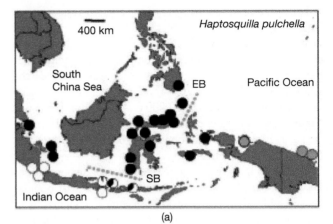

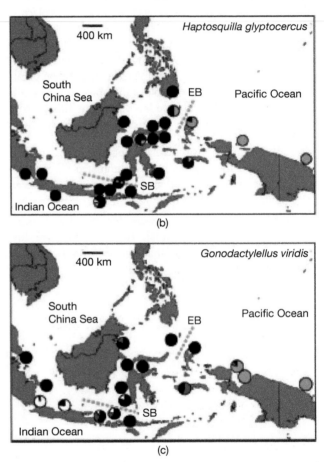

FIG. 20.5 The role of the Indonesian Archipelago as a barrier. Pie diagrams show the relative frequency of highly divergent sequences of mitochondrial *cytochrome c oxidase, subunit 1*, for three stomatopod species: (a) *Haptosquilla pulchella*, (b) *H. glyptocercus*, and (c) *Gonodactylellus viridis*. Dashed red lines highlight the location of the hypothesized Eastern Barrier (EB) and Southern Barrier (SB) across the Maluku and Flores Seas, respectively. (From Barber et al., 2006)

lineages between the Americas and the western Pacific and Indian Ocean species (**Figure 20.6**). A similar break has been noticed in species of the snail genus *Echinolittorina* (Williams and Reid, 2004) and in the cone snail genus *Conus* (Duda and Kohn, 2005). In *Uca* and in *Echinolittorina*, there are completely separated evolutionary lineages

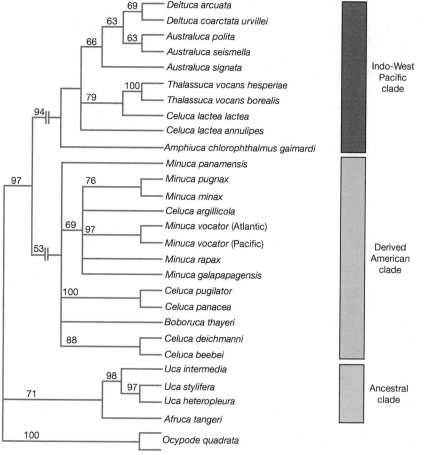

FIG. 20.6 Evolutionary relationships of the fiddler crabs, genus *Uca*. The most ancestral part of the group is found in the American tropics. Another American clade is separated from an Indo-Pacific clade, separated by a large oceanic barrier. Bootstrap values of nodes are a measure of the predictability of nodes of the evolutionary tree. (From Levinton and others, 1996)

in the two regions, but there are some cone shell and coral species that apparently have been able to disperse across the expanse of the Pacific.

■ **Evolutionary lineages are more isolated and regionally diversified in the sea than has been previously thought, especially among islands on archipelagos and newly appearing islands.**

Because many marine species have long-lived plank-totrophic larvae or, like fish, can be transported great distances by currents even as adults, we tend to think of isolation as rare in the ocean. Important exceptions are major barriers, as mentioned earlier, such as open oceans (for coastal benthos), or land barriers, such as the Isthmus of Panama. But many groups have more limited dispersal, and the many oceanic islands and archipelagos throughout the world might provide a strong degree of isolation.

Some surprising evolutionary radiations of species have been found in relatively closely spaced oceanic archipelagos. Pacific islands, though widespread and often isolated from each other, have a number of widespread species, usually with planktotrophic larvae. But many groups have much more limited dispersal, including sea squirts, peracarid crustaceans, and many gastropods. A fascinating

example comes from the predatory venomous snail genus *Conus*. In the Pacific Ocean, species disperse by means of planktotrophic larvae and many occupy very broad geographic ranges. For example, the dangerously venomous *Conus geographus* occurs over a wide range of depth over the entire Indo-Pacific except Hawaii. However, in the eastern Atlantic, *Conus* species disperse by direct release, so dispersal is very limited. This has led to a remarkable subdivision and radiation of species in the Cape Verde Islands, located in the eastern Atlantic at about 16° N latitude (Duda and Rolán, 2005). Nearly 50 species arose, probably from one ancestor, and the level of speciation is so extreme that radiations of species occurred within single islands as small as 25 km long (**Figure 20.7**). We can still see the outcome of past localized, geographically based isolation and speciation. One small-shelled group of species invaded the islands probably at the time of origin of the islands and differentiated into several species, whereas another larger-shelled group evolved and differentiated later (Cunha et al., 2005).

In the western Pacific, the turbinid snail genus *Astralium* has representatives on a broad swath of islands, and these were all thought to belong to probably one variable species. Dispersal is by means of nonplanktotrophic larvae probably of just a few days' duration, so the possibility exists for

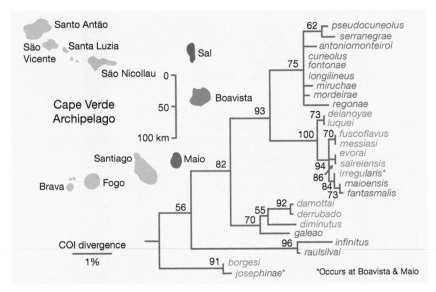

FIG. 20.7 Evolutionary tree of species of *Conus* restricted to the Cape Verde Islands in the eastern Atlantic, showing map of the Cape Verde Archipelago. Geographic distributions of species are indicated with colors of Sal, Boavista, and Maio Islands in map. Bootstrap values shown at nodes are a measure of the predictability of the nodes of the calculated evolutionary tree.

some isolation. Christopher Meyer and colleagues (2005) found two major evolutionary lineages among islands in the western Pacific; but this was even further divided by sublineages, each confined to one of a series of islands that were fairly close to each other. Overall, this suggests that regional differences in biodiversity may be much greater than we might have imagined by first looking at morphology alone.

Sometimes, evolutionary opportunities arise when a new habitat arises. Many islands in the world ocean arise as volcanoes penetrating the surface (as in the Hawaiian Islands) or as geological formations that are thrust upward by tectonic processes. When such islands or archipelagos appear, they may be colonized by species whose spread among the islands interacts with local isolation and strong local natural selection to cause small-scale **adaptive radiations**. Members of the crab family Grapsidae live in shore habitats throughout the world, but a remarkable evolutionary event took place on the island of Jamaica. Sometime in the mid-Tertiary, Jamaica was completely covered by the sea, but in the late Tertiary, sea level declined or uplift occurred, and terrestrial habitats appeared. This has resulted in an evolutionary radiation of terrestrial crabs from one marine ancestral species. The crabs all have a very abbreviated larval stage, and females raise their young in local habitats such as the water retained by leaves. Christoph Schubart and colleagues (1998) collected and sequenced the DNA of Jamaican crabs and found that sequences of all species could be grouped and traced to a single ancestor that must have invaded the island about 4 million years ago. How did they get this age estimate? They used pairs of related crab sister species on either side of the Isthmus of Panama. Using the

independent evidence of the age of the isthmus, they could estimate the number of base pair changes per million years in the gene they studied. This was used to estimate age in the Jamaican group by assuming that the rate of molecular evolution is about the same.

Major Gradients of Species Diversity

■ Biodiversity is an accounting of species within and across habitats.

Diversity can be assessed by a variety of indices of species diversity. Diversity can be described as a function of the number of species and their relative evenness of distribution (see **Going Deeper Box 20.1**), but we will focus on counts of species. The first step in assessing biodiversity is producing a count of species within a specific habitat, such as muddy bottoms of a continental shelf region, often described as alpha diversity. A more complete accounting would be to add more habitat types—for example, rocky subtidal bottoms—and produce comparative species counts for the several habitats within a region, described as beta diversity. The total diversity over a region—for example, a biogeographic province—would refer to gamma diversity. The difference in diversity explained by multiplication of habitats is the **between-habitat component** of species diversity. Any changes in species diversity between regions in a single defined habitat comprise the **within-habitat component** of species diversity. For example, comparison of the number of species living in muddy-bottom shelf subtidal sediments in the Atlantic Ocean versus muddy bottoms in the Pacific Ocean shelf is a within-habitat comparison of species diversity.

GOING DEEPER 20.1

Measuring Biodiversity in Terms of Species Number and Relative Distribution

Most of our attempts to account for biodiversity involve obtaining an accurate count of species, usually within a specific habitat type at a given geographic location. This is often described as alpha diversity. But it is well known that the relative abundance of species is often related to biological processes. In some habitats, strong disturbance leads to strong dominance by just one or a few species that are adapted to colonize rapidly and increase in abundance. Other habitats that are relatively undisturbed accumulate species of a variety of abundances that occupy many different microhabitats and sum up to a more even distribution of species in the habitat. As a result, species diversity indices have been developed to account for a combined species number component and a dominance component of biodiversity.

We will discuss just one example, the Shannon Index, but many others exist.

Shannon's Index, H, is as follows:

$$H = \sum_{i=1}^{S} p_i \ln p_i$$

where S = the number of species, and p_i is the proportion of species from $i = 1$ to S. As H increases, the evenness of distribution, for a given value of S, increases so that if all species are of exactly equal proportions the maximum H, H_{max} is

$$H_{max} = \ln S$$

This allows a calculation of the degree of evenness, known as Equitability, E, to be calculated:

$$E = H/H_{max}$$

In summary, this or any similar index increases with the number of species and the equality of distribution of the proportions of numbers of each species relative to the total sample of individuals. Dominance by one or a small number of species reduces both H and E.

The Latitudinal Diversity Gradient

■ **Species diversity tends to increase with decreasing latitude.**

The best-known species diversity gradient is the **latitude diversity gradient**, which is an increase of species diversity from high to low latitudes in continental shelf benthos, in the plankton in continental shelf regions, and in the open ocean. **Figure 20.8** shows the latitudinal gradient for bivalve mollusks, but the trend applies to other shelf invertebrate groups and to planktonic groups such as copepods. There are some exceptions—for example, nuculid bivalve mollusks are constant in diversity toward higher latitudes. The generalization of increased diversity with decreasing latitude also applies to higher taxonomic categories, such as genera and families (**Figure 20.8**). This generalization applies to species lists of a region, but not necessarily to the number of species living within a small area of, say, a square meter. The regional species list for Costa Rica contains five times as many species as the list for coastal Washington State, yet a square meter of typical tropical beach contains no more species than typical temperate-beach samples. The overall difference in species diversity seems to be due to the relative richness in Costa Rica of the low-shore, cobble-beach habitat. Finally, the latitudinal species diversity gradient is not limited to marine habitats; it is found widely among terrestrial and freshwater biotas.

The present-day latitudinal species diversity gradients correlate with the strong latitudinal gradient of temperature. One should keep in mind some of the conditions that are nearly unique to our time in geological history. At present, there is a strong latitudinal gradient in climate, owing to the recent (last few million years) cooling of world climate and glaciation from the poles. Also, sea-floor spreading and continental drift have produced extensive north–south trending coastlines, a development that distributes shallow-water biotas along a lengthy climatic-geographic gradient.

Comprehensive comparisons of latitudinal gradients in continental shelf species richness are far too scarce throughout the world to permit us to draw general conclusions. One compilation comparing the Atlantic and Pacific coasts of the

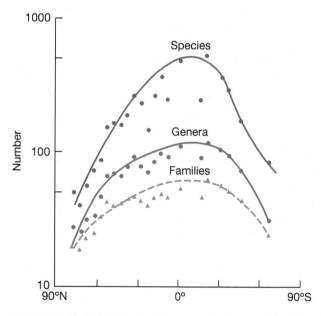

FIG. 20.8 The relationship of bivalve mollusk taxon diversity and latitude. Points are average number of species, genera, and taxonomic families. (After Stehli et al., 1967, courtesy of the Geological Society of America)

Americas gives us an insight into interoceanic comparisons (Roy et al., 1998). A compilation of nearly 4,000 species of marine prosobranch gastropods shows a latitudinal gradient in species richness on both coasts (**Figure 20.9**). It is a bit surprising to those who think of the Pacific as more diverse (which it is, overall) to find that there is not much difference in the number of species between coasts in the tropics. In higher north latitudes, the Pacific coast has about double the number of prosobranch species found in the same latitudes on the Atlantic coast, which may possibly be explained by a more extreme climatic history during the past few million years in the northwest Atlantic.

Can we see any relationships between latitudinal diversity and environmental factors? Macpherson compiled extensive

records of species biodiversity for both benthic and oceanic planktonic groups and found strong evidence supporting an increase of species numbers with decreasing latitude in the North and South Atlantic oceans. Benthic species numbers increased toward the equator in coastal, shelf, and abyssal environments, and species numbers were most closely correlated to average annual temperature. But there were noticeable hotspots of diversity: The Caribbean benthos was noticeably diverse from latitudes 30–10° N latitude and the Mediterranean benthos from latitudes 45–30° N latitude. The overall peak of diversity on both sides of the Atlantic was a bit north of the equator (**Figure 20.10**).

The pattern of a strong peak in biodiversity at the equator was *not* true for water-column species. Benthic species

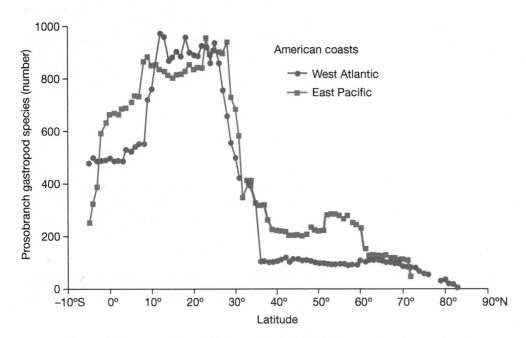

FIG. 20.9 Latitudinal pattern of species richness of eastern Pacific (green-colored squares) and western Atlantic (brown-colored circles) marine prosobranch gastropod species, divided into units of 1 degree of latitude. (After Roy et al., 1998)

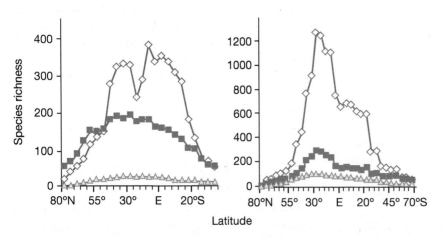

FIG. 20.10 Latitudinal species-richness gradients for benthic taxa along the eastern (left) and western (right) Atlantic Ocean. Diamonds, coastal; green squares, shelf; blue triangles, abyssal. E, Equator. (From Macpherson, 2002)

numbers tended to have a strong peak near the equator, but the highest diversity for water-column species remains constant over a large range of latitude from approximately 50° N to 40° S. This result is very consistent over many different taxonomic groups (**Figure 20.11**). Among species such as open-water gelatinous zooplankton and fishes, species numbers were best correlated with dissolved nitrogen, which reflects ocean productivity. The continuous

and higher productivity in the lower latitudes apparently allowed the production or survival of species to a greater degree than higher latitudes. Areas where productivity is very high and pulsed, such as upwelling systems, do not have high diversity, possibly because the low stability causes species extinction. We must remember that correlation does not necessarily mean causality, and other factors such as climatic fluctuations at higher latitudes may have reduced diversity in water column species in latitudes higher than about 40° N or S.

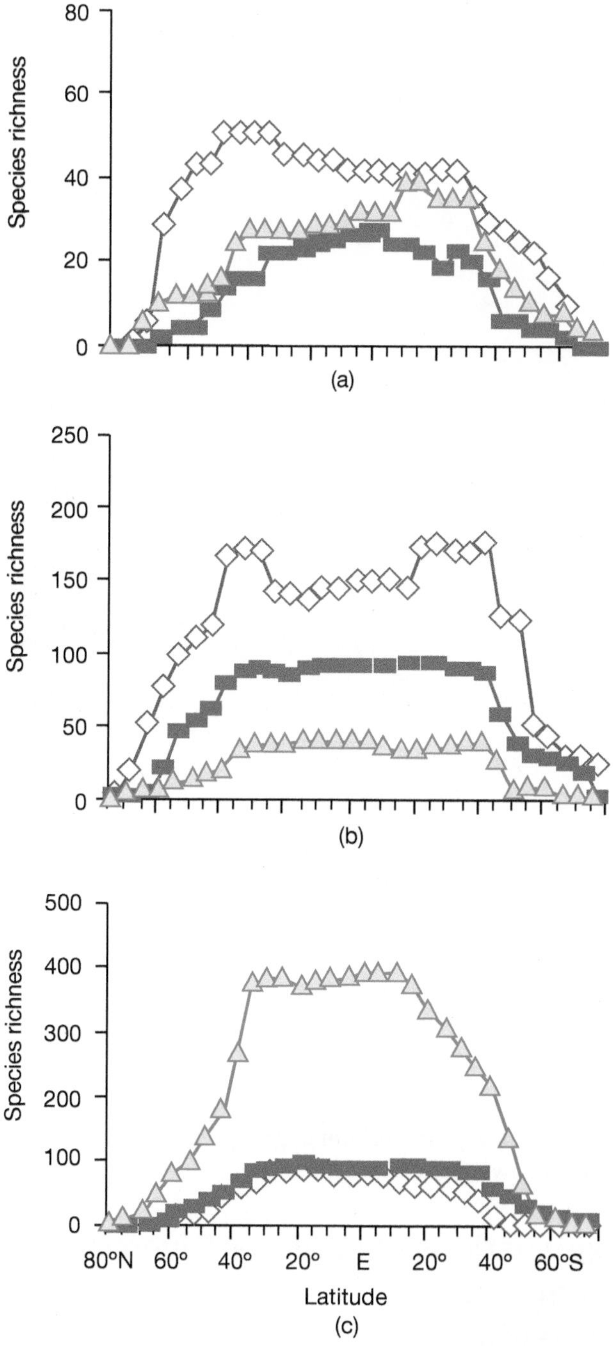

Differences Between and Within Oceans

■ There are differences in species diversity between ocean basins.

Even when latitude is held constant, the Pacific Ocean overall has far more species than the Atlantic. This fact has been documented for a wide variety of invertebrate groups and fishes, especially in coral reef habitats. The contrast is also obvious for many groups on the temperate Atlantic and Pacific sides of North America (**Table 20.1**). In the northwest United States, one can find 19 species of shallow-shelf asteroid starfish, in contrast to only six species in the southern New England region. An interesting exception is the polychaete annelids, which are slightly more species rich in the New England area. Most of the diversity difference, however, is apparent in comparing the tropical Atlantic with the tropical Indo-West Pacific, the latter being far more diverse. The inner-shelf mollusks on the east and west coasts of the tropical Americas appear to be rather similar in species richness.

■ Within the Pacific Ocean, species diversity in coral reefs declines in all directions from an Indo-Pacific diversity maximum.

Although the latitudinal gradient is most prominent in all oceans, the Pacific has a prominent longitude-related peak of diversity, especially in the species associated with coral reef habitats. Diversity reaches a maximum in the southwest Pacific, in the region of the Philippines and Indonesia (**Figure 20.12**). From this center, diversity declines in all directions, although the latitudinal gradient is steeper than the longitudinal gradient. The pattern applies to a wide range of animal groups, including fish and many phyla of invertebrates. We will discuss the possible mechanisms generating this pattern shortly.

Lower Diversity in Inshore and Estuarine Habitats

■ Inshore and estuarine habitats are poorer in species than comparable habitats in the open sea.

In both benthic and water-column assemblages, the open sea tends to have more species than inshore habitats. Estuaries are the extreme cases, where one finds a small number of species, dominated by very few species with high

FIG. 20.11 Latitudinal species-richness gradients for pelagic taxa. (a) Diamonds, Appendicularia; squares, Salpida; triangles, Chaetognatha. (b) Diamonds, Hydromedusae; squares, Siphonophorae; triangles, Euphausiacea. (c) Diamonds, Decapoda; squares, Cephalopoda; triangles, Pisces. E, Equator. (From Macpherson, 2002)

TABLE 20.1 Species Diversity of Various Invertebrate Groups in the Vicinity of Woods Hole, Massachusetts, and Friday Harbor, Washington

TAXONOMIC GROUP	NUMBER OF SPECIES	
	FRIDAY HARBOR, WASHINGTON (LATITUDE 48.5° N)	WOODS HOLE, MASSACHUSETTS (LATITUDE 41.5° N)
Shell-less opisthobranch gastropods	61	23
Shelled gastropods	88	51
Bivalve mollusks	114	55
Asteroid starfish	19	6
Polychaete annelids	174	218
Isopods	42	27
Amphipods	76	47

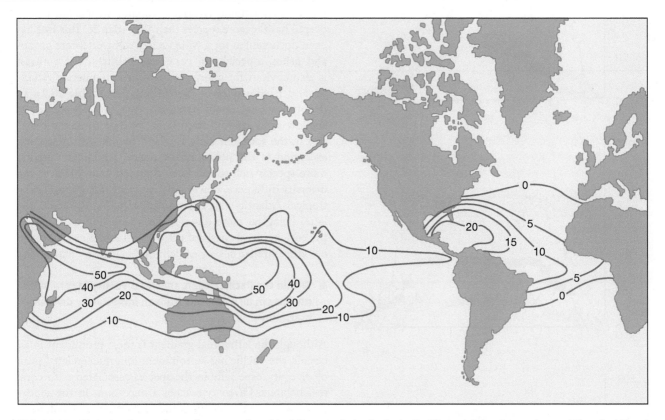

FIG. 20.12 Variation in numbers of genera of reef-building corals in the Indo-Pacific and Atlantic provinces. (After Stehli and Wells, 1971)

abundance. By contrast, open continental shelf communities usually have more species, and the species abundances are more evenly divided. In the plankton, the increasing complexity offshore is correlated with an increase in the number of trophic levels.

In Chapter 18, we discussed the high diversity of the deep-sea benthos. If one considers the muddy-bottom habitat alone, there is a regular change in benthic diversity from the coast to the abyssal plain. Species diversity of macroinvertebrates and fishes increases with depth to a maximum just seaward of the continental rise, and then decreases with increasing distance toward the open abyssal plain. By contrast, the species diversity of

meiofaunal and microfaunal groups seems to increase steadily toward the abyss.

Explanations of Regional Diversity Differences

Effects of Recent Events

■ **Complex recent historical events may explain some current regional differences in species diversity.**

As mentioned earlier, some of the large-scale geographic differences in diversity can be ascribed to recent events. The gradients may not be due to an ongoing equilibrium balance of extinction and speciation. For example, the present-day species pattern along the eastern U.S. coast is due primarily to a rapid large-scale extinction in the Southeast and the Caribbean that probably was followed by the appearance of large numbers of new species. An extinction, followed by an invasion from the northeastern Pacific, strongly influenced the diversity of higher-latitude, shallow-water marine faunas in the northwestern Atlantic. The diversity of the Caribbean has also been shaped by relatively recent events. The large-scale diversity gradient in the tropical American coral reef biota may be due in part to extinctions around the periphery of the province, as the earth's climate became cooler and the latitudinal oceanic temperature gradient steepened strongly in the Atlantic basin during the last few million years in a time of increased polar glaciation. The rise of the Isthmus of Panama a few million years ago was followed by an extinction of many Caribbean species that had close relatives in the Pacific. Despite this extinction, many new species appeared, so the total effect on diversity was not very great. Molluscan diversity apparently even increased after the emergence of the Isthmus (Jackson et al., 1993). The change in diversity in the last few million years, in sum, was very complex and probably consisted of a series of expansions and contractions of species richness.

Effects of Events in Deeper Time

■ Many regional differences arose many millions of years ago and have persisted.

Geographic divergence among groups of species probably results from even more ancient geographic and oceanographic rearrangements that have gone on for many millions of years. Continental drift combined with global rises and falls of sea level have rearranged oceanic realms, isolating major regions with land barriers and combining others as land barriers broke down.

These large-scale processes apply to our understanding of the division of the Atlantic and Pacific marine biotas. Before the Oligocene epoch, approximately 40 million years ago, when the broad, shallow sea known as **Tethys** connected the Atlantic and Pacific, there was greater homogeneity of the marine faunas. Following the rise of landmasses, especially in the Middle East, the Tethyan Sea disappeared and divergence soon followed between Atlantic and Indo-Pacific biotas. The barrier to dispersal not only allowed more local speciation in the two oceans, but was accompanied by strong climatic differentiation, so that species also became adapted to the more variable Atlantic, for example, as opposed to the relatively less environmentally variable Pacific and Indian Oceans. Within the central and eastern Pacific, a great deal of evidence shows that longitudinal distance is accompanied by large differences in species occurrences and also genetic differences within widespread species, which reflects genetic isolation by distance (Palumbi et al., 1997). As we discussed in Chapter 19, the Antarctic benthos today may be the result of isolation from the rest of the global ocean for many millions of years. Such long-term divergence has

been occasionally disrupted by events such as the movement of many species through the trans-Arctic seaway, discussed earlier in this chapter.

Factors Causing High-Diversity Regions

■ The pattern of species appearances within the latitudinal diversity gradient fits a model of tropical origins and spread to higher latitudes.

The high species diversity in the tropics relative to high latitudes is nearly universal. What explains the pattern? Are the tropics a place where more species originate? Do species originate at all latitudes, but disperse and accumulate in the tropics? We can imagine a few alternatives:

Museum hypothesis. A model of geographic isolation would predict that species originate and survive in the tropics, but do not spread to other regions. Instead, regional barriers within the tropics cause speciation but species do not leave the tropics. The high-diversity tropics would, therefore, be a "museum" of diversity, which means that species generally originate in the tropics and stay there in high numbers. Fewer species originate in higher latitudes, which leads to a latitudinal diversity gradient.

Tropical source. Species originate in the tropics and accumulate there but also move toward higher latitudes. This makes for an **out-of-the-tropics hypothesis** to explain the latitudinal diversity gradient.

One approach to this problem is to examine the fossil record to find out where fossil species first appear: tropics or higher latitudes. Unfortunately, it is difficult to identify fossil species using morphology, but there is a good fossil record at the level of genus and subgenus. David Jablonski, Kaustov Roy, and James Valentine (2006) examined many records of fossil appearances of bivalve mollusks in rocks over the past 11 million years and found that genera and subgenera mainly appeared in the tropics and later spread to higher latitudes. These results suggest that the tropics constitute a cradle of origination, but that dispersal to higher latitudes follows, without extinction of the original group in the tropics.

■ On a global scale, temperature is the best explanation for maximum diversity, but the explanation for this is not clear.

The apparent origin of many groups in the tropics begs the question of whether there is a dominant environmental factor that regulates species diversity. The latitudinal gradient appears to suggest that temperature is a cause of diversity, and we discussed earlier some known temperature correlations. A global multivariate study of marine diversity (Tittensor et al., 2010) considered spatial variation in diversity and its correlation with a number of factors, including dissolved oxygen and temperature, and found that temperature is statistically the best regional explanatory variable for diversity. A study of fiddler crabs shows that temperature is closely related to species diversity in three major biogeographic realms (Levinton and Mackie, 2013), even though the exact pattern of latitudinal geographic change differs substantially among the three realms. It is possible that temperature is related to productivity and habitat diversity, or perhaps even a cause of mutation, that increases genetic variation and perhaps the chances of local evolution and speciation. Higher temperatures also may be sites of greater energy at high temperature, where higher metabolic rates might promote higher speciation rates. But so far we have no clear explanation of why temperature matters in diversity.

■ **The maximum of diversity in the Indo-West Pacific is probably the result of a combination of processes.**

The Indo-West Pacific (IWP), in the vicinity of Indonesia, represents a peak of species richness in the enormous Indian Ocean-Pacific realm. As mentioned earlier, this pattern extends to a wide range of groups, including fishes and many invertebrates. It is most noticeable in species associated with coral reefs, but it occurs in other groups as well. Although some set of environmental factors may allow species to originate more frequently in the IWP maximum region, making it a center of origin, other factors may also come into play. Here are the likely components:

1. *Center of origin.* The age and stability of the IWP is conducive to the rise of new species and is a center of origin. Species arise in this area and spread westward and eastward.

2. *Center of accumulation.* Species arise in the periphery of the IWP center and disperse and accumulate as high numbers of species in the central area.

3. *Center of overlap.* The IWP archipelagos at lower levels of sea stands isolated western Pacific from Indian ocean populations and caused species to appear on both sides; these species reentered the IWP center at higher-sea-level stands such as seen today, which increases diversity in the center.

A study by Stehli and Wells (1971) demonstrated that geologically younger reef coral genera are found most abundantly in the IWP center, where diversity today is the greatest. This would seem to support the **center-of-origin theory**. A study of western Pacific sea grasses by Mukai (1993) has also supported the center-of-origin idea. Mukai hypothesized that species should be able to migrate most easily along major current systems. He then examined the number of species of sea grasses from the Indonesian southwestern Pacific center of diversity along the Pacific equatorial current and several other current systems and found that they decreased steadily downcurrent (**Figure 20.13**).

We mentioned earlier the apparent barrier that is created during worldwide glacial advances in the southwestern Pacific, particularly among the archipelago formed by the islands including Indonesia and New Guinea. Donald Potts (1983) first reconstructed the distribution of land and sea following the most recent glacial maximum, when sea level was more than 100 m lower than at present. Most of the IWP archipelago area would be land, which would isolate species between the Pacific and the Indian Ocean.

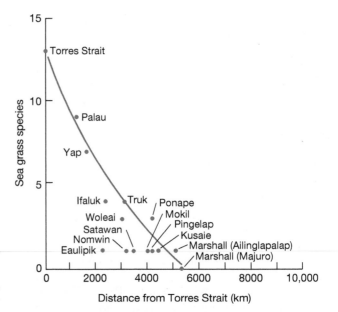

FIG. 20.13 Relationship between species richness of tropical sea grasses on various Indo-Pacific islands and distance from the Torres Strait (northern Australia), eastward along the equatorial countercurrent. (After Mukai, 1993)

Earlier we discussed evidence of regional genetic differences in a number of groups at the species level and at the within-species level, where this region marks a genetic barrier. This would support the idea that the high diversity in the IWP maximum center might be the result of an accumulation of species formed in the high-diversity IWP region of Indonesia and environs during glacial maximum periods. The apparent coral origins in the center, as evidenced by the coral fossil record, fit with an even more ancient center-of-origin theory. Studies of fish lineages and other groups suggest that some of the major splits in diversity predate the Pleistocene, so glacial advances with lowering of sea level, followed by glacial retreats and species immigrations to the center, could not explain the IWP high-diversity area. Even though the pre-Pleistocene center-of-origin theory seems to be the likely explanation for the maximum of species in the IWP, we still do not have a complete set of evidence to substantiate it.

■ **Species diversity increases with increasing habitat area.**

It has long been known that there is a quantitative correlation between geographic area and the number of species that area contains (**Figure 20.14**). This is best described in the form of a simple equation:

$$S = CA^z$$

where S is the number of species, A is the area, z is an exponent (<1) that fits the area term to a logarithmic relationship, and C is a constant that partly accommodates the use of different metrics for area (e.g., km^2 vs. miles^2). The exponent z is usually less than 1, so a proportionate increase of area results in a smaller proportionate increase

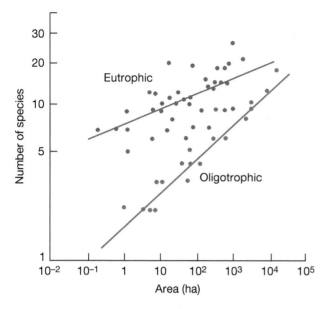

FIG. 20.14 The relationship between area and number of species of freshwater snails in nutrient-rich (eutrophic) and nutrient-poor (oligotrophic) ponds in Denmark. (After Lassen, 1975)

in species; that is, the curve is concave down. For example, this means that the differences in species numbers between the Atlantic and Pacific, for example, can be related to the larger area of the Pacific Ocean. Similarly, the Caribbean coral reef province is far smaller in area than that of the Indo-Pacific, and the latter contains far more species of most invertebrate and vertebrate groups.

A simple theory may explain why larger numbers of species are associated with larger areas. Remember that the number of species in a large region is probably a result of a balance between speciation and extinction. In larger areas, the speciation rate may be greater, owing to the larger diversity of habitats associated with an increased area. Similarly, the rate of extinction may be lower because more refuge habitats are available, and also because larger population sizes and geographic extent in the larger area may make species less prone to extinction. Areas that are richer in food might also support larger populations, which would retard the extinction rate and raise the equilibrium number of species. Thus, food-rich bodies of water might support more species than food-poor ones (**Figure 20.14**).

There is a real problem with the area hypothesis when diversity in the deep sea is considered. It is true that deep-sea diversity is greater in the near-slope abyss. However, diversity decreases with greater distance toward the mid-oceanic bottoms beneath the gyres. These bottoms are greater in area; yet they harbor fewer macrobenthic species than are found in the near-shelf abyssal bottoms. Clearly, reduced food supply has a major influence on diversity in these remote bottoms, perhaps by increasing extinction rate. Area cannot be the direct and only factor that regulates diversity. This fits the results shown in **Figure 20.14**, where lakes poorer in nutrients (oligotrophic) have fewer species than lakes richer in nutrients (eutrophic).

■ Increasing long-term habitat stability may tip the speciation–extinction balance toward higher species diversity.

Physically stable environments may accumulate more species than variable environments. Howard Sanders (1959) proposed that environmental stability influences the flexibility required of the resident species. By environmental stability, we mean the short-term range of environmental variation—for example, the degree of seasonality, seasonal changes in salinity, and frequency of storm events. If the environment varies often from state *a* to state *b*, then a species will become extinct unless it can survive and reproduce in both states. Given that resources may be limiting, the environment may be able to support only relatively few species of such broad adaptability. Unpredictably variable environments may have even more severe effects because species must then have enough flexibility to deal with a wide range of environmental changes. J. W. Valentine (1966) suggested that strong fluctuations in primary production might reduce the ability of consumers to successfully specialize. Stable environments may also accumulate more species owing to reduced extinction rates. In stable environments, major habitat alterations may be less frequent and extinction may, therefore, be less important. This hypothesis also fits the recently discovered latitudinal gradient of benthic diversity in the deep sea of the North Atlantic. Sites in very high latitudes are characterized by greater deposition of detritus from the plankton, and the variation of this deposition is greater than at lower latitudes. It may be that high-diversity centers combine stability with high and stable primary productivity.

Several large-scale patterns of diversity might support the stability hypothesis. The Pacific coral reef province is demonstrably more seasonally constant in temperature variation than the Atlantic. The high diversity of the Pacific coast versus the Atlantic coast of the United States may be explained in the context of the more constant maritime climate of the Pacific as opposed to the continentally dominated climate of the North American Atlantic coast. The deep-sea increase in diversity may be explained by the extreme constancy of temperature and salinity, whereas the drop-off in diversity toward the open-sea abyssal plain may be due to severe food stress. In stressful environments, only a few species capable of evolving tolerance will come to dominate.

Although the deep sea may appear to be more stable today, the difference may be more apparent than real. We have encountered a number of examples in this text of large-scale climatic changes that have ocean-wide if not global effects on the shallow-water marine biota, as occur during El Niño events. One might expect such climatic oscillations to cause changes in global surface and perhaps deep circulation patterns. As we discussed in Chapter 2, the hydrography of the deep sea is very intimately dependent on water motion at the surface.

The deep-sea fossil record of the Pliocene epoch (2.85–2.40 million years ago) demonstrates that the deep sea may not have been rock stable in diversity at all. Deep-sea ostracods apparently fluctuate in diversity by a factor of 3, and fluctuations obey a 41,000-year cycle that correlates with changes in the earth's tilt (Cronin and Raymo, 1997). Diversity tends to be high during warm periods and very low during glacial maxima. This suggests that some form of surface climatic oscillation is affecting deep-sea benthic diversity, perhaps by forcing of climate fluctuation. Our discussion in Chapter 18 of deep-water coral mounds also suggests a strong degree of environmental oscillation. These environments are very diverse, yet they apparently undergo a cycle of buildup during glacial retreats, followed by sedimentary cover and destruction during glacial advances.

Sanders (1968) also suggested that ancient environments might accumulate more species than do young environments simply because of the age of the former. This conclusion was based mainly on observations of the spectacular diversity of many of the rift valley lakes in East Africa and of Lake Baikal in Siberia. All those lakes are millions of years old, unlike the majority of the world's lakes, which are only a few thousand years old at most. The concept is harder to apply to the ocean. The deep-sea biota does not seem to be more ancient on the whole than that of shallow water, despite the presence of a number of famous "living fossils" such as the stalked crinoids and the monoplacophoran *Neopolina*.

Expansion and Extinction in the Geological Past

■ **The fossil record allows us to distinguish periods of origin and extinction. During the Phanerozoic Eon, there have been periods of rapid expansion and several episodes of major extinction.**

Unfortunately, present-day species distributions provide only a kind of snapshot of the current situation, often offering little possibility of insight into the historical events that may have led to current diversity conditions. Although the fossil record is known to be characterized by poor preservation and by marked gaps where there is no record at all, we nevertheless can get an idea of how diversity changed over longer periods of time and whether current conditions might be explained in terms of major changes in recent geological history. Only a brief sketch of some important patterns can be provided here, but such patterns demonstrate that many major changes have occurred, embedded in a history of past climate change and lateral and vertical movements in the earth's crust, which have reoriented coasts and changed current systems.

Figure 20.15 shows the geological time scale, which is broken up into eras, periods, and epochs. (We are living now in the Phanerozoic Eon, the Neogene period, and the Recent epoch.) Although absolute time boundaries have been determined by the use of various radioactive isotopes, most of the relative time scale comes originally from the rock record and the location of fossils in relative positions (older fossils are lower down in the rock record, generally). **Figure 20.16** shows that there was an explosive appearance of new phyla near the beginning of the Paleozoic Era, but that there were two more explosive periods, which led to our current high level of diversity.

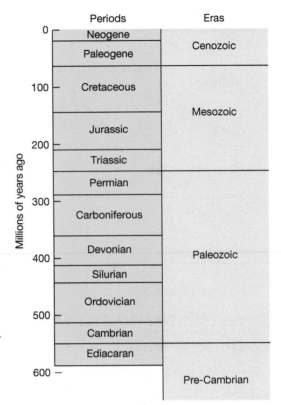

FIG. 20.15 The geological time scale, showing eras, periods, and epochs, and the absolute number of years, determined from radiometric dating.

The general increase, however, was punctuated by so-called **mass extinctions**. The most dramatic occurred at the end of the Permian period, at the end of the Paleozoic Era: Over 95 percent of marine species became extinct at that time (Raup, 1979). More famous is the somewhat less dramatic extinction, at the end of the Cretaceous period, of many marine groups, such as the ammonites and, of course, the dinosaurs. There is still no general theory that satisfactorily and conclusively explains these mass extinctions, but the end-Cretaceous event coincided with a rather large asteroid fall, for which there is evidence in the form of an extraordinary iridium anomaly and a large craterlike structure in Yucatan (Mexico), which suggests a major impact in the Caribbean basin. In the rocks at the end of the Cretaceous, there also are in some places minerals indicative of high pressure (i.e., an impact).

It is still possible, however, that some of the major extinctions were caused by major climate change or even changes in sea level. The Permian mass extinction, about 250 million years ago, coincided with what was probably the greatest fall in sea level in geological history, but a number of other factors have been implicated, including a possible warming period (Erwin, 2006). On the other hand, the large-scale Pleistocene changes in sea level caused by expansion and contraction of glacial ice had little major impact and did not cause global-scale extinctions. The patterns in **Figure 20.16** are biased by strong variation over time in the number of fossils collected and even

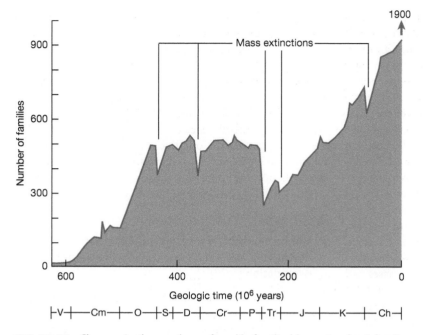

FIG. 20.16 Changes in the numbers of readily fossilizable marine fossil families throughout the Phanerozoic Era (Cambrian times to the present). Major episodes of mass extinction are indicated by downward arrows. V, Vendian; Cm, Cambrian; O, Ordovician; S, Silurian; D, Devonian; Cr, Carboniferous; P, Permian; Tr, Triassic; J, Jurassic; K, Cretaceous; Cn, Cenozoic. Because of subsequent adjustments to the dates since this work was published, the absolute time scale is not linear before 400 million years ago. (After Sepkoski, 1984)

the amount of study done by paleontologists. If there are few fossils preserved at a given time, then we may be getting a poorer sample and may be underestimating diversity. For example, since the Cretaceous, there are far more fossil localities and therefore specimens to count. When corrections are made for such bias, the explosive increase of diversity since the Cretaceous becomes questionable. Diversity may have increased from the Triassic to the Cretaceous and then stabilized until the present day (Alroy et al., 2008).

Conserving Marine Biodiversity

Estimating Diversity

■ **A great deal of the total diversity of marine life is as yet unknown.**

Although many important conservation efforts are under way, most ecologists and systematists agree that there is not yet an adequate understanding of the biodiversity of many marine habitats, particularly in the diverse tropics and in the deep sea. Coral reefs, for example, have hundreds of known species of epibenthic invertebrates, but many species remain undescribed, and probably many more remain undiscovered. Seaweeds are in particular very poorly understood. Application of molecular techniques, such as sequencing of DNA and evaluation of length polymorphisms of mitochondrial DNA, may provide us with important tools for distinguishing unambiguously among marine species that are otherwise very difficult to identify.

An ambitious molecular survey of species worldwide based on sequencing the cytochrome oxidase 1 gene, known as **barcoding**, is now under way.

Value of Biodiversity

■ **Diversity may increase overall biological productivity and the potential for resilience in an ecosystem in the face of environmental change, and also may have the value of providing more sources of drugs and other products.**

There are certainly aesthetic and ethical issues behind the notion that biodiversity should be conserved. We cannot deny our biological heritage, and we benefit in many aesthetic and even emotional ways by conserving the diversity of life. Most of us would agree that a world with only mussels and blue-green algae would be very dull. Some would even argue that it is immoral to allow the destruction of species. A new awareness of the continuity of all of life with humankind has led to a new concern for our fellow species. On the other hand, many see preservation as antithetical to the idea that humanity should have dominion over the earth. This argument is valid only if dominion involves acting purposefully to cause extinctions. Dominion can also involve behaviors to maintain our biological heritage.

There are also practical reasons to be concerned about the loss of diversity. Much of our knowledge of community ecology suggests that many species play crucial roles in elemental cycling and in regulating the distribution and abundance of marine organisms. Several studies in terrestrial systems demonstrate that increased diversity of primary producers tends to increase overall primary production, since more types of species will exploit different combinations of microhabitats and nutrients. In kelp systems, as discussed in Chapter 17, increased predator diversity tends to buffer the effects of removal of particular carnivores that exert important top-down controls on food webs. Top carnivores in food webs are likely to be lowest in population size, and any negative effects of the environment may drive them to extinction. While being most vulnerable, they also may play an important regulatory role in keeping prey populations down, and the presence of alternative predators will buffer against strong cascading effects in food webs. Accidental effects of environmental degradation could eliminate foundation species, such as sea grasses and reef-building corals, that create structural habitats for large numbers of other species, many of which are commercially important (e.g., scallops depend on sea grasses in eastern U.S. coastal habitats). Maintaining diversity allows for the presence of more ecologically similar species that increases resilience. Marine biodiversity may also be a source of innumerable drugs and other products. Unfortunately, it is not predictable where the next important drug will emerge. Diverse ecosystems also provide many sources of insight for design of human products that involve **biologically inspired design**. For example, studies of crab locomotion have proved instrumental in the design of stable autonomous vehicles that can move on beaches in the surf zone and detect explosive mines.

What Is Natural? The Problem of Shifting Baselines

■ **Many marine habitats have been strongly altered by human activity. The reduction of diversity often leads to an incorrect perception of what is natural.**

Consider this experience. You have been living in a city harbor all of your life, watching the shoreline from a boat. The coast is lined with concrete, interspersed by sewer openings and docks. Fish are present but rare, and the docks are made of materials that discourage colonization by intertidal invertebrates. It is likely that most who view this seascape will appreciate that it is not natural and that there once was a softer shoreline, perhaps dominated by marshes and tidal creeks.

Not all seascapes are so obvious. Many coral reefs are dominated by macroalgae and planktivorous fishes with few large predators (Knowlton and Jackson, 2008). Many temperate coastal rocky reefs are low in diversity and covered with red calcareous algae and sea urchins. If we go back just a few human generations, we find reports of abundant large predatory fish, large erect seaweeds, abundant sessile suspension-feeding invertebrates, and generally high diversity. Some investigation results in the conclusion that the shift we observe to the present is related to the overfishing of top predators in a habitat where top-down effects were formerly prominent.

The inability for a person or society to connect the impoverished present with such a rich and biodiverse past is a result of **shifting baselines**.[1] As the environment is disturbed by humans through many decades, successive generations fail to remember what was originally the natural state. This may even result in an inability to establish objectives for ecological restoration that approximate the original undisturbed state. The frightening prospect is that the current generation of citizens has no idea of how rich the marine environment once was because the successive baselines become more and more disconnected from the original pristine state, several human generations ago.

Reduction of Biodiversity

■ **Marine biodiversity may be reduced by habitat destruction, habitat fragmentation, and habitat degradation.**

Biodiversity within a given habitat type can be reduced by three human activities. **Habitat destruction**—such as filling of marshes, human-induced erosion, and destruction of shorelines by the building of seawalls—is one of the major sources of biodiversity reduction because it removes the habitats on which species depend. Removal of one component location of a species' life cycle, such as disruption of a spawning area, may also be very potent. A more subtle, but equally important, source of reduction is **habitat fragmentation**, in which a previously continuous marine habitat is broken up into smaller isolated parcels. Shoreline development in the tropics, for example, tends to have a negative effect on coastal coral reefs, which are broken up into smaller tracts. As a large habitat, such as a sea grass bed, is broken into fragments, colonization of planktonic larvae becomes more difficult. Also, larger predators and other foragers will no longer have a continuous range of habitat over which to feed. A habitat fragment may still contain members of a species, but the population size may be too small to sustain the species over long periods. On the other hand, habitat fragments may have a greater length of habitat edges, which may create new habitats for some species. Finally, **habitat degradation** is probably the most important potential source of loss of biodiversity (see Chapter 22). Nutrient input into estuaries of North America has caused increased phytoplankton density, which in turn has choked off light from sea grasses. Because many fishes and invertebrate species depend on the sea grass as a structural habitat, many species are lost when sea grasses disappear. Degraded marine environments are typically species-poor environments.

■ **The species–area effect might be used to predict the loss of species, but an understanding of habitat and biogeographic effects is crucial as well.**

Earlier in this chapter we introduced the species–area effect, whereby the total number of species in a region can be predicted approximately through a knowledge of habitat area. The proportion of habitat loss might be used to

---

[1] For many resources on the topic of shifting baselines, see www.shifting baselines.org.

predict the loss of species. Destruction of coral reefs, for example, might be used to predict loss of coral-dependent species if one knew the degree of loss of total reef area. Such an approach, however, would have to be modulated by knowledge of effects on habitat loss. Species may not be lost in proportion to area, for example, if some species can be supported on very small habitat fragments. Alternatively, larger top carnivorous fishes—which require large foraging areas and are more sensitive to extinction owing to smaller population size—may be lost even faster than the general species–area relationship would predict.

Conservation Strategies

■ **Conservation remedies can be managed from the perspective of ecosystem function and ecosystem services.**

In recent years, conservationist and government agencies have employed an approach called **ecosystem-based management** to best manage resident species, habitats, and ecosystems. Ecosystem-based management is defined broadly as an approach to conserve ecosystems that includes components of ecosystem function, habitat protection, the social community that interacts with the ecosystem, and economic considerations. Habitat protection approaches are taken by scientists to devise ways to minimize the impact of human activities on natural ecosystems. Sociological approaches involve the engagement of government, nongovernmental organizations, and the public in general to understand how their actions might affect an ecosystem and how ecosystem functioning might affect recreation, food sources, water supply, protection against storms, and other factors important to society.

Economic considerations usually involve attaching a monetary value to a series of **ecosystem services** available when an ecosystem is not severely disturbed by human activities such as development and pollution. The most direct economic calculation can be made by the loss of a fishery, since it is straightforward to evaluate the economic loss of sales of fish in an area and even the impact of a given fish species within a food chain. But other services may also be of great economic value. For example, protection of a salt marsh system might strongly reduce the impact of hurricanes on a coastal community, provide shelter for many valuable resource species, and provide detrital food for coastal species that might in turn be consumed by top predators in coastal food chains. Such complex interactions can be understood only if we study problems of conservation on an ecosystem level.

The connection between ecosystem-based management and biodiversity is important to establish, but we must first understand the relationship between species presence and ecological function. In terms of ecosystem services, several types of species are of special value, including the following:

Foundation species. These species—for example, grasses in marshes, and hermatypic corals in coral reefs—provide the essential structure upon which many other species depend.

Trophic cascading species. Species in different parts of a food web whose loss would result in major effects in the form of a trophic cascade would be targets for protection.

Resource species. Obviously, species important to human food consumption, natural products for drugs, or other products would be important to protect, from an economic perspective.

We might want to conserve a species because of its aesthetic appeal (e.g., a marine mammal), but we also want to understand how that species performs an important ecological function. For example, there might be two abundant benthic suspension feeders in an area. If one became nearly extinct and the other should expand in its place, there would be ecosystem **functional redundancy**, and the ecosystem function performed by suspension feeding would be retained. There may be several species of sea grass in an area, but it might be that the loss of one species (e.g., by disease) might not affect overall the services of sea grasses if another species survives and takes its place. It is, therefore, important to understand whether biodiversity itself has a value in ecosystem resistance to change, or **resilience**.

■ **Biodiversity has a role in ecosystem function and ecosystem services.**

Does diversity matter, or does it take just one species to perform a given ecological function, such as top predator, suspension feeder, or foundation species? In some cases this might be true, but there are reasons to believe that conservation of biodiversity may be important in environmental protection. First, not all species are ecologically exactly alike. Suppose we have an environment with many potential microhabitats. With a large number of species, it is probable that they will each occupy the different microhabitats. The net result will be an occupation of all available food and space resources. On land it has been shown that when many species of plants are grown together, total primary productivity increases because different plant species exploit a more complete range of soil types, moisture conditions, and topographic sites (Tilman, 1999).

We have limited evidence that biodiversity plays a strong role in marine communities; but we would naturally search for this role in foundation species, such as corals and sea grasses. Some very suggestive evidence comes from sea grasses at the species and genotype levels. As we discussed in Chapter 17, sea grasses exert an important effect on shallow-water communities, including provision of shelter against predators and food for many herbivores. Worldwide there are only about 50 species, and higher-latitude eelgrass communities may consist of only one species, which does not allow much room for biodiversity in adding to ecosystem function. In lower latitudes, sea grass meadows often have more than 10 species and consist of a combination of fast-growing and strongly colonizing species that are very available to grazing urchins and vertebrates along with slower-growing climax species that store carbon and are grazed less. Duarte reports that species richness is correlated with community biomass, which suggests a positive role for biodiversity (**Figure 20.17**). Duarte (2000) points out that species combinations in sea grasses often improves total coverage, but that the relationship is not necessarily

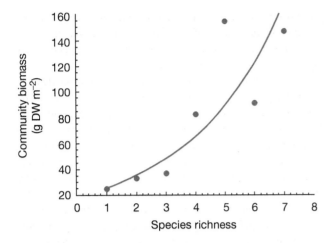

FIG. 20.17 Relationship between biodiversity of sea grass meadows and total sea grass biomass. (From Duarte, 2000)

true for all combinations of species, as suggested by the difference between fast- and slow-growing species.

Diversity of function may be complemented by diversity of population growth patterns among species. John Stachowicz, Robert Whitlatch, and Richard Osman (1999) completed a study using a lovely system of sessile invertebrate species that compete for space on subtidal hard surfaces, such as rock walls. In southern New England, the Pacific Ocean ascidian *Botrylloides diegensis* has invaded and spread in recent years. The native community in any one small area of rock usually consists of three or four different native sessile species, and the question is whether they can resist invasion by virtue of their diversity.

Stachowicz and colleagues hit upon an ingenious and simple experimental design. They allowed species of native sessile suspension-feeding invertebrates in Connecticut shores to settle and dominate small tiles, which were then combined into assemblages of one, two, three, and four species. The experimental "community" consisted of 25 tiles with the different species combinations and five tiles covered by the potential invader *Botrylloides diegensis*. The results were impressive: *B. diegensis* mortality was higher in the tile assemblages with higher species diversity.

The mechanism for resistance to invasion by more sessile species may lie in the compensating effect of having more species that can utilize a resource. Stachowicz and others found that if the tiles had only one or two species, population fluctuations often resulted in large areas of empty space. But with more species, one species might increase in numbers and take the space given up by species in decline. Under these circumstances, an assemblage with low species richness might be more likely to have available resources for an invading population, whereas higher diversity might result in more complete and continuous resource use, which would repel invasions.

Even within a single species, however, there is an interesting role for another level of diversity: genetic variation. North Atlantic sea grass beds consist of just one species, *Zostera marina*. Reusch and colleagues (2005) performed a fascinating experiment using different genotypes of eelgrass.

Plots of eelgrass with multiple genotypes had higher total eelgrass growth than plots with one or few genotypes. This suggests that in species-poor ecosystems, genetic diversity might play the role that species diversity plays in species-rich ecosystems.

The more species present, the more likely that one species might be available in sufficient numbers to perform a given ecosystem function. This is analogous to investing in many stocks: Any single stock may be crashing at a given time, but one of the others that is doing well will likely compensate for this. The complete occupation of resources leads to an obvious prediction: invasive species should have trouble invading high-diversity communities.

■ **Marine protected areas have been established to protect marine biodiversity, but relatively little habitat is under protection, and local practices do not always result in complete protection.**

It may be that the best strategy to conserve biodiversity is to design **marine protected areas** (**MPAs**, also called marine reserves) that allow protection of ecosystems with varying restrictions on fishing all the way to complete closed areas, or **no-take areas** (Gubbay, 1995). Ecosystem-based management cannot succeed without defined areas within which rules of preservation or management can be effectively enforced. One major problem is determining the size required for protection. In ecosystems that have confined boundaries and that are strongly controlled by foundation species, there is a hope of defining areas for control. Such ecosystems include coral reefs, sea grass meadows, kelp forests, and mangrove forests. Reserves that are too small are prone to damage or even local extinction from storms, local pollution, and disease. For example, oyster reefs can be small enough that the gain in oyster shell will be overshadowed by loss of shell through mortality, erosion, and dissolution. A successful oyster reef preserve must therefore be of a minimum area and shell-reef thickness. Groups of preserves might create a metapopulation that guarantees exchange of planktonic larvae and juveniles and also connections of local habitats by mobile carnivores such as large fish.

The most important issues in the success of marine protected areas include stakeholder participation and good local management, especially enforcement of rules against poaching. Larger, older, and relatively isolated MPAs show the greatest success in the increase of large predators such as sharks. MPAs are usually under the jurisdiction of several local political entities, and a number of stakeholder groups (e.g., fishing groups, tourist groups, local residents, local conservation organizations, government organizations) must cooperate in defining the MPA, designing rules for maintenance, and providing personnel for maintenance and enforcement. This is the greatest area of concern because of the widespread variation in types of social structures, governmental agencies, and priorities throughout the geographic range of a habitat type such as coral reefs or mangroves.

MPAs are diverse in the level of protection. For example, the island of Bonaire (Netherlands Antilles) is a marine protected area, focused on coral reefs. Living organisms cannot be taken from the reef, and diving visits are somewhat limited by an array of diving moorings and restrictions against the use of anchored vessels. Conservation takes a three-tiered approach, called zoning. Some areas are completely closed to any visits, others are open to researchers, and the rest can be visited by nature tourists.

A **no-take area** is the strongest level of protection.[2] This is true in many areas managed under the National Marine Sanctuary system in the United States. For example, in the Florida Keys Marine Sanctuary, most areas are accessible to fishing and diving visits, with few restrictions. Only about 1 percent of the sanctuary consists of no-take areas, but this does cover about 65 percent of the coral areas. Most notable are a string of marine-protected areas established mainly on rocky reefs off the coast of California. These areas range from those that are open to recreational and commercial fishing to no-take areas. The Northwestern Hawaiian Islands National Monument, designated in 2006 and consisting of nearly 250,000 km^2 of protected marine environment, is the second-largest marine no-take preserve in the world and has been quadrupled in size in 2016. The Ross Sea preserve in Antarctica, the largest marine reserve, is about 1.5 million km^2.

MPAs have been in place for several decades, and a number of studies have quantified the success of these areas. Benjamin Halpern (2003) examined the impact of 89 marine no-take reserves and was able to quantify the effects on population density, biomass, body size, and biodiversity of 69 reserves, which is shown in **Figure 20.18**. Most of the reserves are 1–10 km^2 in area. Within the reserves, Halpern found positive effects for all of these factors for fishes, and this was also true for invertebrates, with the exception of biomass and body size. Reserve size appeared to have no effect, which was a bit of a surprise, and other studies suggest that MPA size does matter in protecting large predators (Edgar et al., 2014). Still, the results provide convincing evidence that reserves work. A large-scale study demonstrates

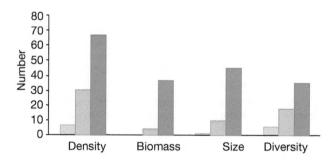

FIG. 20.18 Differences in biological measures (density [no./area], biomass [mass/area], mean size of organism, and diversity [total species richness]) between inside a reserve and outside a reserve (or after vs. before) for all organisms examined. Yellow bars represent lower values inside the reserve, peach bars represent no difference between reserve and nonreserve areas, and orange bars represent higher values inside the reserve. In all cases, biological conditions are better within the reserve at a statistically significant level ($p < 0.001$). (From Halpern, 2003)

that enforcement is crucial in allowing the increase of meso-predators such as large sharks (Kelaher et al., 2015).

At present, coral reefs have been the most consistently targeted for marine protected areas, so some appreciation has developed about the successes and challenges of using this approach to protect and manage ecosystems. But coral reefs exist throughout the world tropics, within the national waters of many countries. The Great Barrier Reef of Australia, probably the largest marine protected area in the world, was established as a protected area in 1975 and is now zoned according to multiple uses (Laffoley, 1995). Zones are (1) for scientific research only, (2) for marine national park (regulated) use by tourists, and (3) for general use (includes commercial and recreational fishing, but with some regulation). Success of conservation efforts has been variable, and further regulations have subdivided the reef into seven conservation, research, tourist, and general-use zones. About one-third of the entire Great Barrier Reef has been established as a group of no-take marine protected areas for fishing. This policy has resulted in strong increases of the leopard coral grouper (*Plectropomus leopardus*) and reductions of their prey, the crown-of-thorns sea star, which has had reduced frequency of sea star outbreaks in reef areas with no-fishing zones (Sweatman, 2008). Unfortunately, the ENSO of 2014–2015 has caused extensive coral bleaching.

Mora and colleagues (2006) have established a worldwide database and kept track of the nearly 1,000 coral reef MPAs around the world. Nearly 19 percent of the approximately 500,000 km^2 of the world's coral reefs are in protected areas, but only 1.4 percent is completely protected. The overwhelming majority of coral reef areas are not protected, especially from the removal by fishers of species such as roving predators and grazing fish that have major impacts on trophic structure of coral reef ecosystems. A special problem is the scale of larval dispersal of coral reef fish and corals themselves, which is commonly in the range of 10–100 km or more. Few areas are being managed effectively on these spatial scales. Mora and colleagues considered a range of attributes, including larval dispersal and poaching of important species, and found adequate protection for only 2 percent of coral reefs. Along with local conditions, global conditions make circumstances grimmer. Increases of sea-surface temperature and concentrations of dissolved carbon dioxide present physiological challenges that are layered on top of local problems of biodiversity protection and protection against overfishing.

Conservation Genetics

■ Genetic approaches can be used to identify species and genetically distinct populations within species.

We have come to appreciate that many marine species have been overlooked completely, partially because morphological similarities often mask complete reproductive separation and independent evolutionary histories. In recent years, a number of molecular techniques have greatly improved our ability to discriminate among species. DNA sequencing is now becoming the routine means of identifying new species.

---

[2] See http://www.bmp.org for more.

It is also important to be able to identify genetically distinct populations that belong to the same species. Consider baleen whales, whose geographic range is often worldwide. If a humpback whale is hunted in Australia, will that affect the New Zealand population? If summering whales in one part of Antarctica are killed, will that affect the population of whales throughout the world or just a section of the population that migrates to a specific place? For example, eastern Pacific humpback whales appear to have a population that migrates between Hawaii and Alaska, and another population that migrates between Mexico and California. DNA differences demonstrate that these two populations are, in fact, genetically distinct. The populations must, therefore, be managed separately.

Genetic markers also have been very useful in proving that the hunting of loggerhead turtles in the western Mediterranean was the cause of decline of the nesting population in the southeastern United States. It is now possible to trace the effects of hunting on populations to distant places that are crucial in the life cycle. This is quite useful for cetaceans and turtles, both of which have species with long-distance migration. Management of the individual populations for the purposes of conservation can thus be based on a detailed knowledge of interchange of individuals among separately migrating populations. A United Nations convention has given the nation in whose waters a species nests or spawns the right to file complaints against those who hunt such species in places removed from the territorial waters of the reproductive site.

Protective Legislation

■ **Laws are now being used to protect both biodiversity and habitats.**

In recent decades, a new concern for conservation has led to a variety of legislative acts that attempt to conserve biodiversity itself and to protect the habitats on which marine species depend. Table 20.2 provides a list of major programs or legislation. Much of the legal effort has been directed toward the identification and protection of endangered species. In the United States, the Endangered Species Act has been the basis for protection of many species. The problem lies with a definition of "endangered." This is a difficult issue, since one cannot always state how much habitat loss or population decline constitutes endangerment, especially because population recovery is difficult to predict. When population density of free spawners becomes very low, as it has for a few abalone species in California, the probability of gamete encounter, let alone larval survival, is very low. On the other hand, one can imagine bringing a vegetatively reproducing species to very low numbers and yet, because recovery is easy, not risking extinction. As a rule of thumb, the International Union for the Conservation of Nature has devised a criterion according to which populations that have experienced 80 percent decline 3 years in a row are considered to be endangered. Although this rule has no real teeth except by general agreement, it at least provides a means of quantifying trends that may be cause for alarm.

Marine Invasions

■ **Many invasions are facilitated by human transport, and introductions of ecologically potent species may cause local extinctions and a homogenization of the world marine biota.**

Increasingly, marine commerce is delivering exotic species over great distances, causing major changes in marine communities around the world. The few excellent colonists have the potential of homogenizing the between-habitat diversity of our coastal environments. In Chapter 16, the tremendous colonization-invasion potential of species of the marsh grass *Spartina* was discussed, as was the disruption of a great many coastal habitats. The arrival in the 1980s of the ctenophore *Mnemiopsis leidyi* in the Black Sea resulted in the loss of a number of zooplankton and fish species

TABLE 20.2 Some Major U.S. Programs and Legislation Now Being Used to Conserve Marine Biodiversity

PROGRAM OR LEGISLATION	MAJOR OBJECTIVES
National Marine Sanctuaries Program, 1972	Identifies and provides management programs to protect important and potentially endangered marine habitats (e.g., Channel Islands off California, Flower Garden Banks in the Gulf of Mexico)
National Estuarine Research Reserve System, 1972	Established as part of the Coastal Zone Management Act in 1972, identifies estuarine sites important in long-term ecological research (e.g., Hudson River Estuarine Reserve)
National Wildlife Refuge System, 1966	Identifies areas where wildlife is especially valuable, especially those in which species or migratory bird sites are threatened by extinction
Endangered Species Act, 1973	Identifies species that are in danger of extinction
Marine Mammal Protection Act, 1972	Intended to halt the decline of marine mammal species and to restore populations to healthy levels
Fisheries Conservation and Management Act, 1976	Intended originally to prevent decline of fisheries within 200 miles of U.S. coast, especially with regard to foreign fisheries (but did not work; see Chapter 21); revised several times since 1976
Coral Reef Conservation Act, 2000	Provides for research, monitoring, and conservation of coral reefs
U.S. Presidential Executive Order, 2000	Declares system of marine protected areas, managed on an ecosystem basis
Magnuson-Stevens Reauthorization, 2007	Expanded regulation, including catch limits, shared fisheries among licensees

(some recovery in recent years has followed efforts at reduction of pollution and overfishing and the invasion of another comb jelly, *Beroe ovata*, that consumed *M. leidyi*). Owing to these transoceanic movements, the world's coastal biota is losing its regional differences. An extreme example: All the dominant hard-bottom intertidal species of San Francisco Bay are now exotics; local species have become rare or extinct.

■ **Invaders probably rarely establish successful populations and usually become extinct; but vectors, high invasion frequency, and ecological suitability of the target habitat all contribute to invasion success.**

Any community is continually colonized by species from other biogeographic provinces or from different habitat types. Most colonizations fail to result in the establishment of a successful invading population in the target area. Imagine a snail larva arriving on a coast after crossing an ocean. Even if the animal is a pregnant female, the chances are good that she and her offspring will not survive. Normal population fluctuations would cause such extinction, but rarity of mates and the likelihood that the habitat is unsuitable make for the poor prospects for such colonists.

Successful invaders must have the following properties:

1. *Dispersal distance.* An invading species must have a means of crossing a long distance, such as an ocean. In recent years, more and more ships use large amounts of **ballast water**, which is rapidly transported across oceans. Long-lived planktonic larvae might survive the journey, as might microbial organisms such as bacteria, protozoa, and phytoplankton species.

2. *Invasion frequency.* It stands to reason that invasion success must be related to frequency of transport. Frequency of ship transport must also matter in human-aided invasions. This factor highlights the special danger of releases from home and professional **aquariums**, where literally thousands of individuals might provide potential invaders. An aquarium release in the Mediterranean led to a takeover of thousands of square kilometers by a species of the alga *Caulerpa taxifolia*, which probably came from the Caribbean but was likely released in Monaco. The invasion of the voracious and spined red lion fish *Pterois volitans* into the Caribbean and eastern United States is causing declines in many prey species. The only likely source for this is a release or series of releases from pet aquaria.

3. *Ecological compatibility.* When a species arrives, it will not survive or reproduce unless it can survive and exploit resources in the target location. Thus, one does not expect tropical species to successfully invade a polar sea. No such newly arrived species would be able to evolve new thermal tolerances within a few generations—assuming that members of a tropical species could survive a polar environment even that long. It would make sense for the **most successful invaders to be ecological generalists**, capable of using whatever resource is available. A specialist is likely to fail to find an appropriate local habitat, even if it did

manage a long journey across the ocean. The Asian clam, *Potamocorbula amurensis*, which tolerates wide ranges of salinity, is a particularly good example of a successful generalist invader that arrived in ballast water from Southeast Asia to San Francisco Bay.

4. *Survival beyond initial population fluctuation.* A newly arrived population is likely to be very small, and random changes in survival and resources will likely drive it to extinction. The population size, therefore, must increase beyond this initial roadblock. There is an interesting trade-off between dispersal ability and the ability to survive the initial population variability at low population size. A poor disperser, once arrived at an invasion site, might be able to build up a large local population, since larvae would not spend a long time in the water column. This would allow buildup of a local population, which would then disperse. By contrast, a species with a long-distance-dispersing larva might become extinct after arriving at a spot because the next generation of larvae, being widely dispersed, would not be able to find mates at the time of reproduction.

An invader may hang on for many years in low abundance until an environmental change allows a local population expansion. For example, the Chinese mitten crab, *Eriocheir sinensis*, probably invaded the United Kingdom from Europe (which was invaded originally from the Far East) at least as far back as the 1930s. Careful surveys in the Thames estuary of southern England reported its presence in low numbers through the 1980s, but it suddenly expanded to a large population with an estuarine migration cycle in 1992 (Attrill and Thomas, 1996). The population flush coincided with a drought period of very low river flow, which probably facilitated retention of larvae and enhanced recruitment dramatically, which allowed the development of an indigenous reproducing population. This case and others suggest that it is not the dispersal ability alone that matters in a successful invasion: **A disturbance may be required to allow the population to pass the threshold from an invading propagule to a locally sustaining population.**

■ **Marine invasions are common, and invaders often come to dominate their newly adopted homes.**

Despite the odds, invaders do often come to dominate and even cause major changes in the communities to which they disperse. The periwinkle *Littorina littorea*, which arrived in northern New England about 100 years ago, is now the most abundant snail on New England rocky shores and exerts major effects on other species, most clearly seaweeds. Brenchley (1982) showed that the invasion of the periwinkle also reduced the ecological distribution of the local eastern mud snail *Tritia obsoleta* by grazing shore cobble habitats, which removed fine material that might be eaten by the mud snails. The periwinkles also preyed upon egg cases laid by mud snails on hard surfaces. An interesting controversy emerged as to when *Littorina littorea* actually arrived from Europe (see Blakeslee, 2012, for details). Molecular data suggested that New England *L. littorea* populations may have arrived much earlier than is apparent from their appearance in Nova Scotia. This means that they may have been present and not noticed in populations north of New England but then spread suddenly southward in the nineteenth century (Cunningham, 2008). Recent evidence, however, comparing European and New England populations shows that genetic diversity of both *L. littorea* and its trematode parasites is a small subset of likely European source populations. A genetic dating technique based on these data suggests a rather recent arrival of *L. littorea*, within the past 500 years. This fits the hypothesis that the periwinkle moved across the Atlantic in the nineteenth century in rock ballast (Blakeslee et al., 2008). Ironically, *L. littorea* has declined severely in southern New England in the past decade owing to predation by the invasive Japanese shore crab, *Hemigrapsus sanguineus*, which is very abundant within cobble shores.

Because inquisitive naturalists now notice invasions, the rate of spread and the ecological effects of an invader are becoming well understood. Because shipping is often the vector, invaders can be traced to a starting point. For example, the feather duster worm *Sabella spallanzanii*, a native to the Mediterranean Sea, established itself in Port Phillip Bay in Victoria, southern Australia, in the late 1980s (**Figure 20.19**). It likely came from an invasion in the 1960s to western Australia. It was found at first in only a small bay in the western part of the larger bay but then spread throughout the bay over the next few years (**Figure 20.20**). Short-distance larval dispersal reduces the rate of spread, but, on the other hand, gregarious settlement allows locally large populations to build up, assuring the probable survival of the population. Their tubes may be 50 cm in height, and suspension feeding must be affecting the local biota, but this has not yet been evaluated.

The shore crab, *Carcinus maenas*, arrived probably in the early nineteenth century on the east coast of the United States from northern Europe and became a major predator (Grosholz and Ruiz, 1996). It was discovered in California

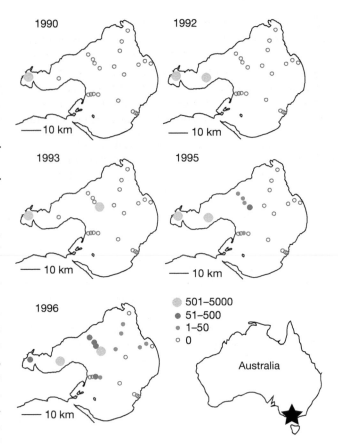

FIG. 20.20   The annual spread of the Mediterranean polychaete *Sabella spallanzanii* into Port Phillip Bay in Australia, from 1990 to 1996. (After Parry et al., 1996)

in 1989. It is now the most abundant carnivorous crab in shallow waters of southern Australia, where it also probably arrived in the early nineteenth century. Its spread to coastal habitats around the world has been documented in terms of arrival time and genetic relatedness with other populations (Geller et al., 1997). In southern Australia, a number of estuaries are intermittently connected with the ocean. The probability of presence of *C. maenas* in such an estuary is positively correlated with the proportion of time it is connected with the open sea (Garside et al., 2014). This relationship suggests that connectivity between nearby marine populations is very high.

Invasions sometimes cause major alterations of a local biota. The predatory nature of *Carcinus maenas* makes it a logical candidate for strong effects on near-shore benthic communities, especially because U.S. west coast populations are of far larger body size than their probable east coast progenitors. The invasion has resulted in a 90 percent decline in populations of the native shore crab *Hemigrapsus oregonensis* and two clam species. It is of interest whether this species is evolving cold-water adaptive capacity in the cold local waters of Oregon to Vancouver Island, Canada, where spread has occurred.

San Francisco Bay is an extreme case of alteration by invasion of marine exotic species. The eastern mud snail *Tritia obsoleta* was introduced into San Francisco Bay and has driven the local mud snail to near extinction. More recently,

FIG. 20.19   The polychaete *Sabella spallanzanii*, a worldwide invader from the Mediterranean. Breadth of crown is about 10 cm. (Photo courtesy of David Paul)

the Asian clam *Potamocorbula amurensis* has invaded the bay and has displaced nearly all other invertebrates (Carlton et al., 1990). Because San Francisco Bay is surveyed regularly, we can be fairly certain that the clam arrived in the northern end in 1986, probably as planktonic larvae in the ballast tanks of ships. It is now the dominant species there and has spread to the rest of the bay. The Asian clam is very tolerant of salinity variation and probably invaded during a flood period, when the salinity was very low. Now it seems to be able to resist the invasion by more typical benthic invertebrates that formerly dominated. It can filter the phytoplankton from the water column and is a major grazer in the southern part of the bay. Currently, San Francisco Bay is dominated by exotic benthic species.

■ **In recent years, the ballast water of ships has greatly accelerated transport of exotic species across oceans. Although transoceanic dispersal of marine larvae certainly occurs occasionally, shipping is currently a far more frequent source of movement.**

Until the last 25 years, invasions facilitated by people probably were due to transport of associated material. The mud snail *Tritia obsoleta* may have arrived in San Francisco Bay with oysters for culture. Adults of many marine benthic species are introduced from Maine with worm bait boxes (Fowler et al., 2016). In the early mid-nineteenth century, invasions with larvae were less likely, given that most larvae could not last long enough to cross an ocean. But now ships are much faster and use seawater for ballast, which may provide a suitable habitat for surviving the journey. In Chapter 16, we discussed the ecological effects of the invasive zebra mussel, *Dreissena polymorpha*, on the Hudson River estuary. It was introduced from Eurasia, probably in a ship's ballast water, and has spread throughout the Great Lakes, in densities as high as 700,000 per square meter! It invaded the Hudson River in 1991, consumed most of the phytoplankton, negatively affected native freshwater mussels, and reduced oxygen levels within the Hudson. Larvae of many species arrive with ballast water into Chesapeake Bay, but they are released in ports of very low salinity; this likely causes the death of most arrivals, which are adapted to more open marine salinities (Smith et al., 1999).

■ **Habitats vulnerable to invasion might include those that are disturbed.**

It is difficult to make rules about the success of invasions, although marginal and disturbed habitats seem to be the most vulnerable. This may be because the best invaders are those adapted to such habitats. It may be no accident that so many invasions have occurred in estuaries such as the Hudson River and San Francisco Bay. While these areas are obviously perfect targets for delivery of propagules by ships, they also may change radically, owing to strong environmental changes from one year to the next.

■ **Invasive species might be more efficiently repelled in tropical environments, owing to high predation rates and high diversity.**

The higher species diversity of the tropics raises strong concerns about invasive species because there is more at stake in the potential loss of biodiversity when an invasive displaces potentially more species. Two possible factors might make such tropical habitats more resistant to invasive species, although we have sparse information. First, higher diversity might result in successful repelling of invasive species (Stachowicz et al., 1999). The more species present, the more potential ways that a native species might be able to outcompete an invasive for a crucial resource such as space. Second, there is evidence suggesting that predation by mobile predators is more intense in tropical shallow bottoms. This conclusion was reached by a simple assay: tethering prey species in a range of habitats at different latitudes. Mobile crustaceans were preyed upon at much higher rates at low-latitude sites (Peterson et al., 2001). Similar results were obtained for the degree of predation and grazing pressure on rocky shores (Menge and Lubchenco, 1981). A. L. Freestone and colleagues (2011) demonstrated that invasive colonial ascidian species were strongly affected by predation in Panama low-latitude sites, but much less in sites in Connecticut higher-latitude shores.

■ **Canals may be a source of large-scale invasions.**

The possibility of such tremendous ecological disruption has always worried biologists contemplating the completion of canals connecting two previously isolated bodies of water. The Suez Canal, connecting the Red Sea with the Mediterranean, has been a source of major invasion and change, especially in the Mediterranean. In the past, plans for a sea-level canal across Panama worried many. Would sea snakes move from the Pacific into the Caribbean? Would the crown-of-thorns sea star, famous for decimating corals throughout the Pacific, make an even bigger splash in the Caribbean? The tremendous ecological importance of invasions is now being appreciated.

■ **CHAPTER SUMMARY**

• Although short-term interactions explain local species diversity, regional patterns are probably due to speciation and extinction. In speciation, relatively isolated populations reach the point where they can no longer interbreed. Habitat change or destruction, diseases, biological interactions, or random fluctuations of population size may cause extinction.

• Geographic isolation and major gradients in temperature and salinity combine to produce statistically distinct groupings of species. The relation of geography to speciation can be determined by comparing evolutionary trees to location.

A geographic barrier may isolate populations of just one species, isolate closely related species, or isolate species from distantly related groups. The biogeography of ocean basins is crucial to patterns of marine biodiversity.

• We must distinguish between diversity *within* and *between* habitats. Species diversity tends to increase with decreasing latitude. Diversity in the Pacific is greater than in the Atlantic. Diversity in coral reefs is greatest in the Coral Triangle in the Indo-Pacific and declines in all directions. Inshore and estuarine habitats are poorer in species than in open-sea coast habitats.

- Complex historical events, including climate change and tectonic events, may explain some regional differences in diversity. Areas of high diversity may be centers of origin of new species, or they may be regions where isolating barriers resulted in past speciation. The latitudinal pattern of species diversity fits a model of tropical origins and spread to higher latitudes. Species diversity increases with increasing habitat area and stability.

- The total diversity of marine life is as yet unknown. Diversity may increase the resilience of an ecosystem. Human activity has caused habitat destruction, fragmentation, and degradation, and the reduction in biodiversity often leads to an incorrect perception of what is natural. The species–area effect might be used to predict species loss, but an understanding of habitat and biogeography is crucial as well.

- Conservation can be viewed from the perspective of individual species, biodiversity, or ecosystem function and value. For example, biodiversity may provide us with drugs and other products. Ecosystems can be evaluated on the services they provide to society. Marine protected areas have been established to protect marine biodiversity, but relatively little habitat is currently under protection, and local practices do not always result in complete protection. Conservation efforts can also be directed toward hotspots of high biodiversity.

- Conservation genetics can be used to identify both species and genetically distinct populations.

- Many marine invasions are facilitated by human transport. The introduction of new species may cause local extinctions and a homogenization of marine biota. Most often, invaders probably become extinct, but vectors, high invasion frequency, and an ecologically suitable target habitat all make an invasion more likely to succeed. Disturbed habitats are also more vulnerable to invasion. In recent years, ballast water from ships has accelerated the transport of exotic species.

■ REVIEW QUESTIONS

1. What major factors enhance speciation in the sea?

2. What has caused groups of coastal marine species to have rather similar biogeographic ranges?

3. What are four consistent gradients or regional differences in species diversity?

4. How might long-term historical factors and shorter-term ecological interactions combine to determine species diversity in a given area?

5. How might environmental stability have contributed to high species diversity?

6. What are the different possible reasons why tropic marine environments are more diverse than those of higher latitudes? How might we test explanatory hypotheses for these differences?

7. Why might regions of greater areal extent tend to harbor greater numbers of species?

8. What does the fossil record reveal about the extent of extinction over long periods of geological time?

9. What are the major processes that contribute to human influences on the loss of biodiversity?

10. How might a more diverse ecosystem change rates and processes, such as productivity, within the ecosystem?

11. How might the species–area relationship allow us to predict the reduction in biodiversity by means of estimates of the extent of habitat destruction? Why might such estimates be inaccurate?

12. Even if a species manages to cross an ocean and invade another region, what factors might result in the failure of the invasion to be successful?

13. What events or conditions in an area might increase the probability that invasive species might successfully colonize that area and expand in population size?

Visit the companion website for *Marine Biology* at www.oup.com/us/levinton where you can find Cited References (under Student Resources/Cited References), Key Concepts, Marine Biology Explorations, and the Marine Biology Web Page with many additional resources.

CHAPTER 21

Fisheries and Food from the Sea

We can get a glimpse of prehistoric fishing practices by examining the reports of anthropologists and explorers who have observed aboriginal societies. When Europeans first encountered them, the aborigines of Australia were using bark boats, from which they threw snail-shell hooks attached to a line fashioned from beaten bark. When a fish was pulled in close to the canoe, it was speared and brought onboard. Native Americans of Pacific coastal Canada dropped lures to the bottom and bobbed them to the surface. The fishes followed the lures and were speared. These natives also used baited hooks, crab traps, and comblike devices with small bone spears to impale fishes.

It is an interesting question as to whether prehistoric natives overfished areas or whether their practices were sustainable. We can't assess motives, but studies of oyster fishing in native America and aboriginal Australia were apparently sustainable and did not drive the fishery to local extinction (see discussion of oyster reefs in Chapter 16). But in the twelfth century, a new type of trawl was so efficient that King Edward II banned its use in the Thames River. Technology has been the main source of rapid overexploitation in the past century. Open-ocean fisheries were essentially limitless until the development of power-driven vessels and gear, aided by modern remote sensing technology. The worldwide search for edible protein has further compounded the problem, and fish landings have doubled every 10 years in recent decades. Fishes are not only consumed directly by people, but are also used as meal for fowl and pigs, and for pet food (which, ironically, is then used as a favored bait for shrimp fishing in the Puget Sound region). These practices have led to many shortages and international disputes.

The Fishery Stock and Its Variability

■ Fish populations are renewable resources.

Fishery populations are a **renewable resource**; nutrients and energy are fueling their regrowth even as we exploit them. This concept of renewable resources serves as the entire basis of fishery management. In some fisheries, even modest fishing depletes the stock much more rapidly than it can be renewed. Such overexploitation can lead to cycles of near extinction of the fishery population, followed by eventual recovery if the fishery is managed. This is the case with many marine mammal populations that have been hunted nearly to extinction. The California sea otter, *Enhydra lutris*, for example, was hunted until only a few populations survived, in the Aleutian Islands and in a few isolated spots on the west coast of North America. The otters were prized for their skins. Hunting has been banned for decades, and population growth, migration, and restocking have resulted in a dramatic recovery of the population.

Stocks and Markers

■ Fishery species are divided into stocks. Various tags and markers can be used to monitor them.

Monitoring is a key to understanding of the trajectory of fisheries. For over a hundred years, many scientific organizations and governmental agencies have been learning about the life histories of important fishery species and have been attempting to monitor the populations and devise management schemes. This began in 1902 with the establishment of the International Council for the Exploration of the Sea, to address fishery problems in the North Sea.

In Chapter 7, we discussed fish migration, which may occur between spawning, nursery, and adult feeding grounds. A fishery species may have a broad geographic range, but, for management purposes, we divide the total range into populations, or **stocks**. Each stock is controlled by separate oceanographic and ecological factors and has nearly complete reproductive separation from all others. This usually involves separation of the spawning grounds, but it may also involve separation of nursery and feeding areas. **Tags** are commonly inserted into fishes to follow their movements but also to delineate stocks (**Figure 21.1**). This allows the recording of time and location of release and capture. Some tags are visible outside the fish, but completely internal tags can be inserted with little mortality. We can detect them by means of magnetic inserts or even by small radio transmitters that are induced by a scanning device. Radio tagging and acoustic reflection can also be used. Tags that communicate with a global positioning system have been used in marine air breathers such as sea turtles, but fish often swim too far beneath the surface for detection, so the tags are designed to disattach and float to the surface, giving a location.

- ■ **Biochemical and molecular markers of various types can also be used to diagnose differences among fish stocks that have become genetically isolated.**

Natural selection or random processes change the genetic composition of stocks after they have been separated for a time. The cod is an example of a species divided into a number of geographically isolated populations that can be diagnosed by a number of biochemical genetic markers. **Hemoglobin** and **transferrin** genotypes were first used to diagnose different cod stocks in the North Atlantic. Other markers, including **enzyme polymorphisms** (genetically distinct variants of enzymes), morphology, and unique parasites can be used to distinguish among stocks. In recent years, **DNA** sequences and markers have been used to delineate stocks.[1] For example, mitochondrial DNA can be fragmented by different types of restriction enzymes, each of which breaks up the DNA at specific points. Such enzyme digestion produces a series of length variants that can be identified by their migration on a stained gel. Different populations tend to exhibit different length variants, and therefore DNA markers are useful in stock identification (**Figure 21.2**). More directly, DNA sequencing can be used to find sequence variants at highly variable genetic loci. At a number of genetic loci, genetic variants are diverse enough that they vary in content between populations and can thus be used to distinguish among stocks. **Microsatellites** are genetic loci where short sequences of nucleotides are extensively repeated. They are quite variable in populations, and the number of certain repeats may vary between species, among populations, or even between individuals. Consequently, they can be used as effective markers at several levels of organization. They have proven valuable, for example, in distinguishing among migrating Pacific salmon stocks from different river systems and

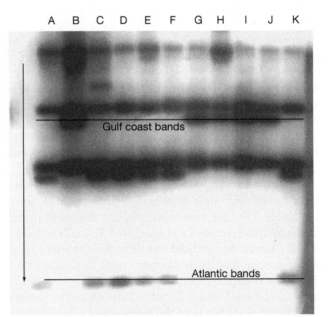

FIG. 21.2 Identification of stocks of the striped bass *Morone saxatilis* by use of DNA markers. An agarose gel, in which restriction enzyme–digested fragments of mitochondrial DNA (labeled with radioactive carbon) have been separated by size, is shown. (Fragments migrate in the direction of the arrow at left, within lanes marked by letters.) Marked fragments in lanes B, G, H, I, and J exhibit a mitochondrial genotype that is seen only in striped bass from the Gulf of Mexico coast, whereas marked fragments in lanes A, C, D, E, F, and K display a mitochondrial DNA genotype diagnostic of striped bass of Atlantic coast ancestry. (Courtesy of Isaac Wirgin)

FIG. 21.1 A striped bass, *Morone saxatilis*, tagged ventrally with an internal tag. (Courtesy of John Waldman)

---

[1] See bonus chapter, "Molecular Tool for Marine Biology," online.

investigating their mixing in the open ocean. It has been possible to cross-check DNA markers with data from tags (Beacham and Wood, 1999). The student should be aware that the use of genetic markers to delineate fisheries stocks is not straightforward (Palsbøll et al., 2007).

■ **Stocks may be isolated from each other on the basis of use of the same drainage but at different times.**

Stocks may differ in geographic occurrence, but they may also use the same geographic space (e.g., river drainage) at different times and thus still be isolated from each other. In the pink salmon, spawning occurs in streams and adult feeding occurs in the adjacent ocean, but there are odd- and even-year populations in the same streams, which are reproductively independent. DNA markers can be used to delineate stocks and help to identify streams that deserve special protection or restoration.

■ **Stock identification can have important societal implications.**

Delineations of stocks according to ecology are often not the main motivations of fishery biologists. Fishermen often have a proprietary interest in their own local fishes, but the fishes also may range over broad areas. For example, many Native American tribes have rights, negotiated by treaty, to specific stocks of Pacific salmon. This requires the ability to distinguish among salmon stocks. To negotiate border disputes between Canadian and U.S. fishers, for example, it is useful to have an accurate means of distinguishing among local stocks and to be able to identify the geographic extents of spawning and feeding grounds. Even when armed with such information, fishery experts may still be frustrated in their efforts to reach an agreement because fishes often swim across international borders during migrations.

Fishing Techniques and Their Effects

■ **Finfishes are mainly caught by (1) hooking fishes individually, (2) entangling fishes in stationary nets, or (3) catching fishes in hauled nets or traps.**

Although the methods may have improved slightly, the basic ways of catching fishes have not changed for centuries. Perhaps spearing is the one technique that has been essentially abandoned by all but aboriginal peoples. It is still quite useful for hunting fishes living on rock and coral reefs. Harpoons are used widely for whales and swordfish. Currently, (a) hooked lines, (b) stationary nets, and (c) hauled nets are the most popular techniques (**Figure 21.3**). Hook

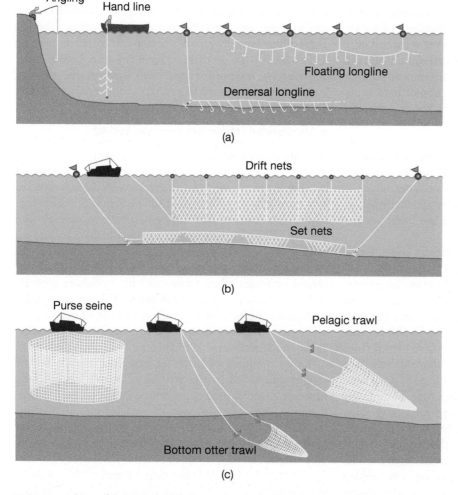

FIG. 21.3   Some fishing techniques: (a) hook-and-line and longline fishing, (b) gill net fishing, and (c) purse seine and trawl netting. (From Cushing, 1988)

and line may appear to be equally primitive, but variants are used extensively. Most popular are **longlines**, from which short leaders with hooks extend at regular intervals. Long-lines can be thousands of meters in length. The line may be paid out and suspended by floats, or it may be trolled behind a boat. Trolling is a common means of fishing for albacore and some species of Pacific salmon.

Many different types of shoreline nets staked into the bottom have been developed that entrap fishes swimming in schools along the shoreline. Such nets are especially effective in trapping migratory fish that have strongly constrained routes, such as American shad, which move upriver to spawn. **Pound nets** often have some sort of leader wire that guides the fish into a blind-ended net. If currents are strong enough, there is no need for a long leader and the fish can be trapped directly. These net arrangements can also be suspended from rigid floating frames and are used to catch some species of salmon. **Fyke nets** are long nets that are usually staked to the bottom. They have leading wings that guide fishes moving with the current into the main bag, from which escape is difficult.

Gill nets have a specified mesh size, chosen to entrap the fishes by their gills, fins, and jaws as they attempt to swim through. These nets could be stationary, but also can be towed from one end through the water, while the other end is attached to a smaller stationary boat or is anchored to the bottom. From both shore and boat, a variety of fine-mesh nets are hauled to entrap large numbers of fishes. One of the laziest techniques is to drop a net stretched over a horizontal frame. The net is raised periodically. This is a common technique used in the Mediterranean to trap small fishes and squid. Many fishing peoples haul seine nets across beach areas, sometimes with the aid of four-wheel-drive trucks on the beach. The top of the net is attached to floats, and the bottom is weighted down. This method catches a large variety of fishes and was used to catch striped bass on the south shore of Long Island, New York. In other areas, a boat offshore hauls the net, which is fixed to the bottom at the beach. The boat gradually brings its end of the net toward the beach, and the fishes are entrapped in the closed portion.

Seine nets are also deployed entirely from boats. **Purse seines** are particularly effective. The end of the net is attached to a small boat, and the net is paid out as a larger boat moves away. The larger boat attempts to encircle a school of fishes with the net, and a rope acts like a purse string as the net is drawn aboard and closes about the fishes. This method is used for a variety of fish species, and is used from inshore waters to the blue ocean. Environmentalists have complained in recent years about the damage to porpoises, which are entrapped in purse seines with tuna. For reasons that are not well understood, schools of tuna congregate beneath groups of porpoises, and fishing boats, therefore, seek out the porpoises, which indicate that tuna are below. As the tuna net is hauled in, the porpoises are often trapped underwater and drown. Many fishing boats now employ divers to help save the porpoises, and there has been strong pressure to eliminate the fishing technique. Many large

fishing fleets in the Pacific have changed fishing practices to avoid trapping porpoises, which enjoy popular support among the public.

Many other nets that are employed are essentially bags towed by a harness of wire. **Bottom trawls** are especially efficient because the harness is attached to two or more panels, called otter boards, that spread the opening of the net efficiently. The bottoms of the boards are weighted in order to drag the net over the bottom. This technique is used widely to bring in bottom fishes such as plaice and flounder along with schools. In midwater, **pelagic trawls** are used in a similar way to net schools of fish. Modern remote-sensing techniques, such as multibeam sonar, permit oceangoing vessels to locate enormous otter trawl gear accurately at depths of hundreds of meters to target large schools of fish.

In recent years, extremely large open-ocean nets were employed to catch squid and other abundant nekton. Fishers from a number of nations used nets over 30 miles long. These nets kill off large numbers of squid and fish and threaten to wipe out a number of different species. A treaty was ratified to ban the nets, which went into effect in 1993.

■ Bottom-associated animals are trapped in baited mesh traps.

A variety of mobile animals, but principally large crustaceans such as crabs and lobsters, are taken from baited traps that are usually marked with surface buoys. Of necessity, these are generally coastal and estuarine fisheries. Traps often have inverted cones at the entrance, making it difficult for the animals to leave the trap. Unlike most net fisheries, trapping is associated with specific areas of bottom, and this has led to a variety of informal and formal divisions of fishing grounds among fishers.

■ Unintended catches are known as bycatch.

Another human-imposed source of fishing mortality is the incidental killing of undesirable species of fishes, known as **bycatch**. As mentioned earlier, dolphins are often trapped in nets during fishing for tuna in the Pacific. U.S. fishery legislation has required that fishing nets have **turtle exclusion devices (TEDs)** consisting of a screen or bars to prevent turtle entry and a trap door, which allows the escape of sea turtles that would otherwise drown in the nets. After some careful estimates of sea turtle mortality by fishery agencies and the realization that TEDs were not large enough for leatherback turtles, current widespread use of inshore and offshore trawlers has dramatically reduced turtle mortality at sea (Finkbeiner et al., 2011).

Large numbers of fishes of inappropriate species are caught and killed in the process of recovering gill nets. Floating longlines used in the central Pacific to catch swordfish may have strong impacts on some species of albatross that feed on fishes near the surfaces. In the northwest Atlantic, swordfish boat longlines catch a large number of blue sharks.

Fisheries biologists have come to realize that oceanographic patterns make bycatch almost inevitable. We have discussed patterns of upwelling and major current systems that cause many species to migrate very long distances and to wind up in high-productivity zones and often in multi-species migration routes (see Chapter 10). Thus, longlines intended for one valuable resource species will likely catch other species attracted to the same high-productivity zone. In such areas in the North Pacific Ocean, longlines laid to catch tuna will likely catch sea turtles or albatrosses attracted to the large populations of forage fish that attract the tuna—and fishers—to the same regions. Peckham and colleagues (2007) describe a fascinating example: A pair of relatively small-scale gill net and longline fisheries near the open Pacific coast of Baja California caught as many endangered North Pacific loggerhead sea turtles, *Caretta caretta*, as the rest of the fisheries in the Pacific! These turtles nest in Japan, but juveniles move throughout a series of distant Pacific Ocean routes to find food. They were snagged by longlines deployed from small boats, but local fishers agreed to change from this fishing technique.

These considerations have led to a variety of techniques to avoid bycatch, many of which have been used in laws and rules set by fisheries agencies (Gilman, 2011; Moore et al., 2009):

Release devices. Turtle exclusion chambers allow sea turtles caught in nets to swim away and escape being trapped in nets.

Avoidance devices. Some nets have devices that make pinging noises, causing dolphins to swim away. We have already mentioned turtle exclusion devices.

Specialty hooks. Because the typical J-shaped hook can also catch turtles and species that are endangered, circular hooks have been adopted in some areas. These hooks efficiently catch fish, but sea turtles cannot easily be ensnared. Some hooks are designed to bend when seized by very large fish, such as Atlantic bluefin tuna, but will retain tuna species of smaller body mass that are acceptable for fishing.

Specialty trawl nets. Nets can reduce bycatch by adjusting net size and structure. For example, the Ruhle net allows haddock to rise as they enter the net and be caught. But Atlantic cod, which require protection, stay to the bottom. For this reason, the trawl net has large openings near the net bottom, allowing the cod to escape.

Timing of fishing and location. Setting of longlines at night is a powerful method of avoiding bycatch of seabirds such as albatrosses, who hunt visually and usually during the day. Undesirable bycatch of threatened species such as sharks and marine mammals can also be reduced by avoiding known hotspots of abundance for potential bycatch species.

Bycatch has become so abundant in some fisheries that seabirds have become dependent on discarded fish. In the Baltic and North Seas, this dependence has produced an ironic linkage. Efforts to decrease bycatch may cause decreases in seabird populations. This is especially true in cases in which normal seabird habitat has been disrupted, making bycatch the only food that is available to some seabird species. When bycatch was reduced, the great skua *Stercorarius skua* shifted to preying on smaller shorebirds, potentially threatening some seabird communities (Votier et al., 2004).

■ **Much of the world's catch consists of bycatch, used for fishmeal, and targeted pelagic fish taken specifically for fishmeal.**

Fishmeal comprises the main marine protein and lipid sources used in animal feed (terrestrial and mariculture food). Some bycatch is set aside for fishmeal, but most fishmeal derives from targeted fisheries (e.g., anchovies), usually smaller pelagic fish such as anchovies, sardines, and capelin. Regulations are beginning to reduce the catch used for this purpose.

■ **Bottom trawling may strongly alter the seabed.**

Bottom trawling very strongly impacts living resources on the seabed (Thrush et al., 1998). In heavily fished areas, just about every square meter of bottom is dragged a few times a year. Many bottoms with gravel and a rich epifauna of bryozoans, hydroids, and echinoderms will be turned over, leaving a smoother soft-bottom interface with smaller soft-bottom invertebrates and much lower benthic diversity (Collie et al., 1997). In places where epifaunal communities are destroyed, the substitute is a mud bottom without the usual benthos (e.g., oysters, sponges, soft corals) that feeds on particles from the water column and deposits them on the sediment–water interface. Instead, the disturbed muddy bottom is easily resuspended, which returns to the water column organic-rich particles, which may be consumed by bacteria and reduce the oxygen content of the water. On the other hand, trawling of sandy bottoms seems to do less damage to the seabed, perhaps because infaunal burrowing organisms are more resilient to sediment overturn. The effects of such strong changes are obvious for the benthic communities, but the feedbacks on the resource species are poorly understood. If a bottom-feeding fish depends on smaller mobile prey, then it might suffer greatly for lack of a variegated cobbly bottom that shelters smaller fish and shrimp. On the other hand, some fish might find a bare bottom to be a more suitable habitat. It will be very important to understand the recovery time, which will help set trawling intensity quotas. On the Grand Banks off Newfoundland, eastern Canada, recovery time may be as long as a year (see Schwinghammer et al., 1998).

Life History and Stock Size

■ **The life history of a fishery species and the size of the stock must be understood before sensible planning can be done regarding fishery management.**

To make sensible decisions about fisheries management, the following must be known: (1) range of temperatures and salinities for maximum growth and survival, (2) location of spawning habitat, (3) location of migration routes, (4) location of feeding grounds, and (5) biological information that minimizes unintended mortality during fishing.

To make proper management decisions, it is necessary to estimate the size of a stock. It is also desirable to know the abundance of the different year and size classes because only fishes of certain sizes are economically important. Certain age classes such as juveniles and reproductive individuals may be important to identify for conservation purposes.

In some cases, fish stocks are assessed by fishery agencies through a carefully designed sampling program that takes into account migration patterns and spatial distribution. Migration over wide geographic regions makes it very difficult to assess a stock size because one cannot easily design a sampling program that adequately follows a fish population over hundreds or thousands of kilometers. The spatial distribution is important because fishes are rarely distributed evenly but usually occur in distinct patches; they may also school. In such cases, small numbers of samples are liable either to miss the population altogether or to hit a dense patch accidentally. In either case, one cannot simply assume that the density estimated from a very few trawls adequately measures the population size. In many cases, we are ignorant of the mechanisms of clumping. For example, leatherback turtles have been found to occur in patches in the open ocean, which are loci of their jellyfish food, which is also very patchy in distribution (Houghton et al., 2006).

It is desirable to use the same types of sampler in different studies because samplers vary in their ability to catch fish. In many cases, population estimates are made with different types of sampling gear, and different results are obtained. When this happens, we cannot be sure whether the difference is due to the gear employed or to actual differences in population size. For purposes of cross-checking, it would be desirable to employ different methods of assessing abundance in the same area at the same time, but this is rarely done. A good example of disparities in stock size estimates is the northern right whale dolphin, *Lisodelphis borealis*, which is often a victim of large drift nets (Mangel, 1993). Estimates from line transects yield far lower population size estimates than do estimates made from accidental catches in drift nets. These differences are crucial because they lead to very different conclusions about the degree of endangerment of this dolphin species in the Pacific Ocean.

Moreover, it is usually not possible to sample eggs, larvae, and adult fishes with the same type of sampling gear. Because each gear type has a different efficiency of catching fish (or eggs), and different life stages have differing spatial distributions, it is often impossible to compare estimates of abundance between very different life stages.

■ **The size of a stock is mainly assessed by quantity of landings by fishers, which is principally a function of the population size, the spatial variability of the fish, and the amount of fishing effort.**

Although fish surveys are done frequently by government agencies, most fishery data do not come from scientific surveys but from data on **landings** by fishing boats. For most important U.S. fisheries, federal and state agencies collect data presented by the fishermen. To some degree, inspectors may verify the data, principally through onboard checks during the fishing. By the end of a year, a fishery agency may know the total number and mass of fishes, and may have a breakdown by size and a record of the number of boats and the time each one spent fishing.

Although the quantity of landings is to some extent a function of population size, landings are difficult to interpret for one of the same reasons applicable to more systematic surveys: The spatial distribution is rarely even, and sampling therefore is biased away from the true density. Moreover, fishermen work in areas that customarily have the highest fish densities, but landings data may ignore other important areas, such as spawning and juvenile feeding grounds. There is also a complication generated by variations in the numbers of boats deployed, fishermen working, type of gear employed, and hours spent fishing. These four factors are collectively known as **fishing effort**. If effort increases, the catch will increase, but not necessarily because there are more fishes in the ocean. If the fishing effort doubles and the catch does not double, however, one may assume that the population is declining. All stock estimates must therefore take into account the **catch per unit effort**. A useful indirect measure of effort is fuel consumed to run the fishing boats involved in a fishery. **Figure 21.4** shows the declining historical trend in catch per unit effort for the blue whale fishery.

Variation in fishing effort often masks the quantitative meaning of landings. If a group of fishers seek a certain catch per unit time, then they might increase effort over time to maintain the catch. Thus, the landings might be constant even if the actual population is declining. If landings decline over time, we might correctly interpret a corresponding decline in the actual population of fish. But fish might learn to avoid fishing boats, or fishing effort might decline over time. Both of these possibilities would give the false conclusion that the size of the stock is declining, when no such thing is occurring. Landings are of course also a function of management goals by government fisheries agencies, so a decline in landings may reflect rules rather than the actual sizes of fish populations.

Stock Health and Production

■ **The health of a stock can be assayed by its production, which is explained in terms of growth of previous year classes and recruitment into the new year class.**

Because a stock has been defined to be a population that is relatively isolated from others of the same species, one can try to evaluate the potential yield of a stock to a fisher. We are interested in the change in the potential yield, measured in pounds of fish per year, from the previous year. To make such an estimate, we have to have a general model for gains and losses in a stock of resource organisms (**Figure 21.5**), which includes recruitment, somatic growth, mortality, and reproduction.

The relationships among these factors can be expressed in an equation that relates them all to the change in population size:

$$\Delta W = W_{t-1} - MW_{t-1} + RW_{t-1} + GW_{t-1}$$

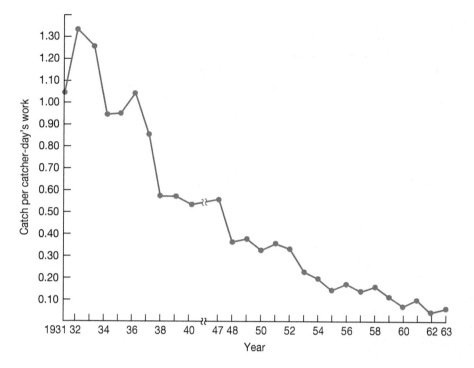

FIG. 21.4 Catch per unit effort of the blue whale through the twentieth century.

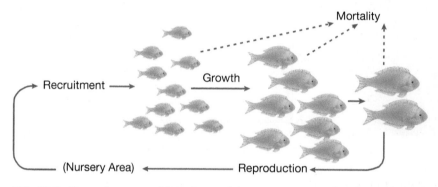

FIG. 21.5 To produce a good fisheries model, we must account for all contributions to reproduction, growth, and mortality throughout the life cycle of the fishery resource species.

where ΔW is the change in mass of the total fish population over 1 year, W_{t-1} is the mass of the population at the same time in the previous year, M is a mortality fraction (varies from 0.0 to 1.0) giving the fraction of the mass of fish that was lost owing to death, R is the fraction by mass that is added from recruitment, and G is the fraction by mass that is added owing to growth of individuals that lived from the previous year to the present. We can neglect immigration and emigration because the stock has been defined to be independent of other populations.

Although we are interested in total mass of fish, fish population models must eventually be expressed in numbers, and it is essential to be able to know the age of the individuals that are surveyed and studied. In some cases, age can be identified by size. This is true in seasonal environments, where birth and rapid growth are often confined to short periods of the year when food is abundant. After several years, the population will consist of a series of **year-size classes**, which may be distinguishable by size

(**Figure 21.6**). Growth slows in later years, and the variation in growth tends to blend the later age classes. Also, fishes growing at different rates may be mixed together in a feeding ground, and the age–size relationship may therefore be too complex to measure. Because of this, fishery biologists have resorted to other techniques. Both **otoliths** (small precipitated spheres used as part of a balancing organ) and scales have rings that record seasonal growth. Otoliths can be used to age larval fish to the day, which is quite important in the understanding of early life histories (**Figure 21.7**). For shellfish, growth rings can also be used, and cross sections of shells often give an accurate age and growth history.

Fishes and shellfish have the capacity to produce thousands to hundreds of thousands of eggs per female. When fertilized, these eggs produce the next generation of larvae and then early stage juveniles. A plankton net can bring in millions of fish eggs and thousands of yolk-sac larvae. As these **young-of-the-year** (0+ year class) feed and grow, one

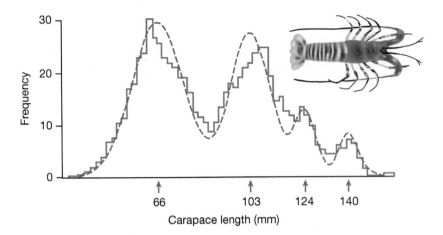

FIG. 21.6 Age classes of the lobster *Panulirus ornatus*. Curved line estimates age classes from the more discontinuous distribution of the histogram. Note the older age classes to the right, which are more indistinct. (Modified from King, 1994)

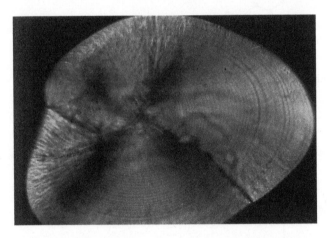

FIG. 21.7 A cross section of an otolith of the bluehead wrasse *Thalassoma bifasciatium* that settled from the plankton at a length of 13 mm. The daily growth record reveals the 41-day mark after hatching, when settlement on the bottom occurred (transition between closely spaced lines and relatively broad band). Otolith is about 300 μm long. (Courtesy of Robert Cowan)

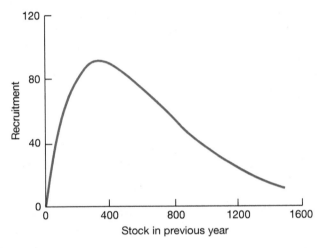

FIG. 21.8 An expected relationship between the size of the stock and recruitment in the following year.

can trace a peak of body size as it increases with time. There is also considerable mortality in the year class as it ages. It is no coincidence that egg production per female is so prodigious. Most eggs and larvae are eaten, starve, or are lost to inappropriate habitats. Few survivors make it to adult age and size, to contribute the next generation.

A major objective of fisheries biology is to project population change into the future. The most common approach is to predict the size of the 0+ year class from the stock size in the previous year. The appearance of the 0+ year class comprises the recruits. A number of different **stock-recruitment models** have been proposed (see Needle, 2002, for an excellent introduction to the different stock recruitment models), but we shall focus here on one that relates reproductive success to population density.

If resources (e.g., food, space for spawning) are limiting, we might expect a relationship between the size of the stock and the recruitment as depicted in **Figure 21.8**. As the stock increases, the number of 0+ recruits in the next

year should increase as well. But past a critical stock size, available food per adult may start to decrease to the point that eggs per female will decrease correspondingly. As the stock increases past this critical threshold, we expect a decline in recruits per reproductive female. Past the critical threshold, resources are limiting and production of recruits into the next year also declines. This reaction of population growth to high density is known as **compensation. Going Deeper Box 21.1** shows one ecological model that might explain compensation in terms of reduced resources per female as population size increases and resource per female decreases. This relationship of stock size to recruitment can be the basis of fishery management, as we discuss shortly.

In practice, often there is a poor correspondence between the size of the stock and the number of recruits surviving into the next year. Postreproductive processes in the first year of the 0+ year class will likely determine population size, but these processes (e.g., predation) may be independent of the parent stock and resource availability to that parent stock. Conditions change rapidly at sea! A simple example may illustrate why such a decoupling between stock size and recruitment of the next 0+ year class

GOING DEEPER 21.1

A Simple Model to Explain the Stock-Recruitment Model and Maximum Sustainable Yield

Presumably, a resource-limited growth rate would also imply the reduction of reproduction as resources become more and more limiting. The following logistic equation* can relate the growth rate of a population to a hypothetical maximum population that available resources can maintain (**Box Figure 21.1a**):

$$\frac{dN}{dt} = rN\frac{(K-N)}{K}$$

where dN/dt is the rate of change of population size or somatic growth, N is the population size, r is a rate of growth when resources are limitless, and K is the maximum population that available resources can support.

Box Figure 21.1b shows dN/dt as a function of N. When $N = 0$, the growth rate must be zero, but the growth rate is also zero when $N = K$. In between these two numbers, growth rate is positive. If we did nothing and resources were limiting, we might expect the population to converge on K. So what might be the effect of fishing if we reduced the population below K? Note that the population growth rate is at a maximum at $K/2$. We would, therefore, gain more fish per unit time if the population were fished down to $K/2$, rather than letting the population grow to the level of K. It may seem counterintuitive at first, but if the population size is greater than $K/2$, eventual fish production would be increased by fishing more and reducing the stock to the size for which growth is maximal. According to the logistic model, the maximum sustainable yield, therefore, is set at $K/2$.

---

*The equation for dN/dt has been used widely in population models, but it is not clear that all populations behave according to it. It is used here only to illustrate how population growth may be maximized at intermediate population size.

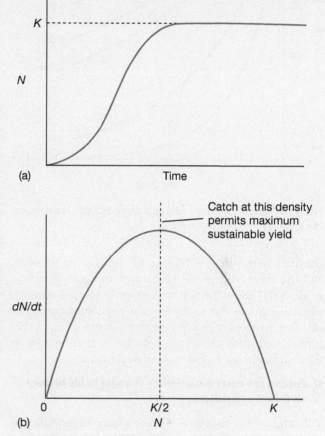

BOX FIG. 21.1 (a) Population change over time according to the logistic model. (b) Population growth rate (dN/dt) as a function of population size N, according to the logistic model.

seems to often exist. Let us take 1,000 eggs as the average produced by a female and assume that all these eggs are fertilized. **Figure 21.9** shows the different adult population sizes obtained when the mortality per day is 1, 1.5, 2, 5, and 10 percent. Note the striking differences in adult population size, even with very slight differences in mortality after only 60 days. It seems quite possible that actual mortality could fluctuate to this degree independent of the density of the previous year class, so it should be no surprise that there could be no systematic relationship between stock size and the recruits of the next generation.

■ **At very low population densities, reproductive output per female might decline with declining population density.**

Because many fishes and invertebrates that are exploited in fisheries produce so many eggs, it is often thought that at very low population density there is no case when recruitment will be limited by the size of the spawning stock. The feeling is that even one male and one female might produce enough eggs and sperm to produce a large population of recruits in the next generation. But some have argued that the reproductive output per female might crash at low population densities. The rapid decline of a population because of the inability to find mates is known as the **Allee effect** (see Chapter 4). In fisheries biology, this relationship has been termed **depensation**. For example, some shellfish populations have become so overexploited by fishing that males and females exist at very low densities and may be too far away from each other for external fertilization to occur efficiently. Some species of abalones on the California coast may be recruitment-limited in just this way. In Great South Bay, Long Island, New York, the northern quahog clam *Mercenaria mercenaria* existed at very high densities until the 1970s, and recruitment was very high. However, by the 1990s, adult population

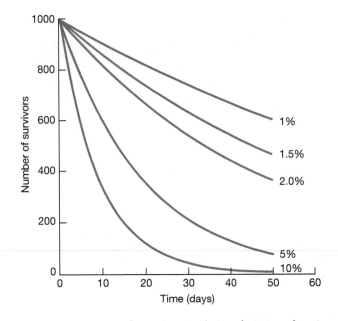

FIG. 21.9 Survivors of a starting population of 1,000 as a function of a differing daily mortality rate.

densities were very low because of high fishing pressure, and the effect of depensation became apparent (Kraeuter et al., 2005). Now the bay has extremely low recruitment, which is partially due to the very low population density of this free spawner with separate sexes. Myers et al. (1995) investigated stock-recruit relationships in many species of fish, but found only a few cases of depensation.

■ **Fishing can exert evolutionary changes in life history on fished populations.**

We can see that fishing is a major source of mortality. If there is genetic variability for life-history traits and if fishing targets certain size classes, human-imposed mortality might be a potent source of natural selection. We saw this in Chapter 7 when discussing fishing on the shrimp *Pandalus borealis*, which is protandrous. Fishing pressure on larger shrimp resulted in the evolution of a smaller size at which sex change occurs from male to female. This is likely to be common in the ocean, but it has not been studied in many species.

Conover and Munch (2002) used a laboratory system to investigate if size-biased fishing might cause an evolutionary change that might have implications for fisheries biology in general. They used laboratory populations of the Atlantic silverside *Menidia menidia* and found genetic variability for growth rate that allowed rapid shifts of body size away from the sizes that they harvested. Harvests of large-sized individuals led to dominance by small fish after just four generations of selection. In effect, this suggests that intense fishing might result in selection for life history traits that impede recovery. This result may be applied directly to fisheries, where size-biased fishing might rapidly select for smaller fish. If smaller females produce fewer eggs, then we would expect an evolutionary change toward lower fish population size. A problem might be that the reverse response to larger size might be slow, even if fishing

were regulated. There is some evidence that intensely exploited fisheries have poor prospects for short-term recovery (Neubauer et al. 2013). This has been the case in the cod fishery of the northwest Atlantic.

On the other hand, most studies from many fisheries show that heavy fishing tends to be correlated with *more rapidly growing fish* instead of the reverse. These selected fish mature at an earlier age (Rochet, 1998). Overall, moderate overfishing tends to be followed by rapid recovery once fishing is relaxed (Neubauer et al. 2013). Selective fishing of larger fish also tends to leave a population of juveniles that tend to respond more to environmental fluctuations.

Population Models Based on Life-History Stages

■ **More effective fishing approaches use models that show the impact of mortality on different life-history stages.**

Models of compensation such as the one shown in **Figure 21.8** have the advantage of building on an important ecological principle: the effect of high population density on further population growth. While this may be important in many cases, it may also be important to understand what parts of the life cycle are most threatened and whether great effort should be expended on protecting different life-history stages. This is especially a problem in many marine vertebrate fisheries because numerous species have life-history stages that live under different ecological conditions and in geographically separate locations. An extreme example is the case of Pacific salmon species: They spawn in freshwater rivers, but adults live and feed in the open North Pacific. In order to understand how to manage such species, we must be able to analyze the contribution of different life-history stages to overall population growth. **Going Deeper Box 21.2** introduces you to how to make such determinations using age-based or life-history stage–based population models so that we can make effective fish management decisions.

The use of life-history stage–based models can very effectively lead to management decisions in the case of some fisheries. For example, all of the American species of sea turtles are classified as threatened or endangered. None constitutes a fishery, but their fate is entwined in the same types of habitat alteration and fishing that affect other species, especially because their nesting beaches are endangered and they are often caught as bycatch. From the 1960s onward, counts of sea turtles uniformly showed severe declines, with expectations of extinction in just a few decades, so policies were badly needed for protection. The numbers of nesting females declined greatly on beaches of the Gulf of Mexico and the Caribbean, but turtles were also being caught by the thousands in shrimp trawl nets. Beach protection has proven important in increasing hatchling survival, but it is crucial to know which life-history stages are contributing most to total population increase because management strategies such as reduction of trawling or rearing of hatchlings can be very expensive and politically difficult to enact.

GOING DEEPER 21.2

Age-Based or Life-History Stage–Based Population Models

When we think of population growth, we think of the factors that cause change in the total numbers of a population per year. But this is an overall number that does not consider two important factors:

1. *Mortality rate differs with age.* Different age classes may have very different mortality rates. For example, juveniles might die at rates that are orders of magnitude higher than adults.

2. *Reproductive contribution of a fish differs with age.* The number of young produced by a fish might vary substantially with age. Immatures obviously produce no young, but later age classes might differ a great deal in their reproductive contribution to population growth. Mortality at a crucial later age might greatly influence future population growth because reproduction at this stage would greatly affect future generations.

This leads to a different approach. We want to know overall population growth, but also want to know how different age classes contribute in terms of reproduction and survival. In many cases, we would want to consider life-history stages, such as larvae, immature juveniles, young mature fish, and old mature fish.

This can be solved with a **Leslie matrix** approach, in which we take account of the survival and reproductive contributions of different age classes or life-history stages. The Leslie matrix we will use considers the change of a population of females. Let's use three age classes: 1 year, 2 years, and 3 years. The 3-year-old individuals reproduce, which adds individuals to the first-year class. We know the survival rate from one age class to the next and the fecundity or number of young produced by the last age class. To know how the population changes, we also must know the initial individual numbers of these three age classes.

The Leslie matrix allows us to project population size and distribution among the age classes in future time periods by a table of fixed survival rates and reproduction rates:

R_i is the reproduction rate if the i th year class
S_i is the survival rate of the i th year class

Suppose we start with a time $(t - 1)$. We create a vector of the initial abundances in the three year classes, X_i where $i = 1, 2,$ or 3 (from top to bottom). In order to project population change, we multiply this vector by a matrix that includes the values of R_i and S_i to get the distribution of the age classes in the next time period t:

$$\begin{pmatrix} x_1^t \\ x_2^t \\ x_3^t \end{pmatrix} = \begin{pmatrix} R_1 & R_2 & R_3 \\ S_1 & 0 & 0 \\ 0 & S_2 & 0 \end{pmatrix} \begin{pmatrix} x_1^{t-1} \\ x_2^{t-1} \\ x_3^{t-1} \end{pmatrix}$$

As the generations proceed, we can find out the population size by taking the initial vector of abundances of fish in each year class and multiplying the matrix of values of R_i and S_i, which is the Leslie matrix, by each successively calculated vector of numbers in the age classes. We will not go into the details of the mathematical manipulation, but clearly the population will increase if the values of R_i are sufficient to increase the population relative to the loss in mortality, represented by values of S_i.

The following is an example:

Let us say that only the individuals in year 3 reproduce. This means that $R_1 = R_2 = 0$. S_1 and S_2 represent survival rates from year 1 to year 2 and from year 2 to year 3, respectively. Let $R_3 = 100$, which means that a female in year class 3 will produce 100 offspring, and S_2 and S_3 both equal 0.5, which means half of the population of the year classes 1 and

2 survive (year class 3 contributes only by reproduction new individuals to year class 1). This produces the following result: with starting populations with 1,000 females in each year class, we get the vector of abundances in the next time period on the left:

$$\begin{pmatrix} 1,000 \\ 0.01 \\ 0.01 \end{pmatrix} = \begin{pmatrix} 0 & 0 & 100 \\ 0.50 & 0 & 0 \\ 0 & 0.50 & 0 \end{pmatrix} \begin{pmatrix} 1,000 \\ 1,000 \\ 1,000 \end{pmatrix}$$

If you add the three numbers in the vector, you get the population size and can therefore follow population size over time. The total population size is now 1,000.02. To get the values of X_i for the next year, we multiply the Leslie matrix by the new vector of X_i at time t. The next vector is, therefore,

$$\begin{pmatrix} 1,000 \\ 1,000 \\ 0.1 \end{pmatrix}$$

You can see the proportions in the three age classes are changing from generation to generation, and population size fluctuates a great deal, mainly because of the starting conditions. **Box Figure 21.2** shows the change in population size after 20 time periods. The population is fairly low for a time but then explodes. The population increase has a lot to do with the combination of the fairly high survival rates (0.5) and the high fecundity (100 offspring per female). Compare this with the green curve on the plot, where we have set survival rates to be 0.1. The population never takes off because low survival offsets high fecundity.

Studies of population dynamics would experiment by varying the starting population sizes, survivorships, and fecundities. Varying fecundity and survivorship allows an investigation of the sensitivity of change of population growth to these parameters. In the text of this chapter you can see a discussion where this is applied to population change in the loggerhead turtle.

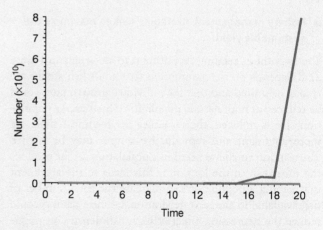

BOX FIG. 21.2 Population change over time in two populations with three age classes under a Leslie matrix model starting with 1,000 individuals in each age class. Red: Reproduction of age 3 individuals is 100 per individual, and survival rates of ages 1 and 2 are each 0.5. Green: Reproduction of age 3 individuals is 100 per individual, and survival rates of ages 1 and 2 are each 0.1.

The life-stage approach discussed in **Going Deeper Box 21.2** shows that different values of survival and reproductive contribution at different ages might have strong effects on life-history stages. One approach, called **elasticity analysis**, measures the relative contributions of survival and reproduction at various life-history stages to overall population change. Crowder and colleagues (1994) used such an approach to analyze the fate of loggerhead turtles in the Gulf of Mexico that suffered high mortality owing to beach disturbance, but also from shrimp trawling nets, which caught and killed loggerheads as bycatch. They used a Leslie matrix approach (**Going Deeper Box 21.2**) to analyze data for loggerheads as hatchlings, small juveniles, large juveniles, subadults, and adults (that were reproductive). The model results demonstrated that large juveniles had the highest contribution to overall population growth. Because large juveniles were suffering high mortality from being caught in fishing nets, it made sense for fishing managers to insist that turtle exclusion devices (TEDs) be used by all shrimp fishers. While protection of hatchlings on beaches is of course still important, this study shows the extreme importance of paying attention to mortality caused by trawling.

Changes in survival of different life-history stages do not immediately result in population increases, but might take a number of years because population growth with several life-history stages is very bumpy, as changes in population effects propagate through the different age classes to the reproductive adult stage (e.g., see **Going Deeper Box 21.2**). Thus, models such as those based on Leslie matrices are crucial in helping fishing managers to understand the impacts of fishing regulations and the expected time for population responses.

Fisheries Impact and Management

The Concept of Maximum Sustainable Yield

■ **Fishery management strategies seek to maximize the sustainable yield.**

The hypothesis relating recruitment to stock size in Figure 21.8 depends on the assumption that adult fish stocks are resource-limited and that the individual growth rate would be reduced at high natural population densities. As population size is reduced, the resources per feeding female are more abundant and reproductive output may be higher. Thus, an intermediate-sized fish population might produce the most fish in the long run. This leads to the argument that fishing a population can increase the total number of fish available in the next generation, because fishing could reduce the depressing effect of high fish density on population growth and reproduction. The level of fishing take that results in a fish density that leads to this maximum yield into the next generation is known as the **maximum sustainable yield (MSY)**.

The MSY model presumes that there is a sweet spot of fishing take. Too much fishing will reduce the stock and may drive it to collapse, which is defined as a reduction

to less than 10 percent of peak stock size. Effective management, therefore, must set a fishing rate that permits removal at a sustainable yield, maximizing recruits into the next generation. The important MSY concept is that some level of fishing will actually enhance fish production.

Body size can be combined in the model. If larger fishes are culled from the population, food will become available to the smaller fishes, which grow more rapidly per unit body weight. We can calculate an intermediate optimal body size for maximum sustainable fishing.

Problems with the Maximum Sustainable Yield Concept

■ **Current thinking suggests that the concept of maximum sustainable yield may be difficult to apply quantitatively in fisheries management, but the principle can be used to create guidelines for catch limits.**

Data supporting the idea of a maximum sustainable yield are limited, and the use of this concept by fisheries biologists in the strict sense has declined over the past few decades. It is still used as a conceptual basis of management, perhaps because the idea of an optimal level of exploitation makes sense and no major approach has replaced this concept. Fisheries managers have to impose limits and have to have some concept upon which to base their decisions.

Maximum sustainable yield depends on the relationship of recruitment to stock, as depicted in Figure 21.8. The decline in recruitment at high stock densities implies that a resource is sufficiently limiting that recruitment will decline as the resource per individual becomes more scarce. As we discussed in Chapter 4, there are two main resource types: space and food. In fish species that require space to breed, as in maintaining a breeding-nesting territory, we can imagine that space might limit recruitment at high adult stock densities. Food might also be limiting, but often it is difficult in fish populations to demonstrate food limitation.

The concept of maximum sustainable yield presupposes a rather precise relationship between population growth and population size. Can this theory be used in practice? First, let's appreciate the problem. Even if the population model is obeyed with precision, we must have accurate measurements of the parameters, and these include population size, reproductive rate, somatic growth rate, and mortality. Larkin (1977) therefore argued that MSY was a concept that was effectively dead for lack of quantitative support. Alternative approaches involve measuring the catch as a function of fishing effort, which can be used indirectly to estimate the optimal fishing pressure.

An equally important problem with the maximum sustainable yield concept involves the range of political pressures that lead to overexploitation (see the discussion on the blue whale fishery that follows). The concept of sustainable yield arose originally as a political effort to protect American fishery interests in international waters (Finley and Oreskes, 2013). In a thoughtful article, Ludwig and colleagues (1993) pointed out that there is a ratchet effect in fishery exploitation. When fish stocks increase, more boats are built and more fishing occurs. However, during periods of constant fish stock, few boats are decommissioned. Then a further period of good fishing years encourages more investment. When the fishery starts to collapse, the societal cost of reducing the workforce or the number of active boats is enormous. Before effective management is possible, the overly large fishery fleet and workforce will cause further collapse of the fishery.

Overfishing and the Food Web

■ Overview: Many fisheries are overfished, but proper management may allow recovery in many cases.

Some recent attempts have been made to understand the big picture of fisheries in the world ocean. This is crucial because fish now constitute about 15 percent of the total protein consumed by humans in the world. A major ongoing report by the Food and Agriculture Organization of the United Nations (FAO, 2016) concluded that most of the stocks of the top 10 finfish species are fully exploited. The proportion of overexploited fish stocks increased from 10 to 29 percent during the period 1974–2008 and at present 89 percent of the world's fisheries are fully or overexploited (FAO, 2016). On the one hand, the results suggest that management could help in recovery of many of the world's fisheries, but on the other hand, the trend is toward overfishing on a global scale. A special concern is the apparent trend toward a plateau of total marine catch in the world, which suggests that we are reaching the limits of exploitation (**Figure 21.10**). Aquaculture production is beginning to surpass the wild catch. These results come from fisheries that are well studied and are also currently being managed. A report that relied on more indirect data, such as trends in catches and studies of the life histories of fishes, suggests that many of the unstudied fisheries also are in danger of severe overexploitation (Costello et al., 2012). The following sections focus on particular groups of fisheries that are more vulnerable because of various ecological traits of the species involved.

■ Open-seas fishing has focused on large carnivores at the apex of food chains.

The size of and extent of fishing fleets coupled with new technologies for locating fish schools have resulted in major reductions of fisheries of whales, certain tuna species, and sharks. We will discuss whales shortly, but large oceangoing predatory fish are often in great danger of collapse by overfishing. Northern bluefin tuna *Thunnus thynnus*, for example, is a highly desirable fish that is used for sashimi throughout the world. As a result, its numbers have plummeted, and catches in the Northeast could not even come close to imposed quotas in recent decades. The use of technologies such as harpoon rifles and hydraulic net lifts has reduced abundant populations of bluefins in the North Atlantic to very low numbers. The worldwide population is believed to have declined by 90 percent since the 1950s, when catches were plentiful. In the North Atlantic, bluefins are important predators of herring, so alterations to the food web are to be expected. A more local example of a smaller predator is the collapse of cod in Atlantic Canada and the Gulf of Maine. In Newfoundland, the collapse was so complete that a moratorium was imposed on fishing in the early 1990s. Overfishing took both large adults and juveniles as bycatch (Myers et al., 1997). There is, therefore, no doubt that some fisheries have been decimated.

■ Overfishing has caused major alterations in the trophic structure of the water column.

We can use our ecological knowledge to predict a fundamental change in water-column communities throughout the world. We have learned that there must, of necessity, be less biomass at the top of a food web, owing to inefficiencies in trophic transfer from trophic levels below (see Chapter 12). Species at the top of the food web, moreover, are usually large in body size, since they are often very mobile predators that must seize and ingest large prey. With low biomass and large body size, top predators are usually few in number.

If fishing focuses on these relatively rare top predators, we should expect fishing collapses to occur much more

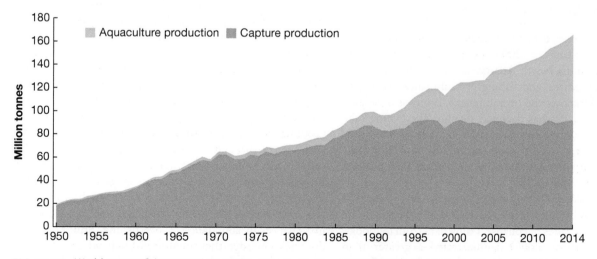

FIG. 21.10   World capture fisheries and aquaculture production from 1950 to 2014. (Data reported by the Food and Agriculture Organization of the United Nations, 2016)

rapidly (too bad we don't prefer to eat bacteria!). Pauly et al. (1998) found just this type of collapse in an analysis of the trophic structure of fisheries. In the past 45 years, fish landings on the global scale have shifted from large piscivorous fishes toward planktivorous fishes and invertebrates lower down in the food web. Pauly has termed this phenomenon **fishing down the food web.** This has led to the development of a **marine trophic index,** a measure of the dominant trophic level, which in turn is a measure of biodiversity (Pauly and Watson, 2005).

Pauly's interpretation makes sense, but an important criticism must also be considered. While it is plausible that removal of top predators leads to inevitable exploitation of lower trophic levels, Essington and colleagues (2006) showed that some of the switch to lower trophic levels has more to do with fishing preferences for certain species at those levels and not just exhaustion of fish stocks at the higher trophic levels; indeed, many tuna stocks are still quite productive. Still, even if many top carnivores might be in better condition than indicated by Pauly's interpretation, there still might be a major ecological trophic cascade effect caused by carnivore reduction.

A focus on one fishery demonstrates the effect of removing a top carnivore. Worm and Myers (2003) examined relationships in the food web of the North Atlantic water column and found a consistent negative relationship between the abundance of cod, *Gadus morhua*, and its major prey, the shrimp *Pandalus borealis*, in nearly all the regions they examined. This result suggests strong connectedness in these food webs. Sometimes cascade effects can be quite unexpected and may come from effects at lower trophic levels. Severe overfishing of sardines in the Benguela upwelling region of southwest Africa resulted in a great increase in jellyfish, which was followed by a population explosion of the bearded goby, *Sufflogobius bibarbatus*, which feeds on jellyfish. The goby switched from feeding on the bottom to the water column. Predatory seabirds, mackerel, and fish that formerly fed on sardines switched to bearded goby (Utne-Palme et al., 2010).

■ Apex predators may be in danger worldwide, with many species facing a major collapse.

How pervasive are declines in apex predatory fish? Christensen and colleagues (2003) used a map-based fishing database from the North Atlantic and reconstructed landings, stock biomass, and fishing pressure since the 1950s. **Figure 21.11** shows the overall results. Biomass of upper-trophic-level fishes has declined steadily. The fishing take increased through to the early 1970s but has declined precipitously ever since. The ratio of catch/biomass is a measure of fishing intensity: It has increased and then reached a plateau since the 1970s.

The late Ransom Myers and colleagues have argued that many tuna and shark fisheries have already collapsed or are within a few decades of collapse. Using data from logbooks, Baum and others (2003) demonstrated a greater than 50 percent decline of most species of sharks in just 15 years and a 75 percent decline for scalloped hammerhead, white,

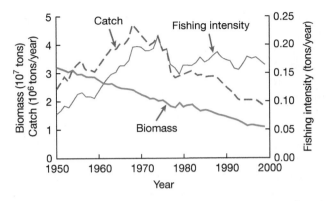

FIG. 21.11 Estimate from a large database of changes of stock biomass, fishing take, and fishing intensity on upper-trophic-level species since the 1950s. (After Christensen et al., 2003)

and thresher sharks (**Figure 21.12**). Overall, 6.4 to 7.9 percent of the total shark population is being taken each year, which is greater than the estimated growth rate of 4.9 percent (Worm et al., 2013). Nearly a third of shark species that have been assessed are in danger of extinction, according to the IUCN. Such evidence of overfishing has led the Convention of International Trades of Endangered Species (CITES) to list five species as requiring permits or being banned for international trade. Much of this decline is due to the harvesting of shark fins for shark fin soup, but new methods of enforcement and changes of cultural practices leave room for hope (see **Hot Topics Box 21.1**).

Myers and Worm (2003) performed a more general analysis of fishing takes of large carnivorous fishes in coastal and open ocean environments and concluded that the biomass of large carnivores exist now at approximately 10 percent of the levels that existed before modern industrial fishing commenced in the 1960s. These are predominantly longline fisheries catching tuna and billfish. The most ominous estimate predicted a complete collapse of ocean fisheries by the year 2048. Worm and colleagues (2006) estimated that by 2003 about 29 percent of all fisheries were at a level of 10 percent of maximums in previous decades. Diverse fisheries might decline more slowly, perhaps because of their ability to switch from one species to another, but species-poor fisheries were collapsing more rapidly. This prediction has been severely criticized, and a more complex view of the state of fisheries is now evolving.

While few fisheries biologists would argue that fisheries are in good condition, there has been some criticism of these collapse scenarios. One question must be directed at the meaning of changes in the size of a catch. Polacheck (2006), for example, examined trends in catches in tuna and found that fishers changed strategies, switching from tuna suitable for canning to tuna suitable for the sashimi market. In other words, the initial decline in catches noticed by Worm and Myers in longline catches might have been largely explained by fishing practices and not necessarily collapses of abundance by overfishing. Still, Polacheck does acknowledge that longline fishing did impose some declines. He argues that the mode of fishing and switches of practice have to be accounted in using landings data to

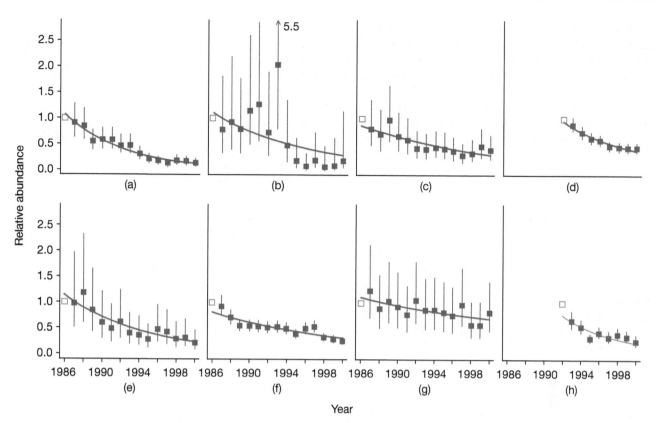

FIG. 21.12 Relative declines among shark species are shown by setting starting abundance at 1. Coastal shark species: (a) hammerhead, (b) white, (c) tiger, and (d) coastal shark species identified from 1992 onward; oceanic shark species: (e) thresher, (f) blue, (g) mako, and (h) oceanic whitetip. Error bars are 95 percent confidence limits. (From Baum et al., 2003)

HOT TOPICS IN MARINE BIOLOGY

Fin-ale for Sharks?   21.1

We discuss in this chapter the great decline of sharks throughout the world in recent decades. Fishing mortality is undoubtedly the main cause of these declines, and they occur in a large number of species, including thresher, blue, oceanic whitetip, and mako sharks. Some of this decline is in parallel with declines in other large fish species, but evidence collected in the past 20 years suggests that a specialized type of fishing for shark fins has been a major driver of shark population collapses.

A major new marketing phenomenon accelerates worldwide fisheries declines and also sometimes thwarts attempts to manage fisheries. There is an old saying, "All politics is local." This can often be said of fisheries: Fisheries management is local! It is tied to population regulation, which is in turn related to spawning grounds, feeding grounds, and other ecological features that are often tied to the conditions of a specific region, such as the shelf of the Gulf of Mexico or the Bering Sea. National law is especially important, particularly since the establishment of large coastal economic zones. Therefore, in order to manage a fishery, we must know fish spawning locations, the ways in which their food is controlled, and other properties that are usually confined to an ocean basin and even sometimes the waters of a coastal nation.

But, once caught, fish may be immediately transported across the globe to a market center that may appear almost overnight. Global sourcing networks bring fish caught all over the world to a remote center,

which may make it very difficult to trace the fish to the source, especially when they are caught illegally. Such contagious spreads to distant marketing centers make it difficult to estimate fishing pressure in any given region, which prevents local fishery agencies from properly managing fishing pressure against what might be a sustainable take.

The market for sharks fits this model very well, perhaps even more so because the marketing center is strongly tied financially and even culturally to the practice of shark finning and the preparation of shark fins for the shark fin soup market. Shark fin soup is a highly valued food, used in major celebrations and often a symbol of the wealth and status of the host. Dried shark fin costs on the order of U.S.$200 per kilogram. Clarke et al. (2006) estimated that in the year 2000, 38 million sharks were taken throughout the world to enter this market.

The fin needles extracted from fins such as the first dorsal fin provide texture to the soup. The three species of oceanic hammerhead sharks have tall first dorsal fins, which have especially high needle content; this makes the species especially valuable for the shark fin trade. Sharks are caught all over the world, usually by enormous longlines with thousands of hooks, and fins are cut from the shark. Often the remaining body is simply tossed back into the ocean because the value of fins is so great. It is extremely hard to regulate shark fishing, much of which occurs in the open sea. For example, the country of Palau in the western Pacific

continues

benefits from tourism by shark-loving divers, and prohibits the taking of oceanic sharks within its fishery zone. But the government of Palau has almost no resources to enforce this law and must patrol a part of the sea about the size of France. Those who fish in such waters and take sharks illegally will rapidly travel to a distant port or transfer their catch to another ship at sea. Then the fins are rapidly transported to marketing centers, but Hong Kong is surely the most important worldwide because of its proximity to the demanding market in east Asia, although wealthy Chinese communities throughout the world spread the demand geographically.

Fishing pressure on sharks is not evenly distributed throughout the world, if only because some species are of far greater value to the shark trade. As a result, some species of oceanic sharks, including three species of hammerheads, oceanic whitetip, and porbeagle, have been especially impacted and are endangered worldwide. So an additional challenge is not only to protect sharks but to pay special attention to those species that are threatened the most. But how do we identify a package of dried shark fin?

In the Hong Kong market, shark fins appear in packages from all over the world. But can we identify the shark species from which the fins originate, and can we develop a technique to locate the shark fin's "postal code" or where it was caught in the world? There are two basic approaches for identification using DNA techniques. First, we can use the DNA barcode, a method using sequencing of the *cytochrome oxidase 1 (CO1)* gene, a mitochondrial gene whose sequences have become a standard for identification of living species. There is an online library of such sequences for nearly all shark species. There are other DNA tests that can also quickly identify shark species. Most important, the use of a number of DNA markers can often be used to identify the geographic region from which the fish came to the central markets. Obviously, it is more effective to intercept the shark fins at the source fishing grounds. There, shark fins might be inspected by local fishing officials or customs agents, before the fins leave the region for a distant market.

The good news is that DNA can be extracted from shark fins at the market, and Demian Chapman and colleagues of Stony Brook University have adapted these techniques to analyze samples for the Hong Kong fin market and even from the needles taken from soup samples in restaurants (see Chapman et al., 2009). While this is theoretically a definitive test for identification of shark species that require protection, it currently seems unlikely that DNA testing will be practiced by fisheries management agencies throughout the world. So it would be highly desirable to identify the fins themselves as they pass into the hands of fish inspectors and customs agents. Debra Abercrombie and colleagues (2013) developed a set of criteria for fin identification that has been converted into

BOX FIG. 21.3 A fin guide developed to allow rapid identification of sharks by combining fin shape, color, and specific traits of the base and forward and trailing edges of the dorsal fin: (a) porbeagle shark; (b) ocean whitetip; (c) hammerhead (Courtesy of Debra Abercrombie).

quick guides, and even cell phone–based apps to be maximally convenient. The web page www.sharkfinid.com provides a set of criteria to identify the three highly endangered species of hammerheads, oceanic whitetips, and porbeagle sharks (**Box Figure 21.3**). This type of guide would allow a customs agent to quickly spot likely declared endangered species under the Convention on International Trade in Endangered Species (CITES) and embargo them for further inspection.

These new techniques raise strong hope that highly endangered shark species can be protected much more effectively than before. DNA testing and fin morphological identification can hopefully be used by fisheries agents in local regions to identify hotspots of illegal fishing for species classified as endangered or threatened under CITES. Even if the fins arrive at the central markets, DNA identifications can still be employed to understand the global pressure on individual species and, occasionally, the local regions from which the fins have been taken. All in all, there is much more hope for the highly endangered species of sharks.

Of course, we can also hope that the culture will change and those who treasure shark fin soup will realize that fishing will eventually collapse the fish at their source. The People's Republic of China banned shark fin soup from public events in 2012 as an austerity measure. Since 2014, news reports suggest that the shark fin trade in China has dropped substantially, owing to a large number of public awareness campaigns. The states of California (2011) and New York (2013) passed a ban on the taking and selling of shark fins and the serving of shark fin soup. All of these changes will prove the most important in reversing the catastrophic decline of these magnificent top predators throughout the world.

estimate population abundance. For example, a switch to fishing using purse seines in shallow water may combine with increased fishing effort to seriously affect the fishery.

■ Fishing pressure at lower levels of the food chain may also produce major ecological impacts.

Although fishing on top predators of large body size is a major concern for fisheries management, fishing targeted at lower levels of the food web may also have serious impacts.

Many fish species—including anchovies, sardines, and menhaden—feed on plankton, and fishing pressure may have strong impacts on the organization of this part of the marine food web. On the Atlantic and Gulf coasts of the United States, the menhaden *Brevoortia tyrannus* is perhaps the largest fishery, whose catch is used mainly for fertilizer (Franklin, 2007). The fish is oily and not regarded as an attractive food for human consumption. Catches off the east coast of the United States increased greatly in the 1940s, and by the 1960s the fishery was strongly impacted, especially in the mid-Atlantic states. Overfishing has resulted in the scarcity of fish older than 5 years, which greatly impacts recruitment, since large fish have disproportionate production of gametes. This species is a planktivore and may have been an ecologically significant consumer before population numbers collapsed. Body size has declined greatly, and mid-Atlantic populations are at a low state. Menhaden also may be the last major prey species available to larger carnivorous fish. Overall, the ecosystem services performed by this species—keeping phytoplankton in check and providing food to top predators—may be worth more than its value as fertilizer. Fishing restrictions have allowed some recovery in northeastern U.S. waters.

Causes and Cures of Stock Reduction

- **Stock reduction can result from random variation as well as from environmental change; fishing would be superimposed on the effects of these factors and appears to cause greater fluctuations than when fishing is absent.**

Fishery managers are concerned with the causes of the great fluctuations in fish populations. Such fluctuations are commonplace in estuarine, shelf, and open-ocean fisheries. One of the most spectacular changes ever recorded illustrates the difficulties in understanding the population changes. In the 1920s and 1930s, the Pacific sardine was landed year after year in the thousands of tons, yet the population declined sharply in the 1940s. After this decline, the anchovy increased greatly in abundance (**Figure 21.13**). A relaxation of fishing pressure did not result in a major recovery of sardines. Competition between the two species, changes in water temperature, and other factors have all been implicated in the switch of dominance, but no factor has been convincingly identified. A similar case can be made for various fish stocks of the North Sea, including herring, cod, and mackerel. Fishing, climate, increased zooplanktonic food for larvae, and pollution have all been suggested as factors, but the one clear fact is the presence of strong fluctuations in population size.

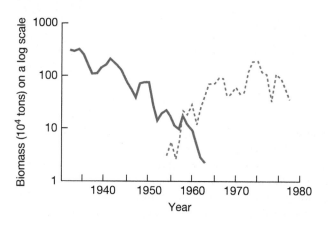

FIG. 21.13 Increase of the anchovy (dashed curve), following the decline of the Pacific sardine (solid curve), off the coast of California.

Fish populations consist of many age classes that grow and reproduce simultaneously. Most species, with the exception of some, such as Pacific salmon, spawn more than once, usually on an annual basis. The individual factors that collectively affect each year class may have a profound effect on population size. Variation in recruitment, for example, may have a great effect on the subsequent age structure. If a given year class is extraordinarily successful, it will appear as a major peak in the size structure of the population and will contribute many more young than will other year classes. Strong fluctuation in recruitment will cause great perturbations in the age structure of subsequent years in the population. Year classes can often be traced through several years as a size peak (as was shown in **Figure 21.6**).

Random fluctuations in recruitment and mortality may be a major background variation in fish populations. Perhaps such fluctuations have specific causes, but these causes may be so complex and varied that they cannot be identified individually, and their effects may be indistinguishable from random variation. Such complex and random fluctuations alone may bring the population down to a very low level. Under such circumstances, the additional imposition of fishing mortality would be quite dangerous for the stock.

Does fishing exacerbate natural population fluctuations? One might expect this to happen, especially when fishing truncates off the larger fish, whose capacity to provide large numbers of young might compensate for failures in success of the new year classes. It was possible to measure this by the use of a long-term data set from the California Cooperative Oceanic Fisheries Investigations, which allowed the comparison of exploited and nonexploited stocks of fishes living in similar environmental conditions. Fished populations clearly were more variable than nonfished populations (Hsieh et al., 2006), which suggests that fishing can cause major instabilities that might lead to local extinctions.

- **Fish stocks characterized by long generation times, small clutches of eggs, and fewer spawnings over time are the most vulnerable to overfishing.**

It is rather easy to see that some fish stocks are potentially much more vulnerable than others, owing to their life-history characteristics. Clearly, fish species with short generation time (which, for our purposes, is the time from birth to age of reproduction), multiple spawnings during adult life, and many offspring produced per female will have a greater chance of rebounding from low population levels. Species with long generation times, few spawnings, and small clutch size may be very vulnerable to combinations of environmental change and fishing pressure. Shark species are clearly vulnerable, as they typically produce very few pups and have typically long generation times, relative to bony fish species (**Table 21.1**). Most species feed at the tops of food chains, and, therefore, relatively small populations can be supported relative to members of lower food chain levels. These general characteristics have resulted in surprisingly rapid declines in shark populations and other top predators such as bluefin tuna.

TABLE 21.1 Life Characteristics of Some Sharks, in Comparison to Atlantic Cod

	WHITE SHARK (*Carcharodon carcharias*)	SANDBAR (*Carcharhinus plumbeus*)	SCALLOPED HAMMERHEAD (*Sphyma lewini*)	SPINY DOGFISH (*Squalus acanthias*)	ATLANTIC COD (*Gadus morhua*)
Age to maturity (y)	M, 9–10; F, 12–14	M, 13–16	M, 4–10; F, 4–15	M, 6–14; F, 10–12	M, 2–4
Litter size	2–10 pups	8–13 pups	12–40 pups	2–14 pups	2–11 million eggs
Reproductive frequency	Biennial (?)	Biennial	?	Biennial	Annual

M, male; F, female.
Source: Data from Klimley, 1999.

■ Fishing technology, boat range, and even fishing policy have initiated or accelerated the decline of many stocks.

Although it may be controversial to predict massive calamity for the fisheries of the United States or the world, there is no doubt that many specific fisheries have suffered greatly from human fishing pressure. In many instances, overfishing has led to drastic depletions of stocks, and fisheries have been closed down as a result. In Great South Bay, New York, a rich fishery of the quahog clam *Mercenaria mercenaria* was overexploited by thousands of clammers who worked the bay with small boats and hand tongs. Owing to this, the landings decreased markedly over the last few decades, and the largest company ceased to fish for clams in 1999. In North America, many fish stocks are now at very low levels, especially on the Atlantic coast. In Newfoundland, cod fishing has essentially ceased, and stocks of bottom-associated fish off New England and Nova Scotia are dangerously low. This pattern has been common in fisheries from the estuaries to the blue ocean. In the past few years, some recovery has been observed in cod on the Nova Scotia continental shelf.

While we have to be cognizant of the difficulties of using landings to estimate population sizes, overfishing can often be identified as a decrease in catch per unit effort. As the stock is overfished, more boats may be deployed, but the fish caught per boat per day decreases. One of the major problems in predicting declines in fisheries is the great increase in fishing effort that occurs in order for a commercial fishery to maintain a given amount of take. Increasing fishing effort over time may result in a continuing stable take, despite an actual underlying decline of stock size. Eventually, no matter how much the effort increases, the take will collapse. This type of decline was found in the Atlantic cod fishery in eastern Canada (Mullon et al., 2005).

One of the most compelling pieces of evidence for depletion of stocks from fishing is the trend of fish landings after World Wars I and II. During both world wars, fishing was understandably reduced. But after each war, the catch increased tremendously. This suggested that the war periods allowed enough time for the fish stocks to recover in numbers.

Overfishing has arisen from the use of long-ranging ships and technological advances that permit efficient catching and preservation. A song captures the old way: "Haul in the nets, same old fisherman, never catches more than he knows he can sell in a day." This pattern ended as large motor-driven trawlers began to ply the seas. After World War II, trawlers threw their nets over the stern. Later the fleets developed the capacity to freeze the fishes at sea, rather than bringing them back on crushed ice. In the years since 1950, a general trend from short-ranging to long-ranging fishing expeditions developed. Off the shores of North America, for example, Japanese and Soviet trawlers represented a dominant part of fishing and certainly took more fish than the small inshore fishing boats that once dominated the coast. Offshore fishing by foreign vessels was concentrated near the highly productive waters just seaward of the continental shelf.

Ironically, protection of U.S. fisheries from foreign exploitation at first did not help at all. In 1976, the passage of the federal U.S. Fishery Conservation and Management Act (also known as the Magnuson-Stevens Act) restricted foreign fishing to the outside of a perimeter greater than 200 miles from U.S. coastlines and established a series of eight regional fishing commissions to help in regulation of domestic fishing. This legislation protected continental shelf fishing grounds from foreign exploitation, and continental shelf species such as haddock began to recover. But no significant limits were imposed on domestic fishers, even as the catches were clearly declining in the 1980s. Some attention was paid to protection of spawning grounds and some size limits were imposed, but the overall effect of management failed to curtail overfishing. The size and number of U.S. fishing boats increased dramatically, and the stocks declined precipitously, to the point that no region of the contiguous lower 48 United States was in good condition, even if some fisheries were on the rise somewhat. In 1994, for example, salmon fishing was curtailed in the Pacific Northwest, and fishing on the Georges Bank of New England was essentially stopped for a few years by federal proclamation in the 1990s. Haddock had declined to very low numbers, but the temporary closure has been followed by a small degree of recovery since 2000. Stocks of yellowtail flounder declined precipitously since the 1970s, but fishing restrictions and recent strong recruitment has resulted in significant recovery (Stone et al., 2004) (**Figure 21.14**).

The Magnuson-Stevens Act has been amended several times with rules designed to establish sustainable fisheries. It was renewed in 2006 and required some more stringent controls and assessment of the health of fish stocks in U.S. waters. Important amendments in 2007 set catch limits for all federally managed fisheries and also established rights-based management approaches, which assigns shares of a

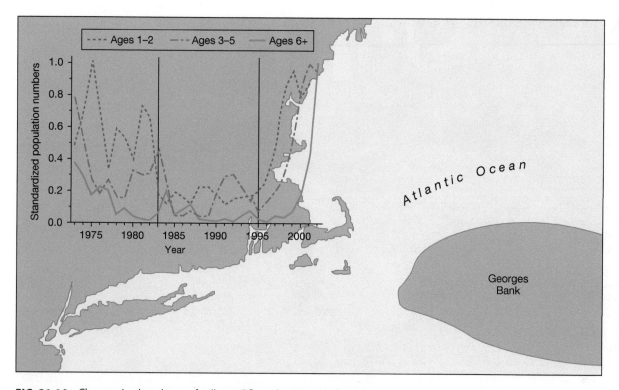

FIG. 21.14 Changes in abundance of yellowtail flounder *Limanda ferruginea*, showing a recovery after restrictions on fishing were imposed in the 1990s. (After Stone et al., 2004)

fishery to past fishery users (see below). As a result, the number of declared severely overfished fisheries in U.S. coastal zones has dropped from 92 to 29 over the period 2000–2015. Management in U.S. waters is finally showing some success.

The trend for coastal zone control of nations has spread throughout the world, and nearly all countries maintain **exclusive economic zones (EEZs)**, within which each country fishes intensively within a border of 200 miles from the coastline. Landings data show that this level of management has not reduced fishing impacts, however; quite the reverse has happened. Despite the difficulties of using landings to estimate fisheries, it is clear that fisheries worldwide, both in the coastal zone and on the high seas, have either reached a plateau or even reduced landings since the 1980s, over a broad range of finfisheries. These results have been accumulated by the Food and Agriculture Organization (FAO) of the United Nations (www.fao.org/fishery) and by the Fisheries Centre at the University of British Columbia (Watson et al., 2005) (www.seaaroundus.org).

■ **Ice retreat caused by anthropogenic climate change is opening up new areas for overexploitation especially in subarctic areas such as the Barents Sea.**

As we discussed in Chapters 3 and 19, anthropogenic climate change is having especially strong effects in high latitudes of the Northern Hemisphere. This has caused retreat of sea ice in many areas especially the Arctic Ocean and parts of the Barents Sea in the North Atlantic region. Cod have exploded (**Figure 21.15**) as their food sources, capelin and krill, have also greatly increased in numbers. The

Norwegian cod catch is shifting steadily toward northern Norway in the Barents Sea as sea ice retreats and waters warm. Agreements between the Russian and Norwegian governments have led to successful regulation of fishing effort on these newly expanded populations of cod and other fish species. Unfortunately, the cod explosion has also led to a greatly expanded bottom trawler fishing effort in the northern Barents Sea from other countries, which may have major negative impacts on the seabed and demersal fish populations. Total benthic biomass is vulnerable to bottom trawling. But epibenthic species—such as *Geodia* sponges, basket stars (*Gorgonocephalus*), sea pens (*Umbellula encrinus*), and sea cucumbers (*Cucumaria frondosa*) (Jørgensen et al., 2015)—are especially vulnerable. Norway has included the island of Svalbard in its exclusive economic zone of 200 km and has some ability to regulate the impending devastation by such intense trawler fishing. Cod fishers have also agreed to curtail, for now, fishing in areas newly opened by the retreat of sea ice. Many major retail purveyors of fish have recently signed an international agreement, stimulated by the efforts of the Greenpeace organization, to refuse to sell cod taken by bottom trawlers in these waters, which may slow down the expansion of bottom-trawling fisheries into this newly opened area where high-latitude sea ice is retreating owing to climate change.

■ **Jellyfish blooms around the world may be the combined result of overfishing, climate change, and pollution.**

The consequences of overfishing can be predicted through an understanding of the role of the missing fish in the food web. This has been one approach in understanding the removal of large sharks from the top of food webs and the

FIG. 21.15 A bottom trawl net of cod being loaded onto a fishing ship near Svalbard, Norway, July 2016. Black steel otter boards that hold the net open as the trawling net moves along the sea bed are visible. (Photograph by Carl Safina)

resulting trophic cascade effects. But many other responses are more complex. First, fish are usually removed by fishers from food webs at several trophic levels. Second, other factors such as ocean warming and pollution must be factored in. An extreme example of this is the invasion of the ctenophore *Mnemiopsis leidyi* into the overfished Black Sea. This species entered a sort of ecological vacuum and its population exploded, causing a large degree of mortality on native fish eggs.

Similar complexities have accompanied the widespread increases of scyphozoan jellyfish populations around the world. For example, jellyfish of various species have increased dramatically in recent decades in the Gulf of Mexico and in the extremely productive Bering Sea near Alaska. In the Mediterranean, very large fluctuations of the jellyfish *Pelagia noctiluca* have been known for many decades, and high population densities are associated with higher sea temperatures and atmospheric pressure. Jellyfish abundance is also known to strongly correlate with climatic fluctuations such as the North Atlantic Oscillation (NAO). In Chesapeake Bay, jellyfish do well when the NAO index is low, which corresponds locally to higher water temperatures and slightly higher salinities. In coastal waters worldwide, the moon jellyfish *Aurelia aurita* has become much more abundant in recent decades. Such increases must impact marine food webs, since jellyfish are efficient predators on zooplankton.

The reasons for such explosions are not so clear, but many fisheries biologists have argued that overfishing has removed planktivorous fishes and jellyfish have replaced them. But jellyfish blooms also are generally correlated with warm temperatures and oscillate in numbers that correspond to climatic oscillations. Abundances may be controlled by regional climate oscillations, and the general global warming of the surface ocean may also contribute strongly to the increases. Finally, pollution may be altering coastal ecosystems to favor jellyfish, but the exact mechanism has not been determined with certainty. Right now, we have only correlations. Ocean warming may also be responsible for shifting the geographic ranges of jellyfish species to higher latitudes (Mills, 2001; Purcell, 2005).

Cures for Overfishing?

CLOSURES AND QUOTAS

■ **Temporary closures and fishing limits allow some fisheries to be sustainable.**

If overfishing occurs, the question is what actions will help. Most fishing nations have laws in place that set limits to the **length of season** and **size of catch** for different fisheries. These limits are often set in the context of strong political pressure to have no limits at all or not to impose very strict limits on the size of the catch. In the United

States, legislation in recent years has prescribed the means to regulate fisheries to be sustainable. Of course, we have discussed a number of cases in which such limits have not forestalled major fishery collapses—or even complete re-organizations of ecosystems. Imposition of limits often occurs only after a major disaster occurs and fish are so scarce that they cannot be recovered economically. Such was the case of the haddock on Georges Bank, where legislation and closures occurred only after major collapses in the 1960s (from foreign fleets) and the 1980s (from U.S. domestic fleets). Many shellfish resources, such as the eastern oyster *Crassostrea virginica* in Chesapeake Bay and the northern quahog in Great South Bay, Long Island, New York, have been fished down to a near-hopeless state, so limits are sometimes too late. On the other hand, several northern Pacific U.S. fisheries, such as the Pacific halibut *Hippoglossus stenolepis* and several species of salmon, have been successfully managed. Pacific halibut in Alaska is divided up in shares, which reduces overfishing, and there are strict limits to catch and bycatch.

■ Catch shares, or individual transferable quotas, might produce a sustainable fishery.

To sustain a fishery is quite difficult, given the strong natural fluctuations of natural populations against which additional fishing pressure must be regulated. As we have mentioned, merely excluding some groups from fishing does not usually solve the problem, as other groups fill in the void. Fishing furthermore is becoming more and more sophisticated, with satellite navigation, sonar detection, and other technologies aiding exploitation.

An interesting philosophy now being used in fisheries is **rights-based management,** in which traditional users of a fishery are declared to have use rights, which can be described as measurable commodities. By assigning these rights, one expects that fishers will adopt long-term sustainable policies, such as catch limits. One interesting approach to this type of management has been the **catch share,** or **individual transferable quota (ITQ) system,** in which a total catch is divided among a series of individual fishers who can sell their personal quotas to other fishers. This approach has been applied in New Zealand, British Columbia (Canada), Australia, Iceland, and Namibia, particularly to invertebrate fisheries, and it might give the economic flexibility needed in deployment of fishing boats and gear (Shotton, 2001). It has also been applied with success to Pacific halibut in Alaska and to ocean clams on the east coast of the United States. Currently, over 15 U.S. federally regulated fisheries are divided into catch shares. Some critics argue that the system is unfair, since it offers shares in a fishery to a group of participants at a specific time, and these participants can make a profit by selling shares to increasingly centralized large fishing corporations, which reduces competition. A recent analysis, however, shows that the approach works well in helping to avert crashes from overfishing (Costello et al., 2008). Participants with shareholder rights have a stake in obeying rules that lead to the sustainability of a fishery if the

participants will benefit from the catch in the long run. This approach is believed to be behind a recent resurgence of a number of fisheries, especially along the Pacific coast of the United States. Regulation has also greatly reduced bycatch in Pacific groundfisheries.

ECOSYSTEM-BASED MANAGEMENT AND MARINE PROTECTED AREAS

■ Ecosystem-based management may allow the environment of a fishery to sustain resource populations.

We discussed ecosystem-based management in Chapter 20. This approach considers ecosystem interactions as the basis for fisheries management. In essence, the protection of an entire ecosystem is believed to maximize biological interactions that have sustained important fishery populations. For example, the diversity of the seabed would be considered because of the role of a variety of species of benthic organisms as food for finfishes. Thus, the damage caused by bottom trawling is not just the removal of fish but the collateral disturbance of the seabed. The negative effects of invasive species must also be considered. For example, a recent invasion of Georges Bank bottoms by an abundant sea squirt *Didemnum* sp. may have strong negative impacts on the availability of benthic invertebrates as food for benthic-feeding fishes. Consideration of the ecosystem as a whole allows a more accurate economic assessment of the ecosystem services provided by any given ecosystem. Marine protected areas can be established to maximize ecosystem functioning. Many have argued that ecosystem-based management should replace management practices based on individual fisheries alone (Crowder et al., 2008).

■ The marine protected area concept can be adapted to fisheries management.

In Chapter 20, we discussed the use of marine protected areas (MPAs) for the conservation of biodiversity. They can also be used for fisheries management. By designating a specific area for protection, one must assign a value to that area for fisheries management. These values include the following:

1. *Crucial habitat for refuge population.* A fishery species might be tightly dependent on a specific habitat. For example, in Chapter 18 we discussed deep-water coral mounds, which are important habitats for a number of species of fish valuable as fisheries. MPAs might allow the preserval of crucial habitat upon which a fishery population might depend. This refuge allows the population to be sustained because a minimal amount of habitat is not destroyed, allowing feeding or reproduction.

2. *Protection of an area where an important life-history stage must live.* An MPA might be erected to allow a resource species to spawn, or for juveniles to live, or for adults to feed. Thinking on this large scale, a network of habitat-related MPAs might have to be

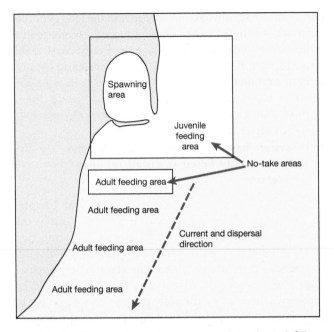

FIG. 21.16 Design of no-take areas in a hypothetical fishery. The spawning and juvenile feeding areas and one adult feeding ground are no-take zones, to allow a minimum population to complete a life cycle. Other areas might be fished, although quotas might be set on the amount of the take.

established to protect a minimal population. This network acknowledges that a metapopulation exists, and the nature of connectivity is important for life-history stages and movement of adults. Such a network would have to take proper habitat types into account, along with water currents for dispersal routes for eggs and larvae (**Figure 21.16**).

3. *Creation of a no-take sanctuary, which individuals might leave and be exploited outside.* If an MPA is large and safe from exploitation, one might expect that mobile species might survive well within the MPA and a **spillover effect** would result from their emigration from the MPA, where they could be fished outside. For example, the coral reefs of the Dry Tortugas National Park ecological reserve in Florida has proven to be a good place for fishing around the periphery of the preserve because individual fish emigrate from a now healthier population within the sanctuary that is closed to commercial fishing (Ault et al., 2013).

4. *Creation of a network of MPAs, among which larvae and adults can disperse.* Along a coast, a network of MPAs may be preferable, to maximize dispersal and successful recruitment as larvae and juveniles are carried by currents. A group of no-take MPAs have been recently established along the California coast. It is crucial to manage both the MPA areas and surrounding areas that are fished, because the establishment of no-take zones might just increase fishing pressure elsewhere, unless the whole ecosystem is managed.

Overexploitation of Whales: A Case History

■ **Whaling began as a shore-based fishery and then developed into an open-ocean fishery.**

Whaling is one of the most romanticized of human endeavors. In our folklore, a sinister and powerful monster appears suddenly on the horizon, and the brave men of the longboats move out after it, perhaps to their death. In the nineteenth century, American whaling ships ranged far and wide, from the coast of Greenland to the antipodes, searching for and killing the profitable beasts. In those days, the sperm whale was sought for its oil for lamps and candles, bones for household implements, and ambergris for perfume. Even today, the waxy substance spermaceti, from the head cavity of sperm whales, is used as a lubricant in outer space, because of its unique stability at low temperature. Other kinds of whale were boiled down for oil, and their flesh might be taken for food. "Whalebone," a bonelike material taken from baleen whales, was prized for women's corsets, buggy whips, and umbrellas.

At first, whalers worked from shore. Fishermen hunted the Biscayan (northern) right whale nearly to extinction as early as the seventeenth century, merely with harpoons thrown from skiffs that quickly returned to shore. On the south shore of Long Island, white settlers and natives harpooned whales from rowboats launched in the surf, and northern right whales were preferred because of their slow swimming speeds (they are baleen whales) and their tendency to keep close to shore. With the advent of large and swift sailing vessels in the early nineteenth century, voyages often lasted for several years, and larger open-ocean whales were hunted and processed at sea. Whales were sighted in the distance as water spouted through the blowhole ("Thar she blows!" was the cry of the spotter, when he saw a whale spouting), and men in longboats rowed out to harpoon the whales by hand. After the kill, the whales were tied along ships and were butchered and cooked on-board. If the killing was a dangerous outing, neither was it exactly safe to stand on the slippery, blubbery carcass and butcher the enormous animal.

We can only imagine the fantastic numbers of whales that the first Antarctic explorers must have encountered, for such numbers exist no more. At first, in the earliest part of the nineteenth century, whales migrating from the Antarctic were caught when they arrived in some of their breeding grounds, south of the Australian continent. The whales were hunted by convicts, who had been transported from England, and also by whalers who had traveled from North America. Shore stations were especially effective in reducing the numbers of the humpback whale, which bred in bays in Australia and New Zealand.

A number of technological advances set the stage for modern whaling in the Antarctic seas. In the 1860s a Norwegian sea captain invented the **cannon-powered harpoon**, and whaling crews began to pump air into carcasses to keep them afloat during butchering. Also, a technique of hunting was developed that was devastating in its efficiency. A series of smaller catcher ships searched out whales. When an individual whale had been harpooned and killed, it was delivered to a larger **factory ship**, equipped with a **stern slipway** for hauling in the whale. The whale was then cut up and processed on-board for oil. (That was the main product obtained from Antarctic whales until after World War II, when whalers began to save the meat for use in pet food and for human consumption.) Even the canning was done on-board the factory ship. This method was first fully developed in about 1925, when the first factory ship equipped with a stern slipway operated in the Antarctic Ocean.

■ Open-ocean fishing technology resulted in the decline of blue whale populations.

The effectiveness of the catcher–factory ship system led to hunting for the blue whale, *Balaenoptera musculus*, in preference to the other species (fin and sei whales), owing to its large size (**Figure 21.17**). By the 1930s, blue whales were already declining. This decline continued until blue whales eventually reached very low numbers. A 1937 agreement set a minimum size limit on hunting blue whales and other species, and prohibited the killing of whales that had calves. A season was also set. No upper limit was placed on whaling in general, however, and the number of catcher ships increased, with no increase in the total blue whale catch. In other words, the total landings remained the same, but the catch per unit effort decreased. Subsequent international whaling conferences set limits on the number of catcher ships, whaling season, size, and number. A blue whale unit (BWU), established in 1944, set the following catch equalities: 1 blue whale = 2 fin whales = 2.5 humpback whales = 6 sei whales. Limits were set in terms of these blue whale units. This was not particularly good for the blue whale itself because whalers included in their limit as many blue whales as possible. The BWU was thought necessary to protect whales in general on the grounds that hunters would not be so impractical as to pass up any whale simply because it was not a blue whale. To do so would have wasted too much ship time.

The **International Whaling Commission (IWC)** was established in 1946. It had the advantage of forcing the representatives of whaling nations to continue to meet to establish quotas and fishing seasons. The charter of the commission permitted nations to ignore the limits, however. This laxity was a political necessity because whaling was a high-seas fishery in the Antarctic and therefore no nation's law regulated fishing. This led to accommodations that were political in nature, and the limits were set above a sustainable yield. The commission failed at first to set limits for each individual nation, although this was later done by means of negotiations outside the formal proceedings of the IWC. The early lack of individual country limits encouraged very intensive fishing effort. The quotas recommended by a scientific committee of the IWC were often ignored by the broader commission in favor of ones that could not even be met by the whalers.

Over time, the catcher ships continued to increase in number and size, and the blue whale stocks declined precipitously through the 1950s. Individual national limits were set in 1962, and fishing for blue whales was stopped in 1965, but blue whales had already ceased to be abundant enough to be taken in significant numbers. By the early 1960s, it had become impossible for the whaling nations to catch as many whales as allowed by the quotas adopted by the IWC. The fin whale also began to decrease greatly, and

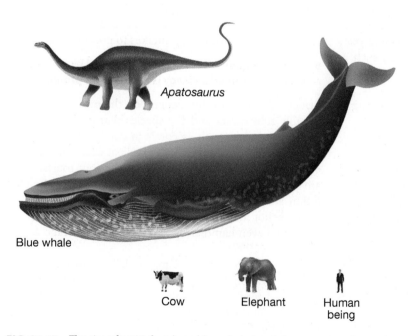

Apatosaurus

Blue whale

Cow   Elephant   Human being

FIG. 21.17   The size of a 100-foot-long blue whale in relation to some other creatures.

fishing effort was shifted to sei whales, which were useful for their meat. Quotas were reduced, but not enough to permit the fin or blue whales to recover very well.

Many pointed out the dangerous position of the blue whale and fin whale stocks, and in 1986 an agreement was finally reached by the IWC that imposed a moratorium on whaling. Of the great whale species, five species are considered endangered, two are vulnerable to declines, and four are at low risk of extinction. The northern right whale is endangered and female survival has declined in recent decades, setting the small remaining population of this species on a course for extinction (Fujiwara and Caswell, 2001). Both Norway and Japan have objections to the moratorium, especially with regard to the minke whale. In 1994, the IWC established a sanctuary throughout much of the Antarctic Ocean, but Japan objected and continues "scientific" whaling, even within the sanctuary. Nevertheless, within the Antarctic minke whales have greatly increased and fin and sei whales are more abundant. In the eastern North Pacific, humpback and blue whales have increased in numbers owing to regional protections.

Other Types of Loss

■ **Fish stocks can be affected by human degradation of water quality or fish habitats, or by killing of fish by means other than direct fishing.**

Although overfishing began to take its toll early in this century, other factors have combined to reduce fish stocks. In many bays and estuaries, the introduction of industrial wastes and sewage has degraded water quality substantially (see Chapter 22 for more details). As a consequence, many fisheries have collapsed, or have become contaminated and therefore unavailable for human consumption. In the most dramatic cases, pesticides have eliminated entire populations. For example, the manufacture and release of the insecticide Kepone into the James River of the Chesapeake Bay region collapsed the blue crab fishery for several years (Schimmel et al., 1979). Other toxic substances, such as heavy metals and various organic compounds, have contaminated fish and shellfish populations. Probably the major water-quality change, however, is in the lowering of dissolved oxygen content (see Chapter 22).

Human disturbance of either spawning grounds or migration routes is also a major cause of the decline of fish populations. In the Pacific Northwest region of North America, dams and deforestation have caused major reductions in salmon stocks (**Figure 21.18**). Dams interrupt the migration route, and juvenile salmon often swim through hydropower turbine intakes, causing extensive damage and mortality. Migrating juveniles often prefer the sluggish parts of lakes behind dams, and these areas are often filled with predators. The extensive clear-cutting of forests in areas such as British Columbia also has drastic effects because trees normally prevent erosion. Following a period of clear-cutting of trees, soil erosion increases greatly, and gravel spawning beds are often ruined by influxes of soft sediment. In many coastal areas, salt marshes have been

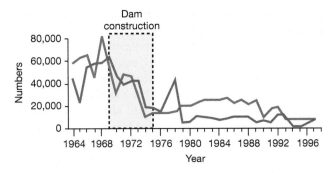

FIG. 21.18 Wild salmon counted on the upper Snake River, Washington: red line, spring–summer Chinook; green line, summer steelhead. (From the Idaho Department of Fish and Game)

filled in to allow coastal construction. The salt marsh creeks are nursery grounds for a variety of fishes, and many of these areas have been eliminated in the eastern United States. Damming of rivers has greatly impeded migration cycles of many species of sturgeon, whose populations have declined throughout the world's estuaries. Climate change may also disrupt migration routes, as shifts of temperature and currents cause changes in coupling of productivity hotspots and the occurrence of forage fish for major oceanic migrating predators (Hazen et al., 2012). A recent removal of the Elwha dams on the Olympic Peninsula of Washington State in 2011–2012 has been followed by increased salmon runs and the reintroduction of marine-derived nutrients into the freshwater system (Tonra et al., 2015).

Many activities cause reductions of fish populations. The intakes of power plants are often the site of extensive larval and adult fish mortality. Water is taken in to cool the turbines, and warm water is piped out into estuaries and coastal areas. The pump intake region entrains fishes and draws them through the intake pipes. When the larvae or adults approach the intake pipes, the sucking force may impinge them on the intakes, although a variety of diverting screens have been placed on intakes to prevent this.

■ **Structural habitats are often endangered by human use. A multitiered strategy is essential for protection.**

Structural habitats often depend on the maintenance of a suitable substratum and the species that construct a biological landscape, such as corals in coral reefs or kelps in kelp forests. Coral reefs are an excellent example. They are endangered from many directions today, but tourist visits are especially worrisome. With tourists come organic pollution and direct disturbance of the reef through diving, boat anchor damage, taking of rare live marine specimens, and even hammering and blasting out coral colonies (Luttinger, 1997).

Temperate rocky reefs are an especially appropriate habitat for conservation, because many fish species use the reefs to breed and feed. They are also targets for sports and commercial fisherman and are therefore highly vulnerable to overfishing. In recent years, a system of protected rocky reefs has been established in California (see Gleason et al., 2013) and along the western coast of the Gulf of California.

By protecting a network of reefs, dispersal among reefs tends to increase the regional population.

Disease as a Major Danger to Coastal Fisheries

Coastal fisheries often consist of species with very high population densities, at least before they are overexploited. Disease is a major factor in fisheries decline and, in some cases in recent years, has proven to be a major challenge in fisheries management, especially in cases where overfishing has occurred. Although a great deal of attention has been paid to diseases of fishes and invertebrates (discussed shortly), we often know very little about diseases that affect natural populations.

In recent decades, bivalve mollusk populations have often been devastated by a variety of diseases that cannot be controlled to any extent. For example, the eastern oyster *Crassostrea virginica* has been infected by two common diseases that exert strong negative effects on populations. MSX is caused by an amoeboid parasite *Haplosporidium nelsoni*, whose complete life cycle is unknown but is believed to occur in some other unknown benthic species. It originated in the western Pacific and was first discovered by Harold Haskin in the late 1950s in Delaware Bay when it caused well over 95 percent mortality. Since that time it has spread throughout the East, including Chesapeake Bay and southern New England. Another oyster disease, Dermo, combines with MSX to be a major factor in the survival of natural oyster reefs. Ocean warming has facilitated the movement of this disease northward along the northeast U.S. coastline.

The interaction of climate change and a parasite is also a factor in strong declines of the blue crab *Callinectes sapidus* in the southeastern United States. For example, landings in the state of Georgia of blue crabs dropped over 75 percent from the 1950s to 2002. Infections of the dinoflagellate *Hematodinium perezi* have greatly increased locally and worldwide (Sheppard et al., 2003). The parasite invades the hemolymph and consumes the crab's blood pigment hemocyanin, which causes oxygen shortage. The infections are prevalent in summer, and the crabs do not have a sufficiently effective immune system to attack the dinoflagellate cells. The parasite does poorly in lower salinities, but the drought in the southeastern United States since 1997 has reduced freshwater flow and allowed the parasite to flourish in salinities over 28 psu, which are now common way up estuaries with low river flow (Lee and Frischer, 2004).

Mariculture

General Principles of Mariculture

■ **In mariculture, the habitats of some natural populations can be simulated, changed for convenience of harvest, or enhanced to increase yields.**

Mariculture, often called aquaculture, is a loose term covering all techniques by which marine organisms are reared for most of their lives under controlled conditions in seawater directly in or connected to the sea. In the crudest form of mariculture, animals are transplanted to habitats that are optimal for growth (e.g., transplantation of hatchery-reared oysters to the seabed). In intermediate cases, organisms are reared throughout their life cycle, but much of the rearing is in open-ocean pens, and conditions are not controlled exactly (e.g., fish farming of salmon). In the most controlled case, the mariculturist controls the rearing environment completely and may attempt to provide completely defined foods (e.g., indoor aquarium hatcheries of striped bass). While the fishing take of wild fish has remained stable or is in decline worldwide, aquacultured fish and invertebrates continue to increase as an important part of total world consumption.

Successful mariculture requires the proper choice of a species for rearing. The following characteristics are desirable:

1. *Desirability as food.* The species should be already known as a desired food item or product.

2. *Uncomplicated reproduction.* The organism should be relatively easy to propagate, or young organisms should be easy to obtain.

3. *Hardiness.* The species should be resistant to handling and changes in environmental conditions.

4. *Disease resistance.* Diseases and parasites should be controllable, to minimize mortality.

5. *High growth rate per unit area.* The organisms should be able to grow rapidly in limited culture areas.

6. *Readily met food requirements.* Feeding the organisms should be easy and cheap. Animals that are high on the food chain are liable to require higher-cost protein foods.

7. *Readily met habitat requirements.* The physical habitat should be easy to duplicate in the mariculture system. There should be relatively low levels of aggressive behavior, and the organisms should be resistant to poisoning by waste products.

8. *Monoculture or polyculture.* It should be considered whether one species is to be grown alone (monoculture) or whether several species will grow most efficiently when placed in the same system (polyculture). For example, some mariculture habitats may be innately complex, and a polyculture, permitting several marketable species to grow in the assemblage of microhabitats, would be more efficient. Monocultures have the disadvantage of the rapid spread of disease that is especially efficient in attacking a single species.

9. *Marketability.* The chosen species should be easy to market, accessible to markets, and of a presentable growth form to consumers.

10. *Minimal ecological side effects.* The mariculture system should have few detrimental effects on the surrounding environment, such as release of cultured organisms into natural habitats.

■ Mariculture may be uneconomical, especially because of some ecologically damaging side effects.

When these requirements are considered, many species turn out to be uneconomical to culture. The New England lobster *Homarus americanus* is a highly prized food, but strong aggression among individuals requires that they be isolated, which increases the expense of rearing them. In contrast, the shrimp *Macrobrachium rosenbergii*, which thrives in both brackish and fresh water, proved ideal for culture in open ponds, as have marine shrimp species. Food is easy to prepare (brine shrimp eggs, fish flesh, etc.), survival is satisfactory, and the animals grow rapidly. Unfortunately, side effects of shrimp farming can be severe. Throughout the tropics, thousands of hectares of mangrove forests and marshes have been destroyed in order to establish shrimp farms. Under crowded conditions, a number of lethal viral diseases may spread and wipe out large shrimp mariculture facilities, which are often permanently abandoned (Kautsky et al., 2000).

In many species, mariculture-reared individuals are released purposefully into natural habitats. In some of these cases, mariculture produces juveniles that escape into coastal fisheries. The large number of Atlantic salmon farms has resulted in a great deal of inadvertent release, and wild salmon fisheries in rivers on both sides of the Atlantic have a considerable number of fish farm–bred salmon (Carr et al., 1997). Salmon raised in pens experience selection for traits that are not conducive to living in a wild environment, so there is a great deal of concern that crosses between farm-bred salmon and wild individuals will produce offspring of lower fitness in the open ocean. Parasites such as sea lice (a type of copepod) spread in fish farms, and planktonic early stages infect wild salmon.

Some Organisms Useful in Mariculture

■ Mollusk mariculture systems enhance the availability of substratum and are located in areas of high phytoplankton supply.

Mollusks, principally bivalves, have been among the most successful mariculture organisms. This has much to do with the rapid growth of some species, combined with the ability to place animals in areas of high phytoplanktonic food supply. All the major cultured species are suspension feeders. A number of species of oysters and mussels have been the mainstay of the industry throughout the world.

The culture of bivalves involves two main steps: (1) rearing of larvae through settlement and metamorphosis into juvenile animals, called spat or seed, and (2) rearing of adults, usually in rafts or attached to suspended poles or ropes. Spat can be collected in natural habitats, but they are often reared in hatcheries.

MUSSELS Mussel culture is best developed in France and became established quite by accident. Patrick Walton, an Irish sailor, was shipwrecked in the year 1235 on the Atlantic coast of France. To snare birds for food, he placed a net attached to poles on a mudflat. Instead of attracting birds, the poles were colonized by settling mussel larvae, which grew far more rapidly than those on the mudflat itself. Today, ropes are still placed near mussel beds, and larvae settle from the plankton on the ropes and grow to a size of 5–10 mm. The ropes are then wrapped in a spiral around poles stuck in the mud, and the mussels then grow rapidly. The bottom of the pole is sheathed in smooth plastic, to prevent predators from climbing the poles. The mussels are thinned out to allow maximum growth, and the harvest is taken about 1 year later.

A variant of this technique, practiced in Japan, Spain, Portugal, and Maine, is to use ropes suspended from rafts. There is some concern that deposition of mussel feces below rafts will cause anoxia on the bottom and negatively affect local marine communities. Mussels are also widely cultured throughout the Mediterranean, especially near estuaries where primary productivity is high.

OYSTERS The oyster is the most profitable bivalve employed in mariculture (**Figure 21.19**). Many different species have been used successfully in a wide variety of climates. Oyster culture is known to go back to the ancient Romans. In the past few decades, oyster culture has been expanded, owing to the high price of oysters and to the decline of many natural oyster beds because of overfishing, disease, and pollution.

Cultured oysters (members of the family Ostreidae) have planktotrophic larvae that swim in the water column for a few weeks. Culture therefore involves the settling and metamorphosis of larvae and the rearing of adults. Spat either are reared in hatcheries or are collected on hard substrata that are placed in natural oyster beds. Dead shells are a common substratum, but plastic, ceramic materials, and wood are also employed. Once spat have been collected, they are transferred to the adult growth area.

Oyster mariculture is complicated by abundant predators, disease, and the problem of maintenance of a natural oyster reef. Oysters can also be grown in cages, which protect them from predators but fail to deliver ecosystem services of increasing biodiversity, among others. The susceptibility of the world bivalve aquaculture industry to ocean acidification is discussed in **Hot Topics Box 21.2**.

OTHER BIVALVES A number of other bivalve species are cultured profitably with aquaculture methods. The common east coast and Gulf coast bivalves *Mercenaria mercenaria* and *M. campechiensis* are popular in Florida, both in the Indian River region on the Atlantic coast and the Cedar Key region of the Gulf coast, and now comprise a multimillion-dollar industry. Another growing industry in the Pacific Northwest focuses on the geoduck, *Panopea generosa*, whose large size (1 kg or more) and use in the sushi industry makes it a profitable species. Juveniles are planted in sediment with a surrounding PVC tube to protect against predators and are allowed to grow for 5–7 years before harvesting, usually in intertidal sites. A most interesting aquaculture industry is farming of the giant clams, a group of species belonging to the genera *Tridacna* and *Hippopus* and found in the

FIG. 21.19 Oysters (*Crassostrea gigas*) are cultured in racks suspended from floats in a tidally flushed bay (top) in San Juan Islands, Washington. Newly settled spat are sorted (lower left) to get the fastest-growing individuals (middle right), and these are raised in suspended plastic trays until ready for market (lower right). (Photographs by Jeffrey Levinton)

HOT TOPICS IN MARINE BIOLOGY

Shellfisheries: Which Will Fall to Ocean Acidification?   21.2

Since about 1750, approximately 337 billion metric tons of carbon have been released by the burning of fossil fuels and the manufacture of concrete. About 35.3 billion tons were released in 2013. The atmosphere's CO_2 concentration has increased from 278 parts per million in 1750, to over 400 today. As discussed in Chapter 3, the totality of this release has had profound effects on our climate, raising air and ocean temperatures in most parts of the earth, and will continue to do so. But another effect of CO_2 emissions also portends profound environmental changes. About 26 percent of all CO_2 emissions enters the ocean. In Chapter 3 we showed how this addition affects the acidity of the ocean, which will lead to areas that are undersaturated with respect to calcium carbonate skeletons—for example, the shells of bivalves, the skeletons of pteropods, and the skeletons of corals. The oceanographic community is now hard at work assessing the potential damage to marine

continues

Shellfisheries: Which Will Fall to Ocean Acidification?   *continued*   21.2

species in the acidic ocean of the future. While pH has declined so far by a small amount on average throughout the global ocean, some marine communities are on the front lines of the threat of increasing ocean acidification.

Put simply, **upwelling systems** are at the head of the brigade of global threats from acidification. We can see this unfortunately already along the west coast of North America, where incidents of upwelling centers approaching the coast are not necessarily more frequent but appear to be more intense (Howes et al., 2015), which will exaggerate episodes of introduction of hypoxic and acidic water into the coastal zone, endangering calcifying organisms.

The process is fairly simple. A general zone of coastal upwelling exists along the west coast of North America from California to Alaska. The upwelling events occur every few years and may be far offshore or sometimes right over the relatively narrow continental shelf. In deep waters of upwelling systems, organic matter accumulates and oxygen levels are low because of microbial consumption of dissolved oxygen during decomposition processes. As the upwelling center approaches the coast, low-pH–low-oxygen water upwells to the surface.

The stress from upwelled water combines with an ever-increasing load of nutrients from the coastal zone, owing to inputs from populated centers and agricultural sources on the coast. The coastal input of nutrients, **eutrophication**, causes increased primary productivity, and decomposition of uneaten phytoplankton further reduces the oxygen concentration in the shallow coastal water column. Thus, we have a dual threat to calcifying organisms: acidification and low oxygen stress. But there is also an additional acidification stress in some areas, since bacteria decay uningested phytoplankton in the enriched coastal waters, producing even more CO_2 and potential acidification.

This dual stress especially has the potential to affect the larval stages of shellfish. Christopher Gobler and colleagues (2014) showed that the combination of low oxygen and high dissolved CO_2 produce both additive and interactive effects that reduce larval survival of east coast bivalves. In an east coast scallop *Argopecten irradians*, both low oxygen and low pH reduced larval survival but their combined effect was additive. But in the quahog *Mercenaria mercenaria*, early stage larvae were affected in the same way, but later stage larvae were not so affected by individual exposures to either low pH or low oxygen. In combination, however, there was some interactive effect that increased mortality. Unfortunately, it is not clear why effects are additive or interactive. Still, we can see that both low dissolved oxygen and low pH in combination are potentially harmful to larvae.

The occurrence of acidic, low-pH water is increasing along the west coast where intense upwelling is an important factor. A study by Booth and colleagues (2012) showed that episodes of intrusions by cold, low-oxygen and low-pH waters into the coastal zone have been increasing in frequency in central California. On the larger scale from California to Washington State waters, such upwelling events have very strong impact on pteropods, a major part of the zooplankton that produce shells made of aragonite. Upwelling events bring acidic water that is undersaturated with regard to aragonite, and the pteropods are frequently very damaged, with obvious evidence of shell corrosion (Bednaršek et al., 2014). Near-shore pteropods are far more damaged in frequency than those from offshore. The large California Current ecosystem has experienced a sixfold increase of undersaturated waters relative to preindustrial conditions (**Box Figure 21.4**).

Along with strong effects on the planktonic pteropods, bivalve shellfisheries in the Pacific Northwest have also been strongly impacted

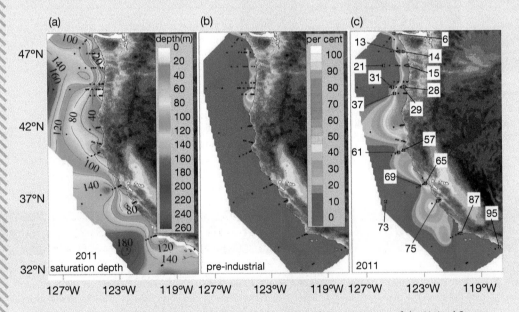

BOX FIG. 21.4 Conditions for aragonite deposition in shells off the west coast of the United States are now poor. (a) Depth of the aragonite saturation horizon. Below this depth, it should not be possible for a pteropod to make an aragonite shell. (b) Percent of the upper 100 m of the water column estimated to be undersaturated in pre-industrial times. (c) Percent of the upper 100 m of the water column estimated to be undersaturated in 2011 [same scale as in (b)]. Numbers refer to sampling localities for pteropods. (From Bednaršek et al., 2014)

HOT TOPICS IN MARINE BIOLOGY

Shellfisheries: Which Will Fall to Ocean Acidification?   *continued*   21.2

by these intrusions of upwelled acidic and hypoxic water. Since 2007, there have been widespread failures in hatcheries of Washington and Oregon. Most shellfish culture is of bivalves, specifically the Asian oyster *Crassostrea gigas*, the Mediterranean oyster *Mytilus galloprovincialis*, and the geoduck, *Panope generosa*. In 2009 the west coast shellfish industry earned about $270 million annually, so losses are quite worrisome. Barton and colleagues did careful studies on oyster larval growth at the Whiskey Creek, Netarts Bay, Oregon, hatchery where intrusions of upwelled water were common. In Chapter 3 we discussed the use of saturation state, or Ω, which best predicts when a mineral will be precipitated as a function of dissolved constituents. **Box Figure 21.5** shows a close positive relationship between Ω and oyster larval production rate. In another hatchery in Dabob Bay, Washington, researchers could sample from relatively saturated shallow waters and deeper, upwelled, more acidic waters and expose larvae to these different waters. Larval oyster shells of larvae placed in the deeper water samples (**Box Figure 21.6**) showed clear evidence of dissolution (Barton et al., 2015).

Upwelling events have been devastating to oyster hatcheries on the outer coasts of Washington and Oregon, but they are widespread in all coastal regions. A recent assessment (Ekstrom et al., 2015) shows that the dangers are widespread but are especially acute from the Pacific Northwest to Alaskan waters. The Alutiiq Pride Shellfish Hatchery in Seward, Alaska, is the only hatchery serving all of Alaska, with larvae of several species of bivalves, all of whose larvae are sensitive to acidification. The largest predictable changes from optimal to suboptimal values of Ω are seasonal, with extended suboptimal waters in fall and winter, when more CO_2 can dissolve into seawater, water-column respiration is elevated, and there is exposure to short-term runoff events (Evans et al., 2015). A survey of shellfish managers suggested that about half in the Pacific Northwest had experienced the effects of ocean acidification in their hatcheries (Mabardy et al., 2015).

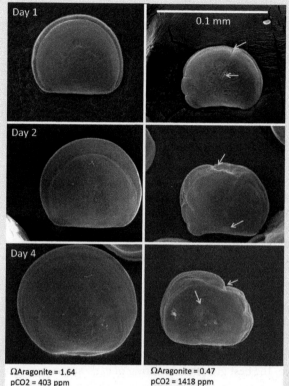

ΩAragonite = 1.64   ΩAragonite = 0.47
pCO2 = 403 ppm   pCO2 = 1418 ppm
pH (total) = 8.00   pH (total) = 7.49

BOX FIG. 21.6 Pacific oyster larvae from the same spawning source were placed in waters of Dabob Bay, Washington from a shallow (left column $\Omega = 1.64$) water source (left column) and deeper water ($\Omega = 0.47$) source (right column). Scanning-electron micrographs of shells show that defects develop in 1-, 2, and 4-day-old larval shells (yellow arrows). (Courtesy of George Waldbusser)

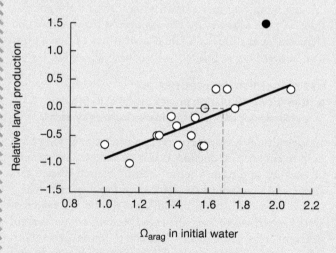

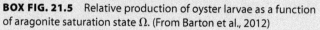

BOX FIG. 21.5 Relative production of oyster larvae as a function of aragonite saturation state Ω. (From Barton et al., 2012)

Is all hopeless? While the imminent and increasing threat of acidification is very worrisome, there are adaptive strategies used by shellfish hatchery managers that might ameliorate the threat to hatcheries, especially because older bivalves are more likely to survive transient acidification events than the larvae. Most important is monitoring, so that production of larvae and juveniles can be timed to avoid the most severe regional upwelling events. As in the case of Alaska we discussed, this may involve a seasonal adjustment to avoid the times of year when Ω is at danger levels. Some hatcheries have also used buffering strategies by adding sodium carbonate into the water, which increases the availability of carbonate ions for precipitation of calcium carbonate. In the future, shellfishery managers will have to maintain a partnership with scientists and continue to develop their own strategies to survive economically at the ever-worsening frontier of climate change.

Indo-Pacific. These species are grown for food, but also for aquarium specimens and for decorative shells. Clams are grown usually in cages to protect against predators such as crabs.

Seaweed Mariculture

■ **Mariculture of seaweeds may be useful in the production of products such as agglutinants, food, and organic matter suitable for methane production.**

Seaweeds are consumed widely as food and are used for a variety of products. The Japanese use a seaweed food (called nori) derived from the red seaweed *Porphyra* in sushi and other foods, and it has become popular in the United States and Canada. In recent years, seaweed has been identified as a "health food." Nori is richer in protein than rice and is more digestible. Seaweeds are also the source of a number of useful substances, including the agglutinant carrageenan, which is used in a number of food products. It has also been suggested that seaweed be grown and converted to fuels such as butanol.

Seaweed culture is well developed in Japan and China. Nori cultivation in Japan goes back to the seventeenth century and involves the collection of spores in nets or on branches of bamboo. The branches are then moved to an area suitable for rapid seaweed growth (e.g., a water body rich in dissolved nutrients such as an estuary). The thalli are harvested frequently and are cut and pressed into sheets for market. The best growth is in spring or early summer, when *Porphyra* grows in the form of a leafy thallus. In June the thalli begin to senesce and gametes are eventually formed, which unite to form a different type of spore. These are the source of a later life stage, the conchocoelis stage, which appears on oyster shells as a red film. This is the stage that produces the conchospores, a haploid stage that settles and germinates to form the leafy thallus that is eventually harvested. The discovery of the conchospore has been used to improve the efficiency of the hatchery rearing of nori. This has greatly increased the efficiency of culture because nori growers can immerse their nets in a tank filled with spores, thus eliminating the uncertainty of recruitment in the natural habitat.

In the eastern Pacific, kelp farming has become a major means of livelihood. The kelps *Macrocystis* and *Nereocystis* grow rapidly, sometimes as much as 15 cm a day. The kelps are harvested mainly for **alginates**, used in a number of foods. Harvesting usually involves boats that have a conveyor belt with cutter blades, which cut the stipes a few feet below the water surface. Kelp populations are enhanced by the culturing of the small gametophyte stage, which produces the sporophytes, whose early stages are cast from boats or applied directly by divers onto the bottom.

FISH FARMING

■ **Fish farming is a major means of rearing finfishes such as salmon.**

The principal advantage of fish farming is the management of a precise area to maximize the productivity and health of all life stages of the food organism. Fish farming has long been well developed in Southeast and east Asia. A variety of freshwater species are reared in ponds and streams. Multispecies culture has been developed in which several distinct microhabitats occur, such as quiet- and running-water habitats. **Salmon farming** is widespread in North America and in the British Isles, Norway, and Ireland. Hatcheries are used to rear fry, which are introduced as young fishes into pens suspended in shallow-water coastal areas. A variety of artificial foods enhance growth, and fishes are raised to market size in the pens. Although keeping the pens free of fouling organisms presents some problems, growth is excellent and the technique is spreading. Popular fish include marine salmonids, marine genetically male *Tilapia*, Branzino (the European marine sea bass *Dicentrarchus labrax*), and the perciform *Cobia*.

Fish farming conflicts with the objective of restoring wild fishes to their former natural levels, before overfishing and water withdrawal take their tolls. There is currently a great controversy over farmed fish, and many marine conservationists have objected to the great expansion of fish farms, which involve principally salmon and European sea bass in the Atlantic and Mediterranean. Escaped salmon are a major component of the populations of "wild salmon" rivers in Norway. Currently, many farms use ground fishmeal as food, which has a great indirect impact on inshore fisheries throughout the world. Another major concern are the diseases that might arise in farmed salmon, since the pens are kept in open water with few precautions taken to prevent release of pathogens. Farmed salmon in British Columbia, Canada, tend to have high occurrences of sea lice, a parasitic copepod that scrapes away at fish skin and causes major lesions, especially in the head area. These parasites are escaping and infecting juvenile wild salmon and causing significant mortality that may drive nearby populations of pink salmon *Oncorhynchus gorbuscha* to extinction (Krkosek et al., 2007). In Chile, a virus has caused a great deal of mortality in fish farms (Naylor et al., 2000).

USE OF GENETIC MANIPULATION

■ **Genetic manipulation proves useful in improving performance of organisms in mariculture systems.**

In agricultural systems, genetic approaches have long been used to enhance production, disease resistance, and general vigor. Such approaches are in their infancy in mariculture systems. At present, there are a large number of hatcheries for fishes and shellfish, and these can be the source of stocks for crossbreeding. The crossbreeding of stocks from different areas, combined with selection for desirable traits, can be used to select a population with high growth and survival characteristics. This approach is useful because one or more desirable traits may initially be fixed in one population and another set of desirable traits may be present in others. It is the combination of such traits that has been a source of vigorous stocks of agricultural plants and stock animals. Modern molecular techniques of gene transfer have also begun to be used to enhance growth of farmed salmon. In the past few years, Dennis Hedgecock has made great strides in producing extensive libraries of genetic markers and linkage maps for oysters (e.g., see Hubert and Hedgecock, 2004).

Drugs from the Sea: Marine Natural Products

■ **Compounds that evolved for a variety of ecological functions—including chemical defense, poisoning prey, and antimicrobial functions—can be adapted for human use as drugs.**

Throughout this text we have mentioned a variety of compounds synthesized by marine organisms whose evolution is related to a variety of ecological functions. For example, many seaweeds have evolved a wide range of compounds ranging from sulfuric acids to polyphenolics, which deter grazers such as herbivorous snails and sea urchins. An amazing variety of conotoxins is synthesized by species of the venomous snail *Conus*, small peptides of about 50 amino acids that have deadly neurotoxic effects on prey. A wide range of organisms, from fungi to polychaetes, synthesize compounds that have strong antibacterial activity. These compounds have a wide range of chemistries and are synthesized in a broad range of biological processes. Many substances, especially toxics found in terrestrial plant leaves, were initially thought be just byproducts of other biosynthetic pathways and were given the misnomer of **secondary compounds**. We now appreciate that the astounding range of such substances has evolved for specific ecological functions, such as chemical defense.

Pharmacologists have discovered an important principle. **Natural products** with strong biological effects may be readily adapted as drugs with applications to human biology, agriculture, and veterinary science. In many cases, the compound's ecological function in the natural species may be adapted to a rather different pharmacological application. For example, the hormone prostaglandin A_2 is synthesized by the Caribbean marine gorgonian *Plexaura homomalla* as an antipredator deterrent, causing a vomiting response in predatory fish (discussed in Chapter 4). But prostaglandins are also active and functional in humans as locally acting cellular signals in a wide variety of functions. Though they were first extracted in quantity from the gorgonian, synthesized prostaglandins can be used in the induction of labor, treatment of ulcers, and many other applications.

■ **Marine natural products are extracted from a wide variety of marine species and may be targeted at a number of specific functional human processes and diseases.**

A wide variety of chemical extraction techniques are now being employed to systematically extract many bioactive molecules from a large number of candidate marine species (for a complete discussion, see Haefner, 2003). Research in this area is extremely active, and literally thousands of new compounds are discovered and researched each year. We can give only a general idea as to the tremendous amount of progress made in several classes of marine compounds.

Ion-Channel Blockers

Species of the carnivorous snail *Conus* employ diverse peptides to poison prey. The general action of the peptides is targeted at **ion channels**, a large array of specialized proteins that allow active transport of ions such as Na and Ca across cell membranes. The diversity of peptide action is correlated with different mechanisms of targeting prey behavior and movement. Literally tens of thousands of conotoxin-like molecules have been synthesized and a number of patented drugs have been developed that target neurologically important Ca channels as pain killers, but do not have the addictive effects of commonly used painkillers such as morphine.

Enzyme Inhibitors

Marine natural products chemists have extracted a large variety of compounds that have strong effects on enzyme function. Many target protein kinases, which are crucial in cell signal transduction and cell division. Because these compounds affect fundamental cell-cycle processes, they are being developed into a wide variety of anticancer drugs in which regulation of cell division is crucial. For example, bryostatin (see Manning et al., 2005) was extracted from the common bryozoan *Bugula neritina*, although the exact ecological function of bryostatin in the bryozoan is obscure, and in fact derives from a bacterium resident in the bryozoan. Bryostatin activity results in the inhibition of protein serine/threonine kinases and has anticancer cell activity. Symbiotic bacteria in some sponges also have related types of activity and their products are being actively investigated as anticancer drugs, via inhibition of protein tyrosine kinases.

Apoptosis Stimulators

Ascidians are sessile and have no hard skeleton. As a result, many species have evolved a wide array of toxic compounds that deter mobile predators such as fishes. Some tropical species are rich in transition metals, and we discussed in Chapter 4 *Phallusia nigra*, which has high concentrations of vanadium and sulfuric acid. The bioactive compound aplidin has been extracted from the ascidian *Aplidium albicans*, found in the eastern North Atlantic and the Mediterranean. Aplidin stimulates the process of apoptosis, or programmed cell death. It is a very promising drug in the treatment of multiple myeloma tumor cells (Mitsiades et al., 2008).

■ **The great diversity of sources of marine bioactive compounds suggests that conservation of biodiversity of species and diverse marine habitats in turn creates a diversity of sources of compounds for drug discovery.**

The previous section demonstrates an important principle. There is a surprising range of sources for new and medically important bioactive compounds. These compounds are diverse because of the existing ecological and taxonomic diversity of biochemical processes that have been discovered in marine organisms. In many cases, compounds have been isolated from single species or groups of closely related species, as in the case of the conotoxins extracted from cone snails. The genus *Conus* is widespread throughout the Indo-Pacific and Atlantic, but the peak of species and therefore biochemical diversity is within the Coral Triangle (see discussion of coral reefs in Chapter 17). Other important compounds come from bacteria that live as symbionts with host invertebrate species such as sponges. Therefore, drug discovery in this area depends on the conservation of mutualistic systems in the ocean. Finally, the complex predator-prey relationships in the ocean have led to the evolution of a broad array of antipredator toxins. This suggests that the conservation of natural and diverse marine food webs will have a direct relationship to our ability to investigate compounds arising from such interactions that might be important in drug discovery. It is therefore no exaggeration to state that marine conservation is crucial to the advancement of human health.

■ CHAPTER SUMMARY

- Fisheries are divided into stocks, usually geographically distinct populations. Tags, as well as biochemical and molecular markers, can help identify and monitor stocks that have become genetically isolated. Because stocks often transcend political borders, they can result in disputes.

- Sensible fishery management is based on an understanding of the life history of species and the size of the stock. Physical variables, spawning and feeding grounds, and migration routes are all important.

- The size of a stock is a function of population size, spatial variability, and the amount of fishing (fishing effort), including numbers of boats, fishing hours, and gear quality. The health of a stock can be assayed by its growth in the previous year and recruitment into production. Stock-recruitment models attempt to predict change in the stock as a function of the stock size in previous years. A popular model predicts that stock increase rate will decline as stock size becomes large, owing to resource limitation.

- Finfishes are caught mainly by hooking fishes individually, entangling them in nets, or catching them in nets or traps. Hooks are used mainly in longline arrays on the surface or at depth. Nets are commonly deployed as trawls, either near the surface or at the bottom, which is greatly disturbed. Unintended catches, or bycatch, are a major source of fish, sea turtle, and mammal mortality.

- In the maximum sustainable yield model, fishing may actually increase productivity, by reducing the effects of high fish density on reproduction and growth. However, this concept is not well supported by data. In recent decades, fishing for top carnivores has severely reduced populations at the apex of food chains and has increased abundance at lower trophic levels.

- Stock reduction results from random variation and environmental change. Overfishing accelerates the decline; stocks characterized by long generation times, small clutches of eggs, and fewer spawnings are the most vulnerable. Many countries have claimed exclusive economic zones of 200 miles from the coast, which allowed local overfishing. Temporary closures, transferable quotas, and fishing limits may allow some fisheries to be sustainable. Ecosystem-based management and marine protected areas are also important tools.

- Once shore based, whaling developed into an open-ocean fishery. New technologies around the turn of the twentieth century—including cannon-powered harpoons, factory ships, and stern slipways—accelerated the decline of whales. A whaling moratorium, molecular detection techniques, and public awareness have helped to arrest the declines of minke and humpback, but other species are still endangered.

- In mariculture, natural habitats are simulated or enhanced, to make harvesting of food fish more convenient or to increase yields. Mariculture systems for mollusks, especially bivalves, are located in areas of high phytoplankton supply, where red tides and other blooms can devastate fisheries. Mariculture of seaweeds aids in the production of agglutinants, food, and organic matter suitable for methane production.

- Fish farming is a major means of rearing finfishes such as salmon, but may produce fish with undesirable traits and spread of parasites. Genetic manipulation can improve the performance of mariculture systems.

- Marine natural products are diverse, probably mainly because of a diversity of evolution of ecological functions. Extraction of many of these compounds has proven useful in the development of drugs for a variety of human ailments.

■ REVIEW QUESTIONS

1. Why is stock identification so crucial in fisheries management?

2. How is the size of a fishery stock usually assessed?

3. What are three factors that might make animals in one fishery more vulnerable than others?

4. Draw a diagram illustrating an expected relationship between stock size and recruitment in the following reproductive season.

5. Why might the expected stock recruitment relationship you have diagrammed fail to develop?

6. What is the basis of the concept of maximum sustainable yield? Why has the effectiveness of this concept as a tool in fisheries management been criticized?

7. How did technology contribute to the decline of open-ocean whale stocks?

8. Why was the blue whale unit adopted? Why did it not represent a good management tool?

9. What characteristics are desirable in choosing a species useful for mariculture?

10. How might harmful algal blooms, discussed in Chapter 8, affect the economy of shellfish mariculture?

11. Why do you think many fishers are resistant to switching from fishing to mariculture?

12. What are some benefits and possible problems connected with fish farming?

13. Why might the culture of just one species of shrimp in crowded shrimp farms accelerate the effects of disease in shrimp population crashes?

14. How can it be that the total landings of a fishery can be constant year after year and then the fishery rapidly collapses?

15. A conservation committee decides to designate a group of subtidal reefs for marine protected areas. What is the value of protecting a group of reefs, as opposed to just the largest one?

16. What is the benefit of removing a dam on a major river connected to the ocean?

17. Suppose that you are given a research budget to find drugs that suppress cell growth to treat cancer. Devise a program of discovery of drugs derived from natural marine species.

18. Why are shellfish farms in special danger on the west coast of the United States?

Visit the companion website for *Marine Biology* **at www.oup.com/us/levinton where you can find Cited References (under Student Resources/Cited References), Key Concepts, Marine Biology Explorations, and the Marine Biology Web Page with many additional resources.**

CHAPTER 22

Environmental Impacts of Industrial Activities and Human Populations

It is hard to believe that many of today's most polluted coastal waters once brimmed over with fish and shellfish. In the early nineteenth century, the waters of metropolitan New York City were a culinary delight. Raritan Bay, now close to the largest landfill in the world, once harbored some of the richest oyster beds in America, and the East River was a paradise for sports fishing. Commercial shell fishing and finfishing were major enterprises. Needless to say, times have changed. The arrival of the Industrial Revolution and the sudden flowering of New York City brought along major habitat deterioration. In New York, the oyster beds are gone, and any remaining shellfish are often dangerous to eat. Fishes are often tainted with toxic substances, and those depending on the bottom are reduced in abundance and diversity. This pattern of decline is typical of urbanized coasts throughout the world. On the other hand, efforts in recent decades have been partially successful in reducing inputs of pollutants. In the Hudson River, mercury in common fishes has declined about two thirds. Still, human impacts are widespread. About 41 percent of the ocean is affected by multiple stressors (Halpern et al., 2008). If toxic substances have been reduced in some areas, they are still on the rise in others. Global climate change adds another layer of stress to the entire oceanic system.

Human Effects on the Marine Environment

■ **Complex interactions of human impacts often make it difficult to understand the role of various pollutants in degrading the marine environment.**

Polluted areas usually are affected by several types of human impacts, which can combine in complex ways to cause biotic degradation, making difficult both the quantification of the degradation and the management decisions on which problem to address. Sediments in most urban harbors are a complex mix of natural sands, silts, and clays, as well as a large number of substances from industrial sources, sewage, and atmospheric deposition. The joint toxic effects of these inputs create difficulties in determining the specific sources of toxicity and effects on organisms.

Human effects on the marine environment may be divided into the following general categories:

1. *Direct habitat destruction.* Alteration of bottom substrate and hydrodynamic conditions through dredging, changing of shoreline structures, and filling.

2. *Toxic substances.* Introduction of chemical substances dangerous either to marine life or to humans.

3. *Sewage, wastewater, and agricultural sources rich in nutrients.* Release of sewage and more generalized sources rich in nutrients used by marine microorganisms.

4. *Heating and water interception.* Effects of power plants, which heat effluent water and capture on screens larval and juvenile marine organisms.

5. *Climate change.* Effects of greenhouse gas emissions on climate and ocean acidification (discussed earlier in this volume).

■ **Pollution may be long term (chronic) or short term (acute).**

It is convenient to divide the effects of pollution into **long-term (chronic) effects** and **short-term (acute) effects.** Chronic pollution involves the introduction of a toxic substance or other anthropogenic factor continuously, causing a degradation of the environment. Year-round inputs of nutrients derived from sewage is a good example. An oil spill is an example of a short-term acute effect. At first, oil often has catastrophic effects on a marine biota, but these effects may gradually be ameliorated as the oil breaks down.

■ Pollution may come from point sources or from a variety of geographic points.

Pollution often comes from a **point source,** such as a single sewer pipe or factory wastewater outfall. In such cases, the concentration of the substance or the intensity of the effect (e.g., temperature near a power plant outfall) should decline with increasing distance from the point source. The nature of the decline depends on the rate of introduction, physicochemical properties of the substance or factor, the water currents, and the sedimentary environment. Such cases are relatively simple in terms of identification and management because a regulatory agency can find the source and monitor the spatial extent of its effects. Mobile organisms may also transport toxic substances by picking them up near a point source and swimming far away. By contrast, **nonpoint source** effects cannot be attributed to any single spot. Runoff following rain is a good example of a nonpoint source: Toxic substances and fertilizer-derived nutrients may then be swept into a basin over a broad extent of the coast, or seep through groundwater into a coastal basin. Such sources are far more difficult to manage because the source cannot be cleaned up as directly as material emanating from a pipe.

Measuring the Impact of Pollutants on Populations and Communities

While it may often be a challenge to identify and measure the concentration of a pollutant, an even greater challenge is to assess the impact of the substance on natural populations and communities. When appropriate baseline data are available, a number of criteria may be used to gauge human impact on a marine environment. These involve effects on single species, effects on communities, such as transfer of toxic substances through food webs, and more general effects on overall community parameters such as biodiversity and trophic structure.

Measuring Effects on Single Species

■ Common species are often chosen as bioassays of pollution effects.

In some cases, certain common or vulnerable species are used as measures of the effects of pollution. The absence of the species, increased mortality, reduced reproduction, or impaired physiological performance may be used as evidence of environmental degradation. For this purpose, much applied biological research has been concentrated on a few common marine species or species used as representative models of pollution impacts. For example, species of the marine mussel genus *Mytilus* have been investigated extensively, and the effects upon it of varying food, temperature, and toxic substances have been measured. In the United States, species of the amphipod *Ampelisca* are used commonly to assay the toxicity of sediments. A method known as *Microtox* employs the effects of a presumed toxic substance or sediment on the light output of a luminescent bacterium growing among the sedimentary grains. Oyster larvae are also used as bioassay organisms because we believe that early developmental stages are more sensitive to pollution than adult stages and mortality of larvae has a great impact on the fate of a marine population.

The use of model species has led to the concept of **bioassay,** which is the measurement of some parameter in a "model" species after being exposed to a stress. The **bioassay species** is usually selected for its natural abundance and ease of rearing. A population is exposed to a range of concentrations of a toxic substance. Mortality rate, uptake rate of the toxic substance, or impairment of physiological function can then be measured. The **LC$_{50}$** is the concentration of a toxic substance that produces 50 percent mortality in an experimental population after a specified time. The reciprocal of the LC$_{50}$ can be used as a measure of tolerance. Estimates of **sublethal responses** are preferred because widespread effects of toxic substances might be operating at concentrations much below the concentration needed to cause mortality. It is useful to measure the relationship between concentration of the toxic substance and oxygen consumption, which is a measure of metabolic rate. One integrating measure of physiological condition is **scope for growth** (see Chapter 5). Increasing exposure to a toxic substance might encumber a metabolic cost at the whole-animal level and reduce scope for growth. On a smaller scale, cellular assays could be usefully employed. One approach is to estimate **cellular energy allocation,** by quantifying cellular energy reserves and measuring mitochondrial activity. On a DNA level, responses of gene expression can be studied through transcriptomic and proteomic assays.

Before laboratory tests are conducted, indicators of pollution effects may be assayed in individuals collected from field populations, such as abnormal body weight or development of abnormal structures. For example, in many gastropods one can find females with extensive development of male anatomical parts including a penis and a vas deferens. Such individuals occur in greater frequency in the presence of toxic metals, and the percentage of females with such male development turns out to be a good indicator of the toxic effects of tributyltin (TBT), a component of antifouling paint once used extensively on boats and still used in U.S. naval vessels. The development of a vas deferens in affected females can block the genital pore and curtail the reproductive output. The decline of the gastropod *Nucella lapillus* in southwestern England has been related to the use of TBT by measurement of the frequency of male pseudogenitalia in field populations (Bryan et al., 1986).

The effects of toxic substances on single species may be measured by constructing models relating toxic substance concentration to population growth.

Effects of toxic substances at the single organism level can be used to predict the fate of populations exposed to pollution. In order to do this, one must be able to estimate the relationship of concentration of a toxic substance to growth, survival, and reproductive output. The life cycle is divided into life-history stages, and then each stage is assayed for survival and reproductive output, when appropriate, from one stage to the next. This approach is analogous to the age-specific approach to population modeling discussed in Chapter 21 in a fisheries context. For example, Diane Nacci and colleagues (2002b) examined the effects of PCBs and dioxins, two toxic substances found in high concentrations in New Bedford Harbor, Massachusetts, and measured the effects on survival and reproduction in the nonmigratory killifish *Fundulus heteroclitus*. Using laboratory data, they predicted significant effects on population growth from exposure to a range of concentrations of the toxic substances (**Figure 22.1**).

The introduction of toxic substances may be related to uptake by individuals in field populations.

Exposure of living organisms to waterborne and sediment-associated toxic substances often results in biological uptake.

The concentration of a toxic substance in an animal at any one time is a function of the environmental concentration of the substance, the biological uptake rate, and the rate of release. Physiological models of uptake and release must include an accurate measure of environmental exposure, uptake and maintenance within various parts of the body, and rates of release by a variety of cellular and exchange methods such as release by excretory organs and gills into the environment. Wang and Fisher (1998) examined uptake and release of metals by copepods and found that processing time was very rapid and metals stayed in a copepod's body only a few days at the most. Zinc and selenium were taken up mainly in the food, but silver, cobalt, and cadmium were taken up from dissolved sources in the water. Benthic animals living within the sediment also derive toxics directly by ingesting sediment but also take up substances from the pore waters.

Gene expression may be an effective means of assaying for effects of toxins.

When an organism is exposed to a toxic substance, a number of genes are induced to express a wide range of defense mechanisms or physiological responses. The degree of expression of genes sensitive to cellular uptake of toxic substances is a very sensitive measure of environmental

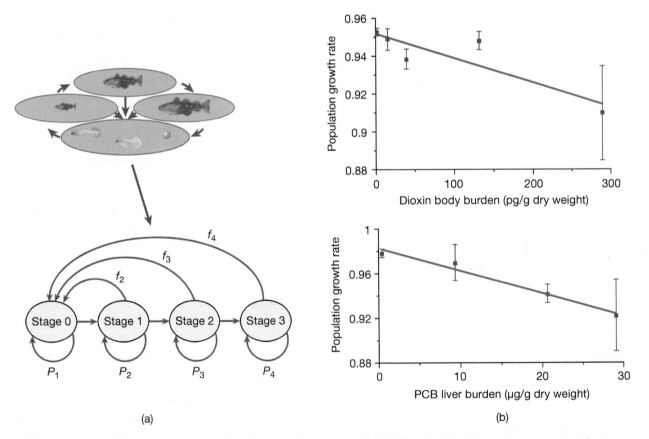

FIG. 22.1 (a) Diagram of a model used to calculate population growth of the killifish *Fundulus heteroclitus*, using four life-history stages (stage 0 is immature; stages 1, 2, and 3 can reproduce). A population projection can be made with the knowledge of stage-specific survival (*P*); reproductive rate (*f*) of stages 1, 2, and 3; and development from one stage into another. (b) Use of data on survival and reproduction to make population rate projections as a function of dioxin and PCB concentrations on the killifish *Fundulus heteroclitus*. (Modified from Nacci et al., 2002b)

exposure.[1] For example, fish are commonly exposed in urbanized estuaries to aromatic hydrocarbons, including dioxins and PCBs. Responses involve arylhydrocarbon receptor (AHR) signaling pathways, and exposure to dioxins and other substances involves increased expression of the *CYP1A* gene, which has been used as a measure of pollutant exposure. AHR signaling and toxicity involve passage of the AHR–dioxin complex into the nucleus. Some metals, such as copper, cadmium, and mercury, are bound and detoxified by metallothionein, a small protein that is rich in metal-binding cysteine amino acid residues. In some cases, increased metals directly induce expression of metallothionein genes, and measures of mRNAs for this protein would be a good assay of pollution exposure. Alternatively, a proteomic approach would estimate the cellular concentration of metallothionein proteins. Also, different genotypes can be shown to produce proteins with differential binding to toxics.

When an organism is exposed to a toxic substance, expression by more than one gene will occur, and complex interactions of gene expression will be employed to allow the organism to respond to the exposure. Studies of multiple gene expression to toxic substances are likely to be complex. **Microarray** techniques are used to develop a DNA sequence library. DNA sequences are synthesized and usually placed on chips. After exposure to a toxic substance, messenger RNA of an organism is isolated and exposed to the chips, and a coupling mechanism to a fluorescent protein system allows a visualization of the expression of multiple genes (**Figure 22.2**). There are two objectives of looking at the responses of groups of genes. First, it would be useful to discover suites of functionally related genes that interact and respond to pollutants. Second, researchers hope to be able to use responses of given suites of genes as indicators of stress by particular pollutants. For example, Venier and colleagues (2006) developed unique sequences for about 1,700 distinct DNA probes in the mussel *Mytilus galloprovincialis* and exposed these mussels in the laboratory to metals and organic toxic substances. They found differences in gene expression and were able to relate these to field-caught mussels in the Bay of Venice, Italy.

■ **With genetic variation in natural populations and differences in fitness among genotypes, evolution of resistance to toxic substances may occur.**

The introduction of a toxic substance may kill off only a fraction of a species population. If there is genetic variability for resistance to the toxic substance, then the population may evolve resistance. After a period of time, the average individual will not be sensitive to the toxic substance. The degree of resistance can, therefore, be used as an index of the biological effect of a substance introduced by human beings. Klerks and Levinton (1989) investigated a population of oligochaetes that had lived exposed to high concentrations of cadmium in the sediment. Worms from a polluted site survived well when exposed to cadmium-rich

FIG. 22.2 Gene expression of a large array of genes of the mussel *Mytilus edulis* when exposed to metals, as shown by a microarray. Each dot represents a sequence on a standard microarray chip, which consists of part of the sequence of DNA of a functioning gene that might prove to be involved in responses to metals. The sequences may or may not bind to messenger RNA sequences from the exposed mussel, which are the result of gene expression (transcription of the gene into mRNAs). Darker green signifies more gene expression for a given sequence type. By analyzing the degrees of expression, the groups of genes that respond to cadmium exposure can be determined, which allows an understanding of the network of genes involved in responses to toxic substances. (Courtesy of Robert Chapman)

sediments, but worms taken from a clean area mostly died when exposed to the same sediments. The ability to survive the cadmium was controlled by genetic variation (**Figure 22.3**). New Bedford Harbor, Massachusetts, is very polluted with toxic PCBs and other substances and should be a location of strong potential natural selection for resistance to toxic substances. New Bedford Harbor killifish had much higher expression of P450 genes than fish from a nearby unpolluted river, suggesting that natural selection had caused the difference (Greytak et al., 2005). Diane Nacci and colleagues (2002a, 2002b) did crossing experiments and demonstrated that resistance of killifish to dioxin-like compounds was genetically based, and that killifish populations were more resistant in areas of New Bedford Harbor where concentrations of PCBs (**Figure 22.4**) and dioxin-like compounds were higher. Resistant genotypes were found in a wide range of New Bedford Harbor habitats, far wider than had been declared to be toxic by the U.S. Environmental Protection Agency.

A molecular mechanism has been found for resistance to PCBs (see later in this chapter for discussion of pollution sources and toxicity of PCBs) of the Hudson River population of the Atlantic tomcod *Microgadus tomcod*. Hudson River tomcod are widely known because of the high occurrence of cancers and skin lesions, an apparent response to toxins in the water. A genetic variant of the AHR protein was commonly found in Hudson River populations but rarely in other less polluted sites. This protein defeats the binding to PCB 126, which is a PCB variant that is toxic to fish. Apparently, rapid natural selection had selected this rare variant, relative to the common one found in most other populations along the northeast coast of the United States and Canada (Wirgin et al., 2011).

---

[1] See bonus chapter "Molecular Tool for Marine Biology," online.

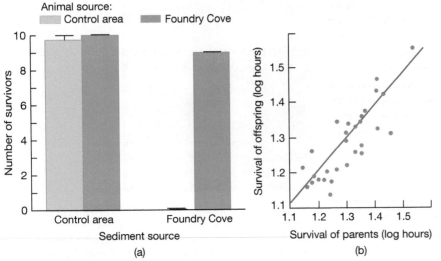

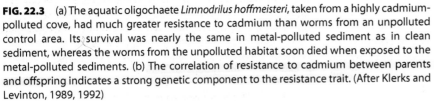

FIG. 22.3   (a) The aquatic oligochaete *Limnodrilus hoffmeisteri*, taken from a highly cadmium-polluted cove, had much greater resistance to cadmium than worms from an unpolluted control area. Its survival was nearly the same in metal-polluted sediment as in clean sediment, whereas the worms from the unpolluted habitat soon died when exposed to the metal-polluted sediments. (b) The correlation of resistance to cadmium between parents and offspring indicates a strong genetic component to the resistance trait. (After Klerks and Levinton, 1989, 1992)

animals are literally evolving to deal with human pollution

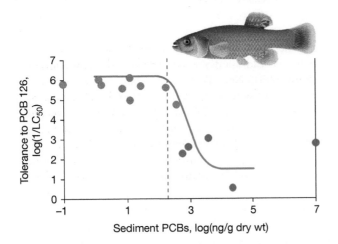

FIG. 22.4   Tolerance for polychlorinated biphenyls (PCB 126) of the nonmigratory killifish *Fundulus heteroclitus* when taken from a range of contaminated areas with different PCB concentrations from New Jersey and Massachusetts, as measured by the reciprocal of LC_{50} after exposure to a given concentration. Low values indicate the fish are more tolerant of PCBs. Green dots signify highly contaminated sites. (After Nacci et al., 2002a)

Measuring Effects on Community Function

■ Many toxic substances are transferred from one trophic level to the next as predators consume prey.

Organisms may take up toxic substances by feeding on other species, resulting in trophic transfer. Some metals, for example, remain in the cytosol, while others become associated with cell membranes and might even be deposited in shells and skeletons. Clearly, the latter will not be as efficiently transferred as toxic substances that remain in the cytosol. William Wallace and Glenn Lopez (1997) found that cadmium was bound to

metal-binding proteins in the oligochaete worm *Limnodrilus hoffmeisteri*. They fed these worms to predatory shrimp and found that trophic transfer was directly correlated with the cadmium that was bound to the metal-binding proteins in the cytosol. Cadmium bound to the cell membrane was not very efficiently transferred to the predator. Cadmium taken up by the shrimp impaired their ability to successfully attack prey.

■ Some toxic substances are magnified in concentration in organisms, as the toxin is transferred through the food web.

Many substances taken up by marine organisms undergo **bioconcentration**, where a substance is taken up from the water directly and is concentrated within the tissues. In some cases, **biomagnification** *increases the concentration of a toxic substance in the tissues as it moves from one trophic level to the next*. A classic case is bird consumption of fish and invertebrate predators contaminated with DDT. DDT is soluble in fat and concentrates in the birds, relative to prey items. However, DDT is usually not biomagnified in other parts of the food web. Methylmercury is usually biomagnified as it moves from animal prey to predator. Methylmercury is also bioconcentrated by large fish (owing to the relatively low loss rate), so large-bodied tuna have relatively high concentrations of methylmercury. In fact, frequent sushi consumers occasionally suffer mercury poisoning, which can cause tremors, impaired vision, and severe neurological damage.

■ When a marine organism is exposed to a toxic substance, the toxic substance may not increase in concentration in the body, or it may indeed continue to increase.

The biological uptake pattern may also differ among species. **Noncumulative toxic substances** do not increase in concentration in the body over time, even if the organism is exposed chronically to the substance. In this case, the organism has an efficient removal mechanism for the toxic substance. By contrast, **cumulative toxic substances** continue to increase in concentration and may be found most abundantly in a single tissue. For example, cadmium tends to increase over time in the digestive gland of crabs. Cumulative uptake occurs only if there is a **specific biological mechanism of tissue concentration** and a **low loss rate** of the toxic substance from the body.

Measuring Effects on Biodiversity

Diversity is said to increase if the number of species increases, or if the abundances of the species are more evenly distributed (see Chapter 20). For example, a low-diversity habitat would consist of very few species, with strong dominance by one species. Although one can clearly find exceptions, diversity tends to decline in strongly polluted habitats (**Figure 22.5**).

Species resistant to pollution are often a small subset of the total species pool that includes species capable of rapid colonization in strongly disturbed habitats; such species are termed **opportunists**. Marine communities subject to pollution often resemble natural assemblages strongly affected by physical disturbance factors. While diversity can be monitored to assess pollution, usually the effects of pollution can be detected with much more sensitivity by using statistical approaches to measure species abundance changes and patterns of association among species. In all cases, it is essential to compare species assemblages in polluted habitats with unpolluted habitats that are otherwise similar in substratum, depth, current strength, and so on. It is also preferable to have data on species occurrences before the impact occurred, to compare to impacted sites and nearby control sites that have not been impacted.

Toxic Substances

Toxic Metals

■ **Metals are often cumulative toxins and have strong effects when consumed by human beings.**

For thousands of years, metals have been released by human industry into the marine environment. Mining is a major source of metals, and mines near estuaries have been a major source of pollution. A wide variety of industrial processes release metals. For example, before regulatory efforts, mercury was released: In emissions from wood pulp processing; as aerosols from coal-burning power plants; to the atmosphere from burning of trash; and in river and atmospheric sources from mercury mines. Metals also are found in great quantities in sewage; in urban areas, this is probably the single greatest source of metal pollution. In agricultural areas, metals are sometimes released as components of insecticides and fungicides. Mercury and copper have been components of antifouling paints, and the toxic effects of tributyltin are recognized widely.

Although the effects of metals on human beings have been studied extensively, relatively little is known about the specific physiological effects on marine organisms. Metals such as zinc and copper are known to denature many proteins and are therefore fundamental poisons. Copper also may bind to blood pigments and impair their function. Many metals are known to increase mortality in a wide variety of species.

Marine organisms produce metal-binding proteins, such as metallothionein, which bind to metals to allow their deposition within cell organelles as **metal-rich granules** in chemically less reactive forms that aid in reducing the exposure of cell constituents to chemical reactions with metals.

Mercury is probably the most notoriously toxic metal. Mercury comes from a number of sources, including atmospheric deposition from often far-distant coal-burning

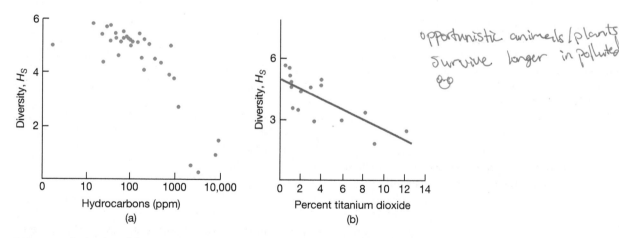

FIG. 22.5   Reduction of diversity of benthic macrofauna (a) along a gradient of increasing sediment hydrocarbon concentration near an oil platform and (b) with increasing concentration of titanium dioxide waste in a Norwegian fjord. H_s is a measure of diversity that increases with the number of species and the relative evenness of abundance of the species. (After Gray, 1989)

power plants, burning of waste materials, industrial sources in factory outfall pipes, mines, and sewage. It is particularly toxic when attached to a short-carbon-chain alkyl group in the form of **methylmercury**. Mercury may be deposited originally in metallic form, but it will be converted into methylmercury when it enters an estuary or a coastal waterway and is exposed to sediments and microbial activity in a bacterially mediated process known as methylation. We do not know enough about the exact microbial mechanism of methylation. Recent evidence suggests the necessity of activity of two genes, *hgcA* and *hgcB*, in the bacteria *Desulfovibrio desulfuricans* and *Geobacter sulfurreducens* (Parks et al., 2013). It is possible that there are other mechanisms of methylation in marine waters and bottom sediments.

The principal effect of methylmercury on human beings is strong disruption of nervous function, especially in the fetus. The odd behavior of the Mad Hatter in *Alice in Wonderland* constitutes an allusion, which would have been obvious to Lewis Carroll's contemporaries, to the effects of the mercury poisoning that was so common among hat makers in mid-nineteenth-century England. In Minamata, Japan, the industrial release of mercury, which was absorbed by fishes and shellfish, led to large outbreaks of nervous disorders and deaths from the 1950s to the 1970s. Unfortunately, mercury passes across the placenta, and many women who had eaten mercury-laden fish bore children who had the affliction that came to be known as Minamata disease. Some other coastal populations have been affected in much the same way. Mercury is also known to cause myocardial infarctions. Mercury in sushi can be a source of toxicity in avid consumers.

One important pathway for mercury to reach human consumers is as follows: atmospheric mercury deposition to watersheds → entry into watershed soil and sediments → dissolved methylmercury in water → methylmercury in fish through a bioaccumulation factor from prey in water → and human consumption of fish. In San Francisco Bay, a considerable amount of mercury arrives from the watershed because of rock formations with high mercury concentrations and also from leakage from mining activities. Mostly because of atmospheric deposition and industrial point sources, there are many hotspots of high mercury concentrations in fish throughout the world. In the Hudson River, anglers who eat locally caught fish have higher tissue mercury concentrations than the rest of the human population. We mentioned earlier in the chapter that methylmercury is biomagnified up the food chain to large carnivores such as tuna. In the past few decades, inputs of mercury from some sources such as burning of wastes have declined. Levinton and Pochron (2008) found that mercury concentrations in five common fish species have declined steadily in the Hudson River since the 1970s. The decline is probably a response to regional regulations that reduced Hg inputs into the local environment. But there is a worry that the need for worldwide energy will result in more burning of coal, which will increase the inputs of mercury into the atmosphere. Such increases are widespread in India and China.

Most mercury exposure to people in the United States is due to consumption of oceanic fish such as various species of tuna, swordfish, and pollock. It is possible that mercury in oceanic fishes, such as North Pacific tuna, is converted to the methylmercury form in open ocean waters. Mercury might be associated with particulate carbon, which sinks through the water column and is methylated by microorganisms during consumption of the POC. This may cause an increase of mercury exposure in the future, as mercury-bearing coal is burned at higher rates in east and south Asia (Sunderland et al., 2009). Mercury has been found in toxic concentrations in the Arctic Ocean in beluga and narwhal, which are hunted by native peoples of Greenland and Canada.

Cadmium is another fairly common metal, although the distribution of its sources is often more variable than is the case for mercury. Cadmium may enter in sewage, but it also comes from outfalls in electroplating factories and battery manufacturing plants. In Foundry Cove, a bay adjacent to the Hudson River estuary, a battery factory outfall caused cadmium concentrations in bottom sediments to be in the range of 1–25 percent! The most notable toxic effect on human beings is on kidney function, but larger accumulations can lead to bone deformities and severe pain. Consumption of cadmium-laden sea turtles and dugongs in northern Australia has had apparent negative effects on kidney function in residents of the Torres Strait. Cadmium is found to be concentrated in rice in certain areas in Japan, but crabs and shellfish also are contaminated. At present, cadmium is highly concentrated in blue crabs in the Hudson River, and frequent crab consumption will produce pathological renal effects. Cadmium has also been found to severely affect eels by impairing lipid accumulation, gamete production, and migration (Pierron et al., 2008), and it may have contributed to the decline, along with overfishing, of the American and European eels in recent decades.

The badly contaminated Foundry Cove of the Hudson River is remarkable for the fact that benthic animals were dense there, despite common sediment cadmium concentrations of 10,000 ppm. The oligochaete *Limnodrilus hoffmeisteri* apparently evolved a resistance to high concentrations of cadmium (**Figure 22.3**). Individuals of this species from other areas died soon after being introduced into Foundry Cove sediments, whereas indigenous worms survived well (Klerks and Levinton, 1989). The resistance is genetically based and may be related to the higher production of a metal-binding protein and the ability to precipitate cadmium sulfide in intracellular organelles. The cove was cleaned up partially in 1994, and local populations lost their resistance in just nine generations. After the cleanup, invasion of nonresistant genotypes from outside the cove probably changed the genetic makeup of the worm population (Levinton et al., 2003).

Lead was used during Roman times in water pipes, pottery, and coins. Since the industrial era, lead has been used extensively in lead storage batteries and paints, and as an additive to motor fuels. Lead is also found in fossil fuels and is therefore emitted during burning in power plants. Lead is most common as an air pollutant, and there has been a movement in recent years to reduce lead in internal combustion engine fuels. The action is the result of knowledge of the strongly toxic effects of lead on the nervous system, and especially on the neurological development of young children.

Lead is now found in relatively high concentrations in estuarine and marine sediments adjacent to urban areas. The most probable sources are release from industrial pipes, dissolution from lead pipe systems, atmospheric deposition, washout of gasoline-ridden wastewater into storm sewers, direct disposal of fuels into seawater, and sewage release.

While other toxic substances are also monitored, metals are a prime target for study in the Mussel Watch Program (an international monitoring study that analyzes mussels in coastal habitats on both sides of the Atlantic). Mussels are analyzed for a variety of metals to judge whether bioavailable toxic substances are entering coastal marine ecosystems. Owing to the differences in chemistry among metals, an increase of concentration must be interpreted carefully.

Pesticides

■ **Pesticides are usually designed to kill terrestrial insects, but they are washed into coastal waters and are often toxic to marine life.**

Pesticides include a wide variety of compounds used typically to kill insects harmful to crops. Because of the large spatial scale of pest infestations, most pesticides are applied in large amounts (nearly 1 billion kg per year in the United States) and in great variety (there are about 1,000 different types). Although some pesticides are very effective, most target species have sufficient genetic variability of resistance to rapidly evolve resistance. At first, the frequency of resistant variants may be very low, but unlike nonresistant individuals, the resistant individuals will survive and reproduce. The evolution of resistance has tended to magnify both the toxicity and the variety of pesticide deployment. Despite this arms race, the insects have been winning by and large, and crop damage has steadily increased over the last few decades. Herbicides are effective in killing undesirable plants, but they also are potentially toxic to human beings.

There is not enough space here to describe the variety of pesticides in any detail, but several have been extremely harmful to marine organisms. To assess the potential harm of a pesticide, the solubility and chemical characteristics related to mobility, toxicity, and rate of degradation must

be known. Substances that are easily mobilized in runoff have generally toxic effects in many organisms, and those that are slow to degrade are liable to be the most dangerous.

Chlorinated hydrocarbons (DDT, dieldrin, chlordane) have the most dangerous combination of harmful properties. DDT's degradation rate is on the order of years, and it washes readily from salt marshes into adjacent shallow estuarine and marine bottoms. **DDT (Figure 22.6)** was used widely as a means of eliminating the *Anopheles* mosquito, which carries malaria. It was later used as a general insecticide on many crops. In the 1960s, the general decline of a large number of seabird and terrestrial predatory species was noticed, especially those at the top of food chains, such as the bald eagle, peregrine falcon, the Bermuda petrel, the brown pelican, and the osprey. DDT and a few related compounds, which are very soluble in fat, were magnified up the food chain to those top predators that ate large numbers of contaminated fishes. DDT and related residues disrupt reproduction and especially eggshell construction, to the degree that shells became too thin to permit normal egg development. Many species declined catastrophically until widespread bans on DDT use were imposed by Western industrialized nations. In recent years, DDT has degraded in the U.S. marine environment. Peregrine falcons, osprey, and bald eagles have all recovered strongly. DDT is no longer used in a number of developed countries but continues to be used to a lesser degree in developing tropical countries because of its continuing effectiveness in dealing with malaria.

Many of the pesticides directed at insects are also toxic to their arthropodan relatives, the crustaceans. Spraying of insecticide on coastal agricultural areas and on marshes for mosquito control may therefore have unfortunate consequences. The spraying of the insecticide kepone caused the closure of the James River (Chesapeake Bay watershed) to fishing and devastated the blue crab population. Other insecticides, such as mirex, harm crabs, especially during larval development. Even when marine invertebrate populations are not affected, they may sequester the toxins, which may pose a danger to human beings. Dioxin, for example, is a

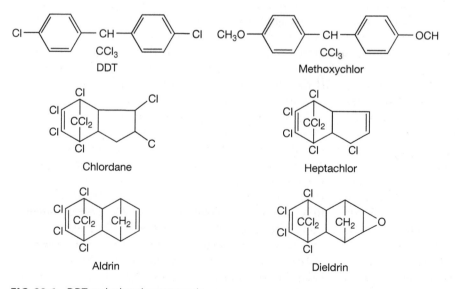

DDT

Methoxychlor

Chlordane

Heptachlor

Aldrin

Dieldrin

FIG. 22.6 DDT and related compounds.

contaminant derivative of some herbicides and is believed to be carcinogenic to people. There have been reports of dioxin in fish and shellfish. Dioxin in the sediment interferes with the reproductive cycle of estuarine fish. The larvicide methoprene is a juvenile hormone mimic that kills *Aedes* mosquito larvae but also strongly affects development of larval stages of estuarine crabs (Christiansen et al., 1977).

Glyphosate is a major component of a variety of herbicides used to control weeds and even invasive plants such as *Spartina* marsh grass species; as a result, glyphosate is found widespread in estuarine environments. It is not yet clear whether there are strong impacts of glyphosate on marine life. Experiments with transcriptomic responses by juvenile oysters demonstrate the moderate increase of gene expression of several candidate response genes such as catalases (Mottier et al., 2014) but no major effects on oyster condition. Effects on larvae and embryos were similarly modest. Direct applications of herbicides with glyphosate are harmful to target organisms such as sprayed invasive *Spartina* but also to seaweeds that might be present.

PCBs

■ **Polychlorinated biphenyls (PCBs) derive from industrial activities and have proven to pose a major toxicity problem in estuarine environments.**

Polychlorinated biphenyls are a class of compounds that have been used extensively as lubricants in various types of industrial machinery. Throughout the world, PCBs have been released into coastal waters and have been found as a contaminant of invertebrates, fishes, and marine mammals. PCBs cause carcinomas in mice and are therefore thought to be a danger to human beings. These substances are particularly a problem because of their very high toxicity and chemical stability. Although marine bacteria can degrade them, the process is very slow.

PCBs have been discovered in a wide variety of commercially captured fishes, such as bluefish and striped bass in the New York and southern New England region. They have been implicated in reproductive failures and reduced populations of seals in the North Sea–Baltic Sea regions. A release of PCBs in the Hudson River from a General Electric Corporation facility resulted in high loads of PCBs in sediments, the contamination of fishes, and the shutdown of the striped bass fishery within the Hudson and even in adjacent waters for a few years. The substances have also turned up in fish caught by Native Americans in Alaska, perhaps owing to exposure of fishes that spawned in industrially polluted rivers in the former Soviet Union. It is not clear whether the fishes take up PCBs primarily from contaminated prey or directly from solution.

In the United States, two large Superfund sites have been declared on the basis of PCB concentration, one in New Bedford Harbor, Massachusetts, and another in the Hudson River Estuary, the largest Superfund site in the country. A plan for dredging the New Bedford Harbor site is already completed, and the Hudson River site dredging and capping has been completed, although many complain that significant PCB sediment areas remain to be dredged.

Plastics

■ **Plastic products and debris enter the ocean from garbage dumps, storm impacts on the shore, and many other sources. The material accumulates in low current areas and breaks down into small pieces that are ingested by a wide variety of marine organisms.**

Approximately 200 million tons of plastic products are produced annually worldwide, and a significant fraction—possibly 10 percent—enters the ocean from oceangoing ships, from dumps near the coastal zone, and from other sources. Oceanic circulation moves material great distances. For example, an enormous region of the central north Pacific has accumulated thousands of square kilometers of plastic-laden surface water. When fresh, this material can entangle mammals and fish (**Figure 22.7**).

Material starting as bags, shoes, toothbrushes, and other products breaks down and is eroded into small particles that enter into biological systems. In surface waters, plastic debris is ingested by fish, seabirds, and sea turtles. An investigation of albatrosses found plastics in 30 of 47 birds, or about 64 percent, recovered as bycatch. Plastics are also found in egested material fed to seabird chicks on nesting grounds (Gray et al., 2012). Plastics also entangle a wide variety of marine organisms. A number of toxic substances, such as PCBs, sorb onto plastic surfaces, making the plastic debris a possible major source of toxicity (Engler, 2012).

Plastics enter the marine environment and immediately become part of the marine ecosystem (**Figure 22.8**): They are transported by currents, deposited on the sea floor, eaten by marine creatures, and even then consumed and transferred through marine food webs. Microplastic particles (plastic particles < 5 mm in size) are now widespread in hand and facial cleansers. They may remain suspended in the water column, where they are ingested by marine suspension feeders or wind up in sediments, eaten by deposit feeders. An investigation near the Great Barrier Reef demonstrated that corals appear to detect plastic particles as prey and ingest as much mass of plastic particles as they ingest of live plankton prey (Hall et al., 2015). The high surface area of the plastic microspheres attracts a number of chemical moieties but especially various metals and organic chemicals such as PCBs that are often toxic. So the question is open: Are microplastics toxic to any degree? If they are, this is a major calamity given the worldwide occurrence.

So far the evidence is very incomplete, and I hope that students can see that this is an area in which more progress in research is needed. For example, a study by Nobre and colleagues (2015) demonstrated that materials leached from microplastic particles appear to cause anomalous development of sea urchin larvae. New plastic pellets appear to be far more toxic than those recovered from a beach, so some of the toxic substances that adsorb to plastic particles may degrade over time, making the plastic particles less effective as vectors of adsorbed toxins. An isopod collected from the North Sea was fed microplastic spheres embedded in food, and no negative effects were found on digestive

FIG. 22.7 Divers free a plastic-entangled Hawaiian monk seal in the northwest Hawaiian Islands. They were successful. (Photograph by Ray Boland, with permission from NOAA)

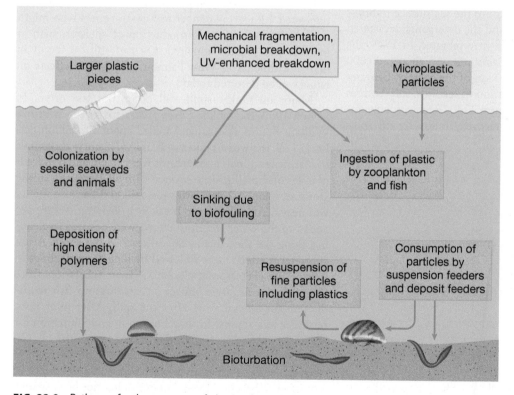

FIG. 22.8 Pathways for the transport of plastics through the ocean and its food webs. (After Wright et al. 2013)

organs, survival, or molting frequency. On the other hand, microplastic spheres were filtered by mussels and taken up and engulfed by cells, and there were clear negative cytological effects (von Moos et al., 2012).

When the Pacific oyster *Crassostrea gigas* (a major species used in aquaculture throughout the world) was fed fresh plastic nanoparticles of about 6 μm, used in facial cleansers, a number of strongly negative effects followed, including reduced oocyte development, sperm velocity, larval survival, and development rate (Sussarellu et al., 2016). This study also gave us great insight on the oyster's responses by following gene expression in the transcriptome. By examining different classes of genes responsible for different reproductive and maintenance functions, it was possible to see a shift of plastic-stressed oysters from reproductive to metabolic activities associated with growth. Apparently the oysters were "interpreting" the presence of the plastic particles as a major stress, which resulted in a shift from reproductive activities to maintenance of the body, in order to maintain survival.

Oil Pollution

■ **Oil pollution can have both short-term and long-lasting effects on communities and individual species.**

In the past 50 years, oil pollution has become a major problem in the coastal zone. Drilling, transport, and burning have all added oil to marine environments. The following are the major sources of oil pollution: (1) leaks from and breakup of oil tankers and barges, (2) leaks from wells drilled offshore, (3) leaks from marine terminals and in harbors, and (4) washout of oil from settled areas into storm drains and direct washout to the shoreline.

The wreck of the tanker *Torrey Canyon* on rocks off the English coast in 1967 was the first oil spill that awakened the international community to the dangers of oil transport. About 80 tons of crude oil was released and 40 tons burned after the Royal Air Force made a bombing run over the site. Detergents were sprayed onto the sea surface to break up the oil slick. Both the oil and the detergents devastated seabird populations and shore invertebrates. In 1978, the wreck of the tanker *Amoco Cadiz* released about 200,000 tons of oil over more than 300 km of the coast of Brittany, France. The result was devastation to seabirds, soft-bottom benthos, and oyster beds (**Figure 22.9**). Recovery now appears complete. In general, recovery following oil spills

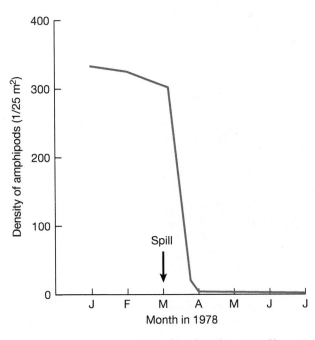

FIG. 22.9 Change in abundance of amphipods in coastal bottom sediments following the wreck of the *Amoco Cadiz* off the coast of Brittany.

takes 2–10 years, but there is strong variation depending on the biological group.

In the United States, an accidental leak in 1969 from an offshore well near Santa Barbara, California, affected marine life in the coastal zone. The effects of the spill were not studied adequately, but the spill had great impact on public opinion concerning environmental issues. In the same year, however, a spill from the relatively small barge *Florida* off Cape Cod, Massachusetts, was studied intensively by a team of benthic ecologists and chemists led by Howard Sanders of the Woods Hole Oceanographic Institution (Sanders et al., 1980). This barge carried 14,000 barrels of oil (1 barrel = 159 liters), a quarter of which was spilled in the area of the relatively small Wild Harbor, Massachusetts. Since the load was number 2 diesel fuel oil, the concentration of relatively toxic aromatic hydrocarbons was high, about 41 percent.

The spill strongly affected the benthic community. A diverse community of clams and polychaetes crashed and was replaced by a very few species. In particular, the polychaete *Capitella capitata* came to dominate the intertidal and subtidal soft bottoms. It took several years for the oil to lose most of its effects. Much of the oil was buried, only to spread during winter storms that eroded the covering sediment. Toxic substances, such as aromatic hydrocarbons, were found in shellfish more than 1 year after the spill. This suggested a lingering danger to human beings who might consume the shellfish. Reproduction of shellfish such as mussels was greatly impaired. Oil can still be found in marsh sediments in West Falmouth, and toxic effects on fauna can be detected today.

Major oil tanker spills still occur, but less often. In March 1989, the tanker *Exxon Valdez* hit a reef in Prince William Sound, Alaska, and spilled about 11 million gallons of oil, the worst tanker spill in U.S. history. Thousands of marine mammals and seabirds were killed, and hundreds of miles of shoreline were covered with oil. Because a strong storm developed 2 days after the spill and because the oil was heavy crude (harder to break up), dispersants were not added to the water and most efforts were devoted to cleaning along the shoreline as the oil reached the shore. Acute short-term effects were followed by long-term declines in marine life (Peterson et al., 2003). Toxic substances from the oil had a negative impact on anadromous fish reproduction for several years after the spill. After pilot tests, fertilizer was applied to shorelines to provide nutrition for bacteria that contributed to the breakdown of the oil. Oil from the spill still remains in the area, especially in the form of large lenses of dried oil beneath the sediment surface in a number of beaches. Even Antarctica has not been spared. Also in 1989, an Argentinean tanker capsized and spilled about 200,000 gallons of oil. Thousands of penguins, seals, and seabirds were killed.

Some of the complex effects of oil spills were well studied by scientists at the Smithsonian Tropical Research Institution following a spill in 1986 on the Caribbean side of Panama (**Figure 22.10**). Bahia Las Minas had been surveyed for years before the spill washed oil ashore, which

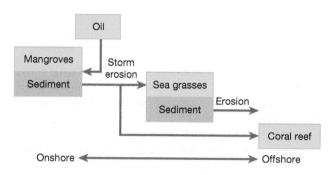

FIG. 22.10 The cascading effects of oil derived from a 1986 spill along the shores of Bahia Las Minas in Panama.

allowed comparisons before and after the spill. Initially the spill destroyed approximately 7 percent of the mangroves on the shoreline. The decay of the mangrove trees caused them to dislodge and roll about the shore, causing damage to the benthos. The oil also damaged sea grass meadows, which caused rhizome mats to disappear. The sediment was no longer held in place: It was transported seaward, which increased the turbidity and sedimentation and reduced the survival of nearby corals. Oil seeped into the mangrove forest sediment, but the rainy season soon came, and storms and freshwater flow caused the oil to be eroded and transported offshore, which caused more damage to sea grasses and corals. Normally, the mangrove ecosystem traps sediment, which helps clarify the water and benefits the nearby coral reefs. The oil spill therefore set off a cascade of ecological effects.

By far, the largest oil releases have come from submarine well drilling. Drilling pipe is extended into a well in

the seafloor from a fixed or floating platform. Most platforms are at shelf depths in relatively shallow water, but in recent years, with the aid of submersible drilling units, some platforms have reached much greater depths, of 1,500 m or more. Wells commonly extend thousands of meters downward from the seafloor surface.

A major blowout occurred in 1979 in a Gulf of Mexico oil well at 60 m depth at Ixtoc, about 100 km offshore in Mexico. Drilling muds failed to balance a back pressure, and the well exploded, releasing about 140 million gallons of oil over a period of 10 months until the well was finally sealed. Because of northward-flowing currents and the long time period of the release, oil spread throughout the shelf waters of the western Gulf of Mexico, reaching shore environments of Texas (**Figure 22.11**). Within Texas, large areas of shoreline were covered with oil, but surface booms (floating barriers) were placed to protect estuaries, and the strategy was largely successful. Although isolated areas are still affected, after 2–3 years, it appeared that most areas had recovered from the blowout.

The Gulf of Mexico experienced an even larger oil release starting April 20, 2010, following an explosion at the British Petroleum (BP) Deepwater Horizon Macondo oil well at 1,525 m depth, approximately 70 km southeast of the Louisiana coast. Following the explosion, 11 platform workers died and a blowout preventer installed at the wellhead failed, resulting in a catastrophic release of oil. By July, over 200 million gallons of oil had been released (Tunnell, 2011; also see OSAT, 2011), making the Deepwater Horizon accident the worst in recorded history. This oil was a lighter crude than that spilled from the *Exxon Valdez*: It

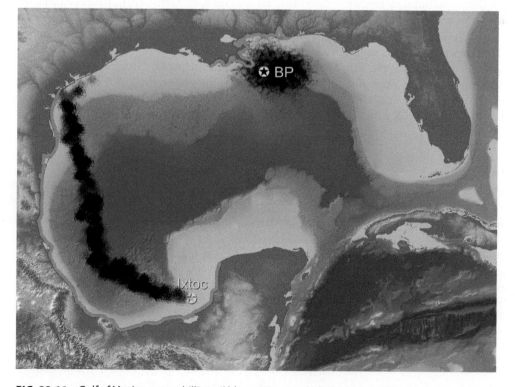

FIG. 22.11 Gulf of Mexico ocean drilling oil blowouts: spread of oil (black area) by currents and extent of the shoreline impacted (red line) from the 1979–1980 Ixtoc well blowout and the 2010 Deepwater Horizon blowout. (Courtesy of I. MacDonald)

contained lower-molecular-mass hydrocarbons, and about half of the hydrocarbons released were methane gas, the remainder being crude oil (King et al., 2015). Thus, the oil was more biodegradable than the oil released in Alaska in 1989. A small minority of the oil was collected from the vicinity of the well. The wellhead was not capped until July 15, 2010, 84 days after the initial event. Toxic drilling muds were also released in the immediate vicinity of the wellhead.

Nearly 2 million gallons of dispersants were added to the water, a large fraction near the wellhead itself. By the beginning of May, the oil reached the Louisiana coast, and authorities used booms to attempt to prevent the oil from reaching shore. The booms were too small and generally failed, however, resulting in concentrated oil-polluted areas in Louisiana. Currents moved the oil to the east and spread as far as Florida, over a period of 3 months. The extent of oil movement and stretch of coastline affected was about 650 km, which was much smaller than the 2,500 km of coast affected during the 10-month spread of the 1979–1980 Ixtoc event in Mexico (**Figure 22.11**).

The coastal environments affected included a large stretch of salt marsh environment. Oil permeated marsh sediments, and a number of seabird and sea turtle nesting grounds were saturated. Marsh sediments impacted with oil killed marsh vegetation, making it likely that such areas will experience erosion in the coming years. Thousands of seabirds and other marine organisms were oiled and killed (see effects of oil on seabirds in next section). Nearly 600 marine mammals, mostly bottlenose dolphins, were found dead ashore, and a few died from an apparent bacterial infection, *Brucella*, that normally afflicts cattle. The longer-term effects for the dolphin are unknown but may involve compromising of the dolphin immune system. While a number of sea turtle eggs were relocated to beaches in Florida, it is likely that most of an entire year class of sea turtles was lost to the Gulf. Economic losses were also severe. About half of the bottlenose dolphins examined showed evidence of direct toxicity from the oil, including lung disease (Schwacke et al., 2014). Most Gulf fisheries

off of Louisiana, Mississippi, Alabama, and the panhandle region of Florida were closed for months, although some began reopening in the fall.

Salt marshes in several areas of Louisiana suffered extensive oiling, which led to plant death and erosion. Seaward edges of marshes were oiled more and suffered erosion, which has caused marsh loss. Even 5 years later, storms erode sediments and wash buried oil onto marshes and shores in Louisiana. Such environments already have been lost at a great rate, so the Gulf oil spill represents a stress that mounts on other problems such as low oxygen concentrations and coastal habitat loss. Still, marsh areas where erosion was slight have been recolonized after 2 years by thick marsh grass (Silliman et al., 2012). Oil was found in a restricted set of marshes in Louisiana, but was not detected in most offshore sediments, except within a few kilometers of the well itself, where polycyclic aromatic hydrocarbons (PAHs) and barium concentrations were believed to be toxic to marine organisms.[2] Both macrobenthos and meiobenthos were severely affected within 3 km of the wellhead, but significant effects could also be detected in a larger area around the wellhead of approximately 150 km^2 (Montagna et al., 2013). Recovery may take several decades. In the short term, Alabama coastal juvenile fishes were not affected within important nursery eel grass beds, and some fish stocks are even larger because of the reduced fishery pressure from the fishing closures after the oil spill began (Fodrie and Heck, 2011). Zooplankton dropped during the first summer but returned to normal a year later. Still, there is early evidence that oil hydrocarbons have been incorporated into zooplankton, which might increase through higher levels of the food web. Longer-term effects or even broad areas of sea bottom that might have been affected in the early days after the blowout are essentially unknown, but after a year, widespread evidence shows that the seabed has dense populations of infaunal benthos and sediment bioturbation seems typical, indicating bottom health. An especially exciting part of the story is the question of breakdown of oil by bacteria and the contribution to recovery within the Gulf, which we discuss in **Hot Topics Box 22.1**.

HOT TOPICS IN MARINE BIOLOGY

Is the Gulf of Mexico Adapted to Oil? 22.1

The Deepwater Horizon oil well blowout dumped about 200 million gallons of crude oil into one of most productive inland seas in the world. The Gulf of Mexico ecosystems are among the most diverse in the world, with over 15,000 species. Most of the known diversity is found in depths shallower than 60 m (**Box Figure 22.1**). These shallow areas have much more habitat diversity, which probably explains the increased species diversity. The shallower areas are therefore the most vulnerable with regard to diversity loss. Shoreline habitats harbor

crucial seabird nesting areas and fish spawning grounds, which makes it likely that other major Gulf-wide ecosystem effects will be detected following a spill as large as the BP accident. Given the size of the Gulf, it seems unlikely that we will ever get a good estimate of the magnitude of larval, benthic invertebrate, and planktonic mortality caused by the release.

When the well was finally capped in July 2010, many expected the worst: a water body with oil that would persist for years, if not decades. So

---

[2] Barium in these sediments derive from oil drilling muds.

HOT TOPICS IN MARINE BIOLOGY

Is the Gulf of Mexico Adapted to Oil? *continued*   **22.1**

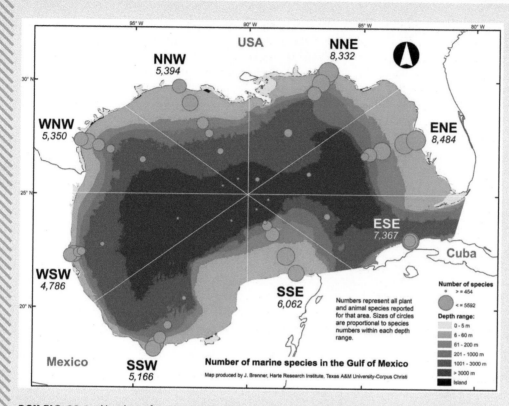

BOX FIG. 22.1   Number of marine species in the Gulf of Mexico as a function of depth. (Courtesy of Harte Research Institute)

it came as a surprise that the volume of oil in the Gulf of Mexico following the BP spill appeared to have been reduced to very low levels in only a matter of months! Some oil was trapped in booms along the shoreline and removed at the wellhead to tankers, but underwater plumes of oil were soon discovered, so not all of the oil was even at the surface to be scooped up or destroyed by surfactants, which were liberally sprayed on the surface from airplanes. Indeed, the National Oceanic and Atmospheric Administration can account for about 25 percent of the total oil as being recovered or burned, so most oil has entered the system in a variety of ways.

Where did the oil go? A previous oil well blowout in Mexico in 1979 resulted in the release of 140 million gallons of oil and, although longer-term data are largely anecdotal, short-term studies in the first 3 years showed that Texas shoreline habitats and biotas seemed to recover rapidly and persistent effects remained only in some isolated Mexican localities. Is there a pattern here?

The Deepwater Horizon well explosion occurred in April 2010, but by the following winter, officials from the U.S. Coast Guard and the National Oceanographic and Atmospheric Administration claimed that most of the oil was gone! This was a great surprise to the scientific community and to the public, who greeted the claim with a measure of appropriate skepticism. But some interesting facts about the oil distribution raised some questions. A December 2010 report from the Operational Science Advisory Team (OSAT), chartered by the U.S. Coast Guard, showed that oil was largely absent in Gulf sediments, with the exception of sites within a few kilometers of the Deepwater Horizon wellhead and some shoreline locations. Extensive sampling failed to

find concentrations of oil that exceeded concentrations of oil or hydrocarbons that might endanger human health. Most important, any samples with oil did not bear the chemical signature of oil from Deepwater Horizon (Operational Science Advisory Team, 2010). As a result, federal agencies announced that more than three-quarters of the oil was gone by the fall of 2010, owing to cleanups and microbial breakdown. PAHs have been found in sediments within a few kilometers of the wellhead, but a recent report suggests that a year later, we have bottom sediments that are burrowed and show no strong evidence of oiled and anoxic sediments. Does this indicate a recovery? Still, a massive underwater plume of hydrocarbons over 35 km long and 1,000 m deep was reported in June. Are these plumes common, and will they disappear over time?

Keep in mind that hydrocarbons have been found in plankton and benthic organisms. Bottlenose dolphins show the effects of oil toxicity. Oil has affected corals but only at sites close to the wellhead. But we are seeing only short-term effects on bottom communities and offshore water column species and apparent recovery. Why? Is it just dilution of oil hydrocarbons?

One possible clue to an important component of the apparent "recovery" can be found in the widespread natural seeps of hydrocarbons that are known throughout the Gulf of Mexico, amounting to a surprising 43 million gallons a year. Most of the known seeps are deeper than the 60 m depth boundary and are found in the central northern part of the Gulf,

continues

Is the Gulf of Mexico Adapted to Oil? *continued* 22.1

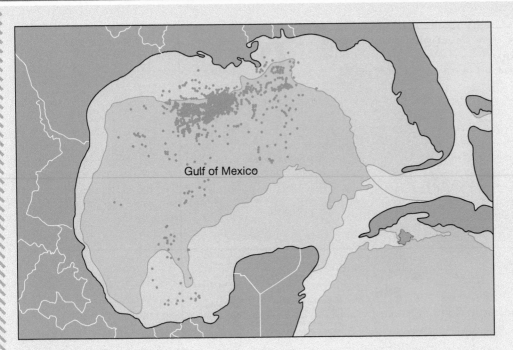

Gulf of Mexico

BOX FIG. 22.2 Distribution of known natural hydrocarbon seeps in the Gulf of Mexico. (Courtesy of Ian MacDonald, after Tunnell, 2011)

many near Louisiana and east Texas (**Box Figure 22.2**). In Chapter 18 we discussed the cold-water hydrocarbon seeps that support local hotspots of benthic diversity, resembling the faunas seen around mid-oceanic hot vents. Natural seeps of oil are very common in the Gulf, so we might expect that microbial organisms have appeared and evolved to break down and derive nutrition from the oil and especially methane gas, which can be broken down by specialized bacteria. If oil seepage suddenly increased, then oil-decomposing bacteria might reproduce rapidly and keep up with oil seepage. Keep in mind, however, that the oil spill added hydrocarbons at seven times the background seep rate from the Gulf, so bacteria may not be powerful enough to keep up with biodegradation.

A second important factor is water temperature: Microbial oil-decomposing activity might be faster in the Gulf than in the colder waters of Alaska, although we should note that the Deepwater Horizon well was about 1,500 m deep, where waters were only about 5°C. Still, currents may have spread the hydrocarbons to shallower and warmer waters. If microbial decomposition happened on the scale of the entire Deepwater Horizon blowout, then microbial breakdown may have produced a spectacular case of fairly rapid natural recovery from a stupendous human error, given the likely longer-term effects on shorebirds, perhaps future year classes of fish species, and the structural integrity of a number of marshes in Louisiana.

Do we have any evidence for such microbial action? Terry Hazen and a large group of colleagues (2010) investigated an undersea oil plume emanating from the area of the wellhead. This oil spill came at a time when next-generation sequencing methods allowed rapid molecular-based identification of bacterial groups specialized to degrade hydrocarbons. Reduced oxygen concentrations within the plume relative to outside of the plume indicated enhanced oil-degrading

microbial activity. Confirmation of increased abundances of bacteria was found within the plume, especially genetic evidence for members of the Oceanospirales within the gamma-Proteobacteria, which are known to break down petroleum hydrocarbons. Molecular biologists could also use state-of-the-art methods of transcriptomics to study degrees of gene expression of the various bacterial groups associated with hydrocarbon degradation. High levels of gene expression of these bacteria was found for *n*-alkane and cycloalkane degradation, although activity for genes involved in breakdown of more resistant but abundant components (e.g., benzene, toluene, xylene) was low (Mason et al., 2012). Oil breakdown of some components was occurring faster than might be expected at the 5°C temperatures found near the bottom at the wellhead. Apparently, propane and ethane were the two main substrates used as energy sources by a very low diversity of bacterial species within the oil plumes (Valentine et al., 2010). Kessler and colleagues (2011) provided evidence that methanotrophic bacteria likely consumed all of the methane found in a large plume that stretched offshore of the Louisiana coast. Later, methane and aromatic hydrocarbons came to dominate the plume (Dubinsky et al., 2013). Hydrocarbons in crude oil were likely also broken down in large measure by bacterial groups (King et al., 2015).

The microbial community of Louisiana beaches also responded directly to the influx of oil, especially in the rapid expansion of bacterial species known to live on oils, as shown by a metagenomic analysis of beach sediment using 16S rRNA sequencing. Species capable of growing on oil as the sole carbon source also appeared. A transcriptomic analysis also demonstrated that expression of oil-degrading genes greatly expanded in beach microbial communities (Lamendella et al., 2014). Oil disappeared faster than in another oil spill studied in a colder and more pristine habitat. Again, the resilience of the Gulf seems notable.

Is the Gulf of Mexico Adapted to Oil? *continued* 22.1

Because studies were undertaken within restricted areas, we do not know what happened on the large scale of the Gulf of Mexico, but we can formulate a hypothesis: Long-term leaks of short-chain hydrocarbons in the Gulf have resulted in evolutionary change that increased the responsiveness of bacteria to hydrocarbons. It is well known that bacteria exposed to petroleum tend to be more capable of breaking down petroleum over time (Leahy and Colwell, 1990). This response could be merely the result of an increase in gene expression as exposure to oil induces certain genes, or it might be the result of natural selection, in which certain bacterial genotypes are selected over others by virtue of their ability to obtain energy and reproduce by breaking down oil more efficiently. As it turns out, there are very few seeps near the Deepwater Horizon site, so it may be the case that natural selection for oil-decomposing bacteria might have occurred throughout the Gulf. At any one place, rare

occurrences of the oil-decomposing genotypes would be selected from low to high abundance when oil appears. It is important to realize that there is no strong evidence currently to prove that bacterial breakdown was responsible for the disappearance of any submarine oil plume. But there is every reason to believe that hydrocarbon-degrading bacteria were present both in deep waters and in coastal marshes before the oil spill. Warm water temperatures also probably accelerated degradation rates. Hydrocarbon-degrading bacteria likely increased in abundance as the oil spill hit, and a form of succession resulted in replacements of bacterial groups by others as specific hydrocarbons became uncommon (Valentine et al., 2012; King et al., 2015). The growth and expansion of hydrocarbon-degrading bacteria likely has had its own disruption on the microbial ecology of Gulf habitats, and it is of great interest to know when a new postspill equilibrium will be reached.

Despite the notoriety of major tanker and offshore drilling accidents, much oil is probably spilled during delivery of oil to harbor terminals. Spills occur when valves malfunction and when workers attempt to pump more oil into a tank than it can hold. U.S. law requires a set of containment booms to surround any marine loading area, but not all countries have legislation like this. Because of the lack of such a precaution, an August 1999 release from open valves of the tanker *Laura D'Amato* in Sydney Harbor, Australia, resulted in a spill of as much as 300,000 liters of Saudi Arabian crude oil along the shores there. The spill oiled thousands of shorebirds but dispersed from the shoreline after a few days. Chronic releases are important in increasing the concentrations of toxic substances, such as polycyclic aromatic hydrocarbons, in marine sediments (see later).

COMPONENTS AND EFFECTS OF OIL

■ The effect of oil varies with chemical composition and the affected organisms.

Oils may have the following components:

1. *Paraffins.* Straight- or branched-chain alkanes that are stable, saturated compounds having the formula C_nH_{2n} 12.

2. *Naphthenes.* Cycloparaffins that are saturated but whose chain ends are joined to form a ring structure.

3. *Aromatics.* Unsaturated cyclic compounds that are based on the benzene ring, with resonating double bonds, and six fewer hydrogen atoms per ring than the corresponding naphthene. Often toxic, aromatics have been implicated in cancers.

4. *Olefins.* Alkenes, or unsaturated noncyclic compounds with two or fewer hydrogen atoms for each carbon atom. Olefins have straight or branched chains; they are not found in crude oil.

5. *Light gases.* Hydrocarbons of very short carbon chains (1–4).

The effect of oil varies with oil chemistry and the organisms affected. Crude oil usually has less than 5 percent aromatics and is widely regarded as the least toxic. Refined oil such as fuel oil may have 40–50 percent aromatic compounds. The toxic compounds in oil are known to impair cell membrane function and may impair behavior in a wide variety of organisms. As mentioned earlier, reproduction can be impaired in invertebrates exposed to these substances. Survival and development of fish eggs and larvae are also affected negatively. Phytoplankton production can also be reduced.

■ Oil affects seabirds via direct toxic effects and by disrupting the mechanical structure of feathers.

Oil has an especially devastating effect on seabirds. Birds maintain a high and constant body temperature, and **feathers** act partially as insulation. The fluffy **down feathers** provide an air space for insulation, and the air is sealed in by **contour feathers**. The **barbules** interlock efficiently, and the hydrophobic surface of the contour feathers helps to keep water from collapsing the downy layer beneath (**Figure 22.12**). Unfortunately, oil readily coats the surface of the contour feathers and collapses their interlock. Seabirds that come into contact with oil, therefore, soon lose their insulation and are likely to die of hypothermia. The oil also impedes flight, and the birds often ingest toxic oil while preening. (Some birds such as puffins are attracted to oil, as if they expect food to be found on the surface.) Both the *Torrey Canyon* and *Amoco Cadiz* spills caused the majority of the affected Atlantic puffins and other diving birds to cease breeding. These are some of the reasons oil spills are usually followed by conservationists' frantic cleanup efforts, but historically these efforts have been in vain (**Figure 22.13**). However, more recent efforts to remove oil from seabirds have been more successful. A study of cleaning efforts of penguins following a 1994 oil spill in South Africa estimated that about 75 percent of the cleaned birds

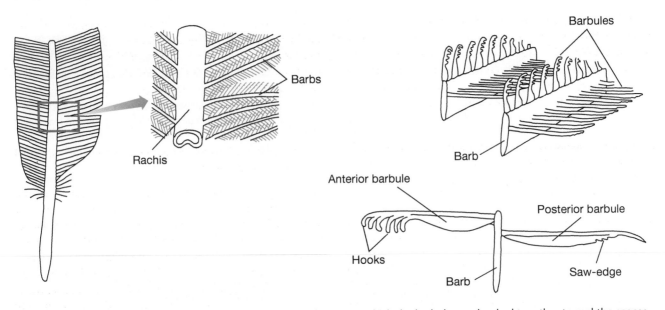

FIG. 22.12 The structure of a bird contour feather, showing the way in which the barbules are hooked together to seal the spaces between barbs. This interlocking system prevents water from penetrating to the downy layer of feathers beneath. (From Nelson-Smith, 1973)

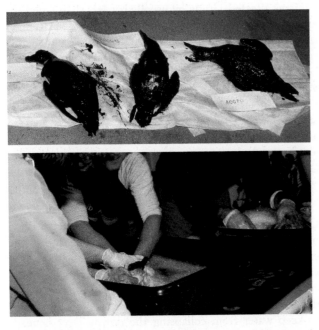

FIG. 22.13 (a) Birds washed but covered with oil from an oil spill. (b) Rescue workers attempting to wash oil from a seabird. (Courtesy of Sam Sadove)

successfully survived in the wild and that cleaning of birds significantly improved population recovery (Underhill et al., 1999). Considerable success was achieved in cleaning seabirds during the 2010 Deepwater Horizon accident.

OIL CONTAINMENT AND DISPERSAL

■ **Oil spills can be contained with floats and are sometimes dispersed with emulsifiers or naturally by storms.**

Because of the devastating effects of oil, a variety of containment and dispersal methods have been developed. In active ports, oil spills may occur every day, and tankers now routinely have floating pens to contain any spilled oil. As mentioned earlier, chemical dispersants have been used extensively, but they often damage marine life further. Surfactants are usually lipophilic (oil-compatible) molecules with a hydrophilic (water-compatible) group at one end. This structure acts to emulsify—that is, to break up—oil. Although these materials are often toxic, they are used in moderation to break up oil slicks, thereby preventing seabird mortality. In some cases, wave action breaks up the oil, so even though it may harm the benthos, it does no damage to diving birds at the surface.

Oil tankers have become larger in the last few decades, and many so-called supertankers exceed 500,000 tons in capacity, moving approximately 2 billion tons of oil by sea each year. Although it is not possible to prevent all collisions, double-walled construction and individual oil compartments help reduce leakage. A series of international agreements has helped the process of compensation for nations victimized by tanker oil spills, but the strongly international nature of oil transport and the concept of freedom of the high seas make for difficulties in regulation and enforcement. The *Torrey Canyon*, for example, was registered in Liberia, owned by an American company, and under charter to a British concern. The wreck was technically in international waters, even though the oil affected both the British and French coasts. Such international complications prevent simple legal redress.

In many cases, the origin of oil spills cannot be isolated. Tankers often spill oil at sea, and minor spills from many small vessels combine to make large-scale oil pollution problems in ports. There has been considerable effort devoted to developing methods of analysis that might fingerprint oil, so that a community could identify the source of the spill. The most promising fingerprinting techniques involve chemical analysis of the hydrocarbons in oil. A technique known as gas chromatography–mass spectrometry is especially effective because it focuses on a number of compounds whose relative proportions can be used to pinpoint a particular oil with respect to its type and origin. Triterpenes and stearanes are the most useful diagnostic compounds. Stable isotopes of carbon are also useful signatures of oil entry into marine food webs.

PAHs

■ **Polycyclic aromatic hydrocarbons (PAHs) are derivatives from fossil fuels. They are known to be carcinogenic in mammals and are major contaminants in coastal marine environments.**

PAHs derive from both point and nonpoint sources, including sewage systems, runoff, various oil spills, and burning of fossil fuels. They range greatly in molecular weight (there are mainly two-, three-, and four-ringed PAHs) and are adsorbed readily onto sedimentary particles, owing to their hydrophobic properties. PAHs therefore have become concentrated in coastal sediments and are widespread in near-shore bottoms in urban zones. PAHs can be toxic to benthic invertebrates, and various tissue abnormalities and cancers in fishes have been related to these compounds. Some PAHs can induce cancers in laboratory mammals, and their occurrence in resource species has troubled environmental managers. In the Hudson River, the tomcod *Microgadus tomcod* accumulates high concentrations of PAHs and other lipophilic contaminants and has high frequencies of liver tumors (Wirgin et al., 2006). Reduced performance has been related to a steady decline in reproductive success over the years. Bacteria in sediments degrade PAHs, but the rate of degradation is much slower for the higher-molecular-weight forms.

■ **PAHs and PCBs severely disrupt endocrine function in vertebrates and may be a major cause of reproductive failure.**

PAHs and PCBs both appear to strongly affect reproductive cycles, particularly in fishes. They act as **endocrine disrupters**, exerting their influence by mimicking the effect of endogenous hormones such as estrogens and androgens, antagonizing the effects of endogenous hormones, altering the pattern of synthesis and metabolism of normal hormones, and modifying hormonal receptor levels. In vertebrates, the secretion of gonadotropin-releasing hormone and follitropin from the hypothalamus and pituitary glands stimulates ovarian follicle growth and estradiol synthesis in the female. In oviparous vertebrates including fish, the release of estradiol from the ovary causes the liver to produce large amounts of vitellogenin, a lipoprotein precursor for egg yolk. Circulating vitellogenin and hormone levels in natural fish populations during the reproductive season have been shown to be a promising indicator of incipient reproductive dysfunction. High vitellogenin concentrations in male fish from contaminated environments have been interpreted as indicative of the estrogenic properties of contaminants.

Induction by PAHs and coplanar PCBs of *CYP1A*, the gene for cytochrome P450–1A, has been found to be one of the most sensitive if not the most sensitive response and has been used extensively in field investigations. The induction of the gene *CYP1A* is known to play a role in both detoxification and activation of toxic compounds and is also significant for health of the fish (Haasch et al., 1993).

Nutrient Input and Eutrophication

■ **Agricultural activities and sewage add nutrients, as well as disease organisms, to the water.**

Agricultural activities and sewage release cause a great deal of damage to marine life and contaminate fishes and shellfish. The major impact is the result of **nutrient release**, which indirectly reduces water quality. With sewage and animal waste, undesirable microorganisms are also released into the marine environment. Pathogens such as hepatitis viruses and the bacterium *Salmonella*, often concentrated by suspension feeders such as clams and mussels, may be the cause of a variety of diseases. Outbreaks of dysentery and other diseases can be common in areas where people collect shellfish near sewage outfalls and in heavily populated areas with septic tanks. Local environmental agencies in the past have counted the number of **coliform bacteria** in seawater, whose source must be the wastewater outlets. These bacteria are associated with the human gut and are counted because they are positively correlated with release of other pathogens. Fecal streptococci, or **enterococci**, have become a substitute indicator of pollution.

Nutrient Sources

■ **Human activities result in large additions of dissolved nutrients to coastal waters.**

Eutrophication is the addition of dissolved nutrients to a water body, resulting in large increases in phytoplankton production and microbial activity. In coastal waters, eutrophication is due to several nutrient sources related to human activity. The following are the principal known sources:

1. Point sources such as sewage treatment outfall pipes
2. Point sources such as storm sewer overflows (SSOs), which may be connected to sewage pipe systems
3. Nonpoint sources stemming from runoff from the watershed, with sources such as general runoff from agricultural and suburban lands owing to the use of commercial fertilizer, animal wastes, and increased supply of nutrients because of disturbed soils
4. Atmospheric deposition

Figure 22.14 shows an estimate of the relative contributions of nitrogen addition to the Chesapeake Bay watershed. Note that **agricultural sources** dominate, but there is a surprisingly large contribution from **atmospheric deposition**, which will be discussed shortly. Municipal wastewater is a considerable source, although septic systems are not. In areas such as New York Harbor and Boston Harbor, point sources such as **combined sewer–storm pipe outfalls and sewage treatment plant outfalls** are the major contributors

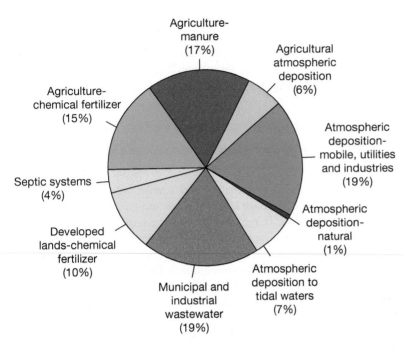

FIG. 22.14   Relative contributions of nitrogen to the Chesapeake Bay watershed. (Courtesy of the Chesapeake Bay Program)

of nitrogen. In New York Harbor, nutrient supply is so plentiful that nitrogen is often not a limiting factor to phytoplankton growth.

■ The atmosphere can be a major source of nutrient addition to coastal bays.

Fossil fuel combustion is a major potential source of nitrogen oxide emissions. These gaseous emissions are eventually returned to the earth as soluble nitrates in wet or dry precipitation. The material becomes part of the now-famous acid rain, whose sulfur components may reduce the pH values of some lake and estuarine waters to the point of toxicity to fishes. The nitrates deposited by the rain are worrisome partially because they may stimulate primary production, which leads to eutrophication. Rainwater in the coastal zone can stimulate phytoplankton growth (Paerl et al., 2002). This effect is likely to be especially strong in more offshore shelf waters, where phytoplankton primary production is severely limited by nitrogen concentration. Along the east coast of the United States, atmospheric deposition accounts for 10–40 percent of the new nitrogen additions to coastal waters.

Effects of Added Nutrients

■ Nutrient stimulation of primary production often results in hypoxia or anoxia.

Nutrient additions of nitrate and ammonia stimulate phytoplankton growth, as discussed in Chapter 11. At modest levels, one might expect higher water column and benthic production. These nutrient additions also stimulate bacterial production, however, especially in water columns where light is reduced. This is a problem in rivers and estuaries, where high particle loads reduce light penetration. As large populations of phytoplankton and bacteria build up, there is little likelihood that a zooplankton population will graze

the phytoplankton. In the shallow waters of southern San Francisco Bay, the benthic suspension feeders can often keep up with the strongly eutrophic waters, and phytoplankton populations do not build up. This is not the case, however, in deeper estuaries with high nutrient loads and sluggish circulation. During the late spring and summer, when the water column stabilizes, much of the phytoplankton may die. Subsequent degradation by aerobic bacteria strongly reduces the level of dissolved oxygen, which is essential for nearly all animals (**Figure 22.15**). **Hypoxia** is the condition of strongly depleted dissolved oxygen. **Anoxia** is the complete absence of oxygen. Both hypoxia and anoxia may be accompanied by significant concentrations of hydrogen sulfide, which is toxic to many marine organisms. The lowered oxygen and the presence of hydrogen sulfide cause mass mortality of fishes and benthos. The rise of bacterial degradation is also exacerbated by the turbidity due to dense phytoplankton, whose shade prevents primary production in deeper water.

Nutrient addition to coastal waters has become one of the major world problems and has resulted in massive **dead zones**, near-surface bodies of waters nearly devoid of oxygen. Such zones have been found in the Black Sea, at the mouth of the Danube River, in the Baltic Sea, in the Gulf of Mexico near the mouth of the Mississippi, and in various sites in shallow seas of east Asia. In the typical case, large amounts of nutrients from a watershed are flushed out of a river into the coastal zone, stimulating primary production, microbial activity, hypoxia, and anoxia. An Oregon case was associated with the California current, usually a source of upwelling and never before associated with hypoxia on the Pacific continental shelf (Chan et al., 2008).

The dead zone off the mouth of the Mississippi (Rabalais et al., 2002) has been quite large: In 2007, it was approximately an area the size of New Jersey! Spring and

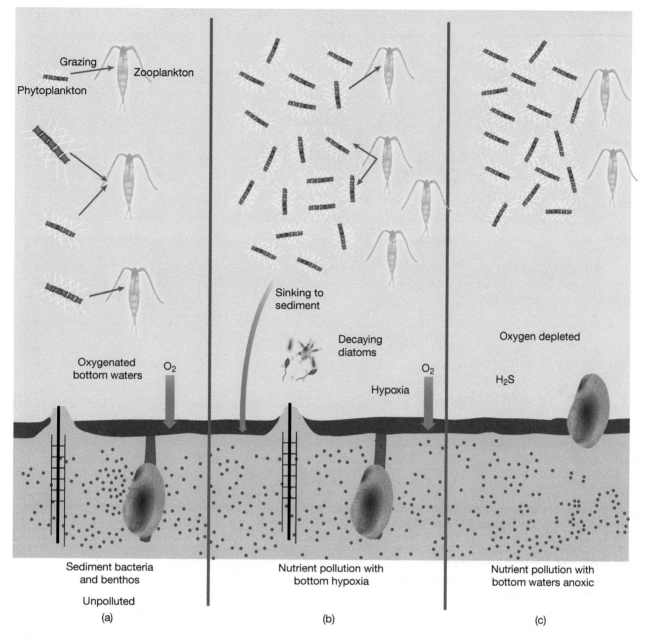

FIG. 22.15 Development of hypoxia in an estuary. (a) Normal situation: Much of the phytoplankton is grazed and bottom waters are oxygenated. (b) Nutrient input from sewage stimulates phytoplankton growth, and some dead phytoplankton sink to bottom waters; bacterial decomposition reduces oxygen, and other material sinks to bottom sediment, where more oxygen is consumed from bottom waters. (c) Oxygen is removed from bottom waters and benthos die.

summer conditions of surface water warming and high freshwater outflow create a strong density stratification, trapping nutrient-rich water in a stable water column near the coastal zone. The high microbial activity in this water body removes most of the oxygen (**Figure 22.16**). Studies from cores show that these conditions started around 1900 but became much more extreme beginning in the 1950s, when the Mississippi watershed was inundated with nitrates deriving from agricultural fertilization. The benthos beneath the dead zone consists of mainly small surface deposit-feeding annelids that respond to the deposition of particulate organic matter. The surface waters reduce fisheries, since fish either die or swim away from this stressful, poorly oxygenated water body.

ABATING EUTROPHICATION

■ **Eliminating ocean dumping of solid sewage waste and better treatment of sewage before wastewaters are released into the coastal zone can abate eutrophication.**

Hypoxia and anoxia events in bays and estuaries have been frequent enough to draw public attention to the problem of nutrient additions, and sewage treatment has generally improved throughout the United States. Sewage treatment plants (**Figure 22.17**) typically practice **primary treatment**, in which solids are intercepted by screens, or **secondary treatment**, in which more toxic nitrogenous organic compounds and colloids are stirred in aerobic tanks so that only phosphates, nitrates, and ammonia will be released into

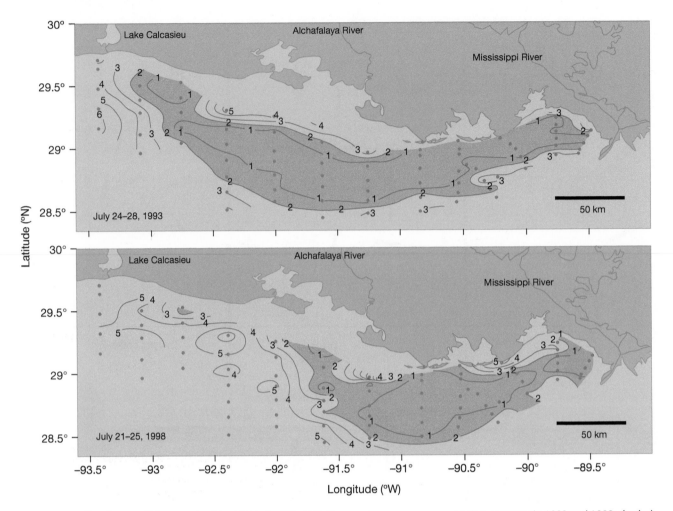

FIG. 22.16 Dead zone off the mouth of the Mississippi River. Bottom-water oxygen concentration contours in 1993 and 1998; shaded areas have concentrations below 2 mg/L. (After Rabalais et al., 2002)

coastal waters. The solid residue must then be disposed of. Very few treatment plants carry out **tertiary treatment**, in which even dissolved phosphates, nitrates, and ammonia are removed. Various anaerobic tanks may be employed to enable microbial removal of the dissolved nitrogen as gas, and iron is used to combine with phosphate; however, such practices are very expensive. These costly tertiary treatment methods, moreover, produce large amounts of sludge, which must be disposed of. Many municipal outfalls throughout the world do not even have primary treatment. The recent increases in hypoxia and anoxia events have stimulated interest in the expansion of secondary and tertiary treatments. Thanks to the improvement of sewage treatment, dissolved oxygen in New York Harbor has improved steadily over the last 50 years (**Figure 22.18**).

Recently, new methods of sewage treatment were introduced to provide a direct and workable alternative or addition to tertiary treatment. The ammonia reduction process (ARP) involves concentration of sewage sludge, removal of ammonia as gas by heating and pressure reduction, reaction with sulfuric acid, and conversion to ammonium sulfate, which can be used directly as fertilizer. This technology is being tested in New York City wastewater treatment plants and is predicted to reduce nitrogen inputs by about half. New York City has adopted a technique in Jamaica Bay in which ammonium is converted to nitrite and nitrate and nitrogen is extracted through the addition of glycerol as a carbon source.

The relative importance of nitrogen and phosphorus as nutrients becomes an important issue in tertiary treatment. Phosphorus is much cheaper to remove than nitrogen, although the ARP process or glycerol addition may prove economical. In most marine waters, nitrogen is generally believed to be the limiting nutrient. In estuaries, there is more room for dispute because phosphorus is known to be limiting in freshwater rivers. An extensive study of lower Chesapeake Bay shows that nitrogen is limiting there. This is a major problem because nitrogen does not come primarily from point sources, such as sewage treatment plants, but from a series of tributaries and general runoff. Control, therefore, will be very difficult.

■ **Reduction of nutrient input into coastal bays and estuaries from point sources has been very successful in reducing phytoplankton in the water column, increasing water clarity, and allowing submerged attached vegetation to recover.**

Reduction of nutrient input in wastewater treatment plants, or point sources, has been a major objective in areas with

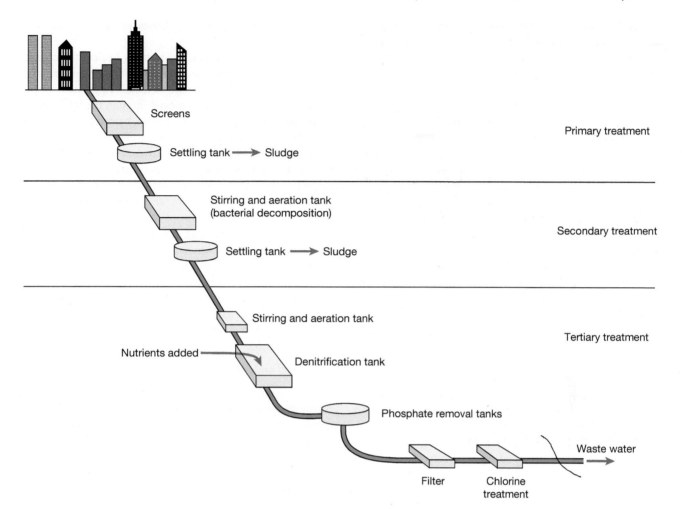

Primary treatment

Secondary treatment

Tertiary treatment

FIG. 22.17   Three types of sewage treatment.

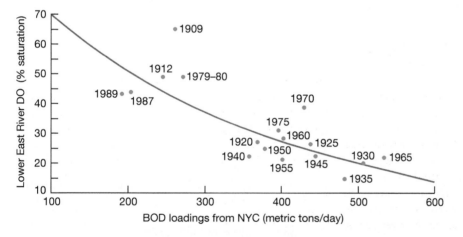

FIG. 22.18   The DO (dissolved oxygen) in the East River, New York Harbor–Estuary, as a function of the particulate organic matter entering the harbor from sewage (measured in terms of the potential biological oxygen demand [BOD] that the material places on the harbor waters). As BOD loading has declined, oxygen in harbor waters has increased. The years 1909 and 1912 probably represent times when material was released into marshlands and not directly through pipes into the harbor—hence, the lower BOD loading. (Courtesy of Dennis Suszkowski, Hudson River Foundation)

large urban and suburban populations. As discussed earlier, nutrients stimulate phytoplankton growth, which adds nutrients to the bottom sediments and causes hypoxia. Increased phytoplankton also intercepts light, which strongly affects and eventually eliminates sea grass growth, as discussed in Chapter 17. Nitrogen reduction is the target in saline waters, but phosphorus is also an important target in freshwater parts of estuaries.

In response to these problems, many coastal areas have instituted plans to reduce nutrient input, often with success.

An example is the Back River estuary near the city of Baltimore. Since the 1980s, a concerted reduction of nitrogen and phosphorus input has resulted in declines of nutrient concentrations and also of phytoplankton. This was especially noticed when wastewater treatment plants were upgraded in the 1990s. The positive response took a few years, probably because a large amount of nutrients remained in the sediment, which continued to fuel phytoplankton growth. Such reductions in phytoplankton also have resulted in increased water quality. Nutrients in Jamaica Bay were reduced from about 22,680 kg d^{-1} to 13,600 kg d^{-1} during the period 2010–2016. In Tampa Bay, a great increase of population growth resulted in heightened nutrient input and strong growths of phytoplankton and the green seaweed *Ulva*, which rotted on shorelines. At least half of the former sea grass beds had been lost because of increased turbidity caused by the phytoplankton. By 1981, upgrades of sewage treatment plants resulted in a 90 percent reduction of nitrogen input. Chlorophyll has been greatly reduced, and sea grass is recovering rapidly (Greening et al., 2011). Reductions of phytoplankton in the water column have also resulted in steady recovery of submerged attached vegetation in the waters of the Potomac River estuary (Ruhl and Rybicki, 2010). Overall in the Chesapeake Bay region, however, water quality has not improved, bottom communities have had reduced biomass diversity and abundance in recent years, and sea grasses are still in great decline. One thing the success stories have in common is that corrections could be made at the wastewater treatment plants because they are point sources of nutrients.

Thermal Pollution and Power Station Fish Mortality

■ **Power-generating stations require water for heating and as a result kill aquatic life, by entraining and impingement.**

Conventional and nuclear power stations require large amounts of water to transport heat from the power-generating system. Two popular means of heat dissipation are used. In the first, cold water is taken from a bay or river and passed through the plant; the heated water is then circulated through large cooling towers, such as those at the famous Three Mile Island nuclear facility in Pennsylvania. The towers then radiate heat into the overlying atmosphere. The relatively cool water is next returned to the river or can be recycled in a closed system between the power plant and the cooling tower. The second major approach involves removing water from a bay or estuary, passing it through the plant, and returning it quite hot to the environment. Typically, the water is returned to a man-made embayment and then passes out to the adjacent estuary or coastline while still approximately 10–15°C over ambient temperature. The released warm-water effluent may also form a plume that moves out into the open water. A warm-water plume may attract fish that should instead migrate away from the otherwise cooling waters in the autumn. If the plume should

dissipate during a storm, the fishes probably would die owing to the cold shock.

Whether or not the returned water is heated, the uptake of large volumes of water at intakes creates the problems of **entrainment** and **impingement** of fishes. Eggs, larvae, and juveniles may be entrained or moved through the intake pipe, passed through the power plant system, heated suddenly, and returned to the open water through an outflow pipe. A significant amount of mortality is often associated with this passage. If the fishes are larger, they may be impinged, or trapped, on the intake screens, which often kills them.

Impingement results from the strong pressure created by water intake pumps. Various screens have been designed, partially to reduce fish kills but mainly to keep pipes from clogging. A barrier net placed in the surrounding area can be helpful, but such nets often become fouled with algae and sometimes cannot be deployed in summer. At intake sites, the angle screen is a popular design because it may divert fish along itself to an exit pipe, supposedly reducing fish mortality, but this has been questioned (Fletcher, 1985).

■ **Thermal emissions may also affect plant production.**

Thermal pollution may inhibit phytoplankton growth and change the character of plant communities in the vicinity of warm-water outfalls. In the Turkey Point nuclear generating station outfall in South Florida, turtle grass was replaced by blue-green cyanobacteria. The hot waters were often too extreme for all but the hardy blue-greens. Similar effects have been observed throughout the world—for example, at Zhanjiang Bay China, where thermal effects reduce phytoplankton production and diversity (Li et al., 2014). Usually, thermal effluents are a certain number of degrees above ambient and therefore exert their strongest effects when summer water temperatures are maximal. These effects, however, are strongly localized, especially when the outfall mixes with coastal waters of high current energy.

Global Environmental Change and the Ocean

Throughout the text we have described the main effects that exist and may continue to develop as a result of global climate change. Here we will summarize these effects and briefly outline their collective possible future impact on marine biological processes, from the level of individual function to the level of biodiversity.

The Effects of Burning of Fossil Fuels and Additions of Greenhouse Gases

■ **Industrial activities have caused the net addition of carbon dioxide and other greenhouse gases to the atmosphere since the nineteenth century. These additions are significant on a geological scale.**

Human activities have reached a point at which they are altering the earth's global climate. Since the Industrial Revolution in the nineteenth century, industrial activity has greatly accelerated the burning of fuels, particularly fossil fuels, such as coal and petroleum products. Mainly as a result of this activity, approximately 35 percent has been added to the storehouse of carbon dioxide in the earth's atmosphere. Measurements from Mauna Loa, Hawaii, and other inferences suggest an increase from 280 ppm in the nineteenth century to the current concentration over 400 ppm. Deforestation and other greenhouse gases are secondary sources of carbon release or heat storage.

■ Carbon dioxide additions to the atmosphere have caused increases of sea-surface temperature through at least the past 150 years.

In the past 150 years, ocean temperature has increased throughout the world, with rather large changes in the Arctic Ocean and in the Antarctic Ocean near the Antarctic Peninsula. Widespread increases have also been recorded in coastal areas of New Zealand, British Columbia, and Massachusetts, and within the Hudson River and Chesapeake Bay. Records at Woods Hole and the Baltic Sea date back to the nineteenth century, and one can see an acceleration of temperature rise in the past 35 years. Worldwide (excluding the Antarctic Ocean), sea-surface temperature has increased approximately 0.6°C from about 1900 to 1990 (Rayner et al., 2003). Sea ice has also been affected, most intensively in the Arctic Ocean, the Barents Sea, and the Antarctic Peninsula.

■ Carbon dioxide additions have resulted in a reduction of seawater pH.

As explained in Chapter 3, carbon dioxide additions to the atmosphere eventually increase dissolved CO_2, which increases ocean acidity. Ocean pH has declined in the surface ocean, although data are only sparsely available. A decline of 0.1 pH unit has been measured ocean-wide, but higher rates have been noticed, for example, in the northeastern Pacific. Within estuaries, pH fluctuates a great deal and low values are to be expected even now, especially when oxygen is depleted, causing microbial activity and increases of dissolved CO_2.

BIOLOGICAL EFFECTS AT THE INDIVIDUAL LEVEL
■ Increases of sea-surface temperature affect physiological function, migration patterns, and geographic range.

Increased sea-surface temperature exerts a variety of physiological functions. Increased temperature will affect cellular oxygen demand, which may be limiting. Species at the low-latitude end of their current geographic range will have their ranges curtailed, as has occurred for the common mussel *Mytilus edulis* on the east coast of the United States. In some cases, this will involve species trapped within estuaries, which will cause local extinction. Increases of ocean temperature at the low-latitude end of a geographic range may also affect migration patterns and foreclose some feeding and spawning areas to some species. At the high-latitude end of a species range, different species may move to cooler

waters. For example, in the North Atlantic, some essential plankton species have moved northward, which has starved sand eels, whose decline has caused reproduction crashes in a number of nesting seabirds in Scotland (Dybas, 2006).

Ocean warming has already exerted some significant changes in reproductive cycles and marine migration patterns. In the North Sea, echinoderm reproductive cycles have been advancing in time over the past few decades, as sea temperature increases (Edwards et al., 2009). Climate change in the northeastern Pacific has caused increases of sea temperatures, and recent work in a warming Alaskan stream demonstrated that migration patterns of pink salmon populations have evolved in the past 4 decades to shift their return to rivers earlier by 2 weeks (Kovach et al., 2012).

Some species or groups will suffer quite specific effects. We discussed coral bleaching in Chapter 18 and the effect of increased temperature, which is well known from studies of effects of ENSO events. A number of predictions suggest widespread increases in bleaching in coming decades. Temperatures in the Caribbean can now be predictably related to coral bleaching. As discussed in Chapter 17, a number of large-scale bleaching events have occurred in recent years, and the northern third of the Great Barrier Reef witnessed unprecedented bleaching in 2016.

■ Increases of sea-surface temperature may affect the impact of disease spread.

The spread and effects of disease may be strongly affected by increases of ocean temperature. Species that are physiologically stressed by high temperature may be more susceptible to disease. Some infectious diseases may spread to higher latitudes as temperatures increase. Disease in oysters and other bivalves has increased in the northeastern United States, and this may continue with regional warming. The oyster disease Dermo is moving northward in the eastern United States as warming occurs.

■ Decreases of pH are influencing calcification.

In certain parts of the ocean, such as the northern Pacific, there already is evidence that increases of ocean acidity will soon affect calcification in organisms ranging from calcareous plankton to corals. Recent evidence suggests that bivalve larval shells are already being affected, relative to presumed levels of atmospheric carbon dioxide in the nineteenth century. Larvae of the bivalve *Mercenaria mercenaria* showed improved survival, growth, and metamorphosis when grown at preindustrial levels of carbon dioxide, relative to those of the present day (Talmage and Gobler, 2010). Species that secrete the less stable form of calcium carbonate, aragonite, are at greater risk than species with skeletons composed of calcite. Because the solubility of calcium carbonate increases with ocean depth, some deep-sea species, such as the corals that live on deep-sea coral mounds, are at risk in depths of 800–1,000 m. As we discussed in Chapter 17, reduced calcification of corals on the Great Barrier Reef in recent years may be related to acidification. Increased CO_2 may increase primary productivity, as it has done in terrestrial experiments; but in the long run, lowered pH will also affect calcifying plankton.

BIOLOGICAL EFFECTS ON BIODIVERSITY

■ **Changes of pH and sea-surface temperature may cause the loss of foundation species for major communities.**

Temperature and associated changes have particularly strong impacts when they affect foundation species in communities. We have discussed a number of communities where dominant living species function as foundation species. Kelps are adapted to cool oceanic waters, and global sea-surface temperature change will likely have strong effects on a number of kelp communities. These impacts are to be expected in both the North Pacific and North Atlantic, and *Laminaria* stands in southern New England are likely now at risk. Temperature-induced bleaching will cause extensive coral mortality, and thousands of species depend on dominant corals in tropical coral reefs. Minimally, there will be strong rearrangements in dominance. Such problems may interact with another predicted associate of global warming, intense storms. Such storms were already instrumental in destroying the typical coral-dominated communities on the north coast of Jamaica, as discussed in Chapter 17.

■ **Changes of sea-surface temperature may cause increases of the success of invasions of alien species and rearrangements of local species abundance.**

Increases of local sea-surface temperature may have a dual impact. On the one hand, native species might suddenly be subjected to temperature stress. But invasive species from warmer water sources might be able to invade and outcompete local species, which were adapted to cooler temperatures. This process appears to be important in the invasion of Long Island Sound, New York, by warmer-water colonial sea squirts. Because local disturbance is an important source of increased risk of invasive success, we can expect global warming to result in increased success of invasive species. We also expect that invasive species will be predominantly warm-adapted or thermally tolerant. The worldwide invasive crab *Carcinus maenas* is notable for its wide thermal tolerance (Tepolt and Somero, 2014), which has allowed it to invade warm and cold waters alike.

■ **Overharvesting of species or habitat destruction may result in complex negative interactions with global climate change impacts.**

The marine realm has been very disturbed by human activities, ranging from habitat destruction to strong overfishing. As temperature increases in some localities, it may become very difficult to restore communities that have been removed by human activities.

■ **Increased temperature may cause sea-level rise and major changes in oceanic circulation.**

Sea-level rise will be caused by a combination of ocean warming, increasing water volume, and melting of glacial ice.

Drowning and rearrangement of estuarine circulation patterns, erosion of marshes, and changes of erosional regimes of sandy shores and barrier bars are all likely outcomes. Storm surges will make some of these effects catastrophic, as in the recent impacts of Hurricane Sandy on the northeast United States. On a more subtle level, melting of Arctic Ocean sea ice has resulted in major changes of productivity patterns and plankton patterns in the northwest North Atlantic continental shelf plankton communities (see Chapter 19 discussion on Arctic climate change).

■ **Sea-level rise and climate change may strongly affect coral reef survival.**

Grigg and Epp (1989) suggested that sea-level rise owing to global warming might outstrip coral growth, which would trigger a worldwide catastrophe. More locally, some reefs are located on sinking oceanic islands and might be vulnerable. Drowned coral reefs have been found throughout the Pacific, but it is not known whether drowning occurred under circumstances similar to today's conditions. Even if the reef were able to grow as sea level rose, increasing sea temperatures might cause extensive coral bleaching and contraction of reef systems. El Niño periods often exceed the thermal limits of hermatypic corals, and general warming could lead to catastrophe. It is not clear, however, that the thermal effects would overwhelm all reef species; some might survive better than others, perhaps coming to dominate a new reef community able to survive warmer temperatures.

■ **Increased temperature and carbon dioxide may increase biological productivity, especially in nutrient-enriched estuaries.**

Any prediction of how global warming might affect production in the ocean must be extremely speculative. Carbon dioxide increases can stimulate plant growth, which may increase primary production in waters in which nitrogen (for example) is not strongly limiting. One can imagine a scenario in which currents increase and bring deep-water nitrogen to the surface while increased carbon dioxide increases photosynthesis. In some polluted estuaries (e.g., New York Harbor), nutrients are not limiting during the year, and primary production increases as a response to increasing temperature. Under such circumstances, global warming might increase primary production, which in turn would cause more waters to be hypoxic, as explained earlier in this chapter. The appearance of dead zones throughout the world in recent years may be related to climate change. As discussed in Chapter 11, some common phytoplankton such as coccolithophores are known to respond positively to carbon dioxide additions to seawater.

■ **Increase of greenhouse gases and global warming could intensify coastal upwelling and increase primary production.**

Any process affected by temperature change may be affected by global warming. One of the most likely factors to be affected is the balance of winds along the coastline. As discussed in Chapter 11, upwelling and enhanced primary productivity occur in regions where wind stress brings nutrient-rich deep water to the surface. Along the California coastline, northerly and northwesterly winds cause the offshore transport of surface water, a process that is in turn balanced by a rise of nutrient-rich deep waters. Global warming can accelerate such a process because daytime temperature on the adjacent land will increase and nighttime cooling will decrease. This overall temperature increase will accentuate the temperature difference between the land and the adjacent cool ocean, which will intensify winds and upwelling. Andrew Bakun (1990) has demonstrated that wind speeds in upwelling regions have steadily increased since the 1950s, when data first began to be collected, and this finding is consistent with the steady increase in greenhouse gases. An increase in primary productivity has not

been measured but likely has occurred because upwelling of nutrient-rich deep water is the limiting factor in these systems. Whether the changes have, or will have, an enhancing effect on fisheries is not yet clear.

■ **Changes in primary production may occur in the open ocean over a few decades, but there is mixed evidence at present that primary production has increased to any degree over the last 70 years or so.**

As discussed in Chapter 11, we might expect sea-surface warming to produce increased stratification, reduced vertical water exchange, and reduced primary production in surface waters of the open sea such as gyre centers. The evidence of change on the scale of 50–100 years is mixed, but reduced primary production in the open ocean near gyre centers since the 1980s has been confirmed.

■ CHAPTER SUMMARY

- Pollution has both chronic (long-term) and acute (short-term) sources that impact the marine environment. Complex interactions among human impacts from many sources can make it difficult to pinpoint the role of particular pollutants.

- Common species are often chosen as bioassays of pollution and its effects on mortality, population growth, physiological condition, and gene expression. Studies may correlate the release of toxic substances with their uptake by individuals. Some populations evolve resistance to toxic substances.

- Many toxic substances transfer from one trophic level to the next as predators consume prey. In the process, the concentration of some substances is biomagnified, reducing biodiversity at the highest trophic levels.

- Heavy metals, including lead, cadmium, and mercury—as well as the deadlier *methylmercury*—are toxic to humans. Pesticides and polychlorinated biphenyls (PCBs) washed into coastal waters are often toxic to marine life. Other major contaminants include oil, particularly refined forms, and derivatives of fossil fuels called polycyclic aromatic hydrocarbons (PAHs). They are known to be carcinogenic in mammals and are major contaminants in coastal marine environments. PAHs and PCBs disrupt endocrine function in vertebrates and may be a major cause of reproductive failure.

- Human activity adds greatly to dissolved nutrients in coastal waters. Sources include sewage, nonpoint coastal sources, and atmospheric deposition. Nutrient stimulation of primary

production often results in hypoxia or anoxia. Most worrisome are coastal dead zones. Eliminating ocean dumping and having better sewage treatment before wastewaters are released can abate eutrophication.

- Industrial activities have been adding greenhouse gases to the atmosphere since the nineteenth century. As a result, increasing sea-surface temperatures have likely increased the spread of disease and affected physiological function, migration patterns, and geographic range. The declining pH of seawater has caused the danger of reduced calcification. These changes may bring about the loss of foundation species for major communities. Sea-level rise and climate change may also directly affect coral reefs, mangroves, and salt marshes.

■ REVIEW QUESTIONS

1. What are the differences between point sources and nonpoint sources of pollution?

2. What are three general approaches that can be used to assess the overall effects of pollution in a given habitat?

3. Why are pesticides, which are designed to kill terrestrial arthropods, often quite dangerous in marine ecosystems?

4. Why is mercury a particular danger?

5. What effects does spilled oil have on marine environments?

6. Why does oil composition and type matter with regard to toxic effects?

7. Why has the treatment of oil spills sometimes been as damaging as the oil spills themselves?

8. Nutrient input increases primary production, which should support more fish. So why are people upset about increasing sewage input throughout the coastal ocean?

9. How does nutrient enrichment lead to hypoxia?

10. In what type of marine environment is atmospheric deposition of nutrients likely to have the greatest effects?

11. Why are plants that generate electric power a potential danger to fish stocks?

12. Which marine environments do you think are more vulnerable to pollution: tropical ones or those in high latitudes? Why?

13. Why has carbon dioxide increased greatly in the atmosphere in the last 100 years or so?

14. How might carbon dioxide increase affect global climate? Oceanic circulation?

15. Why would it be useful to use microarray approaches to studying responses to pollutants?

16. If a toxic substance is removed from an area, what might happen to a

species population that had evolved resistance to that particular pollutant?

17. Why do you think that exposure to two different toxic substances might have very strong effects on an organism, much more than the sum of the individual effects of the two substances?

18. We have shown that in many cases populations can evolve resistance to pollutants. Does that mean that pollution is not really a problem in the marine environment? Under what circumstances to you think that evolution to pollutants might not occur?

Visit the companion website for *Marine Biology* at www.oup.com/us/levinton where you can find Cited References (under Student Resources/Cited References), Key Concepts, Marine Biology Explorations, and the Marine Biology Web Page with many additional resources.

Abiotic interactions. Interactions between an organism and physical and chemical aspects of the environment.

Abyssal plain. The deep ocean floor, an expanse of low relief at depths of 4,000–6,000 m.

Abyssobenthic. Referring to benthic organisms living at abyssal depths.

Abyssopelagic zone. The 4,000–6000 m depth zone, seaward of the shelf-slope break.

Acclimation. Given a change of a single parameter, a readjustment of the physiology of an organism, reaching a new steady state.

Accuracy. The correctness of a measure when comparing to a known standard. *See also* Resolution and Precision.

Action spectrum. Portion of the light spectrum used in photosynthesis.

Active suspension feeders. Suspension feeders that produce a current to actively entrain and capture particles.

Adaptation. An increase in fitness over time as a population evolves in response to a new environment.

Aerobes. Organisms that require and use oxygen in metabolism. *See also* Anaerobes.

Aerobic scope. The range of possible oxidative metabolism from rest to maximal exercise. *See also* Scope for activity.

Aerosol. A variety of substances that may seed clouds and therefore increase cloud cover.

Age structure. The relative abundance of different age classes in a population.

Aggregated spatial distribution. A case in which individuals in a space occur in clusters too frequent to be explained by chance.

Ahermatypic. Non-reef-building (referring to scleractinian corals).

Alginate. A natural polysaccharide that is a major component of many brown seaweeds.

Allee effect. The increase of probability of extinction when the population is so small that population density itself matters. Finding mates in a sparse population is the major mechanism of decline in such a case.

Allele. One of several variants that can occupy a locus on a chromosome.

Allelopathy. The production of a substance by one species that is toxic to another species that may live adjacent to the first species.

Allopatric speciation. The differentiation of geographically isolated populations into distinct species.

Allozyme. A variant of an enzyme type. These may be variants of a specific enzyme (e.g., cytochrome *c*) that are the products of a single genetic locus.

Alternative stable states. Depending on specific circumstances or sequences of events, a community might move toward and then exist in one of two or more states that are stable for long periods of time until a major perturbation occurs.

Amensal. Negatively affecting one or several species.

Amino acids. Basic structural units of proteins.

Amnesic shellfish poisoning. Uptake by shellfish of domoic acid when feeding on certain species of *Pseudonitzschia*, resulting in being poisonous for consumption by humans.

Amphidromic points. Points in the ocean that do not change in tidal height. Tidal currents tend to revolve around these points.

Anadromous fish. Fish species that spends most of its life feeding in the open ocean but migrates to spawn in fresh water.

Anaerobes. Organisms that carry out metabolic processes in the absence of oxygen. *See also* Aerobes.

Anchor dredge. A bottom dredge used to dig into soft sediments down to a specified depth below the sediment–water interface.

Annelida. Phylum characterized by wormlike shape, segmentation, including free-living and parasitic forms, but usually benthic occurrence of free-living forms.

Annual. A species that is born and programmed to die after it reproduces in its first year of life.

Anoxia. Lacking oxygen.

Anthozoa. A class of the phylum Cnidaria, including anemones, corals, and sea fans.

Antibody. A specialized protein produced in response to the introduction of a foreign molecule into the body, usually to destroy the foreign body.

Apoptosis. A form of cell death, in which a series of cellular processes determines or programs the death of the cell, usually without harming the rest of the cells in the organism.

Aposematic coloration. A conspicuous coloration borne by an organism that has evolved to signal to a predator the organism's distastefulness or even poisonous state.

Applied marine biology. The use of marine biology to study practical problems such as pollution, or the use of biological structures to protect against wave surge.

Aragonite. An orthorhombic crystal variant of calcium carbonate that is less stable and more soluble than another variant, calcite. A common constituent of coral skeletons and many snail shells (exception: genus *Conus*).

Aragonite saturation horizon (ASH). Depth below which aragonite-secreting organisms cannot secrete aragonite and live.

Arbuscular mycorrhizae. Fungi that live as mutualists in association with the roots of plants.

Archaea. One of the three major domains of life, including many types of microorganisms that are known mostly from extreme environments, such as those lacking oxygen or at high temperature.

Arctic Oscillation. Temporal variability of climate involving winds circulating counterclockwise around the Arctic at around 55° N latitude.

Aristotle's lantern. A jawlike structure found in sea urchins consisting of a complex series of mobile teeth that can scrape algae or invertebrates from a hard surface.

Arrow worms. Members of the phylum Chaetognatha, a group of planktonic carnivores.

Arthropoda. A phylum of invertebrates characterized by an outer skeleton of chitin, molting, and jointed appendages.

Ascomycetes. Any fungus of the phylum Ascomycota (or class Ascomycetes), including the molds and truffles, characterized by bearing the sexual spores in a sac.

Asexual reproduction. Reproduction of the individual without the production of gametes and zygotes.

Assimilation efficiency. The fraction of ingested food that is absorbed and used in metabolism growth and reproduction.

Assortative mating. The mating of a given genotype with another genotype at a frequency disproportionate to that expected from random encounter.

Asteroidea. A class of carnivorous Echinodermata, the true starfish, having mobile arms, usually in multiples of five, and tube feet.

Atoll. A horseshoe or circular array of islands, capping a coral reef system perched around an oceanic volcanic seamount.

ATP. Adenosine triphosphate, an adenosine nucleotide used universally by organisms to store and transfer energy.

Attenuation (of light). Diminution of light intensity; explained, in the ocean, in terms of absorption and scattering.

Autonomous underwater vehicle (AUV). An underwater robot, equipped with a variety of sensors, that can be released from a ship or dock and is not tethered by a cable.

Autotrophic algae. Algae capable of photosynthesis and growth using only dissolved inorganic nutrients.

Auxospore. In diatoms, a cell that results from successive decreases of cell size during diatom reproduction, resulting in either production of gametes or rejuvenation of size and subsequent renewed cell division.

Auxotrophic algae. Algae requiring a few organically derived substances, such as vitamins, along with dissolved inorganic nutrients for photosynthesis.

Azoic theory. Theory proposed by Edward Forbes in the nineteenth century that no living organisms can be found on the seabed at depths deeper than 300 fathoms.

Bacteria. One of the three major domains of life, including many bacterial groups that decompose organic matter, transmit disease, or may photosynthesize.

Ballast water. Water used as ballast in large ships.

Bar. A linear raised area of bottom parallel to and offshore of the coastline.

Barbule. Hair that extends from a bird's feather barb and interlocks with other barbules.

Barcoding. The identification of a species by means of DNA sequencing, often with the *CO1* gene.

Barrens. A hard bottom dominated by coralline algae, with no kelps (in kelp forest regions) or corals (in coral reef regions).

Barrier island. Elongated offshore island parallel to a soft-sediment shore.

Batesian mimics. A species that is not poisonous that has evolved to resemble a poisonous model species.

Bathyal. A depth zone from 1,000 to 4,000 m.

Bathypelagic zone. The 1,000–4,000 m depth zone in the ocean.

Benthic–pelagic coupling. The cycling of nutrients between the bottom sediments and overlying water column.

Benthos. Organisms that live associated with the sea bottom. Examples include burrowing clams, sea grasses, sea urchins, and acorn barnacles.

Berm. A broad area of low relief in the upper part of a beach.

Between-habitat component. A contrast of diversity in two localities of differing habitat type (e.g., sand vs. mud bottoms).

Bilateria. A group of phyla consisting of bilaterally symmetrical organisms (including Annelida, Chordata, etc.) having a single evolutionary origin.

Bioassay. Use of a biological system; for example, some physiological

parameter of an indicator species, to measure the degree of pollution.

Bioconcentration. The increase of concentration of a substance in a species, relative to its concentration in the external environment. *See also* Biomagnification.

Biodeposition. The deposition on the seabed of material by benthos such as suspension-feeding bivalves, in the form of feces or pseudofeces.

Biodiversity. *See* Species richness.

Bioerosion. Erosion caused by organisms, such as boring sponges, bivalves, and other species.

Biogenic graded bedding. A regular change of sediment median grain size with depth below the sediment–water interface caused by the activities of burrowing organisms.

Biogenically reworked zone. The depth zone, within a sediment, that is actively burrowed by benthic organisms.

Biogeography. The study of the geographic arrangements of populations and species.

Biological pump. The process by which carbon is moved to deeper waters in the ocean as sedimenting biological particles.

Biologically inspired design. The use of natural biological features and phenomena such as morphologies to design novel products for human use.

Bioluminescence. Light emission, often as flashes, by many marine organisms.

Biomagnification. Increase in concentration of a substance as it moves from a species in one trophic level to a species in the next higher trophic level.

Biomass. *See* Standing crop.

Biosphere. The entire set of living things on the earth and the environment in which they interact.

Black band disease. A disease attributed to scleractinian corals, where a black band appears, with abundant cyanobacteria.

Blade. The flattened terminal part(s) of a seaweed.

Blastula. An early stage of embryonic development consisting of a ball of cells.

Bleaching. Referring to corals. The expulsion of zooxanthellae from corals, resulting in coral tissue losing its color and appearing to be the white color of the coral skeleton beneath the living tissue.

Blood pigment. A molecule used by an organism to transport oxygen efficiently, usually in a circulatory system (e.g., hemoglobin).

Bloom (phytoplankton). A population burst of phytoplankton that remains within a defined part of the water column.

Blowhole. A hole at the top of a cetacean's head through which the cetacean breathes air.

Blue whale unit. An equivalency between different whale species, formerly used by the International Whaling Commission in limiting the fishing take of whales.

Bohr effect. When blood pH decreases, the ability of hemoglobin to bind to oxygen decreases. An adaptation to release oxygen in the oxygen-starved tissues in capillaries where respiratory carbon dioxide lowers blood pH.

Boreal zone. Pertaining to the Northern Hemisphere, north temperate zone.

Boring. Capable of penetrating a solid substratum by scraping or chemical dissolution.

Bottom trawls. Fishing by means of an apparatus using boards that hold open a net, which is dragged along the seafloor.

Bottom-up control. Refers to food webs. A control of a population that comes from change lower in a food web (e.g., control of a population of mussels by abundance of phytoplankton food).

Boundary currents. *See* Oceanic boundary currents.

Boundary layer. A layer of fluid near a surface, where flow is affected by viscous properties of the fluid. At the surface, fluid velocity must be zero, and the boundary layer is a thin film that depends on surface texture, fluid velocity in the "mainstream of flow," and fluid mass properties such as salinity.

Bow wave. Water turbulence formed in the front of an object as it moves forward in the water.

Box corer. A ship-deployed benthic sampling device where a box is inserted into soft sediment and a bottom is rotated into place before the sampler is lifted toward the surface.

Brachiopods. A phylum of lophophorate invertebrates with two valves and a suspension-feeding lophophore.

Brackish sea. Semi-enclosed water body of large extent in which tidal stirring and seaward flow of freshwater do not exert enough of a mixing effect to prevent the body of water from having its own internal circulation pattern and lower salinity.

Brevetoxin. Cyclic polyether neurotoxins that bind to sodium channels in nerve cells, produced by some dinoflagellate species.

Broadcast spawning. Production of planktonic gametes that can be transported by water currents.

Brooding. Maintenance of a developing juvenile within the body of the parent.

Bryozoan. A phylum of lophophorate invertebrates, colonial, with individuals living in a flat or upright colony.

Buoyant flow. Flow of lower-salinity surface estuarine water toward the open sea.

Bycatch. The catch of a fishery that is unintended; for example, catching juvenile sturgeon when you were fishing for tuna.

Calanoid copepods. Copepods characterized by large antennae and very abundant in the water column throughout the world.

Calcareous. Made of calcium carbonate.

Calcite. Most common variant of calcium carbonate, a common constituent of limestone and shells and skeletons of many marine species (but not of corals, and most snails).

California Current Large Marine Ecosystem. A series of coastal and offshore environments from the coast of Alaska to Central America that includes a wide series of migration routes, reproductive areas, and upwelling-feeding centers used by a number of long-distance migrating fish and marine mammals.

Cambrian explosion. A hypothetical origin and diversification of all of the Bilateria within the Cambrian geological period.

Canopy. A layer of photosynthetic organisms high off the water column, potentially shading lower layers near the seafloor.

Carbon pump. A mechanism of import of carbon dioxide to great depths by solution and storage in colder deep waters, where solubility is greater than at the surface.

Carnivore. An organism that captures and consumes animals.

Carrying capacity. The total number of individuals of a population that a given environment can sustain.

Catadromous fish. Fish that spawns in seawater but feeds and spends most of its life in estuarine or fresh water.

Catch per unit effort. The fishery take divided by a series of measures such as number of vessels, hours of fishing, and number of fishers. An estimate of how much fishing is done to acquire a given catch.

Catch share. A fisheries management system in which an entire fishery is allocated to a group of fishers who had been exploiting the stock. After the allocation, a member with a share can sell this share to others.

Cementation. The deposition of mineral matter such as calcium carbonate in open spaces that results in the solidification of a habitat such as a reef.

Center-of-origin theory. Theory that high-diversity areas are centers where species originate and then spread to lower-diversity areas.

Cephalopoda. A class of Mollusca characterized by carnivory, well-developed nervous systems, complex behavior, and mobility.

Chaetognaths. *See* Arrow worms.

Character displacement. A pattern in which two species with overlapping ecological requirements differ more when they co-occur than when they do not. The difference is usually in a morphological feature related to resource exploitation, as in the case of head size, which may be related to prey size.

Chemical signaling. Communication by means of production and release of various dissolved substances.

Chemoautotrophic. An organism that depends on inorganic chemicals for its energy and principally on carbon dioxide for its carbon.

Chemolithotrophic bacteria. Bacteria that obtain energy independent of light by chemical modification of inorganic molecules.

Chemophototrophic organisms. Bacteria that obtain energy with the aid of light by chemical modification of inorganic molecules.

Chemosynthesis. Primary production of organic matter, using various substances instead of light as an energy source; confined to a few groups of microorganisms.

Chlorinated hydrocarbons. Compounds containing chlorine, carbon, and hydrogen, including solvents and pesticides, many of which are highly toxic.

Chlorinity. Grams of chloride ions per 1,000 grams of seawater.

Chloroplast. In eukaryotic organisms, the cellular organelle in which photosynthesis takes place.

Choanocytes. Specialized cells found in sponges.

Chondrichthyes. Fishes united by the production of a cartilaginous skeleton.

Chromatophore. A cell on the body surface that contains pigment. Expansion of the cell enlarges the coverage of the pigment and contraction reduces the pigment. This allows organisms such as fish, octopods, and crabs to rapidly change surface coloration and pattern.

Ciliates. Group of protistans characterized by a ciliated surface for locomotion and often a ciliated mouth.

Clade. A group of related species that arose from a single ancestor.

Cladogram. A treelike diagram showing evolutionary relationships. Any two branch tips sharing the same immediate node are most closely related. All taxa that can be traced directly to one node (i.e., they are "upstream of a node") are said to be members of a monophyletic group.

Climax community. A predictable final assemblage of species that arises as the endpoint of succession.

Cline. A regular (usually monotonic) change in gene frequency over a geographic space.

Clone. A group of individuals that have derived from a single individual by nonsexual reproduction.

Cnidaria. A phylum characterized by radially symmetrical organisms with tentacles usually armed with nematocysts.

Coastal reef. A coral reef occurring near and parallel to a coastline.

Coccoliths. Calcium carbonate plates that are the constituents of the spherical skeleton of coccolithophores.

Coelom. A body cavity within mesoderm tissue.

Coevolution. The continual evolutionary response between two species, as one species changes in response to the other species and vice versa.

Cold seeps. Sources of sulfide-rich brines or hydrocarbons from geological structures on the outer continental shelf or continental slope, providing substrates for bacterial growth.

Cold-core rings. A volume of water with warm surface water ringing a core of cooler water, formed when a meander of the Gulf Stream loops and encloses cooler water.

Coliform bacteria. Bacteria that have been used to count in water as an indication of sewage pollution. Bacteria are found in human feces and can ferment lactose.

Collar cells. *See* Choanocytes.

Color polymorphisms. Distinct differences in color of an organism among individuals of a species, usually controlled by genetic variation.

Comb jellies. Members of the phylum Ctenophora, a group of gelatinous forms feeding on smaller zooplankton.

Commensal. Having benefit for one member of a two-species association but neither positive nor negative effect on the other.

Community. A group of species living together and interacting through ecological processes such as competition and predation.

Compensation. In fisheries biology, the reduction of rate of population increase as population density passes an upper threshold.

Compensation depth. The depth of the compensation light intensity.

Compensation light intensity. That light intensity at which oxygen evolved from a photosynthesizing organism equals that consumed in its respiration.

Competent. Larvae capable of settlement and metamorphosis.

Competition. An interaction between or among two or more individuals or species in which exploitation of resources by one affects any others negatively.

Complex life cycle. A life cycle that consists of several distinct stages (e.g., larva and adult).

Compound eyes. Eyes belonging to arthropods consisting of multiple visual units that capture and transmit light signals to the nervous system.

Concentration boundary layer. Boundary layer of water near the bottom characterized by a difference in particle concentration from the overlying water.

Condition index. An index of health in bivalve mollusks determined by the amount of soft tissue relative to shell volume.

Conformer. An organism whose physiological state (e.g., body temperature) is identical to, and varies identically with, that of the external environment.

Connectivity. The degree to which populations at varying localities of a species exchange individuals either by larval dispersal or adult movements.

Continental drift. Horizontal movement of continents located in plates moving via seafloor spreading.

Continental rise. A transition zone between the continental slope and the abyssal plain.

Continental shelf. A broad expanse of ocean bottom sloping gently and seaward from the shoreline to the shelf-slope break at a depth of 100–200 m.

Continental slope. *See* Slope.

Continuity, principle of. The mass flow rate into a pipe is equal to the mass flow rate out, which means that velocity varies inversely with the cross-sectional area of the pipe as the fluid flows through the pipe.

Contour feathers. Feathers that are found on the outside of a bird's body and interlock to reduce heat loss and wetting of feathers beneath.

Control. In designing an experiment, one may have a treatment where an environmental change is imposed. A control is a treatment where there is no change, so a comparison can be made with the environmental change.

Convective heat loss. Loss of heat by movement of fluid across a surface.

Convergence. In oceanography: the contact at the sea surface between two water masses converging, with one plunging below the other. In evolution: the outcome when two unrelated groups evolve a structure that is similar because of evolutionary adaptation to a similar environmental condition.

Copepoda. Order of crustaceans found often in the plankton.

Coprophagy. Feeding on fecal material.

Coral atolls. Oceanic islands topped by a horseshoe or circular array of islands, formed by coral growth.

Coral bleaching. *See* Bleaching.

Coral reef. A wave-resistant structure resulting from cementation processes and the skeletal construction of hermatypic corals, calcareous algae, and other calcium carbonate–secreting organisms.

Coralline algae. Red algae that secrete a skeleton of calcium carbonate.

Coralline algal turfs. A bottom dominated by layers of one or more species of coralline algae.

Corer. Tubular benthic sampling device that is plunged into the bottom in order to obtain a vertically oriented cylindrical sample.

Coriolis effect. The deflection of air or water bodies, relative to the solid earth beneath, as a result of the earth's eastward rotation.

Correlation. A quantitative relationship between two or more variables collected at a series of locations or from a series of individuals.

Corridor. In biogeography, a linear connection between two biogeographically distinct areas, allowing dispersal between them.

Cost of metabolism. The energetic expenditure caused by metabolic activity.

Countercurrent exchange mechanism. Mechanism by which two vessels are set side by side, with fluid flowing in opposite directions, allowing efficient uptake and retention of heat, oxygen, or gas, depending on the type of exchanger.

Countercurrent heat exchanger. A biological device where heat is conserved by moving fluids through adjacent circulatory vessels in opposite directions.

Counterillumination. Having bioluminescent organs that are concentrated on the ventral surface so as to increase the effect of countershading. *See also* Countershading.

Countershading. Condition of organisms in the water column that are dark colored on top but light colored on the bottom.

Crinoidea. A class of echinoderms characterized by long arms covered with tube feet that capture zooplankton.

Critical depth. That depth above which total integrated photosynthetic rate equals total integrated respiration of photosynthesizers.

Critical salinity. A salinity of approximately 3–8 that marks a minimum of species richness in an estuarine system.

Crustacea. A group of arthropods characterized by two pairs of antennae.

Crypsis. Ability to have features that make the organism inconspicuous against an environmental background.

Ctenidium. The gill of a mollusk. Used for respiration and often for feeding.

Ctenophora. *See* Comb jellies.

Cumulative toxic substances. Substances that when absorbed or ingested continue to increase in concentration in the body over time.

Cyanobacteria. A group of photosynthetic and nitrogen-fixing bacteria that live as single cells or chains in both the water column and on the seafloor.

Daily estuary. An estuary in which tidal movements cause substantial changes in salinity at any one location on a daily basis.

Dalton. A measure of molecular mass. One dalton (Da) is equal to the mass of one hydrogen atom. Molecular mass is often measured in kDa, or thousands of daltons.

DDT. A pesticide consisting of chlorinated hydrocarbons, found to accumulate in marine species.

Dead zones. Large areas, usually at the mouths of estuaries, that have low concentrations or the absence of dissolved oxygen.

Deduction. Drawing a conclusion from the logical structure of a set of observations and thoughts.

Deep layer. The layer extending from the lowest part of the thermocline to the bottom.

Deep-scattering layer. Well-defined horizon in the ocean that reflects sonar; indicates a layer usually consisting of fishes, squid, or other larger zooplankton.

Deep-water coral mounds. Communities found in deep water dominated by scleractinian coral colonies.

Demersal. Nektonic, but associated with the seabed.

Demographic. Referring to numerical characteristics of a population (e.g., population size, age, structure).

Denitrifying bacteria. Free-living bacteria that convert nitrates to gaseous nitrogen.

Density (seawater). Grams of seawater in 1 cubic cm of fluid.

Density-dependent factors. Factors, such as resource availability, that vary with population density.

Density-mediated indirect interaction. An indirect action upon the abundance of a species in which a species not directly interacting with the first species changes in abundance. An example would be the increase in abundance of a carnivore, which results in the increase of abundance of a plant because the plant's consumer is reduced in abundance.

Depensation. In fisheries biology, the case in which rate of population increase declines because population size is at a lower threshold. Corresponds to the Allee effect.

Deposit feeder. An organism that derives its nutrition by consuming some fraction of a soft sediment.

Detrended. A fluctuation over time in which the short-term variations are plotted but a longer-term change—for example, increasing temperature—is algebraically subtracted so that the oscillations can be visualized.

Detritus. Particulate material that enters into a marine or aquatic system. If derived from decaying organic matter, it is organic detritus.

Deuterostomes. A group of bilaterian phyla including Echinodermata and Chordata, distinguished by DNA sequence relationships and characteristics of the early embryo, such as formation of the anus at the opening of the gastrular invagination.

Diadromous. Migratory species that move between the open sea and within an estuary.

Diapause. A period of time of year when on organism transitions into a resting mode, usually with no feeding or activity, usually during an unfavorable time of year such as very low temperature.

Diatom. Dominant planktonic algal form with siliceous test, occurring as a single cell or as a chain of cells.

Diel vertical migration. Migration found in many zooplankton and fish where animals rise toward the surface at night and sink to depth during the daytime.

Diet-breadth model. A model of foraging that predicts the array of prey types included in the diet as a function of overall prey density.

Diffusion. The net movement of units of a substance from areas of higher concentration to areas of lower concentration of that substance.

Digestion efficiency. The fraction of living food that does not survive passage through a predator's gut.

Dinoflagellate. Dominant planktonic algal form, occurring as a single cell, often biflagellate.

Direct interception. The interception of particles by suspension feeders from the water directly on threadlike structures such as cilia.

Direct release. The release of juveniles directly into the immediate environment of the parent with little dispersal.

Directional selection. Preferential change in a population, favoring the increase in frequency of one allele over another.

Dispersal. Spread of organisms, usually progeny, from one location to another.

Dissolved organic matter. Dissolved molecules derived from degradation of dead organisms or excretion of molecules synthesized by organisms.

Disturbance. A rapid change in an environment that greatly alters a previously persistent biological community.

Diurnal tide. A tidal cycle where some component (e.g., a low tide) is prominent only once a solar day.

Diversity. A parameter describing, in combination, the species richness and evenness of a collection of species. Diversity is often used as a synonym for species richness.

Diversity gradient. A regular change in species diversity correlated with a geographic space or gradient of some environmental factor.

DNA. A helical, double-stranded pair of macromolecules consisting of two chains of nucleotides whose primary function is to carry of genetic information.

DNA sequence. The precise array of nucleotide bases (A, T, G, C) on a strand of DNA.

Domain. A major evolutionary division of life.

Domoic acid. A toxic substance produced by some phytoplankton responsible for amnesic shellfish poisoning.

Down feathers. Relatively fluffy bird feathers that occupy high volume and enclose air for insulation of the body from the cold external environment.

Drag. A force created on an object because the pressure is different on either side of it.

Dredges. Samplers deployed from ships that drag the bottom and recover sediments with organisms.

Dune. A raised area of sand parallel to shore but significantly upland from the line of low tide on a beach.

Dynamic viscosity. A measure of molecular stickiness of a fluid, or a measure of resistance of a fluid to deform when a force is applied.

Ecdysozoa. A group of protostome phyla, including arthropods and nematodes, united by DNA sequence relationships and an external cuticle.

Echinodermata. Phylum characterized by a spiny skin that encloses a skeleton of interlocking calcium carbonate plates, a water vascular system, and tube feet.

Echinoidea. A class of living Echinodermata, including sea urchins, sand dollars, and heart urchins.

Ecological niche. The range of physical and biological habitats occupied by a species.

Ecology. The study of the interaction of organisms with their physical and biological environments, and how these interactions determine the distribution and abundance of the organisms.

Ecosystem. A group of interdependent biological communities and abiotic factors in a single geographic area that are strongly interactive.

Ecosystem engineer. A species whose activities strongly affect the physical structure of the environment. Often used interchangeably with foundation species.

Ecosystem services. The economic values of a variety of benefits that ecosystems perform for humans, such as the protection by coastal vegetation against storm damage.

Ecosystem-based management. An approach to conserve ecosystems that includes components of environmental protection, the social community that interacts with the ecosystem, and economic considerations.

Ectoparasite. *See* Parasite.

Effect size. In an experiment, the amount of change of a variable (e.g., body temperatures) between one treatment and another.

Ekman transport. Movement of surface water at an angle from the wind, as a result of the Coriolis effect.

El Niño–Southern Oscillation (ENSO). Condition in which warm surface water moves into the eastern Pacific, collapsing upwelling and increasing surface-water temperatures and precipitation along the west coasts of North and South America.

Elasticity analysis. A population analysis in which one examines the effect on population growth by a change of one parameter, such as mortality rate of a single age class.

Emigration. The departure of individuals from a given area.

Endemic. Restricted in geographic range to a particular region.

Endocrine disruptors. A series of pollutants that can disrupt endocrine or hormonal functions in marine organisms.

Endogenous rhythms. A biological rhythm that, at least for a time, is maintained without any outside environmental variation.

Endoparasite. *See* Parasite.

Endosymbiotic. Being symbiotic and living within the body of an individual of the associated species.

Enterococci. Bacteria of the genus *Enterococcus*, which can metabolize lactic acid and are ofen the cause of disease. Indicator of human fecal pollution.

Entrainment. The case when a particle is taken up with the flow of a fluid and moves with the fluid.

Environmental stress. Variously defined as (a) an environmental change to which an organism cannot acclimate and (b) an environmental change that increases the probability of death.

Enzyme polymorphisms. Genetic variation at genetic loci that code for enzymes.

Epibenthic (epifaunal or epifloral). Living on the surface of the seabed.

Epidemic spawning. Simultaneous shedding of gametes by a large number of individuals.

Epifaunal. An animal living on the surface of the seabed.

Epipelagic zone. The 0–150 m depth zone, seaward of the shelf-slope break.

Epiphyte. Microalgal organism living on a surface (e.g., on a seaweed frond).

Estuarine flow. Seaward flow of low-salinity surface water over a deeper and higher-salinity layer.

Estuarine realms. Large coastal water regions that have geographic continuity, are bounded landward by a stretch of coastline with freshwater input, and are bounded seaward by a salinity front.

Estuary. A semi-enclosed body of water that has a free connection with the open sea and within which seawater is diluted measurably with fresh water that is derived from land drainage.

Eubacteria. A group of bacteria often distinguished by cell walls, presence of flagellae, and molecular sequence similarity. One of three major divisions of life.

Eukarya. One of the three major domains of life, including protists, animals, and plants.

Eukaryotes. Organisms distinguished by their cells, which have distinct nuclei and cell organelles, and in which all cells except gametes reproduce by mitosis. *See also* Prokaryotes.

Euphausiacea. A group of shrimplike crustacea, up to a few centimeters long; zooplankton, found usually in upwelling regions.

Euphausiid. Member of an order of holoplanktonic crustacea.

Eusocial. Social organisms where different groups of individuals in a colony serve different purposes and have morphologies determined by a combination of genotype and environment.

Eutrophic. Water bodies or habitats having high concentrations of nutrients.

Eutrophication. Addition of high nutrient concentrations to a water body.

Evaporative cooling. Cooling of an organism by evaporating water from the body surface.

Evenness. The component of diversity accounting for the degree to which all species are equal in abundance, as opposed to strong dominance by one or a few species.

Exclusive economic zone (EEZ). A coastal zone within which a nation has exclusive rights to its fisheries.

Extinction. The complete loss of all members of a taxon, such as a species.

Extremophiles. Microorganisms, often belonging to the major group Archaea, that live in extreme concentrations of metals and often very high temperatures associated with hot vents in the deep sea.

False color. Remote sensing data often are summarized in maps, and the variation in a parameter—for example, sea-surface temperature—is represented in a color code that is not a photographic representation of the original signal but an artificial color code.

Fecal coliform bacteria. Technically, all the facultative anaerobic gram-negative, non-spore-forming, rod-shaped bacteria that ferment lactose in EC medium with gas production within 24 hours at 44.5°C. A measure of bacteria mostly originating from guts that enters waters. Believed to be correlated with disease-causing (pathogenic) bacteria.

Fecal pellets. *See* Pellets.

Fecundity. The number of eggs produced per female per unit time (often: per spawning season).

Fermentation. *See* Fermenting bacteria.

Fermenting bacteria. Bacteria that gain energy by fermentation, the anaerobic breakdown of organic material into end products such as alcohols.

Ferredoxins. Iron-sulfur proteins that mediate electron transfer, as in photophosphorylation reactions in photosynthesis.

Field experiments. Experiments that are designed to manipulate natural communities in the field.

Fish balls. A type of fish aggregation into a ball, where fish move continuously from the surface of the ball to the interior.

Fishing effort. Factors involved in bringing in landings from a fishery, including number of fishing boats, type of fishing gear, and time spent fishing.

Fitness. The rate, relative to other genotypes, at which a genotype reproduces into the next generation. Definition can also be used for alleles.

Flow cytometer. A device that uses a laser light source on particles to analyze for fluorescence and other characteristics that can distinguish the particle from other types (e.g., different types of phytoplankton cells).

Flume. A research device that uses moving water in a contained space to investigate water movement and the response of organisms to water motion.

Flushing time. The time it takes for a parcel of water to leave a confined water body such as a bay.

Foliose coral. A coral whose skeletal form approximates that of a broad, flattened plate.

Food chain. An abstraction describing the network of feeding relationships in a community as a series of links of trophic levels, such as primary producers, herbivores, and primary carnivores; a linear connection of organisms to show the feeding linkages of predators and prey.

Food chain efficiency. Amount of energy or some other quantity extracted from a trophic level, divided by the amount of energy produced by the next-lower trophic level.

Food web. A network describing the feeding interactions of the species in a defined region.

Foraminifera. Protozoan group, individuals of which usually secrete a calcareous test; both planktonic and benthic representatives.

Foraminiferan ooze. A deep-sea sediment composed primarily of skeletons of Foraminifera.

Foundation species. Species that are structurally important for the organization of a community. Includes abundant species like reef-forming corals and sea grasses.

Founder principle. A small colonizing population is genetically unrepresentative of the source of population.

Framework builders. In coral reefs: dominant coral species that comprise most of the calcium carbonate production of the reef.

Free amino acids. Amino acids that are synthesized and kept in free concentration in cells for the purpose of osmoregulation.

Free spawning. Gametes are released directly into the water column.

Free-water techniques. Methods of measuring primary productivity by directly measuring oxygen changes in natural waters, taking into account daily changes in photosynthesis and losses and gains of oxygen between the water column and the atmosphere.

Frequency-dependent selection. Natural selection depends on the frequency of alleles in the population. For example, an allele might be favored more when it is rare than when it is common.

Freshet. An increase of water flow into an estuary during the late winter or spring, owing to increased precipitation and snow melt in the watershed.

Frond. In seaweeds: typically, a flattened structure attached to the stipe, where much of the photosynthesis occurs.

Front. A major discontinuity separating ocean currents and water masses in any combination.

Fugitive species. A species adapted to colonize newly disturbed habitats.

Functional biology. The study of how an organism carries out the basic functions such as reproduction, locomotion, feeding, and the cellular and biochemical processes relating to digestion, respiration, and other aspects of metabolism.

Functional group. A groups of species with similar ecological function (e.g., herbivore species).

Functional redundancy. The case in which the ecological function (e.g., nitrogen fixation) of one species in a community can be replaced by the presence of another ecologically similar species.

Fyke net. Net usually staked to the bottom, sometimes with leading wings that guide fishes moving with the current into the main net bag.

Gametophyte. Haploid stage in the life cycle of a plant.

Gastropoda. A class of mollusks characterized usually by a spiral shell, a muscular foot, and twisting of the embryo during development.

Gelatinous zooplankton. Zooplankton that have a gelatinous support skeleton, including many distantly related phyla such as Cnidaria, Ctenophora, and members of the Chordata.

Gene duplication. A process whereby a gene is duplicated owing to an error in DNA duplication in meiosis or even a duplication of a whole chromosome. Genes may be duplicated by a number of other processes related to DNA replication and repair.

Gene expression. The degree to which a given gene is stimulated to produce a gene product such as a protein.

Gene family. A group of genes that are similar in sequence and usually function and all derive from the same ancestral gene, usually by means of gene duplication and natural selection.

Generation time. The time period from birth to average age of reproduction.

Genetic code. The arrangement of nucleotides to form a code specifying different amino acids.

Genetic drift. Changes in allele frequencies that can be ascribed to random effects.

Genetic locus. A location on a chromosome (possibly of a diploid organism with variants that segregate according to the rules of Mendelian heredity).

Genetic polymorphism. Presence of several genetically controlled variants in a population.

Genome. The total functioning DNA of an organism.

Genotype. The genetic makeup of an organism; with respect to a given genetic locus, the alleles it carries.

Genotype-by-environment interaction. The same genotype may have a different phenotype when raised in different environments.

Genus. (plural: genera) The level of the taxonomic hierarchy above the species but below the family level.

Geostrophic flow. Movement of water in the oceans as a combined response to the Coriolis effect and gravitational forces created by an uneven sea surface.

Geotactic. Moving in response to the earth's gravitational field.

Gill arch. A part of a fish skeleton that supports the gills.

Gill net. Fish net with a mesh size designed to trap by the gills a restricted size class of fish.

Gill rakers. Projections along the gill arch.

Gliders. Autonomous underwater vehicles that can use simple balancing devices and vanes to move the vehicle through the water column.

GIS: Geographic Information System. A system that allows automatic location of information suitable for mapping. Usually involves a software system that takes geographic position data and other data (e.g., type of bottom sediment) in order to create a map. Data on processes (e.g., current speed) can be incorporated to make a geographic model of flow.

Global conveyor belt system. A movement of water currents that couples surface water motion with deep thermohaline water motion.

Global positioning system (GPS). An electronic device that uses positioning signals from satellites in order to locate precisely latitude and longitude. Now used nearly exclusively for locating ship sampling stations at sea, but also useful for locations near and on shore.

Global warming. Predicted increase in the earth's oceanic and atmospheric temperature, owing to additions of carbon dioxide to the atmosphere, often as a result of human activities.

Glycoprotein. A protein having a carbohydrate component.

Gonochoristic. Having separate sexes.

Grab. Benthic sampling device with two or more curved metal plates designed to converge when the sampler hits bottom, grabbing a specified volume of bottom sediment.

Gravity corers. A coring device that drops into the sediment with the aid of a weight, simply by force of gravity.

Grazer. A predator that consumes organisms far smaller than itself (e.g., copepods graze on diatoms).

Greenhouse effect. Carbon dioxide traps solar-derived heat in the atmosphere near the earth.

Greenhouse gases. Gases such as carbon dioxide that enable the greenhouse effect.

Gregarious settling. Settlement of larvae that have been attracted to members of their own species.

Gross primary productivity. The total primary production, not counting the loss in respiration.

Growth efficiency. The efficiency that ingested food is converted into somatic growth.

Guild. A group of species, possibly unrelated taxonomically, that exploit overlapping or similar resources.

Guyot. A submarine oceanic seamount and inactive volcanic mass whose summit is below the sea surface, usually with a flattened top.

Gyre. Major cyclonic surface current systems in the oceans.

Hadal. The depth zone corresponding to oceanic trenches.

Halocline. Depth zone within which salinity changes maximally.

Harmful algal bloom. A bloom of (usually) planktonic microalgae belonging to a strain of a species that has a toxin harmful to marine organisms or humans consuming marine organisms.

Heat of vaporization (water). The amount of heat required to convert a unit mass of water at its boiling point into vapor without an increase in temperature.

Heat-shock proteins. Proteins that are produced under heat stress and reduce the unfolding of functioning enzymes.

Herbivore. An organism that consumes plants.

Heritable character. A morphological character whose given state can be explained partially in terms of the genotype of the individual.

Hermaphrodite. An individual capable of producing both eggs and sperm during its lifetime.

Hermatypic. Reef-building.

Heterocysts. Enlarged cells in cyanobacteria, where nitrogen fixation occurs.

Heterotrophic algae. Algae that take up organic molecules as a primary source of nutrition.

Heterozygote. With respect to a given genetic locus, a diploid individual carrying two different alleles.

Highly stratified estuary. An estuary having a distinct surface layer of fresh or very-low-salinity water, capping a deeper layer of higher-salinity, more oceanic water.

High-nitrogen–low-productivity (HNLP) sites. Areas of the surface ocean that have relatively high concentrations of nitrogen but nevertheless have low primary productivity, probably because of iron limitation.

Hirudinea. The leeches, a class of annelids having a fixed number of segments, lacking parapodia or chaetae, and typically parasitic.

HNLP. *See* High-nitrogen–low-productivity (HNLP) sites.

Holdfast. In seaweeds: a structure that attaches the seaweed to the substratum.

Holobiont. A host organism (animal or plant) and all of its symbiotic microbial organisms.

Holoplankton. Organisms spending all their life in the water column and not on or in the seabed.

Holothuroidea. A class of Echinodermata, the sea cucumbers, characterized by a worm shape, a crown of tentacles with tube feet, and either a deposit- or suspension-feeding ecology.

Homeotherm. An organism that regulates its body temperature despite changes in the external environmental temperature.

Homologous. Having the same evolutionary origin.

Homozygote. With respect to a given genetic locus, a diploid individual carrying two identical alleles.

Hot vents. Openings in oceanic ridge rocks of volcanic origin with hot water emanating with concentrated metals and sulfide.

Hydrogen bonds. A chemical bond with a hydrogen atom between two negatively charged atoms (e.g., oxygen).

Hydrothermal vents. Sites in the deep-ocean floor where hot, sulfur-rich water is released from geothermally heated rock.

Hyperosmotic. Having a higher salt content, or higher content of materials that affect osmosis (e.g., osmolytes), within cells than exists in the surrounding external water environment.

Hypoosmotic. Having a lower salt content, or lower content of materials that affect osmosis, within cells than in the surrounding external water environment.

Hypothesis. A refutable statement about one or a series of phenomena.

Hypoxia. The presence of low concentrations of oxygen in the water that is stressful to marine organisms.

Ice algae. A wide variety of microalgae and macroalgae that live in association with ice floes in polar oceans.

Immunofluorescence. A method of identifying cells by means of using antibodies that are coupled to fluorescent dyes.

Impingement. When a screen in a power plant captures fish or other marine organisms as water is sucked into an intake channel.

Implanted tags. Devices of varying design that are implanted in marine organisms for identification.

Indirect effect. An ecological effect of one species on another that is mediated through changes in abundance of a third species that interacts with the first species.

Individual transferrable quota (ITQ) system. An arrangement in which a fishing entity can transfer its proportional right to a fishery to another fishing entity.

Inducible defense. A defense against predation that grows or develops after an individual is exposed to a predator.

Induction. Reaching a conclusion from an accumulation of facts.

Inertial forces. Forces where an object tends to keep moving after ceasing to apply a force to that object.

Infaunal. Living within a soft sediment and being large enough to displace sedimentary grains.

Inorganic carbon. Carbon in molecules not manufactured by organisms (e.g., as CO_2).

Interference competition. Interspecific competition where individuals of one species directly interfere and prevent individuals of a competing species from access to a limiting resource.

Intermediate predation-disturbance effect. Predation maximizes the number of coexisting and competing species at some intermediate level of predation or disturbance.

International Whaling Commission. An international treaty-based organization that sets policy for whaling practices and quotas, with the objective of conservation and recovery of whale stocks.

Intersexual selection. Selection for traits of one sex that are involved in interactions with the other sex, such as selection for production of bright color in males that attracts mating females.

Interspecific competition. Condition in which one species' exploitation of a limiting resource negatively affects another species.

Interstitial. Living in the pore spaces among sedimentary grains in a soft sediment.

Intertidal zone. The zone between the highest and lowest extent of the tide.

Intrasexual selection. Selection among members of the same sex, for example, selection among body size that gives increases access to mating with females.

Invasive species. A species that successfully disperses to a new region and then increases greatly in abundance and area in the new region.

Isotonic. Having the same overall concentration of dissolved substances as a given reference solution.

Iteroparity. The condition where an individual reproduces more than once.

Kelp. A group of brown seaweeds belonging to the group Laminariales, characterized by rapid growth and occurrence as foundation species of subtidal kelp forests.

Kelp forests. Shallow subtidal communities in relatively cold water, dominated by kelps.

Keystone species. A predator at the top of a food web, or discrete subweb, capable of consuming organisms of more than one trophic level beneath it. Often more broadly defined as any species that has broad impacts on a community.

Kinematic viscosity. The dynamic viscosity divided by water density.

Krill. *See* Euphausiacea.

Laminar flow. The movement of a fluid where movement of the entire fluid is regular and with parallel streamlines.

Laminariales. The group of seaweeds commonly called kelps.

Landings. The catch of a given fishery.

Langmuir circulation. Under steady winds, circulation at the surface is driven into the forms of vertically rotating cells that leave linear traces at the surface, where the cells cause alternating rows of convergence and divergence of water and entrained materials.

Larva. A discrete stage in many species, beginning with zygote formation and ending with metamorphosis.

Larvacea. A group of planktonic tunicates that secrete a gelatinous "house," used to strain unsuitable particles (large particles are rejected). An inner filter apparatus of the house, the so-called food trap or particle-collecting apparatus, is used to retain food particles.

Lateral-line system. In fish: a series of sense organs arranged in a line along the side; used in mechanoreception.

Latitude diversity gradient. The general case where diversity increases from high to low latitude.

LC$_{50}$. The concentration of a toxic substance that produces 50 percent mortality in an experimental population after a specified time.

LD$_{50}$. The value of a given experimental variable required to cause 50 percent mortality.

Leaching. The loss of soluble material from decaying organisms.

Lecithotrophic larva. A planktonic-dispersing larva that lives off yolk supplied via the egg.

Leeward side. The side of an island opposite from the one facing a persistent wind.

Leslie matrix. A matrix of probabilities of reproduction and mortality of difference age classes in a population.

Lichen. A composite symbiosis of filaments of fungi and either algae or cyanobacteria.

Life table. A table summarizing statistics of a population, such as survival and reproduction, all broken down according to age classes.

Light-harvesting complex. The range of photosynthetic pigments that capture light energy and transfer energy for photosynthesis.

Limiting nutrients. Those nutrients that regulate the growth of photosynthetic organisms.

Locus. *See* Genetic locus.

Logistic population growth. Population growth that is modulated by the population size relative to carrying capacity. Population growth declines as population approaches carrying capacity, and is negative when population size is greater than carrying capacity.

Longlines. Lines with hooks at regular intervals used to catch fish.

Longshore current. A current moving parallel to a shoreline.

Lophophorates. A group of phyla, including Brachiopoda, Bryozoa, and Phoronida, united by the presence of a lophophore, which is a horseshoe- or spiral-shaped feeding structure.

Lophophore. A horseshoe- or spiral-shaped feeding structure with tentacles belonging to lophophorate phyla.

Lophotrochozoa. A group of bilaterian phyla within the Protostomia that are evolutionarily related.

Luciferase. A group of oxidative enzymes that catalyze the transformation of luciferin into a light-emitting substance.

Luciferin. A light-emitting compound used by many species in bioluminescence.

Lysis. The breaking open of a cell by destruction of cell membranes resulting from the invasion and reproduction of population of viruses.

Macroalgae. Algae large enough to be detected with the naked eye. Often used as a synonym of seaweeds.

Macrobenthos (macrofauna or macroflora). Benthic organisms (animals or plants) whose shortest dimension is greater than or equal to 0.5 mm.

Macrofauna. Animals whose shortest dimension is greater than or equal to 0.5 mm.

Macrophyte. An individual alga large enough to be seen easily with the unaided eye.

Macroplankton. Planktonic organisms that are 200–2,000 µm in size.

Madreporite. A sievelike plate on the aboral surface of echinoderms that connects the external environment with the internal water vascular system.

Mainstream current. The flow in a part of the fluid (e.g., in a tidal creek) that is well above the bottom or well away from a surface and essentially not under the influence of the boundary layer. *See also* Boundary layer.

Maintenance metabolism. The energy required to maintain an organism at rest.

Male dimorphism. The case where males occur in at least two distinct forms.

Mangel. *See* Mangrove forest.

Mangrove forest. A shoreline ecosystem dominated by mangrove trees, with associated mudflats.

Mantle. A rock layer of the earth beneath the crust.

Marginal seas. Seas located usually in coastal areas where local conditions and past history create distinct local conditions.

Mariculture. The rearing of organisms in confined areas for food or food products.

Marine protected area (MPA). A conservation geographic unit designed to protect crucial communities and to provide reproductive reserves for fisheries that hopefully will disperse over wider areas.

Marine snow. Fragile organic aggregates, resulting from the collision of dissolved organic molecules or from the degradation of gelatinous substances such as larvacean houses. Usually enriched with microorganisms.

Marine trophic index. A measure of the dominant trophic level in a given fishery area.

Mate selection. *See* Intersexual selection.

Maximum sustainable yield (MSY). In fisheries biology, the maximum sustainable catch obtainable per unit time under the appropriate fishing rate.

Mechanical displacement. Burrowing where the animal uses a structure to dig out sediment to form a burrow.

Median grain size. If a sediment sample is divided into a series of size classes by sieves, the median grain size is the size where cumulative addition of sediment of size classes crosses the 50 percent threshold of the total sample.

Megaplankton. Planktonic organisms that are greater than or equal to 2,000 μm in size.

Meiobenthos (meiofauna or meioflora). Benthic organisms (animals or plants) whose shortest dimension is less than 0.5 mm but greater than or equal to 0.1 mm.

Meiofauna. Animals whose shortest dimension is less than 0.5 mm but greater than or equal to 0.1 mm.

Membrane order. The degree of packing of the structural phospholipids in a cell membrane.

Meroplankton. Organisms that spend part of their time in the plankton but also spend time in the benthos (e.g., planktonic larvae of benthic invertebrates).

Merostomata. The class of arthropods including horseshoe crabs.

Mesopelagic. The 200–1,000 m depth zone, seaward of the shelf-slope break.

Messenger RNA (mRNA). RNA molecule that is the template for the amino acid sequence of a protein.

Metabolic rate. The overall rate of biochemical reactions in an organism. Often estimated by rate of oxygen consumption in aerobes.

Metal-rich granules. Small solid deposits of metal compounds found widely in marine species environmentally exposed to metals.

Metamorphosis. Major developmental change as the larva develops into an immature adult.

Metapopulation. A group of interconnected subpopulations, usually of subequal size. The features of individuals now found in one subpopulation might have been determined by conditions affecting them when they were located in another subpopulation.

Metazoa. Equivalent to all of the animals.

Methanogenic bacteria. Anaerobic bacteria that use carbon dioxide as a source of carbon and produce methane as a by-product.

Methylmercury. Organic form of mercury and the form of mercury that is most easily bioaccumulated.

Microarray. An array of sequences, usually attached to a plastic chip, used to test for binding of similar sequences extracted from a species. Used to estimate the degree of gene expression or the presence of given DNA sequences in a given individual.

Microbenthos (microfauna or microflora). Benthic organisms (animals or plants) whose shortest dimension is less than 0.1 mm.

Microbial loop. A part of a marine food web where bacteria is consumed by protistans, which are eventually consumed by larger consumers in the food web.

Microbial stripping hypothesis. When a deposit feeder feeds on sediment, it digests the microbial organisms on particulate organic matter with great efficiency but digests the particulate organic matter itself with very low efficiency.

Microfauna. Animals whose shortest dimension is less than 0.1 mm.

Microsatellites. DNA sequences used as genetic markers. Usually consist of highly repetitive sequences that are quite variable and therefore useful in marking individual populations of a species.

Migration. A directed movement of an organism between specific areas.

Mixed tides. Tides where the vertical extent of the tide is very uneven, usually with two very different alternating low tides.

Mixing depth. The water depth to which wind energy evenly mixes the water column.

Mixoplankton. Planktonic organisms that can be classified at several trophic levels. For example, some ciliates can be photosynthetic but also can ingest other plankton and are heterotrophic or may retain ingested chloroplasts.

Moderately stratified estuary. An estuary in which seaward flow of surface low-salinity water and moderate vertical mixing result in a modest vertical salinity gradient.

Modular. Referring to organisms that consist of repeated connected units that are genetically identical and of similar ecological function (e.g., a coral colony).

Module. A unit in a modular organism.

Molecular clocks. The dating of a biological event (e.g., origin of an evolutionary group) by using the rate at which DNA sequences change over time.

Mollusks. A phylum of protostome mollusks characterized usually by a mantle, calcium carbonate shell(s), and unsegmented body; including snails, bivalves, and squids.

Momentum boundary layer. A layer of water near a surface where physical transport of fluid is affected by the presence of the surface.

Monoclonal antibodies. Antibodies produced by a single immune cell or daughters of that identical cell.

Monophyletic. Refers to a group of species that all have a single common ancestral species.

Monoplacophora. Cap-shaped mollusks, which may include ancestral forms similar to the Mollusca.

Mucous-bag suspension feeder. Suspension feeder employing a sheet or bag of mucus to trap particles nonselectively.

Müllerian mimicry. Mimicry where two species resemble each other and both are toxic.

Multibeam sonar. A sonar signal in a fan shape that can map the spatial arrangement of depth on the sea floor and even detect properties of the seabed.

Mustelidae. A family of mammals including minks and otters.

Mutualism. An interaction between two species in which both derive some benefit.

Mycelium. A tangled mass of hairlike threads of ascomycetes.

Mycorrhizae. Fungi involved in a symbiosis with roots of plants and aiding in plant nutrient uptake.

Mysticeti. Group of Cetacea that use baleen for feeding.

Nanoplankton. Planktonic organisms that are 2–20 μm in size.

NAO. *See* North Atlantic Oscillation.

Natural philosophy. A general approach to studying environmental science and biology that was popular in the nineteenth century, when there were few specialists but scientists who studied diverse subjects.

Natural selection. The differential contribution of genes to the next generation because of fitness differences.

Neap tides. Tides occurring when the vertical range is minimal.

Negative correlation. An inverse relationship between the values of one parameter and the values of another.

Nekton. Organisms with swimming abilities that permit them to move actively through the water column and to move against currents.

Nematoblast. Cells found in Cnidaria, which contain hooks, stingers, or mucus to entrap prey.

Nematocysts. The stinging, hooking, or mucus-producing elements that emerge from nematoblasts.

Nematoda. Free-living and parasitic worms with a cuticle and longitudinal muscles.

Nemertea. Elongate free-living worms, with complete gut; carnivorous, using barbed proboscis to kill prey.

Neritic. Seawater environments landward of the shelf-slope break.

Net primary productivity. Total primary production, minus the amount consumed in respiration.

Neurotoxic shellfish poisoning. Uptake by shellfish of bevetoxin, which is found in dinoflagellate species such as *Gymnodinium breve*. Brevetoxin causes respiratory problems for humans.

Neuston. Planktonic organisms associated with the air–water interface.

Neutral theory of community ecology. Random interactions, combined with occasional extinctions and speciation events, result in indefinite coexistence of many species.

New production. Primary production in a body of water that can be explained by import of usually inorganic nutrients from outside the system, as in upwelling.

Niche. A general term referring to the range of environmental space occupied by a species.

Niche overlap. An overlap in resource requirements by two species.

Niche structure. Any predictable partitioning by coexisting species of a habitat into subhabitats.

Nitrification. A process caused by nitrifying bacteria, where ammonium is oxidized to nitrite or nitrate.

Nitrogen fixation. The conversion of gaseous nitrogen to nitrate or ammonium by specialized bacteria.

Nonconsumptive effect. A response within a food web when one predatory species changes its behavior (perhaps reducing its activity in response to its own predator) and as a consequence reduces its impact on a prey species.

Noncumulative toxic substances. Substances that do not increase in concentration in the body over time.

Nonpoint source. Pollution source that comes to the watershed from many points along a water body, as opposed to from a single source, such as an industrial pipe.

Nonrenewable resource. A resource that, when consumed, is no longer available over the lifetime of the organism.

North Atlantic Oscillation (NAO). A cycle of changing difference in air pressure between a low atmospheric pressure over Iceland at about 64° N latitude and a higher atmospheric pressure over the Azores, at approximately 38° N latitude.

North Pacific Transition Zone. A belt across the Pacific centered at about 35–40° N latitude where frontal transitions often result in high primary productivity.

No-slip condition. The condition where water has zero velocity when in contact with a surface.

No-take area. Geographic area where by law no one is allowed to fish or collect biological specimens. Rules could apply to one or all species.

No-take sanctuary. A marine preserve with a rule that prevents taking of a given species or group of species.

Nucleotides. A building block of a DNA or RNA strand.

Nuisance bloom. A rapid increase of one or only a few species of phytoplankton, resulting in cell densities high enough to cause discoloration of the surface water, possible increase of toxins, and degradation of water quality aspects such as dissolved oxygen.

Null hypothesis. A hypothesis that states an experimental treatment will result in no change, relative to a control or relative to a starting measurement before the experimental treatment is applied.

Nutrient cycling. The pattern of transfer of nutrients between the components of a food web.

Nutrients. Those constituents required by organisms for maintenance and growth (we use this term in this book in application to plants).

Ocean acidification. The decline of pH in the ocean, owing to additions of carbon dioxide in the atmosphere and subsequent dissolution in seawater.

Ocean observatories. Remote sensing systems, on the seafloor or in mid waters, used to collect data and transmit them to land-based laboratories.

Oceanic. Associated with seawater environments seaward of the shelf-slope break.

Oceanic boundary currents. A distinct ocean current that runs parallel to and offshore of the coast usually on the western sides of oceans (e.g., the Gulf Stream).

Oceanic crust. A layer of earth's crust that underlies the ocean basins. Overlies the earth's mantle.

Oceanic ridge. A sinuous volcanic ridge rising from the deep-sea floor.

Odontoceti. A group of Cetacea characterized by reduced appendages, flukes, and teeth used in carnivory.

Oligochaeta. A class of Annelida that are wormlike, usually free living, and have chaetae but lack parapodia; includes earthworms.

Oligotrophic. Refers to water bodies or habitats with low concentrations of nutrients.

Omnivory. Being able to feed in more than one distinct way (e.g., an organism capable of carnivory and herbivory).

Operculum. Hard organic covering of the foot of gastropods, used to protect against predation or desiccation.

Ophiuroidea. Class of Echinodermata that look like sea stars but have very flexible arms.

Opportunists. Species with life-history traits that allow colonization of numerous habitat types and rapid subsequent population growth.

Optimal foraging theory. A theory designed to predict the foraging behavior that maximizes food intake per unit time.

Organic. Deriving from living organisms.

Organic carbon. Carbon derived from organic molecules such as amino acids.

Organic nutrients. Nutrients in the form of molecules synthesized by or originating from other organisms.

Organic osmolytes. Compounds including amino acids and urea that are used in osmoregulation.

Oscillation, climate. A fluctuation in air pressure characteristics, wind systems, sea-surface temperature (SST), or other weather features that occur on the geographic scale of an ocean, such as throughout the North Atlantic Ocean or perhaps even the world ocean.

Osculum. An opening that is used for the exit of wastewater in a living sponge.

Osmoconformer. An organism whose body fluids change directly with a change in the concentrations of dissolved ions in the external medium.

Osmoregulator. An organism that regulates the concentration of dissolved ions in its body fluids irrespective of changes in the external medium.

Osmosis. The movement of pure water across a membrane from a compartment with relatively low dissolved ions to a compartment with higher concentrations of dissolved ions.

Osmotic pressure. A pressure that corresponds to the different osmotic conditions on either side of a biological surface, such as a membrane.

Osteichthyes. The bony fishes.

Otolith. A small mass of calcium carbonate, used in the inner ear of a fish for perception of balance. The mass often grows in increments that can allow us to determine the age of the fish.

Outwelling. The outflow of nutrients from an estuary or salt marsh system to shelf waters.

Overfishing. Case in which high degree of fishing results in decline of a fish population to unsustainable levels.

Oviparous. Reproduction in which eggs are laid and hatch outside the mother's body, with little or no development within the mother. An example would be an unfertilized egg shed into the water that is fertilized by planktonic sperm.

Ovoviparous. Fertilization is internal, but eggs develop within the mother's body with no nutrients supplied. Eggs are eventually released into the environment and hatch.

Oxygen dissociation curve. A curve showing the percent saturation of a blood pigment, such as hemoglobin, as a function of oxygen concentration of the fluid.

Oxygen electrode. An electrode used to measure oxygen concentration in the water.

Oxygen minimum layer. A depth zone, usually below the thermocline, in which dissolved oxygen is minimal.

Oxygen technique (primary productivity). The estimation of primary productivity by the measurement of the rate of oxygen increase.

Pacific Decadal Oscillation. A climatic oscillation in the Pacific of a period of 20–30 years, in which cool and warm waters exchange locations in the eastern and western Pacific north of 20° N latitude.

PAHs. *See* Polycyclic aromatic hydrocarbons.

PAR. *See* Photosynthetically active radiation.

Paradox of the plankton. The coexistence of many species of phytoplankton, despite evidence of resource limitation.

Paralytic shellfish poisoning (PSP). The ingestion of dinoflagellates containing saxitoxin by bivalves, which in turn may be eaten by humans, who are poisoned.

Parapatric speciation. The differentiation into distinct species of populations experiencing some gene flow.

Parapodia. Paired structures on polychaetes, used for locomotion.

Parasite. An organism living on (ectoparasites) or in (endoparasite), and negatively affecting, another organism.

Partially mixed estuary. An estuary or area of an estuary with a gradient of lower-salinity to higher-salinity water as one moves from the surface to the bottom; mixed vertically by wind and tidal motion.

Particulate organic matter (POM). Particulate material in the sea derived from the decomposition of the nonmineral constituents of living organisms.

Patchiness. A condition in which organisms occur in aggregations.

PCR. *See* Polymerase chain reaction.

Peat. A sediment that is rich in organic matter.

Pelagic. Living in the water column seaward of the shelf-slope break.

Pelagic trawls. Fishing by means of a net, which is towed in the water column and kept open by boards at the net opening.

Pellets. Compacted aggregations of particles resulting either from egestion (fecal pellets) or from burrow-constructing activities of marine organisms.

Penetration anchor. In hydraulically burrowing organisms, any device used to penetrate and gain an initial purchase on the sediment so that the body can be thrust in farther.

Pennate diatoms. Diatoms that are bilaterally symmetrical and usually elongate. They tend to occur on the sea bed rather than in the water column.

Peptides. Chains of amino acids; often portions of a protein molecule, or functional molecules in their own right.

Petersen grab. A bottom grab that enters the seabed by gravity and then closes as a chain pulls the sampler upward toward the ship.

pH. Measure of the acidity or basicity of water; pH = $-\log_{10}$ of the activity of hydrogen ions in water.

Phase shift. A predictable response in the form of a change of community composition to a new state in response to an environmental change.

Phenotype. The form of an organism or a trait (as opposed to its genotype).

Phenotypic plasticity. The capacity of an individual of the same genotype to produce different phenotypes under different environmental conditions. Nongenetic potential variability within the range of a single individual.

Phi scale. Scale used for measuring the grain size of sediments; phi = $-\log_2$ (grain diameter).

Phleger corer. A small, gravity-driven bottom corer.

Phoronida. A phylum of wormlike animals, having a lophophore and living infaunally.

Photic zone. The depth zone in the ocean extending from the surface to that depth permitting photosynthesis.

Photoinhibition. Reduction or suppression of photosynthesis due to high light intensity.

Photorespiration. Enhanced respiration of plants in the light relative to dark respiration.

Photosynthate. A substance synthesized in the process of photosynthesis.

Photosynthetic quotient. In photosynthesis, the moles of oxygen produced, divided by the moles of carbon dioxide assimilated.

Photosynthetic rate. The rate of conversion of dissolved carbon dioxide and bicarbonate ion to photosynthetic product.

Photosynthetically active radiation (PAR). That part of the light spectrum that can be used in photosynthesis.

Phototactic. Moving in response to light.

Phylogeography. The study that combines geographic and evolutionary aspects of the distribution of species.

Physiological integration. A general term signifying the degree of coordination of different physiological and biochemical processes within a cell or within an organism.

Physiological race. A geographically defined population of a species that is physiologically distinct from other populations.

Physiological tolerance. The degree to which an organism can survive an extreme environment, by virtue of its physiological traits.

Phytodetritus. Particulate organic matter settling through the water column that derives from dead phytoplankton.

Phytoplankton. The photosynthesizing organisms residing in the plankton.

Pinnipedia. A diverse group of semi-aquatic marine mammals with fin-shaped feet, including seals, sea lions, and walruses.

Planktivorous. Feeding on planktonic organisms.

Plankton. Organisms living suspended in the water column and incapable of moving against water currents.

Planktotrophic larva. Planktonic-dispersing larva that derives its nourishment by feeding in the plankton.

Plant nutrients. Substances required by plants for growth.

Planula. The planktonic larval form produced by scleractinian corals and coelenterates.

Plasticity. *See* Phenotypic plasticity.

Plate. Major section of the earth's crust, bounded by such features as midocean ridges.

Platyhelminthes. A phylum of invertebrates, commonly known as flatworms, with free-living and parasitic representatives.

Pleistocene. Period of time, going back to approximately 2 million years before the present, in which alternating periods of glaciation and deglaciation dominated the earth's climate.

Pleuston. Refers to plankton that have a float protruding above the sea surface, such as the Portuguese man-of-war.

Pneumatocysts. Gas-containing floats found in some brown seaweeds, such as kelps.

Poecilogonic. Species with more than one larval developmental mode.

Pogonophora. A phylum of wormlike animals that are gutless, have a symbiosis with bacteria, and are usually found in deep-sea environments.

Poikilotherm. An organism whose body temperature is identical to that of the external environment.

Point source. A pollution source from a confined spot, such as an industrial pipe.

Polychaeta. A class of annelids characterized by paired parapodia, which are used for locomotion.

Polychlorinated biphenyls (PCBs). Usually very toxic compounds manufactured for insulation, but released into the marine and estuarine environments.

Polycyclic aromatic hydrocarbons (PAHs). Derivatives from fossil fuels that are very toxic and known to be carcinogenic.

Polymerase chain reaction (PCR). A reaction based on fluctuating thermal conditions used to amplify DNA by means of annealing specific strands with nucleotides that bind and amplify the original DNA strand to great abundance.

Polymorphism. The presence of coexisting and distinctly different forms of a species; may be caused by genetic differences or phenotypic plasticity.

Polyp. An individual of a solitary coelenterate or one member of a coelenterate colony.

Polyphyletic. Refers to a group of species that do not have one common ancestor species.

Polyplacophora. A class of Mollusca, comprising the chitons.

Poorly sorted sediments. Sediment consisting of a wide range of mixed groups of differently sized particles.

Population density. Number of individuals per unit area or volume.

Pop-up satellite archival tags. Fish tags that are designed to become detached from fish after a time, so that they can rise to the surface and be detected by a global positioning system.

Porifera. The phylum comprising the sponges.

Positive correlation. An increase in one variable occurs with an increase in another variable, or a decrease in one variable occurs with a corresponding decrease in another variable.

Postmating isolation. Reproductive isolation between species due to mechanisms such as genetic compatibility, despite the fact that mating occurs.

Post-translational modification. Chemical change (e.g., attachment of oligosaccharides) in a protein after it is translated, or after its amino acid sequence is specified.

Pound net. A blind-ended net that captures fish at the blind end.

ppt. A measure of the salt content of seawater in terms of kilograms of salt per kilogram of water, reckoned in parts per thousand. A conductivity version of this measure is the psu, which differs from ppt by very little, on the order of 0.02 psu or less.

Practical salinity unit. *See* psu.

Precision. The repeatability of a measurement. A measurement can be precise, but not accurate. *See also* Accuracy and Resolution.

Predation. The consumption of one organism by another.

Predator. An organism that consumes another living organism (carnivores and herbivores are both predators by this definition).

Premating isolation. Reproductive isolation between species that involves mechanisms such as time of reproduction and mate-recognition signals.

Pressure drag. A difference in pressure upstream and downstream of an object in a flow.

Primary producer. An organism capable of using the energy derived from light or a chemical substance in order to manufacture energy-rich organic compounds.

Primary production. The production of living matter by photosynthesizing organisms or by chemosynthesizing organisms. Usually expressed as grams of carbon per square meter per year.

Primary treatment. Simple screening of organic particulates before sewage is released into the water.

Priority effect. Colonization of an unoccupied site by a species, followed by continued occupation of that site by this species, preventing all other species from invading.

Productivity. The amount of biological material (usually expressed as carbon) produced per unit time (usually expressed per unit of area in the ocean).

Prokaryotes. Organisms distinguished by cells that lack true nuclei or organelles and do not reproduce by means of mitosis.

Protandrous. An animal that, when sexually mature, is first male and then switches sex to female.

Protein. A molecule consisting of one or more chains of amino acids in a specific order.

Protein polymorphism. Presence of several variants of a protein of a given type (e.g., a certain enzyme, such as carboxylase) in a population.

Proteomics. The study of organism response by assaying cellular protein diversity and abundance.

Protista. Group, sometimes considered a kingdom, of mostly unicellular organisms with a true nucleus and chromosomes. Includes ciliates, flagellates, and some macroalgae. This group is not well defined as an evolutionary group of organisms.

Protogynous. An animal that, when sexually mature, is first female and then switches sex to male.

Protostomes. Individuals belonging to a group of Bilateria, including phyla of ecdysozoans (e.g., Arthropoda) and lophotrochozoans (e.g., Annelida).

Province. A geographically defined area with a characteristic set of species or characteristic percentage representation by given species.

Pseudofeces. Material rejected by suspension feeders or deposit feeders as potential food before entering the gut.

psu. Practical salinity unit. A measure of the salt content of seawater (practical salinity), based upon electrical conductivity of a sample relative to a reference standard of seawater at one atmosphere pressure and 15°C temperature. The reference is a set of diluted seawater samples from the North Atlantic of known salt content. *See also* ppt.

Pteropods. Group of holoplanktonic gastropods.

Purple sulfur bacteria. A group of bacteria usually living in stagnant water or on stagnant anoxic sediment surfaces as microbial mats that perform photosynthesis and oxidize hydrogen sulfide.

Purines. The bases adenine and guanine in DNA or RNA.

Purse seine. A seine net that is set in a ring and can be reduced in diameter by pulling a line, trapping fish.

Pycnocline. Depth zone within which seawater density changes.

Pycnogonida. A group of crustacea having long, spiderlike legs.

Pyrimidines. The bases cytosine, thymine, and uracil in DNA and RNA.

Q_{10}. Increase of metabolic rate with an increase of 10°C.

Quantitative genetics. The study of the genetic basis of traits, usually explained in terms of the interaction of a group of genes with the environment.

r. The intrinsic rate of increase of a population.

Radial cleavage. A type of cell division in early embryonic growth in which the cleavage plane is parallel or perpendicular to a single embryonic axis.

Radiance. The amount of electromagnetic radiation (e.g., light energy) arriving at a point on the earth's surface.

Radiocarbon technique (primary productivity). The estimation of primary productivity by the measurement of radiocarbon uptake.

Radiolaria. Protistan phylum whose members are planktonic and secrete an often elaborate siliceous test.

Radiolarian ooze. A deep-sea sediment, composed primarily of radiolarian tests.

Radula. A belt of teeth, found in gastropods and chitons, used for feeding.

Ram feeders. Moving fish that feed by moving rapidly at a prey item with the mouth open.

Random population change. Change in population size over time that has no predictable trend.

Random spatial distribution. Situation in which individuals are randomly distributed in a space; probability of an individual's being located at any given point is the same irrespective of location in the space.

Rarefaction curve. A relationship between number of species collected as a function of the number of individuals collected. As more and more individuals are collected, the number of species collected will eventually reach a plateau. The relationship of species to individuals can be used to predict the expected number of species at different sample sizes of individuals.

Reaction center. Parts of the cell that include chlorophyll and proteins that receive photon energy during photosynthesis.

Real-time PCR. Method of quantifying the amount of DNA as it is being amplified.

Recruitment. The residue of those larvae that have (1) dispersed, (2) settled at the adult site, (3) made some final movements toward the adult habitat, (4) metamorphosed successfully, and (5) survived to be detected by the observer.

Red tide. A dense outburst of phytoplankton (usually dinoflagellates) often coloring water red-brown.

Redfield ratio. Molecular ratio of carbon, nitrogen, and phosphorus found in phytoplankton.

Redox-potential discontinuity (RPD). That depth below the sediment–water interface marking the transition from chemically oxidative to reducing processes.

Refuge. A device by which an individual can avoid predation.

Regeneration production. Primary production that is caused by nutrients that are recycled within a water body from excretions.

Regulator. An organism that can maintain constant some aspect of its physiology (e.g., body temperature) despite different and changing properties of the external environment.

Remote sensing. Acquisition of environmental data with an instrument that is not in contact with the medium to be sampled, often at great distances, as in satellites sensing the earth's surface.

Remotely operated vehicle (ROV). An underwater vehicle connected to the ship by a cable, whose movements are directed from shipboard. ROVs may have sensors to measure salinity, temperature, and other variables, but also may have video cameras and sampling devices.

Renewable resource. A resource that can be regenerated (e.g., a growing diatom population that is being exploited by a copepod).

Reproductive effort. The fraction of assimilated nutrients that are devoted to reproductive behavior and gamete production.

Residence time. The time a unit of a substance spends in a specified location in the environment (e.g., the residence time of sodium in the Antarctic bottom water).

Resilience. The capacity of a community or ecosystem to respond to a major disturbance and recover quickly to its former state.

Resolution. The smallest amount of change that an instrument can

discriminate. *See also* Accuracy and Precision.

Resource. A commodity that is required by an organism and is potentially in short supply.

Respiration. Consumption of oxygen in the process of aerobic metabolism.

Respiratory pigment. A molecule, polymer, or other complex adapted to bind and transport oxygen efficiently, usually in a circulatory system (e.g., hemoglobin).

Respiratory quotient. The ratio of moles of carbon dioxide produced to oxygen consumed in respiration.

Respiratory trees. Paired, branching structures emerging from the cloaca of a sea cucumber.

Rete mirabile. A countercurrent exchange structure of capillaries that allows gas uptake in a fish swim bladder.

Retention time. The time a unit of water remains in a water body such as an estuary before being mixed into an outside water body, such as the shelf.

Reverse Bohr effect. Effect that occurs when lactate builds up in the blood of certain invertebrates and pH decreases, increasing the affinity of hemocyanin for oxygen.

Reynolds number (*Re*). A number that represents the relative importance of viscous forces and inertial forces in a fluid. As *Re* increases, inertial forces become more important. In seawater, *Re* increases with increasing water velocity and with the size of the object in the water.

Rhizome system. A system of runners below the sediment surface that allows sea grasses and salt marsh plants to extend coverage of a plant over large areas and permits transfer of nutrients to new areas where shoots can emerge at the sediment surface.

Rights-based management. Declaration of a fishery as a total commodity, divided among current users who are give the right to fish a percentage of a catch, or to sell that right to others.

Rip current. A concentrated rapid current moving offshore from a beach fronting a longshore current.

Ripple marks. Surface sedimentary structure formed by movement of water over the bottom, resulting in cyclical highs and lows in the sediment.

Rise. Bottom of low relief at the base of the continental slope.

RNA. A macromolecule consisting of a chain of nucleotides, whose primary function is protein synthesis.

ROV. *See* Remotely operated vehicle.

Salinity. Number of kilograms of dissolved salts in 1 kg of seawater, measured in parts per thousand. Actually this definition stands alongside another definition based on water standards whose electrical conductivities are measured. *See also* ppt and psu.

Salps. A group of pelagic tunicates (phylum Urochordata), either colonial or solitary, with buccal and atrial siphons on opposite sides of the body.

Salt glands. In mangroves: in some species, glands in the leaves that excrete salts to the leaf surface.

Salt marsh. A coastal habitat consisting of salt-resistant plants residing in an organic-rich sediment accreting toward sea level.

Saprophytic organisms. Organisms that break down and decay organic matter.

Satellite radiometer. A device in a satellite that measures the amount of electromagnetic radiation over a specified range of wavelengths.

Saturated solution. With respect to a given substance that might precipitate from solution: the concentration of dissolved components are at a maximum before precipitation will occur.

SAV. Submerged attached vegetation.

Saxitoxin. A neurotoxin that blocks sodium channels, produced by certain phytoplankton species.

Scaphopoda. Class of the phylum Mollusca with elongate tusk-shaped conical shell.

Scattering. Interaction of particles in the water column with light, resulting in a decline of light energy with depth.

Scavenger. An organism that feeds on dead or decomposing animals or macrophytes.

Schooling. Fish: movement, usually coordinated, in groups.

Scientific method. Organized means of learning about the natural world, using observation, forming of hypotheses, and hypothesis testing.

Scleractinia. Order of coelenterates, usually producing calcareous skeletons with hexameral symmetry.

Scope for activity. The surplus of energy available for activity, such as swimming, beyond that required for maintenance.

Scope for growth. The surplus of energy available for growth beyond that required for maintenance.

Scyphozoa. The true jellyfish, members of the phylum Cnidaria.

Seafloor spreading. The horizontal movement of oceanic crust.

Seamount. A rise from the bottom that is more than 1 km above the sea floor but whose top does not reach the sea surface.

Seaward. Side of an island that faces the direction of wave action generated either by winds or by currents generated by more indirect forces.

Secondary compounds. Molecules that are manufactured for defense against attack by a predator, parasite, or competitor.

Secondary production. The production of living material per unit area (or volume) per unit time by herbivores. Usually expressed as grams of carbon per square meter per year.

Secondary treatment. Treatment of sewage that encourages breakdown of particulate organic matter but releases dissolved nutrients into the marine environment.

Sedimentation. Deposition of particles and chemical precipitation to form deposits in water.

Seine net. Net placed in the water and pulled along, capturing marine organisms in the mesh.

Semelparity. Reproducing only once.

Semidiurnal tide. A tidal cycle of alternating and subequal high and low tide.

Semi-infaunal. Living partially buried within the sediment but partially projecting into the water column.

Sequential hermaphrodite. An individual that sequentially produces male and then female gametes or vice versa.

Sessile. Immobile because of an attachment to a substratum.

Seston. Particulate matter suspended in seawater.

Setules. Chitinous projections from copepod maxillipeds that trap food particles.

Sex. Combining genetic materials from different types, known as sexes, usually in the production of offspring.

Sexual selection. Selection for traits that are involved in mating success,

such as visual elements (e.g., color) and combat structures (e.g., antlers of deer).

Shear. A force acting parallel to a linear body.

Shelf-slope break. Line marking a change from the gently inclined continental shelf to the much steeper depth gradient of the continental slope.

Shifting baselines. Concept that our perception of the natural environment may change according to how the environment changes over the generations, resulting in a misperception of what was natural several generations before human degradation.

Shoaling. Attraction of individuals, usually fish, as an aggregation to a shallow water area.

Short-term (acute) effects. Immediate response to an environmental change.

Sibling species. Closely related species that are so similar they are nearly indistinguishable morphologically.

Side-scan sonar. A sonar system producing sound energy that bounces off the seafloor and is subsequently picked up by a detector. The signal gives a picture of the seabed surface, revealing a variety of surface sedimentary features. *See also* Multibeam sonar.

Sigma. Parameter expressing the seawater density and equal to 1 minus the density of seawater, measured at a given temperature and at a pressure of 1 atmosphere.

Sill. A raised portion of bottom near the mouth of a water body (e.g., a fjord) that connects to the open sea.

Silt–clay fraction. The particle fraction of a sediment that is less than 62 μm in diameter.

Simultaneous hermaphrodite. An individual capable of producing male and female gametes at the same time.

Siphonophores. A group of specialized hydrozoan cnidarians, consisting of large planktonic polymorphic colonies.

Sled. A benthic sampling device designed to slide along the sediment surface, digging into the bottom to a depth of at most a few centimeters.

Slope. A steep-sloping bottom extending seaward from the edge of the continental shelf and downward toward the rise.

Smith–McIntyre grab. A device that collects a bottom sample by means of spring-loaded sections that close

together and enclose a sediment sample. *See also* Grab.

Snow. *See* Marine snow.

Soft sediment. Sediment composed of separate sedimentary grains.

Solar irradiance. Solar energy that reaches the ocean or earth surface.

Solubility product constant. At saturation, a constant, which is a multiple of the solubilities of the two ionic components (e.g., Na^+ and Cl^-) of the substance that could crystallize in solid phase.

Somatic growth. Growth of the body, exclusive of gametes.

Sorting (of a sediment). The range of scatter of particle sizes about the median grain size of a sediment.

Space limited. Description of a situation in which space is a limiting resource.

Spatial autocorrelation. A situation in which some parameter at any location (e.g., population density) can be predicted through a knowledge of the values of the parameter in other locations.

Spatial distribution. The arrangement of individuals in a space.

Speciation. The process of formation of new species.

Species. A population or group of populations that are in reproductive contact but are reproductively isolated from all other populations

Species richness. The number of species in an area or biological collection.

Species-area effect. A regular logarithmic relationship between the number of species in a confined geographic area (e.g., an island) and the area in which the species occur.

Sperm attractants. Chemicals that sperm use to follow concentration gradients to eggs.

Spicules. Skeletal elements made of silica found in the outer wall of a sponge.

Spillover effect. When a marine no-take area is established, increased population growth within the area might result in resource species moving into an adjacent area, increasing the size of a fishery.

Spongin. The organic material of which the sponge skeleton is composed.

Sporophyte. Diploid stage in the life cycle of a plant.

Spring diatom increase. The major rapid population increase of diatoms, occurring in the spring in temperate-boreal latitudes.

Spring freshet. The increase of flow in an estuary in spring, owing to snow melt and precipitation in the watershed.

Spring tide. The biweekly time, corresponding to full and new moons, of maximum tidal vertical range.

Spur-and-groove topography. Topographic feature of some coral reefs with massive colonies forming an alternation of projections and hollows.

Stability-time hypothesis. Hypothesis that states that higher diversity occurs in habitats that are ancient and stable environmentally.

Standing crop. The amount of living material per unit area or volume; may be expressed as grams of carbon, total dry weight, and so on.

Stern slipway. Large opening in the stern end of a whaling factory ship through which a whale carcass could be dragged onto the ship.

Stipe. In seaweeds: the structure that connects the holdfast to the frond(s) or blades and usually provides mechanical strength to the seaweed in a current.

Stock. A population of a species involved in a fishery whose population dynamics are relatively independent of other stocks (usually geographically separated).

Stock-recruitment model. Fishery model that predicts the amount of juvenile recruitment as a function of the parent stock.

Stokes' law. Mathematical relationship that describes the terminal settling velocity of small particles through a fluid such as water, taking into account the effect of viscosity.

Stratification. In benthos, the presence of different infaunal species at distinct respective horizons below the sediment–water interface.

Subduction zones. Locations where one crustal plate is dragged down below another and eventually is melted within the mantle.

Submarine canyon. An erosional and linear feature found incised in the continental slope, allowing sediment to rapidly move downward to the base of the slope and the continental rise.

Subtidal zone. The depths below the lowest extent of vertical tidal motion in a benthic environment.

Subtropical. Refers to the portion of the temperate zone closest to the equator.

Succession. A predictable ordering of a dominance of a species or groups of species following the opening of an environment to biological colonization.

Sulfate-reducing bacteria. Bacteria that use sulfate as an oxidizing agent, reducing it to sulfide.

Surface browser. Organism that feeds by scraping thin layers of living organisms from the surface of the substratum (e.g., periwinkles feeding on rock-surface diatom films; urchins scraping a thin, filmy sponge colony from a rock).

Surface layer. The layer of the ocean extending from the surface to a depth above which the ocean is homogeneous due to wind mixing.

Surface mixing hypothesis. Explanation for diurnal vertical migrations of zooplankton, arguing that zooplankton go to depth in order to rise to newly mixed surface waters.

Surfactant. A substance that reduces the surface tension between water and various hydrophobic organic compounds (detergent action) and aids in removing these compounds from surfaces. Present in deposit-feeder guts to aid in removing organic compounds from particle surfaces.

Survivorship curve. The curve describing changes of mortality rate as a function of age.

Suspension feeder. An organism that feeds by capturing particles suspended in the water column.

Swarming behavior. Movement of members of a species, usually zooplankton or fish, into a tight aggregation for either protection against predators or for breeding.

Swash rider. Invertebrate that can migrate up and down shore with the rising and falling tide, in order to maintain station at a level that is moist but not overly washed by the waves.

Swim bladder. In fishes, a gas-filled chamber whose volume can be regulated so that fish can regulate their depth.

Tags. *See* Implanted tags.

Teleplanic larva. Larva capable of dispersal over long distances, such as across oceans.

Temperate zone. Pertaining to the latitudinal belt between the tropics (23.5° N latitude) and the Arctic or Antarctic Circle (66.5° S latitude), in the Northern and Southern Hemispheres, respectively.

Tentacle-tube-foot suspension feeder. Suspension feeder that traps particles on distinct tentacles or tube feet (in echinoderms).

TEP. *See* Transparent exopolymers.

Terminal anchor. In hydraulically burrowing organisms: any device used to anchor the leading portion of the burrower, permitting muscular contraction to drag the rest of the body into the sediment.

Territoriality. Defense of a specified location against intruders.

Tertiary production. The production of living material per unit area (or volume) per unit time by organisms consuming the herbivores. Usually expressed as grams of carbon per square meter per year.

Tertiary treatment. Treatment of sewage that removes dissolved nutrients before it enters the environment.

Tethys. An ancient sea that connected the present Indian Ocean with the Mediterranean and Atlantic. The sea was eliminated by a terrestrial uplift in the Miocene.

Thallus. In seaweeds: the life-history form that is usually macroscopic and attaches to a substratum.

Theca (in dinoflagellates). An organic skeleton of dinoflagellates consisting of cellulose plates.

Thermocline. Depth zone within which temperature changes maximally.

Thermohaline circulation. Movement of seawater that is controlled by density differences that are largely explained in terms of temperature and salinity.

Thixotropy. Property of watery sediment in which it liquefies more when a pressure is applied to it.

Tidal current. A water current generated by regularly varying tidal forces.

Tidal wave. *See* Tsunami.

Tides. Periodic movement of water resulting from gravitational attraction between the earth, sun, and moon.

Time-in-patch model. A model that predicts the optimal time that a forager should exploit a patch of food before moving on to another patch.

Top-down control. Refers to food webs where control of a population is mainly explained by consumption by a species or group of species at higher levels of the food chain (e.g., population change of population of mussels controlled by sea star predation).

Toxic algal blooms. Blooms, usually of phytoplankton, that result in toxic effects on other marine organisms or humans.

Trace elements. Elements in the ocean at an average concentration less than 1 part per million.

Trade winds. Persistent winds at low latitudes in both the Northern and Southern Hemispheres, blowing toward the west and the equator.

Trait-mediated indirect interaction. A plastic trait, in which a species changes its behavior or morphology depending on the presence of an interacting species (e.g., a predator), which may lead to different interactions within a community. *See also* Density-mediated indirect interaction.

Transcriptomics. The study of organismic reactions by estimating the degree of gene expression of one or a wide range of genes by assaying RNA content.

Transferrin. Protein in vertebrate blood that binds to iron.

Translation. The process of protein synthesis that determines the amino acid sequence of the protein (primary structure).

Transparent exopolymers (TEP). Large transparent polymers of mainly acidic polysaccharides that enhance aggregation of fine organic particles. TEP are produced mainly by plank-tonic diatoms.

Transverse faults. Large-scale geological faults in oceanic crust.

Trench. Deep and sinuous depression in the ocean floor, usually seaward of a continental margin or an arcuate group of volcanic islands.

Trichomes. Rows of connected cells of cyanobacteria.

Trilobita. Group of extinct arthropods.

Trophic cascade. A strong interaction among trophic levels in a food chain, where changes in density at one level results in indirect effects at a trophic level that does not directly interact with the first level. An example is an increase in carnivores, which indirectly results in the increase in abundance of plants, since herbivores have been reduced.

Trophic group amensalism. Hypothesis of negative effects of deposit feeders in soft sediments on suspension feeders living in the same habitat type.

Trophic level. In a food chain, a level containing organisms of identical feeding habits with respect to the chain (e.g., herbivores).

Trophosome. A part of the body of a vestimentiferan worm that contains symbiotic bacteria.

Tropical. Being within the latitudinal zone bounded by the two tropics of Cancer and Capricorn (23° and ca. 26° N and S latitude).

True jellyfish. Jellyfish belonging to the Cnidarian class Scyphozoa.

Tsunami. A large and fast-traveling ocean wave usually caused by an earthquake or major slide of sediment.

Tube feet. Structures in echinoderms used in locomotion or feeding.

Turbidity. The weight of particulate matter per unit volume of seawater.

Turbulent flow. Movement of water that can be characterized by streamlines moving in a very irregular fashion.

Turtle exclusion device. Device designed to divert turtles from being trapped in a regular fish net.

Ubiquitin. Protein found in all eukaryotic cells that can remove degrading proteins.

Ultraplankton. Planktonic organisms that are less than 2 μm in size.

Uniform spatial distribution. Situation in which individuals are more evenly spread in space than would be expected on the basis of chance alone.

Upper-canopy kelps. Kelps that extend far above the seafloor and have the potential to shade seaweeds below.

Upwelling. The movement of nutrient-rich water from a specified depth to the surface, usually driven by surface winds.

Urochordata. Deuterostome phylum including sea squirts and salps, with larvae that have characters of chordates.

Vadose layer. In mangrove forests, a sediment layer with high saltwater content.

Vents. *See* Hydrothermal vents.

Vertical stratification (in sediment). The occurrence of burrowing species at different levels below the sediment-water interface.

Vertical zonation. The presence of different depth bands dominated by different species in the intertidal zone.

Vertically homogeneous estuary. An estuary in which, at any given location, wind or tidal mixing homogenizes salinity throughout the water column.

Viscosity. As in dynamic viscosity: a measure of the degree that the fluid resists deformation under a force; a measure of "stickiness" of the fluid.

Viscous forces. Forces in a fluid that are explained by viscosity.

Vitamin. Chemical substance required in trace concentrations, acting as a cofactor with enzymes in catalyzing biochemical reactions.

Viviparous. Refers to development of an organism through the juvenile stage within a parent, with live release into the environment.

Warm-core ring. A blob of water formed by a meander of the Gulf Stream, which encloses even warmer water in the ocean, usually on the continental shelf in summer.

Water mass. A body of water that maintains its identity and can be characterized by such parameters as temperature and salinity.

Water vascular system. A system of tubes in echinoderms used to extend and provide suction to tube feet.

Watershed. The land area that is drained by a river or estuary and its tributaries.

Wave height. The vertical distance from the crest to the trough of a wave.

Wave length. The distance between crests in a system of waves.

Westerlies (prevailing westerlies). Persistent eastward-equatorward winds in midlatitudes in both the Northern and Southern Hemispheres.

White band disease. Coral disease, perhaps caused by a bacterium, that results in an advancing white band in the colony.

White plague. A complex of scleractinian coral diseases, sometimes associated with a coccoid bacterium.

Windward side. The side of an island that faces a persistent wind.

Winkler method. A method for measuring dissolved oxygen in water, using a chemical titration technique.

Within-habitat component. A contrast of diversity between two localities of similar habitat type.

Wrack zone. A bank of accumulated litter at the strandline.

Year-class effect. The common domination of a species population by individuals recruited in one reproductive season.

Year-size classes. Size groupings in a size-frequency graph that correspond to year classes.

Young-of-the-year. The new year class of an exploited species, formed usually in a restricted reproductive season.

Zonation. Occurrence of single species or groups of species in recognizable bands that might delineate a range of water depth or a range of height in the intertidal zone.

Zooids. Individuals in a bryozoan colony.

Zooplanktivore. Organism that eats zooplankton.

Zooplankton. Animal members of the plankton.

Zooxanthellae. A group of dinoflagellates living endosymbiotically in association with one of a variety of invertebrate groups (e.g., corals).

Some day you may want to write a research paper in one of many subjects in marine biology. The following is a list of specialty journals that are mainly concerned with marine biological subjects. You may also wish to consult the Marine Biology Web page (http://life.bio.sunysb.edu/marinebio/mbweb.html). There you will find a variety of resources, including reference lists for a number of marine biology subjects.

Marine Biology Journals and Research Papers

Advances in Marine Biology
American Naturalist
Annual Review of Marine Science
Aquaculture
Aquatic Conservation: Marine and Freshwater Ecosystems
Aquatic Toxicology
Archives of Environmental Contamination and Toxicology
Biological Bulletin (Woods Hole)
Biological Conservation
Biological Reviews (Cambridge)
Biology Letters
BMC Genomics
Coral Reefs
Crustaceana
Current Biology
Deep-Sea Research
Ecological Applications
Ecological Monographs
Ecology
Ecology Letters
Ecotoxicology
Environmental Biology of Fishes
Environmental Science and Technology
Environmental Toxicology and Chemistry
Estuaries and Coasts
Estuarine, Coastal & Shelf Science
Evolution
Frontiers in Marine Science
Frontiers in Microbiology

Functional Ecology
Global Change Biology
Global Ecology and Biogeography
Global Ecology and Conservation
Gulf of Mexico Science
Helgolënder wissenschaftliche Meeresuntersuchungen
Hydrobiologia
ICES Journal of Marine Science
Integrative and Comparative Biology
Invertebrate Zoology
Journal of Animal Ecology
Journal of Crustacean Biology
Journal of Experimental Biology
Journal of Experimental Marine Biology and Ecology
Journal of Fish Biology
Journal of Invertebrate Pathology
Journal of the Marine Biological Association of the United Kingdom
Journal of Marine Research
Journal of Molluscan Studies
Journal of Phycology
Journal of Plankton Research
Journal of Shellfisheries Research
Limnology and Oceanography
Limnology and Oceanography Methods
Marine and Freshwater Research
Marine Biology
Marine Biology Research (formerly Sarsia)

Marine Ecology, An Evolutionary Perspective
Marine Ecology—Progress Series
Marine Mammal Science
Marine Pollution Bulletin
Molecular Ecology
Nature
Nature Climate Change
Nature Ecology and Evolution
Oceanography
Oceanography and Marine Biology: An Annual Review
Oceanus
Oecologia
Oikos
Ophelia (Denmark) (now Marine Biology Research)
PLoS Biology
PLoS One
Proceedings of the National Academy of Sciences, USA
Proceedings of the Royal Society of London, Series B
Progress in Oceanography
Quarterly Review of Biology
Sarsia (now Marine Biology Research)
Science
Science Advances
Science of the Total Environment
Systematic Biology
Transactions of the American Fisheries Society
Trends in Ecology and Evolution